Perspectives in Biochemistry

Volume 2

Edited by
Hans Neurath

AMERICAN CHEMICAL SOCIETY
Washington, DC

Library of Congress Cataloging-in-Publication Data

Perspectives in biochemistry, volume 2 / edited by Hans Neurath.

p. cm.

Includes bibliographies and indexes.

ISBN 0-8412-1887-0 (v.2)

1. Biochemistry.

I. Neurath, Hans, 1909-

QP514.2.P47 1989
574.19'2—dc19

89-409
CIP

PRINTED IN THE UNITED STATES OF AMERICA

1990 EDITORIAL ADVISORY BOARD FOR THE JOURNAL *BIOCHEMISTRY*

Contents

*In papers with more than one author, the asterisk indicates the name of the author to whom inquiries about the paper should be addressed.

*In papers with more than one author, the asterisk indicates the name of the author to whom inquiries about the paper should be addressed.

PROTEOLYTIC PROCESSING AND PROTEIN DEGRADATION

CELL GROWTH AND REGULATION

BIOENERGETICS

*In papers with more than one author, the asterisk indicates the name of the author to whom inquiries about the paper should be addressed.

Preface

THIS IS THE SECOND VOLUME of a collection of concise reviews published in *Biochemistry* in 1989 and 1990 under the heading *Perspectives in Biochemistry*. They were solicited from leading experts and are intended to familiarize the general reader with the current status and future trends of topics on the leading edge of biochemical research. This collection should be of special interest to students and research practitioners in the biochemical sciences. The articles are arranged according to generally familiar, descriptive topics.

Authors of these articles contributed their time and effort without financial compensation. We are grateful to them for their generosity and cooperation. Special thanks are due to Beverly J. Brown and the staff in the editorial office of *Biochemistry* and to Robin Giroux of the Books Department.

This book should serve the dissemination of current biochemical knowledge in classrooms, research laboratories, and in institutions that serve basic and applied research in the biological and health-related sciences generally.

HANS NEURATH
Department of Biochemistry
University of Washington
Seattle, WA 98195

November, 1990

STRUCTURE AND FUNCTION OF PROTEINS

Chapter 1

Dominant Forces in Protein Folding

Ken A. Dill
Department of Pharmaceutical Chemistry, University of California, San Francisco, California 94143-1204
Received April 3, 1990; Revised Manuscript Received May 2, 1990

The purpose of this review is to assess the nature and magnitudes of the dominant forces in protein folding. Since proteins are only marginally stable at room temperature,[1] no type of molecular interaction is unimportant, and even small interactions can contribute significantly (positively or negatively) to stability (Alber, 1989a,b; Matthews, 1987a,b). However, the present review aims to identify only the largest forces that lead to the structural features of globular proteins: their extraordinary compactness, their core of nonpolar residues, and their considerable amounts of internal architecture.

This review explores contributions to the free energy of folding arising from electrostatics (classical charge repulsions and ion pairing), hydrogen-bonding and van der Waals interactions, intrinsic propensities, and hydrophobic interactions. An earlier review by Kauzmann (1959) introduced the importance of hydrophobic interactions. His insights were particularly remarkable considering that he did not have the benefit of known protein structures, model studies, high-resolution calorimetry, mutational methods, or force-field or statistical mechanical results. The present review aims to provide a reassessment of the factors important for folding in light of current knowledge. Also considered here are the opposing forces, conformational entropy and electrostatics.

The process of protein folding has been known for about 60 years. In 1902, Emil Fischer and Franz Hofmeister independently concluded that proteins were chains of covalently linked amino acids (Haschemeyer & Haschemeyer, 1973) but deeper understanding of protein structure and conformational change was hindered because of the difficulty in finding conditions for solubilization. Chick and Martin (1911) were the first to discover the process of denaturation and to distinguish it from the process of aggregation. By 1925, the denaturation process was considered to be either hydrolysis of the peptide bond (Wu & Wu, 1925; Anson & Mirsky, 1925) or dehydration of the protein (Robertson, 1918). The view that protein denaturation was an unfolding process was first put forward by Wu (1929, 1931). He proposed that native proteins involve regular repeated patterns of folding of the chain into a three-dimensional network somewhat resembling a crystal, held together by noncovalent linkages. "Denaturation is the breaking up of these labile linkages. Instead of being compact, the protein now becomes a diffuse structure. The surface is altered and the interior of the molecule is exposed" (Wu, 1929). "Denaturation is disorganization of the natural protein molecule, the change from the regular arrangement of a rigid structure to the irregular, diffuse arrangement of the flexible open chain" (Wu, 1931).

Before discussing forces, we ask: Is the native structure thermodynamically stable (the "thermodynamic hypothesis"; Anfinsen, 1973) or metastable, determined, for example, as the protein leaves the ribosome? To prove thermodynamic stability, it is sufficient to demonstrate that the native structure is only a function of state and does not depend on the process or initial conditions leading to that state. By definition, such a state would be at the global minimum of free energy relative to all other states accessible on that time scale. Experiments of Anson and Mirsky (1931) and Anson (1945) showed that hemoglobin folding is reversible as evidenced by similarities in the following properties of native and renatured protein: solubility, crystallizability, characteristic spectrum, binding to O_2 and CO, and inaccessibility to trypsin digestion. The folding of serum albumin and other proteins was shown to be similarly reversible by these coarse measures of native structure (Neurath et al., 1944; Anson, 1945; Lumry & Eyring, 1954). It was then demonstrated that denaturation is also *thermodynamically* reversible for some proteins (Eisenberg & Schwert, 1951; Brandts & Lumry, 1963) and involved large conformational changes (Harrington & Schellman, 1956; Schellman & Schellman, 1958). Recent high-resolution calorimetry experiments show thermodynamic reversibility for many small single-domain globular proteins (Privalov, 1979, 1989; Santoro & Bolen, 1988; Bolen & Santoro, 1988; Pace, 1975) and also for some multidomain and coiled-coil proteins (Privalov, 1982). Reversibility was tested much more specifically by the experiments of Anfinsen et al. (1973) in which the disulfide bonds of bovine pancreatic ribonuclease were

[1] The free energy $\Delta G_{unfold} = G_{denatured} - G_{native}$ is typically 5–20 kcal/mol of protein, less than $(1/10)kT$ per residue, where k = Boltzmann's constant and T is temperature (Pace, 1975; Privalov, 1979).

"scrambled" to random distributions of the 105 possible binding patterns and reacquisition of native structure and activity was observed upon renaturation (Haber & Anfinsen, 1962; Anfinsen, 1973). The advantage of monitoring disulfide bonds is that they are uniquely trappable and identifiable. Similarly, two circularly permuted proteins refold to their original native states (Luger et al., 1989; Goldenberg & Creighton, 1983). Therefore despite often extreme difficulties in the achievement of reversibility, the thermodynamic hypothesis has now been widely established. These experiments do not necessarily imply reversibility is completely general for other conditions, for other proteins, or even for other parts of a given protein than those monitored by the given experiment. It is clear that the folding of some proteins can be catalyzed by other assisting proteins, such as polypeptide binding or "chaperone" proteins (Rothman, 1989; Ostermann et al., 1989; Ellis, 1988; Anfinsen, 1973). Nevertheless, the existence of chaperones bears only on the rate that a protein is folded (provided the chaperone is a true catalyst) and has no bearing on the thermodynamic hypothesis, on the nature of the native state (if the native structure is otherwise reversible on the experimental time scale), or on the driving forces that cause it. The present discussion addresses only those proteins and conditions for which reversibility holds.

In this discussion of the nature of forces, it is useful to distinguish long-ranged and short-ranged forces, on the one hand, from local and nonlocal forces, on the other. The distance dependence defines the range: energies that depend on distance r as r^{-p} are long-ranged if $p \leq 3$ (ion–ion and ion–dipole interactions, for example) or short-ranged if $p > 3$ (Lennard–Jones attractions and repulsions, for example). This inverse third-power dependence is the natural division because for simple pure media the integral that gives the total energy of a system diverges, according to this definition, for long-ranged forces and converges for short-ranged forces (Hill, 1960). For polymer chains such as proteins, segment position in the chain is also important, in addition to the range of force. "Local" interactions are those among chain segments that are "connected" neighbors (i, i+1), or near neighbors, in the sequence (see Figure 1). "Nonlocal" refers to interactions among residues that are significantly apart in the sequence. Local interactions can arise from either long- or short-ranged forces, as can nonlocal interactions.

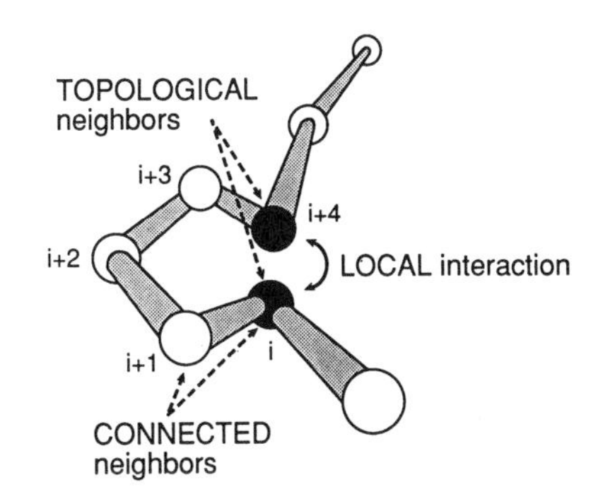

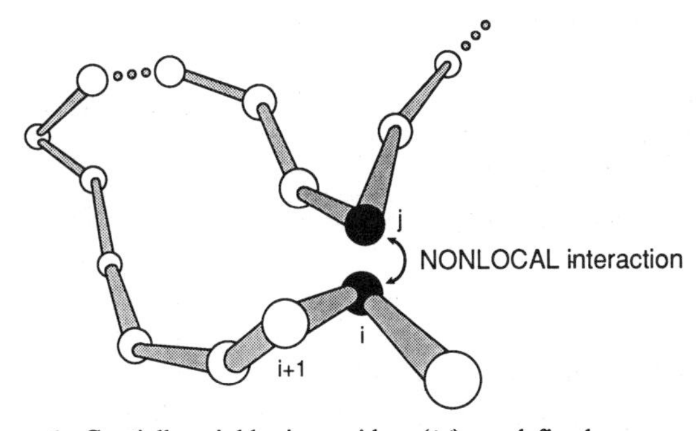

FIGURE 1: Spatially neighboring residues (i,j) are defined as *connected* neighbors if they share a backbone bond, $j = i + 1$; otherwise, they are *topological* neighbors. Interactions are *local* or *nonlocal* depending on the separation along the chain of the interacting residues.

(1) Long-Ranged Interactions: Electrostatics

Because acids and bases were among the earliest known denaturants of proteins, the folding forces were first assumed to be electrostatic in nature. The signature of electrostatically driven processes is a dependence on pH and/or ionic strength. Whereas the pH determines the total charge on the protein, the salt determines the extent of interaction among those charges since salts shield charges. The first quantitative model of electrostatic interactions in native proteins was proposed by Linderstrom-Lang (1924) (when he was 27 years old!). This work appeared less than 1 year after the Debye–Huckel theory on which it was based. By treating a native protein as a multivalent impenetrable spherical particle with its net charge uniformly distributed at the surface, Linderstrom-Lang predicted the number of protons bound and the net charge as functions of the hydrogen ion concentration, i.e., the pH titration curve. The view that protein electrostatics can be represented in terms of charges on a sphere of low dielectric constant in a higher dielectric medium has remained useful. Recent improvements have included (i) the consideration of discrete charges located at specific positions on the spherical native protein (Tanford & Kirkwood, 1957a,b; Matthew & Gurd, 1986; Matthew & Richards, 1982), (ii) modeling native structural deviations from spherical shape (Gilson & Honig, 1988a,b), and (iii) the development of electrostatic theory for the unfolded state and therefore for free energy contributions to stability (Stigter & Dill, 1990; Stigter et al., 1990).

There are two different ways in which electrostatic interactions can affect protein stability. (1) *Classical* electrostatic effects are the nonspecific repulsions that arise when a protein is highly charged, for example, at extremes of pH. The traditional view (Tanford, 1961; Kauzmann 1954; Linderstrom-Lang, 1924) of these effects has been that the electrostatic free energy depends on the square of the net charge. Hence, no electrostatic contribution to protein stability is expected near the isoelectric point. As the net charge on the native protein is increased by increasing acidity or basicity of the solution, the increasing charge repulsion will destabilize the folded protein because the charge density on the folded molecule is greater than on the unfolded molecule. Thus, the process of unfolding leads to a state of lower electrostatic free energy. Hence, acids and bases destabilize native proteins (see Figure 2).

(2) *Specific* charge interactions can also affect stability. For example, ion pairing (salt bridging) occurs when oppositely charged amino acid side chains are in close spatial proximity. Whereas the classical mechanism predicts that increasing the charge could only destabilize the folded state, ion pairing could stabilize it. It has traditionally been held that classical and ion-pairing effects could be distinguished by experiments on the effects of salt concentration (below about 0.1–1.0 M) or the dielectric constant of the solution. Only at low concentrations does salt predominantly affect electrostatic shielding; at higher concentrations the electrostatic shielding is saturated, so that then the dominant effects of salt, like any other additives, are on the solvent properties of the solution (Morrison, 1952). In the traditional view, it is assumed that salts and the dielectric constant of the medium do not affect the net charge on the molecule and that they affect the native state

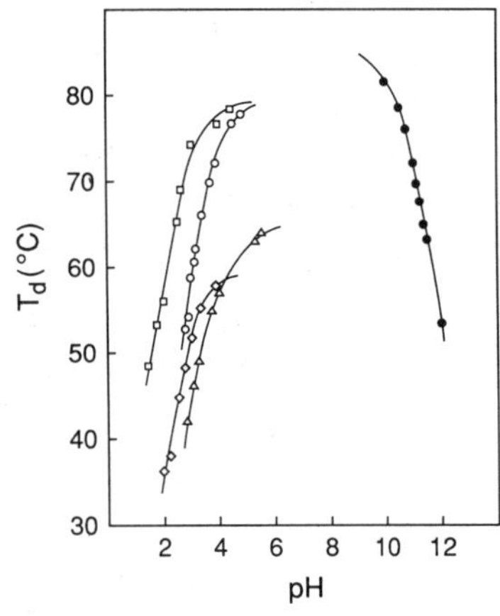

FIGURE 2: Denaturation temperature vs pH for (●) metmyoglobin, (△) ribonuclease A, (○) cytochrome C, (◇) α-chymotrypsin and (□) lysozyme. Increased charge on the protein at extremes of pH (low or high) favors unfolding. Reproduced from Privalov, P. L., & Kechinashvili, N. N. (1974) *J. Mol. Biol. 86*, 665. Copyright (1974) Academic Press Inc. (London) Ltd.

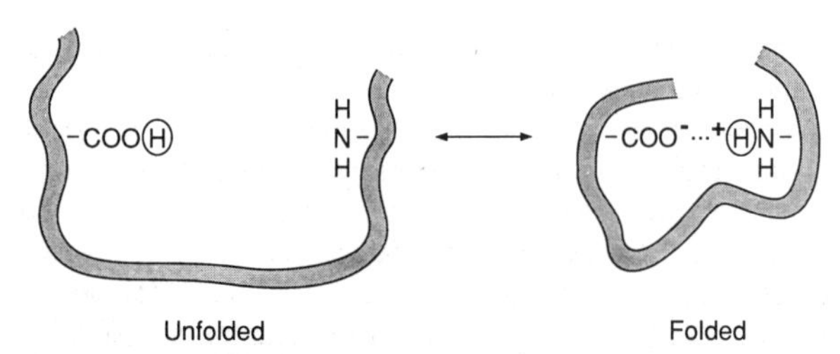

FIGURE 3: Early model in which protein folding was proposed to be driven by ion-paired hydrogen bonding among side chains (Mirsky & Pauling, 1936; Eyring & Stearn, 1939), shown by Jacobsen and Linderstrom-Lang (1949) to be inconsistent with partial molar volumes.

more than the denatured state. It has often been assumed therefore that either adding salt or increasing the dielectric constant of the solution should stabilize proteins if classical effects are dominant or destabilize them if ion pairing is important. It is now clear, however, that effects of neither salt nor dielectric constant can be interpreted so simply, for the following reasons. First, salts strongly affect the unfolded state (see below). Second, ion-pair bonds are generally much shorter than the Debye lengths in salt solutions, so salt should have little effect on ion-pairing stability. [One exception is a Glu-2^-···Arg-10^+ salt bridge in the C-peptide helix (Shoemaker et al., 1990), which is screened by 1 M salt, but this may be a solvent-separated ion pair (R. L. Baldwin, personal communication).] Third, although a decreased dielectric constant will lead to increased charge interactions, it will also decrease the total ionization since charging is energetically more costly in a low-dielectric medium. Moreover, the dielectric constant is also correlated with other solvent properties such as hydrophobicity and is not a simple diagnostic for charge effects alone. Therefore, discriminating between classical and ion-pairing electrostatics contributions to stability has been difficult.

During the 1930s, ion pairing was considered to be the dominant contributor to protein stability (Cohn et al., 1933; Mirsky & Pauling, 1936; Eyring & Stearn, 1939). Mirsky and Pauling suggested that folding was driven by the ion pairing of carboxyl and amino groups on the side chains of the charged amino acids.

If ion pairing is important for protein stability, then such stability must arise from charged pairs at protein surfaces rather than from charged pairs buried in the protein core. The first evidence that few ion pairs are buried was due to Jacobsen and Linderstrom-Lang (1949) on model compounds. An important signature of electrostatic effects in solution is a change in volume: the local volume of water decreases around a molecule of increasing charge. The electrostatic field of the charged molecule orients and orders neighboring water dipoles (electrostriction), decreasing the entropy and volume of the local water molecules. At low pH where only the carboxyl groups are titratable, Jacobsen and Linderstrom-Lang noted that the volume increase upon protonation of COO^- to COOH, of about 10 mL/mol in proteins, is the same as in model carboxyl compounds in water, suggesting that ion pairs in proteins must be exposed. More recent studies of known protein structures show that indeed few ion pairs are buried (on average, only about one ion pair per 150-residue protein is buried) (Barlow & Thornton, 1983). This follows from the very high Born energy required to transfer a charged ion from aqueous solution to the low-dielectric interior of the protein, ranging from 19 kcal/mol for full burial to 4 kcal/mol for a half-exposed ion at the surface, 7 kcal/mol for complete burial of an ion pair (Honig et al., 1986; Honig & Hubbell, 1984). Thus, unless other specific interactions are involved, only surface ion pairs could generally stabilize native states.

It is clear that ion pairing can contribute to protein stability. Studies of X-ray crystal structures of known proteins (Wada & Nakamura, 1981; Barlow & Thornton, 1983) show that ion pairing is common on the surfaces of proteins. Also, variations in sequence that affect ion pairing can change stability by about 1–3 kcal/mol of ion pairs (Fersht, 1972; Perutz & Raidt, 1975); Asp-70–His-31 in T4 lysozyme has recently been found to stabilize by 3–5 kcal/mol (Anderson et al., 1990). Similarly, ion binding sites designed into proteins can affect stability (Pace & Grimsley, 1988). However, it is clear that ion pairing is not the dominant force of protein folding. The first evidence emerged from the pivotal paper of Jacobsen and Linderstrom-Lang (1949). They interpreted the models of Eyring and Stearn (1939) and Mirsky and Pauling (1936) as shown in Figure 3: the hydrogen protonates the carboxyl group in the unfolded state, so both carboxyl and amino groups are uncharged, whereas in the folded state the hydrogen protonates the amino group, so that the carboxyl and amino groups form an ion pair. Jacobsen and Linderstrom-Lang presumed the charges remained solvated upon folding. Model compounds show that the protonation of NH_2 to NH_3^+ leads to an electrostriction of about −4 mL/mol and deprotonation of COOH to COO^- leads to −10 mL/mol, as noted above. Folding should then result in a volume change of −14 mL/mol per ion pair. In contrast to this model, experiments show that folding leads to an *increase* in volume (Jacobsen & Linderstrom-Lang, 1949; Zipp & Kauzmann, 1973; Brandts et al., 1970; Edelhoch & Osborne, 1976).

Also inconsistent with ion pairing as the dominant force of folding is the observation that the stabilities of proteins show little dependence on pH or salt (at low salt concentrations) near the isoelectric point (Tanford, 1968; von Hippel & Schleich, 1968; Hermans & Scheraga, 1961; Acampora & Hermans, 1967). [For some proteins, the pH of maximum stability does not coincide with the isoelectric pH, but this can be accounted for within the classical model by the burial in the hydrocarbon core of some of the titratable groups (often histidines) (Stigter & Dill, 1990).] As further evidence that charge generally contributes only weakly to protein stability, Hollecker and Creighton (1982) found little effect of changing the charges on several different amino groups in three different proteins.

Third, perhaps the most persuasive evidence that ion pairing is not the dominant force of folding comes from the structural studies of Barlow and Thornton (1983). They have observed that ion pairs are not highly conserved in evolution. More importantly, the number of ion pairs in proteins is small. They observe about five ion pairs per 150 residues of protein (about one of which is buried, noted above). It is unlikely that any interaction involving only 10 residues, less than 10% of the molecule, could be the dominant folding force. Using the estimate of 1–3 kcal/mol (Fersht, 1972; Perutz & Raidt, 1975) for the stabilization per ion pair leads to a value of 5–15 kcal/mol stabilization. Even though ion pairing would thus contribute a free energy equal to that of the net stability of the protein, this is still 5–10-fold smaller than the hydrophobic interaction discussed below.

A similar estimate, of about 10 kcal/mol stabilization due to ion pairing, has been made by Friend and Gurd (1979; Matthew & Gurd, 1986). They observed decreased stability of sperm whale ferrimyoglobin with increased salt and interpreted this as evidence for ion pairing. They assumed salt predominantly affects the native state, on the basis of the difference in titration behavior of native and denatured states. However, these results do not necessarily imply the electrostatic stabilization comes from ion pairing. Salt can affect the relative free energies differently than it affects the titration behavior. A recent polyelectrolyte model of proteins (Stigter & Dill, 1990; Stigter et al., 1990) shows instead that increasing salt, by classical effects alone, will reduce the electrostatic free energy of the unfolded state of myoglobin more than the folded state. Increased salt shields the charge repulsions in the unfolded molecule more effectively than in the folded molecule at low pH, probably because of better penetration of the salt solution into the unfolded molecule. The model is consistent with an additional experimental observation that is otherwise difficult to explain on the basis of ion pairing. For β-lactamase, similar to the myoglobin experiments of Friend and Gurd, Goto and Fink (1989) observe that salt destabilizes the native state when the molecule is highly charged at low pH, but they also find that salt *stabilizes* the native structure when the molecule is charged at high pH. It is interesting that a significant fraction of the electrostatic free energy is predicted to arise from the entropy of proton release (Stigter & Dill, 1989, 1990), rather than simply from the charge energetics.

(2) Hydrogen Bonding and van der Waals Interactions

van der Waals attractions arise from interactions among fixed or induced dipoles. A hydrogen bond occurs when a hydrogen atom is shared between generally two electronegative atoms. Hydrogen-bond strength, which depends on the electronegativity and orientation of the bonding atoms, is in the range of 2–10 kcal/mol (Pauling, 1960). For example, the water–water hydrogen bond in the vapor phase is −6.4 kcal/mol (Weiner et al., 1984). A hydrogen bond is primarily a linear arrangement of donor, hydrogen, and acceptor and is comprised of electrostatic, dispersion, charge-transfer, and steric repulsion interactions (Vinogradov & Linnell, 1971). The dominant component of a hydrogen bond is electrostatic (Pauling, 1960; Cybulski & Sheiner, 1989; Vinogradov & Linnell, 1971). In this section we ask: do hydrogen bonds and van der Waals interactions contribute *differently* to folded and unfolded states of proteins, and therefore to stability? While these two types of force are microscopically quite different, there are few simple macroscopic diagnostics that can distinguish between them; hence, in this section we consider them together. The evidence cited below suggests that they may play an important role in protein folding, but the magnitudes, among all the types of force contributing to protein folding, are currently perhaps the most difficult to assess.

Mirsky and Pauling (1936) were the first to suggest that hydrogen bonding was the dominant force of protein folding. Although their focus appears to have been the electrostatic hydrogen bonds arising from ion-paired side chains (see preceding section), they also suggested that hydrogen bonding could occur between the carbonyl C=O and amide NH groups of the peptide backbone. It is the peptide hydrogen bonds we consider here. Their proposal led to the X-ray crystallography studies of amino acid crystals by Pauling et al. begun in 1937, culminating in the discovery of the α-helix and parallel and antiparallel sheets in 1951 (Pauling et al., 1951; Pauling & Corey, 1951a-d). These were first called "secondary structures" by Linderstrom-Lang (1952).

During the 1950s, Doty and his colleagues found a model system, poly(γ-benzyl-L-glutamate), for studying the driving forces in the formation of polypeptide helices in nonaqueous solution (Doty & Yang, 1956; Doty et al., 1954, 1956, 1958). Soon thereafter a theoretical framework emerged for understanding the balance of forces driving the helix–coil transition. The first theoretical model was due to Schellman (1958a). Many other elegant treatments followed, principally based on the one-dimensional Ising model (Peller, 1959; Gibbs & DiMarzio, 1959; Zimm & Bragg, 1959; Zimm & Rice, 1960; Flory, 1969; Poland & Scheraga, 1970). In these models, the intrachain hydrogen bond is considered energetically favorable relative to the hydrogen bond with the solvent. However, to form the first such bond requires overcoming configurational entropy to arrange the immediately adjacent bonds into a helical configuration. At low temperatures with simple solvents, the enthalpic contribution dominates, and the molecule forms a helix; at high temperatures, the entropy dominates and the molecule is configured as a random coil (Shoemaker et al., 1987; Marqusee et al., 1989; Lupu-Lotan et al., 1965; Platzer et al., 1972). A sharp transition between these states results from this subtle balance between the large forces. The entropy is local insofar as it involves the configurations of only immediately neighboring bonds along the chain and thus is assumed to be independent of aspects of the chain configurations more distant. Theoretical helix–coil transition models successfully predict (i) this temperature dependence and (ii) that helices become more stable and that transitions sharpen with increasing chain length (Poland & Scheraga, 1970). The models also predict the influence of pH on the helix–coil transition: greater charge on the molecule destabilizes the helix since the coil has lower charge density and thus lower electrostatic free energy (Peller, 1959; Zimm & Rice, 1960). The helix–coil transition has inverted temperature dependence in some mixed solvents (Zimm et al., 1959; Lupu-Lotan et al., 1965). Solvents that bind to the peptide bond will favor the coil; one example is formic acid, which protonates the bond (Lotan et al., 1967). Consistent with the view that hydrogen bonding is the principal driving force of the helix–coil transition, solvents that form strong hydrogen bonds compete more effectively with the peptide and destabilize the helix relative to the coil. For example, chloroform, dimethylformamide, 2-chloroethanol, trifluoroethanol, and other alcohols favor the helix, relative to formic acid, dichloroacetic acid, or trifluoroacetic acid (Doty & Yang, 1956; Doty et al., 1954, 1956, 1958; Lupu-Lotan et al., 1965; Conio et al., 1970; Nemethy et al., 1981; Nelson & Kallenbach, 1986). Similar theory has been developed for β-sheet formation (Mattice & Scheraga, 1984).

For three reasons, it is natural to assume that hydrogen bonding and van der Waals interactions will be important for the conformational changes of proteins. First, the amino acids that comprise proteins are dipolar and are capable of hydrogen bonding. Second, helices are common features of globular proteins, and the studies cited above show that the helix–coil transition is largely driven by hydrogen bonding. Similarly, intramolecular sheets are also formed by hydrogen bonding (Anufrieva et al., 1968). Third, the conformational forces for nonelectrolyte polymers in nonelectrolyte solvents are short ranged, arising from differences in monomer–monomer attractions of the chain relative to monomer–solvent attractions (Flory, 1953; deGennes, 1979). If monomer and solvent interactions are short ranged, then classical polymer theories would predict that chains should usually be relatively self-attractive, with radius changes characterized by a temperature-independent enthalpy (Flory, 1953). Such a temperature-independent enthalpy has been inferred to contribute to protein folding on the basis of model assumptions about the contribution of the hydrophobic interaction (Baldwin, 1986; Privalov, 1979; Dill et al., 1989). This residual folding force becomes more favorable as the number of polar groups increases (Privalov & Kechinashvili, 1974). For hen lysozyme, the magnitude of this enthalpy is 0.43 kcal/mol of residues (Baldwin, 1986).

Although these short-ranged forces are therefore undoubtedly important, Kauzmann (1954, 1959) concluded that they are probably not the dominant forces that fold proteins in water. A fundamental criterion for a dominant driving force is that it must explain why the folded state is advantageous relative to the unfolded state. He argued that hydrogen bonding would not satisfy this criterion, because there was no basis for believing that the intrachain hydrogen bonds in the folded state would have lower free energy than those of the unfolded chain to water. In support of this view, the distribution of hydrogen-bond angles in proteins is observed to be about the same as in small-molecule compounds (Baker & Hubbard, 1984). It follows however that folded proteins must contain many hydrogen bonds; for otherwise, the protein would denature.

Kauzmann's hypothesis led to model studies on analogues to determine the free energy of peptide hydrogen bonds in water. The several models of the peptide hydrogen bond include urea (Schellman, 1955), valerolactam (Susi, 1969), *N*-methylacetamide (NMA) (Klotz & Franzen, 1962; Kresheck & Klotz, 1969), and cyclic dipeptides, the diketopiperazines (Gill & Noll, 1972). For reasons described below, however, these model studies have not yet yielded definitive estimates for the contribution of hydrogen bonds to protein stability. The dimerization binding constants and their temperature dependences have been measured for these molecules in water, in order to obtain free energies and enthalpies of dimerization. Because the concentration dependences in these experiments are linear at low concentrations, the bound species is presumed to be predominantly in the form of dimers, rather than higher multimers. At 25 °C, dimerization in water is disfavored; the equilibrium ratio of dimers to monomers is only 4.1×10^{-2} for urea and 5.0×10^{-2} for diketopiperazine. Thus, the free energy of dimerization is positive ($\Delta G_{\text{dimerization}} = +1.89$ kcal/mol for urea). However, the enthalpy of dimer formation is negative (−2.1 kcal/mol for diketopiperazine, −2.1 kcal/mol for urea, −2.8 kcal/mol for valerolactam), except for *N*-methylacetamide for which it is approximately zero (Klotz & Franzen, 1962). On the assumption that the hydrogen bond is the only attraction driving dimerization, it has been generally concluded that the hydrogen bond in water is enthalpically favored relative to the monomer–water bond. For *N*-methylacetamide, Jorgensen (1989) has shown by Monte Carlo simulation that this assumption is probably not valid: in water, the amides stack rather than form hydrogen bonds. In contrast, in chloroform, the amides form good hydrogen bonds (Jorgensen, 1989). Susi and Ard (1969) have suggested that ϵ-caprolactam dimerization may also be driven by some mechanism other than hydrogen bonding. Thus in addition to hydrogen bonding, van der Waals and other interactions may also contribute significantly, but their importance for the other model compounds is not yet clear.

An additional problem prevents unequivocal determination of the free energy of hydrogen-bond formation from these model studies. In all the model compounds, there are two ways a dimer can form: singly bonded, wherein one partner in the dimer will have considerable rotational freedom relative to the other, or doubly bonded, with one partner considerably restricted in its rotation relative to the other. The experiments find only the ratio of "complexed" molecules (of singly-bonded plus doubly-bonded dimers) to unbound monomers. To obtain the free energy of hydrogen-bond formation from these data requires additional knowledge of the relative numbers of singly-bonded and doubly-bonded dimers, not currently available for these model compounds. The dilemma is illustrated by the following comparison. Suppose, on the one hand, that the only species in solution was known to be the singly-bonded dimer; then the measured positive free energy implies hydrogen bonding is *disfavored* in water. Suppose alternatively that the only species in solution was known to be the doubly-bonded dimer. Then the binding free energy will include contributions from the two hydrogen bonds and an unfavorable entropy of rotational restriction. If this rotational restriction is sufficiently unfavorable, contributing a large enough positive free energy to the overall dimerization free energy, then the intrinsic free energy of hydrogen-bond formation will be inferred to be negative, implying that hydrogen bonding is *favored* in water. Schellman (1955) estimated this entropic restriction to be 3–6 eu and concluded that the free energy of formation of a hydrogen bond is negative but probably small. Thus the inference as to whether hydrogen-bond formation is favored or disfavored in water depends on (i) which compound is chosen as a model, (ii) the importance of interactions other than hydrogen bonding for the association processes of those model compounds in water, and (iii) estimation of the magnitude of a rotational restriction entropy, presently unknown but probably of about the same magnitude as the free energy of hydrogen bonding itself.

Moreover, this class of experiments has largely been restricted to aqueous solvents. But the peptide hydrogen bond in a globular protein is in a more hydrocarbon-like medium. We are therefore interested in the following equilibrium:

$$\begin{array}{ccc} A_n + B_n & \xrightarrow{K_1} & (AB)_n \\ \uparrow K_2 & \nearrow K_5 & \uparrow K_4 \\ A_w + B_w & \xrightarrow[K_3]{} & (AB)_w \end{array}$$

where A and B are the hydrogen-bond donor and acceptor, w is water, and n is the nonpolar solvent. We aim to determine ΔG_5, the hydrogen-bond contribution to protein stability. We obtain this by using other steps of the thermodynamic cycle. (1) A wide range of hydrogen-bonding species, including NMA, formamide, alcohols, carboxylic acids, and phenols, tend to associate in nonpolar solvents ($K_1 > 1$); for example, for the dimerization of NMA in CCl_4, $\Delta G_1 = -2.4$ kcal/mol

(Vinogradov & Linnell, 1971; Klotz & Farnham, 1968; Kresheck & Klotz, 1969; Roseman, 1988; Sneddon et al., 1989). However, this free energy is solvent dependent. Hydrogen bonding strengthens in nonpolar solvents either if (i) the dielectric constant of the solvent is reduced, with other solvent properties held fixed (Franzen & Stephens, 1963), as expected for Coulomb interactions, or if (ii) the electron- or proton-donating or -accepting capacity of the solvent is varied, with the dielectric constant held fixed (Allen et al., 1966; Krikorian, 1982). (2) Transferring a hydrogen bond into a nonpolar medium is generally disfavored: $\Delta G_2 = +6.12$ kcal/mol has been estimated for NMA from water to CCl_4 (Roseman, 1988). (3) As noted above, AB dimerization is disfavored in water; for NMA, $\Delta G_3 = +3.1$ kcal/mol (Klotz & Farnham, 1968; Roseman, 1988). (4) It follows that $\Delta G_4 = +0.62$ kcal/mol for NMA in CCl_4 (Roseman, 1988); ΔG_4 is also near zero for carboxylic acids in benzene (Klotz & Farnham, 1968) and is predicted to be +2.2 kcal/mol for transferring the formamide dimer from water to CCl_4 (Sneddon et al., 1989). (5) It follows from these estimates that hydrogen bonding opposes folding. For NMA in CCl_4, $\Delta G_5 = +3.72$ kcal/mol; for formamide, $\Delta G_5 = +1.9$ kcal/mol. However, if the transfer of A and B from water into a nonpolar medium is driven by some other force, such as hydrophobicity (see below), then hydrogen-bond formation, process 1, will be strongly favored within the folded structure. Thus whereas hydrogen bonding may not assist the collapse process, it would favor internal organization within the compact protein. Two additional uncertainties make it difficult to estimate the medium effect on the hydrogen-bond strength: (i) the nonpolar core of a protein is not a homogeneous dielectric (Honig et al., 1986; Warshel, 1984), and (ii) hydrogen-bond strength is extremely sensitive to geometric details of bond angles (Scheiner & Hillenbrand, 1985; Scheiner et al., 1986). Since there are so many hydrogen bonds in native proteins, then even small errors in estimating their strength will lead to large errors in determining their effects on protein stability. Only 11% of all C═O groups and 12% of all NH groups have no hydrogen bonds (Baker & Hubbard, 1984). Of all the hydrogen bonds to C═O groups, 43% are to water, 11% are to side chains, and 46% are to main-chain NH groups. Of all the hydrogen bonds to NH groups, 21% are to water, 11% are to side chains, and 68% are to main-chain C═O groups. To reliably estimate the stability of a protein would therefore require model studies more accurate than about $(1/5)kT$ per intrachain peptide bond. For the reasons noted above, this accuracy is not yet available from the current models.

Solvent denaturation studies indicate that hydrogen bonding is not the dominant folding force (Singer, 1962; Edelhoch & Osborne, 1976). If it were, then solvents that form strong hydrogen bonds to the peptide backbone should compete effectively and unfold the protein. Those solvents that do not affect hydrogen bonding should not affect stability. In this light, several observations are of importance. (1) It would be difficult to rationalize the observation that very small concentrations of surfactants (1% dodecyl sulfate, for example) unfold proteins (Tanford, 1968) since they do not destabilize helices (Lupu-Lotan et al., 1965). Moreover, the effectiveness of tetraalkylammonium salts to denature proteins depends on the number of methylene groups, indicative that it is the hydrophobic interaction, rather than hydrogen bonding, which determines stability. (2) Since the C═O group is a strong hydrogen acceptor and the NH group is a weak donor, Singer (1962) has pointed out that the solvents useful for competing with the peptide hydrogen bond would be those which are stronger donors than NH; the peptide bond will generally compete effectively against solvents that are weaker hydrogen acceptors than C═O. Dioxane is only a hydrogen-bond acceptor and therefore should not denature proteins if hydrogen bonding were the dominant folding force. However, dioxane denatures proteins (Singer, 1962). (3) Alcohols are more hydrophobic than water, but they enhance helix formation (Conio et al., 1970). If hydrogen bonding were the dominant folding force, alcohols should stabilize proteins. Hence, the observation that at low concentrations they destabilize proteins is inconsistent with hydrogen bonding as the principal driving force (Hermans, 1966; von Hippel & Schleich, 1969a,b; Parodi et al., 1973). The caveat is that alcohols have complex effects on protein stability, depending on concentration and temperature (Brandts, 1969). Finally, a particularly important comparison involves the effect on protein stability of alcohols (ethanol and propanol) vs the corresponding glycols (ethylene glycol and propylene glycol). The glycols are less hydrophobic and have more hydrogen-bond sites than the corresponding alcohols. The observation that the glycols are worse denaturants is strong evidence that hydrogen bonding is less important than the hydrophobic interaction (Tanford, 1968; Herskovitz et al., 1970; von Hippel & Schleich, 1969a).

Mutation studies show that hydrogen-bonding groups affect stability, but by an amount which can differ considerably depending on the site and nature of the mutation (Bartlett & Marlowe, 1987). Fersht et al. (1985) have estimated from activation free energy measurements of tyrosyl-tRNA synthetase/substrate interactions that breakage of a hydrogen bond increases the free energy by 0.5–1.5 kcal/mol for unpaired uncharged donor and acceptor or about 3.5–4.5 kcal/mol if the donor or acceptor is charged. Site-directed mutagenesis experiments in which nonhelical proline 86 in phage T4 lysozyme is replaced by other amino acids which extend the helix and add new *backbone* hydrogen bonds show marginal *reduction*, rather than increase, in stability (Alber et al., 1988). On the other hand, *side-chain* hydrogen-bonding groups are found to stabilize T4 lysozyme (Alber et al., 1987a; Grutter et al., 1987). However because site-directed mutagenesis experiments measure only the total change in stability upon mutation, $\Delta\Delta G$, and not the individual molecular components of that change, then these mutations may involve more than just hydrogen bonding. For example, for one of them (Thr replacing Val-157), free energy perturbation calculations show that the added stability arises from better van der Waals interactions rather than from the difference in hydrogen-bond strength (Dang et al., 1989).

(3) Local Interactions: Intrinsic Propensities

The term "intrinsic propensity" does not describe any single type of force. Rather it is intended to convey the idea that there are certain conformational preferences of di- or tripeptides, depending on the sequence, which arise from the sum of short- and long-ranged forces that are local among connected neighboring residues. ("Local" may extend to residues three to four monomers distant and may therefore also include hydrogen bonds involved in turns or helices.) Intrinsic propensities have been studied by the measurement of helix/coil equilibria of peptides in water (Sueki et al., 1984; Marqusee et al., 1989; Marqusee & Baldwin, 1987) and turn/coil equilibria (Dyson et al., 1985, 1988a; Wright et al., 1988). The stabilities of long polypeptide helices in aqueous solution can be attributed to intrinsic propensities. Helix stability increases with chain length (Goodman et al., 1969; Zimm & Bragg, 1959; Poland & Scheraga, 1970). Therefore, although the free energy contribution from each individual residue may be small, summed over many residues, the helix can be strongly favored relative to the coil in a long chain.

The traditional view has held that the individual residue helix/coil equilibrium constants are so nearly equal to one, however, and the initiation constants so small that short helices (less than about 15–20 residues) are not stable in aqueous solution. There have been two bases for this view. First, short helices extracted from stable globular proteins have been found to be unstable in isolation in aqueous solution (Epand & Scheraga, 1968; Taniuchi & Anfinsen, 1969; Dyson et al., 1988b). Second, using "guest" amino acids randomly doped into "host" copolymers of hydroxypropyl- and hydroxybutyl-L-glutamine, Scheraga and his colleagues (Sueki et al., 1984) showed that the intrinsic propensities of amino acids to form helices in water are small (with helix/coil equilibrium constants ranging from 0.59 to 1.39 at 20 °C). (Helical propensity can be increased considerably by reducing the temperature to near 0 °C.) However, these equilibrium constants will be universal, in principle, only if the host helix itself is completely inert in its effect on the helix/coil equilibrium of the guest residue. The following recent evidence with other hosts, however, suggests that the helix is not completely inert, i.e., that there are "context" effects. (1) The helix/coil constants differ in other sequences and can be as large as nearly 2 for alanine in alanine-based helices (Marqusee et al., 1989; Padmanabhan et al., 1990). Although the helix/coil constant for uncharged guest residue i appears therefore to depend on the local sequence through residues $i-1$ and $i+1$, it does not appear to further depend on $i-2$ and $i+2$ or otherwise on the position in the chain (Merutka & Stellwagen, 1990). (2) Additional stability results if helix formation leads to burial of nonpolar surface (Tanford, 1968; Chou et al., 1972; Richards & Richmond, 1978). (3) Helices can be stabilized considerably by reducing the helix dipole moment through reduction of the charges at the helical ends (Shoemaker et al., 1985, 1987; Marqusee & Baldwin, 1987). (4) Salt bridges and aromatic interactions can also affect stability (Marqusee & Baldwin, 1987; Shoemaker et al., 1990). In addition, intrinsic propensities can vary with the solvent (Rich & Jasensky, 1980). Recent evidence (Merutka et al., 1990) suggests that context effects may be at least as important as intrinsic propensities.

On the basis of these observations, considerable progress continues to be made in improving stabilities, so that higher helix/coil equilibrium constants are achieved, in shorter chains, and at increasing temperatures up to near room temperature (Marqusee & Baldwin, 1987; Bradley et al., 1990). Nevertheless, intrinsic propensities, in the absence of other forces, appear to be insufficient to account for the full helical stability in globular proteins. Helices in globular proteins are short. The average length is about 12 residues, and the most probable helix length (peak of the distribution) is less than 6 residues (Kabsch & Sander, 1983; Levitt & Greer, 1977; Srinivasan, 1976). Yet protein helices remain 100% helical up to temperatures near the denaturation point. Other forces must therefore also be important for stabilizing helices in globular proteins. One possibility is that there may be additional "context" effects due to the environment provided by the protein interior. For example, charges are distributed in proteins so as to stabilize the helix dipole (Blagdon & Goodman, 1975; Richardson & Richardson, 1988). Helices with modified charges at the ends can affect protein stability (Mitchinson & Baldwin, 1986). In principle, helices could pack in pairs, antialigned, to reduce the net dipole moment; this probably contributes little to stability, however, since the ends of helices are generally found in a high-dielectric medium at protein surfaces (Rogers, 1989; Gilson & Honig, 1989; Presnell & Cohen, 1989). In contrast to these effects, the protein environment may not always stabilize helices: Alber et al. (1988) found that added hydrogen-bonded helix-extending residues in T4 lysozyme had little effect or destabilized the protein.

Other evidence suggests that protein architecture does not arise principally from intrinsic propensities. First, distributions of secondary structures predicted by intrinsic propensities are inconsistent with those in known protein crystal structures. Any model of protein stability based only on local interactions would predict, as is observed in helix/coil equilibria, that longer helices should be more stable, and therefore more probable, than shorter helices. In contrast, studies of the crystal structures of globular proteins (Kabsch & Sander, 1983; Levitt & Greer, 1977; Srinivasan, 1976) show just the opposite: helix probability decreases monotonically with length (see Figure 11). Similarly, longer sheets are observed to be less probable than shorter sheets, for both parallel and antiparallel sheets. These discrepancies are not repaired by local factors alone. For example, it is known that helices can be terminated by "stop" residues (Kim & Baldwin, 1984). Stop signals will not account for the data base trends, however, which are grand averages over residues, positions, and proteins, since it would then follow that most amino acids must be helix destabilizing, in contradiction to the basic premise. Moreover, even given that local interactions impart some stability to helices and turns, it is difficult to rationalize how they would give rise to sheets, which are intrinsically nonlocal. Therefore, alternative explanations for these distributions involve nonlocal factors. For example, at high densities short peptides can form stable helices in crystals (Karle et al., 1990), suggesting the importance of packing effects. It is shown in section 8 that the distributions of internal architecture can be accounted for by steric forces of nonlocal origin.

Second, attempts to predict protein structures by use of only intrinsic propensities have had limited success (Schulz & Schirmer, 1979; Kabsch & Sander, 1984; Argos, 1987; Thornton, 1988; Rooman & Wodak, 1988; Qian & Sejnowski, 1988). Success rates are about 64% when averaged over many proteins (Thornton, 1988; Rooman & Wodak, 1988; Qian & Sejnowski, 1988; Holley & Karplus, 1989). Because this is considerably better than chance, it implies that intrinsic propensities are significant determinants of protein structure. However, according to Qian and Sejnowski (1988), who have used neural net methods, "no method based solely on local information is likely to produce significantly better results for non-homologous proteins." Within a given class of proteins, success rates may be higher (Kneller et al., 1990). Limitations of intrinsic propensities are also found in studies of conformations of identical pentapeptides in different proteins (Kabsch & Sander, 1984; Argos, 1987). Kabsch and Sander found that in 6 of 25 cases one pentamer could be found in a helix whereas the identical sequence in a different protein would be in a sheet, implying that local information alone is not sufficient to fully specify the conformation in the protein. What does the 64% success rate tell us about the magnitude of the nonlocal factors missing from these prediction methods? If we assume that the conformation of any residue can be predicted with a priori success rate $p_o = 1/3$ (Chou & Fasman, 1978), as a rough estimate for classification as a helix, sheet, or other conformation, then a success rate p carries an amount of information, $\langle I \rangle$:

$$\langle I \rangle = p \ln (p/p_o) + (1 - p) \ln [(1 - p)/(1 - p_o)] \quad (1)$$

Thus a success rate of $p = 0.60$–0.70 implies that local factors alone account for only about 15–30% of the total information required to make a perfect prediction. This agrees with estimates (Chan & Dill, 1990b) that the local contributions to the stability of a six-residue helix (based on a 1.05 equilibrium constant) are about 12% of the magnitude of the nonlocal steric free energy that drives helix formation, described in section 8.

(4) HYDROPHOBIC EFFECT

Following the elimination of electrostatics as a principal force of folding by Jacobsen and Linderstrom-Lang (1949), it was suggested that protein folding was driven by the aversion for water of the nonpolar residues (Linderstrom-Lang, 1952; Lumry & Eyring, 1954; Kauzmann, 1954). The same aversion was known to drive micelle formation; it was then assumed to be due to van der Waals interactions (Debye, 1949). In two remarkably insightful papers, Kauzmann (1954, 1959) made the first strong case for the importance of the hydrophobic interaction in protein folding. He reasoned that the formation of one hydrophobic "bond" (which he called an "antihydrogen bond") upon folding involves the gain of a *full* hydrogen bond among water molecules, which should be more important by an order of magnitude than simply a *change of strength* of a hydrogen bond upon folding if hydrogen bonding were the dominant folding force. In support of this view, Kauzmann offered the following evidence. First, nonpolar solvents denature proteins (Singer, 1962; von Hippel & Schleich, 1969a). According to a hydrophobic mechanism, the nonpolar solvent reduces the free energy of the unfolded state by solvating the exposed nonpolar amino acids. Second, experiments of Christensen (1952) had shown an unusual temperature dependence in which stability not only decreases at high temperatures but also decreases at low temperatures. Kauzmann observed that this cold destabilization resembles nonpolar solvation: nonpolar solutes become more soluble in water at low temperatures (Privalov & Gill, 1988).

A considerable body of more recent evidence continues to support the view that hydrophobicity is the dominant force of folding. First, spectroscopic and high-resolution differential scanning calorimetry experiments show the resemblance of the temperature dependence of the free energy of folding and the temperature dependence of the free energy of transfer of nonpolar model compounds from water into nonpolar media (Pace, 1975; Privalov, 1979; Privalov & Gill, 1988). Both involve large decreases in heat capacity. Second, a large number of crystal structures of proteins have become available since Kauzmann's predictions. They show that a predominant feature of globular protein structures is that the nonpolar residues are sequestered into a core where they largely avoid contact with water (Perutz et al., 1965; Chothia, 1974, 1976; Wertz & Scheraga, 1978; Meirovitch & Scheraga, 1980; Guy, 1985). Third, protein stability is affected by different salt species (particularly at high salt concentrations) in the same rank order as the lyotropic (Hofmeister) series (von Hippel & Schleich, 1969a,b; Arakawa & Timasheff, 1984); this is generally taken as empirical evidence for hydrophobic interactions (von Hippel & Schleich, 1969; Collins & Washabaugh, 1985; Morrison, 1952; Morrison & Billet, 1952). From the most stabilizing (for folded ribonuclease) to the most destabilizing, the rank order of anions is found to be SO_4^{2-}, CH_3COO^-, Cl^-, Br^-, ClO_4^-, CNS^- and the rank order of cations is NH_4^+, K^+, Na^+, Li^+, Ca^{2+}. The solubilities of benzene and acetyltetraglycyl ethyl ester in aqueous salt solutions increase in the same rank order. Fourth, accessible surface studies (Richards, 1977) and site-directed mutagenesis

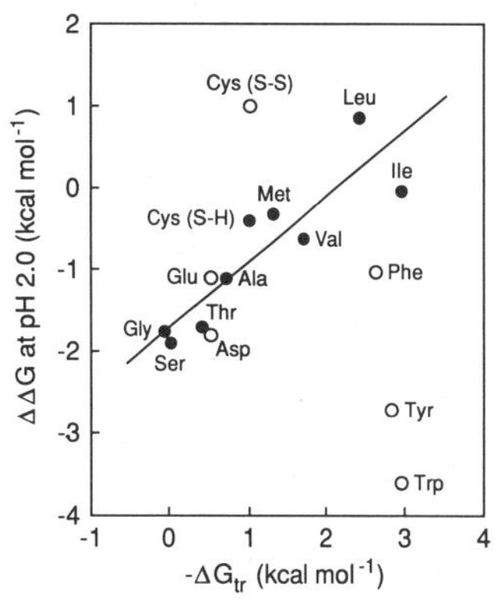

FIGURE 4: Change in free energy of unfolding, $\Delta\Delta G$, of mutant T4 lysozymes at position 3 (wild type is Ile) by substitution of other residues, compared to the corresponding free energy of transfer from water to ethanol, ΔG_{tr}. Reprinted with permission from Matsumura, M., Becktel, W. J., & Matthews, B. W. (1988) *Nature 334*, 406. Copyright (1988) Macmillan Magazines Limited.

experiments involving the replacement of a given residue by other amino acids show that the stability of the protein is proportional to the oil–water partitioning propensity of the amino acid (see Figure 4) (Yutani et al., 1984, 1987; Matsumura et al., 1988a,b; Kellis et al., 1989). Fifth, the hydrophobicities of residues in the cores of proteins appear to be more strongly conserved and correlated with structure than other types of interactions (Lim & Sauer, 1989; Bowie et al., 1990; Kelly & Holladay, 1987; Sweet & Eisenberg, 1983; Bashford et al., 1987). Sixth, computer simulations of incorrectly folded proteins show that the principal diagnostic of incorrect folding of proteins, apart from inappropriate burial of charge, is the interior/exterior distribution of hydrophobic residues (Novotny et al., 1984, 1988; Baumann et al., 1989).

What are "hydrophobic" interactions? There has been some disagreement about the meaning of hydrophobic (effect, force, interactions, etc.) (Hildebrand, 1968, 1979; Nemethy et al., 1968; Tanford, 1979; Ha et al., 1989). At least three different meanings of these terms have been used. (1) Hydrophobic has been used to refer to the transfer of a nonpolar solute to any aqueous solution. (2) Alternatively, it has been used more specifically to refer to transfers of nonpolar solutes into an aqueous solution only when a particular characteristic temperature dependence, described below, is observed. These two meanings describe only experimental observations and make no reference to any particular molecular interpretation. (3) "Hydrophobicity" has also been used to refer to particular molecular models, generally involving the ordering of water molecules around nonpolar solutes. In this review, hydrophobicity will be defined by (2) for reasons discussed below.

What is unusual about the temperature dependence of the hydrophobic interaction? First it is useful to describe "normal" solutions. There are two driving forces relevant to the mixing of simple solutions of A with B, of spherical particles governed by dispersion forces. The tendency to mix is *driven* by an increase in the translational entropy since there are more distinguishable spatial arrangements of the A and B molecules in the mixed system than of the individual pure systems. On the other hand, mixing in simple systems is *opposed* by the enthalpies of interaction; ordinarily, dispersion forces leading to AB attractions are smaller than those leading to the corresponding AA and BB attractions. The latter is captured in the general rule that "like dissolves like" (Hildebrand & Scott, 1950). For simple solutions, the transfer of B from pure B into A is therefore generally opposed by enthalpic interactions. When these interactions are strong, i.e., when A and B are relatively insoluble in each other, then the free energy of transfer is dominated by this opposing enthalpy, which is much larger than the mixing entropy, and transfer is disfavored. This is the ordinary form of incompatibility of two components A and B.

Oil and water are also incompatible at 25 °C. Oil/water incompatibilities tend to be stronger than "normal" incompatibilities. However, the free energy alone (i.e., the solubility or the partition coefficient) is not the principal distinction of normal from hydrophobic processes; in both cases the transfer can be disfavored. The distinction between normal and hydrophobic processes is in the temperature dependence. What was first recognized as unusual about nonpolar transfers to water at 25 °C (Edsall, 1935; Butler, 1937; Frank & Evans, 1945) was that they are not principally opposed by the enthalpy; they are principally opposed by an *excess* (i.e., "unitary") (Gurney, 1953; Tanford, 1970) entropy (Gill & Wadsö, 1976; Tanford, 1980; Privalov & Gill, 1988). The excess entropy is that which remains after the mixing entropy (the $RT \ln x$ term in the chemical potential) is subtracted from the total measured entropy. The enthalpy of mixing oil and water is generally small, sometimes even negative (favorable), at 25 °C (Privalov & Gill, 1988). Because these conclusions derive from experiments with solutes at high dilution, they imply that the excess entropy must arise from the water solvation around the solute rather than from some possible solute–solute interaction. A molecular interpretation of these data, which is supported by computer simulations (Geiger et al., 1979; Pangali et al., 1982; Ravishanker et al., 1982), is that at 25 °C nonpolar solutes are surrounded by ordered waters. Waters surrounding the nonpolar solute prefer to hydrogen bond with other waters rather than to "waste" hydrogen bonds by pointing them toward the nonpolar species (see top of Figure 6) (Stillinger, 1980; Geiger et al., 1979).

However, the aversion of nonpolar solutes for water becomes more ordinary, and less entropy driven, at higher temperatures. This is because there is a second fundamental difference between simple incompatibility on the one hand and oil/water incompatibility on the other. For simple solutions, the heat capacity change upon transfer is small. For nonpolar solutes, the heat capacity change upon transfer from the pure liquid to water is large and positive (Frank & Evans, 1945; Christian & Tucker, 1982; Privalov & Gill, 1988; Edsall & McKenzie, 1983). This means that for simple solutions the transfer of A into B is characterized by an enthalpy and entropy which are temperature independent and a free energy which is constant or linear with temperature. However, the situation is quite different for the transfer of nonpolar solutes into water. A large heat capacity implies the enthalpy and entropy are strong functions of temperature, and the free energy vs temperature is a curved function, increasing at low temperatures and decreasing at higher temperatures. Hence, there will be a temperature at which the solubility of nonpolar species in water is a minimum (Crovetto et al., 1982; Becktel & Schellman, 1987; Edsall & McKenzie, 1983; Privalov & Gill, 1988) (see Figure 5). A striking consequence follows, one which is at variance with the view that water ordering is the principal feature of the aversion of nonpolar residues for water. The free energy of transferring nonpolar solutes into water is extrapolated to be most positive in the temperature range 130–160 °C (Privalov & Gill, 1988). Therefore, the aversion of nonpolar species for water, whatever its molecular nature, is greatest at these high temperatures. Because this maximum aversion, by definition, arises where the free energy of transfer is a maximum, and thus where the entropy (the temperature

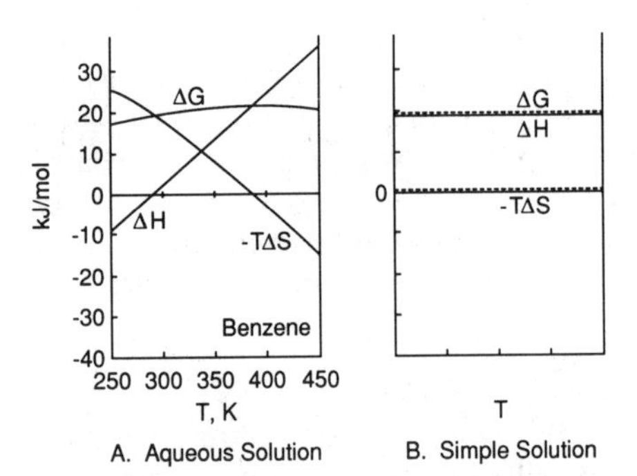

FIGURE 5: Comparison of the enthalpy (ΔH), entropy (ΔS), and free energy (ΔG) of solute transfer from the pure liquid into a hypothetical regular solution and into an aqueous solution. The data in (A) are for benzene, from Privalov and Gill (1988); the entropy at high temperature is extrapolated on the basis of assumed constant heat capacity. The figure on the right represents an idealization according to regular solution theory. Reprinted with permission from Privalov, P. L., & Gill, S. J. (1988) *Adv. Protein Chem. 39*, 191. Copyright (1988) Academic Press.

derivative of the free energy) equals zero, then the maximum aversion of nonpolar solutes for water must be driven by enthalpy (Privalov & Gill, 1988; Baldwin, 1986). In other words, at the temperature for which hydrophobicity is strongest, the entropy of transfer is zero! At those temperatures, the aversion of nonpolar solutes for water is enthalpic, as in simple classical solutions. It is therefore inappropriate to refer to nonpolar solvation processes in water, with large heat capacity changes, as "entropy driven" or "enthalpy driven", since either description is only accurate within a given temperature range. At 25 °C, the hydrophobic effect is entropic; at 140 °C, it is extrapolated to be enthalpic (Privalov & Gill, 1988).

For processes of constant ΔC_p, the enthalpy and entropy of transfer are

$$\Delta H(T) = \Delta H(T_1) + \int_{T_1}^{T} \Delta C_p \, dT = \Delta H(T_1) + \Delta C_p (T - T_1) \quad (2)$$

$$\Delta S(T) = \Delta S(T_2) + \int_{T_2}^{T} \Delta C_p / T \, dT = \Delta S(T_2) + \Delta C_p \ln (T/T_2) \quad (3)$$

and therefore the free energy of transfer is

$$\Delta G(T) = \Delta H - T\Delta S = \Delta H(T_1) - T\Delta S(T_2) + \Delta C_p[(T - T_1) - T \ln (T/T_2)] \quad (4)$$

These quantities are defined in terms of two arbitrary reference temperatures: T_1, for which the enthalpy is known, and T_2, for which the entropy is known. There are three alternative ways to express this free energy in terms of two particularly convenient reference temperatures, T_h, the temperature at which the enthalpy is zero, and T_s, the temperature for which the entropy is zero:

$$T_1 = T_h \qquad T_2 = T_s$$

$$\Delta G(T) = \Delta C_p[(T - T_h) - T \ln (T/T_s)] \quad (5)$$

$$T_1 = T_2 = T_s$$

$$\Delta G(T) = \Delta H(T_s) + \Delta C_p[(T - T_s) - T \ln (T/T_s)] \quad (6)$$

$$T_1 = T_2 = T_h$$

$$\frac{\Delta G(T)}{RT} = -\frac{\Delta S(T_h)}{R} + \frac{\Delta C_p}{R}[(1 - T_h/T) - \ln (T/T_h)] \quad (7)$$

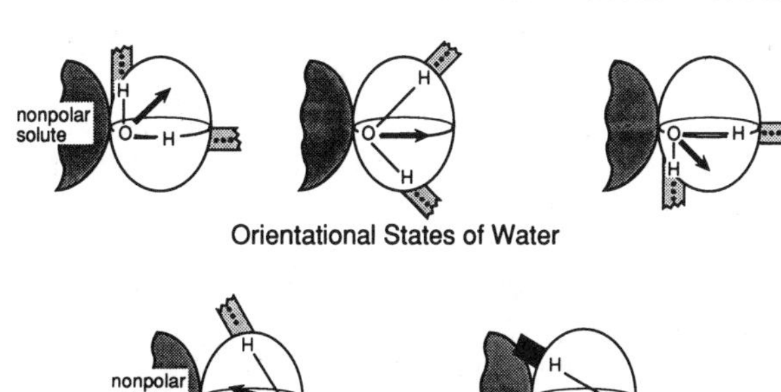

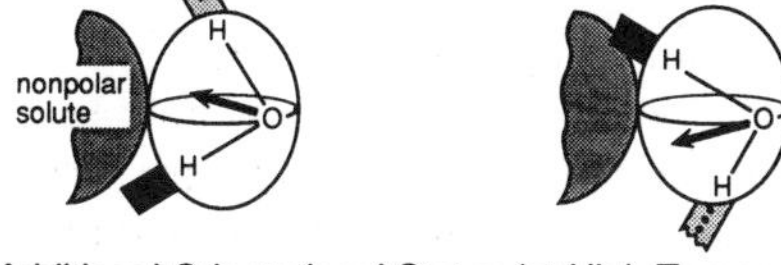

FIGURE 6: "Iceberg" model for the large heat capacity of transfer of nonpolar solutes into water (Frank & Evans, 1945; Gill et al., 1985). At low temperatures (near room temperature for benzene, for example), the water molecules surrounding the nonpolar solute adopt only a few orientations (low entropy), to avoid "wasting" hydrogen bonds; thus, all water configurations are fully hydrogen bonded (low energy). At higher temperatures, more conformations are accessible (higher entropy), but some of them have weaker or unformed hydrogen bonds and/or van der Waals interactions (higher energy). This contributes to the heat capacity because the system energy increases with temperature.

These three expressions have identical content. Which form is used depends on which set of parameters is most convenient: (ΔC_p, T_s, T_h), (ΔC_p, T_s, $\Delta H(T_s)$), or (ΔC_p, T_h, $\Delta S(T_h)$). Ordinary thermodynamic convention is to choose a single reference temperature, rather than two, but the mixed expression is included here because it has been used for protein folding (see below). Equations 5–7 predict the type of temperature dependence shown in Figure 5A when ΔC_p is large or in Figure 5B when $\Delta C_p = 0$.

What is the molecular basis for the large heat capacity of transfer of nonpolar solutes into water? Two observations have contributed significantly to a molecular picture. First, the entropy and heat capacity of transfer are linearly proportional to the surface area of the nonpolar solute (Miller & Hildebrand, 1968; Gill et al., 1985; Jorgensen et al., 1985; Jolicoeur et al., 1986). Second, the large heat capacity of transfer for the simplest solutes (Gill et al., 1985) decreases slightly with increasing temperature. This leads to the view (Frank & Evans, 1945; Nemethy & Scheraga, 1962; Gill et al., 1985; Muller, 1990) that the organization of water molecules in the first shell surrounding the solute is like an "iceberg", a clathrate, or a "flickering cluster"; see Figure 6. At room temperature, the water molecules surrounding the nonpolar solute principally populate a low-energy, low-entropy state: the waters are ordered so as to form good water–water hydrogen bonds. With increasing temperature, the waters surronding the solute increasingly populate a higher energy, higher entropy state: they are less ordered and have weakened attractions. Hence, increased temperature causes "melting" of the surrounding water structure, insofar as the entropy and energy are increased. [This melting process is probably better represented as bent hydrogen bonds than broken bonds, since the bending energy is much smaller than the breaking energy (Lennard-Jones & Pople, 1951).] The reason this results in a large heat capacity is that the two different energetic states of water provide an energy storage mechanism. The higher energy state becomes more populated with temperature. The reason this heat capacity is so large per solute molecule is because each solute molecule is surrounded by a large number (more than 10) of first-shell water molecules, each of which can participate in this energy storage mechanism. Not yet known is the detailed breakdown of the nature of these energies, although they undoubtedly include some combination of solute–water and water–water hydrogen bonding and van der Waals and dipolar electrostatic interactions.

What is the molecular interpretation of T_s and T_h? Equations 2–7 each have temperature-dependent and temperature-independent terms. The slope of the temperature dependence in eqs 5–7 is given by ΔC_p. T_s and T_h can be interpreted as representing a reference enthalpy or entropy at some given temperature. T_s and T_h are diagnostic for liquid-state nonpolar transfer processes (Baldwin, 1986). Sturtevant (1977) observed that several different biomolecular processes at 25 °C have nearly identical values of the ratio $\Delta S/\Delta C_p$. Baldwin (1986) showed that the constancy of this ratio, taken together with $T_2 = T_s$ and $T = 298$ K, substituted into eq 3 implies that these various processes can all be characterized by a single temperature, $T_s = 114$ °C. Using the data of Gill and Wadsö (1976) for ΔS and ΔC_p for the transfer of liquid nonpolar compounds to water at different temperatures, T, substituted into eq 3, Baldwin again found a single characteristic temperature, $T_s = 112.8$ °C, implying that Sturtevant's biomolecular processes resemble nonpolar solvation. Baldwin (1986) preferred the use of eq 5. For eqs 5–7, T_s and T_h are convenient reference quantities because ΔH, ΔS, and ΔC_p depend linearly on solute surface area, so ratios such as $\Delta S/C_p$, and hence T_s, should be independent of solute size.

On the other hand, Murphy, Gill, and Privalov (Murphy et al., 1990; Privalov & Gill, 1988) have preferred the convention given by eq 6. They assumed that at temperature T_s nonpolar solvation is identical with classical solvation and that $\Delta H(T_s)$ is due only to van der Waals forces. They refer to the factors other than $\Delta H(T_s)$ in eq 6 as the "hydration effect". They note that $\Delta H(T_s)$ is a positive enthalpy disfavoring transfer and that the hydration term is always negative (or zero at $T = T_s$). It follows that the hydration effect *favors* nonpolar transfers into water. However, it is not clear that this separation into these molecular factors is warranted. At $T = T_s$, nonpolar solvation is not identical with simple solvation. Even though the entropy of transfer may extrapolate to zero at $T = T_s$, the heat capacity remains large (Muller, 1990). The heat capacity is probably a more fundamental characteristic of nonpolar solvation than the entropy, because the entropy depends on the choice of concentration units, as do the free energy and partition coefficient, whereas that heat capacity and enthalpy do not. In addition, these thermodynamic models predict that for $T > T_s$ the entropy of transfer becomes positive, leading to the questionable prediction that the entropy would then *favor* solvation at high temperatures (see Figure 5A for benzene above 400 K). Also, since water hydrogen bonding persists to beyond 500 K (Crovetto et al., 1982; Franks, 1983), $\Delta H(T_s)$ probably includes water–water hydrogen bonding in addition to van der Waals interactions. Thus, it is not clear that this thermodynamic separation of terms corresponds to a simple molecular picture. Equation 7 provides yet a different view. Whereas ΔG for transferring benzene to water is most positive at $T = T_s$ (near 100 °C), $\Delta G/RT$ is most positive at $T = T_h$ (near room temperature). $\Delta G/RT$ corresponds directly to a solubility, partition coefficient, or Boltzmann population. Benzene is least soluble in water around room temperature. In eq 7, $-\Delta S(T_h)/R$, representing water ordering, is the most positive contribution to $\Delta G/RT$. The remaining terms, whatever their molecular interpretation, favor solvation. Thus at present, T_h and T_s serve to identify nonpolar transfer processes, but their breakdown into molecular components awaits further theory and experiments.

The thermal unfolding of proteins shows important similarities and important differences in comparison with the solvation processes of small nonpolar solutes. The similarity is that protein unfolding involves a large increase in heat capacity,[2] characteristic of nonpolar exposure (Pace, 1975; Privalov, 1979; Baldwin, 1986; Becktel & Schellman, 1987; Schellman, 1987; Privalov & Gill, 1988; Ooi & Oobatake, 1988). Therefore, the enthalpy and entropy of folding strongly decrease with temperature, and the free energy is curved: maximum protein stabilities are in the range of 0–30 °C (see Figure 8). The heat capacity of protein unfolding is itself approximately independent of temperature, although it appears to decrease somewhat at higher temperatures (Becktel & Schellman, 1987; Gill & Privalov, 1988).

The thermal unfolding of proteins differs from nonpolar solvation in the particular values of enthalpy and entropy at any given temperature. For example, as noted above, at 25 °C, the transfer of small nonpolar compounds into water has approximately zero enthalpy change and a large negative excess entropy change. In contrast, at 25 °C, protein unfolding has a positive enthalpy change (except for myoglobin for which the enthalpy is about zero) and a small or positive excess entropy (Baldwin, 1986; Privalov & Gill, 1988). There is an additional positive entropy and enthalpy of unfolding, in excess of that predicted from nonpolar solvation experiments.

What is the origin of the residual enthalpy and entropy of folding? As noted in section 2, the residual enthalpy of unfolding becomes more positive with increased content of polar residues (Privalov, 1979; Privalov & Gill, 1988; Dill et al., 1989). If it is assumed to be approximately temperature independent (Baldwin, 1986), then this result is expected from classical polymer solution behavior wherein chain monomers of type A have normal dispersion-force-driven incompatibilities with solvent B (Flory, 1953). This enthalpy may arise from van der Waals or hydrogen-bonding interactions among the backbone or polar residues. However, thermodynamic models of protein unfolding have generally been based on the assumption that the large heat capacity change is fully attributable to nonpolar solvation. This is open to question. Hydrogen bonding to water weakens with temperature, according to valerolactam dimerization and thymine dissolution experiments (Alvarez & Biltonen, 1973); this would also contribute to a change in heat capacity. Also, increased temperature causes the unfolded states of proteins to expand, reducing nonpolar solvation and contributing additional temperature dependence to the enthalpy and entropy of folding (Dill et al., 1989).

The large residual entropy represents an increased disordering upon unfolding, relative to that expected for nonpolar solvation. In sections 5 and 6, it is suggested that at least a large component of this is due to a difference in the freedom of configurations of the chain backbone, more severely restricted by steric constraints in the folded than in the unfolded states. However, there may also be an *extra* residual entropy due to additional configurational restrictions of the side chains in the folded state. The side chains may be partially frozen if the folded protein resembles a solid-like state (Shakhnovich & Finkelstein, 1989a,b). The residual entropy of protein denaturation is $\Delta S_r(112\ °C) = 18$ J/(K·mol) (Baldwin, 1986; Privalov, 1979; Murphy et al., 1990), similar to that of the dissociation of solid diketopiperazines, $\Delta S_r(112\ °C) = 16$ J/(K·mol), and considerably different than that of liquid hydrocarbon dissolution for which $\Delta S_r(112\ °C) = -0.5$ J/(K·mol) (Murphy et al., 1990). The heat capacity change increment upon dissolution of the solid diketopiperazines is the same as for the liquid-state transfer process (Murphy & Gill, 1989a,b). From those thermodynamic experiments, however, it is not possible to determine how much of the entropy difference originates with the chain expansion and solvation and how much originates from side-chain unfreezing. Bendzko et al. (1986) have suggested that if side-chain freezing is important, then protein denaturation should lead to an increase in partial molar volume of the protein; instead, they and others (see section 5) find a decreased volume upon unfolding. Another test of side-chain restrictions in the folded core is to compare crystal structure distributions of side-chain rotamers (Ponder & Richards, 1987) with computer simulations of the mean position and fluctuations of side chains that are attached to spatially unconstrained backbones. Such comparisons (Janin et al., 1978; Piela et al., 1987) show that equilibrium side-chain positions are predicted relatively well by the unconstrained simulations, but they are constrained somewhat differently in different secondary structures (Piela et al., 1987; McGregor et al., 1987).

Another class of experiment bears on the issue of the solid-like vs liquid-like nature of the protein core. Whereas the core is solid-like in its density and compressibility, it may behave differently insofar as the transfer process is concerned. The experiments involve multiple amino acid substitutions at a given site and the measurement of the change in protein stability, $\Delta\Delta G$, due to each of the replacements (Yutani et al., 1984, 1987; Matsumura et al., 1988a,b; Kellis et al., 1989; Sandberg & Terwilliger, 1989). The slope of $\Delta\Delta G$ vs the free energy of transfer ΔG_{tr} for the corresponding amino acid from water to oil (see Figure 4), provides information about similarities and differences of the protein folding process relative to the simpler process of amino acid transfer from water into liquid oil. This slope depends on a combination of factors: (i) the "deformability" of the native-state cavity, the energetic and entropic constraints affecting the freedom of the cavity wall residues to move to accommodate the mutated amino acid; (ii) the interactions, w, of the residue with the cavity, including both its entropic restrictions (side-chain freezing) and the residue/cavity energetics; (iii) the degree to which the denatured-state environment of the specified residue resembles pure solvent; and (iv) the exposure of the residue in the native state. A slope of 1 is consistent with a process in which the amino acid is exposed in the unfolded state and is transferred into a native cavity that resembles the reference liquid oil. Simple solute transfer to a liquid involves (a) opening the cavity (unfavorable by approximately free energy $w/2$) and then (b) transferring the solute (favorable by a free energy, w). On the other hand, if a solute is transferred instead into an already opened cavity ("preformed"), then only (b) is involved (Kellis et al., 1989). Thus, the free energy of transfer into a simple liquid is only half that of transfer into a preformed cavity. Therefore, a slope of 2 is expected from these experiments if the cavity wall residues are so constrained that they do not move to accommodate the mutant residue. Protein cavities may differ. Some may have much "deformability", as in a liquid, and lead to slopes near 1. Others may have little deformability, if neighboring residues are constrained, and lead to slopes near 2. Cavity deformabilities may also depend on residue size; cavity wall residues may move to accommodate large occupants but not small ones, for example. Overall, slopes ranging from 1 to 2 are expected from factor (i). Slopes smaller than 1 imply that the residue may be exposed in the

[2] One interesting counterexample, however, is a highly stable phosphoglycerate kinase from a thermostable bacterium (Nojima et al., 1977) in which both the enthalpy and heat capacity of folding are small.

native state or may be significantly buried in the denatured state. Slopes considerably greater than 2 must arise at least partly from (ii). A negative slope could arise in principle if the residue is more exposed at the surface of the native structure than it is in the denatured state. A slope of approximately 1 is observed for residue 3 in T4 lysozyme (Matsumura, 1988a). A slope of approximately 2 is observed for residues 88 and 96 in barnase and for a pocket involving residues 35 and 47 of gene V protein of phage f1 (Kellis et al., 1989; Sandberg & Terwilliger, 1989). For two sites normally containing charged residues, Asp-80 in kanamycin nucleotidyl transferase (Matsumura et al., 1988b) and Glu-49 in Trp synthase (Yutani et al., 1984, 1987), the slopes vary with pH, reaching a maximum of about 3.8. These latter experiments are more complex insofar as the proteins have significant populations of intermediates (and do not have two-state transitions) and they have an electrostatic component for the transfer. Although it would appear that current results on noncharged residues can be explained largely by different cavity deformabilities (i), nothing rules out the possibility that the other factors (ii and iii) are important. This type of experiment has not yet established whether side-chain freezing is an important component of protein stability.

Which liquid "oil" best characterizes the native core? While there is some evidence that cyclohexane is good (Radzicka & Wolfenden, 1988), the best correlations of transfer studies with amino acid distributions in proteins appear to involve nonpolar hydrogen-bonding solvents including octanol, ethanol, and dioxane (Fauchere & Pliska, 1983; Nozaki & Tanford, 1971; Kyte & Doolittle, 1982; Rose et al., 1985a). It may be, therefore, that some fraction of the temperature-independent enthalpy, attributed to van der Waals and hydrogen-bonding interactions, is also present in these transfer experiments and that it contributes differently for different oils.

Finally, I return to the meaning of hydrophobicity. I believe the most useful, common (Tanford, 1980; Ha et al., 1989), and unambiguous meaning of this concept is simply in reference to nonpolar transfers from nonaqueous media into aqueous media: (i) that are strongly disfavored and (ii) whenever there is a large associated increase in heat capacity [definition (2) at the beginning of this section]. It was this remarkable feature of nonpolar solvation that was first identified as unusual (Edsall, 1935; Butler, 1937; Frank & Evans, 1945) and which merits special terminology. Definition (1), on the other hand, needs no special term because it otherwise describes ordinary solution processes. There are two problems with (3), hydrophobicity as water ordering or other molecular models: (i) the molecular mechanism is still not fully understood, and (ii) "water ordering" is an appropriate description of the entropic repulsion of nonpolar solutes near room temperature, but not over a broader temperature range. This would lead to unnecessary hairsplitting: benzene insolubility in water would be referred to as hydrophobic at 25 °C but not at 100 °C, where it is even more strongly expelled. This meaning is not subject to the Hildebrand objection (1968, 1979); he noted that hydrophobicity is not an enthalpic disaffinity of nonpolar solutes for water but instead is due to a water–water affinity. Also, because definition (2) describes hydrophobicity in terms of the full transfer process, represented by the total free energies in eqs 5–7 rather than by a particular term in those expressions, it is not subject to difficulties of molecular interpretations (Privalov et al., 1990; Dill, 1990).

(5) What Is Missing?

The dominant force of folding is only half the story. Nearly equal in magnitude is a large opposing force. The structures and stabilities of globular proteins result from the balance between driving and opposing forces. Only recently have the main opposing contributions become better understood. That hydrophobicity is not the whole story becomes immediately apparent from certain puzzles. First, Tanford (1962) and Brandts (1964a,b) showed that hydrophobicity alone would predict protein stability an order of magnitude greater than measured values. They estimated that the free energy of unfolding would be about 100–200 kcal/mol at 25 °C. This estimate is based on free energies of transfer of hydrophobic amino acids from water into ethanol or dioxane, representative of the folded core of the protein, multiplied by the number of nonpolar residues. However, free energies of unfolding are observed to be only about 5–20 kcal/mol (Pace, 1975; Privalov, 1979; Privalov & Gill, 1988). This implies that there must be a large force, of magnitude nearly equal to the hydrophobic driving force, which opposes folding. Second, while the temperature dependence of folding was an important clue for hydrophobicity, Tanford (1962) suggested that it also posed a paradox. If nonpolar components associate more strongly as the temperature is increased, then proteins should fold more tightly with increasing temperature. Just the opposite is observed above room temperature: increasing temperature unfolds proteins. Proteins are typically thermally unfolded in the range of 50–100 °C; the free energies of transfer of hydrocarbons into water have a maximum extrapolated to be in the range of 130–160 °C. Third, if hydrogen bonding is not a dominant force of folding, then what is the origin of the considerable amounts of internal architecture, of secondary and tertiary structures, in proteins? Through what forces do the amino acid sequences so uniquely determine the native structure? The hydrophobic effect would seem to be too nonspecific and an unnatural candidate as the origin of helices and sheets.

Fourth, the pressure dependence of protein stability does not resemble that of model hydrophobic compounds (Brandts et al., 1970; Zipp & Kauzmann, 1973; Kauzmann, 1987). For example, the partial molar volume of methane decreases from 60 to 37.3 mL/mol upon transfer from hexane to water (Masterson, 1954). The decreased volume arises because water molecules pack more efficiently surrounding a nonpolar solute molecule than in its absence. Since the unfolding of a protein leads to increased nonpolar exposure to water, then these model studies would suggest that the partial molar volume of a protein should decrease considerably upon unfolding due to similar contraction of solvating water molecules (by about 20 mL/mol multiplied by a number in the range of 10–40, representing the total hydrophobic exposure upon unfolding). While the volume change of protein unfolding is indeed generally observed to be negative, it is only in the range of −30 to −300 mL/mol (about 0.5% of the total volume) (Brandts et al., 1970; Zipp & Kauzmann, 1973; Edelhoch & Osborne, 1976; Richards, 1977), which is somewhat smaller than the methane model would suggest. Brandts et al. (1970) suggested that other simple factors would not account for this discrepancy; for example, if unfolding leads to exposure of charged groups, then the volume change upon unfolding would be predicted to be even more negative, increasing the discrepancy. There is an additional problem. For model compounds, increasing the pressure leads to more normal water solvation, so the volume change of transfer to water diminishes and ultimately becomes positive at about 1500–2000 atm. For proteins, on the other hand, the negative volume of unfolding does not change much with pressure. Two simple explanations, however, can account for these discrepancies between the model transfer experiments and protein unfolding. First, methane is a poor model amino acid. Better models include alcohols, ketones, and amides, which can form hydrogen bonds; these model compounds have much smaller negative volumes of transfer to water, more closely predictive of the protein experiments (Friedman & Scheraga, 1965; Hvidt, 1975). Second, the pressure dependence is at least qualitatively accounted for by recognizing (i) that model nonpolar solutes in water are less compressible than in the pure liquid (Brandts et al., 1970) and (ii) that the folded state of a protein is much less compressible than the reference liquid hydrocarbon to which it is generally compared. A folded protein is typically 10-fold less compressible than organic liquids (or about half as compressible as ice!) (Kundrot & Richards, 1987; Gavish et al., 1983; Eden et al., 1982; Klapper, 1971; Fahey et al., 1969). Due to (i), the volume change in small-molecule transfers should diminish with increasing pressure; due to (ii), the volume change in protein folding should diminish much less with increasing pressure than in the small-molecule reference experiment. Therefore, it is important to recognize that a protein is not just a sum of transfers of small-molecule model side chains. Proteins are polymers. It is described below how the chainlike nature of proteins and the resultant conformational freedom lead to a strong force that opposes folding.

(6) Principle Opposing Force Is Entropic

Since the 1930s it has been known that the main force opposing protein folding is entropic. Northrop (1932) was the first to observe a sharp thermal denaturation transition; the equilibrium constant depends strongly on temperature. Only more recently has the molecular basis for the opposing entropy become clear. Just as there are translational, rotational, and vibrational entropies of small molecules, depending on the relevant degrees of freedom, likewise there are different possible molecular origins of the entropy gain upon protein unfolding. For example, Mirsky and Pauling (1936) suggested that folding would be opposed by an entropy arising from the proper mating of specific ion pairs. As another example, the helix–coil transition theories (Schellman, 1958a; Zimm & Bragg, 1959; Poland & Scheraga, 1970) showed that local degrees of freedom could be an important source of entropy opposing helix formation.

However, it has long been known that polymers are also subject to another type of configurational entropy, one which is nonlocal (Flory, 1953; deGennes, 1979). It arises from "excluded volume", the impossibility that two chain segments can simultaneously occupy the same volume of space. A chain can occupy a large volume of space in any of a large number of different configurations. However, there are relatively few ways the chain can configure if it is constrained to occupy a small volume of space, simply due to severe steric constraints.

Excluded volume (steric constraints) will play a role in any process that involves a change in the spatial density of polymer segments. These include solution thermodynamic properties of chain molecules, particularly their dependence on concentration, including solubilization and phase behavior, colligative properties, expansion and shrinkage, and virial coefficients and their temperature dependences (Flory, 1953; Munk, 1990). Models for predicting these effects originated with the Flory–Huggins theory (Flory, 1953); others are based on scaling (deGennes, 1979) and renormalization group methods (Freed, 1987).

How can a local entropy be distinguished from a nonlocal entropy? By definition, the entropy is $S/Nk = -\sum_{i=1}^{s} p_i \ln p_i$, where the probabilities p_i describe the distribution of all states, $i = 1, 2, 3, ..., s$, accessible to a system, and N is the number of particles. The issue of local vs nonlocal is the question of what degrees of freedom change in the process of interest. Local entropies arise from the energies responsible for the conformations of connected residues:

$$p_i = z^{-1} \exp[-\epsilon(\phi_i,\psi_i,\chi_i,\chi_{i+1})/kT] \tag{8}$$

where z is the partition function and ϵ is the energy of the dipeptide (or tripeptide, etc.) as a function of ϕ, ψ, and χ angles. These resemble vibrational or internal rotational entropies in small molecules; their contribution to stability is probably small (Karplus et al., 1987). Local entropies are independent of global properties of the chain such as the radius of gyration, or internal/external distributions of nonpolar/polar residues. Local entropies underlie helix/coil processes. There is evidence that changes in local entropies can affect the stabilities of globular proteins. For example, Matthews et al. (1987) have shown that replacement of a glycine for a proline in T4 lysozyme decreases protein stability by increasing the local conformational freedom of one peptide bond.

On the other hand, nonlocal entropies depend on the relative numbers of chain configurations, $N(\rho)$, as a function of the chain segment density, ρ (i.e., the number of monomers divided by the volume occupied by the chain):

$$p(\rho) = \frac{N(\rho)}{\int N(\rho)\, d\rho} \tag{9}$$

Because the folding of a protein involves collapse of the chain from a large volume (the denatured state) to a small volume (the native state), it must lose a considerable amount of this nonlocal entropy in the process. Native states of globular proteins are extremely compact (Klapper, 1971; Richards, 1977). They have as little free volume as small-molecule crystals; they have compressibilities closer to those of glasses than to those of liquids; and their configurational freedom is as restricted as in glasses and crystals of polymers, as evidenced by the existence of virtually unique native configurations of proteins.

One test of the importance of nonlocal entropies in protein stability involves cross-linking the chain. Introducing a cross-link reduces the number of conformations accessible to the unfolded state, making it relatively less favorable, and in that way making the folded state relatively more stable (Kauzmann, 1959; Poland & Scheraga, 1965; Chan & Dill, 1989a, 1990a; Pace et al., 1988; Matsumura et al., 1989). It appears that the most effective way to stabilize a protein at present, with a single amino acid change, is to add a cross-link. Increases of 29 and 25 °C in denaturation temperature have been achieved in hen lysozyme and ribonuclease (Alber, 1989b). For the same reasons, a protein should also be stabilized if it is constrained to be adjacent to an inert surface or contained within a small pore (Dill & Alonso, 1988). Similarly, the restriction of configurations of the unfolded chain may also account for the observation that carbohydrates or other chains appended to proteins appear to stabilize them (Burteau et al., 1989).

(7) Modeling Protein Stability

More quantitative insights into the forces of protein folding can be obtained through theoretical models and experimental tests of them. In principle, if the hydrophobic interaction and conformational entropy are the dominant contributions, then the free energy of folding can simply by calculated as (i) the difference in nonpolar exposure in native and denatured states, multiplied by the free energy of transferring nonpolar surface, and (ii) the difference in conformational entropies. The free energy of folding has been calculated in various studies, on the basis of several of the following simplifying assumptions: (i) in the native state the hydrophobic residues are fully buried in a nonpolar core and (ii) in the unfolded state the hydrophobic residues are fully exposed to solvent. The hydrophobic contributions would then be the sum of the free energies of transfer of the nonpolar residues. If in addition (i) there is only one configuration in the folded state, (ii) the denatured state is an ideal random flight [as in a "theta" solvent (Flory, 1953)], and (iii) the entropy is local, then the difference in configurational entropies will be simply $nk \ln z$, where n is the number of rotational bonds and z is the number of accessible conformations per dipeptide bond. These approaches have not succeeded, perhaps for the following reasons. First, native states have nonpolar exposure. Methods now exist, however (Lee & Richards, 1971; Eisenberg & McLachlan, 1986), to determine the detailed surface exposure and the associated free energies, if the native structure is known. Second, full exposure of the unfolded state is a poor approximation. The ensemble of denatured configurations represents a complex polymeric state, with the following properties. (1) It is often dense, being only 1.3–2-fold greater in volume than the native state, rather than 10–100-fold as in the theta state (Privalov et al., 1986; Ptitsyn, 1987; Goto et al., 1990). (2) The net change in surface exposure to water upon unfolding can be as little as 14% of the maximum possible (Tanford, 1968; Tanford, 1970; Ahmad & Bigelow, 1986; Schrier & Schrier, 1976; M. Hurle, private communication). (3) The radius, and therefore the free energy, of denatured configurations depends on the composition and length of the chain (Tanford, 1968; Goto et al. 1990; Shortle et al., 1988) and on external thermodynamic conditions, including temperature, pH, salt, and denaturants (Privalov et al., 1986). (4) In some cases there are distinctly identifiable different nonnative states and configurational transitions between them (Evans et al., 1987; Goto & Fink, 1989, 1990; Goto et al., 1990; Shortle & Meeker, 1989). (5) Some nonnative states have much secondary structure (Ptitsyn, 1987; Baum et al., 1989; Shortle & Meeker, 1989; Shortle et al., 1988). This is not a class of configurations that can be represented as fully exposed and independent of the solvent and thermodynamic conditions. For these reasons it is also clear that the configurational entropy of folding cannot be due principally to local factors and is not a constant independent of chain and solvent properties. Because the unfolded state contributes to stability on the same footing as does the folded state, satisfactory predictions of stability will require satisfactory models of the unfolded states.

Arguably, the simplest model for polymer chain collapse that treats the chain conformations more realistically is that of homopolymers in poor solvents. The nonlocal conformational entropy favors the open configurations, but when the solvent becomes poor (incompatible), the polymer collapses because polymer–polymer contacts are more favorable than polymer–solvent contacts (Lifschitz, 1968; deGennes, 1975; Post & Zimm, 1979; Sanchez, 1979).

The collapse of a heteropolymer such as a protein is significantly different than that of a homopolymer. In the process of collapse to the folded state, the ensemble of configurations of the molecule must (i) decrease in radius to native compactness and (ii) configure so as to bury much nonpolar surface area (see Figure 7). Hence, there are two degrees of freedom: the radius (or segment density) and the "ordering", the degree of segregation of the nonpolar residues into a core. In contrast, for homopolymers the only degree of freedom is the density. If the free energy is known as a function of these two degrees

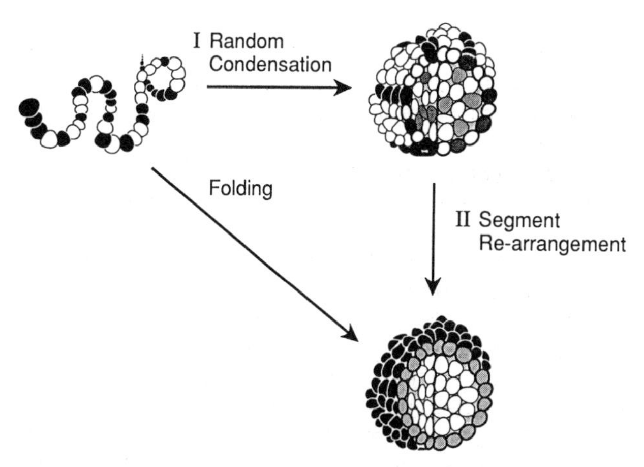

FIGURE 7: Protein folding involves (I) an increase in compactness of the chain and (II) a reconfiguration of nonpolar residues into the core and polar residues to the surface. The collapse of homopolymers only involves (I).

of freedom, then the state of minimum free energy can be found. Theory has recently been developed on this basis for the collapse of heteropolymers (Dill, 1985; Dill et al., 1989). The protein is represented as a chain of bead-like monomers connected by rotatable bonds. Two minima are generally found in this model. One minimum identifies a compact state with nonpolar core, the folded state, and the other minimum identifies a less compact ensemble (with no nonpolar core), the distribution of which changes with solution conditions; this represents the unfolded ensemble. Because there is a free energy barrier between these two states, the model predicts two-state behavior, wherein intermediate states are less populated than native or denatured states. This agrees with experiments, which show that many small single-domain proteins populate predominantly two states (Lumry et al., 1966; Privalov & Kechinashvili, 1974; Privalov, 1979). This approach addresses the problems above: the nature of the unfolded state and its nonpolar exposure are predicted from the balance of forces rather than assumed, and the conformational entropy difference between native and denatured states is calculated as a function of the chain and solution properties. Thus the theory simply provides a procedure for enumerating (i) the relative amount of nonpolar surface buried in the folded and unfolded states (some nonpolar surface is found to be exposed in the folded state, and some is buried in the unfolded state) and (ii) the number of configurations accessible in the folded and unfolded states. Three approximations are used for this counting (Dill, 1985; Dill et al., 1989). First, the Bragg–Williams mean-field approximation (Hill, 1960) is used to count nonpolar contacts by assuming they are uniformly distributed within the chain volume. Second, the Flory (1953) mean-field approximation is used to calculate how the number of conformations diminishes with density due to excluded volume; it too assumes segments are uniformly distributed. Third, only the effects of composition are taken into account (the number of nonpolar residues, and not their sequence along the chain): this is the random copolymer approximation. On this basis, the free energy of folding depends on (i) the chain length, (ii) the number of nonpolar residues, and (iii) the free energy of transferring an amino acid from water to a suitable "protein-like" medium.

The general predictions of the model are at least qualitatively consistent with known protein behavior. For example, the theory predicts that proteins should undergo a (maximally cooperative) first-order transition from the denatured to native state as the solvent becomes poor (Dill, 1985; Dill et al., 1989).

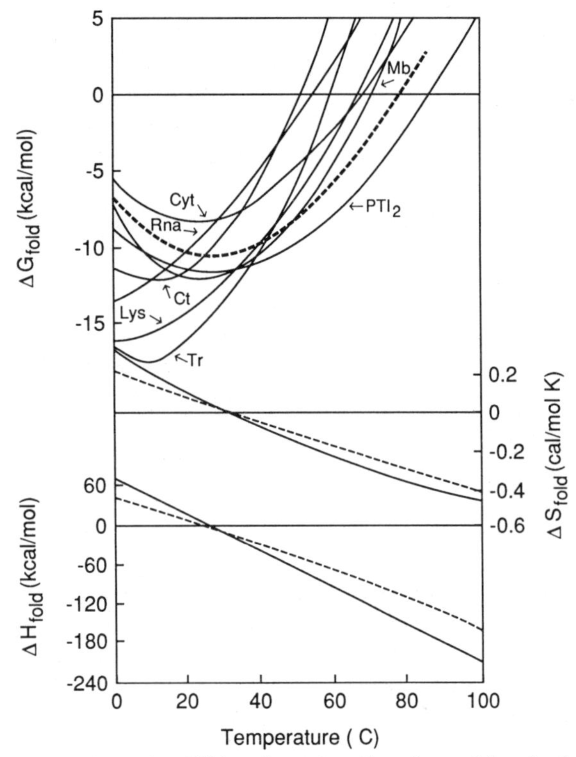

FIGURE 8: Thermal stabilities of proteins. Experimental data for free energies, enthalpies, and entropies of folding taken from Privalov (1979) and Privalov and Kechinashvili (1974) (—). Theoretical predictions are from Dill et al. (1989) (--).

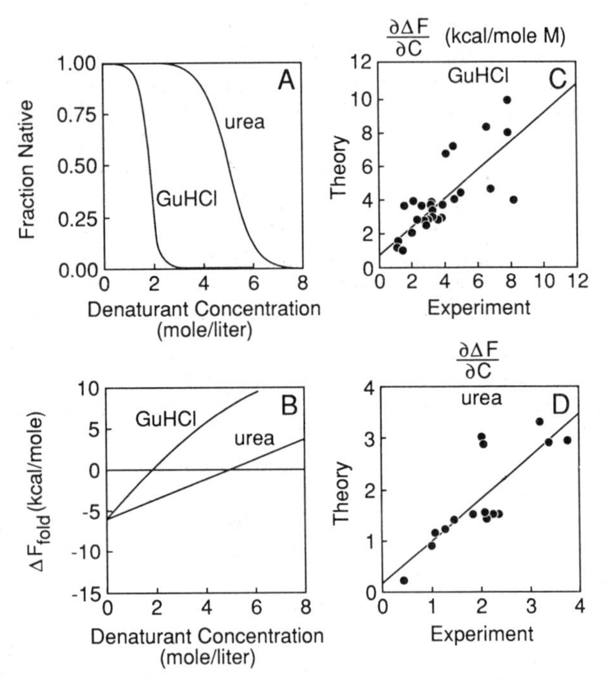

FIGURE 9: Comparison of theory and experiments for denaturation of proteins by guanidine hydrochloride and urea (Alonso and Dill, 1990). (A) Theoretical denaturation curve. (B) Free energy vs denaturant has approximately constant slope for small ranges of denaturant. (C and D) Comparison of theoretical and experimental slopes, $\partial \Delta F_{fold}/\partial c$, at the denaturation midpoint.

If there are too few hydrophobic residues, or if the chain is too short (less than several tens of residues), the molecule will not fold. The hydrophobic free energy is predicted to be about 60 kcal/mol of protein at 25 °C for typical small proteins (Dill et al., 1989). The opposing configurational free energy is predicted to be of nearly the same magnitude as the hydrophobic force: about 50 kcal/mol at 25 °C. This configurational entropy is large because there are tens of orders of magnitude fewer configurations in the folded than in the unfolded states (Chan & Dill, 1990a; Dill, 1985). Thus, the net stability of the protein should be a small difference of these two large free energies. This also resolves a related problem. The number of configurations of the unfolded protein is enormous. On that basis, it has been suggested that a protein would require many ages of the universe to find the native structure by random search, and thus that it could not attain its thermodynamic state of lowest free energy. This theory, on the other hand, predicts that simple collapse will reduce the configurational space enormously, implying that there is no inconsistency with the observation that proteins do attain their states of minimum free energy (Dill, 1985).

The theory also predicts the nature of thermal and solvent stabilities of proteins, in good agreement with experiments. To predict thermal stability, the temperature dependence of the hydrophobic interaction must be put into the model; this is taken (Dill et al., 1989) from the nonpolar small-molecule transfer experiments of Nozaki and Tanford (1971) and Gill and Wadsö (1976). The theory then predicts the existence of two first-order transitions vs temperature. One, a "cold" denaturation, occurs at low temperatures because the dominant temperature dependence of folding at those temperatures is due to the weakening of the hydrophobic interaction with decreasing temperature. Cold denaturation has been observed or has been predicted from extrapolations of experiments in several systems (Christensen, 1952; Hawley, 1971; Privalov et al., 1986; Chen & Schellman, 1989; Brandts, 1964a; Pace & Tanford, 1968; Privalov, 1979; Privalov & Gill, 1988; Griko et al., 1988). The other more familiar thermal denaturation at higher temperatures occurs because the gain in conformational freedom of the chain is more advantageous than the gain of interactions among nonpolar contacts. Comparison of the theoretical and experimental free energies, enthalpies, and entropies of folding is shown in Figure 8.

The theory has also been applied to the prediction of protein stabilities in denaturing solvents, such as urea and guanidine hydrochloride, and in stabilizing solvents (Alonso & Dill, 1990). In general the model can be applied to predicting stability as a function of any external parameter, x, provided that the oil/water partition coefficient for the representative elementary amino acid transfer is known as a function of x. In the case above, x is temperature; in this case, x is the concentration of urea or guanidine in the aqueous solvent. With use of the small-molecule transfer data of Nozaki and Tanford (1970) for this purpose, the resultant protein stability theory predictions are shown in Figure 9, along with experimental results. The theory predicts that the free energy of folding should be a linear function of urea concentration or a slightly curved function of guanidine concentration. These denaturants are predicted to cause unfolding by solvating nonpolar groups better than water does. Similarly, proteins should be stabilized by additives that are worse at solvating nonpolar groups than water is. Stabilizers include a wide range of sugars (Lee & Timasheff, 1981; Arakawa & Timasheff, 1982; Back et al., 1979), glycerol (Gekko & Timasheff, 1981), and polyols, poly(ethylene glycol), and some salts (Back et al., 1979; Arakawa et al., 1990).

(8) WHY DO PROTEINS HAVE INTERNAL ARCHITECTURE?

Globular proteins have internal organization, comprised of a combination of "irregular" structures and "regular" structures such as helices and sheets. This internal organization is uniquely specified by the amino acid sequence. This organization and its uniqueness would appear to be difficult to reconcile with the picture of the nonspecific dominant forces described in the preceding sections. Secondary structures are hydrogen bonded. Therefore, hydrogen bonds must naturally play a significant role in determining internal architecture. Yet according to the arguments above, hydrogen bonding, although prevalent in folded proteins, is a weak driving force. Moreover, hydrogen bonding taken alone cannot readily explain sequence specificity: how the sequence encodes only one native conformation. Most intrachain hydrogen bonds in globular proteins are among backbone C=O and NH groups rather than among side chains (Baker & Hubbard, 1984). From the data base of protein crystal structures, Baker and Hubbard observed that, of all the intrachain hydrogen bonds to C=O groups, 81.3% are with backbone NH groups and 18.7% are with side chains. Of all the intrachain hydrogen bonds to NH groups, 86.2% are with backbone C=O groups and 13.8% are with side chains. However, sequence specificity must arise from differences in side-chain properties, not from differences in backbone properties such as peptide hydrogen bonding. Any one amino acid will have essentially the same backbone interactions as any other. Therefore, if hydrogen bonding were dominant, the native structure of a protein should be essentially independent of the amino acid sequence, and native structures should be regular and periodic, either purely helix or purely sheet. Similarly, to the extent that backbone van der Waals interactions favor folding, they also are probably not very selective for one compact conformation relative to others.

Could it be that there is so much secondary structure in proteins because irregular conformations cannot form hydrogen bonds as well as helices and sheets can? The current limited evidence does not support this view. That the many irregular conformations of proteins can form good intrachain hydrogen bonds is clear from the study of Baker and Hubbard (1984). Of all the hydrogen bonds to C=O groups, 8.9% are in β-structures, 24.6% are in helices, and 5.3% are in turns. In comparison, a minimum of 18.3% of all hydrogen bonds to C=O groups are in "irregular" structures; i.e., they are intrachain (not bonded to water) and not in secondary structures. This is a minium because the Baker and Hubbard study does not itemize bonds to water and in secondary structures as mutually exclusive categories, so only this lower bound can be obtained from their study. Similarly for NH groups, 38.3% are hydrogen bonded in helices, 13.6% in β-structures, and 9% in turns. A minimum of 17.9% of all hydrogen bonds to NH groups are in irregular structures. Hence, hydrogen bonds in irregular structures are not significantly less common than those in helices and sheets. Combined with the observation that 11% of all C=O groups and 12% of all NH groups have no hydrogen bonds (Baker & Hubbard, 1984), this suggests that hydrogen-bonding requirements do not severely constrain the conformations accessible to the chain.

Indeed, the earliest expectations (Wu, 1929; Pauling et al., 1951; Pauling & Corey, 1951a–d; Kendrew et al., 1958; Kendrew, 1961) were that internal architecture in globular proteins would be regular and periodic as in crystals, presumably with all residues in helices or sheets. For example, Pauling et al. (1951) and Pauling and Corey (1951a–d) specifically sought types of structure in which every amino acid was "equivalent", i.e., interchangeable and *not* dependent on the sequence. Therefore, ever since the appearance of the first known structures of globular proteins (Kendrew et al., 1958), the central problem of protein architecture has not been to explain why proteins have so much regular structure. It has been just the opposite: to explain why there is so much irregular structure, and how the sequence uniquely encodes them both. About 53% of all residues in globular proteins are in irregular structures (Kabsch & Sander, 1983; Kneller et al., 1990). Is protein architecture then a consequence of the smaller interactions, including-side chain hydrogen bonding? In the following section, I suggest that sequence-specific internal architecture in globular proteins does not arise principally from these smaller forces but from the dominant forces.

Recent exhaustive simulations of all the possible conformations of chains have shown that protein-like internal architecture is simply a natural consequence of steric constraints in compact polymers (Chan & Dill, 1989a,b, 1990a,b). Any flexible polymer, when made to be compact by *any* driving force, will have much internal architecture composed of helices and sheets. For proteins this driving force is presumably the hydrophobic interaction. There are simply very few possible ways to configure a compact chain, and most of them involve helices and sheets. The existence of internal organization is due to the physical impossibility of steric violation which alternative configurations would require. In other words, consider a pearl necklace. Squeeze it into a ball. One might expect that an ensemble of these compact necklaces would be highly disordered. This is not the case, however; it will have approximately the same distribution of internal architectures as are observed in globular proteins, comprised of helices, sheets, and irregular structures. The evidence for this is the following.

Polymer simulations and theory show that if a chain molecule is to form a single self-contact, it will prefer one that forms the smallest possible loop (Jacobson & Stockmayer, 1950; Poland & Scheraga, 1965; Chan & Dill, 1989a, 1990a). A chain with a local (small) loop has more remaining accessible conformations (and thus greater entropy) than a chain with a nonlocal (large) loop. [Theory and experiment also show that "stiffness", i.e., intrinsic propensities, can further favor or disfavor tight loops (Flory, 1969; Semlyen, 1976; Zhang & Snyder, 1989).] Now consider configurations that have two self-contacts; these provide the most basic description of elements of secondary structure. A most interesting result from the theory and simulations is that if a chain is to form two self-contacts, it will again prefer to form a small loop and to form the second contact as close as possible in sequence to the first, simply because these, among all possible two-contact configurations, have the greatest entropy (Chan & Dill, 1989a, 1990a). The only two ways a chain molecule can form such a loop pair are either as a helix or as the beginning of an antiparallel sheet. Hence given only the conformational freedom and steric restriction in flexible chains, there is a tendency to form helices and sheets, even in the absence of other forces, and even for chain conformations of relatively large radius of gyration. These results are obtained for chains configured in two or three dimensions, on different types of lattices, and by alternative path integral methods and therefore do not appear to be an artifact of the theoretical methods used in these predictions.

Exhaustive simulations further show that as a chain becomes increasingly compact, it develops a considerable amount of secondary structure (Chan & Dill, 1989b) (see Figure 10). Consistent with this result, protein-like organization in compact chains has been observed by Monte Carlo methods for other lattices and chain lengths (Sikorski & Skolnick, 1989; Skolnick et al., 1988, 1989).

The internal distributions of secondary structures obtained in exhaustive simulations of all possible compact conformations (Chan & Dill, 1990b) are in good agreement with the distributions that Kabsch and Sander (1983) have observed in

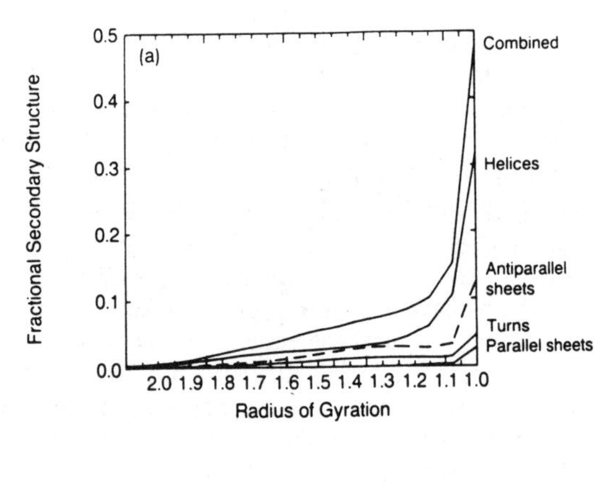

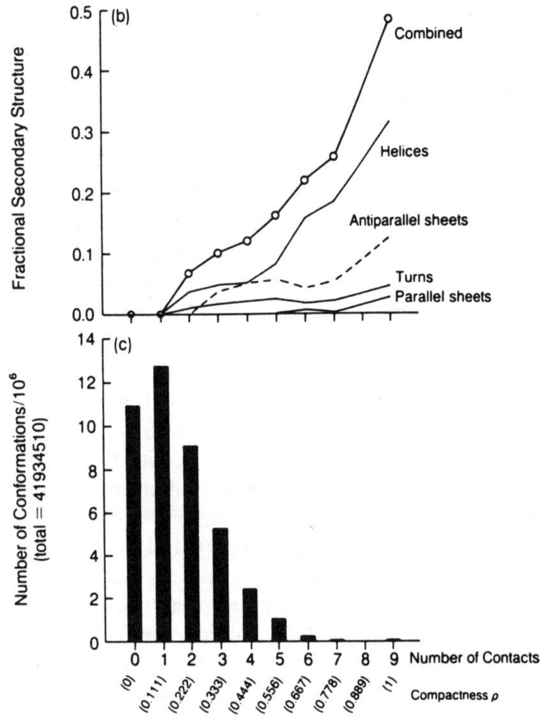

FIGURE 10: Increasing compactness of a chain molecule leads to the formation of secondary structure. With increasing compactness (number of intrachain contacts), exhaustive simulations of all possible configurations of short chains on lattices show that (Chan & Dill, 1989b, 1990b) (c) the number of accessible configurations diminishes rapidly and (b) and the amount of secondary structure increases sharply. (a) Same results as (b), except plotted against radius of gyration instead of compactness.

the data base of protein crystal structures (see Figure 11). For example, theory and experiment agree that the most common helices and sheets will be the shortest ones; that the shortest antiparallel sheets are 3–4-fold more common than the shortest helices, which are about as common as the shortest parallel sheets; and that, among long secondary structures, in decreasing prevalence should be helices, antiparallel sheets, and then parallel sheets. It is interesting then to turn this argument around. Since the known protein architectures are distributed in the same way as in the complete ensemble of all compact conformations, it suggests that the currently known proteins are a reasonably representative sample of all the forms of internal structure that proteins *could* adopt [see also Finkelstein and Ptitsyn (1987)].

These results provide an explanation for certain aspects of protein structure that have otherwise been puzzling. First, they show how the dominant forces due to hydrophobicity and steric constraints give rise to internal architectures. They provide a single framework for comprehending the coexistence of all

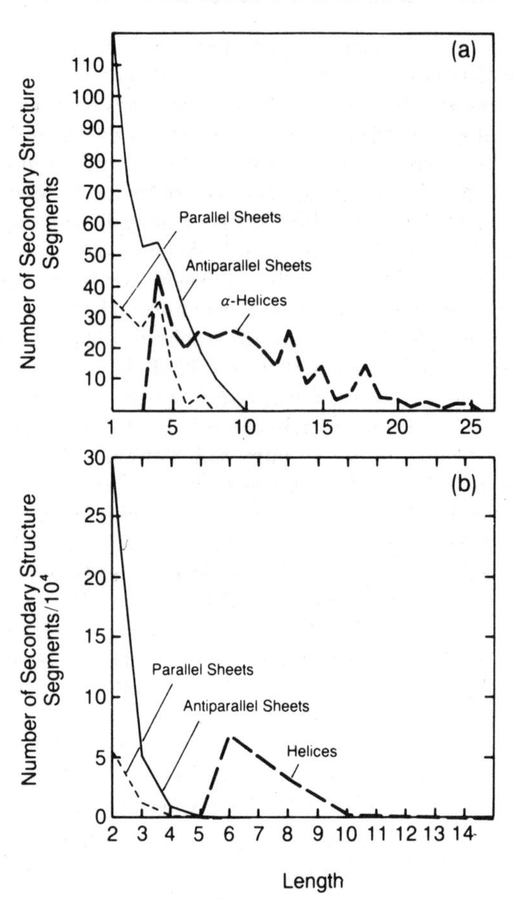

FIGURE 11: Comparison of data base studies of Kabsch and Sander (1983) (a), of the internal distributions of lengths of helices and sheets in globular proteins, with (b) exhaustive lattice simulations of the full compact ensemble (Chan & Dill, 1990b).

the types of internal protein structure, regular and irregular, helices and sheets. Past hypotheses have tended to address a single type of architecture, often principally helices, in which the focus is on factors that are local in the sequence. The present results suggest instead that architectures arise principally from nonlocal factors. That is, "tertiary" forces drive secondary structures rather than "secondary" forces driving tertiary structures. It follows that helices and sheets in globular proteins are only secondarily a consequence of hydrogen bonding. In this regard, it is interesting that of the 176 known crystal structures of different synthetic polymers that are considered to be reliable, 49 are planar zigzags, and 79 are helices of 22 different types (Tadokoro, 1979). Many different crystalline polymers have helical pitches nearly the same as that of the α-helix, between three and four monomers per turn, including polybutadiene, polypropylene, polyvinylnaphthalene, polybutene, and even fibrous sulfur, none of which form hydrogen bonds (Takodoro, 1979). This is not to say that local factors and hydrogen bonds are unimportant. This is only to say that local factors and hydrogen bonding probably contribute to making energetic "decisions" only within an already highly restricted ensemble, one in which the chain will be forced (by the hydrophobic drive to compactness and by the highly selective steric forces) to have considerable stretches of (i, i+3) self-contacts, for example. In that regard, what dictates that these will be specifically α-helices, with 3.6 residues per turn, instead of any of dozens of other types of closely related helices will be local forces uniquely determined by amino acids as monomers.

One aspect of internal architecture that is *not* a consequence of packing forces alone is the spatial distribution of turns. Turns are observed to occur largely at the surfaces of proteins

(Kuntz, 1972). This does not arise from packing constraints (Chan & Dill, 1989b). Rather it appears to be a consequence of the polar nature of turn residues (Rose et al., 1985b). Since the middle residues of turns are geometrically unable to form intrachain hydrogen bonds, then their hydrogen bonding needs can best be met by interacting with water at protein surfaces.

A second puzzle has been why some denatured states of proteins have secondary structure. According to these simulations, in the absence of other interactions, the amount of secondary structure simply depends on the radius of gyration of the chain (Chan & Dill, 1989b, 1990b) (see Figure 10). Therefore, proteins that are denatured only weakly, and thus have small radii, should have some secondary structure. This may account for secondary structures in "molten globule" and other compact denatured states of proteins, with radii only slightly larger than those of native molecules (Ptitsyn, 1987; Goto & Fink, 1989, 1990; Shortle & Meeker, 1989; Kuwajima, 1989; Brems & Havel, 1989; Baum et al., 1989).

Because internal architecture should therefore arise in any compact polymer, it should be possible to design other copolymers, not necessarily comprised of amino acids, which can be driven to compactness in poor solvents and which should then have protein-like architectures. Consistent with this prediction, Rao et al. (1974) have shown that in a solution containing a large number of different sequences of amino acid copolymers roughly half the molecules are highly compact and there is 46% helix observed by circular dichroism. This implies that a large fraction of all possible sequences is capable of coding for large amounts of helix.

(9) Why Is the Native Structure Unique?

A most unusual feature of globular proteins, relative to any other type of isolated polymer molecule, is that they can be found in only a single conformation, the native structure. ("Single" here means that the chain path is identifiable at relatively high resolution as an average over the ensemble of small fluctuations.) What forces, encoded in the amino acid sequence, cause this remarkable uniqueness? As noted above, there is considerable elimination of configurational possibilities simply due to the compactness. Nevertheless, even for compact chain molecules, there are still many accessible conformations; the number of maximally compact conformations increases exponentially with chain length (Chan & Dill, 1989b; Dill, 1985). Within this ensemble of physically accessible compact conformations, all the various types of interactions could play a role, including hydrogen bonding, intrinsic propensities, ion pairing, and hydrophobicity, in causing some compact conformations to be of higher free energy and others to be of lower free energy.

Is there a single one among these types of interaction, by virtue of its encoding within the sequence, that "picks out" the one native structure? The most significant further restrictor of conformational space may be the hydrophobic interaction. By use of a simple model of short self-avoiding flexible chains on lattices, in which the only energetic feature of the sequence is the hydrophobic interaction, every conformation has been explored by exhaustive search in order to determine the native state(s), those at the global minimum of free energy (Lau & Dill, 1989; Chan and Dill, unpublished results). This has been done for many different sequences. In this model, residues are either hydrophobic (H) or polar (P). The free energy of any chain conformation is determined simply by the number of HH topological contacts (topological contacts are defined in Figure 1). Therefore, "native" conformations are those with the maximum number of such HH contacts. How many native structures does any sequence have? Figure 13 shows the

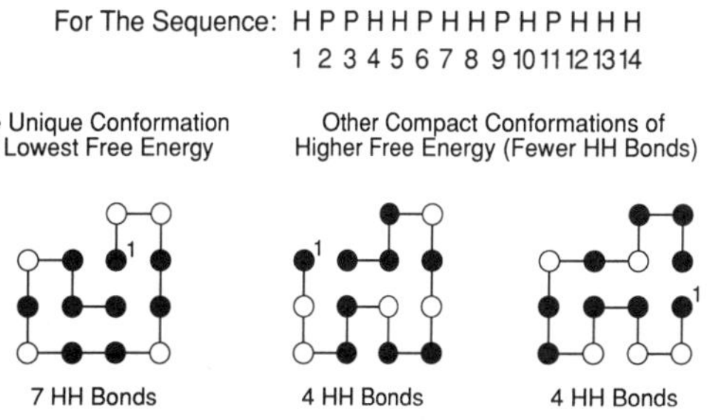

For The Sequence: H P P H H P H H P H P H H H
1 2 3 4 5 6 7 8 9 10 11 12 13 14

The Unique Conformation of Lowest Free Energy

Other Compact Conformations of Higher Free Energy (Fewer HH Bonds)

7 HH Bonds | 4 HH Bonds | 4 HH Bonds

FIGURE 12: Exhaustive lattice simulations of short chains show that any given sequence has many different compact conformations (such as the two on the right) but often only one native state, in which the maximum number of nonpolar contacts is formed. Nonpolar residues (H, ●); polar residues (P, O).

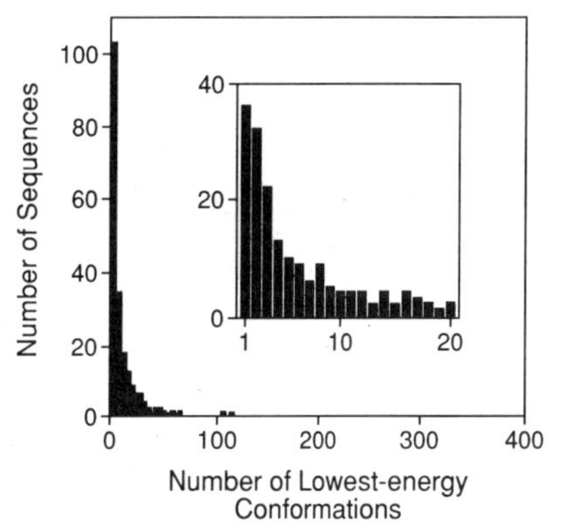

FIGURE 13: How many native structures does a given sequence have? This histogram derives from a 2D lattice model of self-avoiding flexible chains (of length = 24 monomers) subject only to hydrophobic interactions (Lau & Dill, 1990). Exhaustive search permits determination of the structure(s) of global minimum in free energy among the compact ensemble. These native structures are then found for many different sequences. The inset shows finer detail. This decreasing function implies that more sequences have only one native structure than have two, three, etc. Thus for a typical sequence, there are very few configurations that have the maximum possible number of nonpolar contacts.

distribution. The most surprising result is that this is a strongly decreasing function; far more sequences can configure into only one native structure than can configure into 10 or 100 native structures. Hence according to this simulation, for most folding sequences there are exceedingly few ways a chain can configure to form the maximum possible number of HH contacts. One example is shown in Figure 12. This suggests that hydrophobicity is strongly selective and singles out only a very small number of candidate native structures from the compact ensemble. Other evidence also supports this view. Hydrophobicity patterns appear to be more predictive of conformational families than other types of interactions (Sweet & Eisenberg, 1983; Bashford et al., 1987; Bowie et al., 1990). In addition, Covell and Jernigan (1990) have exhaustively explored all conformations of lattice chains confined within known protein shapes using 3D lattices, weighting them using the free energies of Miyazawa and Jernigan (1985), and have compared them with known protein structures. They found the native conformation to be within the best 1.8%. Hence within their shape-restricted ensemble, the transfer free energies, whatever their molecular bases in hydrophobicities, hydrogen bonding, van der Waals, and other interactions, considerably restrict the possible native structures. Thus, the nonlocal forces encoded within the sequence, rather than the local factors involving connected residues, would appear to be largely responsible for the uniqueness of the native structure.

Whatever role hydrophobicity may play in reducing the ensemble to a small set of possibilities, it is also clear that

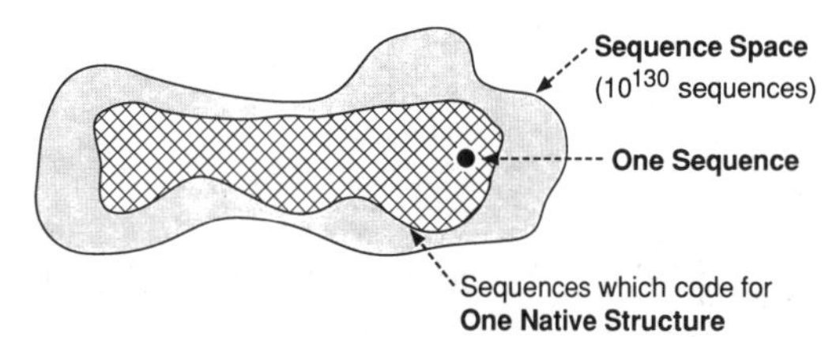

FIGURE 14: For a protein of 100 residues, there are $20^{100} \cong 10^{130}$ different sequences. This is the sequence space. A given protein (such as ribonuclease) has only one sequence. Hence, the probability of drawing the *sequence* of a particular protein by random selection is 10^{-130}. On the other hand, lattice simulations (Lau & Dill, 1990) show that the probability of drawing *any* sequence which will fold to a specified *structure* (the ribonuclease configuration, for example), is estimated to be about 120 orders of magnitude larger than this.

hydrophobicity cannot be *solely* responsible for determining the single native structure. The hydrophobic contact free energies of the native and the next higher free energy level will, in general, differ by no more than one or two nonpolar contacts, less than 2 kcal/mol. At this level of discrimination, the nondominant interactions are important. If they were not, then it would be impossible to account for the existence of hydrogen bonding, the partial successes of the intrinsic propensities, and the observed nonrandom distributions of charges, ion pairs, and aromatic groups (Burley & Petsko, 1985, 1988; Blundell et al., 1986). Thus, hydrophobicity may "select" a relatively small number of compact conformations, from which the native structure is determined by the balance of all types of interactions.

(10) Sequence Space and Origins of Proteins from Random Sequences

That hydrophobicity is the dominant folding force is also consistent with expectations for the nature of mutation-induced changes in proteins. (1) Replacements of hydrophobic residues in the core are more disruptive of stability than other types of substitutions (Perutz & Lehman, 1968; Reidhaar-Olson & Sauer, 1988; Alber et al., 1987b; Alber et al., 1988; Lim & Sauer, 1989; Lau & Dill, 1990). (2) Because hydrophobicity is relatively nonspecific and orientation independent, many different core replacements are tolerated, provided only that they are hydrophobic and not significantly different in size. There is even greater tolerance for replacements of surface residues (Perutz & Lehman, 1968; Lim & Sauer, 1989; Bowie et al., 1990; Alber et al., 1987b; Lau & Dill, 1990). (3) It follows that a large fraction of the molecules in "sequence space" (all the possible sequences) should be able to fold. Two-dimensional lattice simulations predict that more than 50% of all possible sequences will, under native conditions, fold to within 10% of the minimum possible radius of gyration and will therefore have considerable amounts of secondary structure (Lau & Dill, 1989; Chan and Dill, unpublished results). White and Jacobs (1990) have found remarkably little difference between hydrophobicity sequence patterns in more than 8000 known sequences on the one hand and random sequences on the other. (4) There should be an extremely large number of "convergent" sequences: i.e., a given native structure should be encodable in many different sequences (Lau & Dill, 1990; Chan and Dill, unpublished results). This implies a significant probability that a random sequence of amino acids will encode a globular conformation in general and a particular native structure in specific. For example, although there is only one *sequence* of ribonuclease, there should be more than 10^{100} different sequences which will all have the same native backbone *conformation* as ribonuclease (see Figure 14) (Lau & Dill, 1990). Indeed, there has been considerable success in designing proteins based on simple principles (Richardson & Richardson, 1987; DeGrado, 1988; DeGrado et al., 1989). The nonspecific nature of the dominant folding forces may therefore be essential in explaining how functional proteins could have originated from random sequences since only a negligible fraction of sequence space could have been sampled during the origins of life.

Conclusions

More than 30 years after Kauzmann's insightful hypothesis, there is now strong accumulated evidence that hydrophobicity is the dominant force of protein folding, provided that "hydrophobic" is operationally defined in terms of the transfer of nonpolar amino acids from water into a medium that is nonpolar and preferably capable of hydrogen bonding. Other forces are weaker but can affect stability. In acids and bases, electrostatic charge repulsions destabilize native proteins. Near neutral pH, ion pairing can stabilize proteins. There is evidence that hydrogen bonding or van der Waals interactions among polar amino acids may be important, but their magnitude remains poorly understood. An important contributor to protein structure and stability is the dominant opposing force, arising principally from the loss of nonlocal conformational entropy due to steric contraints in the folded state. The marginal stabilities of proteins arise from the small difference between these large driving and opposing forces. Hydrophobicity leads to compact conformations with nonpolar cores, but it is the steric constraints in compact chains that are largely responsible for their considerable internal architecture. The reason that only one native structure is encoded in the amino acid sequence may be largely attributable to the hydrophobic interaction; there are only a small number of ways to configure a chain to maximize the number of nonpolar contacts. These forces are of a nature such that proteins should be tolerant of amino acid substitution, a given native structure should be encodable in many different sequences, and a large fraction of all possible sequences should fold to compact structured native states.

Acknowledgments

I thank Tom Alber, Darwin Alonso, T. Arakawa, R. L. Baldwin, Rod Biltonen, Sarina Bromberg, Hue Sun Chan, Fred Cohen, Linda DeYoung, Gregg Fields, Rick Fine, Stan Gill, Barry Honig, Jack Kirsch, Peter Kollman, Tack Kuntz, Jim Lee, Kip Murphy, Nick Pace, George Rose, Jeff Skolnick, Dirk Stigter, Peter Wright, and Kai Yue for helpful discussions. I thank particularly Wayne Becktel for providing a very helpful list of early references and John Schellman for his unique and insightful perspective. Thanks to the NIH, the DARPA URI Program, and the Pew Scholars Foundation for support.

References

Acampora, G., & Hermans, J. (1967) *J. Am. Chem. Soc. 89*, 1543.

Ahmad, F., & Bigelow, C. C. (1986) *Biopolymers 25*, 1623.

Alber, T. (1989a) *Annu. Rev. Biochem. 58*, 765.

Alber, T. (1989b) in *Prediction of Protein Structure and the Principles of Protein Conformation* (Fasman, G. D., Ed.) p 161, Plenum, New York.

Alber, T., et al. (1987a) *Nature 330*, 41.

Alber, T., et al. (1987b) *Biochemistry 26*, 3754.

Alber, T., et al. (1988) *Science 239*, 631.

Allen, G., Watkinson, J. G., & Webb, K. H. (1966) *Spectrochim. Acta 22*, 807.

Alonso, D. O. V., & Dill, K. A. (1990) *Biochemistry* (submitted for publication).

Alvarez, J., & Biltonen, R. (1973) *Biopolymers 12*, 1815.

Anderson, D. E., Becktel, W. J., & Dahlquist, F. W. (1990) *Biochemistry 29*, 2403.

Anfinsen, C. B. (1973) *Science 181*, 223.

Anson, M. L. (1945) *Adv. Protein Chem. 2*, 361.

Anson, M. L., & Mirsky, A. E. (1925) *J. Gen. Physiol. 9*, 169.

Anson, M. L., & Mirsky, A. E. (1931) *J. Phys. Chem. 35*, 185.

Anson, M. L., & Mirsky, A. E. (1934) *J. Gen. Physiol. 17*, 393.

Anufrieva, E. V., et al. (1968) *J. Polym. Sci. 16*, 3533.

Arakawa, T., & Timasheff, S. N. (1982) *Biochemistry 21*, 6536.

Arakawa, T., & Timasheff, S. N. (1984) *Biochemistry 23*, 5912.

Arakawa, T., Bhat, R., & Timasheff, S. N. (1990) *Biochemistry 29*, 1924.

Argos, P. (1987) *J. Mol. Biol. 197*, 331.

Back, J. F., Oakenfull, D., & Smith, M. B. (1979) *Biochemistry 18*, 5191.

Baker, E. N., & Hubbard, R. E. (1984) *Prog. Biophys. Mol. Biol. 44*, 97.

Baldwin, R. L. (1986) *Proc. Natl. Acad. Sci. U.S.A. 83*, 8069.

Barlow, D. J., & Thornton, J. M. (1983) *J. Mol. Biol. 168*, 867.

Bartlett, P. A., & Marlowe, C. K. (1987) *Science 235*, 569.

Bashford, D., Chothia, C., & Lesk, A. M. (1987) *J. Mol. Biol. 196*, 199.

Baum, J., et al. (1989) *Biochemistry 28*, 7.

Baumann, G., Frommel, C., & Sander, C. (1989) *Protein Eng. 2*, 329.

Becktel, W., & Schellman, J. (1987) *Biopolymers 26*, 1859.

Bendzko, P. I., Pfeil, W. A., Privalov, P. L., & Tiktopulo, E. I. (1988) *Biophys. Chem. 29*, 301.

Blagdon, D. E., & Goodman, M. (1975) *Biopolymers 14*, 241.

Blundell, T., et al. (1986) *Science 234*, 1005.

Bolen, D. W., & Santoro, M. M. (1988) *Biochemistry 27*, 8069.

Bowie, J. U., Reidhaar-Olsen, J. F., Lim, W. A., & Sauer, R. T. (1990) *Science 247*, 1306.

Bradley, E. K., Thomasen, J. F., Cohen, F. E., Kosen, P. A., & Kuntz, I. D. (1990) *J. Mol. Biol.* (submitted for publication).

Brandts, J. F. (1964a) *J. Am. Chem. Soc. 86*, 4291.

Brandts, J. F. (1964b) *J. Am. Chem. Soc. 86*, 4302.

Brandts, J. F. (1969) in *Structure and Stability of Biological Macromolecules* (Timasheff, S. N., & Fasman, G. D., Eds.) p 213, Dekker, New York.

Brandts, J. F., & Lumry, R. (1963) *J. Phys. Chem. 67*, 1484.

Brandts, J. F., Oliveira, R. J., & Westort, C. (1970) *Biochemistry 9*, 1038.

Brems, D. N., & Havel, H. A. (1989) *Proteins 5*, 93.

Burley, S. K., & Petsko, G. A. (1985) *Science 229*, 23.

Burley, S. K., & Petsko, G. A. (1988) *Adv. Protein Chem. 39*, 125.

Burteau, N., Burton, S., & Crichton, R. R. (1989) *FEBS Lett. 258*, 185.

Butler, J. A. V. (1937) *Trans. Faraday Soc. 33*, 229.

Chan, H. S., & Dill, K. A. (1989a) *J. Chem. Phys. 90*, 492.

Chan, H. S., & Dill, K. A. (1989b) *Macromolecules 22*, 4559.

Chan, H. S., & Dill, K. A. (1990a) *J. Chem. Phys. 92*, 3118.

Chan, H. S., & Dill, K. A. (1990b) *Proc. Nat. Acad. Sci. U.S.A.* (in press).

Chen, B., & Schellman, J. A. (1989) *Biochemistry 28*, 685.

Chick, H., & Martin, C. J. (1911) *J. Physiol. 43*, 1.

Chothia, C. (1974) *Nature 254*, 304.

Chothia, C. (1976) *J. Mol. Biol. 105*, 1.

Chou, P. Y., & Fasman, G. D. (1978) *Adv. Enzymol. 47*, 45.

Chou, P. Y., Wells, M., & Fasman, G. D. (1972) *Biochemistry 11*, 3028.

Christensen, J. K. (1952) *C. R. Trav. Lab. Carlsberg, Ser. Chim. 28*, 37.

Christian, S. D., & Tucker, E. E. (1982) *J. Solution Chem. 11*, 749.

Cohn, E. J., et al. (1933) *J. Biol. Chem. 100*, 3.

Collins, K. D., & Washabaugh, M. W. (1985) *Q. Rev. Biophys. 18*, 323.

Conio, G., Patrone, E., & Brighetti, S. (1970) *J. Biol. Chem. 245*, 3335.

Covell, D. G., & Jernigan, R. L. (1990) *Biochemistry 29*, 3287–3294.

Crovetto, R., Fernandez-Priri, R., & Japas, M. L. (1982) *J. Chem. Phys. 76*, 1077.

Cybulski, S., M., & Scheiner, S. (1989) *J. Phys. Chem. 93*, 6565.

Dang, L. X., Merz, K. M., & Kollman, P. A. (1989) *J. Am. Chem. Soc. 111*, 8505.

Debye, P. (1949) *Ann. N.Y. Acad. Sci. 51*, 575.

deGennes, P. G. (1975) *J. Phys. (Paris) 36*, L-55.

deGennes, P. G. (1979) in *Scaling Concepts in Polymers Physics*, Cornell University Press, Ithaca, NY.

DeGrado, W. F. (1988) *Adv. Protein Chem. 39*, 51.

DeGrado, W. F., Wasserman, Z. R., & Lear, J. D. (1989) *Science 243*, 622.

Dill, K. A. (1985) *Biochemistry 24*, 1501.

Dill, K. A. (1990) *Science* (in press).

Dill, K. A., & Alonso, D. O. V. (1988) in *Colloquium Mosbach der Gessellschaft fur Biologische Chemie: Protein Structure and Protein Engineering* (Huber, T., & Winnacker, E. L., Eds.) Vol. 39, p 51, Springer-Verlag, Berlin.

Dill, K. A., Alonso, D. O. V., & Hutchinson, K. (1989) *Biochemistry 28*, 5439.

Doty, P., & Yang, J. T. (1956) *J. Am. Chem. Soc. 78*, 498.

Doty, P., et al. (1954) *J. Am. Chem. Soc. 76*, 4493.

Doty, P., Bradbury, J. H., & Holtzer, A. M. (1956) *J. Am. Chem. Soc. 78*, 947.

Doty, P., et al. (1958) *Proc. Natl. Acad. Sci. U.S.A. 44*, 424.

Dyson, H. J., et al. (1985) *Nature 318*, 480.

Dyson, H. J., et al. (1988a) *J. Mol. Biol. 201*, 161.

Dyson, H. J., et al. (1988b) *J. Mol. Biol. 201*, 201.

Edelhoch, H., & Osborne, J. C. (1976) *Adv. Protein Chem. 30*, 183.

Eden, D., et al. (1982) *Proc. Natl. Acad. Sci. U.S.A. 79*, 815.

Edsall, J. T. (1935) *J. Am. Chem. Soc. 57*, 1506.

Edsall, J. T., & McKenzie, H. A. (1983) *Adv. Biophys. 16*, 53.

Eisenberg, D., & McLachlan, A. D. (1986) *Nature 319*, 199.

Eisenberg, M. A., & Schwert, G. W. (1951) *J. Gen. Physiol. 34*, 583.

Ellis, R. J. (1988) *Nature 328*, 378.

Epand, R. M., & Scheraga, H. A. (1968) *Biochemistry 7*, 2864.

Evans, P. A., et al. (1987) *Nature 329*, 266.

Eyring, H., & Stearn, A. E. (1939) *Chem. Rev. 24*, 253.

Fahey, P. F., Krupke, D. W., & Beams, J. W. (1969) *Proc. Natl. Acad. Sci. U.S.A. 63*, 548.

Fauchere, J.-L., & Pliska, V. E. (1983) *Eur. J. Med. Chem.-Chem. Therm. 18*, 369.

Fersht, A. R. (1972) *J. Mol. Biol. 64*, 497.
Fersht, A. R. (1985) *Nature 314*, 235.
Finkelstein, A. V., & Ptitsyn, O. B. (1987) *Prog. Biophys. Mol. Biol. 50*, 171.
Flory, P. J. (1953) in *Principles of Polymer Chemistry*, Cornell University Press, Ithaca, NY.
Flory, P. J. (1969) *Statistical Mechanics of Chain Molecules*, Wiley, New York.
Frank, H. S., & Evans, M. W. (1945) *J. Chem. Phys. 13*, 507.
Franks, F. (1983) *Water*, The Royal Society of Chemistry, London.
Franzen, J. S., & Stephens, R. E. (1963) *Biochemistry 2*, 1321.
Freed, K. F. (1987) in *Renormalization Group Theory of Macromolecules*, Wiley, New York.
Friedman, M. E., & Scheraga, H. A. (1965) *J. Phys. Chem. 69*, 3795.
Friend, S. H., & Gurd, F. R. N. (1979) *Biochemistry 18*, 4612.
Gavish, B., Gratton, E., & Hardy, C. J. (1983) *Proc. Natl. Acad. Sci. U.S.A. 80*, 750.
Geiger, A., Rahman, A., & Stillinger, F. H. (1979) *J. Chem. Phys. 70*, 263–276.
Gekko, K., & Timasheff, S. N. (1981) *Biochemistry 20*, 4667.
Gibbs, J. H., & DiMarzio, E. A. (1959) *J. Chem. Phys. 30*, 271.
Gill, S. J., & Noll, L. (1972) *J. Phys. Chem. 76*, 3065.
Gill, S. J., & Wadsö, I. (1976) *Proc. Natl. Acad. Sci. U.S.A. 73*, 2955.
Gill, S. J., et al. (1985) *J. Phys. Chem. 89*, 3758.
Gilson, M. K., & Honig, B. H. (1988a) *Proteins 3*, 32.
Gilson, M. K., & Honig, B. H. (1988b) *Proteins 4*, 7.
Gilson, M. K., & Honig, B. H. (1989) *Proc. Natl. Acad. Sci. U.S.A. 86*, 1524.
Goldenberg, D. P., & Creighton, T. E. (1983) *J. Mol. Biol. 165*, 407.
Goodman, M., et al. (1969) *Proc. Natl. Acad. Sci. U.S.A. 64*, 444.
Goto, Y., & Fink, A. L. (1989) *Biochemistry 28*, 945.
Goto, Y., & Fink, A. L. (1990) *J. Mol. Biol.* (in press).
Goto, Y., Calciano, L. J., & Fink, A. L. (1990) *Proc. Natl. Acad. Sci. U.S.A. 87*, 573.
Griko, Y. V., Privalov, P. L., Sturtevant, J. M., & Venyamenov, S. Y. (1988) *Proc. Natl. Acad. Sci, U.S.A. 85*, 3343.
Grutter, M. G., et al. (1987) *J. Mol. Biol. 197*, 315.
Gurney, R. W. (1962) in *Ionic Processes in Solution*, Dover, New York.
Guy, H. R. (1985) *Biophys. J. 47*, 61.
Ha, J. H., Spolar, R. S., & Record, M. T., Jr. (1989) *J. Mol. Biol. 209*, 801.
Harrington, W. F., & Schellman, J. A. (1956) *C. R. Lab. Carlsberg, Ser. Chim. 30*, 21.
Haschemeyer, R. H., & Haschemeyer, A. E. V. (1973) in *Proteins*, Wiley, New York.
Hawley, S. A. (1971) *Biochemistry 10*, 2436.
Hermans, J. (1966) *J. Am. Chem. Soc. 88*, 2418.
Hermans, J., & Scheraga, H. A. (1961) *J. Am. Chem. Soc. 83*, 3283.
Hildebrand, J. H. (1968) *J. Phys. Chem. 72*, 1841.
Hildebrand, J. H. (1979) *Proc. Natl. Acad. Sci. U.S.A. 76*, 194.
Hildebrand, J. H., & Scott, R. L. (1950) in *The Solubility of Nonelectrolytes*, Reinhold, New York.
Hill, T. L. (1960) in *Introduction to Statistical Thermodynamics*, Addison-Wesley, Reading, MA.
Hollecker, M., & Creighton, T. E. (1982) *Biochim. Biophys. Acta 701*, 395.
Holley, L. H., & Karplus, M. (1989) *Proc. Natl. Acad. Sci. U.S.A. 86*, 152.
Honig, B., & Hubbell, W. (1984) *Proc. Natl. Acad. Sci. U.S.A. 81*, 5412.
Honig, B., Hubbell, W., & Flewelling, R. F. (1986) *Annu. Rev. Biophys. Biophys. Chem. 15*, 163.
Hvidt, A. (1975) *J. Theor. Biol. 50*, 245.
Jacobsen, C. F., & Linderstrom-Lang, K. (1949) *Nature 164*, 411.
Janin, J., et al. (1978) *J. Mol. Biol. 125*, 357.
Jolicoeur, C., et al. (1986) *J. Solution Chem. 15*, 109.
Jorgensen, W. L. (1989) *Acc. Chem. Res. 22*, 184.
Jorgensen, W. L., Geo, J., & Ravimohan, C. (1985) *J. Phys. Chem. 89*, 3470.
Kabsch, W., & Sander, C. (1983) *Biopolymers 22*, 2577.
Kabsch, W., & Sander, C. (1984) *Proc. Natl. Acad. Sci. U.S.A.* 1075.
Karle, I. L., et al. (1990) *Proteins 7*, 62.
Karplus, M., Ichiye, T., & Pettitt, B. M. (1987) *Biophys. J. 52*, 1083.
Kauzmann, W. (1954) in *The Mechanism of Enzyme Action* (McElroy, W. D., & Glass, B., Eds.) p 70, Johns Hopkins Press, Baltimore, MD.
Kauzmann, W. (1959) *Adv. Protein Chem. 14*, 1.
Kauzmann, W. (1987) *Nature 325*, 763.
Kellis, J. T., Nyberg, K., & Fersht, A. R. (1989) *Biochemistry 28*, 4914.
Kelly, L., & Holladay, L. A. (1987) *Protein Eng. 1*, 137.
Kendrew, J. C. (1961) *Sci. Am. 205*, (6), 96.
Kendrew, J. C., et al. (1958) *Nature 181*, 662.
Kim, P. S., & Baldwin, R. L. (1984) *Nature 307*, 329.
Klapper, M. H. (1971) *Biochim. Biophys. Acta 229*, 557.
Klotz, I. M., & Franzen, J. S. (1962) *J. Am. Chem. Soc. 84*, 3461.
Klotz, I. M., & Farnham, S. B. (1968) *Biochemistry 7*, 3879.
Kneller, D. G., Cohen, F. E., & Langridge, R. (1990) *J. Mol. Biol.* (in press).
Kresheck, G. C., & Klotz, I. M. (1969) *Biochemistry 8*, 8.
Krikorian, S. E. (1982) *J. Phys. Chem. 86*, 1875.
Kundrot, C. E., & Richards, F. M. (1987) *J. Mol. Biol. 193*, 157.
Kuntz, I. D. (1972) *J. Am. Chem. Soc. 94*, 4009.
Kuwajima, K. (1989) *Proteins 6*, 87.
Kyte, J., & Doolittle, R. F. (1982) *J. Mol. Biol. 157*, 105.
Lau, K. F., & Dill, K. A. (1989) *Macromolecules 22*, 3986.
Lau, K. F., & Dill, K. A. (1990) *Proc. Natl. Acad. Sci. U.S.A. 87*, 638.
Lee, B. K., & Richards, F. M. (1971) *J. Mol. Biol. 55*, 379.
Lee, J. C., & Timasheff, S. N. (1981) *J. Biol. Chem. 256*, 7193.
Lennard-Jones, J., & Pople, J. A. (1951) *Proc. R. Soc. London A205*, 155.
Levitt, M., & Greer, J. (1977) *J. Mol. Biol. 114*, 181.
Lifschitz, I. M. (1968) *Zh. Eksp. Teor. Fiz. 55*, 2408.
Lim, W. A., & Sauer, R. T. (1989) *Nature 339*, 31.
Linderstrom-Lang, K. U. (1924) *C. R. Trav. Lab. Carlsberg, 15*, 70.
Linderstrom-Lang, K. U. (1952) *Lane Medical Lectures, Vol.* 6, p 53, Stanford University Press, Stanford, CA.
Lotan, N., Bixon, M., & Berger, A. (1967) *Biopolymers 5*, 69.
Luger, K., et al. (1989) *Science 243*, 206.
Lumry, R., & Eyring, H. (1954) *J. Phys. Chem. 58*, 110.
Lumry, R., Biltonen, R., & Brandts, J. F. (1966) *Biopolymers 4*, 917.

Lupu-Lotan, N., Yaron, A., Berger, A., & Sela, M. (1965) *Biopolymers 3*, 625.
Marqusee, S., & Baldwin, R. L. (1987) *Proc. Natl. Acad. Sci. U.S.A. 84*, 8898.
Marqusee, S., Robbins, V. H., & Baldwin, R. L. (1989) *Proc. Natl. Acad. Sci. U.S.A. 86*, 5286.
Masterson, W. L. (1954) *J. Chem. Phys. 22*, 1830.
Matsumura, M., Becktel, W. J., & Matthews, B. W. (1988a) *Nature 334*, 406.
Matsumura, M., et al. (1988b) *Eur. J. Biochem. 171*, 715.
Matsumura, M., Matthews, B. W., Levitt, M., & Becktel, W. J. (1989) *Proc. Natl. Acad. Sci. U.S.A. 86*, 6562.
Matthew, J. B., & Richards, F. M. (1982) *Biochemistry 21*, 4489.
Matthew, J. B., & Gurd, F. R. N. (1986) *Methods Enzymol. 130*, 413.
Matthews, B. W. (1987a) *Harvey Lect. 81*, 33.
Matthews, B. W. (1987b) *Biochemistry 26*, 6885.
Matthews, B. W., Nicholson, A., & Becktel, W. J. (1987) *Proc. Natl. Acad. Sci. U.S.A. 84*, 6663.
Mattice, W. L., & Scheraga, H. A. (1984) *Biopolymers 23*, 1701.
McGregor, M. J., Islam, S. A., & Sternberg, M. J. E. (1987) *J. Mol. Biol. 198*, 295.
Meirovitch, H., & Scheraga, H. A. (1980) *Macromolecules 13*, 1406.
Merutka, G., & Stellwagen, E. (1990) *Biochemistry 29*, 894.
Merutka, G., et al. (1990) *Biochemistry* (submitted for publication).
Miller, K. W., & Hildebrand, J. H. (1968) *J. Am. Chem. Soc. 90*, 3001.
Mirsky, A. E., & Pauling, L. (1936) *Proc. Natl. Acad. Sci. U.S.A. 22*, 439.
Mitchinson, C., & Baldwin, R. L. (1986) *Proteins 1*, 23.
Miyazawa, S., & Jernigan, R. L. (1985) *Macromolecules 18*, 534.
Morrison, T. J. (1952) *J. Chem. Soc. 3*, 3814.
Morrison, T. J., & Billett, F. (1952) *J. Chem. Soc. 3*, 3819.
Muller, N. (1990) *Acc. Chem. Res. 23*, 23.
Murphy, K. P., & Gill, S. J. (1989a) *J. Chem. Thermodyn. 21*, 903.
Murphy, K. P., & Gill, S. J. (1989b) *Thermochim. Acta 139*, 279.
Murphy, K. P., Privalov, P. L., & Gill, S. J. (1990) *Science 247*, 559.
Nelson, J. W., & Kallenbach, N. R. (1986) *Proteins 1*, 211.
Nemethy, G., & Scheraga, H. A. (1962) *J. Chem. Phys. 36*, 3401.
Nemethy, G., Scheraga, H. A., & Kauzmann, W. (1968) *J. Phys. Chem. 72*, 1842.
Nemethy, G., Peer, W. J., & Scheraga, H. A. (1981) *Annu. Rev. Biophys. Bioeng. 10*, 459.
Neurath, H., et al. (1944) *Chem. Rev. 34*, 157.
Nojima, H., et al. (1977) *J. Mol. Biol. 116*, 429.
Northrup, J. H. (1932) *J. Gen. Physiol. 16*, 323.
Novotny, J., Bruccoleri, R. E., & Karplus, M. (1984) *J. Mol. Biol. 177*, 787.
Novotny, J., Rashin, A. A., & Bruccoleri, R. E. (1988) *Proteins 4*, 19.
Nozaki, Y., & Tanford, C. (1970) *J. Biol. Chem. 245*, 1698.
Nozaki, Y., & Tanford, C. (1971) *J. Biol. Chem. 246*, 2211.
Ooi, T., & Oobatake, M. (1988) *J. Biochem. (Tokyo) 103*, 114.
Ostermann, J., et al. (1989) *Nature 341*, 125.
Pace, C. N. (1975) *CRC Crit. Rev. Biochem. 3*, 1.
Pace, C. N., & Tanford, C. (1968) *Biochemistry 7*, 198.
Pace, C. N., & Grimsley, G. R. (1988) *Biochemistry 27*, 3242.
Pace, C. N., Grimsley, G. R., Thomson, J. A., & Barnett, B. J. (1988) *J. Biol. Chem. 263*, 11820.
Padmanabhan, S., et al. (1990) *Nature 344*, 268.
Pangali, C., Rao, M., & Berne, B. J. (1982) *J. Chem. Phys. 71*, 2982.
Parodi, R. M., Bianchi, E., & Ciferri, A. (1973) *J. Biol. Chem. 248*, 4047.
Pauling, L. (1960) in *The Nature of the Chemical Bond*, 3rd ed., Cornell University Press, Ithica, NY.
Pauling L., & Corey, R. B. (1951a) *Proc. Natl. Acad. Sci. U.S.A. 37*, 235.
Pauling, L., & Corey, R. B. (1951b) *Proc. Natl. Acad. Sci. U.S.A. 37*, 251.
Pauling, L., & Corey, R. B. (1951c) *Proc. Natl. Acad. Sci. U.S.A. 37*, 272.
Pauling, L., & Corey, R. B. (1951d) *Proc. Natl. Acad. Sci. U.S.A. 37*, 729.
Pauling, L., Corey, R. B., & Branson, H. R. (1951) *Proc. Natl. Acad. Sci. U.S.A. 37*, 205.
Peller, L. (1959) *J. Phys. Chem. 63*, 1194.
Perutz, M. F., & Lehmann, H. (1968) *Nature 219*, 902.
Perutz, M. F., & Raidt, H. (1975) *Nature 255*, 256.
Perutz, M. F., Kendrew, J. C., & Watson, H. C. (1965) *J. Mol. Biol. 13*, 669.
Piela, L., Nemethy, G., & Scheraga, H. A. (1987) *Biopolymers 26*, 1273.
Platzer, K. E. B., et al. (1972) *Macromolecules 5*, 177.
Poland, D. C., & Scheraga, H. A. (1965) *Biopolymers 3*, 379.
Poland, D. C., & Scheraga, H. A. (1970) *Theory of the Helix-Coil Transition*, Academic Press, New York.
Ponder, J. W., & Richards, F. M. (1987) *J. Mol. Biol. 193*, 775.
Post, C. B., & Zimm, B. H. (1979) *Biopolymers 18*, 1487.
Presnell, S. R., & Cohen, F. E. (1989) *Proc. Natl. Acad. Sci. U.S.A. 86*, 6592.
Privalov, P. L. (1979) *Adv. Protein Chem. 33*, 167.
Privalov, P. L. (1989) *Annu. Rev. Biophys. Biophys. Chem. 18*, 47.
Privalov, P. L., & Kechinashvili, N. N. (1974) *J. Mol. Biol. 86*, 665.
Privalov, P. L., & Gill, S. J. (1988) *Adv. Protein Chem. 39*, 191.
Privalov, P. L., et al. (1986) *J. Mol. Biol. 190*, 487.
Privalov, P. L., Gill, S. J., & Murphy, K. P. (1990) *Science* (in press).
Ptitsyn, O. B. (1987) *J. Protein Chem. 6*, 273.
Qian, N., & Sejnowski, T. J. (1988) *J. Mol. Biol. 202*, 865.
Radzicka, A., & Wolfenden, R. (1988) *Biochemistry 27*, 1664.
Rao, S. P., Carlstrom, D. E., & Miller, W. G. (1974) *Biochemistry 13*, 943.
Ravishanker, G., Mezei, M., & Beveridge, D. L. (1982) *Faraday Symp. Chem. Soc. 17*, 79.
Rich, D. H., & Jasensky, R. D. (1980) *J. Am. Chem. Soc. 102*, 1112.
Richards, F. M. (1977) *Annu. Rev. Biophys. Bioeng. 6*, 151.
Richards, F. M., & Richmond, T. (1978) in *Molecular Interactions and Activity in Proteins*, (Wolstenholme, G. E., Ed.) p 23, Ciba Foundation Symposium 60, Excerpta Medica, Amsterdam.

Richardson, J. S., & Richardson, D. C. (1987) in *Protein Engineering* (Oxender, D. L., & Fox, C. F., Eds.) p 149, Alan R. Liss, New York.
Richardson, J. S., & Richardson, D. C. (1988) *Science 240*, 1648.
Robertson, T. B. (1918) *The Physical Chemistry of the Proteins*, Longmans, Green and Co., New York.
Rogers, N. K. (1989) in *Prediction of Protein Structure and the Principles of Protein Conformation* (Fasman, G. D., Ed.) p 359, Plenum, New York.
Rooman, M. J., & Wodak, S. J. (1988) *Nature 335*, 45.
Rose, G. D., et al. (1985a) *Science 229*, 834.
Rose, G. D., Gierasch, L. M., & Smith, J. A. (1985b) *Adv. Protein Chem. 37*, 1.
Roseman, M. A. (1988) *J. Mol. Biol. 201*, 621.
Rothman, J. E. (1989) *Cell 59*, 591.
Sanchez, I. C. (1979) *Macromolecules 12*, 980.
Sandberg, W. S., & Terwilliger, T. C. (1989) *Science 245*, 54.
Santoro, M. M., & Bolen, D. W. (1988) *Biochemistry 27*, 8063.
Scheiner, S., & Hillenbrand, E. A. (1985) *Proc. Natl. Acad. Sci U.S.A. 82*, 2741.
Scheiner, S., Redfern, P., & Hillenbrand, E. A. (1986) *Int. J. Quant. Chem. 29*, 817.
Schellman, C., & Schellman, J. A. (1958) *C. R. Lab. Carlsberg, Ser. Chim. 30*, 463.
Schellman, J. A. (1955) *C. R. Trav. Lab. Carlsberg, Ser. Chim. 29*, 223.
Schellman, J. A. (1958a) *J. Phys. Chem. 62*, 1485.
Schellman, J. A. (1958b) *C. R. Trav. Lab. Carlsberg, Ser. Chim. 30*, 450.
Schrier, M. Y., & Schrier, E. E. (1976) *Biochemistry 15*, 2607.
Schulz, G. E., & Schirmer, R. H. (1979) in *Principles of Protein Structure*, Springer-Verlag, New York.
Semlyen, J. A. (1976) in *Advances in Polymer Science* (Cantow, H. J., et al., Eds.) Vol. 21, Springer, Berlin.
Shakhnovich, E. I., & Finkelstein, A. V. (1989a) *Biopolymers 28*, 1667.
Shakhnovich, E. I., & Finkelstein, A. V. (1989b) *Biopolymers 28*, 1681.
Shoemaker, K. R., et al. (1985) *Proc. Natl. Acad. Sci. U.S.A. 82*, 2349.
Shoemaker, K. R., et al. (1987) *Nature 326*, 563.
Shoemaker, K. R., et al. (1990) *Biopolymers 29*, 1.
Shortle, D., & Meeker, A. K. (1989) *Biochemistry 28*, 936.
Shortle, D., Meeker, A. K., & Freire, E. (1988) *Biochemistry 27*, 4761.
Sikorski, A., & Skolnick, J. (1989) *Proc. Natl. Acad. Sci. U.S.A. 86*, 2668.
Singer, S. J. (1962) *Adv. Protein Chem. 17*, 1.
Skolnick, J., Kolinski, A., & Yaris, R. (1988) *Proc. Natl. Acad. Sci. U.S.A. 85*, 5057.
Skolnick, J., Kolinski, A., & Yaris, R. (1989) *Proc. Natl. Acad. Sci. U.S.A. 86*, 1229.
Sneddon, S. F., Tobias, D. J., & Brooks, C. L., III (1989) *J. Mol. Biol. 209*, 817.
Srinivasan, R. (1976) *Indian J. Biochem. Biophys. 13*, 192.
Stigter, D., & Dill, K. A. (1989) *J. Phys. Chem. 93*, 6737.
Stigter, D., & Dill, K. A. (1990) *Biochemistry 29*, 1262.
Stigter, D., Alonso, D. O. V., & Dill, K. A. (1990) (submitted for publication).
Stillinger, F. H. (1980) *Science 209*, 451.
Sturtevant, J. M. (1977) *Proc. Natl. Acad. Sci. U.S.A. 74*, 2236.
Sueki, M., et al. (1984) *Macromolecules 17*, 148.
Susi, H. (1969) in *Structure and Stability of Biological Macromolecules* (Timasheff, S. N., & Fasman, G. D., Eds.) Marcel Dekker, New York.
Susi, H., & Ard, J. S. (1969) *J. Phys. Chem. 73*, 2440.
Sweet, R. M., & Eisenberg, D. (1983) *J. Mol. Biol. 171*, 479.
Tadokoro, H. (1979) in *Structure of Crystalline Polymers*, Wiley, New York.
Tanford, C. (1961) in *Physical Chemistry of Macromolecules*, Chapter 7, Wiley, New York.
Tanford, C. (1962) *J. Phys. Chem. 84*, 4240.
Tanford, C. (1968) *Adv. Protein Chem. 23*, 121.
Tanford, C. (1970) *Adv. Protein Chem. 24*, 1.
Tanford, C. (1979) *Proc. Natl. Acad. Sci. U.S.A. 76*, 4175.
Tanford, C. (1980) in *The Hydrophobic Effect*, 2nd ed., Wiley, New York.
Tanford, C., & Kirkwood, J. G. (1957a) *J. Am. Chem. Soc. 79*, 5333.
Tanford, C., & Kirkwood, J. G. (1957b) *J. Am. Chem. Soc. 79*, 5340.
Taniuchi, H., & Anfinsen, C. B. (1969) *J. Biol. Chem. 244*, 3864.
Thornton, J. (1988) *Nature 335*, 10.
Vinogradov, S. N., & Linnell, R. H. (1971) in *Hydrogen Bonding*, Van Nostrand Reinhold, New York.
von Hippel, P. H., & Schleich, T. (1969a) in *Structure and Stability of Biological Macromolecules* (Timasheff, S., & Fasman, G., Eds.) Vol. II, p 417, Dekker, New York.
von Hippel, P. H., & Schleich, T. (1969b) *Acc. Chem. Res. 2*, 257.
Wada, A., & Nakamura, H. (1981) *Nature 293*, 757.
Weiner, S., et al. (1984) *J. Am. Chem. Soc. 106*, 765.
Wertz, D. H., & Scheraga, H. A. (1978) *Macromolecules 11*, 9.
White, S. H., & Jacobs, R. E. (1990) *Biophys. J.* (in press).
Wright, P. E., Dyson, H. J., & Lerner, R. A. (1988) *Biochemistry 27*, 7167.
Wu, H. (1929) *Am. J. Physiol. 90*, 562.
Wu, H. (1931) *Chin. J. Physiol. 5*, 321.
Wu, H., & Wu, D. (1931) *Chin. J. Physiol. 5*, 369.
Yutani, K., et al. (1984) *J. Biol. Chem. 259*, 14076.
Yutani, K., et al. (1987) *Proc. Natl. Acad. Sci. U.S.A. 84*, 4441.
Zhang, R., & Snyder, G. H. (1989) *J. Biol. Chem. 264*, 18,472.
Zimm, B. H., & Bragg, J. K. (1959) *J. Chem. Phys. 31*, 526.
Zimm, B. H., & Rice, S. A. (1960) *Mol. Phys. 3*, 391.
Zimm, B. H., et al. (1959) *Proc. Natl. Acad. Sci. U.S.A. 45*, 1601.
Zipp, A., & Kauzmann, W. (1973) *Biochemistry 12*, 4217.

Chapter 2

The Mechanism of Protein Folding. Implications of in Vitro Refolding Models for de Novo Protein Folding and Translocation in the Cell†

Gunter Fischer‡ and Franz X. Schmid*,§

Enzymologie, Sektion Biowissenschaften, Martin-Luther-Universität, Halle-Wittenberg, Domplatz 1, DDR-4020 Halle/Saale, GDR, and Laboratorium für Biochemie, Universität Bayreuth, D-8580 Bayreuth, FRG

Received September 6, 1989

The in vitro refolding of purified, denatured proteins is a spontaneous process. It is driven by a small, but significant, difference in Gibbs free energy under native conditions and generally requires neither the presence of additional factors nor the input of energy (Tanford, 1968, 1970; Kim & Baldwin, 1982; Jaenicke, 1987; Tsou, 1988). In the past few years, however, an increasing number of proteins have been discovered, which were proposed to be involved in the maturation and folding of nascent protein chains in the cell (Pelham, 1986, 1988; Rothman & Kornberg, 1986; Freedman, 1987). These proteins are found in the cytoplasm as well as in the endoplasmic reticulum (ER),[1] the mitochondria, or the chloroplasts. They are thought to be involved in functions such as catalyzing protein folding, keeping protein chains in a nonnative state, or "guiding" proteins to their cellular association partners. When recombinant proteins are overexpressed in foreign hosts, the polypeptide chains frequently do not fold correctly, but they are deposited in the cell in an insoluble, nonnative form (Marston, 1986), which could mean that the correct "folding helpers" are missing. We review here current concepts for the mechanism of in vitro protein folding and evaluate their significance for folding processes in vivo and the role that could be ascribed to cellular components in this process.

Physicochemical Data on the in Vitro Folding and Stability of Proteins

(*1*) *Thermodynamic Stability of Proteins*. Folded proteins are usually stable in a thermodynamic sense at ambient temperature and at neutral pH; however, the difference in Gibbs free energy between the native and the unfolded state is generally small. Frequently, values in the range of −40 kJ/mol are found. Consequently, small shifts in the solvent conditions, such as changes in pH, temperature, or composition, can lead to large changes in the net stability of folded proteins. Also, associations with ligands, other proteins, or subunits may lead to major alterations in the stability of a protein or even promote its unfolding. The thermal or denaturant-induced unfolding transitions of several small proteins are described well by a simple two-state model, which involves only native and unfolded protein at equilibrium (Privalov, 1979). In other words, folding of these proteins is a cooperative process, and molecules where only part of the stabilizing interactions are missing are not populated at equilibrium. Large proteins consist of folding domains, which can behave as independent structural entities with individual unfolding transitions (Privalov, 1982). A number of proteins unfold irreversibly in in vitro experiments. The reasons for the lack of reversibility are not always clear. In some cases posttranslational covalent modifications, such as proteolytic processing, can lead to a loss of reversibility. Generally, small changes of the covalent structure, such as the removal of a few residues from either end of the chain, often result in large changes in the overall stability of a protein (Wetlaufer & Ristow, 1973).

(*2*) *The Unfolded State and Equilibrium Intermediates*. Unfolding of proteins in the presence of high concentrations of "strong" denaturants, such as GdmCl or urea, leads to a state that shows a number of properties that are expected for a randomly coiled polypeptide. The titratable groups are normalized and the aromatic residues are exposed to solvent (Tanford, 1968, 1970). As judged by amide circular dichroism, the secondary structure has disappeared and the NMR spectrum gives no indication for ordered structure. The thermally unfolded state is more difficult to assess. The thermodynamic properties of thermally and denaturant-un-

† Writing of this paper was supported by a grant from the Volkswagenstiftung.

* To whom correspondence should be addressed.

‡ Martin-Luther-Universität.

§ Universität Bayreuth.

[1] Abbreviations: ER, endoplasmic reticulum; BiP, immunoglobulin heavy-chain binding protein; GdmCl, guanidinium chloride; PDI, protein disulfide isomerase; PPIase, peptidyl-prolyl cis–trans isomerase; hsp, heat shock protein.

folded proteins are very similar or identical; i.e., in both cases cooperative unfolding transitions with a similar increase in heat capacity are observed for small proteins. However, some residual structure appears to be present in proteins denatured at high temperature. Exposure to solvent of the aromatic residues is not complete, and residual secondary structure is found in some cases (Tanford, 1968).

For a few proteins, such as α-lactalbumin, myoglobin, or carbonic anhydrase, folding intermediates have been found, which are stable in the presence of intermediate concentrations of denaturant and/or in the acid or alkaline pH region (Kuwajima, 1977; Wong & Tanford, 1973; Dolgikh et al., 1984; Brems et al., 1985). Such intermediates are frequently observed for proteins with a low overall stability, i.e., with weak long-range interactions that are abolished under "mildly denaturing" conditions. The addition of extra stabilizing interactions, such as binding of Ca^{2+} ions to α-lactalbumin, leads to an increase in the stability of the native protein and to the disappearance of the partially folded intermediate (Kuwajima et al., 1989). Such intermediates are sometimes called "molten globules". They have some nativelike properties, such as a high amount of secondary structure and a compact shape. On the other hand, there appears to be only little or no well-defined tertiary structure and the aromatic residues are mostly unordered, suggesting that the flexibility of the polypeptide chain is very high in this state. In its enthalpy and heat capacity the molten globule still resembles the unfolded protein (Kuwajima, 1989).

(*3*) *Short Peptides as Models for Local Structure*. Contrary to earlier views that small peptides are essentially devoid of ordered structure in aqueous solution, it has become evident in the last few years that certain peptides as short as four or five residues can adopt significant nonrandom conformations under "native" aqueous conditions. Early evidence for such ordered structure came from circular dichroism measurements by Brown and Klee (1971) on the amino-terminal fragment of ribonuclease A, the S-peptide, which is 20 residues long. A significant amount of helical structure was detected in this peptide in the presence of salt near 0 °C. Extensive studies of the S-peptide and a large number of derivatives by Baldwin and his co-workers demonstrated that the helix in the S-peptide is similar to the helix found in the respective chain region of intact ribonuclease A (Kim & Baldwin, 1984). Furthermore, they showed that the helical content could be increased by amino acid substitutions, which favored salt bridge formation and stabilized the helix dipole (Shoemaker et al., 1987). Also, alanine residues were found to increase the helical content (Marqusee et al., 1989). Dyson et al. (1988a,b) searched for nonrandom structure in a large number of short peptides by NMR. A remarkable finding was that the short pentapeptide YPGDV exists to about 50% in a β-turn conformation at low temperature. Other short peptides contained significant amounts of β-turns in aqueous solution as well, albeit at a lower percentage. They found a correlation between the turn-forming potential of short peptides and the turn probabilities of the same sequences in proteins, as derived from an analysis of known crystal structures. This suggests that formation of some secondary structures can be specified by strictly local sequences. Such chain segments might thus serve as initiation sites for folding (Wright et al., 1988) and lead to a very fast restriction of the conformational space early in the folding process. The results on the sequence variations of the ribonuclease S-peptide show that the naturally occurring sequence is not the one with the highest helical content. This indicates (i) that protein secondary structures have clearly not evolved to maximal stability and (ii) that a high stability of local structures is not desirable for early stages of the folding process. Local ordered structures should indeed be fairly unstable, as required from theoretical considerations. To avoid formation of stable, but incorrect, structures, it is necessary to use the long-range interactions (which are formed late in folding) with their high resolving power to select and further stabilize those early local structures that are on the correct folding pathway. Incorrect local structures are not stabilized and consequently convert back into the ensemble of unfolded conformations (Go, 1983). Thus the finite, but very low, stability of local structures could be the kinetic "proofreading" mechanism of protein folding. In this context it is interesting to note that the stability of ordered structure in natural peptides is always small and that not all turns observed in peptides are also turns in the respective folded proteins, from whence the sequences were derived (Dyson et al., 1988a,b). Introduction of a defined long-range contact, such as a disulfide bond, strongly increases the stability of secondary structure (Oas & Kim, 1988).

(*4*) *Structure and Stability of Protein Fragments*. The question of how much polypeptide chain is necessary for the stability and the correct folding was investigated for a number of proteins. In the case of ribonuclease A, it was found that removal of either 20 residues from the amino terminus or four residues from the carboxy terminus both resulted in the inability of the truncated protein chains to re-form all correct disulfide bonds after denaturation and reduction (Haber & Anfinsen, 1961; Taniuchi, 1970). This defect was "healed" by adding back the S-peptide (1–20) or the C-terminal 110–124 peptide, respectively, to the reoxidation mixture (Kato & Anfinsen, 1969; Andria & Taniuchi, 1978). Large fragments of staphylococcal nuclease were found to be devoid of specific structure; however, folding was possible when overlapping peptides were incubated together (Andria et al., 1971). These examples suggest that for such small single domain proteins essentially the entire chain is necessary for stability and folding. On the other hand, chain extensions, such as a transit peptide, could easily inhibit the formation of correctly folded structures by contributing a few "bad" contacts.

In the case of larger proteins, stable and autonomously folding domains have been detected. Examples include the domains of the immunoglobulin light chain (Tsunenaga et al., 1987), of γ-crystallin (Siebendritt, 1989), and of tryptophan synthase (Zetina & Goldberg, 1980; Matthews & Crisanti, 1981). An interesting system is provided by the carboxy-terminal region of thermolysin. A number of proteolytic fragments of that region can be prepared which show autonomous folding behavior. The smallest of these stable and cooperative fragments, comprising residues 255–316, is highly helical; it is equivalent to a three-helix motif in native thermolysin (Dalzoppo et al., 1985).

(*5*) *Kinetics of Protein Folding Reactions*. The rates of protein folding reactions vary extensively. Folding can be complete within a few milliseconds or can require several hours. To a first approximation, small proteins tend to fold more rapidly than large ones; however, many exceptions from that rule exist. In addition, a single protein can refold on different fast- and slow-folding pathways (Garel & Baldwin, 1973). The occurrence of such fast- and slow-folding species in unfolded proteins is explained by the proline hypothesis (Brandts et al., 1975), which assumes that slow-folding molecules contain incorrect isomers of Xaa–Pro peptide bonds. The presence of such incorrect isomers does not block refolding, as assumed originally; however, it leads to a significant destabilization of folding intermediates and hence to a strong decrease of folding rates. Fast-folding molecules have all important prolyl peptide bonds in the native conformation, and therefore they refold rapidly (in many cases in the millisecond time range). Presumably not all proline residues are important for folding; their role for folding may depend on the location within the polypeptide chain and also on the conditions employed for refolding (Schmid & Blaschek, 1981). Additional slow, rate-determining folding steps exist. Examples are the formation of disulfide bonds (see below), the association of prefolded monomers in the case of oligomeric proteins (Jaenicke, 1987), or slow reshuffling reactions of almost nativelike intermediates (Vaucheret et al., 1987). In general, the slowest, rate-limiting step of unfolding and refolding appears to be close to the native state in terms of ordered structure (Segawa & Sugihara, 1984; Goldenberg & Creighton, 1985). In other words, the activated state of folding with the highest energy is a distorted form of the native protein. The refolding of many proteins is not completely reversible in vitro, and reactivation yields of less than 100% are frequently observed. This holds in particular for experiments at high protein concentration. Competing aggregation of unfolded or partially refolded protein molecules may be the major reason for a lowered reversibility (Jaenicke, 1987). This could reflect simply the lack of success of finding the correct solvent conditions and protein concentrations for reversible refolding, but it could also be a genuine problem of in vitro folding.

The slow rate-limiting steps of folding, such as proline isomerization or other processes, are generally preceded by the rapid formation of folding intermediates. These intermediates are frequently compact molecules and contain most of the secondary structure of the native protein (Kuwajima, 1989). They are still flexible and show little defined tertiary structure. In some aspects they resemble the molten globule intermediates that were found in the equilibrium unfolding transitions of some proteins (see above). High-resolution NMR experiments showed that most of the chain is involved in the formation of these early intermediates and that extensive secondary structure has already formed and is fairly stable (Udgaonkar & Baldwin, 1988; Roder et al., 1988). For cytochrome *c* the interaction of two helices in the amino- and carboxy-terminal regions, respectively, is important for the stabilization of this intermediate. Similar results were obtained for α-lactalbumin, where a sizable part of the molecule is already organized in a nativelike manner in the early intermediate (Kuwajima et al., 1989). These findings demonstrate that information encoded in distant regions of the chain is required already for early steps in folding. These processes are much faster than the biosynthesis of protein chains. On the other hand, chain regions, such as *cis*-proline containing β-turns, can be tolerated in a nonnative state until a late stage in refolding (Lang & Schmid, 1990; Kiefhaber et al., 1990b).

Kinetic intermediates of folding are sometimes sensitive to aggregation (Brems, 1988), because they still have an increased amount of exposed hydrophobic surface. This transient low solubility of incompletely folded chains may also be a problem for the de novo folding of nascent proteins. There is no unique folding mechanism for a particular protein, but the kinetic mechanism may depend strongly on the refolding conditions. The stability of folding intermediates is generally small; therefore, variations in the solvent conditions can strongly affect the formation of critical intermediates. For a review on early steps in protein folding, see Baldwin (1989).

(*6*) *Formation of Disulfide Bonds during Refolding.* Unfolded proteins with reduced disulfide bonds can spontaneously regain the native set of disulfide bonds under suitable conditions. Re-formation of disulfides requires the presence of a redox partner, such as oxygen, glutathione, or dithiothreitol (Creighton, 1978). The redox partner has to be present in both the oxidized and the reduced forms to allow for the reshuffling of incorrect nonnative disulfides, which are generated rapidly at the beginning of reoxidation (Saxena & Wetlaufer, 1970). Incorrect disulfides are generally observed on the in vitro reactivation pathways of reduced proteins. Formation and stabilization of the correct disulfides depend on the local concentrations of the respective sulfhydryl groups, which in turn depend on the conformation of the protein chain (Creighton, 1983). For the complete re-formation of the disulfides, it is essential that the respective sulfhydryl groups remain accessible for the disulfide reagent. Hence partially refolded and reoxidized molecules may not have a compactly folded structure, which could block the access of the redox partner. In the case of pancreatic trypsin inhibitor, a fraction of the refolded molecules contained two buried sulfhydryl groups. They were able to form a disulfide bridge only after heating, which increased the mobility of the chain, thus allowing oxygen to approach and oxidize these two cysteine residues (States et al., 1984).

Proteins That Are Involved in Cellular Protein Folding Events

(*1*) *Protein Disulfide Isomerase.* The first protein that was suggested to be involved in the de novo folding of nascent polypeptide chains is protein disulfide isomerase (PDI). Anfinsen's group showed that reduced and unfolded ribonuclease A can regain its native state with correct disulfide bonds in a spontaneous, but very slow, reoxidation reaction (Sela et al., 1957). Soon after that, an enzyme activity was discovered in liver extracts that was capable of accelerating the reoxidation of reduced proteins, and it was immediately suggested that it might be involved in the formation of disulfide bonds in nascent proteins (Goldberger et al., 1963; Venetianer & Straub, 1963). Generally, only secreted proteins contain disulfide bonds; therefore, PDI should be localized in cell compartments that form the secretory path. Indeed, it is found in the endoplasmic reticulum (ER) at a very high concentration. In rat liver microsomes, PDI amounts to about 2% of the entire protein content (Freedman, 1984, 1989). It carries a carboxy-terminal KDEL sequence, which is typical for proteins resident in the ER (Munro & Pelham, 1987).

PDI catalyzes the in vitro reoxidation of many different proteins. Its activity may depend on the presence of a low molecular weight oxidant, such as molecular oxygen or a mixture of oxidized and reduced glutathione. It is a true catalyst; the product of its action depends on the redox potential of the solution. Accordingly, the formation, reduction, or isomerization of disulfide bonds can be catalyzed by PDI. The conformation of the final product depends on the stability of the target protein under the employed solvent and redox conditions (Freedman, 1984).

Conclusive evidence for an important role of PDI for in vivo folding was provided by Bulleid and Freedman (1988). They depleted dog pancreas microsomes of soluble proteins, including PDI. These microsomal preparations were still able to translocate and process nascent protein chains, but they were defective in the formation of correct disulfide bonds. Addition of purified PDI to these microsomes restored the capacity to generate correctly disulfide-bonded protein.

The role of PDI apears to be fairly clear now. It facilitates the formation of the correct set of disulfide bonds during de novo folding of secreted proteins. It does not determine the folding pathway (Creighton et al., 1980), but catalyzes slow steps, presumably by rapid reshuffling of incorrect disulfide bonds in the presence of a low molecular weight thiol compound, which may be reduced and oxidized glutathione. The direction of folding and the end product are determined by the protein itself, i.e., by the stable, native set of disulfide bonds and by the suitable solvent and redox conditions. The formation of the correct disulfides can be very rapid. In some cases, such as in the immunoglobulin chains, it is a cotranslational event: as soon as an entire domain is translocated into the lumen of the endoplasmic reticulum, the single intradomain disulfide is formed (Bergman & Kuehl, 1979). Whether PDI is responsible for this particular reaction and whether disulfide bond formation is a cotranslational event for other, more extensively cross-linked proteins as well remain to be elucidated.

PDI is identical with the β-subunit of prolyl-4-hydroxylase (Koivu et al., 1987) and very similar to a protein that is presumably involved in the oligosaccharide transferase system (Geetha-Habib et al., 1988). These recent findings are discussed by Freedman (1989).

(*2*) *Peptidyl-Prolyl Cis–Trans Isomerase.* Some conformational steps in protein folding can be slow; therefore, enzymatic catalysis of folding should be of advantage. An acceleration of crucial folding steps would decrease the risk of proteolytic degradation of partially folded chains, suppress competing unproductive pathways, such as aggregation, and hence "keep the protein on the correct, productive folding pathway". The isomerization of incorrect Xaa–Pro peptide bonds is one of the slow, rate-determining steps in in vitro refolding (Brandts et al., 1975; Schmid & Baldwin, 1978; Lang et al., 1987). In 1984 an enzyme was found by Fischer et al. that accelerates efficiently the cis–trans isomerization of prolyl peptide bonds in short oligopeptides. Accordingly, this protein was named peptidyl-prolyl cis–trans isomerase (PPIase). The catalyzed reaction is a 180° rotation about the C–N linkage of the peptide bond preceding proline, which involves neither net cleavage nor net formation of covalent bonds. Thus, PPIases are "conformases" with a very high efficiency. The porcine 17-kDa PPIase shows a k_{cat}/K_M value that is estimated to be higher than $10^6\ M^{-1}\ s^{-1}$ for the isomerization of short oligopeptides. PPIase activities are widely distributed; they are found in virtually all tissues and organisms, ranging from mammals to bacteria. They apparently represent a new class of enzymes that are diverse in molecular weight, in cellular location, and in substrate specificity. Surprisingly, it was found that PPIase from porcine kidney is identical with bovine cyclophilin, a protein that binds the immunosuppressant cyclosporin with high affinity (Takahashi et al., 1989; Fischer et al., 1989a).

In addition to their activity on small proline-containing peptides, PPIases also catalyze the slow in vitro refolding of several small proteins for which from independent experiments proline isomerization was assumed to be a rate-limiting step (Lang et al., 1987; Lin et al., 1988). The efficiency of catalysis varies strongly, depending on the target protein. One particular slow step in the folding of ribonuclease T_1 is accelerated more than 100-fold by prolyl isomerase. Catalysis depends on both PPIase and ribonuclease T_1 concentrations (Fischer et al., 1989a; Kiefhaber et al., 1990a,b), and it is abolished in the presence of low concentrations of cyclosporin A, which binds with high affinity to the active site. The folding of some proteins, such as bovine ribonuclease A, chymotrypsinogen, or thioredoxin, is not catalyzed significantly, although good evidence exists that proline isomerization is important for their refolding mechanism (Lang et al., 1987; Lin et al., 1988; Lang, 1988). Apparently, catalysis of in vitro refolding depends on the sequence surrounding the respective proline residues and also on their steric accessibility in partially structured folding intermediates. The importance of accessibility was demonstrated for the folding of ribonuclease T_1, where the efficiency of catalysis depends on the structural environment of two proline peptide bonds (Kiefhaber et al., 1990a,b). The folding of collagen (types III and IV) is also accelerated in the presence of PPIases (Bächinger, 1987; Davis et al., 1989). At this point, however, it is important to keep in mind that proline isomerization is not the only slow process in protein folding. Other conformational steps, unrelated to proline isomerization, can be very slow as well. This may in particular hold for multidomain and oligomeric proteins, which frequently refold in the time range of minutes to hours.

The influence of the flanking sequences around the target proline on the catalysis of protein folding by PPIase has not yet been investigated. Data on small peptides indicate that at least the four amino acids around the rotating bond are important for catalysis (Fischer et al., unpublished).

Little is known at present about the enzymatic mechanism of PPIases. A single thiol group, which is near the active site, appears to be essential for the activity of the porcine 17-kDa PPIase (Fischer et al., 1989a). Inverse secondary kinetic isotope effects in substrates, in which the C_α-H that precedes proline is substituted by deuterium, indicate that a covalent intermediate, possibly a hemi ortho thioamide, is formed transiently (Fischer et al., 1989b).

The role of PPIases for in vivo protein folding or other transconformational reactions in the cell is not yet known, and the effect of the immunosuppressant cyclosporin A is not understood. As shown by immunohistochemical staining, actively growing cells show a higher content of 17-kDa PPIase (cyclophilin) than resting cells and an increased staining of vesicular structures (Harding & Handschumacher, 1988). In *Neurospora crassa* PPIase is found in the cytosol and in the mitochondria; the two respective proteins appear to be encoded by the same gene (Tropschug et al., 1988). The product of the *ninaA* gene of *Drosophila* is homologous with the 17-kDa PPIase. It carries an additional carboxy-terminal hydrophobic extension, which could serve as a membrane anchor. Mutant flies with a defect in the *ninaA* gene show a 10-fold reduction in rhodopsin level, although the opsin mRNA is produced normally. Possibly the *NinaA* protein is important for the maturation of rhodopsin (Shieh et al., 1989; Schneuwly et al., 1989). A nucleotide binding motif was suggested to be present in 17-kDa PPIase (Gschwendt et al., 1988); however, several features typical for such proteins (Wierenga & Hol, 1983) are lacking in the sequence. Also there is no experimental evidence that the activity of PPIases is influenced by the presence of nucleotides.

(*3*) *Proteins Involved in Polypeptide Transport across Membranes.* A number of proteins are translocated during or after biosynthesis across the plasma membrane of procaryotes or across organelle membranes in eukaryotes. It is clear now that these proteins are not transported in a compact globular state but that a more or less unfolded state is required for efficient transfer (Schleyer & Neupert, 1985; Randall & Hardy, 1986; Zimmermann & Meyer, 1986; Chen & Douglas, 1987; Vestweber & Schatz, 1988; Wickner, 1988). Structure formation can be avoided easily by a tight coupling of transport to protein biosynthesis, i.e., by the cotranslational transfer of the nascent chains across membranes, as is observed normally for the transport of proteins into the ER. However, post-translational transport into the ER (Watts et al., 1983), into the mitochondria (Harmey et al., 1977), and across the procaryotic plasma membrane (Zimmermann & Wickner, 1983; Randall, 1983) is possible as well.

A distinctive feature of proteins that are targeted to different cellular locations is the presence of an amino-terminal leader sequence, which, after transport, is usually removed from the mature protein. The simplest way to keep the preprotein in a "membrane-transport competent" form after the completion of biosynthesis would be to ensure the presence of the leader sequence, thus preventing the formation of stable three-dimensional structure in the preprotein. A few unfavorable interactions of the leader sequence with the remaining polypeptide chain should be sufficient to block or at least decelerate the formation of rigid, folded structure. Such a deceleration of refolding in the presence of the leader sequence was indeed observed for the precursors of maltose- and of ribose-binding protein from *Escherichia coli* (Park et al., 1988). Also, the interaction of the presequence with (negatively charged) membrane lipids may lead to partial unfolding of the precursor prior to transport (Endo & Schatz, 1988; Endo et al., 1989).

Such ways of keeping precursor proteins competent for translocation may not be sufficient in all cases. Also, incompletely folded precursors are presumably sensitive to aggregation. Therefore, the posttranslational import of polypeptide chains to mitochondria or the export from *E. coli* requires the presence of proteins, which, in an ATP-dependent manner, confer or retain transport competence to these polypeptides and keep them in solution (Pfanner et al., 1987; Crooke et al., 1988). The requirement for this ATP-dependent step depends on the folded state of the "passenger" polypeptide. Import of unfolded precursor proteins required less or no ATP (Verner & Schatz, 1987; Pfanner et al., 1988; Crooke et al., 1988). An inverse correlation exists between the stability of the folded state of a precursor and the efficiency of its transport across the mitochondrial membrane. The transfer into the mitochondria of mouse dihydrofolate reductase fused to the signal sequence of yeast cytochrome oxidase subunit IV is facilitated by urea denaturation and impaired by binding the inhibitor methotrexate, which strongly stabilizes the folded state of this protein (Eilers & Schatz, 1986). The implicated ATP-dependent unfolding proteins have not yet been isolated. However, there are indications that members of the "heat shock" protein family such as hsc70 (in cooperation with additional proteins) can stimulate import into microsomes (Chirico et al., 1988; Deshaies et al., 1988; Zimmermann et al., 1988). The precursor of an outer membrane protein from *E. coli*, proOmpA, is held in a membrane assembly competent form by a 63-kDa protein, the trigger factor (Crooke et al., 1988). The difference in Gibbs free energy between the native and the unfolded state of a protein is small, and therefore, the hydrolysis of ATP should be sufficient to drive such a transition to a less ordered conformation.

(*4*) *Proteins That Recognize Unfolded or Incompletely Folded Polypeptide Chains.* Other proteins that have been implicated in folding and unfolding processes in the cell apparently recognize misfolded or aggregated molecules or bind transiently to partially folded or nonassembled polypeptide chains, to prevent them from premature aggregation. Similar to the proteins that facilitate transport through membranes, these proteins belong to the family of the heat shock proteins [for reviews see Pelham (1986, 1988) and Ellis and Hemmingsen (1989)].

The immunoglobulin heavy-chain binding protein (BiP), which is identical with the glucose-regulated protein GRP78, is one of these proteins. BiP was suggested to be a "helper" in the folding and assembly of oligomeric proteins in the ER, such as the immunoglobulins (Munro & Pelham, 1986; Bole et al., 1986; Gething et al., 1986), or alternatively to be a binding protein for misfolded, unassembled, or otherwise aberrant proteins in the endoplasmic reticulum (Kassenbrock et al., 1988; Hurtley et al., 1989). Most of the presently available evidence supports the assumption that the major function of BiP is the recognition of incorrectly folded or aggregated proteins and their ultimate removal from the ER (Kassenbrock et al., 1989; Hurtley et al., 1989). Also, treatments of cells that lead to the accumulation of incomplete or misfolded proteins in the ER induce an increase in BiP–mRNA (Normington et al., 1989; Rose et al., 1989). Whether BiP is also involved in "normal" protein folding and association in the ER is still not entirely clear.

Protein molecules that are necessary for the correct folding of monomeric proteins and for the formation of oligomeric assemblies have been localized in *E. coli* (the GroEL and GroES proteins), in chloroplasts of higher plants (the Rubisco subunit binding protein), and in mitochondria from *Saccharomyces cerevisiae* and *N. crassa* (hsp60). Since these molecules "guide" other proteins to their final native functional state, they were named "molecular chaperones" (Hemmingsen et al., 1988). These chaperones are highly abundant, essential proteins, they show a M_r of about 60000, and they are strongly homologous to each other in sequence (Hemmingsen et al., 1988; Cheng et al., 1988). Related proteins exist in human leucocytes (Waldinger et al., 1988). All three chaperones occur as multimers of 12–16 identical or different subunits, associated probably with additional proteins such as GroES. Rubisco binding protein is necessary for the correct assembly of the eight large and eight small subunits to form the active complex in the chloroplasts (Roy et al., 1988; Hemmingsen et al., 1988). Rubisco assembly can also be promoted by the GroE proteins, after expression in *E. coli* (Goloubinoff et al., 1989). The mitochondrial hsp60 protein was found to be essential for the correct association of several oligomeric enzymes in the mitochondria (Cheng et al., 1988), as well as for the correct folding of monomeric imported proteins. It is suggested that folding occurs at the surface of the oligomeric hsp60 in an ATP-dependent reaction. Unlike in the case of the BiP protein, it was shown unequivocally that association with hsp60 is transient and is on the pathway of folding. ATP depletion led to an arrest of folding in the hsp60-bound state. When ATP was added back, folding went to completion, and the product dissociated from hsp60 (Ostermann et al., 1989). Hsp60 and GroEL are constitutively expressed (GroEL accounts for about 1% of total cell protein), but the amounts of both proteins increase upon thermal stress. The molecular mechanism of folding assistance by chaperones is unknown. Also, it is not certain whether these proteins are real enzymes. It appears possible that the extended surfaces of the oligomeric chaperones offer sites for the reversible binding of partially folded target molecuels and thereby suppress aggregation. ATP hydrolysis is necessary for the assisted folding. The predominant energy requirement may be related to the dissociation of the folded or nearly folded protein from the chaperone. Whether this is the only energy-requiring step and whether ATP is needed by the chaperone itself or another protein are not yet known [cf. Ostermann et al. (1989)]. The GroEL protein can also interact transiently with newly synthesized unfolded proteins. This association is decreased in the presence of other unfolded proteins, such as heat-denatured apomyoglobin, and it is abolished by ATP hydrolysis (Bochkareva et al., 1988).

Correlation of in Vitro and in Vivo Results

Several key results have emerged from in vitro folding experiments. (i) The structure of the folded state and the folding mechanism are both encoded in the amino acid sequence; hence folding is a spontaneous process under suitable conditions. (ii) The stabilization energy of a native protein is very small. (iii) Partially folded states can be populated rapidly, and the slow rate-determining steps occur late in folding. (iv) Aqueous medium is a poor solvent for unfolded protein chains in the absence of urea or GdmCl. (v) The yields of refolding are frequently smaller than 100% (particularly for large proteins), because aggregation can compete with proper folding. (vi) The structure of prefolded monomers is fairly unstable and different from the structure in the native oligomeric protein.

From point i it is immediately evident that proteins can gain their native structure without any help and most of them (in particular monomeric single-domain proteins) can achieve this in a biologically reasonable time. Times estimated for the in vivo maturation of proteins lie in the range of a few minutes [Goldenberg & King, 1982; for a review, see Wetlaufer (1985)] around 30 °C. Many in vitro folding reactions are similar in rate, or even faster.[2] Nevertheless, assistance from "folding helpers" is useful (i) to accelerate some slow rate-limiting steps, (ii) to suppress competing aggregation reactions, and (iii) if necessary, to reverse or suppress premature and incorrect folding. Examples for all three tasks have been described. Protein disulfide isomerase catalyzes the slow formation of the correct set of disulfide bonds; prolyl isomerase could accelerate the intrinsically slow isomerization of incorrect prolyl peptide bonds. This may be important for prolines that become cis in the native protein, because the product of protein biosynthesis is presumably an all-trans polypeptide chain. Whether catalysis of folding is indeed the in vivo function of the prolyl isomerases is still unknown. Proline isomerization during refolding can also be strongly accelerated by prefolding in the vicinity of the incorrect proline (Schmid & Blaschek, 1981), so that enzymatic catalysis might not be necessary. An alternative in vivo function of prolyl isomerases could be the recognition and/or interconversion of certain surface-exposed turn regions of folded proteins that contain proline residues. Such interactions could be used to trigger and to time slow activation/deactivation processes. PPIases might thus "liquify" the polypeptide backbone in certain regions of folded or partially folded proteins.

The solubility of "unfolded" or partially folded chains in aqueous solvents is poor; therefore, the premature aggregation of proteins prior to the final slow-folding or assembly steps is a major problem in in vitro folding experiments. It can be overcome in some cases by working at high dilution or by following certain time-dependent assembly protocols. By such procedures it was possible to reassemble functional *E. coli* ribosomes in vitro (Nierhaus & Dohme, 1974). In cells, the same problems are apparent, aggravated by the high concentrations of folded and partially folded polypeptide chains. Under cellular conditions newly formed chains will never adopt an extended conformation, but presumably collapse into a nonspecific globule with still a high tendency to aggregate. This state is presumably characterized (i) by packing defects in the hydrophobic core and (ii) by a suboptimal hydrogen bonding of the main chain. Of course, some proteins can complete folding rapidly to the soluble native state. For others, assistance from folding helpers or binding proteins is necessary to cope with the aggregation problem.

The task of keeping polypeptide chains in an incompletely folded state or of converting folded proteins to such a state is managed by several proteins, which are involved in protein transport. These proteins need energy to prevent folding to the native state. The energy requirement is small, since native and unfolded states do not differ strongly in energy. ATP hydrolysis should be sufficient in this respect to convert the binding proteins from a high-affinity state (for incompletely folded proteins) to a state with a low affinity for folded proteins. Little is known about the folding assistance given to cytoplasmic proteins. It is conceivable that the major function of the hsc70-like proteins is the guidance of their folding or even the protection of folded molecules from aggregation. Evidence for such a role is provided by Nguyen et al. (1989).

The marginal stability of folded proteins is the result of a delicate balance between opposing forces. This guarantees a high flexibility of the polypeptide chain, a prerequisite for the proper functioning of a protein, e.g., as a biocatalyst. In the present context it becomes evident that this marginal stability of folded structures is necessary for another vital property as well, i.e., the facile transient unfolding of a protein with a small input of energy to render or to keep it competent for transport or assembly.

As first pointed out by Bychkova et al. (1988), the incompletely folded state that is involved in protein translocation may resemble the "molten globular" state. This state was detected in the folding of several proteins at equilibrium as well as during refolding under native conditions. States such as the molten globule can be regarded as the "unfolded state" under native solvent conditions (as in cells). It is a globular state that is less compact than the native protein with a large amount of secondary and some tertiary structure. However, a number of important long-range interactions are missing and the flexibility of the polypeptide chain is high (Kuwajima, 1989). With respect to cellular protein folding and transport, two properties are important. (i) The difference in stability between the flexible intermediate and the native state is small. Therefore, a small perturbation, such as binding to the ribosome, to the signal-recognition particle, or to a binding protein, will be sufficient either to arrest a folding protein in that state or to convert a folded protein back to a molten globular conformation. This may be particularly easy for precursor proteins, which are additionally destabilized by the presence of the signal peptide. (ii) In addition to this thermodynamic aspect, the kinetic properties of the molten globule are very important. Unfolding of this partially folded intermediate is rapid; further folding to the native state, however, is a slow reaction. These kinetic properties ensure that a passenger protein can rapidly be further destabilized as soon as it leaves the cytoplasm and enters the translocation machinery. In addition, unwanted refolding immediately after dissociation from the binding protein is blocked, since the reaction to the native state is slow.

The last considerations can be described in more general terms. Unfolded molecules, molecules with local structures, and even intermediates with a fairly high amount of ordered structure can occur in very rapid pseudoequilibria at the beginning of folding. This ensemble of structures is separated from the completely folded state by a high activation barrier, which makes the final steps of folding slow (Segawa & Sugihara, 1984; Goldenberg & Creighton, 1985). So the ultimate function of binding proteins and chaperones would be 2-fold. First, they prevent the nascent or prefolded polypeptide chains

[2] The reader should keep in mind that the majority of in vitro folding experiments are carried out at low temperature, such as at 0–10 °C, where folding is slow, in order to resolve individual steps of the folding kinetics.

from crossing the barrier to the native state, e.g., by providing some of their binding energy to stabilize the partially folded state. Second, they protect these incompletely folded, flexible molecules from premature aggregation by binding at their surface. This binding could be to exposed hydrophobic patches. Alternatively, the chaperones might represent a class of "polypeptide binding proteins", which (similar to single-strand DNA binding proteins) recognize sequence-independent properties of their target molecules, i.e., the peptide backbone. They will not bind to completely folded proteins, because the peptide bonds of their backbone are buried in the interior of the molecule. Binding of BiP and hsc70 to various peptides has indeed been observed by Flynn et al. (1989). In summary, these considerations suggest that many of the emerging molecular features of protein folding and transport in the cell and the properties of the involved proteins can be rationalized in the light of the existing models and concepts that resulted from in vitro studies on the stability and folding of mature proteins.

References

Andria, G., & Taniuchi, H. (1978) *J. Biol. Chem. 253*, 2262–2270.

Andria, G., Taniuchi, H., & Cone, J. L. (1971) *J. Biol. Chem. 246*, 7421–7428.

Baechinger, H. P. (1987) *J. Biol. Chem. 262*, 17144–17148.

Baldwin, R. L. (1989) *Trends Biochem. Sci. 14*, 291–294.

Bergman, L. W., & Kuehl, W. M. (1979) *J. Biol. Chem. 254*, 8869–8876.

Bochkareva, E. S., Lissin, N. M., & Girshovich, A. S. (1988) *Nature 336*, 254–257.

Bole, D. G., Hendershot, L. M., & Kearney, J. F. (1986) *J. Cell Biol. 102*, 1558–1566.

Brandts, J. F., Halvorson, H. R., & Brennan, M. (1975) *Biochemistry 14*, 4953–4963.

Brems, D. N. (1988) *Biochemistry 27*, 4541–4546.

Brems, D. N., Plaisted, S. M., Havel, H. A., Kauffman, E. W., Stodola, J. D., Eaton, L. C., & White, R. D. (1985) *Biochemistry 24*, 7662–7668.

Brown, J. E., & Klee, W. A. (1971) *Biochemistry 10*, 470–476.

Bulleid, N. J., & Freedman, R. B. (1988) *Nature 335*, 649–651.

Bychkova, V. E., Pain, R. H., & Ptitsyn, O. B. (1988) *FEBS Lett. 238*, 231–234.

Chen, W. J., & Douglas, M. G. (1987) *Cell 49*, 651–658.

Cheng, M. J., Hartl, F.-U., Martin, J., Pollock, R. A., Kalousek, F., Neupert, W., & Horwich, A. L. (1989) *Nature 337*, 620–625.

Chirico, W. J., Waters, M. G., & Blobel, G. (1988) *Nature 332*, 805–809.

Creighton, T. E. (1978) *Prog. Biophys. Mol. Biol. 33*, 231–297.

Creighton, T. E. (1983) *Biopolymers 22*, 49–58.

Creighton, T. E., Hillson, D. A., & Freedman, R. B. (1980) *J. Mol. Biol. 142*, 43–62.

Crooke, E., Guthrie, B., Lecker, S., Lill, R., & Wickner, W. (1988) *Cell 54*, 1003–1011.

Dalzoppo, D., Vita, C., & Fontana, A. (1985) *J. Mol. Biol. 182*, 331–340.

Davis, J. M., Boswell, B. A., & Baechinger, H. P. (1989) *J. Biol. Chem. 264*, 8956–8962.

Deshaies, R. J., Koch, B. D., Werner-Washburne, M., Craig, E. A., & Schekman, R. (1988) *Nature 332*, 800–805.

Dolgikh, D. A., Kolomiets, A. P., Bolotina, I. A., & Ptitsyn, O. B. (1984) *FEBS Lett. 165*, 88–92.

Dyson, H. J., Rance, M., Houghton, R. A., Lerner, R. A., & Wright, P. E. (1988a) *J. Mol. Biol. 201*, 161–200.

Dyson, H. J., Rance, M., Houghton, R. A., Wright, P. E., & Lerner, R. A. (1988b) *J. Mol. Biol. 201*, 201–207.

Eilers, M., & Schatz, G. (1986) *Nature 322*, 228–232.

Ellis, R. J., & Hemmingsen, S. M. (1989) *Trends Biochem. Sci. 14*, 339–342.

Endo, T., & Schatz, G. (1988) *EMBO J. 7*, 1153–1158.

Endo, T., Eilers, M., & Schatz, G. (1989) *J. Biol. Chem. 264*, 2951–2956.

Fischer, G., Bang, H., & Mech, C. (1984) *Biomed. Biochim. Acta 43*, 1101–1111.

Fischer, G., Wittmann-Liebold, B., Lang, K., Kiefhaber, T., & Schmid, F. X. (1989a) *Nature 337*, 268–270.

Fischer, G., Berger, E., & Bang, H. (1989b) *FEBS Lett. 250*, 267–270.

Flynn, G. C., Chappell, T. G., & Rothman, J. E. (1989) *Science 245*, 385–390.

Freedman, R. B. (1984) *Trends Biochem. Sci. 9*, 438–441.

Freedman, R. B. (1987) *Nature 329*, 196–197.

Freedman, R. B. (1989) *Cell 57*, 1069–1072.

Garel, J.-R., & Baldwin, R. L. (1973) *Proc. Natl. Acad. Sci. U.S.A. 70*, 3347–3351.

Geetha-Habib, M., Noiva, R., Kaplan, H. A., & Lennarz, W. J. (1988) *Cell 54*, 1053–1060.

Gething, M. J., McCammon, K., & Sambrook, J. (1986) *Cell 46*, 939–950.

Go, N. (1983) *Annu. Rev. Biophys. Bioeng. 12*, 183–210.

Goldberger, R. F., Epstein, C. J., & Anfinsen, C. B. (1963) *J. Biol. Chem. 238*, 628–635.

Goldenberg, D. P., & King, J. (1982) *Proc. Natl. Acad. Sci. U.S.A. 79*, 3403–3407.

Goldenberg, D. P., & Creighton, T. E. (1985) *Biopolymers 24*, 167–182.

Goloubinoff, P., Gatenby, A. A., & Lorimer, G. H. (1989) *Nature 337*, 44–47.

Gschwendt, M., Kittstein, W., & Marks, F. (1988) *Biochem. J. 256*, 1061.

Haber, E., & Anfinsen, C. B. (1961) *J. Biol. Chem. 236*, 422–424.

Harding, M. W., & Handschumacher, R. E. (1988) *Adv. Inflammation Res. 12*, 283–293.

Harmey, M. A., Hallermayer, G., Korb, H., & Neupert, W. (1977) *Eur. J. Biochem. 81*, 533–538.

Hemmingsen, S. M., Woolford, C., van der Vies, S. M., Tilly, K., Dennis, D. T., Georgopoulos, C. P., Hendrix, R. W., & Ellis, J. (1988) *Nature 333*, 330–334.

Hurtley, S. M., Bole, D. G., Hoover-Litty, H., Helenius, A., & Copeland, C. S. (1989) *J. Cell Biol. 108*, 2117–2126.

Jaenicke, R. (1987) *Prog. Biophys. Mol. Biol. 49*, 117–237.

Kassenbrock, C. K., Garcia, P. D., Walter, P., & Kelly, R. B. (1988) *Nature 333*, 90–93.

Kato, I., & Anfinsen, C. B. (1969) *J. Biol. Chem. 244*, 1004–1007.

Kiefhaber, T., Quaas, R., Hahn, U., & Schmid, F. X. (1990a) *Biochemistry* (in press).

Kiefhaber, T., Quaas, R., Hahn, U., & Schmid, F. X. (1990b) *Biochemistry* (in press).

Kim, P. S., & Baldwin, R. L. (1982) *Annu. Rev. Biochem. 51*, 459–489.

Kim, P. S., & Baldwin, R. L. (1984) *Nature 307*, 329–334.

Koivu, J., Mylyllä, R., Helaakoski, T., Pihlajaniemi, T., Tasanen, K., & Kivirikko, K. I. (1987) *J. Biol. Chem. 262*, 6447–6449.

Kuwajima, K. (1977) *J. Mol. Biol. 114*, 241–258.

Kuwajima, K. (1989) *Proteins: Struct., Funct., Genet. 6*, 87–103.

Kuwajima, K., Mitani, M., & Sugai, S. (1989) *J. Mol. Biol. 206*, 547–561.

Lang, K. (1988) Thesis, Regensburg, West Germany.

Lang, K., & Schmid, F. X. (1990) *J. Mol. Biol.* (in press).

Lang, K., Schmid, F. X., & Fischer, G. (1987) *Nature 329*, 268–270.

Lin, L.-N., Hasumi, H., & Brandts, J. F. (1988) *Biochim. Biophys. Acta 956*, 256–266.

Marqusee, S., Robbins, V. H., & Baldwin, R. L. (1987) *Proc. Natl. Acad. Sci. U.S.A. 86*, 5286–5290.

Marston, F. A. O. (1986) *Biochem. J. 240*, 1–12.

Matthews, C. R., & Crisanti, M. M. (1981) *Biochemistry 20*, 784–792.

Munro, S., & Pelham, H. R. B. (1986) *Cell 46*, 291–300.

Munro, S., & Pelham, H. R. B. (1987) *Cell 48*, 899–907.

Nguyen, V. T., Morange, M., & Bensaude, O. (1989) *J. Biol. Chem. 264*, 10487–10492.

Nierhaus, K. N., & Dohme, F. (1974) *Proc. Natl. Acad. Sci. U.S.A. 71*, 4713–4717.

Normington, K., Kohno, K., Kozutsumi, Y., Gething, M. J., & Sambrook, J. (1989) *Cell 57*, 1223–1236.

Oas, T. G., & Kim, P. S. (1988) *Nature 336*, 42–48.

Ostermann, J., Horwich, A. L., Neupert, W., & Hartl, F.-U. (1989) *Nature 341*, 125–130.

Park, S., Lin, G., Topping, T. B., Cover, W. H., & Randall, L. L. (1988) *Science 239*, 1033–1035.

Pelham, H. R. B. (1986) *Cell 46*, 959–961.

Pelham, H. R. B. (1988) *Nature 332*, 776–777.

Pfanner, N., Tropschug, M., & Neupert, W. (1987) *Cell 49*, 815–823.

Pfanner, N., Hartl, F.-U., & Neupert, W. (1988) *Eur. J. Biochem. 175*, 205–212.

Privalov, P. L. (1979) *Adv. Protein Chem. 33*, 167–241.

Privalov, P. L. (1982) *Adv. Protein Chem. 35*, 1–104.

Randall, L. L. (1983) *Cell 33*, 231–240.

Randall, L. L., & Hardy, S. J. S. (1986) *Cell 46*, 921–928.

Roder, H., Elöve, G. A., & Englander, S. W. (1988) *Nature 335*, 700–704.

Rose, M. D., Misra, L. M., & Vogel, J. P. (1989) *Cell 57*, 1211–1221.

Rothman, J. E., & Kornberg, R. D. (1986) *Nature 322*, 209–210.

Roy, H., Cannon, S., & Gilson, M. (1988) *Biochim. Biophys. Acta 957*, 323–334.

Saxena, V. D., & Wetlaufer, D. B. (1970) *Biochemistry 9*, 5015–5023.

Schleyer, M., & Neupert, W. (1985) *Cell 43*, 339–350.

Schmid, F. X., & Baldwin, R. L. (1978) *Proc. Natl. Acad. Sci. U.S.A. 75*, 4764–4768.

Schmid, F. X., & Blaschek, H. (1981) *Eur. J. Biochem. 114*, 11–117.

Schneuwly, S., Shortridge, R. D., Laarrivee, D. C., Ono, T., Ozaki, M., & Pak, W. L. (1989) *Proc. Natl. Acad. Sci. U.S.A. 86*, 5390–5394.

Segawa, S.-I., & Sugihara, M. (1984) *Biopolymers 23*, 2473–2488.

Sela, M., White, F. H., & Anfinsen, C. B. (1957) *Science 125*, 691–692.

Shieh, B.-H., Stamnes, M. A., Seavello, S., Harris, G. L., & Zuker, C. S. (1989) *Nature 338*, 67–70.

Shoemaker, K. R., Kim, P. S., York, E. J., Stewart, J. M., & Baldwin, R. L. (1987) *Nature 326*, 563–567.

Siebendritt, R. (1989) Thesis, Regensburg, FRG.

States, D. J., Dobson, C. M., Karplus, M., & Creighton, T. E. (1984) *J. Mol. Biol. 174*, 411–418.

Takahashi, N., Hayano, T., & Suzuki, M. (1989) *Nature 337*, 473–475.

Tanford, C. (1968) *Adv. Protein Chem. 23*, 121–282.

Tanford, C. (1970) *Adv. Protein Chem. 24*, 1–95.

Taniuchi, H. (1970) *J. Biol. Chem. 245*, 5459–5468.

Tropschug, M., Nicholson, D. W., Hartl, F.-U., Köhler, H., Pfanner, N., Wachter, E., & Neupert, W. (1988) *J. Biol. Chem. 263*, 14433–14440.

Tsou, C. L. (1988) *Biochemistry 27*, 1809–1812.

Tsunenaga, M., Goto, Y., Kawata, Y., & Hamaguchi, K. (1987) *Biochemistry 26*, 6044–6051.

Udgaonkar, J. B., & Baldwin, R. L. (1988) *Nature 335*, 694–699.

Vaucheret, H., Signon, L., Le Bras, G., & Garel, J. R. (1987) *Biochemistry 26*, 2785–2790.

Venetianer, P., & Straub, F. B. (1963) *Biochim. Biophys. Acta 67*, 166–168.

Verner, K., & Schatz, G. (1988) *Science 241*, 1307–1313.

Vestweber, D., & Schatz, G. (1988) *J. Cell Biol. 107*, 2037–2043.

Waldinger, D., Eckerskorn, C., Lottspeich, F., & Cleve, H. (1988) *Biol. Chem. Hoppe-Seyler 369*, 1185–1189.

Watts, C., Wickner, W., & Zimmermann, R. (1983) *Proc. Natl. Acad. Sci. U.S.A. 80*, 2809–2813.

Wetlaufer, D. B. (1985) *Biopolymers 24*, 251–255.

Wetlaufer, D. B., & Ristow, S. (1973) *Annu. Rev. Biochem. 42*, 135–158.

Wickner, W. (1988) *Biochemistry 27*, 1081–1086.

Wierenga, R. K., & Hol, W. G. J. (1983) *Nature 302*, 842–844.

Wong, K.-P., & Tanford, C. (1973) *J. Biol. Chem. 248*, 8518–8523.

Wright, P. E., Dyson, H. J., & Lerner, R. A. (1988) *Biochemistry 27*, 7167–7175.

Zetina, C., & Goldberg, M. E. (1980) *J. Mol. Biol. 137*, 401–414.

Zimmermann, R., & Wickner, W. (1980) *J. Biol. Chem. 255*, 7973–7977.

Zimmermann, R., & Meyer, D. I. (1986) *Trends Biochem. Sci. 11*, 512–515.

Zimmermann, R., Sagstetter, M., Lewis, J. M., & Pelham, H. R. B. (1988) *EMBO J. 7*, 2875–2880.

Chapter 3

Structural Characteristics of α-Helical Peptide Molecules Containing Aib Residues†

Isabella L. Karle*,‡ and Padmanabhan Balaram§

Laboratory for the Structure of Matter, Naval Research Laboratory, Washington, D.C. 20375-5000, and Molecular Biophysics Unit, Indian Institute of Science, Bangalore 560 012, India

Received March 30, 1990

The α-aminoisobutyryl residue $-NHC(CH_3)_2C(O)-$ (Aib or U), although not one of the 20 amino acid residues found in proteins, is a common residue that occurs in peptides produced by microbial sources. Examples of such peptides that possess antibiotic properties are chlamydocin, a peptide with a cyclic backbone (Closse & Huguenin, 1974; Flippen & Karle, 1976), and linear peptides composed of 15–20 residues such as alamethicin, antiamoebin, emerimicin, and zervamicin (Rinehart et al., 1979) that produce voltage-gated ion channels in lipid membranes (Mueller & Rudin, 1968; Mathew & Balaram, 1983). Each of the linear peptides contains 5–8 Aib residues in addition to L-residues that occur in proteins. The preliminary results of a crystal structure of alamethicin (Fox & Richards, 1982) showed that the 20-residue alamethicin molecule folds into a single helix that is predominantly α-helical. Recently, a published abstract concerning the crystal structure of trichorzianine (Le Bars et al., 1988) reports that its conformation is very similar to that of alamethicin.

The replacement of the proton on the C^α atom in an alanine residue with another methyl group severely restricts the possible rotations about the $N-C^\alpha$ and $C^\alpha-C'$ bonds. The torsion angles about these bonds are designated by ϕ and ψ, respectively. In the manner of Ramachandran et al. (1963, 1968) for calculating allowable ϕ and ψ space, Marshall and Bosshard (1972) and Burgess and Leach (1973) showed that the allowable ϕ and ψ angles for the Aib residue occurred in two very restricted regions near −57°, −47° and +57°, +47°. These two regions correspond to a right-handed α-helix or 3_{10}-helix and a left-handed α-helix or 3_{10}-helix, respectively. Since the Aib residue does not have an asymmetric C^α atom, either the L- or D-configuration is equally possible. The handedness of a helix containing Aib residues is fixed by the presence of other residues in the sequence that have the L- or D-handedness.

The implications of the ϕ, ψ map for the Aib residue have indeed been borne out. A review of 28 crystal structures of synthetic tri-, tetra-, and peptapeptides containing at least one Aib residue has shown that all but one contain an incipient 3_{10}-helix (Toniolo et al., 1983). Crystal structure analyses of longer (6–20 residues) Aib-containing peptides, listed in Table I, have shown that these peptides fold into predominantly α-helices, although some have 3_{10}-helices or mixed $3_{10}/\alpha$-helices. Among the very few exceptions to helix formation by the Aib residue are the structure of dihydrochlamydocin (Flippen & Karle, 1976), where the Aib residue is part of an all-trans cyclic tetrapeptide backbone with ω values near 162° instead of 180°, and the structure of

Boc-Cys-Val-Aib-Ala-Leu-Cys-NHMe
(Cys1–Cys6 linked: S-S)

where the Aib residue is part of a β-bend in the antiparallel β-hairpin structure (Karle et al., 1988a).

Many of the peptides shown in Table I have been modeled on fragments of the naturally occurring membrane-active peptides. Others have been designed to examine the effects of the positioning of the Aib residue and of special sequences. All are composed of apolar residues and all yield good crystals from various organic solvents by slow evaporation.

The crystals of many of the longer peptides are stable only when surrounded by mother liquor or water and must be mounted in thin glass capillaries for X-ray data collection. Crystals of other similar peptides are quite stable in the dry state. Better diffraction data, with less attenuation of intensity with increasing scattering angle, can often be obtained by cooling the crystal with a stream of liquid nitrogen to ∼−70 °C. However, this technique has not always been successful, since some of the peptide crystals shatter upon cooling. X-ray scattering data are usually obtained to resolutions of 0.90 Å

†This research was supported in part by the Office of Naval Research, by National Institutes of Health Grant GM30902, and by a grant from the Department of Science and Technology, India.

‡Naval Research Laboratory.
§Indian Institute of Science.

Table I: Crystal Structures of Aib-Containing Peptides[a]

No.	Sequence	Helix Type[b]	Reference
Ia)	Ac-Aib-Pro-Aib-Ala-Aib-Ala-Gln-Aib-Val-Aib-Gly-Leu-Aib-Pro-Val-Aib-Aib-Glu-Gln-Pho	A	Fox & Richards 1982
Ib)	Boc- Ala-Aib-Ala-Aib-Ala-Glu(OBzℓ)-Ala-Aib-Ala-Aib-Ala-OMe	A	Bosch et al, 1985a
Ic)	Boc -Leu-Aib-Pro-Val-Aib-Aib-Glu(OBzℓ)-Gln-Pho	A	Bosch et al, 1985b
Id)	Boc-Aib-Pro-Val-Aib-Val-Ala-Aib-Ala-Aib-Aib-OMe	T	Francis et al, 1983
IIa)	Ac-Phe-Aib-Aib-Aib-Val-Gly-Leu-Aib-Aib-OBzℓ	A†	Marshall et al, 1990
IIb)	pBrBz -Aib-Aib-Aib-Val-Gly-Leu-Aib-Aib-OMe	T	Bavoso et al, 1986
IIc)	pBrBz -Aib-Aib-Aib-Aib-Aib-Leu-Aib-Aib-OMe (2 conformers)	T T	Bavoso et al, 1988
IId)	pBrBz -Aib-Aib-Aib-Aib-Leu-Aib-Aib-OMe	T	Bavoso et al, 1988
IIe)	pBrBz -Aib-Aib-Aib-Aib-Aib-Aib-Aib-Aib-OMe	T	Bavoso et al, 1986
IIIa)	Boc-Val-Aib-Val-Aib-Val-Aib-Val-OMe	T	Francis et al 1985
IIIb)	Boc-Aib-Val-Aib-Aib-Val-Val-Val-Aib-Val-Aib-OMe (2 conformers)	A† A/T	Karle et al, 1989a
IIIc)	Boc-Leu-Leu-Leu-Aib-Leu-Leu-Leu-Aib-OBzℓ	A	Okuyama et al, 1988
IIId)	Boc-Val-Val-Aib-Pro-Val-Val-Val-OMe	T	Karle et al, 1990a
IIIe)	Boc-Val-Ala-Leu-Phe-Aib-Val-Ala-Leu-Phe-OMe	A	To be published
IVa)	Boc-Trp-Ile-Ala-Aib-Ile-Val-Aib-Leu-Aib-Pro-Ala-Aib-Pro-Aib-Pro-Phe-OMe	A,T,B	Karle et al, 1987
IVb)	Boc-Trp-Ile-Ala-Aib-Ile-Val-Aib-Leu-Aib-Pro-OMe·2H_2O	A†	Karle et al, 1986
IVc)	Boc-Trp-Ile-Ala-Aib-Ile-Val-Aib-Leu-Aib-Pro-OMe·isopropanol	A	Karle et al, 1990b
IVd)	Boc-Trp-Ile-Ala-Aib-Ile-Val-Aib-Leu-Aib-Pro-OMe (anhydrous)	T/A	Karle et al, 1988b
IVe)	Ac-Trp-Ile-Ala-Aib-Ile-Val-Aib-Leu-Aib-Pro-OMe (anhydrous)	A/T	Karle et al, 1990b
Va)	Boc-Val-Ala-Leu-Aib-Val-Ala-Leu-----Val-Ala-Leu-Aib-Val-Ala-Leu-Aib-OMe	A†	Karle et al, 1990c
Vb)	Boc-Val-Ala-Leu-Aib-Val-Ala-Leu-Aib-Val-Ala-Leu-Aib-Val-Ala-Leu-Aib-OMe	A	Karle et al, 1990c
Vc)	Boc -Aib-Val-Ala-Leu-Aib-Val-Ala-Leu-Aib-Val-Ala-Leu-Aib-OMe	A	Karle et al, 1989b
Vd)	Boc -Aib-Val-Ala-Leu-Aib-Val-Ala-Leu-Aib-OMe	A	Karle et al, 1988c
Ve)	Boc -Val-Ala-Leu-Aib-Val-Ala-Leu-Aib-OMe (2 conformers)	A A/T	Karle et al, 1990d
Vf)	Boc -Val-Ala-Leu-Aib-Val-Ala-Leu- OMe (2 conformers)	A A(W)	Karle et al, 1990d
VIa)	Boc-Aib-Ala-Leu-Ala-Leu-Aib-Leu-Ala-Leu-Aib-OMe·isopropanol	A	Karle et al, 1990e
VIb)	Boc-Aib-Ala-Leu-Ala-Leu-Aib-Leu-Ala-Leu-Aib-OMe·methanol	A†	Karle et al, 1990e
VIc)	Boc-Aib-Ala-Leu-Ala-Aib-Aib-Leu-Ala-Leu-Aib-OMe (anhydrous)	A†	Karle et al, 1990f
VId)	Boc-Aib-Ala-Leu-Ala-Aib-Aib-Leu-Ala-Leu-Aib-OMe·isopropanol	A†	Karle et al, 1990g
VIe)	Boc-Aib-Ala-Aib-Ala-Leu-Ala-Leu-Aib-Leu-Aib-OMe (2 conformers)	A† A	Karle et al, 1990f
VIIa)	Boc -Ala-Leu-Aib-Ala-Leu-Aib-OMe	T(W)	Karle et al, 1989c
VIIb)	Boc-Aib-Ala-Leu-Aib-Ala-Leu-Aib-Ala-Leu-Aib-OMe·2H_2O·CH_3OH	A(W)	Karle et al, 1988d
VIIc)	Boc-Aib-Ala-Leu-Aib-Ala-Leu-Aib-Ala-Leu-Aib-OMe (anhydrous)(2 conformers)	A† A	To be published

[a] With six or more residues. [b] Helix types: A†, totally α-helix; A, predominantly α-helix; T, 3_{10}-helix; A/T or T/A, mixed $\alpha/3_{10}$-helix; A, T, B, α-, 3_{10}-, and β-ribbon helix; (W), water inserted into backbone.

or better. Anisotropic least-squares refinements result in *R* factors ranging from 4 to 10% for 2000–9000 carefully measured X-ray intensities. The results of these single-crystal analyses of linear peptides not only establish the precise helical conformation of each crystalline peptide but also provide valuable information on hydrogen bonding, helix aggregation, solvation, and conformational heterogeneity.

The stabilization of helical peptide conformation by incorporation of Aib residues into peptide sequences is a recognized and well-established phenomenon (Balaram, 1984; Prasad & Balaram, 1984; Bosch et al., 1985a; Bavoso et al., 1988; Uma & Balaram, 1989). Crystallization may, in fact, be a consequence of appreciable conformational homogeneity in solution, suggesting that such sequences may be ideal candidates for design and construction of stereochemically rigid modules in a "molecular Meccano (or Lego) set" approach to the synthesis of protein mimics. The use of apolar (hydrophobic) sequences permits characterization of conformations in poorly solvating organic solvents, under conditions where peptide folding is largely controlled by nonbonded interactions, intramolecular hydrogen bonding, and electrostatic effects. Such a situation is less complex and more tractable at the present stage of synthetic design than studies in aqueous solutions, where hydrophobic effects generally play a major but incompletely understood role in dictating the nature of polypeptide folding (Creighton, 1985). Aib-containing sequences have therefore been chosen for the construction of stereochemically rigid helical segments in a modular approach to synthetic protein design (Karle et al., 1989b).

Factors Governing Type of Helix

The conformational preference of Aib-containing peptides has been addressed recently, considering such factors as helix length, the inherent stability of the α-helix as compared to the 3_{10}-helix in long sequences, the additional NH···O=C hydrogen bond in a 3_{10}-helix for the same number of residues, solvent polarity, intermolecular interactions, packing efficiency in crystals, and sequence dependence (Marshall et al., 1990). A significant number of additional crystal structure analyses of longer helical peptides have become available since the preparation of the above paper. Accordingly, additional observations can be made.

Table I contains a list of helical peptides with 6–20 residues whose crystal structures are known presently. They are divided into seven groups, with each group having similar sequences. Those labeled 2 conformers have two independent molecules per asymmetric unit of a unit cell in a crystal. Some others with the same sequence (but not necessarily the same cocrystallized solvent) occur in different crystal forms (polymorphs) with different packing motifs. The helix type denoted by the symbol A† indicates a completely α-helical backbone, and A indicates an α-helical backbone except for a 3_{10}-helix-type hydrogen bond at one or both termini. The symbol T indicates a complete 3_{10}-helix, whereas the symbol T/A denotes a mixed 3_{10}/α-helix. The 16-residue peptide IVa has three Pro residues near the C-terminus, which are involved in three successive β-bends that twist into an approximate 3_{10}-helix. The helix types from Table I for peptides with six residues or more are represented in Figure 1.

Figure 1, showing the distribution of the types of helices with a plot of the total number of residues in a peptide vs the number of Aib residues (the type of helix is indicated by ▲ = T, ● = A, and Ⓐ = T/A), provides a simple graphic delineation between helix type. The 3_{10}-helix is preferred for the shorter peptides ($n \sim 8$ or less), and the α-helix is preferred for the longer peptides (n = 9–20). Furthermore, the presence of a large fraction of Aib residues in a peptide induces a 3_{10}-helix. These results reaffirm the suggestions from earlier observations when fewer structures of the longer peptides were known (Bosch et al., 1985b; Toniolo et al., 1985; Karle et al., 1986).

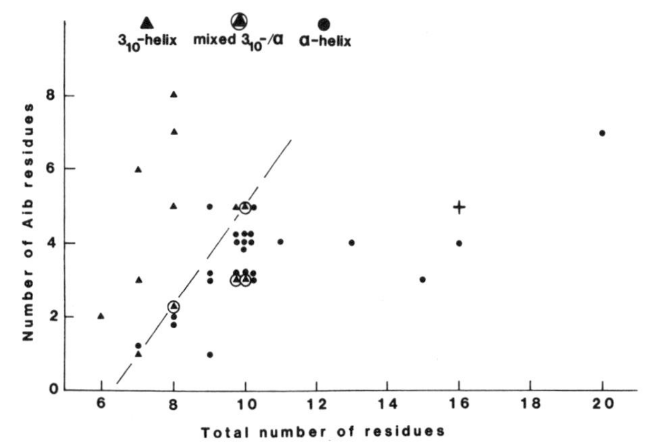

FIGURE 1: Helix type as a function of the total number of residues in peptide and total number of Aib residues. The dashed line separating the 3_{10}-helices from the α-helices has been placed arbitrarily. Peptides containing Aib residues, but with five or fewer total residues, overwhelmingly form a 3_{10}-helix or an incipient 3_{10}-helix and are omitted from the plot. The + symbol indicates a 16-residue, totally helical peptide with three Pro residues.

At the interface between 3_{10}-helices and α-helices in Figure 1, there is some overlap in helix types. For the most part, they are pairs of peptides of the same sequence that undergo a facile 3_{10}/α transition when they are crystallized in different polymorphs or when they occur in the same crystal with more than one independent molecule (i.e., not related by crystal symmetry). For example, the structure of Boc-Trp-Ile-Ala-Aib-Ile-Val-Aib-Leu-Aib-Pro-OMe (IVb–e in Table I and points at 3 Aib/10 residues in Figure 1) has been analyzed in four different crystal forms, with different cocrystallized solvents or a different end group (Ac replacing Boc) (Karle et al., 1986, 1988b, 1990b). In crystals IVb and IVc, both in space group *P*1 with parallel packing of helices, the helix is the α-type, except for the helix reversal at Aib,[9] which occurs relatively often at an Aib residue if it is in an ultimate position or occasionally in the penultimate position. In crystal IVd, in space group $P2_1$ with parallel packing, the helix has switched to a primarily 3_{10}-type with only two 5→1-type hydrogen bonds in the middle. Figure 2 shows a superposition of the two helix types with the predominantly 3_{10}-helix being longer than the predominantly α-helix. In crystal IVe, where the Boc end group has been replaced by Ac, the helices pack in an antiparallel fashion in space group $P2_1$. In this crystal, there is one 3_{10}-type hydrogen bond in the middle of the α-helix. Usually in mixed 3_{10}/α-helices, the 3_{10}-type hydrogen bonds are only at the termini. The 3_{10}/α-helix transitions in this group of four polymorphs cannot be ascribed to the differences in environment between parallel and antiparallel packing of helices nor to differences evolving from the exchange of Ac and Boc end groups at the N-terminus. The cocrystallized solvents, or lack thereof, do change the environment of the peptides somewhat and may contribute to the helix transitions.

In the second case of 3_{10}/α-helix transitions in the same peptide, Boc-(Val-Ala-Leu-Aib)$_2$-OMe (Ve in Table I and points at 2 Aib/8 residues in Figure 1), the two helices alternate in the same crystal. One helix is completely α-helical; the other is mixed with three 4→1-type hydrogen bonds interspersed with three 5→1 hydrogen bonds (Karle et al., 1990d). Similarly, in the crystal of Boc-Aib-Val-Aib-Aib-Val-Val-Val-Aib-Val-Aib-OMe (IIIb in Table I and points

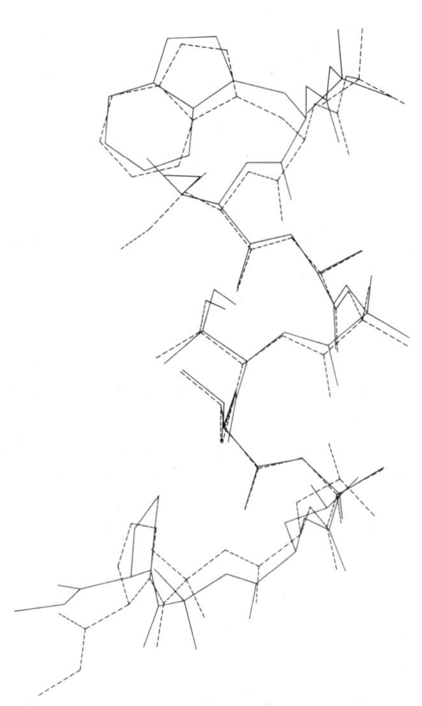

FIGURE 2: Superposition (with a least-squares fit of the backbone atoms) of Boc-Trp-Ile-Ala-Aib-Ile-Val-Aib-Leu-Aib-Pro-OMe in two different crystal forms, solid line in space group $P2_1$ (Karle et al., 1988b) and dashed line in space group $P1$ (Karle et al., 1986). The backbone of the conformer shown by the dashed line is predominantly α-helical, while the backbone in the conformer shown by the solid line has only two α-helical-type hydrogen bonds near the middle and the remainder at both ends are the 3_{10}-helix type.

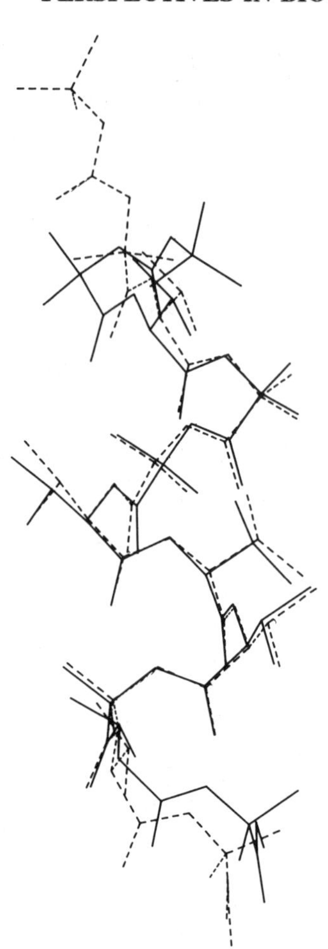

FIGURE 3: Superposition of the two conformers of Boc-Aib-Val-Aib-Aib-Val-Val-Val-Aib-Val-Aib-OMe occurring side by side in the same crystal. The conformer shown by the solid line is completely α-helical. The one shown by the dashed line has 3_{10}-type hydrogen bonds at both ends plus an unwinding of the backbone at the N-terminus (Karle et al., 1989a).

at 5 Aib/10 residues in Figure 1), one helix that is completely α-helical alternates with another that has 3_{10}-helical segments at either end plus an unwinding of the backbone at the N-terminus. Figure 3 shows a superposition of the two conformers (Karle et al., 1989a). In these cases, the crystallizing conditions are identical since only one crystal is involved. Finally, the completely α-helical Ac-Phe-Aib-Aib-Aib-Val-Gly-Leu-Aib-Aib-OBzl (IIa and points at 5 Aib/9 residues in Figure 1) (Marshall et al., 1990) occupies a somewhat anomalous position in Figure 1, considering the high Aib content. Previous experience suggests that a different polymorph of this peptide could have a mixed $3_{10}/\alpha$-helix. An explanation for the easy $3_{10}/\alpha$ transitions in different molecules of the same peptide, as well as the partial unwinding of the backbone, as in Figure 3, must rest on the near equality of the stability of the two forms and on other factors concerning subtle changes in environment that are not well understood.

NUMBER OF AIB RESIDUES

The question of how many Aib residues are needed to induce helix formation in longer peptides has been answered partially by the structures of three peptides: Boc-Val-Val-Aib-Pro-Val-Val-Val-OMe, IIId (Karle et al., 1990a); Boc-Val-Ala-Leu-Aib-Val-Ala-Leu, Vf (Karle et al., 1990d); and Boc-Val-Ala-Leu-Phe-Aib-Val-Ala-Leu-Phe-OMe, IIIe (to be published). These hepta- and nonapeptides have only one Aib residue, located in the middle in each sequence. The first peptide forms a 3_{10}-helix, and the latter two are predominantly α-helical. The high propensity of Aib for helix formation is demonstrated particularly by peptide IIId, in which one Aib residue overcomes the effects of five Val residues and one Pro residue, which are known to be poor helix formers (Chou & Fasman, 1974). The longest sequence of non-Aib residues in a helix, among the known structures, occurs as a string of six residues in the middle of Boc-Val-Ala-Leu-Aib-*Val-Ala-Leu-Val-Ala-Leu*-Aib-Val-Ala-Leu-Aib-OMe, Va (Karle et al., 1990c). This peptide is completely α-helical, as shown in Figure 4.

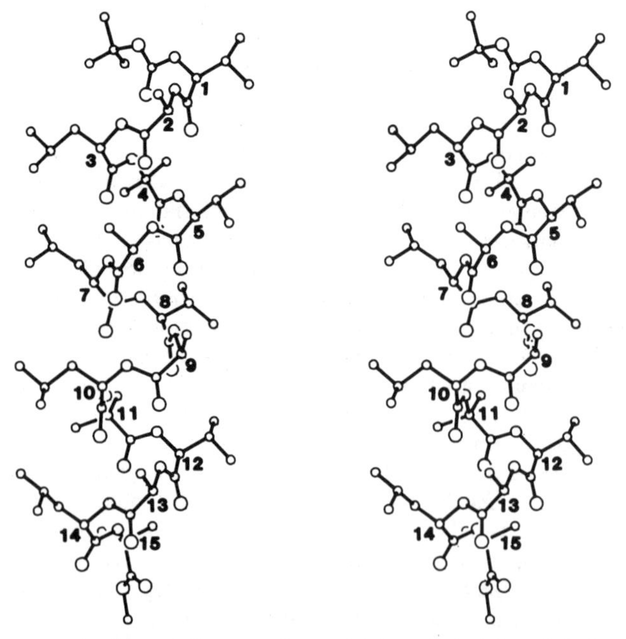

FIGURE 4: Stereo diagram of completely α-helical Boc-Val-Ala-Leu-Aib-Val-Ala-Leu-Val-Ala-Leu-Aib-Val-Ala-Leu-Aib-OMe (Karle et al., 1990c). This helical peptide contains six contiguous non-Aib residues.

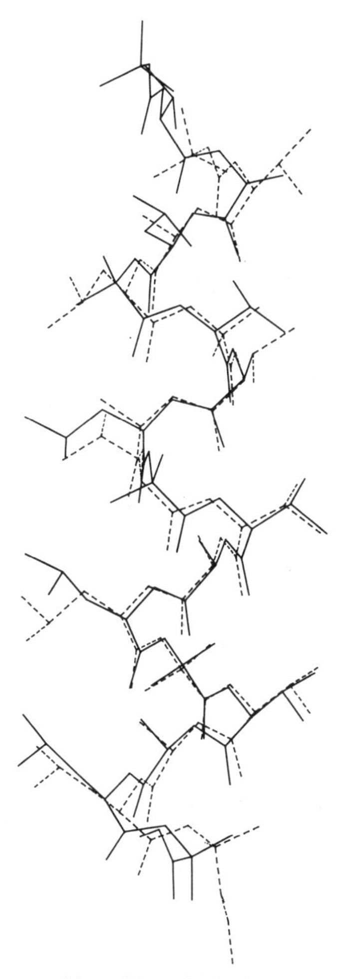

FIGURE 5: A superposition of Boc-(Val-Ala-Leu-Aib)$_4$-OMe (solid line) and the 15-residue peptide with Aib^8 deleted (dashed line, also shown in Figure 4). The sequence of the last eight residues is the same, and the sequence in the top half has a shift of one due to the omission of Aib^8 from the 15-residue peptide; nevertheless, the helical nature of the backbone is almost undisturbed (Karle et al., 1990c).

Position of Aib Residue in Sequence

Two series of peptides were designed and analyzed to evaluate the effects of the presence of Aib residues at the C- or N-termini of a sequence, the removal of Aib from the middle of a sequence, and the exchange of Aib with Leu or Ala residues. In the first series, Va–f in Table I, 7–16-residue peptides were synthesized by using the tetrad (Val-Ala-Leu-Aib)$_n$ with Aib residues added or removed. In the second series, VIa–e and VIIa–c in Table I, 6- and 10-residue peptides were synthesized by using the triad (Ala-Leu-Aib)$_n$, or permutations of Ala, Leu, and Aib, with Aib added to the N-terminus of the peptide.

The 15-residue peptide Va differs from the 16-residue peptide Vb only by the deletion of Aib^8 from the middle of Vb. The helical structure of the backbones in the two peptides is entirely similar. A least-squares fit of backbone atoms N(3)–O(14) in Vb with N(2)–O(13) in Va yields a rms deviation of 0.28 Å. A superposition of the two molecules is shown in Figure 5, where the final eight residues are the same but the initial residues are shifted by one because of the deletion of the middle Aib in Va (Karle et al., 1990c).

The decapeptides VIa and VIc differ only by the exchange of Leu^5 in VIa by Aib^5 in VIc. The effect of such a replacement on conformation is minuscule. The rms deviation of the backbone atoms in the two peptides is only 0.12 Å (Karle et al., 1990f). The isomeric decapeptides VIc and VIe differ by the interchange of Leu and Aib at residues 3 and 5 and the interchange of Ala and Aib at residues 6 and 8. Even with changes at four sites, the conformations of the two peptides are very similar, with a rms deviation of only 0.26 Å for the backbone atoms (Karle et al., 1990f). The conclusion that can be drawn from these structure analyses is that an Aib residue can replace an apolar residue, or vice versa, without disturbing the α-helix. The precise positioning of Aib residues in these cases is without effect on the α-helical nature of the backbone.

All but two of the 18 peptides in groups V, VI, and VII, as well as several others in Table I, terminate with a Leu-Aib-OMe sequence; peptides Id and IIa–e terminate with an Aib-Aib-OMe (or OBzl) sequence and two others with an Ala-Aib-OMe sequence. Only in three of the peptides does the terminal Aib residue continue the right-handed helix with ϕ and ψ values near −54° and −47°, respectively. In 14 of the peptide molecules there is a helix reversal at the final Aib residue to a left-handed helix with ϕ and ψ values near +54° and +45°, respectively. The helix reversal at the final Aib residue had already been noted in the 2–5-residue peptides (Toniolo et al., 1983). However, another conformation for the terminal Aib residue appeared in some of the longer peptides. The final Aib residue is semiextended with ϕ and ψ values near −57° and +150°, respectively, in peptides Va, Vc, and VIa, near −57° and +180° in peptides Vd, Ve, and VIb, and near +57° and −150° in peptide VIIb. Conformational energy maps for the Aib residue (Marshall, 1972; Burgess & Leach, 1973; Uma & Balaram, 1989) do show minor elongated minima in the ϕ, ψ space corresponding to the observed ϕ and ψ values.

The presence or absence of an Aib residue at the N-terminus does not appear to affect the helical nature of peptides; see, e.g., peptides Va–f.

The Spread of ϕ, ψ Conformational Angles for α-Helices

The torsional angles ϕ (torsion about the N–C$^\alpha$ bond) and ψ (torsion about the C$^\alpha$–N bond) have been plotted in a Ramachandran-type plot in Figure 6 for each residue separately for the helical peptides in groups VI and VII in Table I, which contain only Aib, Ala, and Leu residues, and for helical peptides in group V, which contain Aib, Ala, Val, and Leu residues, in panels a and b, respectively. The values for the Aib residues at the C-terminus have been omitted because of the helix reversal that occurs often. A previous plot, using a different choice of peptide helices (Karle et al., 1990f), separated only the Aib residues from the remainder. In both the previous plot and the plots in Figure 6, the ϕ and ψ values for Aib (●) occur in a concentrated region near −55° and −45°, respectively (at the right-hand side of the distribution), a region that coincides with the original calculations of conformational energy maps for the Aib residue. In the present maps, the ϕ and ψ values for individual residues are plotted separately for each to determine whether the large spread in the observed ϕ and ψ values is a function of the type of residue. The clusters of points for the Ala residue (+) are much larger than those for Aib. In both panels a and b, the center of the Ala clusters is near −65° and −35°. In Figure 6b, the cluster for the Val residues (/) is longer than that for Ala residues (along the diagonal direction where ϕ and ψ are related inversely), and the center of the cluster appears to have moved to the left slightly with ϕ and ψ near −67° and −35°. Finally, the distribution for the Leu residues (▲) not only overlaps the distributions for Ala and Val but also extends far to the left with some points at $\phi < -100°$. Many of the Leu residues that fall on the far left in the distribution are the penultimate residue

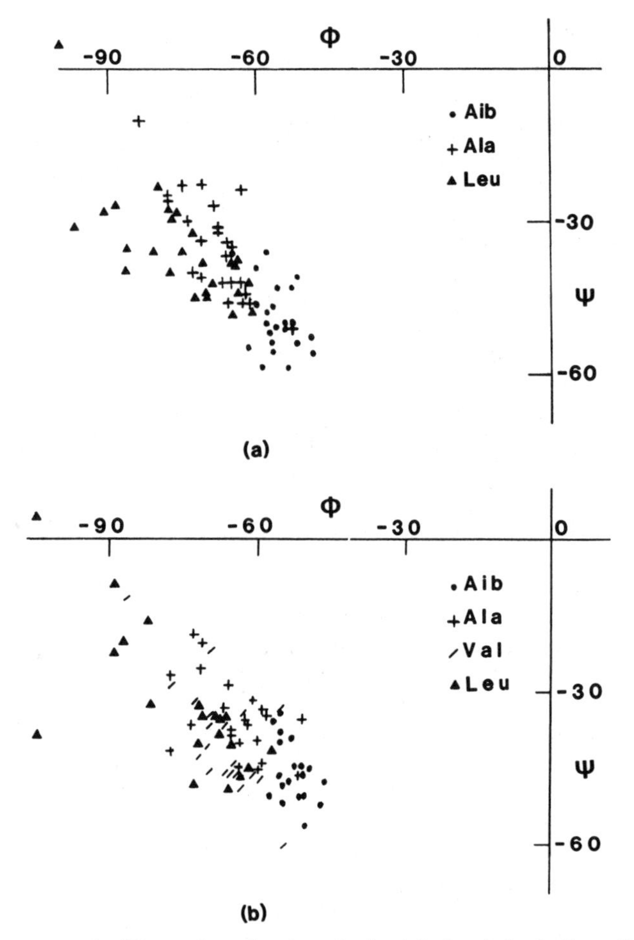

FIGURE 6: Observed torsional angles in right-handed α-helices of Aib-containing peptides where ϕ is the rotation about the N–C^{α} bond and ψ is the rotation about the C^{α}–C′ bond. (a) Data from crystal structures of decapeptides containing Aib, Ala, and Leu residues in various sequences. (b) Data from crystal structures of 7–16-residue peptides containing the tetrad $(\text{Val-Ala-Leu-Aib})_n$ and the addition or deletion of Aib residues.

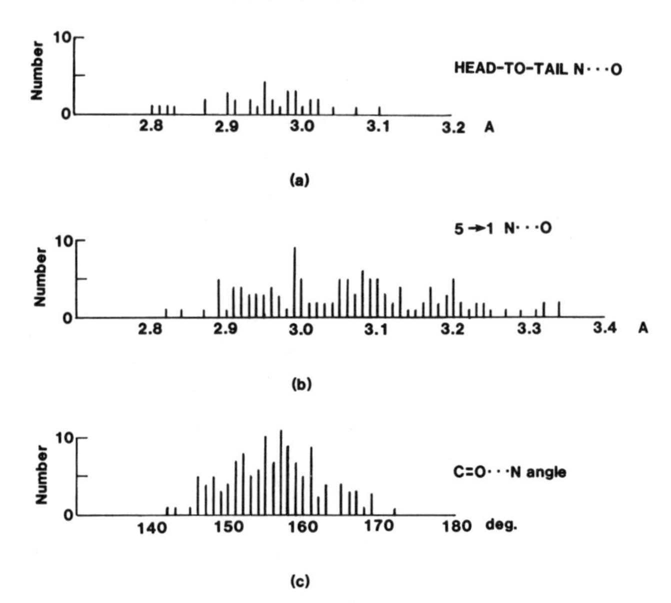

FIGURE 7: Observed values of hydrogen bonds in α-helices of Aib-containing peptides. (a) Distribution of N···O distance values in hydrogen bonds between helices. (b) Distribution of N···O distance values in 5→1-type hydrogen bonds in α-helices. (c) Distribution of C═O···N angle values in α-helix-type hydrogen bonds.

in the peptide sequence and followed by an Aib that reverses the helix direction. However, other Leu residues with aberrant ϕ and ψ values occur in the middle of a peptide sequence. The wide range of ϕ and ψ values actually observed is larger than previously anticipated. Furthermore, there is a suggestive dependence upon type and/or size of residue, since the clusters of observed ϕ and ψ values in α-helices increase in spread and move leftward and upward in Figure 6 for the Ala, Val, and Leu residues, respectively.

Hydrogen Bonds

5→1 Hydrogen Bond in α-Helix. The spread of ϕ and ψ angles, discussed above, contributes to minor irregularities in the α-helix. These irregularities are manifested primarily by the spread of N···O distances in the 5→1 hydrogen bonds. Criteria for accepting a 5→1 hydrogen bond (or choosing between a 4→1 and a 5→1 hydrogen bond in a distorted helix) include not only the N···O distance but also the NH···O distance and the direction of the NH bond toward the O atom (with the NH···O angle 150°–180°). The distribution of the N···O distances in 5→1 hydrogen bonds in peptides IIIb, IVa–c, Va–f, VIa–e, and VIIb–c from Table I is shown in Figure 7b. There is a continuous distribution from 2.89 to 3.25 Å, with several outliers on either side and a median value of 3.06 Å. A plot of C═O···N angles for the same 5→1 hydrogen bonds, Figure 7c, is approximately Gaussian with a median value of 156°. Plots of N···O distances and C═O···N angles in 12 α-helices of refined proteins (resolution 2.0 Å or better) indicate about the same median value for the angle and a somewhat smaller median value for the N···O distance, ~3.00 Å (Barlow & Thornton, 1988), although the spread of N···O values is quite comparable for α-helices in proteins and in Aib-containing helical peptides.

Head-to-Tail Hydrogen Bonds. All the peptides listed in Table I form continuous rods in crystals by head-to-tail hydrogen bonding. Since almost all the side chains in these peptides are hydrocarbons (except for group I in Table I), there are no direct lateral hydrogen bonds between helices. In an isolated ideal α-helix, N(1)H, N(2)H, and N(3)H at the top of the helix, with the NH moieties directed upward, and C═O(f), C═O(f − 1), and C═O(f − 2) (where f is the final residue at the bottom of the helix), with all C═O moieties directed downward, do not participate in the intramolecular 5→1 hydrogen bonds. These NH and C═O moieties participate in head-to-tail hydrogen bonds in various modes: (a) It is possible to form three direct NH···O═C hydrogen bonds in perfect register, as illustrated by peptide IIIb in Figure 8a. (b) Two direct NH···O═C bonds may be formed by utilizing N(1)H and N(2)H, while N(3)H remains unsatisfied. The head-to-tail region of peptide VIe is shown in Figure 8b. The nearest atom in any neighboring molecule to N(3) is O(10) at 3.9 Å. It is not unusual for an N(2)H or N(3)H moiety not to participate in any hydrogen bonding owing to a lack of acceptor. Inefficient crystal packing appears to be the cause. (c) There may be only one direct NH···O═C hydrogen bond and the remaining head-to-tail hydrogen bonds are mediated by several water molecules, as shown by the 15-residue peptide Va in Figure 8c. (d) There are no direct NH···O═C hydrogen bonds in peptide IVd in Figure 8d. Rather, the NH in a side chain of the first residue (Trp) forms a bifurcated hydrogen bond with two carbonyl oxygens from the upper molecule, and the OH of a 2-propanol solvent molecule mediates hydrogen bonding between N(1)H and O(7) of the upper molecule. A 4→1 type of hydrogen bond is formed between N(3)H···O(0), where O(0) is part of the Boc end group. Further, N(2)H remains unsatisfied. The closest atom in any neighboring molecule to N(2) is O(10) (in the upper molecule) at 5.5 Å.

The above examples show the adaptability of helical peptides, even with end groups present, to a variety of head-to-tail

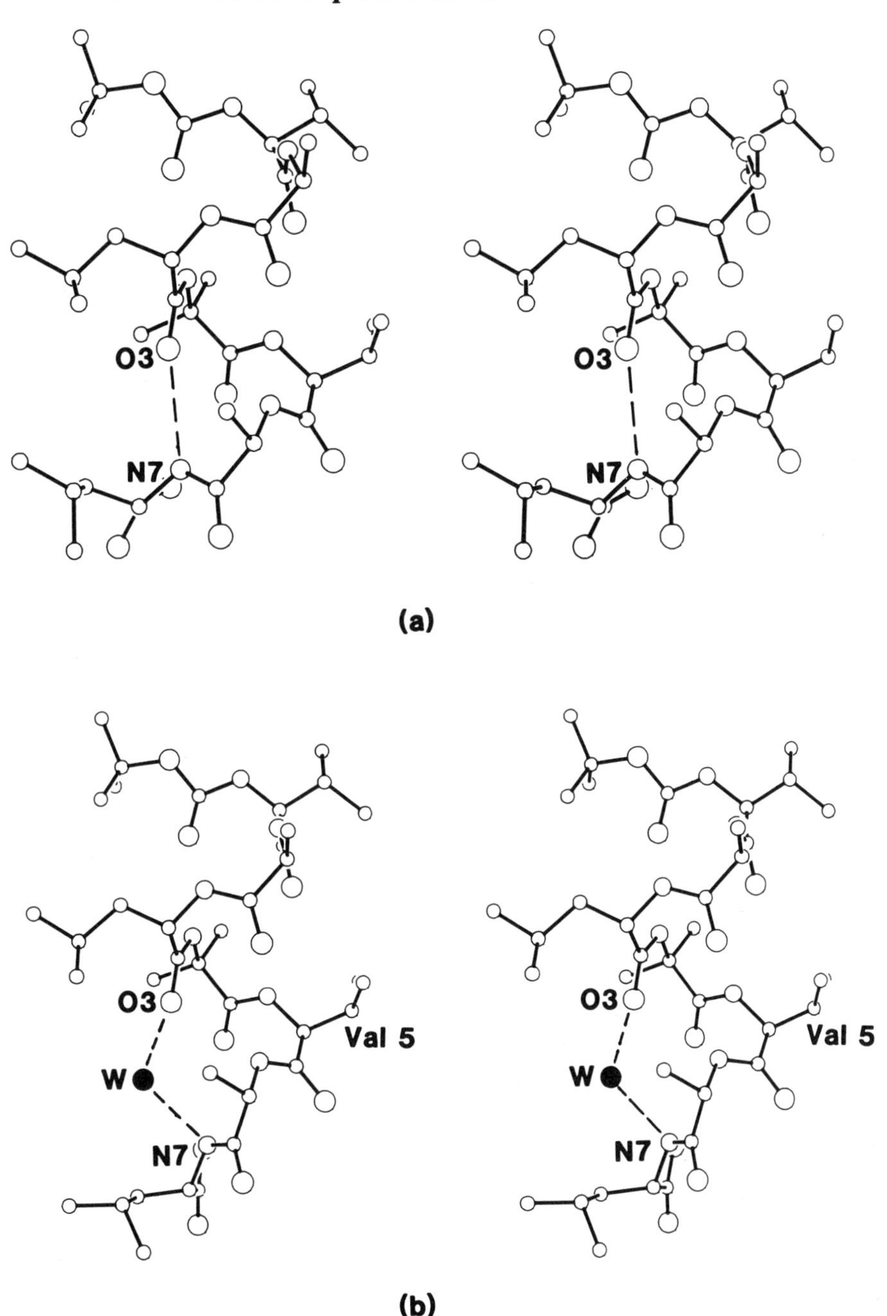

FIGURE 9: Stereo diagrams of two peptide molecules that occur side by side in a crystal of Vf. (a) The backbone is in the form of a normal helix. (b) The N(7)H···O(3) hydrogen has been broken by the insertion of a water molecule W. Two new hydrogen bonds have been formed, W···O(3) and N(7)H···W (Karle et al., 1990d).

Bifurcated 5→1 Hydrogen Bonds. The side chains in apolar helices often pack in an inefficient manner in a crystal lattice, leaving voids predominantly surrounded by hydrophobic moieties. These voids can be occupied by water molecules that make a hydrogen bond to a carbonyl group in the helix that is already involved in a 5→1 hydrogen bond. Such bifurcated hydrogen bonds involving water molecules (or 2-propanol) have been observed in the apolar peptides Va, Vc, Vd, VIc, and VIe. In peptides Vc and VIe, a water molecule links neighboring helices by hydrogen bonds to carbonyl oxygens and may be instrumental in stabilizing the structure, which otherwise lacks the possibility for any lateral hydrogen bonds (Karle et al., 1989b). Similar bifurcated hydrogen bonds in proteins between helix and water or a side-chain group have been reported and analyzed (Blundell et al., 1983; Barlow & Thornton, 1988; Sundaralingam & Sekharudu, 1989). In proteins, such bifurcated bonds generally occur in polar rather than in apolar environments. A notable exception occurs in the structure of the photosynthetic reaction center of the purple bacterium *Rhodopseudomonas viridis*, where a water molecule cross-links two hydrophobic transmembrane helices by forming hydrogen bonds with carbonyl oxygens that also are involved in α-helix hydrogen bonds (Deisenhofer & Michel, 1989).

Water Insertion into Backbone. The most interesting mode of hydration is that of a rupture of a 5→1 hydrogen bond in a helix and the spreading of the NH and CO moieties to incorporate a water molecule into the helix backbone. In peptide Vf (Karle et al., 1990d) there are two independent peptide molecules side by side in the crystal. One peptide molecule is totally helical (Figure 9a). In the other peptide molecule, the α-helical N(3)···O(7) hydrogen bond has been severed by the entry of water molecule W, which in turn makes two new hydrogen bonds, W···O(3) and N(7)···W, 2.77 and 2.97 Å, respectively. The opening of the helix has been accomplished by a minimum of motion, that is, by ϕ and ψ rotations at $C^{\alpha}(5)$ of Val^5, −87°, −11° before hydration and

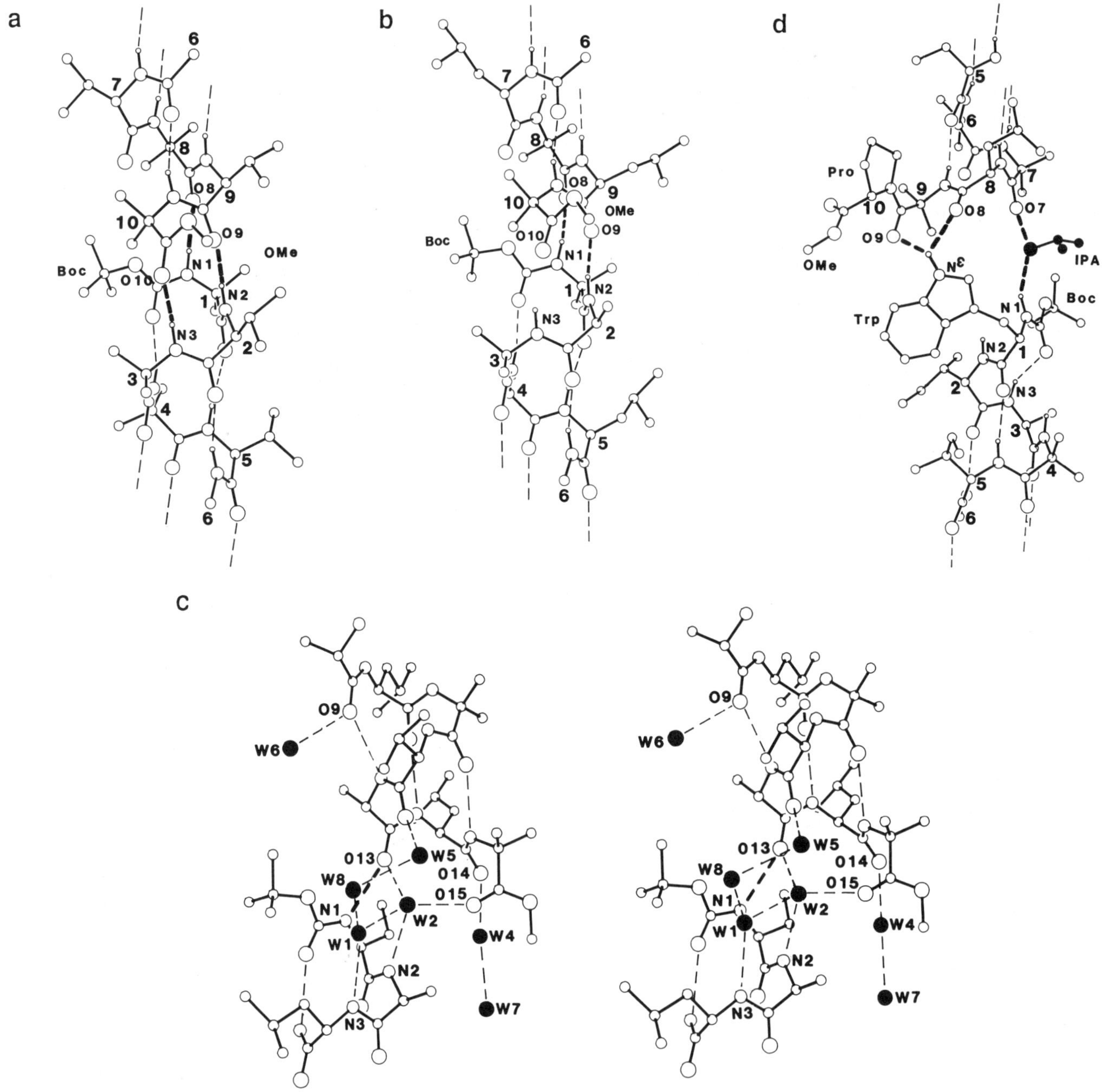

FIGURE 8: Modes of head-to-tail hydrogen bonding between helices. The lower part of one helical molecule is shown over the upper part of a neighboring helical molecule. The head-to-tail hydrogen bonds are emphasized by heavy dashed lines. (a) Three direct intermolecular NH···O=C bonds in perfect register in peptide IIIb (Karle et al., 1989a). (b) Two direct NH···O=C bonds in peptide VIe. The N(3)H moiety does not participate in any hydrogen bonding (Karle et al., 1990f). (c) One direct N(1)H···O(13) bond in peptide Va. Water W(2) mediates hydrogen bonds between N(2)H and O(13) and O(15). Water molecules W(1), W(8), and W(5) mediate hydrogen-bond connections between N(3)H and O(12) (Karle et al., 1990c). A stereo diagram is shown. (d) No direct NH···O=C backbone head-to-tail hydrogen bonds in peptide IVc. The NH in the Trp side chain makes a bifurcated hydrogen bond to O(8) and O(9). The OH in 2-propanol mediates hydrogen bonds between N(1) and O(7). The N(2)H moiety does not participate in any hydrogen bonding (Karle et al., 1989b).

hydrogen-bonding schemes. The distribution of N···O distances in direct head-to-tail bonding is shown in Figure 7a, where the median N···O distance is 2.95 Å.

WATER IN HYDROPHOBIC HELICAL STRUCTURES

An example of the intimate association of water with hydrophobic α-helices has been demonstrated in the head-to-tail hydrogen-bonding region (Figure 8c). Both ends of the α-helices or 3_{10}-helices are quite polar around the extended NH and CO moieties that are free to form direct NH···O=C hydrogen bonds between helices or hydrogen bonds mediated by water molecules (or alcohol molecules). The occurrence of water near the middle of helices where the apolar side chains create an apparently totally hydrophobic environment is more unusual. Three different modes of hydration have been found among Aib-containing helical peptides.

Exposed Carbonyls. The presence of a Pro residue in the middle of a sequence that forms a helix results in the loss of an intramolecular hydrogen bond due to the lack of a proton on the N in the Pro. The bulk of the pyrrolidine ring causes the helix to bend ~30° and exposes the free C=O group at the bend to the outside environment. The 16-residue apolar peptide IVa affords such an example, in which the bend in the helix caused by Pro^{10} results in the extension of $C{=}O^{7}$ away from the nearby hydrocarbon side chains from Leu and Aib. The $C{=}O^{7}$ attracts a water molecule, which in turn makes a hydrogen bond to another water molecule, resulting in a minipolar area on the hydrophobic helix (Karle et al., 1987).

−91°, +2° after hydration. None of the other torsional angles has been changed significantly. It should be noted that, except for the hydrogen bonds, the water molecule is entirely surrounded by hydrophobic side chains of its own molecule as well as from neighboring molecules.

Another example of a very similar insertion of water into a helix is afforded by the hydrophobic peptide VIIb (Karle et al., 1988d) and its anhydrous polymorph VIIc. In this case, the normal helix for VIIc and the hydrated helix for VIIb occur in different crystals grown from ethylene glycol and methanol/water, respectively. The opening of the helix to accommodate the water molecule occurs at Leu^3 with major changes of ϕ and ψ values from −75°, −42° for the normal helix to −102°, +15 for the hydrated helix. Contrary to the predominantly hydrophobic environment for the water in the peptide in Figure 9, peptide VIIb acquires amphiphilic character by the insertion of water into the helix and subsequent attraction of additional water (Karle et al., 1988d). The six-residue peptide VIIa (a fragment of VIIb) also has a water molecule inserted into the helix in a mode identical with that of VIIb (Karle et al., 1989c).

The occurrence of both the normal helix and internally hydrated helix next to each other in the same crystal of Vf and in polymorphs VIIb and VIIc grown in the same manner (but with a different solvent) points to the facile transformation between the unhydrated and hydrated helices and to the probable role of water in helix folding.

Similar water insertion into the helix of troponin C (Satyshur et al., 1988) prompted a search that identified an additional 19 proteins with internally hydrated helices (Sundaralingam & Sekharudu, 1989c).

Implications for Peptide Design

The crystal structures of Aib-containing peptides described in this paper establish that stable helical conformations, predominantly or totally α-helical, can be maintained in 10–16-residue peptides containing triads, tetrads, and even hexads of non-Aib residues. Helical rods having three to more than four turns of $3_{10}/\alpha$-helix can be readily constructed. The precise positioning of Aib residues, or their total number, appears to be without effect on the backbone. The helical conformation of diverse sequences, with 10–50% Aib residues, suggests that stable helical segments can be constructed reproducibly and that Aib can be substituted for other residues to promote helix formation. Such segments can provide an important element in a modular approach to synthetic protein design, as, for example, in piecing together a hydrophobic four-helix bundle.

Acknowledgments

We are deeply indebted to our colleagues K. Uma and M. Sukumar (India) and Judith L. Flippen-Anderson (Washington) for their crucial contributions to the investigations reviewed here.

Registry No. Aib, 62-57-7.

References

Balaram, P. (1984) *Proc.—Indian Acad. Sci. 93*, 703–717.

Barlow, D. J., & Thornton, J. M. (1988) *J. Mol. Biol. 201*, 601–619.

Bavoso, A., Benedetti, E., Di Blasio, B., Pavone, V., Pedone, C., Toniolo, C., & Bonora, G. M. (1986) *Proc. Natl. Acad. Sci. U.S.A. 83*, 1988–1992.

Bavoso, A., Benedetti, E., Di Blasio, B., Pavone, V., Pedone, C., Toniolo, C., Bonora, G. M., Formaggio, F., & Crisma, M. (1988) *J. Biomol. Struct. Dyn. 5*, 803–817.

Blundell, T. L., Barlow, T. J., Borkakoti, N., & Thornton, J. M. (1983) *Nature (London) 306*, 281–283.

Bosch, R., Jung, G., Schmitt, H., & Winter, W. (1985a) *Biopolymers 24*, 961–978.

Bosch, R., Jung, G., Schmitt, H., & Winter, W. (1985b) *Biopolymers 24*, 979–999.

Burgess, A. W., & Leach, S. J. (1973) *Biopolymers 12*, 2599–2605.

Chou, P. Y., & Fasman, G. D. (1974) *Biochemistry 13*, 222–245.

Closse, A., & Huguenin, R. (1974) *Helv. Chim. Acta 57*, 533–545.

Creighton, T. E. (1985) *J. Phys. Chem. 89*, 2452–2459.

Deisenhofer, J., & Michel, H. (1989) *EMBO J. 8*, 2149–2170.

Flippen, J. L., & Karle, I. L. (1976) *Biopolymers 15*, 1081–1092.

Fox, R. O., Jr., & Richards, F. M. (1982) *Nature (London) 300*, 325–330.

Francis, A. K., Iqbal, M., Balaram, P., & Vijayan, M. (1983) *FEBS Lett. 155*, 230–232.

Francis, A. K., Vijayakumar, E. K. S., Balaram, P., & Vijayan, M. (1985) *Int. J. Pept. Protein Res. 26*, 214–223.

Karle, I. L., Sukumar, M., & Balaram, P. (1986) *Proc. Natl. Acad. Sci. U.S.A. 83*, 9284–9288.

Karle, I. L., Flippen-Anderson, J. L., Sukumar, M., & Balaram, P. (1987) *Proc. Natl. Acad. Sci. U.S.A. 84*, 5087–5091.

Karle, I. L., Kishore, R., Raghothama, S., & Balaram, P. (1988a) *J. Am. Chem. Soc. 110*, 1958–1963.

Karle, I. L., Flippen-Anderson, J. L., Sukumar, M., & Balaram, P. (1988b) *Int. J. Pept. Protein Res. 31*, 567–576.

Karle, I. L., Flippen-Anderson, J. L., Uma, K., & Balaram, P. (1988c) *Int. J. Pept. Protein Res. 32*, 536–543.

Karle, I. L., Flippen-Anderson, J. L., Uma, K., & Balaram, P. (1988d) *Proc. Natl. Acad. Sci. U.S.A. 85*, 299–303.

Karle, I. L., Flippen-Anderson, J. L., Uma, K., Balaram, H., & Balaram, P. (1989a) *Proc. Natl. Acad. Sci. U.S.A. 86*, 765–769.

Karle, I. L., Flippen-Anderson, J. L., Uma, K., & Balaram, P. (1989b) *Biochemistry 28*, 6696–6701.

Karle, I. L., Flippen-Anderson, J. L., Uma, K., & Balaram, P. (1989c) *Biopolymers 28*, 773–781.

Karle, I. L., Flippen-Anderson, J. L., Uma, K., Balaram, H., & Balaram, P. (1990a) *Biopolymers* (in press).

Karle, I. L., Flippen-Anderson, J. L., Sukumar, M., & Balaram, P. (1990b) *Int. J. Pept. Protein Res.* (in press).

Karle, I. L., Flippen-Anderson, J. L., Uma, K., & Balaram, P. (1990c) (submitted for publication).

Karle, I. L., Flippen-Anderson, J. L., Uma, K., & Balaram, P. (1990d) *Proteins: Struct., Funct., Genet. 7*, 62–73.

Karle, I. L., Flippen-Anderson, J. L., Uma, K., & Balaram, P. (1990e) *Biopolymers* (in press).

Karle, I. L., Flippen-Anderson, J. L., Uma, K., & Balaram, P. (1990f) *Biopolymers* (in press).

Karle, I. L., Flippen-Anderson, J. L., Uma, K., & Balaram, P. (1990g) *Curr. Sci.* (submitted for publication).

Le Bars, M., Bachet, B., & Mornon, J. O. (1988) *Z. Kristallogr. 185*, 588.

Marshall, G. R., & Bosshard, H. E. (1972) *Circ. Res. 30/31 (Suppl. II)*, 143–150.

Marshall, G. R., Hodgkin, E. E., Langs, D. A., Smith, G. D., Zabrocki, J., & Leplawy, M. T. (1990) *Proc. Natl. Acad. Sci. U.S.A. 87*, 487–491.

Mathew, M. K., & Balaram, P. (1983) *Mol. Cell. Biochem. 50*, 47–64.

Mueller, P., & Rudin, D. O. (1968) *Nature (London) 217*, 713–719.

Okuyama, K., Tanaka, N., Doi, M., & Narita, M. (1988) *Bull. Chem. Soc. Jpn. 61*, 3115–3120.

Prasad, B. V. V., & Balaram, P. (1984) *CRC Crit. Rev. Biochem. 16*, 307–348.

Ramachandran, G. N., & Sasisekharan, V. (1968) *Adv. Protein Chem. 23*, 284–437.

Ramachandran, G. N., Ramakrishnan, C., & Sasisekharan, V. (1963) *J. Mol. Biol. 7*, 95–99.

Rinehart, K. L., Jr., Pandey, R. C., Moore, M. L., Tarbox, S. R., Snelling, C. R., Cook, J. C., Jr., & Milberg, R. H. (1979) in *Peptides. Proceedings of the Sixth American Peptide Symposium* (Gross, E., & Meienhofer, J., Eds.) pp 59–71, Pierce Chemical Co., Rockford, IL.

Satyshur, K. A., Rao, S. T., Pyzalska, D., Drendel, W., Greaser, M., & Sundaralingam, M. (1988) *J. Biol. Chem. 263*, 1628–1647.

Sundaralingam, M., & Sekharudu, Y. C. (1989) *Science 244*, 1333–1337.

Toniolo, C., Bonora, G. M., Bavoso, A., Benedetti, E., Di Blasio, B., Pavone, V., & Pedone, C. (1983) *Biopolymers 22*, 205–215.

Toniolo, C., Bonora, G. M., Bavoso, A., Benedetti, E., Di Blasio, B., Pavone, V., & Pedone, C. (1985) *J. Biomol. Struct. Dyn. 3*, 585–598.

Uma, K., & Balaram, P. (1989) *Indian J. Chem. 28B*, 705–710.

Chapter 4

Progress with Laue Diffraction Studies on Protein and Virus Crystals[†]

Janos Hajdu[‡] and Louise N. Johnson*[,‡]

Laboratory of Molecular Biophysics, University of Oxford, The Rex Richards Building, South Parks Road, Oxford OX1 3QU, U.K.

Received July 21, 1989; Revised Manuscript Received September 27, 1989

X-ray diffraction studies have made outstanding contributions to structural molecular biology. The resulting image of the molecule is time averaged over the period needed to make the measurements and spatially averaged over all the molecules within the volume of the crystal. Until recently, the measurements from protein crystals took days or weeks with conventional X-ray sources. Now, the ability to obtain atomic information on short-lived structures that may accumulate transiently during a reaction in the crystal is within our grasp. These new opportunities in macromolecular crystallography, which has previously been considered a static technique, require new developements for initiating and monitoring events in protein crystals.

Several approaches have demonstrated that proteins can exhibit dynamic properties in the crystal despite the constraints of the crystal lattice. Studies on catalysis have indicated that many enzymes are active in the crystal with thermodynamic properties similar to those shown in solution but often with reduced rate constants [Quiocho & Richards, 1966; reviewed by Makinen and Fink (1977)]. In the crystal, as in solution, each structural state of a protein represents a subset of closely related structures that undergo thermal fluctuations around the mean structure. Analysis of temperature factors of refined protein crystal structures [Artymiuk et al., 1979; reviewed by Petsko and Ringe (1984)] has given indications of mobility and restraints on atoms in protein molecules, and many ligand binding studies have shown the ability of proteins to respond with conformational changes in the crystal. The structural states that a molecule can adopt are restrained by lattice forces, and conformational changes incompatible with the lattice break up the crystal.

In a diffraction experiment, for a novel structural state to be detected and its structure determined, there must be a transient buildup of the intermediate before it disappears during the reaction. There is the expectation that if X-ray data collection rates can be made commensurate with the time scale of the dynamic events, then transient structural states that are simultaneously present in most of the unit cells could be analyzed. High-intensity synchrotron radiation sources and the revival of the Laue method (white X-radiation, stationary crystal) have made fast X-ray crystallographic data collection a reality. Many full diffraction data sets have been collected with overall data acquisition times ranging from seconds to milliseconds, and recent developments suggest that data collection on the picosecond time scale may be possible in the near future. The Laue method has been applied to the study of enzyme structures, ligand–enzyme interactions, viruses, and viral drug complexes (Table I). The advantages of the method have been demonstrated for small crystals (Harding et al., 1988), for radiation-sensitive crystals (Hedman et al., 1985), and for ligand binding studies where data from the native and ligand-bound structures are collected from the same crystal (Hajdu et al., 1987b). So far the Laue method has yielded acceptable diffraction data for structural studies, but full exploitation of the method for the study of dynamic events in crystals requires developments in the physics, chemistry, and biochemistry of protein crystal systems, similar to those that have been achieved for the time-resolved studies on muscle [see, e.g., Kress et al. (1986)]. The time scales of some of these dynamic events are given in Figure 1.

In this review we summarize the physical basis of Laue diffraction with synchrotron radiation and the results that have been obtained so far. We then assess the progress toward time-resolved studies with protein crystals and the problems that remain. Time-resolved macromolecular crystallography has been reviewed recently with emphasis on the biophysical principles (Moffat, 1989). Time-resolved diffraction studies on noncrystalline biological materials have also been described (Gruner, 1987; Potschka et al., 1988) and are not discussed here.

Diffraction Studies with Synchrotron Radiation

Synchrotron radiation refers to the electromagnetic radiation produced when charged particles are accelerated at relativistic

[†]This work has been supported by the Medical Research Council and the Science and Engineering Research Council, U.K.

[‡]Member of the Oxford Centre for Molecular Sciences.

Table I: Laue Diffraction Studies on Protein Crystals

protein	exposure time per photo	experiment	ref
(1) calcium binding protein	30 s	test photograph	Moffat et al., 1984
(2) pea lectin	45 s (single bunch, 1.8 GeV, 15 mA)	test photograph; data set leading to 6982 out of 12 906 reflections subsequently collected in four such photographs with merging $R = 0.082$	Helliwell, 1984; Helliwell et al., 1988
(3) gramicidin A	50 s (20-μm^3 crystal)	demonstration of Laue diffraction from very small crystals and investigation of radiation damage	Hedman et al., 1985
(4) hen egg white lysozyme	64 ms	monitoring of time-dependent variation in intensities following thermally induced structural changes	Moffat et al., 1986
(5) glycogen phosphorylase *b*	1 s	three-dimensional data set collected in three photographs to 2.4-Å resolution; difference electron density map showed oligosaccharide bound to enzyme	Hajdu et al., 1987b
(6) xylose isomerase	1 s	three-dimensional data set collected in three photographs; difference electron density map showed Eu binding site	Faber et al., 1988
(7) insulin	3 s	demonstration of transformation of four Zn insulin crystals to two Zn insulin crystals induced by contact with vapor from two Zn crystallization buffer	Reynolds et al., 1988
(8) tomato bushy stunt virus	24 s	three-dimensional data collected for native virus and for virus in which Ca^{2+} has been removed by EDTA; data to 3.5-Å resolution contained 38 720 reflections ($I > 2\sigma$) representing 31% of the data with $R_m = 0.14$ from one photograph; Ca^{2+} binding sites identified in difference electron density maps	Hajdu et al., 1989
(9) γ-chymotrypsin	1 s	three-dimensional data collected from three photographs (total exposure 3 s) to yield a data set of 5–2.5-Å resolution with $R_m = 0.078$; structure refined to $R = 0.21$ with Brookhaven Data Bank coordinates as starting model; results revealed the presence of exogenous peptide at the catalytic site	Almo et al., 1989
(10) turkey lysozyme	1 s	three-dimensional data collected from two photographs (total exposure 2 s) to yield 55% of data between 5- and 2.5-Å resolution ($I > 2\sigma$) with $R_{sym} = 0.05$; structure solved by molecular replacement and refined to $R = 0.19$	P. L. Howell et al., unpublished results

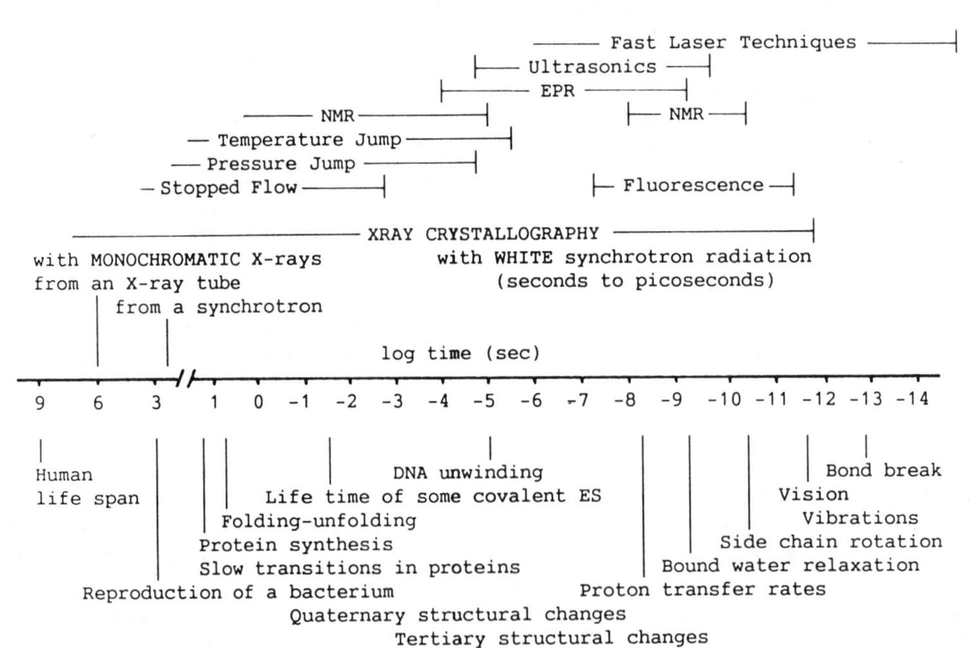

FIGURE 1: Approximate time scales for biological events (lower lines) and for physical methods for monitoring these (upper lines).

velocities and constrained to a curved trajectory by an external magnetic field. The energy the particle emits as electromagnetic radiation per unit time increases as the fourth power of the particle's energy (E^4) and decreases with the fourth power of its rest mass ($1/m^4$) (Schwinger, 1949). Hence, electrons or positrons are most useful, having low mass and being capable of acceleration to high energies. In a typical machine electrons or positrons are injected from a linear accelerator into a booster synchrotron and from there into a storage ring that is run at energies between 1 and 20 GeV with circulating currents of up to several hundred milliamps. In the storage ring, the particles are grouped into small bunches (a few millimeters in length), giving rise to short (10–500 ps long) and intense bursts of synchrotron radiation whenever they pass through the intense fields of the magnets. At relativistic speeds, the radiation is confined to a narrow cone around the instantaneous direction of flight tangential to the curved particle orbit. The radiation emitted is polychromatic, is extremely intense, has a pulsed nature, and is highly polarized. It can span a spectral range from γ-rays to radio frequencies (Figure 2). With insertion devices, such as wiggler magnets or undulator magnets, a local reduction in bending radius is achieved, and the spectral distribution of the radiation is altered (Figure 2).

The first demonstration of the applicability of synchrotron radiation to biological specimens came with the work of Ro-

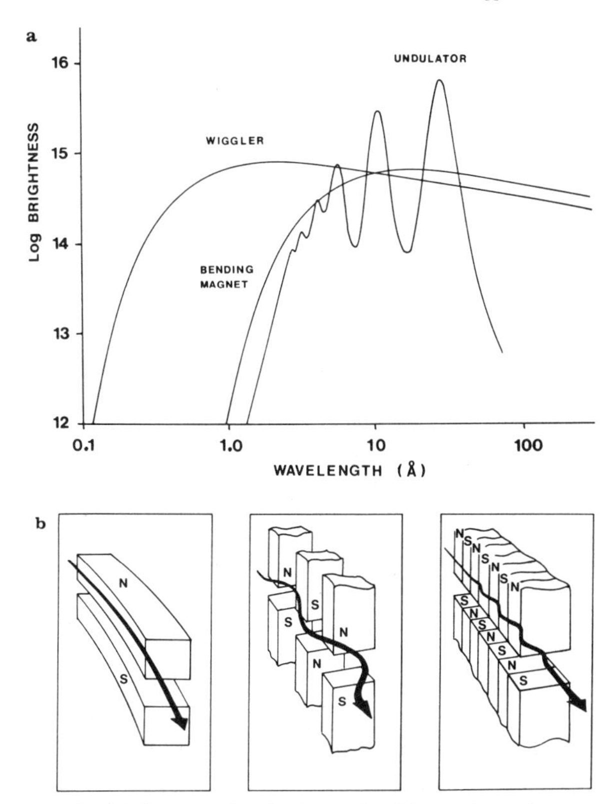

FIGURE 2: (a) Spectra of radiation emitted from the main storage ring (bending magnet) and from two special magnets, a wiggler and an undulator, at a synchrotron radiation source. The insertion of such devices into the straight sections of the storage ring allows the operation of a beam line with altered spectral properties. (b) Electron paths in the magnets. Wiggler magnets operate at very high field strength and force the circulating electron beam onto a tightly bent (and usually) periodic path on a short section. Undulators are multipole wigglers with moderate field strengths. The departure of the electron beam from a straight path is very small. The radiation produced with undulators has a spectrum with enhanced intensities at particular wavelengths produced by interference effects.

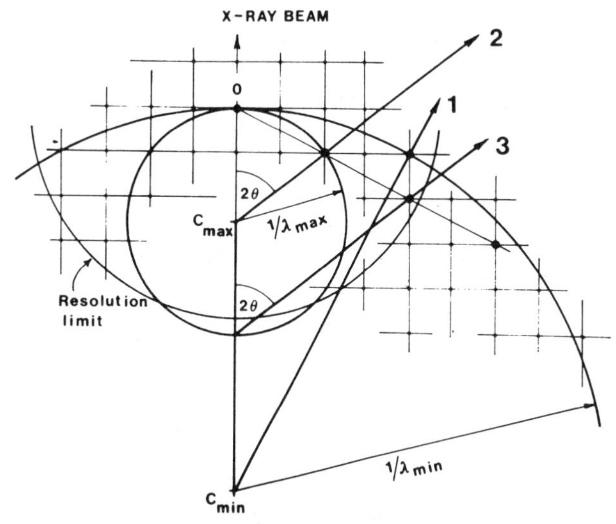

FIGURE 3: Laue geometry of diffraction. The diagram shows the Ewald construction, a geometrical representation of Bragg's law. A sphere is drawn with center *C* and a radius equivalent to the reciprocal of the wavelength. The reciprocal lattice is placed with origin *O* at the point where the straight through the beam cuts the sphere. Diffraction occurs whenever a reciprocal lattice point cuts the sphere. With the Laue method the reciprocal lattice is illuminated with a spectrum of X-rays with wavelengths continously varying from λ_{min} to λ_{max} between the two limiting spheres with radii $1/\lambda_{min}$ and $1/_{max}$. All reciprocal lattice points between the limits will lie on an appropriate sphere and will diffract (for example, rays 1, 2, and 3). Note that in reality the resolution limit and the wavelength limits are smoothly varying soft limits.

senbaum et al. (1971) on muscle. These authors used monochromatized X-rays. Throughout the 1970s protein crystallographers were parasitic users of synchrotron radiation, using sources designed for elementary particle physics. In 1981 the first synchrotron dedicated to the production of radiation came on line at SRS, Daresbury. There are now nine protein crystallographic facilities available at sources around the world: LURE, Orsay, France; DESY, Hamburg, West Germany; SRS, Daresbury, U.K.; NSLS, Brookhaven, NY; CHESS, Ithaca, NY; SPEAR and PEP, Stanford, CA; PHOTON Factory, Tsukuba, Japan. Further sources (ESRF, Grenoble, France; APS, Argonne, IL; ALS, Berkeley, CA; ELETTRA, Trieste, Italy) are also planned. The application of synchrotron radiation studies to biological macromolecules has been reviewed (Greenhough & Helliwell, 1983).

Diffraction with Monochromatic X-rays. In order to solve a protein crystal structure, an almost complete diffraction data set from a crystal to a given resolution must be measured. With monochromatic X-rays, only a small proportion of the lattice planes diffract at any particular orientation of the crystal, and the crystal has to be rocked through a small angle (1–2°) in order to record the full intensity of the reflection. To bring other planes into the diffracting position, the crystal has to be rotated to a new setting and the procedure repeated with a new film. A monochromatic data set to 2.4-Å resolution for glycogen phosphorylase *b* required 32 exposures each with a 1.5° oscillation. At the Daresbury synchrotron, the very high intensity permitted the collection of such data sets within about 20 min (in the best case). This is an increase of almost 3 orders of magnitude compared to the data acquisition rates possible with a conventional rotating-anode X-ray source in the home laboratory. Even so, a new crystal was required for each data set, and the time resolution was still relatively coarse. In spite of this, a sequence of such data sets has been used to observe a very slow catalytic reaction in tetragonal crystals of phosphorylase *b* (Hajdu et al., 1986a, 1987a).

Diffraction with Polychromatic X-rays (*Laue Diffraction*). Laue diffraction refers to the method used by Friedrich, Knipping, and von Laue to record the first X-ray diffraction image from a crystal of copper sulfate (Friedrich et al., 1912). The method utilizes the whole polychromatic spectrum instead of a single wavelength. It fell into disuse because conventional X-ray sources did not give a satisfactory spectrum and because of difficulties in unraveling the complicated diffraction patterns. The broad spectral range (Figure 2) and the high intensity of synchrotron radiation have been the prime driving force behind the revival of the Laue method (Wood et al., 1983) and its application to biological macromolecules (Moffat et al., 1984; Helliwell, 1984), along with new developments in computing techniques to process the data (Machin, 1985, 1987; Campbell et al., 1986, 1987; Rabinovich & Lourie, 1987). With white X-radiation, a large number of lattice planes diffract simultaneously as the Bragg condition is satisfied for each of these planes by at least one wavelength of the spectrum (Figure 3). Many reflections can thus be recorded in a short time with a single exposure. The diffracting position of a stack of lattice planes is determined by their orientation (relative to the direction of the incident beam), their spacing, and the wavelength range applied; the wider the wavelength range the greater the number of planes in the diffracting position. With crystals of high symmetry, a large proportion of the unique data set may be recorded with a single photograph. For example, under certain experimental conditions, almost 98% of the unique data set may be recorded in a single exposure from a cubic crystal while, under similar conditions, only about 55% of the unique set is accessible from a monoclinic crystal. This requires more than one photograph to be taken in order to complete the data set in the latter case (I. A. Clifton et al., unpublished results).

White synchrotron X-radiation was first used for recording topographs from inorganic crystals by Tuomi et al. in 1974. The first biological application of this very intense radiation came in a study of collagen fibers with energy-dispersive X-ray scattering by Bordas et al. (1976). This was done at the old NINA storage ring in Daresbury (U.K.). In 1977, Steinberger et al. (1977) recorded Laue photographs from zinc sulfide crystals there, and in the same year, Bordas and Kam obtained Laue diffraction images from a crystal of lysozyme at the DESY storage ring in Hamburg, West Germany (J. Bordas and Z. Kam, unpublished results). The first successful structural work applying the broad band path Laue diffraction technique was the refinement of a known inorganic crystal structure (aluminium phosphate) from Laue diffraction data by Wood, Thompson and Matthewman at the Daresbury Laboratory in 1982 (Wood et al., 1983). Moffat et al. (1984) showed that Laue diffraction could be applied to proteins and proposed that a narrow band path of the white radiation be used in order to reduce the number of harmonic overlaps (see below). Later that year, Helliwell and colleagues (Helliwell, 1984; Helliwell et al., 1989) performed experiments in Daresbury with the complete white spectrum of radiation, which gives greater efficiency of data collection than the narrow band path technique. Our own work on Laue diffraction started that year. Various feasibility studies followed (Hedman et al., 1985; Clifton et al., 1985). The first experimental results in macromolecular crystallography (Table I) provided information on the glycogen storage site in crystals of rabbit muscle phosphorylase (Hajdu et al., 1987b), located a metal binding site in crystals of xylose isomerase (Farber et al., 1988), gave detailed structural information on the divalent cation binding sites in crystals of tomato bushy stunt virus (Hajdu et al., 1989), and led to the solution of two new protein structures: lysozyme from turkey egg white (P. L. Howell et al., unpublished results) and glyceraldehyde-3-phosphate dehydrogenase (form I) from *Trypanosoma brucei* (F. M. D. Vellieaux et al., unpublished results). These experiments were performed at station 9.7 of the Daresbury synchrotron by utilizing the full white spectrum of the wiggler magnet with a single optical element (a pinhole) in the beam. Data were processed with software written by the Computing Systems and Application Group in Daresbury under the leadership of the late Pella A. Machin.

There are several inherent problems that complicate measurements from Laue photographs:

(*1*) *Harmonic Overlaps* (*Multiplets*). The diffracted rays of polychromatic X-rays with wavelengths of λ, $\lambda/2$, $\lambda/3$, ..., and λ/n from parallel lattice planes with spacings of d, $d/2$, $d/3$, ..., and d/n (where n is a positive integer) will have the same diffracting angle (θ), according to Bragg's law ($2d \sin \theta = n\lambda$). Thus, these reflections (e.g., rays 2 and 3 in Figure 3) will produce a single spot containing more than one reflection on the film. The number of reflections buried in harmonic spots (or "multiplets") increases with increasing wavelength range, reaching an upper limit of only about 17% at an infinite wavelength range (Cruickshank et al., 1987). At realistic wavelength ranges, they form an even smaller part of the data. Harmonic reflections require separation during processing. Methods to deconvolute harmonic spots into component reflections are available (Zurek et al., 1985; Helliwell et al., 1989a,b) with some promising results for doublet spots.

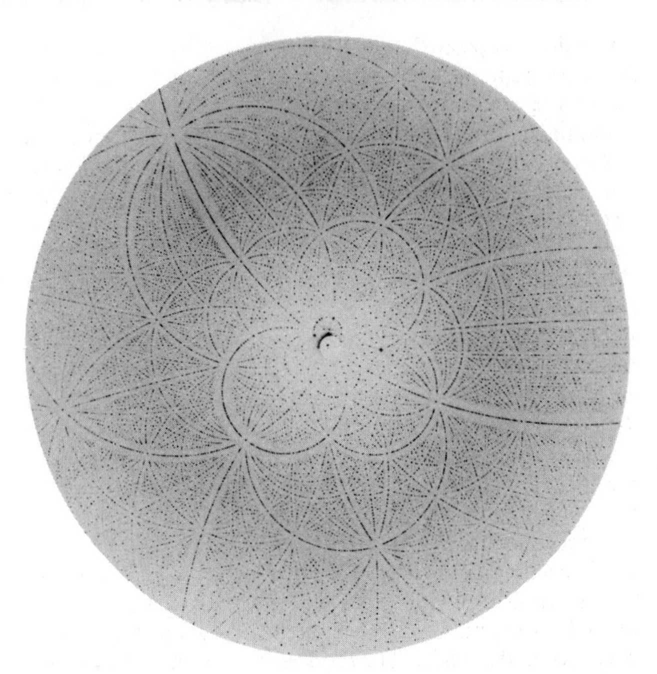

FIGURE 4: Laue photograph of ribulosebisphosphate carboxylase: space group $C222_1$, a = 157.2 Å, b = 157.2 Å, c = 201.3 Å. Wavelength range 0.2–2.1 Å; exposure time 1 s; crystal to film distance 142.3 mm; number of reflections predicted on photograph 126 270. Photograph recorded at SRS, Daresbury, U.K. (I. A. Andersson et al., unpublished results).

(*2*) *Spatial Overlaps*. As is apparent in a typical Laue photograph (Figure 4) the space on the film is crowded with spots. The photograph from a crystal of ribulosebisphosphate carboxylase [space group $C222_1$, a = 157.2 Å, b = 157.2 Å, c = 201.3 Å (Andersson & Branden, 1984)] contains 126 270 predicted reflections. In the early studies with Laue photography [e.g., Hajdu et al. (1987b)], intensities could only be obtained if the spots were separated by more than 0.2 mm. For the photograph in Figure 4 some 66 753 reflections (over 50% of the data) would be rejected as spatially overlapped by this criterion. Recently, profile-fitting procedures have been implemented that allow the spatial deconvolution of spots, and measurements of spots that are separated by more than 0.1 mm (2 raster units) are now possible (T. J. Greenhough et al., unpublished results). Thus for the photo of Figure 4, only 21 948 reflections are rejected as spatially overlapped out of 126 270 predicted on the film. This development has allowed considerably increased numbers of reflections to be measured for crystals of large unit cells including viruses.

(*3*) *Wavelength Normalization*. Techniques are needed to take into account the effects of wavelength-dependent factors on intensity measurements. The intensity of the incident radiation (Figure 2) modulated by the optical elements used to focus the beam, the interaction of the radiation with the crystal, and the interaction of the radiation with the detector all vary with wavelength. Two experimental approaches have been developed for the treatment of wavelength-dependent effects. In the first method, a wavelength normalization curve is deduced. This can be done in a number of ways. Wood et al. (1983) used a standard Si crystal (i.e., an external reference) to record the variation in intensity of a single reflection while rotating the crystal in the white X-ray beam. Another technique is based on the comparison of intensities of symmetry-related reflections stimulated by X-rays of different wavelengths (Campbell et al., 1986). The technique uses an internal reference and is fairly robust but may not be applicable where significant anomalous scattering is present. It has been used in the determination of an unknown crystal structure of an organometallic compound (Harding et al., 1988) where only very small crystals were available. A recent comparison of structure determination of a small molecule ($C_{10}H_{11}NOClF$) has shown that a structure based on data recorded with Laue diffraction in 6.5 s is of comparable precision in atomic coordinates to a structure based on data recorded in 24 h with standard monochromatic methods (Gomez de Anderez et al., 1989). With proteins this approach was used in the location of heavy atom sites in crystals of xylose isomerase (Farber et al., 1988) and in the pea lectin data processing (Helliwell et al., 1989a).

The second method devised to compensate for the wavelength-dependent terms is designated the difference method and is used for the analysis of structural changes relative to a known starting structure. A sequence of nearly identical Laue photographs are recorded before, during, and after the initiation of the reaction or ligand binding. Data sets in the sequence are scaled to the starting native set, and the fractional difference between the structure factor amplitudes of the initial and intermediate data sets is multiplied by the appropriate structure factor amplitude from a reference data set recorded with monochromatic radiation [see Moffat et al. (1986)]. This method was used to produce the first interpretable Laue difference Fourier of a ligand bound to an enzyme, glycogen phosphorylase *b*, where the total exposure time per data set was 3 s (Hajdu et al., 1987b).

(*4*) *Low-Resolution Data.* Figure 3 reveals that the Laue geometry allows a large proportion of medium- to high-resolution reflections to be recorded but is much less efficient for low-resolution reflections. This is due to two factors: (i) the narrowing of the reciprocal space swept between the two limiting Ewald spheres at low diffraction angles (Figure 3) and (ii) the fact that many of the low-resolution reflections are buried in harmonic overlaps (Amoros et al., 1975; Cruickshank et al., 1987). In principle, some of these reflections could be retrieved through deconvolution of the harmonic overlaps. In a recent structural study on ribulosebisphosphate carboxylase (Andersson & Branden, 1984; Andersson et al., 1989; I. A. Andersson et al., unpublished results) data to 1.8-Å resolution from these orthorhombic crystals were measured with the Laue method by using four different orientations of the crystal. The low-resolution data to 5-Å resolution were recorded separately with monochromatic radiation. Such an approach allows rapid data collection from only a few crystals and is of considerable value when crystals are rare or severely radiation sensitive but would not be possible for time-resolved experiments. The low-resolution terms contribute to the definition of the boundary between protein and solvent in protein electron density maps and to the localization of less well ordered regions. In a trial study with glycogen phosphorylase (K. R. Acharya et al., unpublished results) comparison of electron density maps based on all the data to 1.9-Å resolution and that based on data from 5-Å resolution to 1.9-Å resolution showed essentially no difference in those parts of the molecule that were well localized, such as the internal cofactor pyridoxal phosphate, but those regions which were less well ordered and which exhibited high thermal factors ($B > 60$ Å^2) were significantly better defined in the maps containing all the data than those for which the low-resolution terms were ommitted, as might be expected. Lack of low-resolution terms did not affect the structure determination of a small molecule (Gomez de Anderez et al., 1989). In an experiment with Laue data the structure of a γ-chymotrypsin–tetrapeptide inhibitor complex has been analyzed. The lack of low-resolution data did not lead to any problems in the refinement of the structure, and good density for the inhibitor was observed at the catalytic site (Almo et al., 1989) in agreement with the work of Dixon and Matthews (1989). This density was also observed in the initial structure determination of native γ-chymotrypsin (Cohen et al., 1981), but since there was no reason to expect the presence of an exogenous peptide in the structure, the density was modeled by a string of water molecules. Current experience suggests that ommission of low-resolution terms does not affect structure determination by molecular replacement methods or difference electron density maps but may interfere with determination of heavy atom positions by Patterson methods (S. C. Almo et al., unpublished results).

SOME RECENT EXAMPLES OF STRUCTURAL RESULTS FROM LAUE DIFFRACTION

The increasing number of applications of Laue diffraction methods to proteins and viruses is summarized in Table I. Some of the achievements are described below, and these demonstrate that the method is now proving successful as a reliable method for data collection with dramatically reduced data collection times.

Determination of the Structure of Turkey Egg White Lysozyme. Hexagonal turkey lysozyme crystals (Bott & Sarma, 1976; Sarma & Bott, 1977) are of potential interest for crystal studies on lysozyme catalysis. The catalytic site cleft may be more accessible to substrates in this crystal lattice than in the well-studied tetragonal crystals of hen egg white lysozyme (HEWL; Blake et al., 1965) where the lower part of the catalytic site (subsites E and F) is blocked by symmetry-related molecules. The structure of turkey egg white lysozyme (TEWL), which differs from HEWL in only six amino acid positions, has been solved simultaneously in two laboratories by two different methods. Parsons and Phillips (Parsons, 1989) have used monochromatic diffraction data and the technique of isomorphous replacement. P. L. Howell et al. (unpublished results) used Laue diffraction data and molecular replacement based on the structure of hen egg white lysozyme. The structures agree with each other but differ from the structure published by Sarma and Bott (1977). In the molecular replacement solution of P. L. Howell et al. (unpublished results) Laue data between 5- and to 2.5-Å resolution were used. The starting model determined by the molecular replacement method with MERLOT (Fitzgerald, 1988) gave an *R* factor of 0.39. The model was refined with the programs XPLOR (Brunger et al., 1988) and PROLSQ (Hendrickson, 1985) to a final crystallographic *R* factor of 0.18. The bound solvent was not modeled. This is the first macromolecule whose structure has been solved from Laue diffraction data alone. The results show that the structure and the positions of the key catalytic residues (Figure 5) are closely similar to those of HEWL. The packing of the molecules in the crystal lattice leaves the lower part of the catalytic cleft accessible although the upper part (subsite A and part of subsite B) is blocked by an adjacent molecule.

Tomato Bushy Stunt Virus. Laue diffraction studies have been extended to systems with very large unit cells. The structure of tomato bushy stunt virus (TBSV) ($T = 3$ icosahedral, 180 subunits in the capsid) has been solved to 2.9-Å resolution (Harrison et al., 1978). The subunits have identical sequence but are present in three different conformations in the icosahedral asymmetric unit. A reversible conformational change occurs when two structurally bound calcium ions are removed from each of the trimer interfaces. At the pH of the cell, this is followed by an expansion of the virion [see Robinson

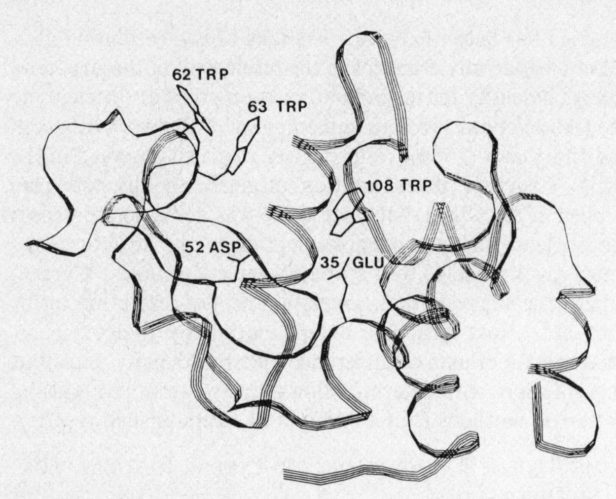

FIGURE 5: Ribbon diagram of turkey egg white lysozyme based on data from Laue diffraction. The diagram shows the chain topology and the positions of some key residues (P. L. Howell et al., unpublished results).

and Harrison (1982)], leading to uncoating.

Crystals of the virus are cubic (*I*23), allowing the recording of an almost complete data set on a single Laue photograph. A 24-s exposure was recorded from a crystal that had been soaked in EDTA to remove the bound Ca^{2+} ions. This was predicted to represent 92.7% of the unique data set to 3-Å resolution and yielded 31% of the data set with intensities greater than two standard deviations of the measurements between 6- and 3.5-Å resolution. A difference Fourier synthesis calculated with this subset of data clearly showed the three pairs of Ca^{2+} binding sites related by quasi-symmetry (Figure 6) (Hajdu et al., 1989). The extent and quality of data obtained from a single Laue photograph were sufficient to detect clearly a small alteration, i.e., the replacement of a Ca^{2+} ion by a water molecule and the conformational changes induced in the capsid. It is suggested that drug binding as well as molecular processes associated with infectivity may be studied with this technique. Following this work, Laue data on human rhino virus, canine parvo virus, mengo virus, and the ϕX174 phage as well as drug–virus complexes have been recorded (M. G. Rossmann et al., unpublished results).

Kinetic Laue Crystallography

A major limitation to the kinetic approach in X-ray crystallography is the fact that structures derived from diffraction images represent mixtures of all structural states in the crystal. There will be a time-dependent moving average of various structures in the crystal along the reaction coordinate. Even with the shortest possible data collection times (a few picoseconds), the major problem is the lack of "synchronization" among molecules in various parts of the crystal. With our present toolbox, for a novel structural state to be detected and its structure determined, there must be a transient buildup of this intermediate before it disappears during the reaction. This means a population inversion leading to one conformation largely predominant over the whole volume of the crystal (Hajdu et al., 1988). The structures observable are set by the system and not by the experimenter, and the favorable case of a single conformation (subject to some thermal motion) predominant in the crystal may not always be obtained. In general, the concentrations of some interesting intermediates may be low and masked by more predominant structures. Analysis of these mixed structures requires the development of reliable techniques to detect and to deconvolute the average into components. Further, the detailed chemical state of a bound molecule may not be identifiable even by high-resolution protein crystallography (such as the oxidation state of a cofactor, for example).

Thus, methods are needed that allow the amount and concentration of various intermediates present in the crystal at the time of the X-ray exposure to be measured. This can be accomplished if there is a suitable diagnostic such as a distinct spectral signal (fluorescence, UV or visible absorption, infrared

FIGURE 6: Difference Fourier map of tomato bushy stunt virus (*I*23, a = 383 Å) based on monochromatic data for the virus in the presence of Ca^{2+} and Laue data for the Ca^{2+} free virus. The location of the Ca^{2+} ions at each of the trimer interfaces is apparent (Hajdu et al., 1989).

absorption) associated with at least some of the components. In the past, spectroscopic methods have been used to establish the nature of the components in protein crystals (e.g., the semiquinone form of flavodoxin in the crystal; Eaton et al., 1975) and to follow reactions such as the transamination reaction catalyzed by aspartate aminotransferase (Eichele et al., 1978; Mozzarelli et al., 1979), formation of an acyl-enzyme intermediate with glyceraldehyde-3-phosphate dehydrogenase (Mozzarelli et al., 1982), and electron transfer between flavocytochrome b_2 and cytochrome *c* (Tegoni et al., 1983). With alcohol dehydrogenase, single-crystal microspectroscopic measurements were used to detect NADH present and to test for NADH/NAD conversion in crystals of complexes used for X-ray data collection (Bignetti et al., 1979). With tryptophan synthase chromophoric intermediates formed between the pyridoxal phosphate and substrates at the catalytic site of the β subunits and the effects of ligands bound to the α subunits have been monitored by polarized absorption spectroscopy for this fascinating bifunctional enzyme (Mozzarelli et al., 1989). Fourier transform infrared (FTIR) spectroscopy has been applied to crystals of the photosynthetic reaction center (Gerwert et al., 1988). The crystals were grown on the CaF_2 windows to dimensions of 1.0 mm × 0.5 mm × 6.5 μm, and the difference spectra between the dark and steady-illumination states showed that the intramolecular processes which took place in the chromophores, protein side chains, and protein backbone on light illumination in the crystal were similar to those observed for the reaction center in reconstituted lipid vesicles. These studies indicate the ease with which formation of intermediates can be monitored in the crystal with spectroscopic measurements although the need for very thin samples in the direction of the radiation beam means that FTIR may not be easily applicable to X-ray-size protein crystals.

In order to coordinate X-ray diffraction and spectroscopic studies, a diode array microspectrophotometer is being developed for use at the X-ray station. The hardware is nearing completion, and the software is at present being developed in collaboration with the Daresbury Synchrotron Laboratory (A. Hadfield et al., unpublished results). The instrument utilizes optical fibers to relocate the beam of light in the spectrophotometer onto the crystal in the confined space available in the Laue camera. A reflecting objective is then used to focus the light onto the crystal and another one to collect the light after its passage through the crystal. In addition to the advantage of being able to record spectra during X-ray exposures, it is hoped that the instrument might also be used to monitor a reacting system so that the X-ray camera shutter can be triggered at the right moment as determined by changes in the spectrum.

Initiation of Reactions in Enzyme Crystals. Protein crystals contain a large amount of water. Typically between 30 and 80% of the volume of a protein crystal may be solvent of crystallization. This is fortunate for it means that the environment of the protein in the crystal is similar to that in solution and that ligand binding studies may readily be accomplished by diffusion of ligands into preformed crystals. Many such studies have been carried out, and information on binding sites, intermolecular interactions, and conformational responses have been determined. The diffusion method is not appropriate where large conformational changes are likely to be hindered by the lattice forces of the crystal, although, surprisingly, recent studies suggest that aspartate carbamoyltransferase can accomplish a significant part of the T to R allosteric response in the crystal (Gouaux & Lipscomb, 1989).

In the time-resolved studies on the conversion of substrate to product in phosphorylase crystals using monochromatic data collection methods (Hajdu et al., 1986a, 1987a), the flow cell technique (Wyckoff et al., 1967; Hajdu et al., 1986b) was used to initiate the reaction. Solutions containing the substrate were flowed past the crystal, and the reaction was started by diffusion of the substrate into the enzyme crystal. Such an approach is possible where the diffusion and binding are not rate limiting and the reaction is slower than the time resolution of the experiment (in this instance 30 min). Diffusion of a ligand such as glucose 1-phosphate into a phosphorylase crystal of dimensions of 0.4 mm × 0.4 mm × 1.6 mm had a mean half-saturation binding time of 1.7 min, a time that is consistent with approximate calculations based on knowledge of the free diffusion coefficient of the ligand, the pore radii in the crystal, the dimensions of the crystal, and the external concentration and the dissociation constant of the ligand (Johnson & Hajdu, 1989). Laue photographs of phosphorylase crystals show an unexpected order–disorder–reorder phenomenon as substrates are diffused into the crystal. The diffraction pattern of a native crystal shows sharp spots; 40 s after the start of diffusion of glucose 1-phosphate into the crystals, the pattern exhibits disorder; 20 min later the crystal has reordered, and the pattern is sharp again. Similar effects were observed on conversion of the Zn insulin crystals (Reynolds et al., 1988). The transient disorder means that the diffraction pattern is lost just at the time it would be most desirable to monitor the structure. Thus for Laue diffraction studies with time resolution of the order of seconds or milliseconds, it is necessary to prediffuse the ligand into the crystal under inactive conditions and to initiate the reaction by (say) temperature jump, pH jump, pressure jump, or photodissociation. Indeed, photodissociation was used in one of the earliest time-resolved studies in protein crystallography. The changes in certain reflection intensities were measured following laser illumination of a CO–myoglobin complex with a time resolution of 0.5 ms (Bartunik et al., 1981). Temperature jump has been used (Moffat et al., 1986) with lysozyme crystals, and the changes in intensities in the Laue diffraction pattern were measured with a streak camera on a 0.2–6-s time scale. The interpretation of the results was complicated by crystal movement, changes in unit cell, and nonuniform heating.

Caged compounds (Kaplan et al., 1978) offer one of the most promising approaches for the synchronization of the start of the reaction with the start of data collection. The substrate is made biologically inert through covalent attachment of a photolabile protecting group, most commonly a nitrophenyl ester. Subsequent illumination by a laser or xenon flash lamp results in photodissociation of the protecting group and liberation of the substrate. The nitrophenyl group has the advantage that it absorbs strongly in the near-UV (300–360 nm) in a region where many biological molecules are optically transparent. The physiological and the chemical and physical aspects of the use of caged compounds have been reviewed (Gurney & Lester, 1987; McCray & Trentham, 1989). A number of relevent caged biologically important molecules have been synthesized: caged ATP (Kaplan et al., 1978; McCray et al., 1980; Hibberd et al., 1985); caged cyclic nucleotides (Karpen et al., 1988); caged *myo*-inositol triphosphate (Walker et al., 1987); caged calcium (Adams et al., 1988; Kaplan & Ellis-Davies, 1988); caged protons (McCray & Trentham, 1985); caged neurotransmitters such as carbamoylchloride (Milburn et al., 1989). Rate constants for the liberation of the cage vary from 100 to 100 000 s^{-1} depending on the precise chemical nature of the cage, substrate, and

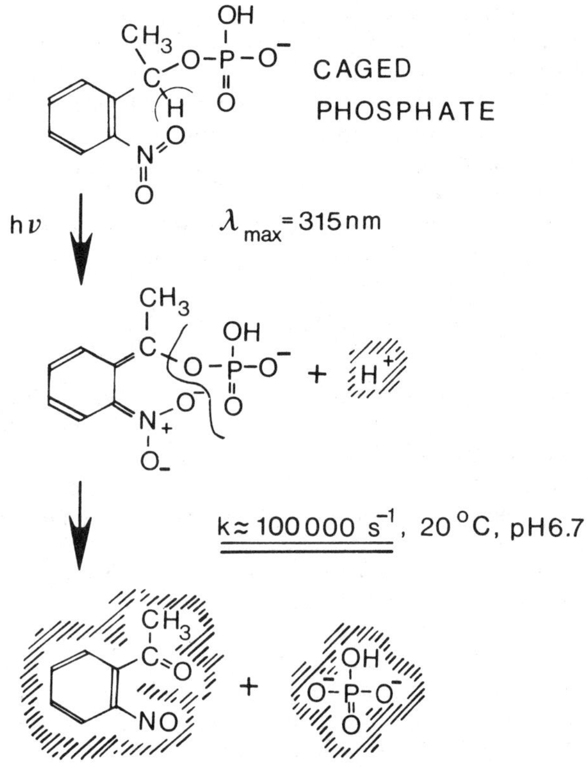

CAGED PHOSPHATE

$h\nu$ λ_{max} = 315 nm

$k \approx 100000\ s^{-1}$, 20 °C, pH6.7

FIGURE 7: Scheme for the release of phosphate from caged 2-nitrophenyl phosphate (McCray & Trentham, 1989).

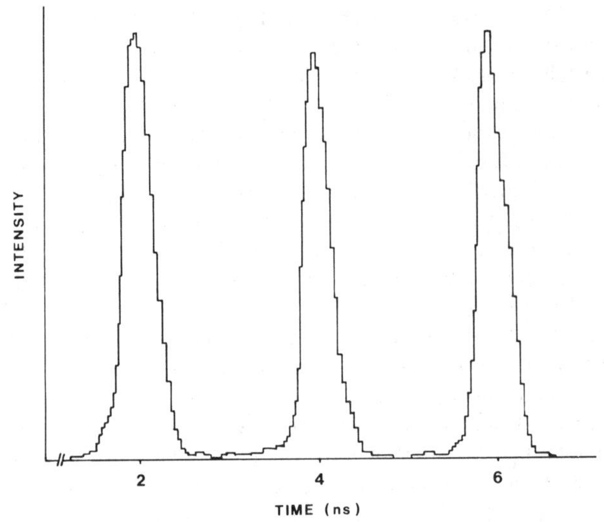

FIGURE 8: Time characteristics of pulsed radiation from the Daresbury Synchrotron Radiation Source run in the multibunch mode.

external conditions. Liberation of the cage results in an almost instantaneous change in the concentration of the biologically active form of the substrate. McCray and Trentham (1989) report that 20 mJ of 347-nm irradiation is sufficient to liberate 2 mM ATP from 5 mM caged ATP (the P^3-[(2-nitrophenyl)ethyl] ester of ATP) over an area of 10 mm^2 and 0.1 mm thickness within a period of a few milliseconds. A scheme for the liberation of phosphate from caged phosphate is shown in Figure 7.

Crystallographic experiments with tetragonal phosphorylase *b* crystals were performed in 1984 (J. Hajdu, D. R. Trentham, D. I. Stuart, K. R. Acharya, and L. N. Johnson, unpublished results), utilizing "caged" phosphate compounds similar to those that have been developed in the study of ATP utilization in muscle contraction (Hibberd et al., 1985). Full monochromatic data sets to 2.8-Å resolution were collected before and after the photolysis of 25 mM caged phosphate in the crystal. The binding sites for the intact caged phosphate have been established, and the structure of the photolyzed products has been determined. These results revealed a recognized problem (Kaplan et al., 1978; Goldman et al., 1984) with the use of saturating concentrations of 2-nitrobenzyl caged substrates. The photolabel when released (a nitroso ketone in this case) reacts with thiol and other nucleophilic groups, extensively modifying the enzyme. These modifications were observed in the electron density difference maps, and chemical modification may lead to an inhibition of the enzyme reaction. The complication of the nitroso ketone reaction is usually overcome by inclusion of equimolar amounts of scavenging thiols. Phosphorylase requires a high concentration of substrate phosphate for reaction when the enzyme is in the T state, and the crystals do not tolerate high concentrations of thiol reagents. As an alternative route to alleviate this problem, it is planned to cage the enzyme. Glycogen phosphorylase contains the essential cofactor, pyridoxal phosphate, which plays an obligatory role in catalysis through its phosphate group. The enzyme concentration in the crystal is about 7 mM, so lower concentrations of cage will be liberated that can be scavenged by suitable concentration of thiols. Caged pyridoxal phosphate has been synthesized, and kinetic, reconstitution, and crystallization studies are underway. Caged ATP has been used in a recent study with hexokinase (Bartunik et al., unpublished results) and caged GTP with the ras oncogene P21 protein (Schlichting et al., 1989).

Finally, there is a need to consider the detector technology. Photographic film has served well, but there are indications that Fuji image plates (Amemiya et al., 1987) and Kodak storage phosphor plates (Whiting et al., 1988) may have advantages in terms of increased absorption efficiency and increased dynamic range, thus partly removing the need to record data with multifilm packs. In a recent development at the Photon Factory a rapid image plate exchanger has been constructed so that 40 images with minimum exposures of 0.1 s and intervals of 0.2 s can be recorded (Amemiya et al., 1989). The use of charge coupled devices is being explored with some promising results, but the small geometric size of these systems is a limitation (Strauss et al., 1987; Allinson et al., 1989).

Picosecond Laue Diffraction at High-Energy Storage Rings. Exposure times of 1 s to 1 ms are exciting to a protein crystallographer, but to a kineticist these times are still very slow. What is the shortest exposure that might be achieved with Laue diffraction? Calculations suggest that about 10^{13} photons are needed to produce an interpretable Laue photograph. In the Daresbury synchrotron a 1-s exposure is sufficient to produce such a picture. Due to the bunch structure of the radiation, 5×10^8 X-ray pulses (each about 120 ps long) will pass through the sample during that time. The source brilliance of high-energy storage rings like the PEP synchrotron in Stanford, CA, may be 8–10 orders of magnitude than that in Daresbury (Wiedemann, 1987) and under similar geometric conditions offers the unique possibility of recording a full Laue diffraction pattern from a crystal by a single X-ray pulse from a single electron bunch (Figure 8). This would mean an overall exposure time on the picosecond scale. Some work toward this goal has already been achieved. In the summer of 1988 a group at Cornell university obtained a 100-ps photograph of a lysozyme crystal using an X-ray station on the undulator magnet at CHESS, a remarkable achievement based on careful calculations of flux and ingenious instrumentation [Szebenyi et al., 1988; also reported by Poole (1988)]. The photograph was recorded with a single bunch with flux of about 10^6 photons and showed 50–200 spots. Only a narrow radiation band-pass was possible with the undulator (compare Figure 2). This meant that a limited number of lattice planes diffracted, a limitation that can be overcome by using the broad-band Laue technique possible with wiggler magnets at synchrotron radiation sources. Single-bunch exposure experiments in the picosecond time scale using broad-band radiation may open up a completely new era in crystallography and structural chemistry. The very short (picosecond) exposure times could result in increased data quality since chemical processes associated with radiation damage may not have enough time to take place, although, due to these processes and the absorbed radiation energy, the crystal may explode in a few microseconds following the exposure. It will be interesting to see whether or not a nearly complet lack of radiation damage (on the photograph only!) could be achieved this way.

ACKNOWLEDGMENTS

The work described here is a collaborative effort with scientists from a number of laboratories from various countries. We are greatly indebted to all of them: Ian J. Clifton, David I. Stuart, K. Ravi Achavya, and Andrea Hadfield (Laboratory of Molecular Biophysics, Oxford University, Oxford, U.K.), P. Lynne Howell, Steven C. Almo, and Gregory A. Petsko (Department of Chemistry, Massachusetts Institute of Technology, Cambridge, MA 02139), Trevor J. Greenhough (Department of Physics, University of Keele, Keele, U.K., and SERC Daresbury Laboratory, Daresbury, U.K.), Annette K. Shrive (Department of Physics, University of Keele, Keele, U.K.), John W. Campbell and Susan Bailey (SERC Daresbury Laboratory, Daresbury, U.K.), Mark Parsons (Department of Biophysics, Leeds University, Leeds, U.K.), Robert M. Stroud (Department of Biochemistry, University of California, San Francisco, CA 94143), Steven C. Harrison and Robert C. Liddington (Department of Biochemistry and Molecular Biology, Harvard University, Cambridge, MA 02139), Michael G. Rossmann and Michael Chapman (Department of Biological Sciences, Lilly Hall of Life Sciences, Purdue University, West Lafayette, IN 47907), David R. Trentham (Medical Research Council, Mill Hill, London, U.K.), and Inger A. Andersson (Swedish University of Agricultural Sciences, Uppsala Biomedical Center, Sweden).

REFERENCES

Adams, S. R., Kao, J. P. Y., Grynkiewicz, G., Minta, A., & Tsien, R. Y. (1988) *J. Am. Chem. Soc. 110*, 3212–3220.

Allinson, N. M., Brammer, R., Helliwell, J. R., Harrop, S., Magorrian, B. G., & Wan, T. (1989) *J. X-ray Sci. Technol.* (in press).

Almo, S. C., Howell, P. L., Petsko, G. A., & Hajdu, J. (1989) *Proc. Natl. Acad. Sci. U.S.A.* (submitted for publication).

Amemiya, Y., Wakabayashi, K., Tanaka, H., Ueno, Y., & Miyahara, J. (1987) *Science 237*, 164–168.

Amemiya, Y., Matsushita, T., Nakagawa, A., Kishimoto, S., Ando, M., Chikawa, J., Wakabayashi, K., Iwamoto, H., & Kobayashi, T. (1989) in *Biophysics and Synchrotron Radiation* (Hasnain, S., Ed.) Ellis Horwood, Chichester (in press).

Amoros, J. L., Buerger, M. J., & Canut de Amoros, M. (1975) *The Laue Method*, Academic Press, New York.

Andersson, I. A., & Branden, C.-I. (1984) *J. Mol. Biol. 172*, 363–366.

Andersson, I. A., Knight, S., Schneider, G., Lindqvist, Y., Lundqvist, T., Branden, C.-I., & Lorimer, G. H. (1989) *Nature 337*, 229–234.

Artymiuk, P. J., Blake, C. C. F., Grace, D. E. P., Oatley, S. J. Phillips, D. C., & Sternberg, M. J. E. (1979) *Nature 280*, 563–568.

Bartunik, H. D., Jerzembek, E., Press, D., Huber, G., & Watson, H. C. (1981) *Acta Crystallogr. A37*, C-51.

Bignetti, E., Rossi, G. L., & Zeppezauer, E. (1979) *FEBS Lett. 100*, 17–22.

Blake, C. C. F., Koenig, D. F., Mair, G. A., North, A. C. T., Phillips, D. C., & Sarma, V. R. (1965) *Nature 206*, 757–763.

Bordas, J., Munro, I. H., & Glazer, A. M. (1976) *Nature 262*, 541–545.

Bott, R., & Sarma, R. (1976) *J. Mol. Biol. 106*, 1037–1046.

Brunger, A. T., Karplus, M., & Petsko, G. A. (1989) *Acta Crystallogr. A45*, 50–61.

Campbell, J. W., Habash, J., Helliwell, J. R., & Moffat, K. (1986) *Inf. Q. Protein Crystallogr., Daresbury Lab. No. 18*, 23–31.

Campbell, J. W., Clifton, I. J., Elder, M., Machin, P. A., Zurek, S., Helliwell, J. R., Habash, J., Hajdu, J., & Harding, M. M. (1987) in *Springer Series in Biophysics*, Vol. 2, Biophysics and Synchrotron Radiation (Bianconi, A., & Congiu Castellano, A. Eds.) pp 52–60, Springer-Verlag, Berlin, Heidelberg, New York, London, Paris, and Tokyo.

Clifton, I. A., Cruickshank, D. W. J., Diakun, G., Elder, M., Habash, J., Helliwell, J. R., Liddington, R. C., Machin, P. A., & Papiz, M. Z. (1985) *J. Appl. Crystallogr. 18*, 296–300.

Cohen, G. H., Silverton, E. W., & Davies, D. R. (1981) *J. Mol. Biol. 148*, 449–479.

Cruickshank, D. W. J., Helliwell, J. R., & Moffat, K. (1987) *Acta Crystallogr. A43*, 656–674.

Dixon, M. M., & Matthews, B. W. (1989) *Biochemistry 28*, 7033–7038.

Eaton, W. W., Hofrichter, J., Makinen, M. W., Anderson, R. D., & Ludwig, M. L. (1975) *Biochemistry 14*, 2146–2151.

Eichele, G., Karabelnik, D., Halobrenner, R., Jansonius, J. N., & Christen, P. (1978) *J. Biol. Chem. 253*, 5239–5242.

Farber, G. K., Machin, P. A., Almo, S. C., Petsko, G. A., & Hajdu, J. (1988) *Proc. Natl. Acad. Sci. U.S.A. 85*, 112–115.

Fitzgerald, P. M. (1988) *J. Appl. Crystallogr. 21*, 274–278.

Friedrich, W., Knipping, P., & von Laue, M. (1912) *Sitzungsber. Math.—Phys. Kl. Bayer. Akad. Wiss. Muenchen*, 303–322.

Gerwert, K., Hess, B., Michel, H., & Buchanan, S. (1988) *FEBS Lett. 232*, 303–307.

Goldman, Y. E., Hibberd, M. G., & Trentham, D. R. (1984) *J. Physiol. 354*, 577–604.

Gomez de Anderez, D., Helliwell, M., Habash, J., Dodson, E. J., Helliwell, J. R., Bailey, P. D., & Gammon, R. E. (1989) *Acta Crystallogr. B45*, 482–488.

Gouaux, J. E., & Lipscomb, W. N. (1989) *Proc. Natl. Acad. Sci. U.S.A. 86*, 845–848.

Greenhough, T. J., & Helliwell, J. R. (1983) *Prog. Biophys. Mol. Biol. 41*, 67–164.

Gruner, S. M. (1987) *Science 238*, 305–312.

Gurney, A. M., & Lester, H. A. (1987) *Physiol. Rev. 67*, 583–617.

Hajdu, J., Acharya, K. R., Stuart, D. I., McLaughlin, P. J., Barford, D., Klein, H., & Johnson, L. N. (1986a) *Biochem. Soc. Trans. 14*, 538–541.

Hajdu, J., McLaughlin, P. J., Helliwell, J. R., Shelden, J., & Thompson, A. W. (1986b) *J. Appl. Crystallogr. 18*, 528–532.

Hajdu, J., Acharya, K. R., Stuart, D. I., McLaughlin, P. J., Barford, D., Klein, H. W., Oikonomakos, N. G., & Johnson, L. N. (1987a) *EMBO J. 6*, 539–546.

Hajdu, J., Machin, P. A., Campbell, J. W., Greenhough, T. J., Clifton, I. J., Zurek, S., Gover, S., Johnson, L. N., & Elder, M. (1987b) *Nature 329*, 115–116.

Hajdu, J., Acharya, K. R., Stuart, D. I., & Johnson, L. N. (1988) *Trends Biochem. Sci. 13*, 104–109.

Hajdu, J., Greenhough, T. J., Clifton, I. J., Campbell, J. W., Shrive, A. K., Harrison, S. C., & Liddington, R. C. (1989) in *Synchrotron Radiation in Structural Biology* (Sweet, R. M., Ed.) pp 331–339, Plenum Press, New York.

Harding, M. M., Maginn, S. J., Campbell, J. W., Clifton, I., & Machin, P. A. (1988) *Acta Crystallogr. B44*, 142–146.

Harrison, S. C., Olson, A. J., Schutt, C. E., Winkler, F. K., & Bricogne, G. (1978) *Nature 276*, 368–373.

Hedman, B., Hodgson, K., Helliwell, J. R., Liddington, R. C., & Papiz, M. Z. (1985) *Proc. Natl. Acad. Sci. U.S.A. 82*, 7604–7606.

Helliwell, J. R. (1984) *Rep. Prog. Phys. 47*, 1403–1497.

Helliwell, J. R., Habash, J., Cruickshank, D. W. J., Harding, M. M., Greenhough, T. J., Campbell, J. W., Clifton, I. J., Elder, M., Machin, P. A., Papiz, M. Z., & Zurek, S. (1989a) *J. Appl. Crystallogr. 22*, 483–497.

Helliwell, J. R., Harrop, S., Habash, J., Magorrian, B. G., Allinson, N. M., Gomez, D., Helliwell, M., Derewenda, Z., & Cruickshank, D. W. J. (1989b) *Rev. Sci. Instrum. 60*, 1531–1536.

Hendrickson, W. A. (1985) *Methods Enzymol. 115*, 252–270.

Hibberd, M. G., Dantzig, J. A., Trentham, D. R., & Goldman, T. E. (1985) *Science 228*, 1317–1319.

Johnson, L. N., & Hajdu, J. (1989) in *Biophysics and Synchrotron Radiation* (Hasnain, S., Ed.) Ellis Horwood, Chickester (in press).

Kaplan, J. H., & Ellis-Davies, G. C. R. (1988) *Proc. Natl. Acad. Sci. U.S.A. 85*, 6571–6575.

Kaplan, J. H., Forbush, B., & Hoffman, J. F. (1978) *Biochemistry 17*, 1929–1935.

Karpen, J. W., Zimmerman, A. L., Stryer, L., & Baylor, D. A. (1988) *Proc. Natl. Acad. Sci. U.S.A. 85*, 1287–1291.

Kress, M., Huxley, H. E., Faruqi, A. R., & Hendrix, J. (1986) *J. Mol. Biol. 188*, 325–342.

Machin, P. A. (1985) *Inf. Q. Protein Crystallogr., Daresbury Lab. No. 15*, 1–16.

Machin, P. A. (1987) in *Computational Aspects of Protein Crystal Data Analysis* (Helliwell, J. R., Machin, P. A., & Papiz, M. Z., Eds.) DL/SCI/R25, pp 75–83, Daresbury Laboratory, Daresbury, U.K.

Makinen, M. W., & Fink, A. L. (1977) *Annu. Rev. Biophys. Bioeng. 6*, 301–342.

McCray, J. A., & Trentham, D. R. (1985) *Biophys. J. 47*, 406a.

McCray, J. A., & Trentham, D. R. (1989) *Annu. Rev. Biophys. Biophys. Chem. 18*, 239–270.

McCray, J. D., Herbette, L., Kihara, T., & Trentham, D. R. (1980) *Proc. Natl. Acad. Sci. U.S.A. 77*, 7237–7241.

Milburn, T., Matsubara, N., Billington, A. P., Udgaonkar, J. B., Walker, J. W., Carpenter, B. K., Webb, W. W., Marque, J., Denk, W., McCray, J. A., & Hess, G. P. (1989) *Biochemistry 28*, 49–55.

Moffat, K. (1989) *Annu. Rev. Biophys. Biophys. Chem. 18*, 309–332.

Moffat, K., Szebenyi, D. M. E., & Bilderback, D. H. (1984) *Science 223*, 1423–1425.

Moffat, K., Bilderback, D., Schildkamp, W., & Volz, K. (1986) *Nucl. Instrum. Methods A246*, 627–635.

Mozzarelli, A., Ottonello, S., Rossi, G. L., & Fasella, P. (1979) *Eur. J. Biochem. 98*, 173–179.

Mozzarelli, A., Berni, R., Rossi, G. L., Vas, M., Bartha, F., & Keleti, T. (1982) *J. Biol. Chem. 257*, 6739–6744.

Mozzarelli, A., Peracchi, A., Rossi, G. L., Ahmed, S. A., & Miles, E. W. (1989) *J. Biol. Chem. 264*, 15774–15780.

Parsons, M. (1989) Ph.D. Thesis, University of Leeds.

Petsko, G. A., & Ringe, D. (1984) *Annu. Rev. Biophys. Bioeng. 13*, 331–371.

Poole, R. (1988) *Science 241*, 295.

Potschka, M., Kock, M. H. J., Adams, M. L., & Schuster, T. M. (1988) *Biochemistry 27*, 8481–8491.

Quiocho, F. A., & Richards, F. M. (1966) *Biochemistry 5*, 4062–4076.

Rabinovich, D., & Lourie, B. (1987) *Acta Crystallogr. A43*, 774–780.

Reynolds, C. D., Stowell, B., Joshi, K. K., Harding, M. M., Maginn, S. J., & Dodson, G. G. (1988) *Acta Crystallogr. B44*, 512–515.

Robinson, I. K., & Harrison, S. C. (1982) *Nature 297*, 563–568.

Rosenbaum, G., Holmes, K. C., & Witz, J. (1971) *Nature 230*, 129–131.

Sarma, R., & Bott, R. (1977) *J. Mol. Biol. 113*, 555–565.

Schlichting, I., Rapp, G., John, J., Wittinghofer, A., Pai, E. F., & Goody, R. S. (1989) *Proc. Natl. Acad. Sci. U.S.A. 86*, 7687–7690.

Schwinger, J. (1949) *Phys. Rev. 75*, 1912–1925.

Steinberger, I. T., Bordas, J., & Kalman, Z. H. (1977) *Philos. Mag. 35*, 1257–1267.

Strauss, M. G., Naday, I., Sherman, M. R., Westbrook, E. M., & Zaluzec, N. J. (1987) *Nucl. Instrum. Methods A266*, 563.

Szebenyi, D. M. E., Bilderback, D., LeGrand, A., Moffat, K., Schildkamp, W., & Teng, T.-Y. (1988) *Trans. Am. Crystallogr. Assoc. 24*, 167–172.

Tegoni, M., Mozzarelli, A., Rossi, G. L., & Labeyrie, F. (1983) *J. Biol. Chem. 258*, 5424–5427.

Tuomi, T., Naukkarinen, K., & Rabe, P. (1974) *Phys. Status Solidi A25*, 93–98.

Walker, J. W., Somlyo, A. V., Goldman, Y. E., Somlyo, A. P., & Trentham, D. R. (1987) *Nature 327*, 249–252.

Whiting, B. R., Owen, J. F., & Rubin, B. H. (1988) *Nucl. Instrum. Methods A266*, 628.

Wiedemann, H. (1987) in *Proceedings of the Workshop on PEP as a Synchrotron Radiation Source*, Oct 20–21, Stanford CT, pp 18–38.

Wood, I. G., Thompson, P., & Matthewman, J. C., (1983) *Acta Cryst B39*, 543–547.

Wyckoff, H. W., Doscher, M., Tsernoglou, D., Inagami, T., Johnson, L. N., Hardman, K. D., Allewell, N. N., Kelly, D. M., & Richards, F. M. (1967) *J. Mol. Biol. 27*, 5372–5382.

Zurek, S., Papiz, M. Z., Machin, P. A., & Helliwell, J. R. (1985) *Inf. Q. Protein Crystallogr., Daresbury Lab. No. 16*, 37–40.

Chapter 5

Implications of the Three-Dimensional Structure of α_1-Antitrypsin for Structure and Function of Serpins

R. Huber*,‡ and R. W. Carrell§

Max-Planck-Institut für Biochemie, D-8033 Martinsried, FRG, and Department of Haematology, University of Cambridge MRC Centre, Cambridge CB2 2QH, U.K.

Received June 6, 1989

There is now much interest in a newly recognized superfamily of proteins, the serpins (Carrell & Travis, 1985; Carrell et al., 1987a,b). More than 40 members of the family have been identified in viruses and plants as well as higher organisms. The serpins have developed by divergent evolution over a period of some 500 million years (Hunt & Dayhoff, 1980), most of the members retaining the presumed function of the original ancestral protein as serine proteinase inhibitors. Some, however, have lost this function and developed specialized roles as carriers of lipophilic molecules (thyroxine- and cortisol-binding globulins) or as peptide hormone precursors (angiotensinogen) or have no recognized function (ovalbumin).

The best studied members are those in human plasma where there is a diversity of inhibitory specialization that illustrates the way in which the serpins have evolved in parallel with their cognate proteases: antithrombin with thrombin, C_1-inhibitor with C_1-esterase, antiplasmin with plasmin, and so on. A key plasma serpin is α_1-antitrypsin; this is an efficient inhibitor of trypsin, but its prime physiological role is as an inhibitor of the elastase released by leukocytes. Interest focused on α_1-antitrypsin because its common genetic deficiency is associated with the development of premature lung degeneration (Laurell & Eriksson, 1963).

The establishment of α_1-antitrypsin as the archetype of the serpins was strengthened by the determination of its crystallographic structure, in a modified form, by Löbermann et al. (1984). It seems timely now to show how it can act as a general template for the other serpins. We look here at the common structural features of the family: the location of insertions and deletions and their compatibility with the three-dimensional template, the conserved amino acid residues and their relevance for the integrity of the spatial structure, the location of cysteine residues and disulfide bridges, and the sites of glycosylation. We well also discuss binding sites of functional modulators of some serpins (i.e., heparin in antithrombin) and ligand binding sites in serpins with carrier function. Finally, we demonstrate the overall validity of the α_1-antitrypsin model by showing how it provides a general explanation of the molecular pathology associated with diverse variants of the human serpins.

Molecular Structure of α_1-Antitrypsin

Fortunately, α_1-antitrypsin has turned out to be a typical member of the serpin family (Carrell et al., 1982). It is a glycoprotein of 394 residues with MW 51 000 and functions by forming a tight complex with its target protease. The serpins are believed to function as ideal substrates with association rates of the order of 10^4 M^{-1} s^{-1} or more and negligible dissociation rates (Travis & Salvesen, 1983). In particular, the sequence at the reactive center helps define specificity by providing a putative cleavage site for the target proteinase. Thus, the methionine 358 reactive center residue of α_1-antitrypsin provides a cleavage site of choice for leukocyte elastase, whereas in antithrombin the homologously aligned reactive center arginine 393 provides a specific cleavage site for thrombin. The critical role of the reactive center residues was highlighted by the finding of a pathological variant of α_1-antitrypsin in a child with a bleeding disorder in which methionine 358 had been substituted by an arginine, thus converting the protein from an inhibitor of elastase to a highly effective inhibitor of thrombin (Owen et al., 1983).

The molecular structure of α_1-antitrypsin as reported by Löbermann et al. (1984) is based on that of the cleaved molecule subsequent to release from the complex with chymotrypsin. It crystallizes in three different crystal forms which have been analyzed and found to be based on very similar molecular structures (Löbermann et al., 1984; Engh et al., 1989). The surprising feature of the structure was the separation of methionine and serine at the cleaved 358–359 reactive

‡Max-Planck-Institut für Biochemie.
§University of Cambridge MRC Centre.

Table I: Secondary Structural Elements in α_1-Antitrypsin[a]

helixes	sheets	turns	bulges
hA: 20–44 (kink at 28 Pro)	s6B: 49–53	thAs6B: 45–48	169–172
	s5B: 380–389	thBhC: 68–70	171–174
hB: 53–68	s4B: 369–378	thChD: 81–88 (lh: 81)	173–176 (series of overlapping bulges)
hC: 69–81	s3B: 247–255		
hC1: 83–87	s2B: 236–245	thDs2A: 105–110	
hD: 88–105	s1B: 228–233	ts2AhE: 122–127	
hE: 127–139	s6A: 290–299	thEs1A: 139–140 (lh: 139)	bs5B: 382–385
hF: 149–166	s5A: 326–342		
hF1: 200–203 (one open turn)	s4A: 343–356	ts1AhF: 146–149	bs5A: 329–332
	s3A: 181–194	thFs3A: 166–181 (lh: 166) (series of bulges)	
hF2: 232–236 (one open turn)	s2A: 109–121		
	s1A: 140–146		
hG: 259–264	s4C: 203–212	ts3AhF1: 194–199	
hH: 268–278	s3C: 213–226	ts4Cs3C: 211–214	
hI: 299–306	s2C: 283–289	ts3Cs1B: 226–228	
hI1: 309–312 (one open turn)	s1C: 362–367	ts1Bs2B: 233–236 (lh: 236)	
hI2: 376–380 (one open turn)		ts2Bs3B: 244–248	
		ts3BhG: 256–259	
hI3: 390–393 (one open turn)		thHs2C: 278–283	
		thI1s5A: 318–325	
		ts5As4A: 341–344	
		ts4Bs5B: 377–380 (lh: 380)	
		ts5Bc-ter: 389–394	

[a] Residues at the termini of helixes are included if at least one of their main-chain conformational angles is canonical; strands of sheets are defined similarly; appropriate hydrogen bonds are not always made by these residues. hX, helix X; sXY, strand X in sheet Y; thXhY, turn between helix X and helix Y; bsXY, bulge in strand X of sheet Y; lh, left-handed helical conformation.

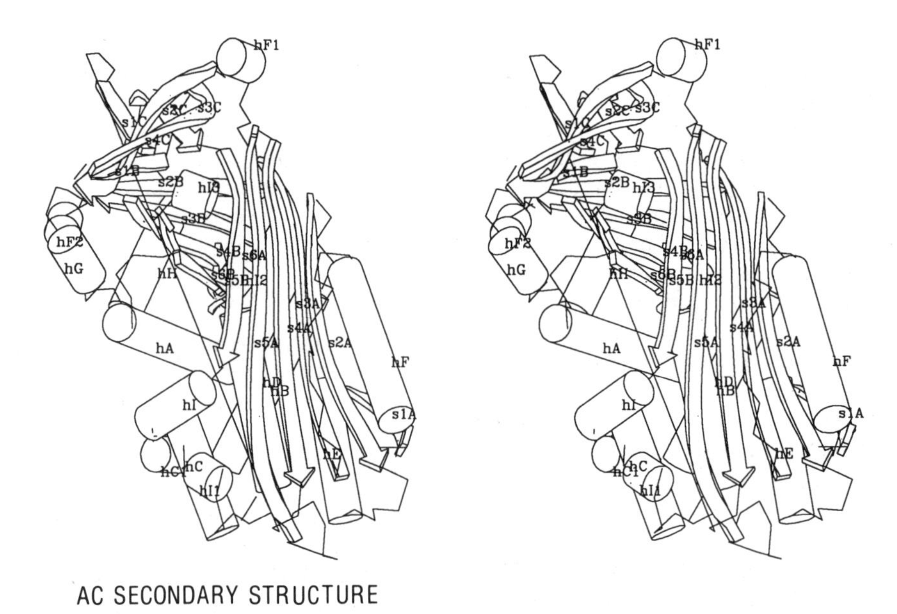

FIGURE 1: Polypeptide chain folding of α_1-antitrypsin with the secondary structure elements represented by arrows (sheet strands) and cylinders (helixes) and marked according to Table II. Residue numbers refer to α_1-antitrypsin and the alignment in Table II. Residues C-terminal to 358 have 100 added to their numbers in the plot.

center to opposite poles of the molecule with a distance of 69 Å. The significance of this change and the reconstruction necessary to give the native, uncleaved, protein are discussed in a subsequent section.

Table I gives a concise account of the secondary structural elements. The structure as a whole is remarkably ordered, with 80% of the amino acids in eight well-defined helixes (A–H) and three large β-sheets (A–C), as shown in Figure 1.

The α-helixes are regular, but hA, hB, and hC have 3_{10} geometry at their N- and C-termini. These helixes are fully or partly buried. The short helixes hF2 and hI2 are helical turns between β-strands. The β-sheets are antiparallel, except for short parallel strands added to sheet A (s1A) and C (s1C). They are regular, and their strands are only twice interrupted by bulges bs5B and bs5A. Because of these bulges, a proline residue, P382,[1] is accommodated in s5B and a lysine residue, K331, is turned to the surface of s5A, respectively.

The segment 169–176, which is antiparallel to α-helix hF, forms a sequence of bulges, by which the peculiar series of apolar and polar residues D171, L172, V173, K174, E175, and L176 are appropriately positioned internally and externally, respectively. It resembles the sequence of reverse turns in lysozyme (residues 17–22) but lacks the $O_j \rightarrow N_{j+3}$ hydrogen bonds characteristic for type I and II turns. The conserved N158 in hF makes hydrogen bonds to the main chain of V173 which may be crucial for the stabilization of this segment. The

[1] The one-letter code for amino acids is used. The amino acid numbers are based on the α_1-antitrypsin sequence and the alignment of Table II. When two numbers are given, the first refers to the protein discussed and the second to the homologous residue in α_1-antitrypsin.

turns are β-hairpins (ts4Cs3C, ts1Bs2B, ts2Bs3B, ts5As4A, ts4Bs5B), other sharp turns (ts1AhF, ts3BhG, thEs1A, thFs3A, thDs2B, thI1s5A), or wide excursions (ts2AhE, thFs3A, ts3AhF1, thHs2C). Of the β-hairpins only ts1Bs2B and ts5As4A have a regular $O_j \rightarrow N_{j+3}$ hydrogen bond.

Consistently in Ramachandran plots of all α_1-antitrypsin crystal structures, six non-glycine residues are outside the favorable region of conformational space, of which five have the slightly unfavorable left-handed α-helical conformation. This conformation is infrequent but clearly established in some proteins where it occurs in turns (Huber et al., 1987). Also, in α_1-antitrypsin the left-handed helical residues mediate turns: N81, at the end of hC, abruptly changes the chain direction and buries L80 and F82 in the interior of the protein. H139 is at the C-terminus of hE and allows an abrupt change of chain direction into s1A. Similarly, Q166 ends hF and leads into the antiparallel segment and series of bulges 169–176. It is adjacent to invariant T165. S236 ends the helical turn hF2 to lead into s2B. K380 ends the helical turn hI2 and leads into s5B so that the side chain of K380 projects to the surface.

A70 is in a high-energy conformation in all crystal forms. It is located in the turn thBhC and well defined. The carbonyl oxygen of the preceding K69 is hydrogen bonded to O^{γ} of T72 and to N of H73. These interactions enforce a strained conformation of A70, as observed similarly in other proteins [see, e.g., Bode et al. (1989)]. Eight out of 18 sequences in Table II have A or G at this position.

In the completely buried and apolar segments (hB, s4B, s5B, s6B), which are strongly conserved within the serpin superfamily, all hydrogen bonds are satisfied, mainly by interactions of main-chain atoms. Some irregularities in secondary structure occur where neutral polar amino acid side chains participate in hydrogen bonds to main-chain atoms: N49 by its O^{δ_1} and N^{δ_2} side-chain atoms forms hydrogen bonds to main-chain O and N atoms of N390, to N of V389, and to N^{ϵ_2} of Q393. E376 is at the N-terminus of hI2 and forms hydrogen bonds to the main-chain N atoms of N378 and T379 and also balances the charge of the helical dipole. N158 is hydrogen bonded by N^{δ_2} and O^{δ_1} to the main-chain N and O atoms of 173 and anchors hF to the series of bulges. O^{γ} of S53 initiates hB and is hydrogen bonded to N of S56. It also stabilizes bs5B by bonding to O of L383. T165 O^{γ} is hydrogen bonded to the main-chain O atom of V161 and N atom of I169 and acts as a clamp for the helical turn thFs3A. N186 is hydrogen bonded by its side chain to O and O^{γ} of S56 and to O^{γ} of T59. W194 has its N^{ϵ} hydrogen bonded to the main-chain O of E341. Y297 O^{ϵ} is hydrogen bonded to N of 51Phe. H334 has its N^{δ} hydrogen bonded to O of A350 and its N^{ϵ} to N of S53. These internal and polar residues are rather conserved or conservatively replaced in the serpin superfamily.

Family Relationships

Table II shows the alignment of 20 members of the superfamily modified from that of Bock et al. (1986) and Carrell et al. (1987a,b) and reevaluated on the basis of three-dimensional modeling using FRODO (Jones, 1978). Some 20 other serpins are not included because of either incomplete sequences or their recent elucidation as with the vaccinia virus serpins [e.g., Kotwal and Moss (1989)].

There is considerable variation in length due to long additions at the N- (C1 inhibitor) and C-termini (antiplasmin) and a 33-residue insertion between hC and hD (placental PAI). Other species have shorter insertions or deletions in the same area. Both N- and C-termini of modified α_1-antitrypsin protrude into solution so that insertions can be accommodated. The shortest species begin with residue 23 and end with residue 391, which are indeed the first and the last residues buried in the globular structure. hC and hD are connected by a wide loop including short helical segment hC1 so that long insertions can be incorporated without conflict with the globular fold of the molecule. Also, deletions between hC and hD are possible by eliminating hC1 as in endothelial PAI and antiplasmin. Short insertions or deletions occur in various members of the superfamily usually in turns between secondary structural elements: One- or two-residue insertions or deletions occur in thAs6B, a surface turn and also a carbohydrate attachment site. Up to three-residue deletions or two-residue insertions occur in thDs2A where they can be accommodated without affecting the helix and sheet elements. Similarly, two-residue insertions are found in ts1AhF where they can be accommodated between the series of bulges (ending with L176) and the start of s3A (T181). One- or two-residue insertions occur in ts3Cs1B. Up to five-residue insertions are found in ts3BhG (256–259). P255 and L260 are strongly conserved and in contact. It is likely that the insertions are arranged as a protruding loop. A two-residue insertion occurs in thHs2C, which is a wide exposed loop. It is unlikely to affect the adjacent secondary structure A one-residue insertion occurs between s6A and hI and can be accommodated by extending hI by one residue. A two-residue insertion is found near A325 at the beginning of s5A, which is exposed. The variations in lengths near the site of specific cleavage (358–359) are likely to be tolerable because it must be exposed, as discussed later.

Heparin cofactor II, antithrombin III, and C_1-inhibitor have extensions of 81, 25, and 93 residues at their N-termini, respectively. In the latter two cases they are linked by disulfide bonds to the rest of the molecule so that their general locations are defined and will be discussed later.

In Table III we note conserved residues and indicate their locations and environments to provide clues as to their importance for the structural integrity. All conserved residues are internal or in surface niches. They are clustered. Four segments, hB, s3A, s4B, and s5B, have more than four conserved residues. hB, s4B, and s5B constitute the core of the molecule, while s3A represents the spine (Figure 2).

Disulfide Linkages

Disulfide bonds absent in α_1-antitrypsin but present in family members provide direct evidence for structural similarity, but very few of them have been chemically defined. Two occur in conserved parts: C216–C392 in antithrombin III replacing V and T in α_1-antitrypsin (Petersen et al., 1979) and C27–C381 in placental PAI which replace T and S (U. Kiso, A. Henschen, I. Leander, and B. Astedt, unpublished results). Figure 3 shows that these disulfide bonds can be formed by substitution in the α_1-antitrypsin structure without significant structural perturbations. This observation is relevant also with regard to the structural change upon limited proteolysis.

Other disulfide linkages are in areas with deletions or insertions: C87h–C133 in ovalbumin. The C^{α} atoms of residues 90 and 133 in α_1-antitrypsin are 11 Å apart so that the disulfide bond can be easily made if the nine-residue insertion between hC and hD is in an appropriate conformation, perhaps extending hC1. Defined cystine residues are also in antithrombin III and C_1-inhibitor in the long N-terminal extensions, which they link to the globular domain. They serve to define the general course of the N-termini in these molecules as shown in Figure 3. The N-terminus of C_1-inhibitor is longer by 93 residues compared to that of α_1-antitrypsin. It is constrained from residue 9 on by the disulfide bonds 9–319 and 14b–88. In the α_1-antitrypsin structure, H20 is the first spatially defined residue. Its distance to residue 88 is 16.5 Å,

Table II: Alignment of Amino Acid Sequences of 20 Members of the Serpin Family[a]

Protein	No.	Sequence
		-90
C1-INHIBITOR	5.	NPNATSS
		-80 -70 -60 -50 -40
secondary structure (Table 1)		
ANTITRYPSIN HUMAN	1.	
ANTITRYPSIN BABOON	2.	
HEPARIN COFAC II	3.	GSKGPLDQLEKGGETAQSADPQWEQLNNKNLSMPLLPADFHKENTV
ANTITHROMBIN	4.	mysnvigtvtsgkrkvyllsll
C1-INHIBITOR	5.	SSQDPESLQDRGEGKVATTVISKMLFVEPILEVSSLPTTNSTTNSATKIT
ENDOTHELIAL PAI	6.	
THYROXINE B G	7.	
ANTICHYMO MOUSE	8.	
ANTICHYMO HUMAN	9.	
ANTIPLASMIN	10.	
OVALBUMIN	11.	
GENE Y PROTEIN	12.	
PLACENTAL PAI	13.	
ANGIOTENS RAT	14.	mtptgaglkatifciltwvsltagDRVYIHPFHLLY
ANGIOTENS HUMAN	15.	mrkrapqsemapagvslratilcllawaglaagDRVYIHPFHLVI
BARLEY Z PROTEIN	16.	
CORTICOSTEROID BG	17.	
PROTEIN C INHIB.	18.	
PROTEASE NEXIN	19.	
RAB ORF1	20.	

Protein	No.	Sequence
		-30 -20 -10 -11 10
secondary structure		
ANTITRYPSIN HUMAN	1.	mpssvswgilllaglcclvpvslaEDPQGDAAQKTDTS-
ANTITRYPSIN BABOON	2.	lllaglccllpgslaEDPQGDAAQKTDTP-
HEPARIN COFAC II	3.	TNDWIPEGEEDDDYLDLEKIFSEDDDYIDIVDSLSVSPTDSDVSAGNIL-
ANTITHROMBIN	4.	ligfwdcvtcHGSPVDICTAKPRDIPMNPMCIYRSPEKKATEDEGSEQK-
C1-INHIBITOR	5.	ANTTDEPTTQPTTEPTTQPTIQPTQPTTQLPTDSPTQPTTGSFCPGPVTL
ENDOTHELIAL PAI	6.	mqmspaltclvlglalvf
THYROXINE B G	7.	mspflylvllvlglhatihcASPEGKVTA
ANTICHYMO MOUSE	8.	
ANTICHYMO HUMAN	9.	mermlpllalgllaagfcpavlchpNSPLDEENLTQE-
ANTIPLASMIN	10.	NQEQVSPLTLLKLGNQEPGGQTALKSPPGV
OVALBUMIN	11.	
GENE Y PROTEIN	12.	
PLACENTAL PAI	13.	
ANGIOTENS RAT	14.	YSKSTCAQLENPSVETLPEPTFEPVPIQAKTSPVDEKTLRDKLVLATEK-
ANGIOTENS HUMAN	15.	HNESTCEQLAKANAGKPKDPTFIPAPIQAKTSPVDEKALQDQLVLVAAK-
BARLEY Z PROTEIN	16.	
CORTICOSTEROID BG	17.	mplllytcllwlptsglwtvqaMDPNA-
PROTEIN C INHIB.	18.	mqlflllclvllspqgaslHRHHPREMKK-
PROTEASE NEXIN	19.	MNWHLPLFLLASVTL-
RAB ORF1	20.	

Protein	No.	Sequence
		20 30 40 50 60
secondary structure		^^^^^^^^^^^^hA^^^^^^^^^^^ -s6B-^^^^^^hB^
ANTITRYPSIN HUMAN	1.	-HHDQDHPTFNKITPNLAEFAFSLYRQLAHQSNS-TNI-FFSPVSIATAF
ANTITRYPSIN BABOON	2.	-PHDQNHPTLNKITPSLAEFAFSLYRQLAHQSNS-TNI-FFSPVSIATAF
HEPARIN COFAC II	3.	-QLFHGKSRIQRLNILNAKFAFNLYRVLKDQVNTFDNI-FIAPVGISTAM
ANTITHROMBIN	4.	-IPEATNRRVWELSKANSRFATTFYQHLADSKNDNDNI-FLSPLSISTAF
C1-INHIBITOR	5.	CSDLESHSTEAVLGDALVDFSLKLYHAFSAMKKVETNM-AFSPFSIASLL
ENDOTHELIAL PAI	6.	gegsaVHHPPSYVAHLASDFGVRVFQQVAQASKD-RNV-VFSPYGVASVL
THYROXINE B G	7.	CHSSQPNATLYKMSSINADFAFNLYRRFTVETPD-KNI-FFSPVSISAAL
ANTICHYMO MOUSE	8.	
ANTICHYMO HUMAN	9.	-NQDRGTHVDLGLASANVDFAFSLYKQLVLKALD-KNV-IFSPLSISTAL
ANTIPLASMIN	10.	CSRDPTPEQTHRLARAMMAFTADLFSLVAQTS-TCPNL-ILSPLSVALAL
OVALBUMIN	11.	MGSIGAASMEFCFDVFKELKVHHAN-ENI-FYCPIAIMSAL
GENE Y PROTEIN	12.	MDSISVTNAKFCFDVFNEMKVHHVN-ENI-LYCPLSILTAL
PLACENTAL PAI	13.	MEDLCVANTLFALNLFKHLAKASPT-QNL-FLSPWSISSTM
ANGIOTENS RAT	14.	-LEAEDRQRAAQVAMIANFMGFRMYKMLSEARGVASGA-VLSPPALFGTL
ANGIOTENS HUMAN	15.	-LDTEDKLRAAMVGMLANFLGFRIYGMHSELWGVVHGATVLSPTAVFGTL
BARLEY Z PROTEIN	16.	
CORTICOSTEROID BG	17.	-AYVNMSNHHRGLASANVDFAFSLYKHLVALSPK-KNI-FISPVSISMAL
PROTEIN C INHIB.	18.	-RVEDLHVGATVAPSSRRDFTFDLYRALASAAPS-QNI-FFSPVSISMSL
PROTEASE NEXIN	19.	-PSICSHFNPLSLEELGSNTGIQVFNQIVKSRP-HDNI-VISPHGIASVL
RAB ORF1	20.	MKYLVLVLCLTSCACRDIGLWTFRYVYNES---DNV-VFSPYGLTSAL

Protein	No.	Sequence
		70 80
secondary structure		^^^^^^^^^^^^^hC^^^^^. ^ hC1^
ANTITRYPSIN HUMAN	1.	AMLSLGTKADTHDEILEGLNFN-LTEI-----------------------
ANTITRYPSIN BABOON	2.	AMLSLGTKADTHSEILEGLNFN-LTEI-----------------------
HEPARIN COFAC II	3.	GMISLGLKGETHEQVHSILHFKDFVNA-----------------------
ANTITHROMBIN	4.	AMTKLGACNDTLQQLMEVFKFDTISEK-----------------------
C1-INHIBITOR	5.	TQVLLGAGQNT-KTNLESILSYPKDFT-----------------------
ENDOTHELIAL PAI	6.	AMLQLTTGGETQQQIQAAMGFK----------------------------
THYROXINE B G	7.	VMLSFGACCSTQTEIVETLGFN-LTDT-----------------------
ANTICHYMO MOUSE	8.	
ANTICHYMO HUMAN	9.	AFLSLGAHNTTLTEILKASSSP-HGDL-----------------------
ANTIPLASMIN	10.	SHLALGAQNHTLQRLQQVLH------------------------------
OVALBUMIN	11.	AMVYLGAKDSTRTQINKVVRFDKLPGF-----------------------
GENE Y PROTEIN	12.	AMVYLGARGNTESQMKKVLHFDSITGA-----------------------
PLACENTAL PAI	13.	AMVYMGSRGSTEDQMAKVLQF-NEVGANAVTPMTPENFTSCGFMQQIQKG
ANGIOTENS RAT	14.	VSFYLGSLDPTASQLQVLLGVPVKEGD-----------------------
ANGIOTENS HUMAN	15.	ASLYLGALDHTADRLQAILGVPWKDKN-----------------------
BARLEY Z PROTEIN	16.	
CORTICOSTEROID BG	17.	AMLSLGTCGHTRAQLLQGLGFN-LTER-----------------------
PROTEIN C INHIB.	18.	AMLSLGAGSSTKMQILEGLGLN-LQKS-----------------------
PROTEASE NEXIN	19.	GMLQLGADGRTKKQLAMVMRY-----------------------------
RAB ORF1	20.	SVLRIAAGGNTKREIDVPESVVED--------------------------

Table II (Continued)

```
                                 90       100         110       120
secondary structure           ^^^^^^^hD^^^^^^^^^^     -  ----s2A-----
ANTITRYPSIN HUMAN  1. ----------PEAQIHEGFQELLRTLNQPDSQ--LQLTTDGGLFLSEGLK
ANTITRYPSIN BABOON 2. ----------PEAQVHEGFQELLRTLNKPDSQ--LQLTTGNGLFLNKSLK
HEPARIN COFAC II   3. ------SSKYEITTIHNLFRKLTHRLFRRNFG--YTLRSVNDLYIQKQFP
ANTITHROMBIN       4. ----------TSDQIHFFFAKLNCRLYRKANK-SSKLVSANRLFGDKSLT
C1-INHIBITOR       5. ----------CVHQALKGFTTKG-------------VTSVSQIFHSPDLA
ENDOTHELIAL PAI    6. ---------IDDKGMAPALRHLYKELMGPWNK--DEISTTDAIFVQRDLK
THYROXINE B G      7. ----------PMVEIQHGFQHLICSLNFPKKE--LELQIGNALFIGKHLK
ANTICHYMO MOUSE    8.
ANTICHYMO HUMAN    9. ----------LRQKFTQSFQHLRAPSISSSDE--LQLSMGNAMFVKEQLS
ANTIPLASMIN       10. --------AGSGPCLPHLLSRLCQDLG-PG-----AFRLAARMYLQKGFP
OVALBUMIN         11. -GDSIEAQCGTSVNVHSSLRDILNQITKPND--VYSFSLASRLYAEERYP
GENE Y PROTEIN    12. -GSTTDSQCGSSEYVHNLFKELLSEITRPNA--TYSLEIADKLYVDKTFS
PLACENTAL PAI     13. SYPDAILQAQAADKIHSSFRSLSSAINASTGD--YLLESVNKLFGEKSAS
ANGIOTENS RAT     14. -CTSRLDGH-KVLTALQAVQGLLVTQGGSSSQTPLLQSTVVGLFTAPGLR
ANGIOTENS HUMAN   15. -CTSRLDAH-KVLSALQAVQGLLVAQGRADSQAQLLLSTVVGVFTAPGLH
BARLEY Z PROTEIN  16.
CORTICOSTEROID BG 17. ----------SETEIHQGFQHLHQLFAKSDTS--LEMTMGNALFLDGSLE
PROTEIN C INHIB.  18. ----------SEKELHRGFQQLLQELNQPRDG--FQLSLGNALFTDLVVD
PROTEASE NEXIN    19. ----------GVNGVGKILKKINKAIVSKKNK--DIVTVANAVFVKNASE
RAB ORF1          20. --------------------------------S--DAFLALRELFVDASVP

                         130      140         150       160       170
secondary structure    ^^^^^^hE^^^^^-- s1A--  ^ ^^^^^^hF^^^^^^^^^
ANTITRYPSIN HUMAN  1. LVDKFLEDVKKLYHSE-AFTVNFGD-TEEAKKQINDYVEKGTQGKIVDLV
ANTITRYPSIN BABOON 2. VVDKFLEDVKNLYHSE-AFSVNFED-TEEAKKQINNYVEKGTQGKVVDLV
HEPARIN COFAC II   3. ILLDFKTKVREYYFAE-AQIADFSD--PAFISKTNNHIMKLTKGLIKDAL
ANTITHROMBIN       4. FNETYQDISELVYGAK-LQPLDFKENAEQSRAAINKWVSNKTEGRITDVI
C1-INHIBITOR       5. IRDTFVNASRTLYSSS-PRVLSNN--SDANLELINTWVAKNTNNKISRLL
ENDOTHELIAL PAI    6. LVQGFMPHFFRLFRST-VKQVDFSE-VERARFIINDWVKTHTKGMISNLL
THYROXINE B G      7. PLAKFLNDVKTLYETE-VFSTDFSN-ISAAKQEINSHVEMQTKGKVVGLI
ANTICHYMO MOUSE    8.
ANTICHYMO HUMAN    9. LLDRFTEDAKRLYGSE-AFATDFQD-SAAAKKLINDYVKNGTRGKITDLI
ANTIPLASMIN       10. IKEDFLEQSEQLFGAK-PVSLT--GKQEDDLANINQWVKEATEGKIQEFL
OVALBUMIN         11. ILPEYLQCVKELYRGG-LEPINFQTAADQARELINSWVESQTNGIIRNVL
GENE Y PROTEIN    12. VLPEYLSCARKFYTGG-VEEVNFKTAAEEARQLINSWVEKETNGQIKDLL
PLACENTAL PAI     13. FREEYIRLCQKYYSSE-PQAVDFLECAEEARKKINSWVKTQTKGKIPNLL
ANGIOTENS RAT     14. LKQPFVESLGPFTPAIFPRSLDLSTDPVLAAQKINRFVQAVTGWKMNLPL
ANGIOTENS HUMAN   15. LKQPFVQGLALYTPVVLPRSLDF-TELDVAAEKIDRFMQAVTGWKTGCSL
BARLEY Z PROTEIN  16.
CORTICOSTEROID BG 17. LLESFSADIKHYYESE-VLAMNFQDW-ATASRQINSYVKNKTQGKIVDLF
PROTEIN C INHIB.  18. LQDTFVSAMKTLYLAD-TFPTNFRD-SAGAMKQINDYVAKQTKGKIVDLL
PROTEASE NEXIN    19. IEVPFVTRNKDVFQCE-VRNVNFED-PASACDSINAWVKNETRDMIDNLL
RAB ORF1          20. LRPEFTAEFSSRFNTS-VQRVTFN--SENVKDVINSYVKDKTGGDVPRVL

                            180       190       200        210       220
secondary structure            ------s3A-----   ^hF1-- -s4C-------s3C-
ANTITRYPSIN HUMAN  1. KELDRD--TVFALVNYIFFKGKWERPFEVKDTEE-EDFHVDQVTTVKVPM
ANTITRYPSIN BABOON 2. KELDRD--TVFALVNYIFFKGKWERPFEVEATEE-EDFHVDQATTVKVPM
HEPARIN COFAC II   3. ENIDPA--TQMMILNCIYFKGSWVNKFPVEMTHN-HNFRLNEREVVKVSM
ANTITHROMBIN       4. PSEAINELTVLVLVNTIYFKGLWKSKFSPENTRK-ELFYKADGESCSASM
C1-INHIBITOR       5. DSLPSD--TRLVLLNAIYLSAKWKTTFDPKKTRM-EPFHFKNSV-IKVPM
ENDOTHELIAL PAI    6. GKGAVDQLTRLVLVNALYFNGQWKTPFPDSSTHR-RLFHKSDGSTVSVPM
THYROXINE B G      7. QDLKPN--TTMVLVNYIHFKAQWANPFDPSKTEDSSSFLIDKTTTVQVPM
ANTICHYMO MOUSE    8.           VVLVNYIYFKGKWKISFDPQDTFE-SEFYLDEKRSVKVPM
ANTICHYMO HUMAN    9. KDP--DSQTMMVLVNYIFFKAKWEMPFDPQDTHQ-SRFYLSKKKWVMVPM
ANTIPLASMIN       10. SGLPED--TVLLLLNAIHFQGFWRNKFDPSLTQR-DSFHLDEQFTVPVEM
OVALBUMIN         11. QPSSVDSQTAMVLVNAIVFKGLWEKAFKDEDTQA-MPFRVTEQESKPVQM
GENE Y PROTEIN    12. VSSSIDFGTTMVFINTIYFKGIWKIAFNTEDTRE-MPFSMTKEESKPVQM
PLACENTAL PAI     13. PEGSVDGDTRMVLVNAVYFKGKWKTPFEKKLNGL-YPFRVNSAQRTPVQM
ANGIOTENS RAT     14. EGVSTD--STLFFNTYVHFQGKM-RGFSQ-LTGL-HEFWVDNSTSVSVPM
ANGIOTENS HUMAN   15. MGASVD--STLAFNTYVHFQGKM-KGFSL-LAEP-QEFWVDNSTSVSVPM
BARLEY Z PROTEIN  16.
CORTICOSTEROID BG 17. SGLDS--PAILVLVNYIFFKGTWTQPFDLASTRE-ENFYVDETTVVKVPM
PROTEIN C INHIB.  18. KNLDS--NAVVIMVNYIFFKAKWETSFNHKGTQE-QDFYVTSETVVRVPM
PROTEASE NEXIN    19. SPDLIDGVTRLVLVNAVYFKGLWKSRFQPENTKK-RTFVAADGKSYQVPM
                             L
RAB ORF1          20. DASLDRD-TKMLLLSSVRMKTSWRHVFDPSFTTD-QPFYSGNV-TYKVRM

                               230        240        250            260
secondary structure   ------   ----^^hF2^----s2B--     ----s3B--     ^^^
ANTITRYPSIN HUMAN  1. MKRLGMF--NIQHCKK-LSSWVLLMKYL-GNANAIFFLPD-----EGKLQ
ANTITRYPSIN BABOON 2. MRRLGMF--NIYHCEK-LSSWVLLMKYL-GNATAIFFLPD-----EGKLQ
HEPARIN COFAC II   3. MQTKGNF--LAANDQE-LDCDILQLEYV-GGISMLIVVPHK----MSGMK
ANTITHROMBIN       4. MYQEGKF--RYRRVAE--GTQVLELPFKGDDITMVLILPKP----EKSLA
C1-INHIBITOR       5. MNSKKYP-VAHFIDQT-LKAKVGQLQLS-HNLSLVILVPQNL--KHRLED
ENDOTHELIAL PAI    6. MAQTNKFNYTEFTTPDGHYYDILELPYHGDTLSMFIAAPYE---KEVPLS
THYROXINE B G      7. MHQMEQY--YHLVDME-LNCTVLQMDYS-KNALALFVLPK-----EGQME
ANTICHYMO MOUSE    8. MKMKLL-TTRHFRDEE-LSCSVLELKYT-GNASALLILPD-----QGRMQ
ANTICHYMO HUMAN    9. MSLHHL-TIPYFRDEE-LSCTVVELKYT-GNASALFILPD-----QDKME
ANTIPLASMIN       10. MQARTYP-LRWFLLEQ-PEIQVAHFPFK-NNMSFVVLVPTH---FEWNVS
OVALBUMIN         11. MYQIGLF--RVASMAS-EKMKILELPFASGTMSMLVLLPDE----VSGLE
GENE Y PROTEIN    12. MCMNNSF--NVATLPA-EKMKILELPYASGDLSMLVLLPDE----VSGLE
PLACENTAL PAI     13. MYLREKL--NIGYIED-LKAQILELPYA-GDVSMFLLLPDEIADVSTGLE
ANGIOTENS RAT     14. LSGTGNF--QHWSDAQ-NNFSVTRVPL-GESVTLLLIQPQ----CASDLD
ANGIOTENS HUMAN   15. LSGMGTF--QHWSDIQ-DNFSVTQVPF-TESACLLLIQPH----YASDLD
BARLEY Z PROTEIN  16.        YISSSDNLK-VLKLPYAKGHDKRQFSMYILLPG----AQDGLW
CORTICOSTEROID BG 17. MLQSSTI--SYLHDSE-LPCQLVQMNYV-GNGTVFFILPD-----KGKMN
PROTEIN C INHIB.  18. MSREDQY--HYLLDRN-LSCRVVGVPYQ-GNATALFILPS-----EGKMQ
PROTEASE NEXIN    19. LAQLSVFRCGSTSAPNDLWYNFIELPYHGESISMLIALPT---ESSTPLS
RAB ORF1          20. MNKIDTL-KTETFTLRNVGYSVTELPYKRRQTAMLLVVP------DDLGE
```

Table II (Continued)

```
                                   270          280            290          300
secondary structure        hG^    ^^^^^hH^^^^          --s2C-----s6A---^^ ^hI^^^
ANTITRYPSIN HUMAN   1. HLENELTHDIITKFLENEDR--RSASLHLPKLSITGTYDLK-SVLGQLGI
ANTITRYPSIN BABOON  2. HLENELTHDIITKFLENENR--RSANLHLPKLAITGTYDLK-TVLGHLGI
HEPARIN COFAC II    3. TLEAQLTPRVVERWQKSMTN--RTREVLLPKFKLEKNYNLV-ESLKLMGI
ANTITHROMBIN        4. KVEKELTPEVLQEWLDELEE--MMLVVHMPRFRIEDGFSLK-EQLQDMGL
C1-INHIBITOR        5. MEQALSPSVFKAIMEKLEMSKFQPTLLTLPRIKVTTSQDML-SIMEKLEF
ENDOTHELIAL PAI     6. ALTNILSAQLISHWKGNMTR--LPRLLVLPKFSLETEVDLR-KPLENLGM
THYROXINE B G       7. SVEAAMSSKTLKKWNRLLQK--GWVDLFVPKFSISATYDLG-ATLLKMGI
ANTICHYMO MOUSE     8. QVEASLQPETLRKWRKTLFPS-QIEELNLPKFSIASNYRLEEDVLPEMGI
ANTICHYMO HUMAN     9. EVEAMLLPETLKRWRDSLEFR-EIGELYLPKFSISRDYNLN-DILLQLGI
ANTIPLASMIN        10. QVLANLSWDTLHPPLVWE----RPTKVRLPKLYLKHQMDLV-ATLSQLGL
OVALBUMIN          11. QLESIINFEKLTEWTSSNVMEERKIKVYLPRMKMEEKYNLT-SVLMAMGI
GENE Y PROTEIN     12. RIEKTINFDKLREWTSTNAMAKKSMKVYLPRMKIEEKYNLT-SILMALGM
PLACENTAL PAI      13. LLESEITYDKLNKWTSKDKMAEDEVEVYIPQFKLEEHYELR-SILRSMGM
ANGIOTENS RAT      14. RVEVLVFQHDFLTWIKNPPP--RAIRLTLPQLEIRGSYNLQ-DLLAQAKL
ANGIOTENS HUMAN    15. KVEGLTFQQNSLNWMKKLSP--RTIHLTMPQLVLQGSYDLQ-DLLAQAEL
BARLEY Z PROTEIN   16. SLAKRLSTEPEFIENHIPKQTVEVGRFQLPKFKISYQFEAS-SLLRALGL
CORTICOSTEROID BG  17. TVIAALSRDTINRWSAGLTS--SQVDLYIPKVTISGVYDLG-DVLEEMGI
PROTEIN C INHIB.   18. QVENGLSEKTLRKWLKMFKK--RQLELYLPKFSIEGSYQLE-KVLPSLGI
PROTEASE NEXIN     19. AIIPHISTKTIDSWMSIMVP--KRVQVILPKFTAVAQTDLK-EPLKVLGI
RAB ORF1           20. IVRALDLSLVRFWIRNMRK---DVCQVVMPKFSVESVLDLR-DALQRLGV

                       310          320            330        340         350
secondary structure     ^hI1                      -------s5A----------s4A-------
ANTITRYPSIN HUMAN   1. TKVFSNGAD-LSGVTEEA--PLKLSKAVHKAVLTIDEKGTEAAGAMFLEA
ANTITRYPSIN BABOON  2. TKVFSNGAD-LSGVTEDA--PLKLSKAVHKAVLTIDEKGTEAAGAMFLEA
HEPARIN COFAC II    3. RMLFDKNGN-MAGISDQR---IAIDLFKHQGTITVNEEGTQATTVTTVGF
ANTITHROMBIN        4. VDLFSPEKSKLPGIVAEGRDDLYVSDAFHKAFLEVNEEGSEAAASTAVVI
C1-INHIBITOR        5. FD-FSYDLN-LCGLTEDP--DLQVSAMQHQTVLELTETGVEAAAASAISV
ENDOTHELIAL PAI     6. TDMFRQFQADFTSLSDQE--PLHVAQALQKVKIEVNESGTVASSSTAVIV
THYROXINE B G       7. QHAYSENAD-FSGLTEDN--GLKLSNAAHKAVLHIGEKGTEAAAVPEVEL
ANTICHYMO MOUSE     8. KEVFTEQAD-LSGIIETK--KLSVSQVVHKAVLDVAETGTEAAAATGVIG
ANTICHYMO HUMAN     9. EEAFTSKAD-LSGITGAR--NLAVSQVVHKAVLDVFEEGTEASAATAVKI
ANTIPLASMIN        10. QELF-QAPD-LRGISEQ---SLVVSGVQHQSTLELSEVGVEAAAATSIAM
OVALBUMIN          11. TDVFSSSAN-LSGISSAE--SLKISQAVHAAHAEINEAGREVVGSAEAGV
GENE Y PROTEIN     12. TDLFSRSAN-LTGISSVD--NLMISDAVHGVFMEVNEEGTEATGSTGAIG
PLACENTAL PAI      13. EDAFNKGRANFSGMSERN--DLFLSEVFHQAMVDVNEEGTEAAAGTGGVM
ANGIOTENS RAT      14. STLLGAEAN-LGKMGDTN--PRVGEVLNSILLELQAGEEEQPTESAQQPG
ANGIOTENS HUMAN    15. PAILHTELN-LQKLSNDR--IRVGEVLNSIFFELEA-DEREPTESTQQLN
BARLEY Z PROTEIN   16. QLPFSEEAD-LSEMVDSS-QGLEISHVFHKSFVEVNEEGTEAGAATVAMG
CORTICOSTEROID BG  17. ADLFTNQAN-FSRITQDA--QLKSSKVVHKAVLQLNEEGVDTAGSTGVTL
PROTEIN C INHIB.   18. SNVFTSHAD-LSRISNHS--NIQVSEMVHKAVVEVDESGTRAAAATGTIF
PROTEASE NEXIN     19. TDMFDSSKANFAKITTGSE-NLHVSHILQKAKIEVSEDGTKASAATTAIL
RAB ORF1           20. RDAFDPSRADFGQASPSN--DLYVTKVLQTSKIEADERGTTASSDTAITL

                             360              370        380         390 394
secondary structure     -         --        s1C- ---s4B-^hI2^---s5B---^hI3
ANTITRYPSIN HUMAN   1. IP-MSIPPE-----VKFNKPFVFLMIEQNTKSPLFMGKVVNPTQK
ANTITRYPSIN BABOON  2. IP-MSIPPE-----VKFNKPFVFLMIEQNTKSPLFIGKVVNPTQK
HEPARIN COFAC II    3. MP-LSTQVR-----FTVDRPFLFLIYEHRTSCLLFMGRVANPSRS
ANTITHROMBIN        4. AG-RSLNPN--RVTFKANRPFLVFIREVPLNTIIFMGRVANPCVK
C1-INHIBITOR        5. A--RTLLV------FEVQQPFLFVLWDQQHKFPVFMGRVYDPRA
ENDOTHELIAL PAI     6. SA-RMAPEE-----IIMDRPFLFVVRHNPTGTVLFMGQVMEP
THYROXINE B G       7. SD-QPENTFLHPI-IQIDRSFMLLILERSTRSILFLGKVVNPTEA
ANTICHYMO MOUSE     8. GIRKAILPA-----VHFNRPFLFVIYHTSAQSILFMAKVNNPK
ANTICHYMO HUMAN     9. TL-LSALVETRTI-VRFNRPFLMIIVPTDTQNIFFMSKVTNPKQA
ANTIPLASMIN        10. S--RMSLSS-----FSVNRPFLFFIFEDTTGLPLFVGSVRNPNPSAPREL
OVALBUMIN          11. DA-ASVS-EE----FRADHPFLFCIKHIATNAVLFFGRCVSP
GENE Y PROTEIN     12. NIKHSLELEE----FRADHPFLFFIRYNPTNAILFFGRYWSP
PLACENTAL PAI      13. TG-RTGHGG---PQFVADHPFLFLIMHKITKCILFFGRFCSP
ANGIOTENS RAT      14. SP--------EVLDVTLSSPFLFAIYERDSGALHFLGRVDNPQNVV
ANGIOTENS HUMAN    15. KP--------EVLEVTLNRPFLFAVYDQSATALHFLGRVANPLSTA
BARLEY Z PROTEIN   16. VA-MSMPLKVDLVDFVANHPFLFLIREDIAGVVVFVGHVTNPLISA
CORTICOSTEROID BG  17. NL-TSKPII-----LRFNQPFIIMIFDHFTWSSLFLARVMNPV
PROTEIN C INHIB.   18. TF-RSARLN--SQRLVFNRPFLMFIV---DNNILFLGKVNRP
PROTEASE NEXIN     19. IA-RSSPPW-----FIVDRPFLFFIRHNPTGAVLFMGQINKP
RAB ORF1           20. IP-RNALTA-----IVANKPFMFLIYHKPTTTVLFMGTITKGEKVIYDTE

                       400        410        420         430        440    446
ANTITRYPSIN HUMAN   1.
ANTITRYPSIN BABOON  2.
HEPARIN COFAC II    3.
ANTITHROMBIN        4.
C1-INHIBITOR        5.
ENDOTHELIAL PAI     6.
THYROXINE B G       7.
ANTICHYMO MOUSE     8.
ANTICHYMO HUMAN     9.
ANTIPLASMIN        10. KEQQDSPGNKDFLQSLKGFPRGDKLFGPDLKLVPPMEEDYPQFGSPK
OVALBUMIN          11.
GENE Y PROTEIN     12.
PLACENTAL PAI      13.
ANGIOTENS RAT      14.
ANGIOTENS HUMAN    15.
BARLEY Z PROTEIN   16.
CORTICOSTEROID BG  17.
PROTEIN C INHIB.   18.
PROTEASE NEXIN     19.
RAB ORF1           20. GRDDVVSSV
```

Table II (Continued)

[a] The secondary structural elements are indicated in the headline. Most amino acid sequences were taken from MIPSX Database. *α_1-Antitrypsin precursor—Human.* Bollen, A., Herzog, A., Cravador, A., Herion, P., Chuchana, P., Vander Straten, A., Loriau, R., Jacobs, P., & Van Elsen, A. (1983) *DNA 2*, 255–264. Rosenberg, S., Barr, P. J., Najarian, R. C., & Hallewell, R. A. (1984) *Nature 312*, 77–80 (this sequence differs from that shown in having His 125, Gly 139, Asn 140, Thr 273, and Ile 326). Carrell, R. W., Jeppsson, J.-O., Laurell, C.-B., Brennan, S. O., Owen, M. C., Vaughan, L., & Boswell, D. R. (1982) *Nature 298*, 329–334 (sequence of residues 25–418; this sequence differs from that shown in having Gly 139, Asn 140, and Thr 273). *α_1-Antitrypsin—Baboon.* Kurachi, K., Chandra, T., Degen, S. J. F., White, T. T., Marchioro, T. L., Woo, S. L. C., & Davie, E. W. (1981) *Proc. Natl. Acad. Sci. U.S.A. 78*, 6826–6830. *Heparin cofactor II precursor—Human.* Blinder, M. A., Marasa, J. C., Reynolds, C. H., Deaven, L. L., & Tollefsen, D. M. (1988) *Biochemistry 27*, 752–759. *Antithrombin III precursor—Human.* Bock, S. C., Wion, K. L., Vehar, G. A., & Lawn, R. M. (1982) *Nucleic Acids Res. 10*, 8113–8125. Chandra, T., Stackhouse, R., Kidd, V. J., & Woo, S. L. C. (1983) *Proc. Natl. Acad. Sci. U.S.A. 80*, 1845–1848. Prochownik, E. V., Markham, A. F., & Orkin, S. H. (1983) *J. Biol. Chem. 258*, 8389–8394 (this sequence differs from that shown in having Arg 97). Petersen, T. E., Dudek-Wojciechowska, G., Sottrup-Jensen, L., & Magnusson, S. (1979) *The Physiological Inhibitors of Coagulation and Fibrinolysis* (Collen, D., Wiman, B., & Verstraete, M., Eds.) pp 43–54, Elsevier/North-Holland Biomedical Press, Amsterdam. *Complement C_1-inhibitor precursor—Human.* Carter, P. E., Dunbar, B., & Fothergill, J. E. (1988) *Eur. J. Biochem. 173*, 163–169. *Plasminogen activator inhibitor 1 precursor—Human.* Pannekoek, H., Veerman, H., Lambers, H., Diergaarde, P., Verweij, C. L., van Zonneveld, A. J., & van Mourik, J. A. (1986) *EMBO J. 5*, 2539–2544. Ginsburg, D., Zeheb, R., Yang, A. Y., Rafferty, U. M., Andreasen, P. A., Nielsen, L., Dano, K., Lebo, R. V., & Gelehrter, T. D. (1986) *J. Clin. Invest. 78*, 1673–1680. *Thyroxine-binding globulin precursor—Human.* Flink, I. L., Bailey, T. J., Gustafson, T. A., Markham, B. E., & Morkin, E. (1986) *Proc. Natl. Acad. Sci. U.S.A. 83*, 7708–7712. *Contrapsin—Mouse (fragment).* Hill, R. E., Shaw, P. H., Boyd, P. A., Baumann, H., & Hastie, N. D. (1984) *Nature 311*, 175–177. *α_1-Antichymotrypsin precursor—Human.* Chandra, T., Stackhouse, R., Kidd, V. J., Robson, K. J. H., & Woo, S. L. C. *Biochemistry 22*, 5055–5061 (partial sequence derived from the 3′ half of the m-RNA). Hill, R. E., & Hastie, N. D. (1987) *Nature 326*, 96–99 (this sequence is reported wherever there are differences). Morii, M., & Travis, J. (1983) *J. Biol. Chem. 258*, 12749–12752 (inhibitory site). *α_2-Antiplasmin precursor—Human.* Holmes, W. E., Nelles, L., Lijnen, H. R., & Collen, D. (1987) *J. Biol. Chem. 262*, 1659–1664. *Ovalbumin—Gallus gallus.* Woo, S. L. C., Beattie, W. G., Catterall, J. F., Dugaiczyk, A., Staden, R., Brownlee, G. G., & O'Malley, B. W. (1981) *Biochemistry 20*, 6437–6446. Benoist, C., O'Hare, K., Breathnach, R., & Chambon, P. (1980) *Nucleic Acids Res. 8*, 127–142. Breathnach, R., Benoist, C., O'Hare, K., Gannon, F., & Chambon, P. (1978) *Proc. Natl. Acad. Sci. U.S.A. 75*, 4853–4857. Gannon, F., O'Hare, K., Perrin, F., LePennec, J. P., Benoist, C., Cochet, M., Breathnach, R., Royal, A., Garapin, A., Cami, B., & Chambon, P. (1979) *Nature 278*, 428–434. McReynolds, L., O'Malley, B. W., Nisbet, A. D., Fothergill, J. E., Givol, D., Fields, S., Robertson, M., & Brownlee, G. G. (1978) *Nature 273*, 723–728. Robertson, M. A., Staden, R., Tanaka, Y., Catterall, J. F., O'Malley, B. W., & Brownlee, G. G. (1979) *Nature 278*, 370–372. *Gene Y protein (ovalbumin-related)—Chicken.* Heilig, R., Muraskowsky, R., Kloepfer, C., & Mandel, J. L. (1982) *Nucleic Acids Res. 10*, 4363–4382. *Placental plasminogen activator inhibitor, type II—Human.* Ye, R. D., Wun, T. C., & Sadler, J. E. (1987) *J. Biol. Chem. 262*, 3718–3725. *Angiotensinogen precursor—Rat.* Ohkubo, H., Kageyama, R., Ujihara, M., Hirose, T., Inayama, S., & Nakanishi, S. (1983) *Proc. Natl. Acad. Sci. U.S.A. 80*, 2196–2200. Bouhnik, J., Clauser, E., Strosberg, D., Frenoy, J. P., Menard, J., & Corvol, P. (1981) *Biochemistry 20*, 7010–7015 (sequence of residues 25–41). *Angiotensinogen precursor—Human.* Kageyama, R., Ohkubo, H., & Nakanishi, S. (1984) *Biochemistry 23*, 3603–3609 (it is uncertain whether Met-1 or Met-10 is the initiator). Tewksbury, D. A., Dart, R. A., & Travis, J. (1981) *Biochem. Biophys. Res. Commun. 99*, 1311–1315 (sequence of residues 34–58; this sequence differs from that shown in having Ser 51 and Asp 58, residue 47 was not determined). *Protein Z—Barley (fragment).* Hejgaard, J., Rasmussen, S. K., Brandt, A., & Svendsen, I. (1985) *FEBS Lett. 180*, 89–94. Nielsen, G., Johansen, H., Jensen, J., & Hejgaard, J. (1983) *Barley Genet. Newslett. 13*, 55–57 (map position). *Human corticosteroid binding globulin mRNA, complete cds.* Hammond, G. L., Smith, C. L., Goping, I. S., Underhill, D. A., Harley, M. J., Reventos, J., Musto, N. A., Gunsalus, G. L., Bardin, C. W. (1987) *Proc. Natl. Acad. Sci. U.S.A. 84*, 5153–5157. *Protein C inhibitor precursor—Human.* Suzuki, K., Deyashiki, Y., Nishioka, J., Kurachi, K., Akira, M., Yamamoto, S., & Hashimoto, S. (1987) *J. Biol. Chem. 262*, 611–616. *Glia-derived neurite promoting factor precursor—Human.* Gloor, S., Odink, K., Guenther, J., Nick, H., & Monard, D. (1986) *Cell 47*, 687–693 (sequence). *Hypothetical protease inhibitor—Rabbit plasmid.* Upton, C., Carrell, R. W., & McFadden, G. (1986) *FEBS Lett. 207*, 115–120. MIPSX: F. Pfeiffer, Martinsried Institute for Protein Sequences (MIPS), unpublished.

which is 21 Å away from residue 319. Both disulfide bonds can therefore be made if the segments 14b–20 and 9–14b have appropriate conformations. The N-terminus of C_1-inhibitor is rich in prolines and may prefer the conformation of a polyproline II helix with a rise of 3.1 Å per residue to span considerable distances. In antithrombin III two disulfide bonds link the N-terminus to the globular molecule. The segment from residue −5 to −18 is linked to 69 and 101, respectively, which are 29.7 Å apart. It is rich in prolines (3 out of 15) and can easily span the long distance. It is obvious that in both molecules the N-termini are located close to helix hD.

Carbohydrate Attachment Sites

α_1-Antitrypsin has three branched carbohydrate chains linked to asparagines 46, 83, and 247. These are located in turns thAs6B, thChD, ts2Bs3B, respectively. Most of the members of the serpin family are glycoproteins but only in a few of them are the sites of attachment chemically defined. In others the presence of consensus amino acid sequences, Asn-X-Ser/Thr, in relation to the content of carbohydrate determined chemically suggests sites of attachment.

These are shown in Figure 4, projected onto the α_1-antitrypsin structure except for the glycosylation sites at the N-terminal extension in C_1-inhibitor. Carbohydrate is apparently distributed over most of the surface of the molecule with no obvious preference. All glycosylation sites are external with the asparagine side chains as sites of attachment projecting into solution. Most of them are located in turns; a few are on the hydrophilic side of peripheral α-helixes.

Structure of the Active Inhibitor

The crystallographic structure obtained by Löbermann et al. (1984) is that of antitrypsin cleaved at the reactive center 358 methionine. Although the separation of the cleaved ends to opposite poles of the molecule suggests that the modified molecule has undergone a major conformational change, other evidence indicates that the changes are relatively limited. In the first place, the known structure of the cleaved form puts constraints on the structural transition, and as described here, the cleaved structure is compatible with the requirement of the native protein for the placement of oligosaccharide attachments and for the heparin binding site. The cleaved structure also provides a satisfying explanation for the functional changes observed in mutants of native serpins. In particular, the disulfide linkages observed in well-defined conserved areas of antithrombin and placental PAI, as in Figure 3, indicate that the central part of the molecule is not involved in significant conformational shifts.

What then is the change that occurs at the reactive center on cleavage? There is good evidence that a reconstruction of the native reactive center involves an extraction of much of the central 4A strand of the A sheet in which methionine 358 forms the C-terminus. Exposure of the native serpins to a range of proteases shows that they are consistently vulnerable to cleavage at sites within s4A (Carrell et al., 1987a,b). This is compatible with the model of Löbermann et al. (1984) in which the reconstructed reactive center is situated near the position of serine 359 in the cleaved molecule with removal of s4A followed by annealing of the A sheet. The evidence

Table III: Conserved Residues in the Serpin Family

Element	Residue	Comments
hA	F33	internal; close to conserved I57
hA	Y/F38	internal; O^{ϵ} hydrogen bonded to O^{ϵ}E264(conserved); close to conserved F52,M385
s6B	N49	internal; side chain hydrogen bonded to main chain of V380 and N390; close to conserved I293,V388
s6B	S53	internal; initiates hB; O^{γ} hydrogen bonded to main chain of S56 and L383
hB	P54	internal; initiates hB; close to conserved P382,F52
hB	S56	internal; O^{γ} hydrogen bonded to side chain of conserved, internal N186
hB	I57	internal; close to conserved F33,H334; in pocket formed by these residues and by F61,M351,L37
hB	T59	internal; O^{γ} close to side chain of conserved N186
hB	L66	in surface niche; close to conserved F130,Y138
hB	G67	terminates hB; a side chain would interfere with conserved F130,G320
	T72	internal; O^{γ} hydrogen bonded to O K69, N H73, O N317
s2A	F/Y119	close to conserved I157; in pocket formed by I157,Y160, F143
hE	F/Y130	close to G67; in pocket formed by L118,V321
hE	Y138	internal; ends hE; O^{ϵ} hydrogen bonded to N^{δ}H93
hF	I157	internal tight pocket; close to conserved F(Y)119
hF	N158	internal; side chain hydrogen bonded to main chain of V173; holds hF to antiparallel segment of bulges 169-176
hF	T165	O^{γ} hydrogen bonded to main chain of K168,I169,V161 which are conserved; the adjacent residue Q166 is in left handed helical conformation
	I169	internal; in tight pocket
s3A	L184	internal;
s3A	N186	internal; side chain hydrogen bonded to side chain of conserved S56; close to T59 and N116
s3A	I188	internal pocket; close to conserved F384,S56
s3A	F/Y190	internal; close to conserved F384,M374
s3A	G192	a side chain would interfere with conserved F190
s3A	W194	internal; tightly packed, close to conserved F198,Y244; N^{ϵ} hydrogen bonded to main chain D341 close to conserved E342

Table III (Continued)

-	F198	internal; close to conserved W194
	T203	ends hF1; O^{γ} is hydrogen bonded to main chain O of V200 and to the side chain of the invariant E342
s4C		
	F208	internal pocket formed by conserved P391,F370
	V218	internal; close to conserved P391, F208, M220
	M220	internal; close to conserved F208
s3C		
	M221	internal; close to conserved T203,E342,K290,F198
-	P255	ends s3B and initiates ts3BhG; close to conserved L260
	L260	in niche; close to conserved P369
hG		
	E264	salt link to K387 in hydrophobic surface pocket; hydrogen bonded to O^{ε} of conserved Y38
-	P289	internal; close to conserved F370, M220, F208
	Y297	O^{ε}hydrogen bonded to main chain of F51
hI		
	L303	ties hI to molecule; close to conserved F312, F33
-	F312	ties hI1 to molecule
-	L318	
-	G320	see G67
	H334	internal; side chain hydrogen bonded to main chain S53 and A355
s5A		
	E342	salt link to K290 and hydrogen bonded to side chain of conserved T203
s4A		
-	G344	a side chain would interfere with conserved W194 and Y244
	P369	turns s1C into s4B; close to conserved F208,P255,K387
	F370	
s4B		close to opposite strand s5B with conserved F384,G386,V388; also close to Y38
	V/L371	
	F372	
	F384	see F370,V371,F372
	G386	
s5B		
	V388	internal; close to conserved F370,F372,F208
	N390	$O^{\delta 1}$ hydrogen bonded to N, O^{γ} T392; initiates hI3
	P391	internal; close to conserved F208; initiates one turn C-terminal helix (hI3)

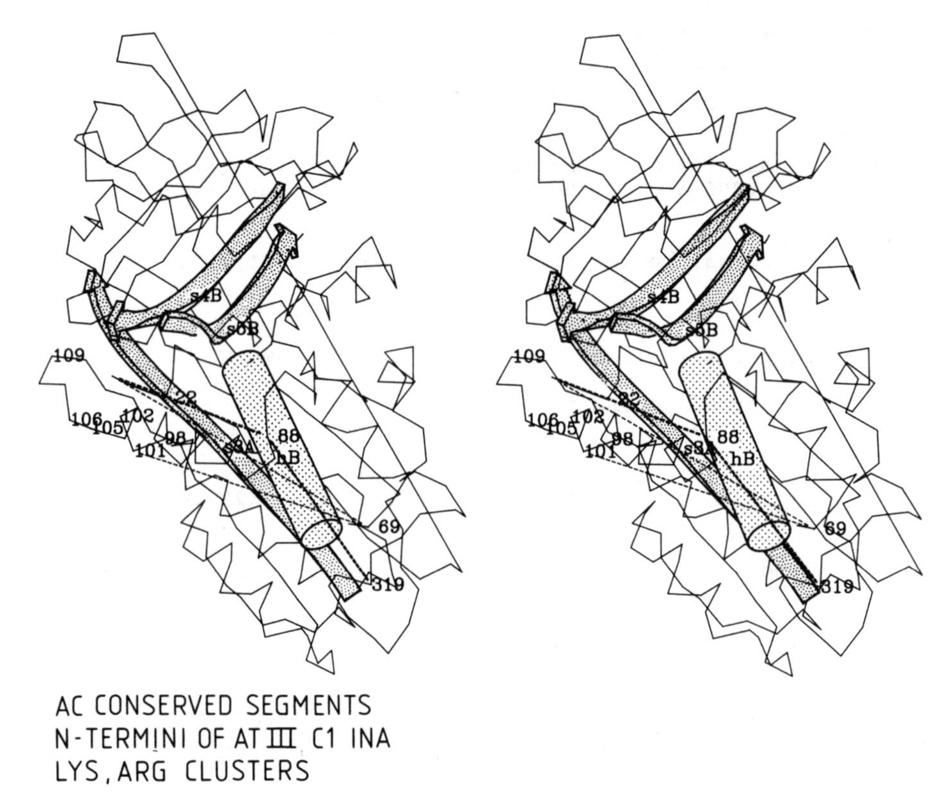

FIGURE 2: Polypeptide chain folding of α_1-antitrypsin with the locations of the most conserved secondary structure elements drawn as dotted arrows and cylinders (hB, s3A, s4B, s5B). The broken lines indicate the general course of the N-termini in antithrombin III (thin line) and C_1-inhibitor (thick line) as derived from the disulfide linkages made to 69 and 101 and to 88 and 319, respectively. Residue numbers 22, 98, 102, 105, 106, and 109 mark the cluster of lysine and arginine residues in antithrombin III. Residue numbers refer to α_1-antitrypsin and the alignment in Table II. Residues C-terminal to 358 have 100 added to their numbers in the plot.

from proteolysis is clearly incompatible with the alternative reconstruction of Toma et al. (1987) involving a movement of serine 359 along with strand s1C to methionine 358 fixed in the A sheet.

Confirmation of the Löbermann et al. model comes from their prediction that the cleaved structures will have increased conformational stability over that of the native structures. Experimental evidence confirms this and shows that the serpins in general undergo a striking change in thermal stability on cleavage, with an increase in denaturation temperature from 58 °C in the native inhibitors to 80 °C or above in the cleaved inhibitors (Carrell & Owen, 1985). This so-called stressed (S) to relaxed (R) conformational change on cleavage at various sites within the exposed s4A segment has been demonstrated in α_1-antitrypsin, C_1-inhibitor, α_1-antichymotrypsin, antithrombin, and cortisol- and thyroxine-binding globulins (Pemberton et al., 1988) and quantitatively characterized in some of these by proton NMR and circular dichroism studies (Gettins & Horton, 1988; Bruch et al., 1988). It does not occur with the noninhibitors ovalbumin and angiotensinogen (Gettins, 1989; Stein et al., 1989); this is compatible with our proposal that the stressed conformation has been conserved in an evolutionary sense as a requirement for the inhibitory function of the reactive center.

A problem intimately related to the structure of the active, intact inhibitor concerns the docking to its cognate enzyme, leukocyte elastase. This is related to specific interactions of the substrate binding area of leukocyte elastase (Bode et al., 1986) with the primary binding segment but probably also to electrostatic attractions acting between more extended surface areas of the molecules. Calculations of the electrostatic potential of α_1-antitrypsin and human leukocyte elastase (HLE) using the Atanasov and Karshikov (1985) approach based on the Tanford and Kirkwood (1957) theory confirm the very dipolar character of the inhibitor with the positive pole at the S359 and the negative pole at the M358 end. Conversely, HLE has a negative electrostatic potential in the substrate binding area. This suggests that docking takes place at the S359 end of the inhibitor and is in accord with the view that s4A is removed from sheet A to approach S359 in the intact inhibitor.

The proposed docking area encompasses the site of access to the barrel formed by residues 190–300 and 359–394, i.e., β-sheets B and C. The barrel is the proposed ligand binding site in serpins with carrier function. Bound ligands may therefore modulate protease binding. Reciprocally, limited proteolysis is expected to affect ligand binding.

Carrier Function

Plasma proteins which function as carriers of thyroxine or corticosteroids have been identified as members of the serpin family (Flink et al., 1986; Hammond et al., 1987). There is direct information on the binding site of corticosteroid ligands in the carrier protein. Corticosteroid binding protein has two cysteine residues (69 and 237, α_1-antitrypsin numbering), one of which is affinity labeled by a ligand analogue but has not been identified (Defye et al., 1980; Kahn & Rosner, 1977). C69 is located in thBhC with no obvious ligand binding cavity, while C237 is in s2B at the mouth of a deep pocket formed by a barrel of β-strands, characteristic of other ligand binding sites (Katsunuma et al., 1980; Ragg, 1986). The binding site of thyroxine in thyroxine-binding globulin has been labeled by the chemical cross-linking of ligand and receptor. A labeled residue is K256 (D in α_1-antitrypsin) (Tabachnik & Perret, 1987) at the mouth of the proposed pocket of the β-barrel very close to residue 237 (Figure 5). Protein families which exhibit general ligand binding properties analyzed so far are immunoglobulins (Huber, 1984), retinol and bilin binding proteins

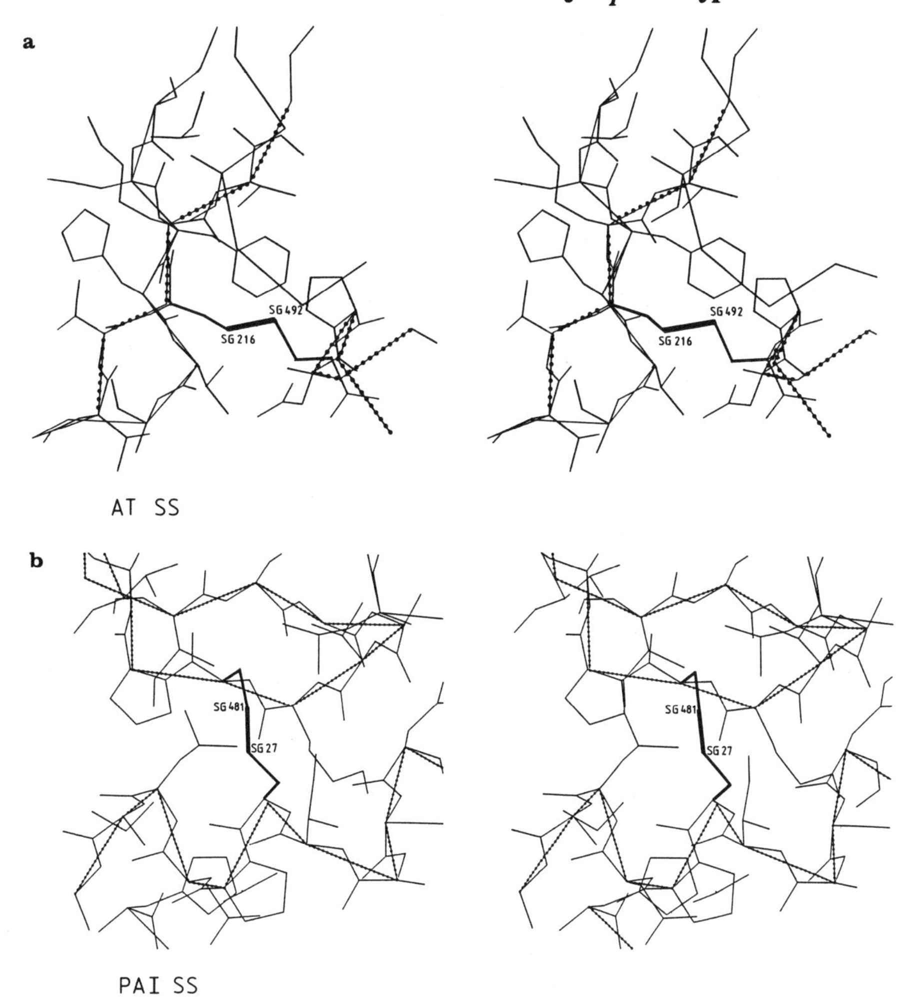

FIGURE 3: Model of antithrombin III and placental PAI at the (a) C216–C392 and the (b) C27–C381 disulfide groups, respectively. The dotted lines mark the positions of the C^{α} atoms in α_1-antitrypsin. Residue numbers refer to α_1-antitrypsin and the alignment in Table II. Residues C-terminal to 358 have 100 added to their numbers in the plot.

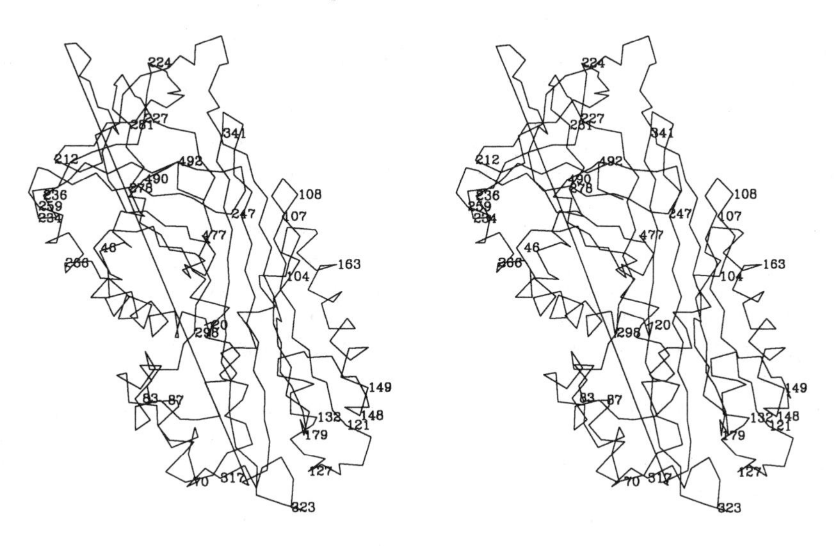

FIGURE 4: Potential glycosylation sites in members of the serpin superfamily projected onto the α_1-antitrypsin structure. Residue numbers refer to α_1-antitrypsin and the alignment in Table II. Residues C-terminal to 358 have 100 added to their numbers in the plot.

(Newcomer et al., 1984; Huber et al., 1987), and prealbumin (Blake & Oatley, 1977). Prealbumin deserves particular attention as it is also a thyroxine binding protein. Though seemingly unrelated in detailed folding topology and probably also in evolution, all three protein classes have β-barrel structures in common, which in the latter two cases are the ligand binding sites. There is a similar structural motif in α_1-antitrypsin in the S359 end of the molecule consisting of residues 190–300 and 359–394, as shown in Figure 5 in a comparison with bilin binding protein. In α_1-antitrypsin the entrance to the barrel as a putative ligand binding site is blocked by bulky W238, which in thyroxine-binding globulin

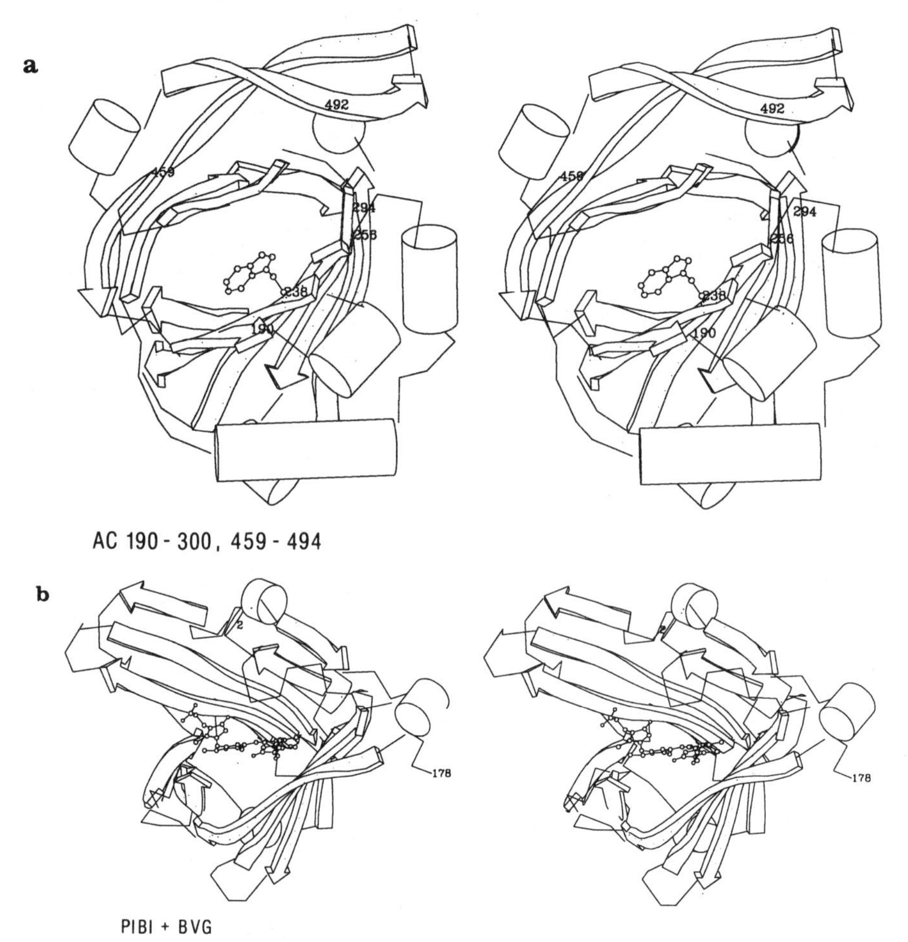

FIGURE 5: Bilin binding protein oriented with a view onto the bound biliverdin IXγ and residues 190–300 and 359–394 of α_1-antitrypsin with a similar view into the putative ligand binding pocket. Residue numbers refer to α_1-antitrypsin and the alignment in Table II. Residues C-terminal to 358 have 100 added to their numbers in the plot.

is T and in corticosteroid binding protein is Q. These small residues may permit access to the barrel.

Heparin Binding Site

Four of the human serpins, antithrombin III, heparin cofactor II, protease nexin I, and protein C inhibitor, are relatively inefficient inhibitors until they are activated by specific sulfated polysaccharides. The best studied of these inhibitors is antithrombin III, which undergoes a 1000-fold increase in its association constant with thrombin in the presence of heparin. The binding function of heparin is dependent on a pentasaccharide sequence with four sulfate groups (Choay et al., 1983; Lindahl et al., 1983; Beetz & Van Boeckel, 1986). Evidence as to the binding site of the pentasaccharide on antithrombin has come primarily from human variant antithrombins with decreased heparin affinity. Three of these variants were shown to be due to mutations of a single arginine at position 47/22 to cysteine, to histidine, and to serine (Koide et al., 1984; Owen et al., 1987; Borg et al., 1988). This finding together with other chemical evidence (Peterson et al., 1987) implicating K125/98 supported the inference that the binding of heparin to antithrombin was primarily due to salt bridging between the sulfates of the heparin and basic residues on the protein. Carrell et al. (1987a,b) compared the aligned serpins to determine which basic residues were uniquely conserved in antithrombin, in heparin cofactor II, and, subsequently, in protease nexin I. When these mutually conserved arginines and lysines are projected on the three-dimensional α_1-antitrypsin template, they are seen to form a band of positive charge stretching from the base of the A helix and across the underside of the D helix (Figure 6). Further support comes from the identification of other human heparin affinity variants. In the variant P41/16 → L (Chang & Tran, 1986) heparin binding is affected. This mutation is spatially closed to the suggested binding site. Also, Brennan et al. (1987, 1988) showed that the heparin affinity of antithrombin was decreased by the presence of oligosaccharide side chains at either the upper or lower ends of the proposed site. The change at the upper end of the site occurs due to aberrant glycosylation of N135/108, to give the high-affinity β-antithrombin normally present as a minor component in plasma. The change at the lower end of the site occurs in a mutant where I7/−19 is replaced by an asparagine which consequently is subject to glycosylation. The position of I7 can be approximately fixed on the antitrypsin model since it is adjacent to C8/−18, which is linked to C128/101. This places asparagine 7, and hence its bulky attached oligosaccharide, near the base of the proposed heparin site where it would predictably interfere with the binding of the heparin pentasaccharide.

On the basis of these results, we suggest that the primary heparin binding site on these three serpins is as shown in Figure 6, involving in the case of antithrombin arginines 47, 129, and 132 and lysines 125 and 133. Support for this conclusion is given by Smith and Knauer (1987) and Griffith et al. (1985), who show that the isolated antithrombin peptide 114–159 is preferentially bound by heparin. A further confirmation has been given by the recent identification of a low heparin affinity variant of heparin cofactor II in which the residue homologous to antithrombin's R129 (R189 in heparin cofactor) has been replaced by a histidine (Blinder et al., 1989).

The story is of course more complex than this because heparin activates these serpins as well as binding to them. The

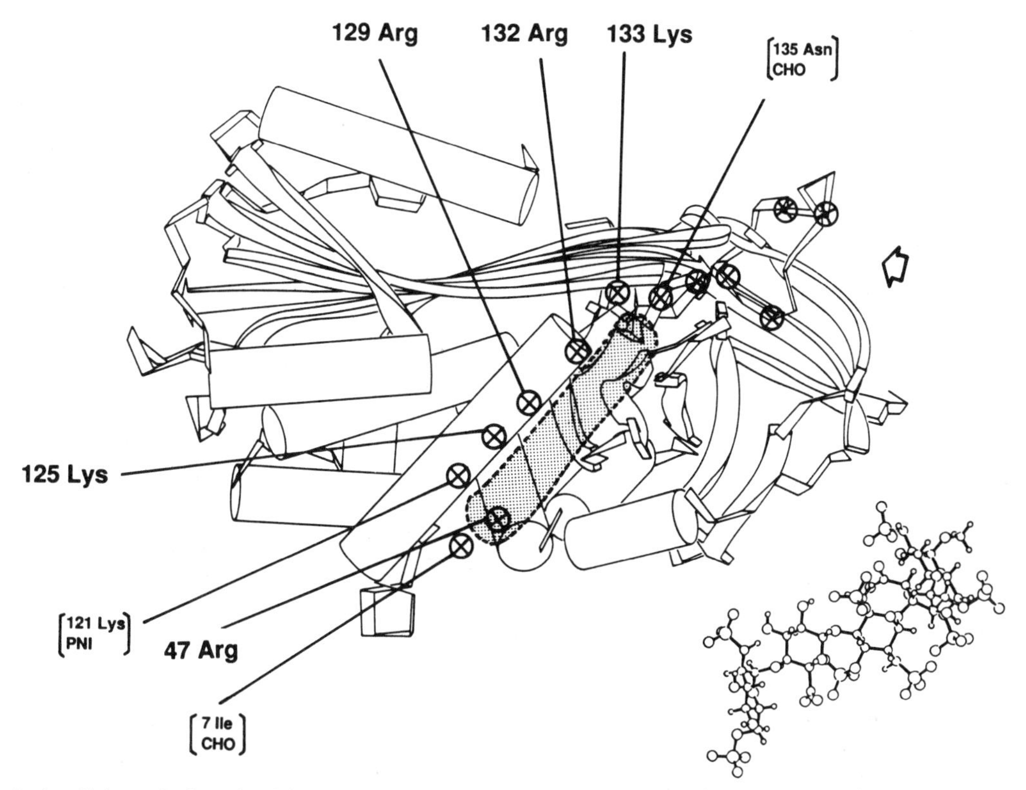

FIGURE 6: Heparin site. Schematic view of antithrombin projected onto the structure of cleaved α_1-antitrypsin, showing the prime (pentasaccharide) binding site (shaded) formed by the side chains of lysines and arginines 47, 125, 129, 132, and 133. The site extends around the molecule to the reactive center with K136, K228, R235, and K236 (shown but not labeled). The estimated region of the reactive center is arrowed. The prime site is flanked by oligosaccharide attachment points 135 and 7 (Asn in variant Rouen-III). In protease nexin I the equivalent of R47 is absent, but at position 121 is a lysine whose side chain overlaps that of 47. The size of the pentasaccharide is indicated in the lower right. Residue numbers refer to antithrombin III. Residues C-terminal to 358 have 100 added to their numbers in the plot.

mechanism of activation of antithrombin is uncertain, but it is likely to involve both a conformational change on the binding of heparin, as occurs with the activation by the pentasaccharide of factor Xa inhibition, and also a direct influence on the reactive site as probably occurs with the large heparins (12–15 units) necessary for the activation of thrombin inhibition. The latter type of activation is compatible with the observation from the model that the primary basic binding extends, in a less well-defined way, to the reactive center pole of the molecule—hence, the longer heparins may influence the electrostatic environment of the apposition site of thrombin and antithrombin. More appropriate to this paper is the conformational change that occurs on the binding of the pentasaccharide (Villanueva & Danishefsky, 1977). Antithrombin has an initial low affinity for the pentasaccharide with a subsequent high-affinity state (Olson et al., 1981). The binding constant for the low-affinity state is the same as the binding constant of cleaved antithrombin (Björk & Fish, 1982). We have also noted that some reactive center mutants of antithrombin are locked in the high-affinity state (Owen et al., 1988). Our deduction from this (with our colleagues M. C. Owen, P. B. Christey, and J.-Y. Borg) is that antithrombin has two conformational states, on the basis of changes at the primary heparin site, which influence the reactive center. The native state is the low-affinity conformation, as present in the locked form in cleaved antithrombin and represented sterically by the Löbermann et al. antitrypsin structure. The high-affinity state results from a conformational change at the binding site induced by the approach of heparin and transmitted reciprocally to the reactive center, hence the high-affinity state of the reactive center variants.

A detailed discussion of the likely mechanisms is not appropriate here, but we believe that the model of antithrombin, derived from cleaved antitrypsin, provides prospects of determining the likely conformational contribution to heparin-induced activation.

NATURAL SERPIN VARIANTS

A confirmation of the structure–function deductions made from the α_1-antitrypsin model is provided by natural variants of the serpins. Some 40 of these have now been characterized (Table IV, listing and references), and the mutations in each can be assessed in terms of the model. Overall there is a convincing correlation between the structural consequences predicted and the actual functional changes as observed in the affected individuals. Where no such functional changes are apparent, as in the physiologic polymorphisms, the underlying mutations are indeed seen to involve minimal structural perturbations on the model as in the substitutions of amino acids with similar properties in sterically noncritical areas. Examples in antitrypsin are the major common polymorphism V213 → A in s3C, the polymorphism E376 → D in s4B, the M2 polymorphism R101 → H in hD, and the nonpathological variants E204 → K in s4C and Christchurch E363 → K in s1C.

The pathological variants, i.e., those associated with significant abnormalities, are due to mutations that fall into three groups: those that directly affect the expression of the protein (which are not relevant to this discussion); those that affect a critical functional site in the molecule; those that affect the integrity of the overall structure. Examples of variants affecting critical functional sites include the mutations at the reactive center. These have been identified and show predicted functional consequences, in antitrypsin, antiplasmin, antithrombin, and C_1-inhibitor; they include alterations in the s4A reactive site loop as well as at the reactive center. The other major functional site is that of heparin binding; as previously

Table IV: Natural Serpin Variants[a]

structure	serpin and name	mutation	consequence	disease[b]	ref
prehA	antithrombin Rouen-III	7 Ile → Asn	new carbohydrate; heparin affinity decreased	thrombosis	Brennan et al., 1988
prehA	antithrombin Basel	41 Pro → Leu	perturbation of heparin site	thrombosis	Chang & Tran, 1986
hA	antithrombin Rouen-I	47 Arg → His	loss of heparin binding	thrombosis	Owen et al., 1987
hA	antithrombin Toyama	47 Arg → Cys	loss of heparin binding	thrombosis	Koide et al., 1984
hA	antithrombin Rouen-II	47 Arg → Ser	loss of heparin binding	thrombosis	Borg et al., 1988
hA	antitrypsin I	39 Arg → Cys	loss of salt bridge; deficiency	emphysema	Kalsheker et al., 1989
hA	antitrypsin M Procida	41 Leu → Pro	helix distortion; unstable	emphysema	Takahashi et al., 1987
s6B	antitrypsin I Malton	52 Phe deleted	misfolding; nonexpression	emphysema	Kalsheker et al., 1989
hD	thyroxine BG Gary	96 Ile → Asn	impaired T4 binding	?	Mori et al., 1986
hD	antitrypsin M2	101 Arg → His	apparently normal	polymorphism	
hD	heparin cofactor II	189 Arg → His	decreased dermatan sulfate binding		Blinder et al., 1989
s2A	antitrypsin Null Newport	115 Gly → Ser	steric distortion?; deficiency	emphysema	Kalsheker et al., 1989
hDs2A	antithrombin β	135 Asn	loss of carbohydrate; decreased heparin affinity	physiological	Brennan et al., 1987
hF	antitrypsin Null Granite Falls	160 Tyr → Stop	nonexpression	emphysema	Nukiwa et al., 1987
s4C	antitrypsin X	204 Glu → Lys	apparently normal	nil	Jeppsson & Laurell, 1988
s3C	antitrypsin polymorphism	213 Val → Ala	apparently normal	nil	Carrell et al., 1982
s3C	antitrypsin Null Bellingham	217 Lys → stop	nonexpression	emphysema	Satoh et al., 1988
ts3BhG	antitrypsin Null Cardiff	256 Asp → Val	impaired folding	emphysema	Kalsheker et al., 1989
hG	antitrypsin S	264 Glu → Val	loss of salt bridge; deficiency	emphysema	Owen & Carrell, 1976
thI1s5A	antitrypsin Null Hong Kong	318 Leu, TC deleted	nonexpression	emphysema	Sifers et al., 1988
s5A	antitrypsin Z	342 Glu → Lys	loss of salt bridge; incomplete processing	emphysema cirrhosis	Jeppsson, 1976
s4A	antithrombin Hamilton	382 Ala → Thr	loop variant; nonfunctional	thrombosis	Devraj-Kisuk et al., 1988
s4A	antitrypsin Null Mattawa	insert 353	termination position 376	emphysema	Curiel et al., 1988
s4A	antiplasmin Enschede	Ala insert 353–357	? loss of loop stress; nonfunctional	bleeding	Holmes et al., 1987
reactive center	antitrypsin Pittsburgh	358 Met → Arg	changed inhibitory specificity	bleeding	Owen et al., 1983
reactive center	C_1-inhibitor	444 Arg → His	nonfunctional	angioedema	Aulak et al., 1988b
reactive center	C_1-inhibitor	444 Arg → Cys	nonfunctional	angioedema	Aulak et al., 1988a
reactive center	antithrombin Glasgow	393 Arg → His	nonfunctional	thrombosis	Lane et al., 1987
reactive center	antithrombin Northwick Park	393 Arg → His	nonfunctional	thrombosis	Erdjument et al., 1988
reactive site	antithrombin Denver	394 Ser → Leu	reduced activity	thrombosis	Stephens et al., 1987
s1C	antitrypsin Christchurch	363 Glu → Lys	normal function	nil	Brennan & Carrell, 1986
s1Cs4B	antitrypsin M Heerlen	369 Pro → Leu	turning-point instability	emphysema	Hofker et al., 1987
s1Cs4B	antithrombin Utah	407 Pro → Leu	as above; disrupts turn	thrombosis	Bock et al., 1985
s4B	C_1-inhibitor	458 Met → Val	apparently normal	polymorphism	Bock et al., 1986
s4B	antitrypsin M3	376 Glu → Asp	normal function	polymorphism	Jeppsson & Laurell, 1988

[a] Adapted from Carrell et al. (1989). The sequence numbers refer to the residue numbers of the given protein. [b] Disease, i.e., predisposition to.

discussed, the variants of antithrombin and heparin cofactor II have helped to define this site but, conversely, also demonstrate the validity of the antitrypsin structure for the other serpins.

Further support for the model is given by the third group of variants, in which the overall integrity of the molecule is affected. Two good examples are the much-studied S and Z deficiency variants of antitrypsin which occur commonly in Northern Europeans. Both are due to substitutions of residues involved in key salt bridges—as seen on the model and as indicated by the consistent conservation of the residues involved. The Z variant has replacement of a glutamate in the bridge E342–K290 by a lysine; the S variant has replacement of the glutamate in the bridge E264–K387 by a valine. This variant has been studied crystallographically (Engh et al., 1989). In both variants there is evidence of decreased structural stability and in the Z variant the additional problem of a failure in transport/solubility at the final stage of the vesicular synthesis pathway. A second mutation of K290 → E in the Z variant appears to correct the defect (Brantley et al., 1988).

Another deficiency mutant, the I variant of antitrypsin, also has a similar consequence to that seen with the S mutation. This is in keeping with the structural change of the replacement, by cysteine, of the arginine at position 39 that contributes to the stabilization of the turn between helixes G and H and donates hydrogen bonds to the carbonyl oxygens projecting from the C-terminus of hG. The correlation between observed changes and the conformational predictions is also seen with other variants that affect the overall structure. The mutation of P369 → L, at the beginning of s4B, occurs in variants of both antitrypsin (M Heerlen) and antithrombin III (Utah, 407 in antithrombin numbering) with consequent gross dysfunction leading to a predisposition to emphysema with one and thrombosis with the other. A final example is that of two antitrypsin null variants—so-called because there is no circulating gene product. DNA analysis shows that one variant (Null Malton) has F52 deleted from an internal strand, s6B; the other (Null Cardiff) has the mutation D256 → V. In the first example β-sheet B would be obstructed; in the second a salt bridge between D256 and H231 was eliminated and turn ts3BhG destabilized. In both, misfolding of the molecule may result to explain the apparent nonexpression of the protein. In conclusion, experience in correlating and predicting the changes consequent on single amino acid substitutions supports our confidence that the Löbermann et al. model of α_1-antitrypsin provides a generally valid model of the native molecule in vivo.

References

Atanasov, B. P., & Karshikov, A. D. (1985) *Stud. Biophys. 105*, 11–22.

Aulak, M. S., Pemberton, P. A., Rosen, F. S., Carrell, R. W., Lachmann, P. J., & Harrison, R. A. (1988a) *Biochem. J. 253*, 615.

Aulak, M. S., Lachmann, P. J., Rosen, F. S., & Harrison, R.

A. (1988b) *Complement 5*, 181.

Beetz, T., & Van Boeckel, C. A. A. (1986) *Tetrahedron Lett. 27*, 5889–5892.

Björk, I., & Fish, W. W. (1982) *J. Biol. Chem. 257*, 9487–9493.

Blake, C. C. F., & Oatley, S. J. (1977) *Nature 268*, 115–120.

Blinder, M. A., Andersson, T. R., Abildgaard, U., & Tollefsen, D. M. (1989) *J. Biol. Chem.* (in press).

Bock, S. C., Harris, J. F., Schwartz, C. E., Ward, J. H., Hershgold, E. J., & Skolnick, M. H. (1985) *Am. J. Hum. Genet. 37*, 32.

Bock, S. C., Skriver, K., Nielsen, E., Thogersen, H. C., Wiman, B., Donaldson, V. H., Eddy, R. L., Marrinan, J., Radziejewska, E., Huber, R., Shows, T. B., & Magnusson, S. (1986) *Biochemistry 25*, 4292–4301.

Bode, W., Wei, An-Zhi, Huber, R., Meyer, E., Travis, J., & Neumann, S. (1986) *EMBO J. 5*, 2453–2458.

Bode, W., Greyling, H. J., Huber, R., Otlewsky, J., & Wilusz, T. (1989) *FEBS Lett. 242*, 285–292.

Borg, J. Y., Owen, M. C., Soria, C., Caen, J., & Carrell, W. R. (1988) *J. Clin. Invest. 81*, 1292–1296.

Brantley, M., Courtney, M., & Crystal, R. G. (1988) *Science 242*, 1700–1702.

Brennan, S. O., & Carrell, R. W. (1986) *Biochim. Biophys. Acta 873*, 13.

Brennan, S. O., George, P. M., & Jordan, R. E. (1987) *FEBS Lett. 219*, 431–436.

Brennan, S. O., Borg, J.-Y., George, P. M., Soria, C., Soria, J., Caen, J., & Carrell, R. W. (1988) *FEBS Lett. 237*, 118.

Bruch, M., Weiss, V., & Engel, J. (1988) *J. Biol. Chem. 32*, 16626–16630.

Carrell, R. W., & Owen, M. C. (1985) *Nature 317*, 730–732.

Carrell, R. W., & Travis, J. (1985) *Trends Biochem. Sci. 10*, 20.

Carrell, R. W., Jeppson, J.-O., Laurell, C.-B., Brennan, S. O., Owen, M. C., Vaughan, L., & Boswell, D. R. (1982) *Nature 298*, 329–334.

Carrell, R. W., Pemberton, P. A., & Boswell, D. R. (1987a) *Cold Spring Harbor Symp. Quant. Biol. 52*, 527–535.

Carrell, R. W., Christey, P. B., & Boswell, D. R. (1987b) in *Thrombosis and Haemostasis* (Verstraete, M., Vermylen, J., Lijnen, H. R., & Arnout, J., Eds.) pp 1–15, Leuven University Press, Leuven, Belgium.

Carrell, R. W., Aulak, K. S., & Owen, M. C. (1989) *Mol. Biol. Med.* (in press).

Chang, J.-Y., & Tran, T. H. (1986) *J. Biol. Chem. 261*, 1174–1176.

Choay, J., Petitou, M., Lormeau, J. C., Sinay, P., Cosu, B., & Gatti, G. (1983) *Biochem. Biophys. Res. Commun. 116*, 492–499.

Curiel, D., Brantly, M., Curiel, E., Stier, L., & Crystal, R. G. (1988) *Am. Rev. Respir. Dis. 137*, 210.

Defye, G., Basset, M., Monnier, N., & Chambaz, E. M. (1980) *Biochim. Biophys. Acta 623*, 280–294.

Devraj-Kizuk, R., Chui, D. H. K., Prochownik, E. V., Carter, C. J., Ofosu, F. A., & Blajchman, M. A. (1988) *Blood 72*, 1518.

Engh, R., Löbermann, H., Schneider, M., Wiegand, G., Huber, R., & Laurell, C.-B. (1989) *Protein Eng. 2*, 407–415.

Erdjument, H., Lane, D. A., Panico, M., Di Marzo, V., & Morris, H. R. (1988) *J. Biol. Chem. 263*, 5589.

Eriksson, U., Rask, I., & Peterson, D. A. (1984) *EMBO J. 3*, 1451–1454.

Flink, I. L., Bailey, T. J., Gustafson, T. A., Markham, B. E., & Markin, E. (1986) *Proc. Natl. Acad. Sci. U.S.A. 83*, 7708–7712.

Gettins, P. (1989) *J. Biol. Chem. 264*, 3781.

Gettins, P., & Horton, B. (1988) *Biochemistry 27*, 3634–3639.

Griffith, M. J., Noyes, C. M., & Church, F. C. (1985) *J. Biol. Chem. 260*, 2218–2225.

Hammond, G. L., Smith, C. L., Goping, I. S., Underhill, D. A., Harley, M. J., Revetos, J., Musto, N. A., Gunsalus, G. L., & Bardin, C. W. (1987) *Proc. Natl. Acad. Sci. U.S.A. 84*, 5153–5157.

Hofker, M. H., Nukiwa, T., Van Paassen, H. M. B., Nelen, M., Frants, R. R., Klasen, E. C., & Crystal, R. G. (1987) *Am. J. Hum. Genet. 41*, A2200.

Holmes, W. E., Lijnen, H. R., Nelles, L., Kluft, C., Nieuwenhuis, H. K., Rijken, D. C., & Collen, D. (1987) *Science 238*, 209.

Huber, R. (1984) *Behring Inst. Mitt. 76*, 1–14.

Huber, R., Schneider, M., Mayr, I., Müller, R., Deutzmann, R., Suter, F., Zuber, H., Falk, H., & Kayser, H. (1987) *J. Mol. Biol. 198*, 499–513.

Hunt, L. T., & Dayhoff, M. O. (1980) *Biochem. Biophys. Res. Commun. 95*, 864–871.

Jeppsson, J.-O. (1976) *FEBS Lett. 65*, 195.

Jeppsson, J.-O., & Laurell, C.-B. (1988) *FEBS Lett. 231*, 327.

Jones, T. A. (1978) *J. Appl. Crystallogr. 11*, 268–272.

Kahn, M. S., & Rosner, W. (1977) *J. Biol. Chem. 252*, 1895–1900.

Kalsheker, N. A., Newton, C., Graham, A., Bamforth, F. J., Powell, S., & Markham, A. (1989) *J. Med. Genet.*, Abstract (in press).

Katsunuma, T., Tsuda, M., Kusumi, T., Ohkubo, T., Mitoni, T., Nakasaki, H., Tajima, T., Yokoyama, S., Kamiguchi, H., Kobayashi, K., & Shinoda, H. (1980) *Biochem. Biophys. Res. Commun. 93*, 552–557.

Koide, T., Odani, S., Tokahashi, K., Ono, T., & Sakuragawa, N. (1984) *Proc. Natl. Acad. U.S.A. 81*, 289–293.

Kotwal, G. J., & Moss, B. (1989) *J. Virol. 63*, 600.

Lane, D. A., Lowe, G. D. O., Flynn, A., Thompson, E., Ireland, H., & Erdjument, H. (1987) *Br. J. Haematol. 66*, 523.

Laurell, C.-B., & Eriksson, S. (1963) *Scand. J. Clin. Lab. Invest. 15*, 132–140.

Lindahl, U., Bäckström, G., & Thunberg, L. (1983) *J. Biol. Chem. 258*, 9826–9830.

Löbermann, H., Tokuoka, R., Deisenhofer, J., & Huber, R. (1984) *J. Mol. Biol. 177*, 531–556.

Mori, Y., Refetoff, S., Seino, S., Flink, I. I., & Murata, Y. (1986) *N. Engl. J. Med. 314*, 694 (Abstract).

Newcomer, M. E., Jones, T. A., Åqvist, J., Sundelin, J., Eriksson, U., Rask, I., & Peterson, P. A. (1984) *EMBO J. 3*, 1451–1454.

Nukiwa, T., Takahashi, H., Brantly, M., Courtney, M., & Crystal, R. G. (1987) *J. Biol. Chem. 262*, 11999.

Olson, S. T., Srinivasan, K. R., Björk, I., & Shore, J. D. (1981) *J. Biol. Chem. 256*, 11073–11079.

Owen, M. C., & Carrell, R. W. (1976) *Br. Med. J. 1*, 130.

Owen, M. C., Brennan, S. O., Lewis, J. H., & Carrell, R. W. (1983) *N. Engl. J. Med. 309*, 694.

Owen, M. C., Borg, J. Y., Soria, C., Soria, J., Caen, J., & Carrell, R. W. (1987) *Blood 69*, 1275–1279.

Owen, M. C., Beresford, C. H., & Carrell, R. W. (1988) *FEBS Lett. 231*, 317–320.

Pemberton, P. A., Stein, P. E., Pepys, M. B., Potter, J. M., & Carrell, R. W. (1988) *Nature 336*, 257–258.

Petersen, T. E., Dudek-Wojciechowska, G., Sottrup-Jensen, L., & Magnusson, S. (1979) in *Physiological Inhibitors of Coagulation and Fibrinolysis* (Collen, D., Wiman, B., & Verstraete, M., Eds.) pp 43–54, Elsevier/North-Holland Biomedical Press, Amsterdam.

Peterson, C. B., Noyes, C. M., Pecon, J. M., Church, F. C., & Blackburn, M. N. (1987) *J. Biol. Chem. 262*, 8061–8065.

Ragg, H. (1986) *Nucleic Acids Res. 14*, 1073–1088.

Satoh, K., Nukiwa, T., Brantly, M., Garver, R. I., Jr., Hofker, M., Courtney, M., & Crystal, R. G. (1988) *Am. J. Hum. Genet. 42*, 77.

Sifers, R. N., Brashears-Macatee, S., Kidd, V. J., Muensch, H., & Woo, S. L. C. (1988) *J. Biol. Chem. 263*, 7330.

Smith, J. W., & Knauer, J. (1987) *J. Biol. Chem. 262*, 11964–11972.

Stein, P. E., Tewkesbury, D. A., & Carrell, R. W. (1989) *Biochem. J.* (in press).

Stephens, A. W., Thalley, B. S., & Hirs, C. H. W. (1987) *J. Biol. Chem. 262*, 1044.

Tabachnik, M., & Perret, V. (1987) *Biochem. Int. 15*, 409–417.

Takahashi, H., Nukiwaa, T., Ogushi, F., Brantly, M., Courtney, M., & Crystal, R. G. (1987) *Am. Rev. Respir. Dis. 135*, A292.

Tanford, C., & Kirkwood, J. G. (1957) *J. Am. Chem. Soc. 79*, 5333–5339.

Toma, K., Yamaomoto, S., Deyashiki, Y., & Suzuki, K. (1987) *Protein Eng. 1*, 471–475.

Travis, J., & Salvesen, G. S. (1983) *Annu. Rev. Biochem. 52*, 655–709.

Villanueva, G. B., & Danishefsky, I. (1977) *Biochem. Biophys. Res. Commun. 74*, 803–809.

Chapter 6

Zinc Coordination, Function, and Structure of Zinc Enzymes and Other Proteins†

Bert L. Vallee* and David S. Auld

Center for Biochemical and Biophysical Sciences and Medicine and Department of Pathology, Harvard Medical School and Brigham and Women's Hospital, 250 Longwood Avenue, Boston, Massachusetts 02115

Received March 30, 1990; Revised Manuscript Received April 18, 1990

Zinc is an essential component of many enzymes involved in virtually all aspects of metabolism. It has been found to be an integral component of nearly 300 enzymes in different species of all phyla, indispensable to their functions, which encompass the synthesis and/or degradation of all major metabolites. Advances in the isolation and characterization of enzymes and analysis of metals were basic to the rapid growth of the field. The capacity to measure picograms of zinc in nanograms of proteins is now widespread, and such analyses of biological matter have ceased to be a problem (Vallee, 1988; Riordan & Vallee, 1988).

About a decade ago, the role of zinc in gene expression began to attract interest (Vallee, 1977a,b; Vallee & Falchuk, 1981; Hanas et al., 1983) and this field has recently gained wide attention. In addition to its roles in catalysis and gene expression, zinc stabilizes the structure of proteins and nucleic acids, preserves the integrity of subcellular organelles, participates in transport processes, and plays important roles in viral and immune phenomena (Vallee & Auld, 1990a). Its nutritional essentiality has focused attention on the pathology and clinical consequences of both its deficiency and toxicity (Vallee, 1986).

A perspective on the biological occurrence and role of zinc should not only seek out the generalizations that are now apparent but should also call attention to the participation of zinc in metabolic processes, the bases of which have yet to be explained. We shall therefore highlight the salient accomplishments of the past two decades, achieved primarily through the study of zinc enzymes, while calling attention to the urgent but unanswered questions regarding the functions of zinc in metallothionein and DNA-binding proteins.

We will first consider the 12 zinc enzymes whose crystal structures are known and are now the standards of reference for catalytically active and structural zinc sites. In all of these a catalytic zinc atom is coordinated to three amino acid residues of the protein and an activated water molecule, whereas structural zinc atoms are coordinated to four cysteine residues. In members of their respective families whose crystal structures and/or sequences are known, the situation is identical. We will further examine the predictive capacity of such crystal structures for other zinc enzymes whose functions may be related and whose sequence is known but whose three-dimensional structure is not. This perspective will also deal with the induction of enzymatic activity in the procollagenases, which is accomplished through alterations in zinc coordination chemistry. It will conclude with the zinc cluster structure of the metallothioneins and its potential relevance to the DNA-binding zinc proteins and the problem of zinc coordination in those proteins in general.

(A) Crystal Structures of Zinc Enzymes as Standards of Reference

The realization that the affinity of zinc for nitrogen and oxygen is nearly equal to that for sulfur ligands has brought about a major change of viewpoint regarding zinc coordination chemistry and its manifestation in biological systems. Present perceptions indicate that there is as great a need to inspect the chemical properties of zinc complex ions containing combinations of nitrogen, oxygen, and sulfur ligands as well as the properties of those that only contain sulfur ligands.

X-ray crystallographic analysis of 12 zinc enzymes has identified the zinc ligands and defined the modes of coordination at both their active and structural sites. The primary and tertiary structures of these zinc enzymes provide standards of reference for the zinc sites in the sequences of protein families (Vallee & Auld, 1989, 1990b). These 12 zinc enzymes represent examples from classes I, II, III, and IV of the six classes of enzymes (Table I). X-ray crystallographic analyses of zinc enzymes in classes V and VI are not yet on record. The ligands are now known with certainty, and they define the features of the catalytic and structural zinc binding sites of zinc enzymes.

† This work was supported by grants from the Endowment for Research in Human Biology, Inc. (Boston, MA).

* To whom correspondence should be addressed.

		97		100		103		111		Ref.
Horse E	P Q	C	G K	C	R V	C	K H P E G N F	C	L K	1
Human α	P Q	C	G K	C	R I	C	K N P E S N Y	C	L K	2
Human β	P Q	C	G K	C	R V	C	K N P E S N Y	C	L K	3
Human γ	P Q	C	G K	C	R I	C	K N P E S N Y	C	L K	4
Human π	P L	C	R K	C	K F	C	L S P L T N L	C	G K	5
Human χ	P Q	C	G E	C	K F	C	L N P K T N L	C	Q K	6
Baboon*	P Q	C	G K	C	R V	C	K S P E G N Y	C	V K	7
Horse χ	P Q	C	G E	C	K F	C	L N P Q T N L	C	Q K	8
Rat χ	P Q	C	G E	C	K F	C	L N P K T N L	C	Q K	9
Rat	P Q	C	G K	C	R I	C	K H P E S N L	C	C Q	10
Mouse*	P Q	C	G E	C	R I	C	K H P E S N F	C	S R	11
Chicken*	P Q	C	G E	C	R S	C	L S T K G N L	C	I K	12
Quail*	P Q	C	G E	C	R S	C	L S T K G N L	C	I K	13
Maize 1*	G E	C	K E	C	A H	C	K S A E S N M	C	D L	14
Maize 2*	G E	C	K E	C	A H	C	K S E E S N M	C	D L	15
Barley*	G E	C	K E	C	P H	C	K S A E S N M	C	D L	16
Potato*	G E	C	K D	C	A H	C	K S E E S N M	C	S L	17
Pea*	G E	C	G E	C	P H	C	K S E E S N M	C	D L	18
Arab.*	G E	C	G D	C	R H	C	Q S E E S N M	C	D L	19
Asper.*	G S	C	L S	C	E M	C	M Q A D E P L	C	P H	20
Yeast 1*	G S	C	M A	C	E Y	C	E L G N E S N	C	P H	21
Yeast 2*	G S	C	M A	C	E Y	C	E L G N E S N	C	P H	22
Yeast 3*	G S	C	M T	C	E F	C	E S G H E S N	C	P D	23
Yeast 4*	S S	C	G N	C	E Y	C	M K A E E T I	C	P H	24

FIGURE 1: Structural-site zinc ligands of alcohol dehydrogenase. Lightly stippled boxes denote the enzyme(s) X-ray standard of reference for each family. An asterisk denotes those for which zinc was not measured directly. Black vertical columns indicate the proposed metal-binding ligands based on the structure of the standard of reference. Tetrameric ADH enzymes are shown at the bottom. Arab. and Asper. refer to *Arabidopsis thaliana* and *Aspergillus nidulans*, respectively. Yeasts 1 and 2 are respectively *Saccharomyces cerevisiae* cytosolic isozymes 1 and 2, yeast 3 is a *S. cerevisiae* mitochondrial enzyme, and yeast 4 is a *Schizosaccharomyces pombe* enzyme. References: (1) Jörnvall, H. (1970) *Eur. J. Biochem. 16*, 41–49; (2) von Bahr-Lindström, H., Höög, J.-O., Heden, L.-O., Kaiser, R., Fleetwood, L., Larsson, K., Lake, M., Holmquist, B., Holmgren, A., Hempel, J., Vallee, B. L., & Jörnvall, H. (1986) *Biochemistry 25*, 2465–2470; (3) Duester, G., Smith, M., Bilanchone, V., & Hatfield, G. W. (1986) *J. Biol. Chem. 261*, 2027–2033; (4) Höög, J. O., Hedēn, L.-O., Larsson, K., Jörnvall, H., & von Bahr-Lindström, H. (1986) *Eur. J. Biochem. 159*, 215–218; (5) Höög, J.-O., von Bahr-Lindström, H., Hedēn, L.-O., Holmquist, B., Larsson, K., Hempel, J., Vallee, B. L., & Jörnvall, H. (1987) *Biochemistry 26*, 1926–1932; (6) Kaiser, R., Holmquist, B., Hempel, J., Vallee, B. L., & Jörnvall, H. (1988) *Biochemistry 27*, 1132–1140; (7) Trezise, A. E. O., Godfrey, E. A., Holmes, R. S., & Beacham, I. R. (1989) *Proc. Natl. Acad. Sci. U.S.A. 86*, 5454–5458; (8) H. Jörnvall, personal communication; (9) Julia, P., Pares, X., & Jörnvall, H. (1988) *Eur. J. Biochem. 172*, 73–83; (10) Crabb, D. W., & Edenberg, H. J. (1986) *Gene 48*, 287–291; (11) Edenberg, H. J., Zhang, K., Fong, K., Bosron, W. F., & Li, T. K. (1985) *Proc. Natl. Acad. Sci. U.S.A. 82*, 2262–2266; (12) H. Jörnvall, personal communication; (13) *ibid.*; (14) Brandēn, C. I., Eklund, H., Cambillau, C., & Pryor, A. J. (1984) *EMBO J. 3*, 1307–1310; (15) Dennis, E. S., Sachs, M. M., Gerlach, W. L., Finnegan, E. J., & Peacock, W. J. (1985) *Nucleic Acids Res. 13*, 727–743; (16) Good, A. G., Pelcher, L. E., & Crosby, W. L. (1988) *Nucleic Acids Res. 16*, 7182; (17) D. P. Matton and M. Brisson, in GenBank/Los Alamos; (18) Llewellyn, D. J., Finnegan, E. J., Ellis, J. G., Dennis, E. S., & Peacock, W. J. (1987) *J.* Mol. *Biol. 195*, 115–123; (19) Chang, C., & Meyerowitz, E. M. (1986) *Proc. Natl. Acad. Sci. U.S.A. 83*, 1408–1412; (20) McKnight, G. L., Kato, H., Upshall, W., Parker, M. D., Saari, G., & O'Hara, P. J. (1985) *EMBO J. 4*, 2093–2099; (21) Jörnvall, H. (1977) *Eur. J. Biochem. 72*, 425–442; (22) Russell, D. W., Smith, M., Williamson, V. M., & Young, E. T. (1983) *J. Biol. Chem. 258*, 2674–2682; (23) Young, E. T., & Pilgrim, D. (1985) *Mol. Cell. Biol. 5*, 3024–3034; Young, E. T., & Pilgrim, D. (1986) *Mol. Cell. Biol. 6*, 2284; (24) Russell, P. R., & Hall, B. D. (1983) *J. Biol. Chem. 258*, 143–149.

Searches of the protein and gene sequence data banks and literature have ascertained the structural homology or identity and functional similarity of zinc enzyme families related by common ancestry.[1] Their characteristics and the identity as well as the conformations of their zinc ligands will likely prove critical to the specificity and mechanisms of action of these enzymes. On the basis of the reference structures, a combination of three His, Glu, Asp, or Cys residues creates a tridentate active zinc site (Table I). "Activated" H_2O fills and completes the coordination sphere in all such enzymatically active zinc sites, which contrasts with tetradentate structural zinc sites in which four cysteines are coordinated to zinc. Histidine is the most frequent ligand in active sites. In non-coenzyme-dependent zinc enzymes, a short spacer of 1–3 amino acids, intervening between the first two ligands L_1 and L_2, provides a zinc-binding nucleus. The third ligand, L_3, separated from L_2 by a long spacer of ~20 to ~120 amino acids, completes the coordination sphere with a long polypeptide loop that further aligns protein residues with the zinc and thereby also brings about interactions with the substrate. In our efforts to delineate the role of zinc in catalysis, we have centered on the chemistry of zinc and its effect on interacting proteins and vice versa rather than on the details of the participation of zinc in the mechanism of action of any particular zinc enzyme.

Carboxypeptidases A and B of bovine pancreas, the neutral proteases of *Bacillus thermoproteolyticus* and *Bacillus cereus*, carbonic anhydrases I and II of human red blood cells, and the dimeric alcohol dehydrogenase of equine liver are the standards of reference for those members of their respective families whose sequence is known but whose three-dimensional structure is not. Specifically, the identities of the zinc ligands and the amino acid sequences in their immediate vicinities are compared. The findings may have general implications for the elucidation of the mechanisms of related enzymes and the design of enzymatically active model systems.

(B) STRUCTURAL ZINC ATOMS OF ENZYMES

(1) Class I: Alcohol Dehydrogenase. In alcohol dehydrogenase (ADH)[2] (EC 1.1.1.1), zinc can have either a

[1] The National Biomedical Research Foundation and GenBank/Los Alamos database files of the Molecular Biology Computer Research Resource at Harvard Medical School were used for these searches.

Table I: Zinc Ligands and Their Spacing for the Catalytic and Structural Zinc[a]

enzyme	L_1	X	L_2	Y	L_3	Z	L_4	ref[b]
Class I								
alcohol dehydrogenase	Cys	20	His	106	Cys (C)		H_2O	1
alcohol dehydrogenase[c]	Cys	2	Cys	2	Cys (C)	7	Cys (C)	1
Class II								
aspartate transcarbamoylase[c]	Cys	4	Cys	22	Cys (C)	2	Cys (C)	2
Class III								
carboxypeptidase A	His	2	Glu	123	His (C)		H_2O	3
carboxypeptidase B	His	2	Glu	123	His (C)		H_2O	4
thermolysin	His	3	His	19	Glu (C)		H_2O	5
B. cereus neutral protease	His	3	His	19	Glu (C)		H_2O	6
DD carboxypeptidase	His	2	His	40	His (N)		H_2O	7
β-lactamase	His	1	His	121	His (C)		H_2O	8
phospholipase C	His	3	Glu	13	His (N)		H_2O	9
alkaline phosphatase	Asp	3	His	80	His (C)		H_2O	10
Class IV								
carbonic anhydrase I	His	1	His	22	His (C)		H_2O	11
carbonic anhydrase II	His	1	His	22	His (C)		H_2O	12

[a] X is the number of amino acids between L_1 and L_2; Y is the number of amino acids between L_3 and its nearest zinc ligand neighbor; Z is the number of amino acids between L_3 and L_4. L_3 is contributed by either the amino (N) or the carboxyl (C) portion of the protein. [b] References: (1) Brandén et al., (1975; (2) Honzatko et al., 1982; (3) Rees et al., 1983; (4) Schmid & Herriott, 1976; (5) Matthews et al., 1972; (6) Pauptit et al., 1988; (7) Dideberg et al., 1982; (8) Sutton et al., 1987; (9) Hough et al., 1989; (10) Kim & Wyckoff, 1989; (11) Kannan et al., 1975; (12) Liljas et al., 1972. [c] Structural zinc site; all others are in catalytic zinc sites.

structural or a catalytic (see below) role. The structural zinc atom is bound tetrahedrally to the four sulfur atoms of Cys-97, -100, -103, and -111 (Eklund & Brandén, 1983), all close to one another in the primary sequence and separated by 2, 2, and 7 amino acid residue spacers, respectively (Figure 1). Remarkably, this zinc atom, inaccessible to solvent, is close to the surface of the molecule. Its cysteine ligands are part of a lobe that projects out of the catalytic domain and has only a few side-chain interactions with the remainder of the subunit. These circumstances together with energy calculations lead to the inference that this zinc atom primarily affects local structure and conformation.

(*2*) *Class II*: *Aspartate Transcarbamoylase*. Aspartate transcarbamoylase (ATCase) (EC 2.1.3.2) contains but a single structural zinc atom located in the regulatory subunit, bound tetrahedrally to four cysteines, Cys-109, -114, -137, and -140, separated by 4, 22, and 2 intervening amino acid residue spacers (Honzatko et al., 1982). The polypeptide chain in this region forms two loops, which the zinc atom holds together. Stabilization of these loops, which form part of the interface between the regulatory and catalytic subunits, is thought to be responsible for stabilizing the quaternary structure. Zinc likely influences the local conformation and structure of the regulatory subunit to fine-tune its interaction with the catalytic subunit, as is apparent from the entropic contribution to the activation energy of zinc binding at such sites (Keating et al., 1988). Similar considerations and conclusions could pertain to the role of zinc atoms in DNA-binding proteins (see below). While in these two enzymes such circumstances may well affect enzymatic activity indirectly, there is no evidence of direct involvement of the zinc atom in their enzymatic—or any other—function; there is, of course, no zinc atom at the active site of the catalytic subunits of ATCase. In both instances cited, ADH and ATCase, the *structural* zinc atoms are *fully coordinated tetrahedrally* to cysteines. Neither water nor substrate has access to an open coordination site of that metal atom in these molecules, which apparently maintains the local structure in its immediate vicinity.

The prominence of the sulfur atoms of cysteine in the coordination chemistry of structural sites of these enzymes coincides with earlier views from inorganic chemistry and geochemistry of a predilection of zinc for sulfur ligands. The interaction of zinc with proteins that play critical roles in the regulation of gene transcription has revived such views (Klug & Rhodes, 1987).

(C) Functional Zinc Atoms of Enzymes

(*1*) *Class I*: *Alcohol Dehydrogenases*. The ligands of the active-site zinc atom of alcohol dehydrogenase (EC 1.1.1.1) differ significantly from those encountered thus far in all other zinc enzymes examined by X-ray crystallography. This is the only coenzyme-dependent zinc enzyme whose three-dimensional structure is known. The involvement of both zinc and NADH in the catalytic process requires suitable alignment of amino acid residues that can provide for both metal chelation and coenzyme binding sites. In ADH this has been accomplished by the use of residues 46 and 47 as zinc- and NADH-binding ligands, respectively, with two cysteines and one histidine as ligands to the active-site zinc, and by extending the short spacer between L_1 and L_2 to 20 amino acids.

The peptide backbones of all three zinc ligands are firmly anchored in secondary structural elements; none is part of a flexible loop region (Eklund & Brandén, 1983). Cys-46 (L_1) is simultaneously the first residue of a short α-helix, comprising residues 46–55, and the last residue of a β-strand. His-67 (L_2) is the first residue of a β-strand, while Cys-174 (L_3) is in the middle of an α-helix. The variation observed in residue 47, one of the two sites to which the coenzyme NAD(H) binds through its phosphate, contrasts markedly with the invariant nature of a number of other amino acids around both Cys-46 (L_1) and His-67 (L_2) (Figure 2).

A "long" spacer of 106 amino acids separates His-67 (L_2) from Cys-174 (L_3). Considering that the known sequences of dimeric and tetrameric ADHs span a remarkable range of evolution, the homology around this Cys residue is notable. In all of these ADHs a H_2O molecule completes the zinc coordination sphere. Among the enzymes whose structure is known so far, alcohol dehydrogenase is the only one in which cysteine is an active-site zinc ligand.

(*2*) *Class IV*: *Lyases*. The crystal structure of carbonic anhydrases I (Kannan et al., 1975) and II (Liljas et al., 1972) from human erythrocytes (EC 4.2.1.1) are known. The ac-

[2] Abbreviations: ADH, alcohol dehydrogenase; ATCase, aspartate transcarbamoylase; collag, collagenase; L_1, L_2, L_3, and L_4, the first, second, third, and fourth zinc-binding ligand, respectively; MC, mast cell; NEP, neutral endoprotease; TFIIIA, transcription factor IIIA; TL, thermolysin.

		46			67			174		Ref.
Horse E	T G I	C	R S D	I A G	H	E A A	C L I G	C	G F S T	1
Human α	V G I	C	G T D	I L G	H	E A A	C L I G	C	G F S T	2
Human β	V G I	C	R T D	I L G	H	E A A	C L I G	C	G F S T	3
Human γ	A G I	C	R S D	I L G	H	E A A	C L I G	C	G F S T	4
Human π	T S L	C	H T D	I V G	H	E A A	C L L G	C	G F S T	5
Human χ	T A V	C	H T D	I L G	H	E G A	C L L G	C	G I S T	6
Baboon*	V G I	C	R T D	I L G	H	E A A	C L I G	C	G F S T	7
Horse χ	T A V	C	H T D	I L G	H	E G A	C L L G	C	G V S T	8
Rat χ	T A V	C	H T D	I L G	H	E G A	C L L G	C	G I S T	9
Rat	T G V	C	R S D	V L G	H	E G A	C L I G	C	G F S T	10
Mouse*	T G V	C	R S D	V L G	H	E G A	C L I G	C	G F S T	11
Chicken*	T G I	C	R S D	I L G	H	E A A	C L I G	C	G F S T	12
Quail*	T G I	C	R S D	I L G	H	E A A	C L I G	C	G F S T	13
Maize 1*	T S L	C	H T D	I F G	H	E A G	C V L S	C	G I S T	14
Maize 2*	T A L	C	H T D	I L G	H	E A G	C I L S	C	G I S T	15
Barley*	T S L	C	H T D	I F G	H	E A G	C V L S	C	G I S T	16
Potato*	T S L	C	H T D	I L G	H	E A A	C V L S	C	G I S T	17
Pea*	T S L	C	H T D	I F G	H	E A G	C I L S	C	G I C T	18
Arab.*	T S L	C	H T D	I F G	H	E A G	C I V S	C	G L S T	19
Asper.*	S G V	C	H T D	I G G	H	E G A	A P I L	C	A G I T	20
Yeast 1*	S G V	C	H T D	V G G	H	E G A	A P V L	C	A G I T	21
Yeast 2*	S G V	C	H T D	V G G	H	E G A	A P I L	C	A G I T	22
Yeast 3*	S G V	C	H T D	V G G	H	E G A	A P I L	C	A G V T	23
Yeast 4*	T G V	C	H T D	I G G	H	E G A	A P I M	C	A G I T	24

FIGURE 2: Active-site zinc ligands for alcohol dehydrogenases. For definition of abbreviations, references, and key, see Figure 1.

		94		96			119		Ref.
Human I	F Q F	H	F	H	W G S	S A E L	H	V A H W	1
Human II	I Q F	H	F	H	W G S	A A E L	H	L V H W	2
Bovine I	F Q F	H	F	H	W G I	S A E L	H	L V H W	3
Mouse I*	T Q F	H	F	H	W G N	S G E L	H	L V H W	4
Horse I*	V Q F	H	F	H	W G S	S A E L	H	L V H W	5
Rabbit I*	S Q F	H	F	H	W G K	S A E L	H	L V H W	6
Monkey*	F Q F	H	F	H	W G S	S S E L	H	I V H W	7
Bovine II	V Q F	H	F	H	W G S	A A E L	H	L V H W	8
Mouse II*	I Q F	H	F	H	W G S	A A E L	H	L V H W	9
Rabbit II*	I Q F	H	F	H	W G S	A A E L	H	L V H W	10
Sheep II*	V Q F	H	F	H	W G S	A A E L	H	L V H W	11
Chicken II*	V Q F	H	I	H	W G S	D A E L	H	I V H W	12
Human III	R Q F	H	L	H	W G S	A A E L	H	L V H W	13
Bovine III	R Q F	H	L	H	W G S	A A E L	H	L V H W	14
Horse III*	R Q F	H	L	H	W G S	A A E L	H	L V H W	15

FIGURE 3: Zinc ligands of carbonic anhydrases. For key to figures, see Figure 1. References: (1) Barlow, J. H., Lowe, N., Edwards, Y. H., & Butterworth, P. H. W. (1987) *Nucleic Acids Res. 15*, 2386; (2) Henderson, L. E., Henriksson, D., & Nyman, P. O. (1976) *J. Biol. Chem. 251*, 5457–5463; (3) Sciaky, M., Limozin, N., Filippi-Foveau, D., Gulian, J.-M., & Laurent-Tabusse, G. (1976) *Biochimie 58*, 1071–1082; (4) Fraser, P. J., & Curtis, P. J. (1986) *J. Mol. Evol. 23*, 294–299; (5) Jabusch, J. R., Bray, R. P., & Deutsch, H. F. (1980) *J. Biol. Chem. 255*, 9196–9204; (6) Konialis, C. P., Barlow, J. H., & Butterworth, P. H. W. (1985) *Proc. Natl. Acad. Sci. U.S.A. 82*, 663–667; (7) Henriksson, D., Tanis, R. J., & Tashian, R. E. (1980) *Biochem. Biophys. Res. Commun. 96*, 135–142; (8) Tashian, R. E., Hewett-Emmett, S. K., Stroup, M., Goodman, M., & Yu, Y.-S. L. (1980) in *Biophysics and Physiology of Carbon Dioxide* (Bauer, C., Gros, G., & Bartels, H., Eds.) p 165, Springer, Berlin, FRG; (9) Curtis, P. J., Withers, E., Demuth, D., Watt, R., Venta, P. J., & Tashian, R. E. (1983) *Gene 25*, 325–332; Venta, P. J., Montgomery, J. C., Hewett-Emmet, D., Wiebauer, K., & Tashian, R. E. (1985) *J. Biol. Chem. 260*, 12130–12135; (10) Ferrell, R. E., Stroup, S. K., Tanis, R. J., & Tashian, R. E. (1978) *Biochim. Biophys. Acta 533*, 1–11; (11) Tanis, R. J., Ferrell, R. E., & Tashian, R. E. (1974) *Biochim. Biophys. Acta 371*, 534–548; (12) Roger, J. H. (1987) *Eur. J. Biochem. 162*, 119–122; (13) Lloyd, J., McMillan, S., Hopkinson, D., & Edwards, Y. H. (1986) *Gene 41*, 233–239; (14) Tashian, R. E., Hewett-Emmett, S. K., Stroup, M., Goodman, M., & Yu, Y.-S. L. (1980) in *Biophysics and Physiology of Carbon Dioxide* (Bauer, C., Gros, G., & Bartels, H., Eds.) p 165, Springer, Berlin, FRG; (15) Wendorff, K. M., Nishita, T., Jabusch, J. R., & Deutsch, H. F. (1985) *J. Biol. Chem. 260*, 6129–6132.

tive-site cavity is situated in the middle of a large twisted β-sheet of the protein (Liljas et al., 1972). A β-strand encompassing residues 88–108 supplies two of the zinc ligands, His-94 (L_1) and His-96 (L_2), while another β-sheet extending from residues 113 to 126 contributes His-119 (L_3). A H_2O molecule occupies the fourth coordination site, and the resulting geometry about the zinc is a distorted tetrahedron.

The shortest possible spacer, a single amino acid, separates His-94 (L_1) from His-96 (L_2); in 15 different carbonic anhydrases, five of the seven amino acids surrounding these ligands are 95% invariant (Figure 3). A long spacer of 22 amino acids supplies His-119 (L_3). In these 15 carbonic anhydrases, four of the eight amino acids surrounding His-119 (L_3) are invariant and the other four are highly similar.

	69 72	196	Ref.
Bovine A	L G - I H S R E W I T	F - L S - - I H S Y S Q	1
Bovine B	C G - F H A R E W I S	Y - L T - - I H S Y S Q	2
Rat A1 *	T G - I H S R E W V T	F - I S - - I H S Y S Q	3
Rat A2 *	A G - I H A R E W V T	F - I T - - L H S Y S Q	4
Mouse MC A *	C G - I H A R E W I S	Y - I T - - F H S Y S Q	5
Human MC A *	C G - I H A R E W V S	Y - I T - - F H S Y S Q	6
Crayfish B *	G G - I H A R E W I A	Y - L T - - F H S Y S Q	7
Rat B *	C G - F H A R E W I S	Y - L T - - I H S Y S Q	3
Bovine E *	I G N M H G N E A V G	F V L S A N L H G G D L	8
Human N *	V G N M H G N E A L G	F V L S A N L H G G A V	9
Human M *	V A N M H G D E T V G	F V L S A N L H G G A L	10

FIGURE 4: Zinc ligands of carboxypeptidases. For key to figures, see Figure 1. References: (1) Bradshaw, R. A., Ericsson, L. H., Walsh, K. A., & Neurath, H. (1969) *Proc. Natl. Acad. Sci. U.S.A. 63*, 1369–1394; (2) Titani, K., Ericsson, L. H., Walsh, K. A., & Neurath, H. (1975) *Proc. Natl. Acad. Sci. U.S.A. 72*, 1666–1670; (3) Clauser, E., Gardell, S. J., Craik, C. S., MacDonald, R. J., & Rutter, W. J. (1988) *J. Biol. Chem. 263*, 17837–17845; (4) Gardell, S. J., Craik, C. S., Clauser, E., Goldsmith, E. J., Stewart, C.-B., Graf, M., & Rutter, W. J. (1988) *J. Biol. Chem. 263*, 17828–17836; (5) Reynolds, D. S., Stevens, R. L., Gurley, D. S., Lane, W. S., Austen, K. F., & Serafin, W. E. (1989) *J. Biol. Chem. 264*, 20094–20099; (6) Reynolds, D. S., Gurley, D. S., Stevens, R. L., Sugarbaker, D. J., Austen, K. F., & Serafin, W. E. (1989) *Proc. Natl. Acad. Sci. U.S.A. 86*, 9480–9484; (7) Titani, K., Ericsson, L. H., Kumar, S., Jakob, F., Neurath, H., & Zwilling, R. (1984) *Biochemistry 23*, 1245–1250; (8) Fricker, L. D., Evans, C. J., Esch, F. S., & Herbert, E. (1986) *Nature 323*, 461–464; (9) Gebhard, W., Schube, M., & Eulitz, M. (1989) *Eur. J. Biochem. 178*, 603–607; (10) Tan, F., Chan, S. J., Steiner, D. F., Schilling, J. W., & Skidgel, R. A. (1989) *J. Biol. Chem. 264*, 13165–13170.

(*3*) *Class III*: *Hydrolases*. (*a*) *Metalloexoproteases*. Carboxypeptidase A (EC 3.4.17.1) is a member of a large family of zinc proteases (Vallee et al., 1983; Auld & Vallee, 1987, and references cited therein) and is one of those best characterized. The crystal structures of the bovine A and B enzymes show His-69 (L_1), Glu-72 (L_2), and His-196 (L_3) to be the three zinc ligands (Quiocho & Lipscomb, 1971; Schmid & Herriott, 1976). Two amino acid residues constitute the short amino acid spacer between His-69 (L_1) and Glu-72 (L_2), while the long spacer between Glu-72 (L_2) and His-196 (L_3) consists of 123 amino acids (Vallee & Auld, 1989). In eight carboxypeptidases A and B from human, cow, rat, mouse, and crayfish, all three ligands to the zinc atom are conserved, and very few changes occur in the residues adjacent to them (Figure 4).

His-69 (L_1) and Glu-72 (L_2) are at the ends of a reverse turn, while His-196 (L_3) is the last residue in a β pleated sheet structure extending from amino acids 191 to 196 (Rees et al., 1983). In addition, structure determination, chemical modification, and kinetic studies have identified Arg-145 and Glu-270 as functionally essential residues in the active site (Lipscomb et al., 1970; Vallee et al., 1970).

The sequences of a number of other carboxypeptidases have recently been established by means of DNA technology. Carboxypeptidase E or enkephalin convertase (Fricker et al., 1986) processes prohormones in secretory granules; carboxypeptidase N or kinase I (Gebhard et al., 1989), a blood enzyme, regulates peptide hormone activity at neutral pH; and carboxypeptidase M, a plasma membrane bound enzyme placed strategically at local tissue sites, also acts on peptide hormones (Tan et al., 1989). The sequence identity among these three proteins is approximately 41% but decreases to 15% when compared to carboxypeptidase A or B. However, suitable alignment of residues shows the preservation of all residues binding to the active-site zinc (Figure 4) as well as of Glu-270 and Arg-145 (Tan et al., 1989), which are essential to enzymatic function of carboxypeptidases A and B. A short spacer of 2 amino acid residues between L_1 and L_2 is common to all carboxypeptidases A, B, E, M, and N. The long spacer consists of 126, 130, and 138 amino acid residues for the human N, bovine E, and human M enzymes, respectively.

The most notable differences in the sequences of carboxypeptidases E, M, and N from those of carboxypeptidases A and B, which serve as standards, are the replacements of Tyr-198 and Arg-71, considered to have roles in substrate binding but not in catalysis, by Gly and Asn, respectively. Further, Ala or Thr replaces Trp-73 (Figure 4).

(*b*) *Metalloendoproteases*. The pH optimum of thermolysin (TL) (EC 3.4.24.4) from *B. thermoproteolyticus*, M_r 34 000, is near neutrality, representative of a number of neutral metalloproteinases. It contains one catalytic zinc and several structural calcium atoms[3] (Holmquist & Vallee, 1974) and has been studied extensively both in solution and in crystals. The zinc atom is bound to His-142 (L_1), His-146 (L_2), and Glu-166 (L_3) (Matthews et al., 1974). These residues are located around the zinc atom in such a manner that a water molecule completes a distorted tetrahedral coordination (Matthews et al., 1972). A short three amino acid spacer separates L_1 from L_2. Both are part of an α-helical domain behind the zinc atom extending from residues 137 to 150, approximately parallel to the cleft and traversing the entire center of the molecule. Another long internal helix, residues

[3] All four calcium sites—in which Glu and Asp are the principal ligands—are distant from the zinc atom. The calcium atoms, arranged in distorted octahedral coordination geometry, generate a complex network that may stabilize the enzyme structurally and thermally (Matthews et al., 1974).

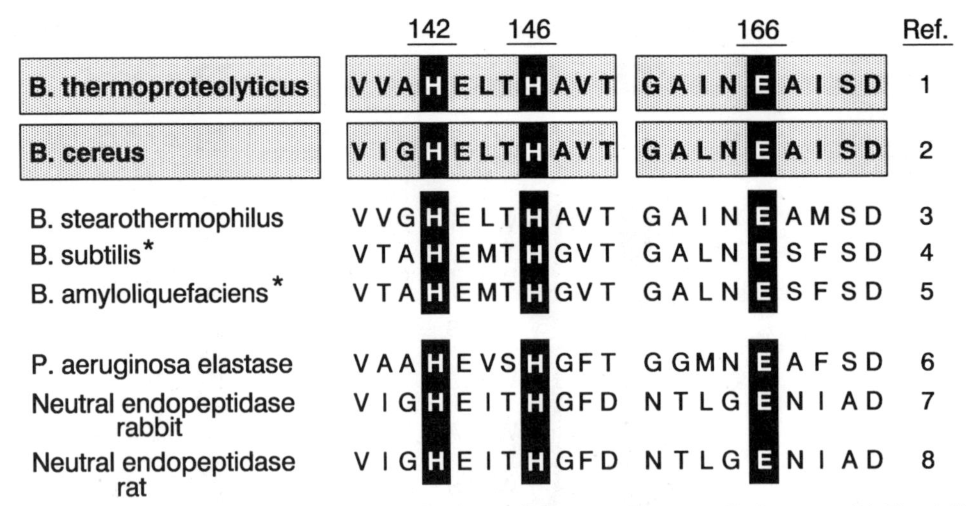

FIGURE 5: Zinc ligands of thermolysin and neutral proteases. For key to figures, see Figure 1. References: (1) Titani, K., Hermodson, M. A., Ericsson, L. H., Walsh, K. A., & Neurath, H. (1972) *Nature, New Biol. 238*, 35–37; (2) Sidler, W., Niederer, E., Suter, F., & Zuber, H. (1986) *Biol. Chem. Hoppe-Seyler 367*, 643–657; (3) Takagi, M., Imanaka, T., & Aiba, S. (1985) *J. Bacteriol. 163*, 824–831; (4) Levy, P. L., Pangburn, M. K., Burstein, Y., Ericsson, L. H., Neurath, H., & Walsh, K. A. (1975) *Proc. Natl. Acad. Sci. U.S.A. 72*, 4341–4345; (5) Vasantha, N., Thompson, L. D., Rhodes, C., Banner, C., Nagle, J., & Filpula, D. (1984) *J. Bacteriol. 159*, 811–819; (6) Beuer, R. A., & Iglewski, B. H. (1988) *J. Bacteriol. 170*, 4309–4314; (7) Devault, A., Lazure, C., Nault, C., Le Moual, H., Seidah, N. G., Chrêtien, M., Kahn, P., Powell, J., Mallet, J., Beaumont, A., Roques, B. P., Crine, P., & Boileau, G. (1987) *EMBO J. 6*, 1317–1322; (8) Malfroy, B., Schofield, P. R., Kuang, W.-J., Seeburg, P. H., Mason, A. J., & Henzel, W. J. (1987) *Biochem. Biophys. Res. Commun. 144*, 59–66.

160–180, underlying the central helix and placed at about 135° to it, contributes L_3, which is separated from L_2 by a long spacer of 19 amino acids. Thus, the secondary protein structure that supplies the zinc ligands in thermolysin differs markedly from that in the carboxypeptidases (see above). Moreover, among these two enzyme families both the length of the spacers and the order of the histidines in the sequence (His, His, Glu in the carboxypeptidases versus His, Glu, His in thermolysin) differ significantly (Vallee & Auld, 1989).

The crystal structure of the closely related neutral protease from *B. cereus* has recently been defined at 3.0-Å resolution (Pauptit et al., 1988). Its sequence is 73% identical with that of thermolysin and is particularly notable for all the residues flanking the zinc-binding ligands (Figure 5). Thermolysin and three bacterial neutral proteases from *Bacillus stearothermophilus*, *Bacillus subtilis*, and *Bacillus amyloliquefaciens* share the same specificity, zinc ligands His-142 (L_1), His-146 (L_2), and Glu-66 (L_3), near identity of the proximal amino acids, and the lengths of their short and long spacers (Figure 5). Glu-143, which has been thought to function as a general base, is present in all of the above neutral proteases.

The sequence of the elastase from *Pseudomonas aeruginosa*, an extracellular enzyme, is quite similar to that of thermolysin. While the overall amino acid identity between the mature elastase and thermolysin is 49%, their respective amino and carboxyl termini differ. Identity within the central portions of the two molecules is much higher, i.e., 67%. On that basis, His-140, His-144, and Glu-164 of this elastase have been predicted to be the equivalents of L_1, L_2, and L_3, respectively, in thermolysin (Beuer & Iglewski, 1988) (Figure 5). The long and short spacers would therefore be identical with those of thermolysin. The amino acids flanking these putative zinc ligands also closely resemble those of other members of the thermolysin family.

The specificity of neutral endopeptidase (NEP) (EC 3.4.24.11) from rabbit kidney brush border is similar to that of thermolysin; it contains 1.1 mol of zinc/mol of 93 000 protein (Kerr & Kenny, 1974) and is thought to be identical with the enkephalinase in brain membrane fractions that cleaves the Gly–Phe bond in Met- and Leu-enkephalins (Auld, 1987, and references cited therein). The amino acid sequence of NEP (749 amino acids) (Devault et al., 1987; Malfroy et al., 1987) differs greatly from that of thermolysin (316 amino acids). There are only two short, albeit highly homologous, regions of similarity (Devault et al., 1987). Yet, as judged by a hydrophobic clustering analysis (Benchetrit et al., 1988), all amino acids involved in the catalytic sites of thermolysin appear to be conserved in NEP. The His-583 and His-587 of one of these short homologous regions could correspond to His-142 (L_1) and His-146 (L_2) of thermolysin (Figure 5). Their substitution by Phe in NEP completely abolishes both the activity and interaction of the recombinant enzymes with a neutral endopeptidase metal-binding inhibitor (Devault et al., 1988). Similarly, the substitution of Val or Asp for Glu-646 results in the loss of both enzymatic activity and binding of the inhibitor (Le Moual et al., 1989), a result that could be consistent with Glu-646 being the third zinc ligand.

The amino acid sequences flanking the proposed His-583 (L_1) and His-587 (L_2) are very similar to those of thermolysin, and the 3 amino acid spacer between them is identical. However, adjacent to Glu-646, the similarity of amino acids in NEP and the thermolysin family decreases greatly. Further, the 58 amino acid long spacer of the NEP differs significantly from that of the thermolysin family (Figure 5).

(D) ZINC ENZYME FAMILIES FOR WHICH CRYSTAL STRUCTURE STANDARDS OF REFERENCE DO NOT EXIST AS YET

(*1*) *Class III*: *Hydrolases*. The structures of thermolysin and the neutral protease of *B. cereus*, which are the only ones known for the neutral proteases, have served for comparison with sequences of other metalloproteinases (Stöcker et al., 1988; Jongeneel et al., 1989). Compared with the above examples, in these instances it is less clear whether or not these belong to the same family of enzymes that the structural models represent. Such comparisons may or may not be warranted but are presented here as current viewpoints, clearly subject to reinterpretation once more specific structural standards become available. A persuasive case has been made for the prediction of the active zinc-binding sites in the neutral proteases from the structures, though a good many uncertainties remain. Only structure determination of a member

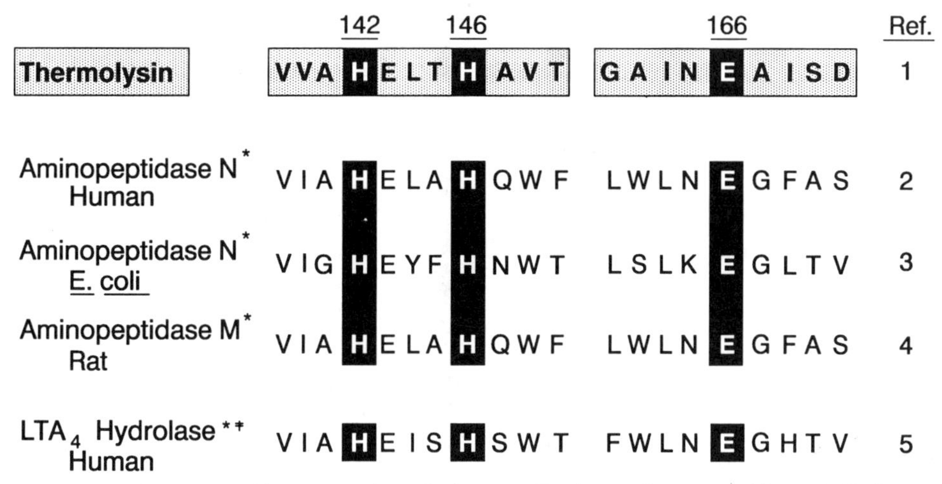

FIGURE 6: Putative zinc ligands of aminopeptidases. For ‡, see footnote 4. For key to figures, see Figure 1. References: (1) see Figure 5; (2) Olsen, J., Cowell, G. M., Kønigshøfer, E., Danielsen, E. M., Møller, J., Laustsen, L., Hansen, O. C., Welinder, K. G., Engberg, J., Hunziker, W., Spiess, M., Sjöström, H., & Norén, O. (1988) *FEBS Lett. 238*, 307–314; (3) McCaman, M. T., & Gabe, J. D. (1986) *Gene 48*, 145–153; Foglino, M., Gharbi, S., & Lazdunski, A. (1986) *Gene 49*, 303–309; (4) Watt, V. M., & Yip, C. C. (1989) *J. Biol. Chem. 264*, 5480–5487; Malfroy, B., Kado-Fong, H., Gros, C., Giros, B., Schwartz, J.-C., & Hellmiss, R. (1989) *Biochem. Biophys. Res. Commun. 161*, 236–241; (5) Funk, C. D., Rådmark, O., Fu, J. Y., Matsumoto, T., Jörnvall, H., Shimizu, T., & Samuelsson, B. (1987) *Proc. Natl. Acad. Sci. U.S.A. 84*, 6677–6681.

of the family of enzymes can render a final judgment.

(*a*) *Metalloexopeptidases*. Aminopeptidases (EC 3.4.11.2) catalyze the hydrolysis of N-terminal amino acid residues of proteins, peptides, and amino acid amides. Several have been isolated from a wide range of tissues and bacteria and fall into two categories, but structure determinations are not on record. Human liver aminopeptidase (Garner & Behal, 1974), aminopeptidase M (Wacker et al., 1971), two bacterial enzymes, from *Bacillus licheniformis* (Rodriguez-Absi & Prescott, 1978) and *B. subtilis* (Wagner et al., 1979), and that from porcine kidney (Van Wart & Lin, 1981) each contain one zinc atom. In contrast, the bovine lens (Thompson & Carpenter, 1976) and Aeromonas aminopeptidases (Prescott et al., 1985) each contain two zinc atoms. In all of these, zinc is essential for catalytic activity.

Human intestinal aminopeptidase contains 967 amino acids (Olsen et al., 1988) and, in a domain of ~300 amino acids, is remarkably similar both to an equivalent region in the *Escherichia coli* aminopeptidase N (McCaman & Gabe, 1986; Foglino et al., 1986) and to rat kidney aminopeptidase M (Watt & Yip, 1989; Malfroy et al., 1989). In particular, a short segment in these domains contains two histidines and a glutamic acid in a linear arrangement that corresponds closely to the active zinc-binding site of thermolysin (Figures 5 and 6). If this comparison to thermolysin were to prove valid, the short spacer between His-388 (L_1) and His-392 (L_2) for the intestinal aminopeptidase would consist of 3 amino acids, identical with that of thermolysin, and the long spacer between His-392 (L_2) and Glu-411 (L_3) would be 18 instead of 19 amino acids (Vallee & Auld, 1989). If correct, this seemingly would be the first instance in which comparison of sequence identities and order of active-site zinc ligands gained from structure analysis of one enzyme family serves to predict that of another.[4]

(*b*) *Metalloendopeptidases*. The bacterial collagenases (EC 3.4.24.7) and neutral proteinases share similar zinc and calcium contents and pH activity optima, but their substrate specificity differs markedly and that of the collagenases is unique. In higher vertebrates, collagen is the most abundant protein, accounting for about a third of the total, but the triple helix of collagen renders it inert to most proteinases, including the neutral proteases. Rearrangement, synthesis, and degradation of connective tissues, as occurs in growth and development, arthritis, emphysema, lupus, tumor metastasis, osteomalacia, wound healing, bone resorption, uterine involution, etc., have all been said to involve the action of collagenases at some stage.

The purification and characterization of six collagenases from *Clostridium histolyticum* finally established them as zinc enzymes, confirming earlier speculations. Their molecular weights range from 68 000 to 125 000 (Bond & Van Wart, 1984a). All six enzymes (α, β, γ, δ, ϵ, and ζ) contain from 0.8 to 1.1 mol of zinc/mol of monomeric protein, and the calcium content varies from 1.9 to 6.8 mol/mol of protein (Bond & Van Wart, 1984b).

In one particular domain of the sequences of all known collagenases, transins, stromelysins, and human pump 1 proteinases, a short spacer of 3 amino acids separates two histidines from one another (Figure 7). In all matrix metalloproteases, a Glu is conserved juxtaposed to the potential equivalent of His-142 (L_1) of thermolysin. They have been suggested to correspond to those of the zinc-binding site of thermolysin (Figure 5) (Birkedal-Hansen, 1990, and references cited therein). The similarity of the three residues preceding and succeeding L_1 and L_2 lends weight to such a deduction, much as it does not prove it (Figure 7).

In the thermolysin family, His-146 (L_2) is separated from Glu-166 (L_3) by a 19 residue long spacer (Figure 5). However, in the matrix metalloproteinases there is no Glu at that position or anywhere near it (Figure 7). Among the other two known active-site ligands, histidines are 5, 19, 32, and 47 amino acids removed from the nearest proposed L_1 (His-218) or L_2 (His-222), and conserved aspartic acids are found after spacers of 15, 17, 21, 31, 32, 40, 45, 57, 84, and 91 amino acids, all in conserved sequences. Thus, the location and identity of L_3 in the matrix metalloproteases will remain speculative until the structure of one of the members of this family can serve as a standard of reference for the active-site zinc ligands of the others. It is not even a foregone conclusion that the two

[4] Leukotriene A_4 hydrolase, whose specificity is completely different, exhibits 20% sequence identity with aminopeptidase N (Malfroy et al., 1989). Its potential zinc binding site displays remarkable similarity to that of the aminopeptidases (Figure 6). However, neither the metal content nor the esterase or peptidase activity of this enzyme has been reported.

	142 146	166	Ref.
Thermolysin	V V A H E L T H A V T	G A I N E A I S D	1
	218 222	242	
Human Collag. *	V A A H E L G H S L G	P S Y T F S G D V	2
Rabbit Collag. *	V A A H E L G H S L G	P N Y M F S G D V	3
Rat Collag. *	V A A H E L G H S L G	P V Y K S S T D L	4
Rat Transin *	V A A H E L G H S L G	P V Y K S S T D L	5
Rat Transin 2 *	V A A H E L G H S L G	P V Y R F S T S Q	6
Human Stromelysin *	V A A H E I G H S L G	P L Y H S L T D L	5
Human Stromelysin 2 *	V A A H E L G H S L G	P L Y N S F T E L	7
Human Pump 1 *	A A T H E L G H S L G	P T Y G N G D P Q	8
Human Collag. IV (72 kDa)*	V A A H E F G H A M G	P I Y T Y T K N F	9
Human Collag. IV (92 kDa)*	V A A H E F G H A L G	P M Y R F T E G P	10

FIGURE 7: Putative zinc ligands of collagenase. For key to figures, see Figure 1. The second sequence begins at 16 amino acids beyond the putative second zinc ligand. References: (1) see Figure 5; (2) Goldberg, G. I., Wilhelm, S. M., Kronberger, A., Bauer, E. A., Grant, G. A., & Eisen, A. Z. (1986) *J. Biol. Chem. 261*, 6600–6605; (3) Fini, M. E., Plucinska, I. M., Mayer, A. S., Gross, R. H., & Brinckerhoff, C. E. (1987) *Biochemistry 26*, 6156–6165; (4) Matrisian, L. M., Glaichenhaus, N., Gesnel, M. C., & Breathnach, R. (1985) *EMBO J. 4*, 1435–1440; (5) Whitham, S. E., Murphy, G., Angel, P., Rahmsdorf, H. J., Smith, B. J., Lyons, A., Harris, T. J. R., Reynolds, J. J., Herrlich, P., & Docherty, A. J. P. (1986) *Biochem. J. 240*, 913–916; (6) Breathnach, R., Matrisian, L. M., Gesnel, M. C., Staub, A., & Leroy, P. (1987) *Nucleic Acids Res. 15*, 1139–1151; (7) Muller, D., Quantin, B., Gesnel, M. C., Millon-Collard, R., Abecossis, J., & Breathnach, R. (1988) *Biochem. J. 253*, 187–192; (8) Quantin, B., Murphy, G., & Breathnach, R. (1989) *Biochemistry 28*, 5327–5334; (9) Collier, I. E., Wilhelm, S. M., Eisen, A. Z., Marmer, B. L., Grant, G. A., Seltzer, J. L., Kronberger, A., He, C., Bauer, E. A., & Goldberg, G. I. (1988) *J. Biol. Chem. 263*, 6579–6587; (10) Wilhelm, S. M., Collier, I. E., Marmer, B. L., Eisen, A. Z., Grant, G. A., & Goldberg, G. I. (1989) *J. Biol. Chem. 264*, 17213–17221.

particular His residues that have been thought to be involved in binding zinc are, indeed, those that do. Clearly there are several additional His residues that might be the ones; in addition, there are Asp residues that could be involved as well.

(E) "Velcro" Mechanism of the Activation of the Matrix Prometalloproteinases

The metalloproteinases (Figure 7) that catalyze the hydrolysis of the major components of the extracellular matrix are synthesized as zymogen-like inactive or latent precursors and are converted subsequently to the active form. They are another example of an ever-growing number of physiological processes initiated by selective enzymatic cleavage of peptide bonds in enzymes and hormone precursors, vasoactive products, proteins involved in growth and development, blood coagulation, fibrinolysis, digestion, complement activation, and yet others (Neurath, 1986, 1989). Linderstrøm-Lang (1952) proposed long ago that limited proteolysis proceeds by either a "one-by-one" or a "zipper" mechanism.

The activation of procollagenases and progelatinases proceeds by a different mechanism, which we choose to call the "Velcro" mechanism. Their propeptides contain a solitary cysteine residue at position 92 (fibroblast collagenase numbering) (Figure 8), in a highly conserved region, PRCGVPDV (Witham et al., 1986; Sanchez-Lopez et al., 1988). It seems established that this forms a mercaptide with the sole zinc atom of what will become the mature enzyme. A number of studies on fibroblast procollagenase show that it can be activated by trypsin (Vaes, 1972; Birkedal-Hansen, 1976; Stricklin et al., 1983), organomercurials (Werb & Burleigh, 1974; Sellers et al., 1977; Grant et al., 1987), salts such as NaI and NaSCN (Shinkai & Nagai, 1977), detergents (Birkedal-Hansen & Taylor, 1982), and thiol exchange reactions (Macartney & Tschesche, 1983; Springman et al., 1990). These results suggest that it is the dissociation and/or displacement of that cysteine from the zinc atom that results in activity by zymogen activation (Springman et al., 1990). The cysteine, sticking to the zinc atom through its SH group and acting like Velcro,

	92	Ref.
Human Collag.	M K Q P R C G V P D V A	2
Rabbit Collag.	M K Q P R C G V P D V A	3
Rat Collag.	M H K P R C G V P D V G	4
Rat Transin	M H K P R C G V P D V G	5
Rat Transin 2	M H K P R C G V P D V G	6
Human Stromelysin	M R K P R C G V P D V G	5
Human Stromelysin 2	M R K P R C G V P D V G	7
Rabbit Stromelysin	I R K P R C G V P D V G	5
Human Pump 1	M Q K P R C G V P D V A	8
Human Collag. IV (72 kDa)	M R K P R C G N P D V A	9
Human Collag. IV (92 kDa)	M R T P R C G V P D L G	10

FIGURE 8: Propeptide of matrix metalloproteinase precursors. For key to figures, see Figure 1. For references, see Figure 7.

prevents the zinc atom from becoming enzymatically active until the cysteine is removed.

While the details of the process still require definition, there seems no doubt that the principal chemical event which induces activity is the removal of the SH ligand of cysteine from the zinc (Figure 9). This is seemingly the first instance in which a fully coordinated, tetradentate (i.e., structural type) zinc atom is converted into a tridentate (i.e., enzymatically functional) one through the displacement of one ligand, i.e., cysteine, which is then replaced by water. However, the identity of the remainder of the ligands is not yet known (see above).

The dissociation of the zinc monothiolate complex of the inactive matrix prometalloproteinases into its constituents, i.e., cysteine plus a tridentate zinc complex of the active enzyme forms, represents a new activation mechanism that is based on zinc coordination chemistry. Since activated water is only found in relation to tridentate sites, while the cysteine is coordinated to the zinc, the protein will be inactive. The sole cysteinyl residue of the activation peptide apparently blocks the active-site zinc atom, thereby preventing its participation in catalysis. Its removal through physiological or pathological processes constitutes the activation process, allowing the entry of H_2O or substrate. This mechanism, hitherto unknown,

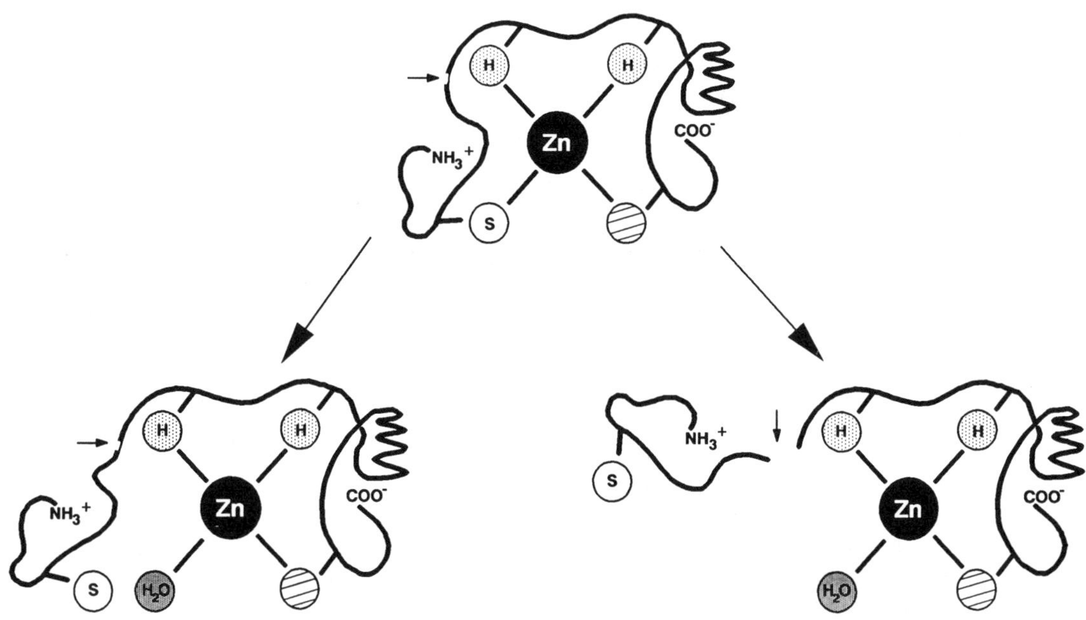

FIGURE 9: Schematic of the Velcro mechanism for the activation of the matrix prometalloproteinases. Activation occurs through the conversion of the zinc coordinated to four amino acid residues (tetradentate) into one that is coordinated to only three residues (tridentate) through the removal of one ligand, cysteine, S, which is then replaced by water. The displacement of cysteine is induced either by proteolytic cleavage and/or by conformational changes of the propeptide. The small arrow indicates the peptide region of the presumably autocatalytic cleavage site. Ligands to the functional zinc are thought to be two histidines, H; a third ligand is as yet unidentified and is symbolized by the crosshatched circle.

apparently represents yet another example of the versatility of zinc chemistry in biological reactions.

(F) Zinc Coordination Chemistry Relevant to the Structure and Function of Zinc Enzymes

Cysteine residues of zinc enzymes form tetradentate zinc complexes with very high stability constants, which ensure both overall structures and local conformations akin to those provided by disulfides. Tridentate combinations of histidine, glutamic and aspartic acid, and cysteine side chains bind zinc firmly and have proved to be characteristic of catalytic function by providing coordination sites open to water and/or substrate complexes and their transition-state intermediates. Which features of zinc chemistry may account for these findings?

The inherent chemical potential and reactivities of zinc are not exceptional. Oxidoreduction, characteristic of the neighboring transition elements, is a major source of changes in coordination geometries, rates of ligand substitution, and amphoteric properties of these elements. Importantly, however, zinc is both stable and inert to oxidoreduction. Indeed, it has been emphasized that generally the divalent state is more stable than are higher oxidation states (Cotton & Wilkinson, 1988).

The lack of redox changes makes zinc stable in a biological medium whose potential is in constant flux. Furthermore, some of its physical–chemical qualities present important advantages in biology. Zinc is amphoteric and exists in both metal hydrate and hydroxide forms at pH values near neutrality. The coordination sphere has proven exceptionally flexible. The stereochemical adaptability of zinc coordination complexes in enzymes is unusual and constitutes one of the striking features of its coordination complexes; the multiplicity of coordination numbers and geometries denotes that zinc submits readily to the demands of its ligands. In fact, it is through these properties that proteins and other biological macromolecules alter the reactivities of zinc. Thus, proteins and enzymes affect the chemistry of zinc, much as it, in turn, adapts to these macromolecules. Collectively, these physicochemical features are important means for the translation of chemical structure into multiple biological functions. Zinc thereby becomes a versatile interactant for different donor groups of varying ligand types resulting in a broad range of stability constants, reactivities, and functions.

(G) Zinc Thiolate Clusters: Metallothioneins

Zinc enzymes and metallothionein were discovered almost contemporaneously. While efforts to understand the role of zinc in enzyme mechanisms rapidly became an important biochemical topic, the existence of metallothionein went virtually unnoticed for many years, and its precise function(s) remain(s) unknown to the present. It required 25 years of work to establish its structure, which revealed a remarkable coordination complex that may well prove as important to an extension of zinc coordination chemistry itself and to its implications for biochemistry as turned out to be the case for the examination of zinc enzymes in relation to the mechanism of enzyme action.

Metallothionein, first isolated from equine kidney cortex in 1957 (Margoshes & Vallee, 1957), has been found in the animal kingdom wherever it has been sought (Kägi & Kojima, 1987). It has a molecular weight of 6 700 and is composed of 62 (or 61) amino acids, including 20 cysteines, but cystine and heterocyclic and aromatic amino acids are absent. It contains 7 mol of zinc and/or cadmium/mol of protein.[5] In some instances copper, iron, or mercury has also been detected.

In multiple species, a series of metals, hormones, and other organic molecules induce its formation, but the functions of metallothionein itself remain unknown (Kägi & Kojima, 1987; Kägi & Schäffer, 1988). In native mammalian metallothionein, the 7 mol of zinc and/or cadmium/mol of protein, together with the presence of 20 cysteinyl residues, absorption maxima at 215 and 248 mm, and the resultant CD and MCD

[5] Plants and fungi contain cadmium(II) thiolate polypeptides with the primary structure $(\gamma\text{-Glu-Cys})_n$ or $(\gamma\text{-Glu-Cys})_n\text{-}\beta\text{-Ala}$, where $n = 2\text{–}11$. They have variously been given trivial names, i.e., phytochelatins, homophytochelatins, or cadystins. These polypeptides are mentioned to complete the inventory of Zn and Cd chemistry in biology even though native zinc complexes of this type have not been reported thus far (Rauser, 1990).

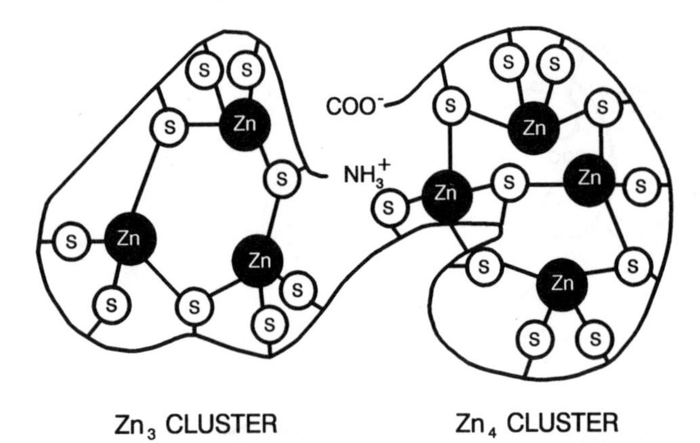

FIGURE 10: Zinc thiolate clusters of metallothioneins (Kägi & Kojima, 1987).

spectra are consistent with an average ratio of approximately 3 SH/Me atom. This was thought to reflect a negatively charged complex owing to the formation of three or four mercaptide bonds (Kägi & Vallee, 1961). Subsequent structural studies of metallothionein by ^{113}Cd NMR and 2D ^{1}H NMR and X-ray diffraction have demonstrated that cadmium and zinc are the nuclei of metal clusters in metallothioneins (Otvos & Armitage, 1980; Braun et al., 1986; Furey et al., 1986; Kägi & Schäffer, 1988).

The seven zinc atoms form two distinct and separate metal clusters, Zn_4Cys_{11} (residues 33–60) and Zn_3Cys_9 (residues 5–29), with five and three cysteine residues acting as bridging ligands between two metal ions in each cluster, respectively (Figure 10). Such zinc thiolate clusters had neither been anticipated inductively nor could they have been deduced, since at that time neither complex ions nor proteins were known to form them. They are reminiscent of the ferredoxin, rubredoxin, or molybdenum clusters, much as those participate in oxidoreductive reactions. Awareness of the existence of zinc and cadmium clusters is timely given the structural role of zinc in some zinc enzymes and its role in the expression and transcription of the genetic message. Whatever the function(s) of metallothionein will turn out to be, it (they) will likely involve its interactions with other molecules. Hence, this zinc and cadmium thiolate cluster structure, until now unique to metallothionein, may ultimately be found to telltale specific metabolic functions. This view is supported by the fact that the zinc thiolate cluster structure has most recently been observed in DNA-binding proteins (see below). Furthermore, discovery of such a structure in a biochemical system has provided the impetus to attempt to design, synthesize, and inspect the properties of simpler inorganic adamantane structures formed with the phenylthiols, $Zn_4(SR)_{10}$ (Dance, 1981; Hencher et al., 1981; Hagen et al., 1982).

(H) Zinc in the Control of Gene Expression

(*1*) *Xenopus Transcription Factor IIIA*. Evidence that zinc is crucial to DNA and RNA synthesis and cell division emerged in the 1970s (Vallee, 1977a,b; Auld, 1979; Vallee & Falchuk, 1981). It was reinforced by Wu's demonstration that transcription factor IIIA (TFIIIA), which activates the transcription of the *Xenopus* 5S RNA gene, is a zinc protein containing from 2 to 3 mol of zinc/mol of protein (Hanas et al., 1983). Subsequently, TFIIIA was shown to consist of 9 repeat sequences of about 30 amino acids, each containing 2 Cys residues separated by a short spacer of 2–5 amino acid residues and 2 His residues separated by two short spacers of 3–4 amino acid residues. A long spacer between the putative inner Cys and His ligands consists of 12 amino acid residues (Ginsberg et al., 1984; Brown et al., 1985). Each 7S particle was reported to contain 7–11 zinc atoms/mol (Miller et al., 1985).

It was proposed that 2 Cys and 2 His residues of each repeat unit form a tetrahedral coordination complex with each of nine zinc atoms, thereby generating a peptide domain postulated to interact with DNA. The results of limited proteolytic degradation, EXAFS measurements (Miller et al., 1985), organic synthesis of small peptide domains and NMR of their zinc complexes (Lee et al., 1989), and absorption spectroscopy of their cobalt complexes (Frankel et al., 1987; Green & Berg, 1989) were thought to be consistent with the above findings and interpretation of their functional significance.

However, a very recent extension of the earlier studies renewedly claims and affirms that the isolated protein and its RNA complex (the 7S particle) contain only two—not nine—firmly bound, intrinsic zinc atoms essential to the transcription factor IIIA activation of the transcription of the 5S RNA gene. No evidence for the presence of additional loosely bound zinc ions was found. Addition of zinc to the protein containing only the two intrinsic zinc atoms enhanced neither the affinity of protein binding to DNA or 5S RNA nor transcription activation (Shang et al., 1989).

(*2*) *Other DNA-Binding Putative Zinc Proteins*. Subsequent to the initial reports regarding zinc in TFIIIA, a large number and variety of sequences of DNA-binding proteins were shown to contain Cys and His residues thought to be analogous and separated by short and long intervening amino acid spacer sequences (Rosenberg et al., 1986; Vincent et al., 1985; Hartshorne et al., 1986; Schuh et al., 1986; Chowdhury et al., 1987). They were identified by computer searches intended to single out proteins, loosely defined as not homologous with but containing TFIIIA-like sequences, that bind to nucleic acids. It was inferred further that these reflected putative metal-binding domains which might participate in DNA binding and, hence, gene regulation. Such searches have now revealed well in excess of 150 such proteins, suspected—but not shown—to contain zinc. The initial postulate was that one pair each of spatially juxtaposed Cys and His ligands in such sequences, each of them separated by short amino acid spacers, would prove characteristic of zinc sites in the domains of DNA-binding proteins. On the basis of analytical findings demonstrating the presence of zinc in the glucocorticoid receptor protein (Freedman et al., 1988), this proposition was subsequently amended to accommodate an alternative ligand arrangement consisting of two pairs of Cys zinc ligands but no His. A total of six classes of proteins with putative metal-binding sites and participating in binding to nucleic acids have been enumerated: (1) low molecular weight nucleic acid binding or gene regulatory proteins, (2) adenovirus E1A gene products, (3) aminoacyl-tRNA synthetases, (4) large T antigens, (5) bacteriophage proteins, and (6) hormone receptors (Berg, 1986, 1988). Additional computer searches have resulted in an increasing number of such sequences featuring cysteine and histidine residues in locations analogous to and homologous with those stipulated above (Sunderman & Barber, 1988).

In these, the evidence for the presence of zinc-binding sites and their relevance to these hypotheses is quite variable but may imply that the "putative" zinc-binding sites are tantamount to both the presence and a biological function of zinc in a given instance. A large number of such studies ignored the fact that once proteins which contain putative metal-binding domains are recognized, the presence, mode of binding, and role of zinc ions in such systems must still be verified. In some instances, the induction or enhancement of an activity

Table II: Transcription Proteins Established To Contain Zinc by Quantitative Analysis

protein	MW × 10^{-3}	Zn (mol/mol of protein)	ligands	ref[a]
TFIIIA	40	2	2 Zn { 4 Cys, 4 His	1
TFIIIA	40	7–11	9 Zn { 18 Cys, 18 His	2
Glu Rec	19	2	2 Zn { 9 Cys, 1 His	3
GAL4	17	2	2 Zn { 6 Cys	4
g32P	35	1	1 Zn { 3 Cys, 1 His	5

[a] References: (1) Hanas et al., 1983; Shang et al., 1989; (2) Miller et al., 1985; (3) Freedman et al., 1988; (4) Hollenberg et al., 1987, and Pan & Coleman, 1989, 1990; (5) Giedroc et al., 1986.

by addition of Zn^{2+} ions and/or its diminution or abolition by chelating agents are cited to either confirm or prove this. However, in the past, the results of such experiments have not proven to be reliable criteria for either the presence or potential function of zinc or other metals (Vallee & Wacker, 1970; Vallee & Auld, 1990a).

Considering the vast number of articles whose titles refer to "zinc fingers", it is important to realize that the presence of zinc has been confirmed analytically in only four instances (Table II). Moreover, in each of those, the combination and identity of the putative zinc ligands differ. The 2 Cys and 2 His of TFIIIA have been mentioned already. In the glucocorticoid receptor, there are 9 Cys and 1 His together with at least two zinc atoms (Freedman et al., 1988). The transcription factor of GAL4 contains 6 Cys and two zinc atoms (Hollenberg et al., 1987; Pan & Coleman, 1989), and the gene 32 protein required for DNA replication in bacteriophage T4 contains one zinc as well as 3 Cys and 1 His ligands (Giedroc et al., 1986). Thus, in these DNA-binding proteins where the presence of zinc is documented, the relevant ligands are very variable and seemingly not predictable on the basis of the hypotheses that have been suggested (Berg, 1986). Yet other combinations of ligands in such binding sites may exist, but first and foremost, the presence of zinc should be established for each. X-ray crystallographic or NMR data, once available as structural standards of reference for members of these protein families, will no doubt answer most of the questions yet remaining.

(*3*) *GAL4*: *Zinc Thiolate Cluster*. The DNA-binding domain of the transcription factor GAL4 consists of the 62 N-terminal residues, denoted by GAL4(62*). Its study by Cd NMR was timely (Pan & Coleman, 1990). ^{1}H–^{113}Cd heteronuclear multiple quantum NMR spectroscopy and phase-sensitive double-quantum-filtered ^{1}H COSY of the ^{112}Cd- and ^{113}Cd-substituted GAL4(62*) derivatives provide direct evidence that the two bound ^{113}Cd ions are coordinated by *six* cysteines, two of which form bridging ligands between the ^{113}Cd ions (Figure 11). The overall arrangement is that of a zinc thiolate cluster structure akin to that of metallothionein (Kägi & Kojima, 1987), not that of a zinc finger. The highly conserved arrangement of Cys in GAL4 and other fungal transcription factors almost certainly predicts the occurrence of such binuclear zinc clusters in other similar DNA-binding proteins (Pan & Coleman, 1990).

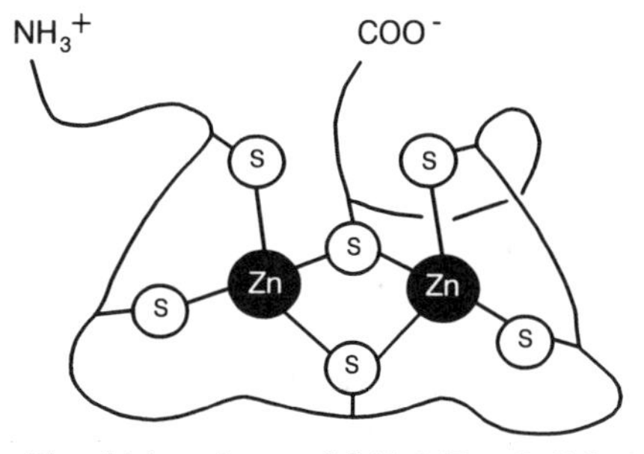

FIGURE 11: Zinc thiolate cluster of GAL4 (Pan & Coleman, 1990).

SUMMARY

The last decade has seen remarkable developments in establishing zinc-dependent interrelationships between enzyme activities, protein structures and folding, and products of gene expression. Different permutations of zinc chemistry seem pivotal both for the expression of catalytic function of zinc enzymes and their local structure and for DNA-binding proteins and their interactions that result in gene expression. Conversely, the structure, conformation, and folding of the protein determine the response of the metal. This reciprocity expresses itself in particular through the number and nature of the protein ligands forming the coordination complex. Tridentate zinc sites (plus activated water) are characteristic of catalysis. In enzymes, tetradentate zinc sites are inaccessible to solvent, critically controlling local protein folding, structure, and conformation. The extent to which they may regulate protein–protein and protein–DNA interactions remains to be defined. In enzymes, histidines predominate in catalytic zinc sites while cysteines are the exclusive zinc ligands in structural sites. Tetradentate zinc coordination in procollagenase is convertible and becomes tridentate in collagenase through the removal of the cysteine ligand of the propeptide from the zinc atom. This conversion yields a catalytically active zinc enzyme site in which a water molecule replaces the cysteine ligand.

Biology has availed itself of zinc coordination chemistry for specific biological objectives. The zinc clusters of metallothionein seem to go beyond this, representing an example of "inorganic natural products chemistry", since such a zinc structure had never been observed in nonbiological zinc chemistry prior to discovery of metallothionein. While the function of metallothionein is still unknown, its zinc thiolate cluster, a structural motif unique to biology, is perhaps trying to telltale function. Ultimately such zinc cluster structures, if proven to exist widely, might, in fact, be characteristic of and synonymous with functions yet to emerge—much as is the case for active zinc site enzymes. The zinc thiolate cluster of the transcription factor GAL4 may just be the first of its kind in DNA-binding proteins.

REFERENCES

Auld, D. S. (1979) *Adv. Chem. Ser. 172*, 112–133.
Auld, D. S. (1987) in *Enzyme Mechanisms* (Page, M. I., Ed.) Royal Society of Chemistry, Letchworth, U.K.
Auld, D. S., & Vallee, B. L. (1987) in *Hydrolytic Enzymes* (Neuberger, A., & Brocklehurst, K., Eds.) pp 201–255, Elsevier, New York.
Benchetrit, T., Bissery, V., Mornon, J. P., Devault, A., Crine, P., & Roques, B. P. (1988) *Biochemistry 27*, 592–596.
Berg, J. M. (1986) *Science 232*, 485–487.
Berg, J. M. (1988) *Proc. Natl. Acad. Sci. U.S.A. 85*, 99–102.
Beuer, R. A., & Iglewski, B. H. (1988) *J. Bacteriol. 170*, 4309–4314.
Birkedal-Hansen, H. (1990) in *Matrix Metalloproteinases and Inhibitors* (Birkedal-Hansen, H., Werb, Z., Welgus, H., &

Van Wart, H., Eds.) Gustav Fischer Verlag, Stuttgart, West Germany (in press).

Birkedal-Hansen, H., & Taylor, R. E. (1982) *Biochem. Biophys. Res. Commun. 107*, 1173–1178.

Birkedal-Hansen, H., Cobb, C. M., Taylor, R. E., & Fullmer, H. W. (1976) *J. Biol. Chem. 251*, 3162–3168.

Bond, M. D., & Van Wart, H. E. (1984a) *Biochemistry 23*, 3077–3085.

Bond, M. D., & Van Wart, H. E. (1984b) *Biochemistry 23*, 3085–3091.

Brandēn, C. I., Jörnvall, H., Eklund, M., & Furugren, B. (1975) in *Enzymes, 3rd Ed.* (Boyer, P. D., Ed.) Vol. 11, p 103, Academic Press, New York.

Braun, W., Wagner, G., Wörgötter, E., Vasâk, M., Kägi, J. H. R., & Wüthrich, K. (1986) *J. Mol. Biol. 187*, 125–129.

Brown, R. S., Sander, C., & Argos, P. (1985) *FEBS Lett. 186*, 271–274.

Chowdhury, K., Deutsch, U., & Gruss, P. (1987) *Cell 48*, 771–778.

Cotton, F. A., & Wilkinson, G. (1988) in *Advanced Inorganic Chemistry*, 5th ed., p 599, Wiley-Interscience, New York.

Dance, I. G. (1981) *Inorg. Chem. 20*, 2155.

Devault, A., Lazure, C., Nault, C., Le Moual, H., Seidah, N. G., Chrētien, M., Kahn, P., Powell, J., Mallet, J., Beaumont, A., Roques, B. P., Crine, P., & Boileau, G. (1987) *EMBO J. 6*, 1317–1322.

Devault, A., Nault, C., Zollinger, M., Fournie-Zaluski, M.-C., Roques, B. P., Crine, P., & Boileau, G. (1988) *J. Biol. Chem. 263*, 4033–4040.

Dideberg, O., Charlier, P., Dive, G., Joris, B., Frēre, J. M., & Ghuysen, J. M. (1982) *Nature 299*, 469–470.

Eklund, H., & Brandēn, C.-I. (1983) in *Zinc Enzymes* (Spiro, T. G., Ed.) pp 123–152, Wiley-Interscience, New York.

Foglino, M., Gharbi, S., & Lazdunski, A. (1986) *Gene 49*, 303–309.

Frankel, A. D., Berg, J. M., & Pabo, C. O. (1987) *Proc. Natl. Acad. Sci. U.S.A. 84*, 4841–4845.

Freedman, L. P., Luisi, B. F., Korszun, Z. R., Basavappa, R., Sigler, P. B., & Yamamoto, K. R. (1988) *Nature 334*, 543–546.

Fricker, L. D., Evans, C. J., Esch, F. S., & Herbert, E. (1986) *Nature 323*, 461–464.

Furey, W. F., Robbins, A. H., Clancy, L. L., Winge, D. R., Wang, B. C., & Stout, C. D. (1986) *Science 231*, 704–710.

Garner, C. W., Jr., & Behal, F. J. (1974) *Biochemistry 13*, 3227–3233.

Gebhard, W., Schube, M., & Eulitz, M. (1989) *Eur. J. Biochem. 178*, 603–607.

Giedroc, D. P., Keating, K. M., Williams, K. R., Konigsberg, W. H., & Coleman, J. E. (1986) *Proc. Natl. Acad. Sci. U.S.A. 83*, 8452–8456.

Ginsberg, A. M., King, B. O., & Roeder, R. G. (1984) *Cell 39*, 479.

Grant, G. A., Eisen, A. Z., Marmer, B. L., Roswit, W. T., & Goldberg, G. I. (1987) *J. Biol. Chem. 262*, 5886–5889.

Green, L. M., & Berg, J. M. (1989) *Proc. Natl. Acad. Sci. U.S.A. 86*, 4047–4051.

Hagen, K. S., Stephan, D. W., & Holm, R. H. (1982) *Inorg. Chem. 21*, 3928.

Hanas, J. S., Hazuda, D. J., Bogenhagen, D. F., Wu, F. Y.-H., & Wu, C.-W. (1983) *J. Biol. Chem. 258*, 14120–14125.

Hartshorne, T. A., Blumberg, H., & Young, E. T. (1986) *Nature 320*, 283–287.

Hencher, J. L., Khan, M., Said, F. F., & Tuck, D. G. (1981) *Inorg. Nucl. Chem. Lett. 17*, 187.

Hollenberg, S. M., Giguēre, V., Segui, P., & Evans, R. M. (1987) *Cell 49*, 39–46.

Holmquist, B., & Vallee, B. L. (1974) *J. Biol. Chem. 249*, 4601–4607.

Honzatko, R. B., Crawford, J. L., Monaco, H. L., Ladner, J. E., Edwards, B. F. P., Evans, D. R., Warren, S. G., Wiley, D. C., Ladner, R. C., & Lipscomb, W. N. (1982) *J. Mol. Biol. 160*, 219–263.

Hough, E., Hansen, L. K., Birknes, B., Jynge, K., Hansen, S., Horvik, A., Little, C., Dodson, E., & Derewenda, Z. (1989) *Nature 338*, 357–360.

Jongeneel, C. V., Bouvier, J., & Bairoch, A. (1989) *FEBS Lett. 242*, 211–214.

Kägi, J. H. R., & Vallee, B. L. (1961) *J. Biol. Chem. 236*, 2435–2442.

Kägi, J. H. R., & Kojima, Y., Eds. (1987) *Metallothionein II*, Proceedings of the Second International Meeting, Zurich, Switzerland, Birkhäuser Verlag, Basel, Switzerland.

Kägi, J. H. R., & Schäffer, A. (1988) *Biochemistry 27*, 8509–8515.

Kannan, K. K., Notstrand, B., Fridborg, K., Lövgren, S., Orlsson, A., & Petef, M. (1975) *Proc. Natl. Acad. Sci. U.S.A. 72*, 51–55.

Keating, K. M., Ghosaini, L. R., Giedroc, D. P., Williams, K. R., Coleman, J. E., & Sturtevant, J. M. (1988) *Biochemistry 27*, 5240–5245.

Kerr, M. A., & Kenny, A. J. (1974) *Biochem. J. 137*, 489–495.

Kim, E. E., & Wyckoff, H. W. (1989) *Clin. Chim. Acta 186*, 175–188.

Klug, A., & Rhodes, D. (1987) *Trends Biochem. Sci. 12*, 464–469.

Lee, M. S., Gippert, G. P., Soman, K. V., Case, D. A., & Wright, P. E. (1989) *Science 245*, 635–637.

Le Moual, H., Crine, P., & Boileau, G. (1989) in *Abstract of Matrix Metalloproteinase Conference* (Birkedal-Hansen, H., Werb, Z., Welgus, H., & Van Wart, H., Eds.) Destin, FL.

Liljas, A., Kannan, K. K., Bergstēn, P. C., Waara, I., Fridborg, B., Strandberg, B., Carlbom, U., Järup, L., Lövgren, S., & Petef, M. (1972) *Nature, New Biol. 235*, 131–137.

Linderstrøm-Lang, K. (1952) in Lane Medical Lectures, *Stanford Univ. Publ., Univ. Ser., Med. Sci. 6*, 1–115.

Lipscomb, W. N., Reeke, G. N., Hartsuck, J. A., Quiocho, F. A., & Bethge, P. H. (1970) *Philos. Trans. R. Soc. London, B 257*, 177–214.

Macartney, H. W., & Tschesche, H. (1983) *Eur. J. Biochem. 130*, 71–78.

Malfroy, B., Schofield, P. R., Kuang, W.-J., Seeburg, P. H., Mason, A. J., & Henzel, W. J. (1987) *Biochem. Biophys. Res. Commun. 102*, 206–214.

Malfroy, B., Kado-Fong, H., Gros, C., Giros, B., Schwartz, J.-C., & Hellmiss, R. (1989) *Biochem. Biophys. Res. Commun. 161*, 236–241.

Margoshes, M., & Vallee, B. L. (1957) *J. Am. Chem. Soc. 79*, 4813.

Matthews, B. W., Jansonius, J. N., Colman, P. M., Schoenborn, B. P., & Dupourque, D. (1972) *Nature, New Biol. 238*, 37–41.

Matthews, B. W., Weaver, L. H., & Kester, W. R. (1974) *J. Biol. Chem. 249*, 8030–8044.

McCaman, M. T., & Gabe, J. D. (1986) *Gene 48*, 145–153.

Miller, J., McLachlan, A. D., & Klug, A. (1985) *EMBO J. 4*, 1609–1614.

Neurath, H. (1986) *Chem. Scr. 27B*, 221–229.

Neurath, H. (1989) *Trends Biochem. Sci. 14*, 268–271.

Olsen, J., Cowell, G. M., Kønigshøfer, E., Danielsen, E. M., Møller, J., Laustsen, L., Hansen, O. C., Welinder, K. G., Engberg, J., Hunziker, W., Spiess, M., Sjöström, H., & Norën, O. (1988) *FEBS Lett. 238*, 307–314.

Otvos, J. D., & Armitage, I. M. (1980) *Proc. Natl. Acad. Sci. U.S.A. 77*, 7094–7098.

Pan, T., & Coleman, J. E. (1989) *Proc. Natl. Acad. Sci. U.S.A. 86*, 3145–3149.

Pan, T., & Coleman, J. E. (1990) *Proc. Natl. Acad. Sci. U.S.A. 87*, 2077–2081.

Pauptit, R. A., Karlsson, R., Picot, D., Jenkins, J. A., Niklaus-Reimer, A.-S., & Jansonius, J. N. (1988) *J. Mol. Biol. 199*, 525–537.

Prescott, J. M., Wagner, F. W., Holmquist, B., & Vallee, B. L. (1985) *Biochemistry 24*, 5350–5356.

Quiocho, F. A., & Lipscomb, W. N. (1971) *Adv. Protein Chem. 25*, 1–58.

Rauser, W. E. (1990) *Annu. Rev. Biochem. 59*, 61.

Rees, D. C., Lewis, M., & Lipscomb, W. N. (1983) *J. Mol. Biol. 168*, 367–387.

Riordan, J. F., & Vallee, B. L. (1988) *Methods Enzymol. 158*, 3–6.

Rodriguez-Absi, J., & Prescott, J. M. (1978) *Arch. Biochem. Biophys. 186*, 383–391.

Rosenberg, U. B., Schröder, C., Preiss, A., Kienlin, A., Côtê, S., Riede, I., & Jäckle, H. (1986) *Nature 319*, 336–339.

Sanchez-Lopez, R., Nicholson, R., Gesnel, M.-C., Matrisian, L. M., & Breathnach, R. (1988) *J. Biol. Chem. 263*, 11892–11899.

Schmid, M. F., & Herriott, J. R. (1976) *J. Mol. Biol. 103*, 175–190.

Schuh, R., Aicher, W., Gaul, U., Côtê, S., Preiss, A., Maier, D., Seifert, E., Nauber, U., Schröder, C., Kemler, R., & Jäckle, H. (1986) *Cell 47*, 1025.

Sellers, A., Cartwright, E., Murphy, G., & Reynolds, J. J. (1977) *Biochem. J. 163*, 303.

Shang, Z., Liao, Y.-D., Wu, F. Y.-H. & Wu, C.-W. (1989) *Biochemistry 28*, 9790–9795.

Shinkai, H., & Nagai, Y. (1977) *J. Biochem. (Tokyo) 81*, 1261–1268.

Springman, E. B., Angleton, E. L., Birkedal-Hansen, H., & Van Wart, H. W. (1990) *Proc. Natl. Acad. Sci. U.S.A. 87*, 364–368.

Stöcker, W., Wolz, R. L., Zwilling, R., Strydom, D. J., & Auld, D. S. (1988) *Biochemistry 27*, 5026–5032.

Stricklin, G. P., Jeffrey, J. J., Roswit, W. T., & Eisen, A. Z. (1983) *Biochemistry 22*, 61–68.

Sunderman, F. W., Jr., & Barber, A. M. (1988) *Ann. Clin. Lab. Sci. 18*, 267–288.

Sutton, B. J., Artymiuk, P. J., Cordero-Borboa, A. E., Little, C., Phillips, D. C., & Waley, S. G. (1987) *Biochem. J. 248*, 181–188.

Tan, F., Chan, S. J., Steiner, D. F., Schilling, J. W., & Skidgel, R. A. (1989) *J. Biol. Chem. 264*, 13165–13170.

Thompson, G. A., & Carpenter, F. H. (1976) *J. Biol. Chem. 251*, 53–60.

Vaes, G. (1972) *Biochem. J. 126*, 275–289.

Vallee, B. L. (1977a) *Experientia 33*, 600.

Vallee, B. L. (1977b) in *Biological Aspects of Inorganic Chemistry* (Dolphin, D., Ed.) pp 37–70, Wiley, New York.

Vallee, B. L. (1986) in *Zinc Enzymes* (Bertini, I., Luchinat, C., Maret, W., & Zeppezauer, M., Eds.) pp 1–15, Birkhäuser, Boston, MA.

Vallee, B. L. (1988) *Biofactors 1*, 31–36.

Vallee, B. L., & Wacker, W. E. C. (1970) *Proteins (2nd Ed.) 5*, 1–192.

Vallee, B. L., & Falchuk, K. F. (1981) *Philos. Trans. R. Soc. London, B 294*, 185–197.

Vallee, B. L., & Auld, D. S. (1989) *FEBS Lett. 257*, 138–140.

Vallee, B. L., & Auld, D. S. (1990a) in *Matrix Metalloproteinases and Inhibitors* (Birkedal-Hansen, H., Werb, Z., Welgus, H., & Van Wart, H., Eds.) Gustav Fischer Verlag, Stuttgart, West Germany (in press).

Vallee, B. L., & Auld, D. S. (1990b) *Proc. Natl. Acad. Sci. U.S.A. 87*, 220–224.

Vallee, B. L., Riordan, J. F., Auld, D. S., & Latt, S. A. (1970) *Philos. Trans. R. Soc. London, B 257*, 215–230.

Vallee, B. L., Galdes, A., Auld, D. S., & Riordan, J. F. (1983) in *Zinc Enzymes* (Spiro, T. G., Ed.) pp 26–75, Wiley, New York.

Van Wart, H. E., & Lin, S. H. (1981) *Biochemistry 20*, 5682–5689.

Vincent, A., Colot, H. V., & Rosbash, M. (1985) *J. Mol. Biol. 186*, 149–166.

Wacker, H., Lehky, P., Fischer, E. H., & Stein, E. H. (1971) *Helv. Chim. Acta 54*, 473–484.

Wagner, F. W., Ray, L. E., Ajabnoor, M., Ziemba, P. E., & Hall, R. L. (1979) *Arch. Biochem. Biophys. 197*, 63–72.

Watt, V. M., & Yip, C. C. (1989) *J. Biol. Chem. 264*, 5480–5487.

Werb, Z., & Burleigh, M. C. (1974) *Biochem. J. 137*, 373.

Witham, S. E., Murphy, G., Angel, P., Rahmsdorf, H. J., Smith, B. J., Lyons, A., Harris, T. J. R., Reynolds, J. J., Herrlich, P., & Docherty, A. J. P. (1986) *Biochem. J. 240*, 913–916.

Chapter 7

Eye Lens ζ-Crystallin Relationships to the Family of "Long-Chain" Alcohol/Polyol Dehydrogenases. Protein Trimming and Conservation of Stable Parts[†]

Teresa Borrás,[‡] Bengt Persson,[§] and Hans Jörnvall*,[§]

Department of Chemistry I, Karolinska Institutet, S-104 01 Stockholm, Sweden, and Laboratory of Mechanisms of Ocular Diseases, National Eye Institute, National Institutes of Health, Bethesda, Maryland 20892

Received March 21, 1989

ABSTRACT: ζ-Crystallin of guinea pig lens is distantly related to the family of zinc-containing alcohol/polyol dehydrogenases. The amino acid residues binding the catalytic zinc atom in the alcohol dehydrogenase are exchanged in ζ-crystallin, explaining lack of known enzyme activity, and those residues binding the noncatalytic zinc in the dehydrogenase are located in a segment absent from the crystallin. Mammalian alcohol dehydrogenase, polyol dehydrogenase, and ζ-crystallin therefore constitute a series of proteins exhibiting successive changes in subunit metal content, from two to one and probably zero zinc atoms, respectively. In common with tetrameric dehydrogenases, the crystallin lacks a loop structure present in the dimeric dehydrogenases. Significantly, the crystallin is tetrameric, and a correlation between extra subunit interactions and lack of the loop segment is indicated. The lacking segment in crystallin is extended, encompassing a second loop in the dehydrogenase. The greatest conservation corresponds to the coenzyme-binding domain of the dehydrogenases, the central parts of which are remarkably similar to those in the crystallin. Glycine is by far the most conserved residue and corresponds to positions at bends in the conformation of the alcohol dehydrogenase. The conservation of the stable parts of the fold, the absence of the loop structure, the lack of the metal atoms, and the presence of only a small proportion of oxidation-sensitive cysteine residues in crystallin (5 versus 15 in the β_1 dehydrogenase subunit) suggest an increased stability of the lens protein and a derivation from the alcohol dehydrogenase family. This is compatible with the recruitment of stable enzyme structures for lens crystallin functions, with trimming of protein structures through these dehydrogenases or a yet unknown enzyme, and with multiple changes in the dehydrogenase family.

Analysis of mammalian ζ-crystallin in comparison with zinc-containing dehydrogenases reveals similarities and variations that are of interest for evaluation of properties of protein structures. The results show remarkable correlations between conserved residues and conformational properties. The variations in metal binding, absence of known enzyme activity, extreme stability of basic folds, and trimming of a parent protein through losses/nonacquirements of structurally nonessential parts are outlined after an account of the crystallins and alcohol/polyol dehydrogenases.

ζ-Crystallin and Other Crystallins

Crystallins in General. Crystallins constitute structural proteins of the eye lens and are present at high concentration in this organ, which has very little resynthesis capacity. Stable conformations, giving long half-lives, are essential. Apparently, nature has acquired these proteins by repeated recruitments of globular protein structures for incorporation as crystallins in the lens (Wistow & Piatigorsky, 1987; Wistow et al., 1987). Some crystallins, such as α, β, and γ, are present in the lenses of all vertebrates; α-crystallin shows a distant relationship to small heat-shock proteins, while β- and γ-crystallins belong to another family. A different group of crystallins appear to be taxon-specific and frequently involve common enzymes of

[†]This work was supported by Grant 03X-3532 from the Swedish Medical Research Council.

* Address correspondence to this author.

[‡]National Eye Institute, NIH.

[§]Karolinska Institutet.

well-known metabolic pathways [review in Wistow and Piatigorsky (1987)]. Presumably, the ancient origins and widespread occurrence of these enzymes or derivatives of them resulted in stable conformations, which made them suitable for functions also in the lens.

Taxon-Specific Mammalian Crystallins and the Present Case. Recently, two taxon-specific crystallins were purified from mammalian lenses, λ-crystallin from rabbit (Mulders et al., 1988) and ζ-crystallin from guinea pig (Huang et al., 1987). They are related to hydroxyacyl coenzyme A dehydrogenase (Mulders et al., 1988) and alcohol dehydrogenase (Rodokanaki et al., 1989), respectively. The ζ-crystallin/alcohol dehydrogenase case illustrates an apparent conservation of stable parts as presently described. The consequences are of interest for evolution of protein structures in general.

Nonmammalian Crystallins. Characterized relationships between enzymes and lens proteins in nonmammals involve duck ε-crystallin, identical with a lactate dehydrogenase (Wistow et al., 1987), and the following crystallins more or less different[1] from the parent proteins: frog ρ-crystallin and aldose reductase or prostaglandin synthase (Carper et al., 1987; Watanabe et al., 1988), reptile/avian δ-crystallins and argininosuccinate lyase [cf. R. F. Doolittle as quoted in Wistow and Piatigorsky (1987); Piatigorsky et al. (1988)], turtle τ-crystallin and enolases (Wistow & Piatigorsky, 1987), or invertebrate S_{III} crystallin and cytosolic glutathione transferase (Siezen & Shaw, 1982; Tomarev & Zinovieva, 1988; Wistow & Piatigorsky, 1987). These and the presently discussed dehydrogenase relationships indicate that the recruitments of old protein structures to serve new functions have frequently occurred via gene duplication(s), accumulation of mutational differences, and acquisition of different genetic regulatory mechanisms resulting in altered tissue expression (Wistow & Piatigorsky, 1987).

Alcohol Dehydrogenase

Multiple Forms. The finding of ζ-crystallin as a distant relative of an alcohol dehydrogenase answers questions concerning which properties of a parent protein have been conserved in the new function and which have been altered. However, alcohol dehydrogenase, like crystallins, is not a single protein of simple origin. At least three separate types of alcohol dehydrogenase exist, with largely unrelated structures and mechanisms, the "long-chain" and "short-chain" alcohol dehydrogenases, with and without zinc at the active site, respectively (Jörnvall et al., 1981), and an iron-activated alcohol dehydrogenase (Scopes, 1983; Neale et al., 1986).

The Zinc-Enzyme, "Long-Chain" Type. The best known type, that of the zinc-containing, long-chain enzymes, is the one first characterized; it includes the common liver and yeast alcohol dehydrogenases and is the type related to ζ-crystallin. This type exhibits at least three levels of gene duplication (Jörnvall et al., 1987b). They have led to (1) different enzymes, alcohol and polyol dehydrogenases (Jörnvall et al., 1981; (2) different alcohol dehydrogenase classes, I, II, and III (Vallee & Bazzone, 1983); and (3) different intraclass isozymes [review in Jörnvall et al. (1987a)]. These separate levels of change in the zinc-enzyme type affect many properties, including metal content [different for polyol and alcohol dehydrogenases (Jeffery et al., 1984)], quaternary structure [also different (Jörnvall et al., 1987c)], enzymatic activities, and rates of evolutionary divergence [different among the alcohol dehydrogenase classes (Kaiser et al., 1989)].

[1] Some of these differences may turn out to represent close similarities or even identities although not yet established as such, since the crystallins and the enzymes have been characterized from separate species. Consequently, species variations and enzyme/crystallin variations are presently superimposed in the structures characterized.

Table I: ζ-Crystallin Relationships toward Each of the Zinc Dehydrogenase Lines Characterized[a]

line	residue identity with crystallin	spread
mammalian ADH	86 (27)	74–86 (23–27)
plant ADH	70 (22)	67–72 (21–23)
yeast ADH	67 (22)	60–68 (19–22)
mammalian SDH	64 (21)	64 (21)

[a] Column 2 gives the residue identity in number of positions and in percent (in parentheses) between crystallin and the alcohol dehydrogenase structures in Figure 1 (human β_1 in the case of the mammalian alcohol dehydrogenase line), while column 3 gives the values for crystallin compared with all characterized alcohol dehydrogenase structures within each line (only one alternative thus far reported for mammalian sorbitol dehydrogenase). ADH, alcohol dehydrogenase; SDH, sorbitol dehydrogenase.

Alignment

An alignment of the primary structure of ζ-crystallin with those of alcohol/polyol dehydrogenases is given in Figure 1. The analysis included 20 alcohol/polyol dehydrogenases [17 summarized in Jörnvall et al. (1987c), plus three class III structures (Kaiser et al., 1989)] and one ζ-crystallin structure (Rodokanaki et al., 1989), all compared in spans of variable size (Jörnvall et al., 1981). The zinc-containing dehydrogenases have been subgrouped into four lines: two dimeric (mammalian and plant alcohol dehydrogenase) and two tetrameric [yeast alcohol dehydrogenase and mammalian polyol dehydrogenase (Jörnvall et al., 1987c)]. The alignment given is based on maximal residue identities and minimal gap introductions and, in a manner similar to that for the dehydrogenase alignments (Jörnvall et al., 1978, 1987c; Jeffery et al., 1984; Brändén et al., 1984; Eklund et al., 1985, 1987), is compatible with the tertiary structure of horse liver alcohol dehydrogenase directly analyzed by X-ray crystallography [Eklund et al., 1976; recent summary in Eklund and Brändén (1987)]. At a few segments of the alignment, additional identities could have been obtained by further gap introductions, especially in the terminal parts where similarities are weak. Since variations within each dehydrogenase line are small (Table I), only representative structures discussed are shown. Figure 1 lists one dehydrogenase of each of the major lines (mammalian alcohol, plant alcohol, yeast alcohol, and mammalian polyol dehydrogenase), with the addition in the mammalian alcohol dehydrogenase line of alternatives to include all characterized classes of this enzyme line.

ζ-Crystallin exhibits the greatest overall similarity to the mammalian alcohol dehydrogenase line. Regarding the separate classes within this line, the extent of similarities does not differ much but is greatest with class I, intermediate with class III, and least with class II. Class I is the traditional and abundant liver type of the enzyme. The somewhat closer similarity of crystallin to this class than to the other two classes is noteworthy because the class I structure is the one for which most variation has been characterized thus far [about 3-fold more variable than class III (Kaiser et al., 1989)].

However, the crystallin structure also shows particular similarities with the structures of the tetrameric dehydrogenase lines (yeast alcohol and mammalian polyol dehydrogenases). Thus, many residues are unique to ζ-crystallin and these dehydrogenases (56 positions, versus 55 unique to the dimeric dehydrogenases and ζ-crystallin). Furthermore, a segment lacking in the crystallin has an identically positioned border with a segment lacking in yeast alcohol dehydrogenase and

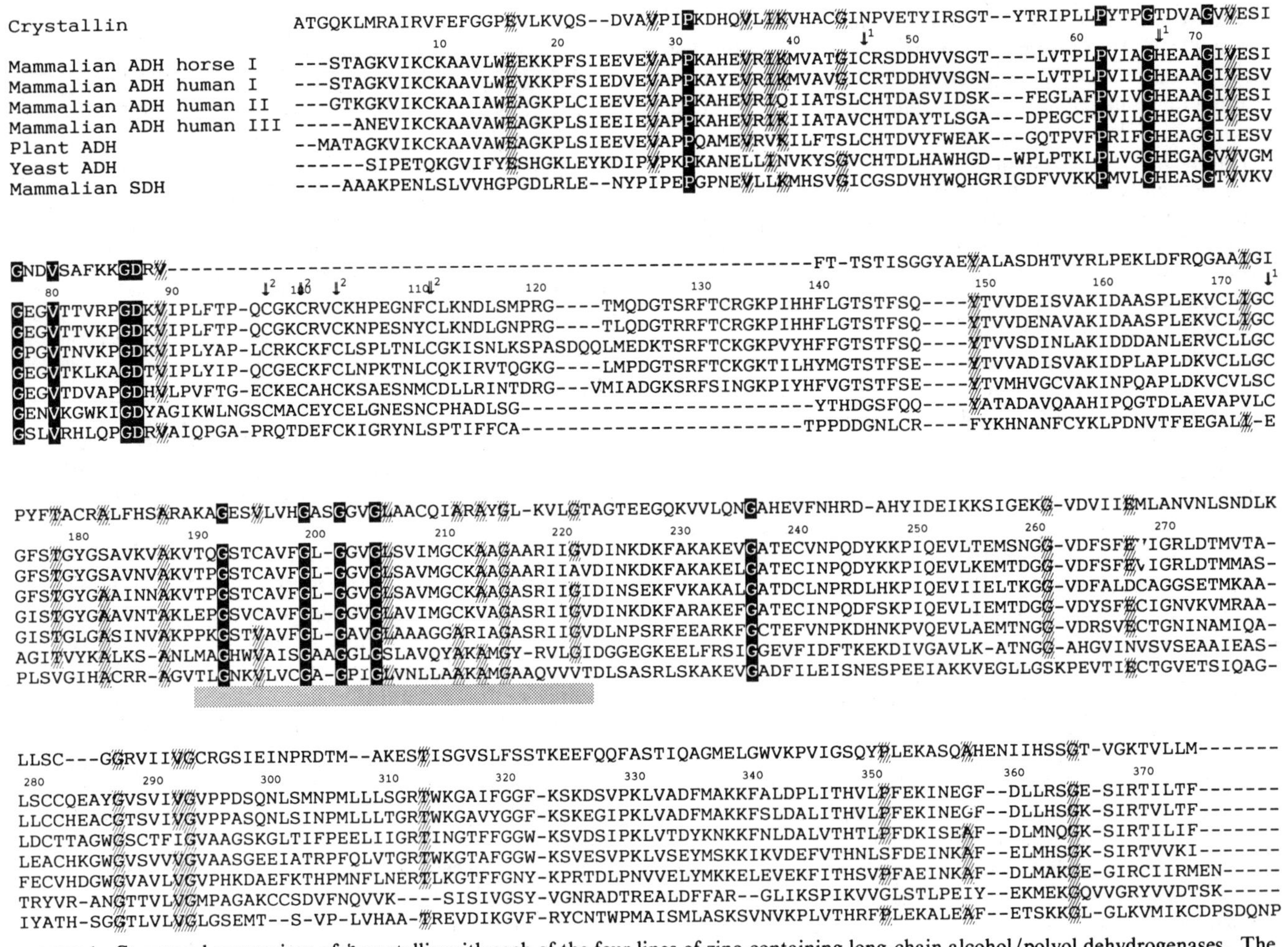

```
Crystallin                     ATGQKLMRAIRVFEFGGPEVLKVQS--DVAVPIPKDHQVLIKVHACGINPVETYIRSGT--YTRIPLLPYTPGTDVAGVVESI
                                         10        20        30        40   ↓1  50           60    ↓1  70
Mammalian ADH horse I          ---STAGKVIKCKAAVLWEEKKPFSIEEVEVAPPKAHEVRIKMVATGICRSDDHVVSGT----LVTPLPVIAGHEAAGIVESI
Mammalian ADH human I          ---STAGKVIKCKAAVLWEVKKPFSIEDVEVAPPKAYEVRIKMVAVGICRTDDHVVSGN----LVTPLPVILGHEAAGIVESV
Mammalian ADH human II         ---GTKGKVIKCKAAIAWEAGKPLCIEEVEVAPPKAHEVRIQIIATSLCHTDASVIDSK---FEGLAFPVIVGHEAAGIVESI
Mammalian ADH human III        -----ANEVIKCKAAVAWEAGKPLSIEEIEVAPPKAHEVRIKIIATAVCHTDAYTLSGA---DPEGCFPVILGHEGAGIVESV
Plant ADH                      --MATAGKVIKCKAAVAWEAGKPLSIEEVEVAPPQAMEVRVKILFTSLCHTDVYFWEAK---GQTPVFPRIFGHEAGGIIESV
Yeast ADH                      ------SIPETQKGVIFYESHGKLEYKDIPVPKPKANELLINVKYSGVCHTDLHAWHGD--WPLPTKLPLVGGHEGAGVVVGM
Mammalian SDH                  ----AAAKPENLSLVVHGPGDLRLE--NYPIPEPGPNEVLLKMHSVGICGSDVHYWQHGRIGDFVVKKPMVLGHEASGTVVKV

GNDVSAFKKGDRV-----------------------------------------------FT-TSTISGGYAEYALASDHTVYRLPEKLDFRQGAAIGI
   80        90      ↓2 1↓0↓2 ↓2    110↓2     120             130          140          150          160          170 ↓1
GEGVTTVRPGDKVIPLFTP-QCGKCRVCKHPEGNFCLKNDLSMPRG----TMQDGTSRFTCRGKPIHHFLGTSTFSQ----YTVVDEISVAKIDAASPLEKVCLIGC
GEGVTTVKPGDKVIPLFTP-QCGKCRVCKNPESNYCLKNDLGNPRG----TLQDGTRRFTCRGKPIHHFLGTSTFSQ----YTVVDENAVAKIDAASPLEKVCLIGC
GPGVTNVKPGDKVIPLYAP-LCRKCKFCLSPLTNLCGKISNLKSPASDQQLMEDKTSRFTCKGKPVYHFFGTSTFSQ----YTVVSDINLAKIDDDANLERVCLLGC
GEGVTKLKAGDTVIPLYIP-QCGECKFCLNPKTNLCQKIRVTQGKG----LMPDGTSRFTCKGKTILHYMGTSTFSE----YTVVADISVAKIDPLAPLDKVCLLGC
GEGVTDVAPGDHVLPVFTG-ECKECAHCKSAESNMCDLLRINTDRG---VMIADGKSRFSINGKPIYHFVGTSTFSE----YTVMHVGCVAKINPQAPLDKVCVLSC
GENVKGWKIGDYAGIKWLNGSCMACEYCELGNESNCPHADLSG-------------------------YTHDGSFQQ----YATADAVQAAHIPQGTDLAEVAPVLC
GSLVRHLQPGDRVAIQPGA-PRQTDEFCKIGRYNLSPTIFFCA-----------------------TPPDDGNLCR----FYKHNANFCYKLPDNVTFEEGALI-E

PYFTACRALFHSARAKAGESVLVHGASGGVGLAACQIARAYGL-KVLGTAGTEEGQKVVLQNGAHEVFNHRD-AHYIDEIKKSIGEKG-VDVIIEMLANVNLSNDLK
    180       190       200        210        220        230        240        250        260        270
GFSTGYGSAVKVAKVTQGSTCAVFGL-GGVGLSVIMGCKAAGAARIIGVDINKDKFAKAKEVGATECVNPQDYKKPIQEVLTEMSNGG-VDFSFE'IGRLDTMVTA-
GFSTGYGSAVNVAKVTPGSTCAVFGL-GGVGLSAVMGCKAAGAARIIAVDINKDKFAKAKELGATECINPQDYKKPIQEVLKEMTDGG-VDFSFEVIGRLDTMMAS-
GFSTGYGAAINNAKVTPGSTCAVFGL-GGVGLSAVMGCKAAGASRIIGIDINSEKFVKAKALGATDCLNPRDLHKPIQEVIIELTKGG-VDFALDCAGGSETMKAA-
GISTGYGAAVNTAKLEPGSVCAVFGL-GGVGLAVIMGCKVAGASRIIGVDINKDKFARAKEFGATECINPQDFSKPIQEVLIEMTDGG-VDYSFECIGNVKVMRAA-
GISTGLGASINVAKPPKGSTVAVFGL-GAVGLAAAGGARIAGASRIIGVDLNPSRFEEARKFGCTEFVNPKDHNKPVQEVLAEMTNGG-VDRSVECTGNINAMIQA-
AGITVYKALKS-ANLMAGHWVAISGAAGGLGSLAVQYAKAMGY-RVLGIDGGEGKEELFRSIGGEVFIDFTKEKDIVGAVLK-ATNGG-AHGVINVSVSEAAIEAS-
PLSVGIHACRR-AGVTLGNKVLVCGA-GPIGLVNLLAAKAMGAAQVVVTDLSASRLSKAKEVGADFILEISNESPEEIAKKVEGLLGSKPEVTIECTGVETSIQAG-

LLSC---GGRVIIVGCRGSIEINPRDTM--AKESTISGVSLFSSTKEEFQQFASTIQAGMELGWVKPVIGSQYPLEKASQAHENIIHSSGT-VGKTVLLM-------
 280       290       300        310        320        330        340        350        360        370
LSCCQEAYGVSVIVGVPPDSQNLSMNPMLLLSGRTWKGAIFGGF-KSKDSVPKLVADFMAKKFALDPLITHVLPFEKINEGF--DLLRSGE-SIRTILTF-------
LLCCHEACGTSVIVGVPPASQNLSINPMLLLTGRTWKGAVYGGF-KSKEGIPKLVADFMAKKFSLDALITHVLPFEKINEGF--DLLHSGK-SIRTVLTF-------
LDCTTAGWGSCTFIGVAAGSKGLTIFPEELIIGRTINGTFFGGW-KSVDSIPKLVTDYKNKKFNLDALVTHTLPFDKISEAF--DLMNQGK-SIRTILIF-------
LEACHKGWGVSVVVGVAASGEEIATRPFQLVTGRTWKGTAFGGW-KSVESVPKLVSEYMSKKIKVDEFVTHNLSFDEINKAF--ELMHSGK-SIRTVVKI-------
FECVHDGWGVAVLVGVPHKDAEFKTHPMNFLNERTLKGTFFGNY-KPRTDLPNVVELYMKKELEVEKFITHSVPFAEINKAF--DLMAKGE-GIRCIIRMEN-----
TRYVR-ANGTTVLVGMPAGAKCCSDVFNQVVK----SISIVGSY-VGNRADTREALDFFAR--GLIKSPIKVVGLSTLPEIY--EKMEKGQVVGRYVVDTSK-----
IYATH-SGGTLVLVGLGSEMT--S-VP-LVHAA-TREVDIKGVF-RYCNTWPMAISMLASKSVNVKPLVTHRFPLEKALEAF--ETSKKGL-GLKVMIKCDPSDQNP
```

FIGURE 1: Structural comparison of ζ-crystallin with each of the four lines of zinc-containing long-chain alcohol/polyol dehydrogenases. The four evolutionary lines are from Jörnvall et al. (1987c), and in each case except the mammalian dehydrogenase line, the representative has been chosen that corresponds to the structure previously utilized for conformational comparisons, i.e., the horse E-type alcohol dehydrogenase subunit actually analyzed by X-ray crystallography (Eklund et al., 1976; Eklund & Brändén, 1987), the maize isozyme 1 structure (Brändén et al., 1984), the yeast (*Saccharomyces cerevisiae*) isozyme 1 structure (Jörnvall et al., 1978), and the sheep liver sorbitol dehydrogenase structure (Eklund et al., 1985). In addition, all three classes of human alcohol dehydrogenase have been included in the mammalian line (human β_1 subunit of class I, human π subunit of class II, and human χ subunit of class III) in order to allow judgment on class distinctions within the mammalian enzyme. Residues strictly conserved in all these lines are given against a black background, whereas residues conserved between the majority of the alternatives within at least three of the four dehydrogenase lines (column 4 in Table II) are given against a hatched background. Numbers given refer to the positions of the class I horse and human liver alcohol dehydrogenase subunits (Jörnvall et al., 1987c). ADH, alcohol dehydrogenase; SDH, sorbitol dehydrogenase. The stippling below the mammalian SDH line indicates the segment of maximal conservation and corresponds to the stippling in Figure 2B. Arrows marked 1 indicate the ligands to the active-site zinc atom, while arrows marked 2 indicate the ligands to the second zinc atom.

mammalian polyol dehydrogenase (Figure 1). Consequently, the overall residue identities with the mammalian alcohol dehydrogenase line and some of the particular similarities with the tetrameric lines emphasize different relationships. Both the similarities and differences can be further interpreted as discussed below.

Conformational and Functional Consequences

The consequences of the alignment in Figure 1 are of special interest in relation to the characteristic properties of alcohol dehydrogenases, i.e., overall conformation, quaternary structure, active site, and coenzyme-binding fold.

Overall Conformation. The crystallin is homologous with the alcohol dehydrogenases over extensive parts of the protein chains (60–86 residues are identical; Table I). Furthermore, the large gap segment in the crystallin overlaps a gap segment in two of the alcohol dehydrogenase lines (and coincides at one end; cf. Figure 1, top and bottom two lines around positions 90/120–140). Consequently, the crystallin structure is concluded to exhibit an overall conformation related to those of the alcohol dehydrogenases. The latter have previously been suggested to be related in tertiary structure (Brändén et al., 1984; Eklund et al., 1985, 1987; Jörnvall et al., 1978), and the extent of residue identity between the four separate alcohol dehydrogenase lines was similar to those now observed with the crystallin (Table I).

The type and positions of the residue identities are of significance. Thus, glycine is by far the residue most strictly conserved (Table II). Such a distribution is a property typical of distantly related proteins with similar overall conformations. Furthermore, the actual positions of these glycine residues in the tertiary structure of the alcohol dehydrogenase show that a large proportion of them are located at reverse turns or other bends in the enzyme conformation (Figure 2A). This fact supports a conservation of the general characteristics of the dehydrogenase tertiary structures in that of the crystallin, with largely similar folding patterns. This conclusion applies to overall properties only, and localized deviations could well occur as detected below.

Deviating Patterns, Quaternary Structure. The overall similarity between crystallin and the mammalian alcohol dehydrogenase line (Table I) does not apply to the crystallin gap

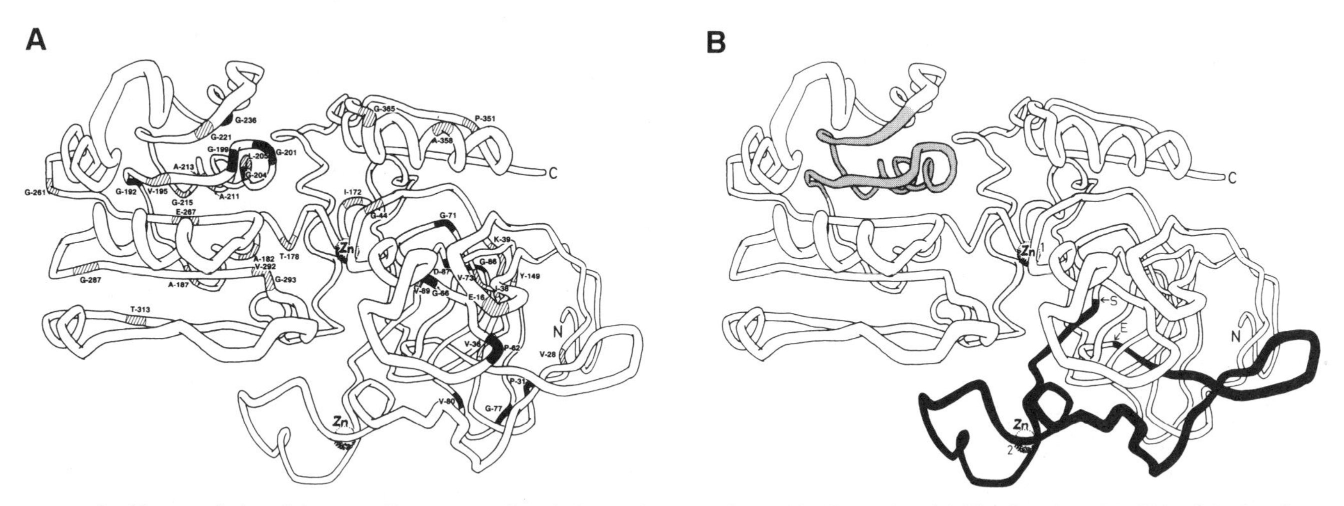

FIGURE 2: Characteristics of the crystallin structure in relation to the crystallographically analyzed (Eklund et al., 1976; Eklund & Brändên, 1987) horse liver alcohol dehydrogenase structure. (A) Positions of conserved residues. Residues strictly conserved (corresponding to column 2 in Table II) are shown black, and those conserved between the majority of the alternatives within at least three of the four dehydrogenase lines (column 4 in Table II) are hatched. The corresponding residues are in all cases also indicated by the one-letter code and the position in the horse enzyme. Data for the conserved residues are taken from Figure 1, while the conformation of the model enzyme is from Eklund et al. (1976) and Eklund and Brändên (1987). (B) Regions with the most conserved and variable segments. The long black region constitutes the gap segment in the crystallin of Figure 1, showing it to correspond exactly to surface loops of the dehydrogenase catalytic domain, including the structural zinc atom. A shortcut of the protein chain at the arrows marked S (for gap start) and E (for gap end) illustrates a possibility for direct continuity without large alterations of remaining chain folds. The stippled segment corresponds to the most conserved region shown in Table III at the center of the six β-pleated sheet strands of the stable conformation of the coenzyme-binding domain. The catalytic zinc atom is numbered 1 and the second (structural) zinc atom 2.

Table II: Conserved Residues between ζ-Crystallin and the Four Dehydrogenase Lines[a]

		conserved between crystallin and over half of all DHs in	
residue	strictly conserved	each of the four DH lines	at least three of the four DH lines
Asp	1	1	1
Thr			2
Glu			2
Pro	2	2	3
Gly	9	13	16
Ala		1	5
Val	1	3	7
Ile			2
Leu			1
Tyr			1
Lys			1
sum	13	20	41

[a] Residues in columns 2 and 4 are black and hatched, respectively, in Figure 2A. DH, dehydrogenase.

region covering positions 90–139 of the dehydrogenase. Instead, the crystallin gap segment coincides with a similarly positioned but smaller gap in the yeast alcohol and mammalian sorbitol dehydrogenases (Figure 1). The latter gap has previously been shown to correlate with the absence of one loop at a surface of the catalytic domain (Eklund et al., 1985; Jörnvall et al., 1978) and has been suggested as a possible reason (because of exposure of a novel subunit surface) why the two dehydrogenase lines lacking this loop represent tetrameric enzymes, while the two enzyme lines having this loop (mammals, plants) give dimeric enzymes. Significantly, the ζ-crystallin quaternary structure is also tetrameric (Huang et al., 1987; Rodokanaki et al., 1989), further supporting a correlation between this surface loop and dimer–dimer interactions.

In addition, the segment lacking from ζ-crystallin is considerably extended toward the N-terminal part, compared with the segment lacking from the yeast alcohol and mammalian sorbitol dehydrogenase lines (Figure 1). However, this extension corresponds to positions 90–118 in the mammalian alcohol dehydrogenase structure (Figure 1), which exactly constitutes a second loop, adjacent to the first and also located at the surface of the catalytic domain in the dehydrogenase structure (Figure 2B). Consequently, the extra segment lacking from the crystallin structure, though increasing the difference between crystallin and dehydrogenases, still confines the variations to the same surface of the subunit. The extended gap segment therefore supports the conclusion that this surface area may be involved in determining the quaternary structure and does not violate the conclusion about overall related conformations. In fact, the protein overall conformation can be obtained even with a "shortcut" between adjacent chain parts at the arrows indicated in Figure 2B, accounting for the absence of the two loops without large effects on remaining parts. As a consequence, the second zinc atom (the structural zinc atom), characteristic of the mammalian alcohol dehydrogenases, is absent in ζ-crystallin, exactly as in sorbitol dehydrogenase, where the lack has been shown by actual zinc analysis (Jeffery et al., 1984).

Active Site, Catalytic Properties. The three residues binding the catalytic zinc atom in all the dehydrogenase structures (Eklund et al., 1976, 1985, 1987; Jörnvall et al., 1978, 1987c; Jeffery et al., 1984; Brändên et al., 1984; Eklund & Brändên, 1987; Kaiser et al., 1989) are all absent in the ζ-crystallin structure (Figure 1). Thus, Cys-46 is replaced by Asn, His-67 by Thr, and Cys-174 in the alcohol dehydrogenases (Glu-154 in the sorbitol dehydrogenase) by Ile. None of these crystallin residues is typical of zinc-binding residues in proteins. Consequently, it can be concluded that ζ-crystallin lacks the catalytic zinc atom, thereby explaining the absence of alcohol/sorbitol dehydrogenase activity, which is in agreement with the direct enzymatic measurements (Zigler, unpublished results). Importantly, this change appears to be absolute. Thus, there is no Cys, His, or Glu residue that could easily replace the residues lost (Figure 1). Furthermore, the segment around the His-67 zinc-binding residue in the dehydrogenases, although being the most conserved region among the variable

Table III: Residue Identities between ζ-Crystallin and the Dehydrogenases for the Central Part of the Dehydrogenase Coenzyme-Binding Domain and the Whole Domain in Relation to Overall Protein Identities[a]

region	residue identity (%)
dehydrogenase positions 190–222	34 (31–52)
dehydrogenase coenzyme-binding domain	29 (19–29)
whole protein chains	27 (19–27)

[a] Values shown relate to the crystallin/human β_1 alcohol dehydrogenase subunit similarity and, within parentheses, those for the comparison of crystallin and any of the dehydrogenases.

dehydrogenases and the only segment where they have three consecutive residues strictly conserved (Jörnvall et al., 1987c), is also largely exchanged in the crystallin structure (having only Gly-66 conserved; Figure 1).

Strict Conservation, Coenzyme-Binding Fold. Despite the many exchanges between the crystallin and dehydrogenase structures (Table I), one segment is especially well conserved, the one corresponding to positions 190–222 in the dehydrogenase structure (Figure 1). This constitutes the center of the coenzyme-binding fold and encompasses central strands of the six-strand β-pleated sheet structure forming the backbone of the coenzyme-binding domain of the dehydrogenases (Eklund et al., 1976; Eklund & Brändén, 1987). The maximally conserved strands are highlighted in Figures 1 and 2B. This 33-residue segment is far more conserved than the whole protein chain and has many alcohol dehydrogenase typical residues, covering the βA–αB–βB center of the coenzyme-binding fold [Figure 2B; cf. Eklund et al. (1976) and Eklund and Brändén (1987)]. Consequently, the most characteristic dehydrogenase structure, constituting an evolutionarily old and heavily hydrogen-bonded structure (Rossman et al., 1978), is extensively conserved in the crystallin structure. Furthermore, this conservation is large (Table III), independent of whether the crystallin is compared to one or the other of the four dehydrogenase lines, and is largest (36–52%) toward the yeast enzyme, whereas overall residue similarities are most clearly visible toward the mammalian alcohol dehydrogenase (Table I).

Consequences of the Relationships

The comparisons clearly show that mammalian ζ-crystallin and alcohol/polyol dehydrogenases are structurally related. The homology extends over the whole protein chains (Figure 1); the most conserved residues are glycine residues (Table I), largely confined to bends in the tertiary structure (Figure 2A). The single long segment lacking in the crystallin corresponds to surface loops and is compatible with an otherwise largely unchanged chain fold through a polypeptide "shortcut" (Figure 2B). The absence of the structural zinc atom coincides with lack of this segment and the absence of the catalytic zinc atom with the lack of known enzyme activity. Finally, the most extensively conserved structure is represented by a central segment in one domain, corresponding to stable secondary structures with β-strands (Figure 2B). Naturally, the overall strict agreements do not exclude further localized dissimilarities than those regarding the zinc atoms and the surface loop. Nevertheless, the overall agreement establishes the common origin through an ancestral gene duplication and places ζ-crystallin as a deviating member of the alcohol/polyol long-chain dehydrogenase superfamily (Jörnvall et al., 1981, 1987c). The inclusion of the crystallin in this protein group has consequences for conclusions on both the crystallin structure and the dehydrogenase structures.

Consequences for the Crystallin Protein. The fact that ζ-crystallin is related to alcohol dehydrogenase strongly supports the recruitment theory (Wistow & Piatigorsky, 1987; Wistow et al., 1987) for the taxon-specific crystallins of the eye lens. These recruitments have occurred repeatedly, explaining why different crystallin forms are taxon-specific and showing separate relationships between crystallin forms and lactate dehydrogenase, argininosuccinate lyase, enolase, glutathione transferase, prostaglandin synthase, aldose reductase, hydroxyacyl coenzyme A dehydrogenase, and alcohol dehydrogenase [Carper et al., 1987; Doolittle as quoted in Wistow and Piatigorsky (1987); Huang et al., 1987; Mulders et al., 1988; Piatigorsky et al., 1988; Rodokanaki et al., 1989; Siezen & Shaw, 1982; Tomarev & Zinovieva, 1988; Watanabe et al., 1988; Wistow et al., 1987; Wistow & Piatigorsky, 1987]. In addition, the present case illustrates that the recruitment apparently involves conservation of the most stable structural characteristics, but lack of structures little stabilized. Thus, in the case of the alcohol dehydrogenase/ζ-crystallin pair, the crystallin has conserved the hydrogen-bonded most central part of the coenzyme-binding domain that constitutes a very old and common region of many dehydrogenase and kinase structures. Similarly, the crystallin structure lacks the two zinc atoms per alcohol dehydrogenase subunit and six zinc-binding cysteine residues, which, when present, are more or less reactive (Johansson et al., unpublished results) and hence sensitive to oxidation or other reactions. The crystallin structure has only 5 thiols per subunit, versus 15 in human alcohol dehydrogenase. Finally, the loop segment absent in crystallin corresponds to the only segment in the entire alcohol dehydrogenase structure that lacks stabilization by an extensive secondary structure (Eklund et al., 1976; Eklund & Brändén, 1987). In summary, the crystallin structure has all properties of a stable conformation with much secondary structure and with few sensitive residues, as expected from efficient utilization of a gene product recruited from a metabolic enzyme structure.

Regarding the question from which line the recruitment occurred, different possibilities seem to exist. Since ζ-crystallin is quite different from alcohol/sorbitol dehydrogenase, one possibility is that it represents recruitment from yet another line, hitherto unknown, of this enzyme superfamily. If so, that enzyme could already have lost/never acquired the zinc atoms and have a different enzymatic mechanism and metabolic role. No such enzyme is known and its possible activity or existence is hypothetical. However, it should be noted that a novel protein similar to ζ-crystallin has been detected in liver from several mammalian species (Zigler & Du, 1989). Also, the time of guinea pig taxon-specific divergence appears short in relation to the large differences observed. Consequently, a recruitment from such a hypothetical enzyme line is possible, provided the demonstration of the novel protein in liver does not reflect an ectopic expression of the ζ-crystallin gene.

The other alternative is that ζ-crystallin has been recruited from the alcohol/sorbitol dehydrogenase line. If so, the properties of the crystallin are noteworthy. Thus, the ζ-crystallin shows partly different similarities. The lacking segment and one set of unique residue identities suggest relationships with the tetrameric alcohol/polyol dehydrogenases, while overall residue identities and another set of positions suggest relationships with the dimeric enzymes. This mixed pattern appears to indicate the existence of both divergence and convergence in the evolution of ζ-crystallin: divergence from an original line and convergence toward the other(s), presumably because of structural requirements. Since the ζ-crystallin is hystricomorphic and guinea pig proteins may deviate to a considerable extent from those of other mammalian forms [cf. Persson et al. (1989) for another enzyme], it would be of interest to know the structures of guinea pig alcohol and sorbitol dehydrogenases. However, guinea pig alcohol dehydrogenase has been purified and its properties, including the presence of different enzyme classes, resemble those of other mammalian alcohol dehydrogenases (Keung & Fong, 1988). Consequently, guinea pig alcohol dehydrogenase has not evolved differently than other alcohol dehydrogenases and most likely does not constitute the novel hypothetical enzyme eluded to above. Thus, the two possibilities remain, both equally unexpected, demonstrating recruitment involving either an exceptionally large trimming of a parent alcohol/sorbitol dehydrogenase structure or the existence of a hitherto uncharacterized liver enzyme with an unknown substrate.

Consequences for Alcohol Dehydrogenase Relationships. Three levels of duplicatory events have been traced in the evolution of the zinc-containing long-chain alcohol dehydrogenases (Jörnvall et al., 1987b,c). At the first level, the alcohol dehydrogenase and polyol dehydrogenase lines separated; these two types of structure now exhibit overall residue identities of only about 25% (Jörnvall et al., 1987b). At the most recent level, the separate isozymes of mammalian alcohol dehydrogenases became apparent, presently exhibiting residue identities of more than 90%. The intermediate level is the one giving rise to the three classes of mammalian zinc-containing alcohol dehydrogenases. These three classes constitute the abundant, classical liver-type (class I) enzyme and the separate alcohol dehydrogenases of classes II and III, which differ in structure, substrate specificity, and organ distribution (Vallee & Bazzone, 1983; Jörnvall et al., 1987b,c). Species variations within the classes suggest that class III is highly constant in structure, whereas class I exhibits a variation about 3-fold more divergent (Kaiser et al., 1989). Furthermore, the three classes differ at the active site, the area for subunit interactions, and in the segment around the second zinc atom, suggesting that they constitute separate enzymes rather than simple isozymes (Jörnvall et al., 1987b). Noticeably, the same types of variation, i.e., variations affecting areas of subunit interactions and the loop around the second zinc atom, are exactly those now found to affect the crystallin structure in relation to those of the alcohol dehydrogenases. Thus, the crystallin shows that the interclass dehydrogenase variability is not an isolated phenomenon for just the alcohol dehydrogenases, but affects similarly other members of this protein superfamily.

Furthermore, the crystallin relationship can illustrate the origin of the separate classes. This question is important, because the three classes differ in substrate specificity (Vallee & Bazzone, 1983) and neither class has a clearly defined, known metabolic role. It is now apparent that the crystallin relationship starts to show a distinction between class II and the other classes of alcohol dehydrogenase (23% residue identity toward class II versus 26–27% toward class I and 25–26% toward class III). Thus, class II appears to be the class least similar to crystallin. It is the only class for which species variations have thus far not been characterized and for which the extent of divergence is therefore unknown. In contrast, classes I and III (Kaiser et al., 1989) reveal different extents of changes and separate properties. Separate properties are now also extended to class II, showing its relation to crystallin not to resemble those of either of the other two classes.

Finally, the successive changes in metal content within the alcohol dehydrogenase/polyol dehydrogenase/crystallin family are noteworthy. Thus, the corresponding subunits exhibit two zinc atoms, one zinc atom, and zero zinc atoms, respectively, as judged from previous analyses of the dehydrogenases (Jeffery et al., 1984) and the present comparisons for the crystallin. The successive differences in metal content were noteworthy already when only the alcohol and polyol dehydrogenases were characterized, but are now even more apparent, showing a lineage of related proteins with three different stages regarding metal content.

Conclusions

The following properties are of particular interest for conclusions on protein structures in general.

(1) The relationships add the "long-chain" zinc-containing dehydrogenase family to the group of protein families that contain highly divergent members involving both enzymes and other proteins. The early cases of lysozyme/α-lactalbumin (Hill & Brew, 1975) and serine proteases/haptoglobin (Kurosky et al., 1974), together with recent additions of, for example, lipases/vitellogenins (Bownes et al., 1988; Persson et al., 1989) and enzymes/crystallins (Wistow & Piatigorsky, 1987), with the present case in particular, show that many protein families are large with wide relationships across conventional borders. Apparently, protein structures exhibit limited variability. The crystallin member now detailed of the "long-chain, zinc-containing dehydrogenase" family has neither the zinc ligands (hence not the zinc) nor the same substrates if at all any enzyme activity. Successive variations in metalloprotein metal content are well illustrated by alcohol dehydrogenase/sorbitol dehydrogenase/ζ-crystallin. The lack of known enzyme activity can be ascribed to the absence of single residues (here active-site zinc ligands) in the same manner as for the loss of Glu-35 in lysozyme, Ser-195 in chymotrypsin, and a putative functional serine in lipases regarding the other examples of enzymes/nonenzymes mentioned above. Thus, small protein changes can have large effects on important properties at the same time as extreme divergence does apparently not alter basic folds.

(2) The most conserved segments are central regions of an evolutionarily old domain, particularly involving glycine residues at critical positions, as shown early for cytochrome *c* (Smith & Margoliash, 1964) and known within the alcohol/polyol dehydrogenases themselves (Jörnvall et al., 1984). Apparently, such structures are stable, making them suitable also for recruitments as structural proteins.

(3) Natural protein trimming occurs and suggests the removal (if ζ-crystallin is derived from alcohol dehydrogenase) or nonacquirement (if derived from a novel, hitherto unknown protein already having these characteristics) of unstable parts, such as oxidation-sensitive cysteine residues, superficial segments, and loops nonstabilized by elements of secondary structure. Obviously, observations on the naturally occurring variants complement those possible to obtain from directed mutagenesis. Native variability exhibits successive changes, and as presently shown, extensive differences between enzymes and structural proteins can be correlated with particular residues or segments in critical regions. Combined, observations on both natural variants and directed mutants facilitate delineation of basic folds. Regarding differences in metal content, quaternary structure, and biological activity, the crystallin/alcohol dehydrogenase/polyol dehydrogenase protein family illustrates exceptionally well several types of relationship, emphasizing stability and multiplicity.

Added in Proof

A report has recently appeared, extending the family of long-chain zinc-containing dehydrogenases to include also prokaryotic threonine dehydrogenase (Aronson et al., 1989).

One further line can therefore be added to Figure 1 and Table I, and the family can now be summarized as alcohol/polyol/threonine dehydrogenases/ζ-crystallin.

ACKNOWLEDGMENTS

We are grateful to Carina Palmberg and Eva Andersson for figures and to Bo Furugren for the original drawing of the schematic chain fold of the horse enzyme crystallographically analyzed by Carl-Ivar Brändén, Hans Eklund, and co-workers.

Registry No. ADH, 9031-72-5; SDH, 9028-21-1; Zn, 7440-66-6; polyol dehydrogenase, 80448-98-2.

REFERENCES

Aronson, B. D., Somerville, R. L., Epperly, B. R., & Dekker, E. E. (1989) *J. Biol. Chem. 264*, 5226–5232.

Bownes, M., Shirras, A., Blair, M., Collins, J., & Coulson, A. (1988) *Proc. Natl. Acad. Sci. U.S.A. 85*, 1554–1557.

Brändén, C.-I., Eklund, H., Cambillau, C., & Pryor, A. J. (1984) *EMBO J. 3*, 1307–1310.

Carper, D., Nishimura, C., Shinohara, T., Dietzchold, B., Wistow, G., Craft, C., Kador, P., & Kinoshita, J. H. (1987) *FEBS Lett. 220*, 209–213.

Eklund, H., & Brändén, C.-I. (1987) in *Biological Macromolecules and Assemblies* (Jurnak, F., & McPherson, A., Eds.) Vol. 3, pp 74–142, Wiley, New York.

Eklund, H., Nordström, B., Zeppezauer, E., Söderlund, G., Ohlsson, I., Boiwe, T., Söderberg, B.-O., Tapia, O., Brändén, C.-I., & Åkeson, Å. (1976) *J. Mol. Biol. 102*, 27–59.

Eklund, H., Horjales, E., Jörnvall, H., Brändén, C.-I., & Jeffery, J. (1985) *Biochemistry 24*, 8005–8012.

Eklund, H., Horjales, E., Vallee, B. L., & Jörnvall, H. (1987) *Eur. J. Biochem. 167*, 185–193.

Hill, R. L., & Brew, K. (1975) *Adv. Enzymol. Relat. Areas Mol. Biol. 43*, 411–490.

Huang, Q.-L., Russell, P., Stone, S. H., & Zigler, J. S., Jr. (1987) *Curr. Eye Res. 6*, 725–732.

Jeffery, J., Chester, J., Mills, C., Sadler, P. J., & Jörnvall, H. (1984) *EMBO J. 3*, 357–360.

Jörnvall, H., Eklund, H., & Brändén, C.-I. (1978) *J. Biol. Chem. 253*, 8414–8419.

Jörnvall, H., Persson, M., & Jeffery, J. (1981) *Proc. Natl. Acad. Sci. U.S.A. 78*, 4226–4230.

Jörnvall, H., von Bahr-Lindström, H., & Jeffery, J. (1984) *Eur. J. Biochem. 140*, 17–23.

Jörnvall, H., Hempel, J., & Vallee, B. L. (1987a) *Enzyme 37*, 5–18.

Jörnvall, H., Höög, J.-O., von Bahr-Lindström, H., & Vallee, B. L. (1987b) *Proc. Natl. Acad. Sci. U.S.A. 84*, 2580–2584.

Jörnvall, H., Persson, B., & Jeffery, J. (1987c) *Eur. J. Biochem. 167*, 195–201.

Kaiser, R., Holmquist, B., Vallee, B. L., & Jörnvall, H. (1989) *Biochemistry* (in press).

Keung, W.-M., & Fong, W.-P. (1988) *Comp. Biochem. Physiol. 89B*, 85–89.

Kurosky, A., Barnett, D. R., Rasco, M. A., Lee, T.-H., & Bowman, B. H. (1974) *Biochem. Genet. 11*, 279–293.

Mulders, J. W. M., Hendriks, W., Blankesteijn, W. H., Bloemendal, H., & de Jong, W. W. (1988) *J. Biol. Chem. 263*, 15462–15466.

Neale, A. D., Scopes, R. K., Kelley, J. M., & Wettenhall, R. E. H. (1986) *Eur. J. Biochem. 154*, 119–124.

Persson, B., Bengtsson-Olivecrona, G., Enerbäck, S., Olivecrona, T., & Jörnvall, H. (1989) *Eur. J. Biochem. 179*, 39–45.

Piatigorsky, J., O'Brien, W. E., Norman, B. L., Kalumuck, K., Wistow, G. J., Borrás, T., Nickerson, J. M., & Wawrousek, E. F. (1988) *Proc. Natl. Acad. Sci. U.S.A. 85*, 3479–3483.

Rodokanaki, A., Holmes, R., & Borrás, T. (1989) *Gene 78*, 215–224.

Rossmann, M. G., Liljas, A., Brändén, C.-I., & Banaszak, L. J. (1978) *Enzymes (3rd Ed.) 11*, 61–102.

Scopes, R. K. (1983) *FEBS Lett. 156*, 303–306.

Siezen, R. J., & Shaw, D. C. (1982) *Biochim. Biophys. Acta 704*, 304–320.

Smith, E. L., & Margoliash, E. (1964) *Fed. Proc., Fed. Am. Soc. Exp. Biol. 23*, 1243–1247.

Tomarev, S. I., & Zinovieva, R. D. (1988) *Nature 336*, 86–88.

Vallee, B. L., & Bazzone, T. J. (1983) *Isozymes: Curr. Top. Biol. Med. Res. 8*, 219–244.

Watanabe, K., Fujii, Y., Nakayama, K., Ohkubo, H., Kuramitsu, S., Kagamiyama, H., Nakanishi, S., & Hayaishi, O. (1988) *Proc. Natl. Acad. Sci. U.S.A. 85*, 11–15.

Wistow, G., & Piatigorsky, J. (1987) *Science 236*, 1554–1556.

Wistow, G. J., Mulders, J. W. M., & de Jong, W. W. (1987) *Nature 326*, 622–624.

Zigler, J. S., Jr., & Du, X.-y. (1989) *Invest. Ophthalmol. Visual Sci. 30 (Suppl.)*, 266.

Chapter 8

Probing Protein Structure and Dynamics with Resonance Raman Spectroscopy: Cytochrome *c* Peroxidase and Hemoglobin[†]

Thomas G. Spiro,[*,‡] Giulietta Smulevich,[§] and Chang Su[‡]

Department of Chemistry, Princeton University, Princeton, New Jersey 08562, and Dipartimento di Chimica, Università di Firenze, Via G. Capponi 9, 50121 Firenze, Italy

Received October 31, 1989; Revised Manuscript Received December 14, 1989

Resonance Raman (RR) spectroscopy is increasingly being applied to biological molecules (Spiro, 1988). The technique involves laser scattering from a light-absorbing sample. The scattering spectrum contains peaks that report the frequencies of vibrational modes of the molecules being irradiated. Resonance occurs when the laser wavelength matches that of an electronic transition. Because of coupling between electronic and nuclear motions, certain vibrational modes are enhanced, those which mimic the distortion of the molecule in its resonant excited state (Spiro & Stein, 1977). For example, resonance with π–π^* transitions enhances modes in which π bonds of the chromophore are stretched, while resonance with ligand–metal charge-transfer transitions in metal complexes enhances modes in which the metal–ligand bonds are stretched. This selectivity in the enhancement means that RR spectroscopy can be used as a probe for chromophoric sites in complex biological systems. Hemes, chlorophyll, flavins, the retinylidene cofactor of visual pigments and bacteriorhodopsin, and a variety of iron and copper metalloprotein sites are among the biological chromophores that have been examined by RR spectroscopy (Spiro, 1988). Recently the technique has been extended to aromatic protein residues (Hudson & Mayne, 1988) and to the purine and pyrimidine bases of nucleic acids (Tsubaki et al., 1988), thanks to the advent of practical UV laser sources (Ziegler & Hudson, 1983; Asher et al., 1983; Fodor et al., 1986; Jones et al., 1987).

Of course, selective enhancement also means a loss of information since many sites of molecular interest are nonchromophoric, although they can sometimes be labeled with extrinsic chromophores (Carey, 1988). Complete vibrational spectra can be obtained with nonresonance Raman spectroscopy, in which the laser wavelength is in a transparent region of the spectrum, or with infrared spectroscopy. The complete vibrational spectrum is very crowded for complex biological molecules, however, and it is difficult to associate a spectral feature with a particular site, although impressive results can be obtained with well-controlled difference spectra (Derguini et al., 1986; Chen et al., 1987). In those cases where the site of interest is chromophoric, RR spectra provide much more detailed structural information than nonresonance Raman or IR spectra.

The structural information in a vibrational spectrum is complementary to that produced by X-ray crystallography. Occasionally an error in a crystal structure determination can be detected by vibrational spectroscopy, as in the case of the 3-Fe center of *Azotobacter vinelandii* ferredoxin I, for which the original report of a hexagonal Fe_3S_3 ring (Ghosh et al., 1982) was shown to be incompatible with the RR spectrum (Johnson et al., 1983) [other spectroscopic probes also threw doubt on this structure (George et al., 1984)]; the structure was later redetermined (Stout et al., 1988; Stout, 1988) and shown to be a Fe_4S_4 cube with a missing corner, as had earlier been proposed on the basis of chemical evidence (Beinert & Thomson, 1983). Only X-ray crystallography, or, for relatively small macromolecules, 2-D NMR spectroscopy, can produce the full three-dimensional arrangement of all the atoms, however. What vibrational spectroscopy can provide is a monitor of the strengths of bonds and the conformations of molecular fragments, as they interact with their environments. The vibrational frequencies are highly sensitive to these molecular parameters, which are at the heart of biological activity.

In this review the kinds of information available from RR spectroscopy are illustrated with new results on two heme proteins, cytochrome *c* peroxidase (CCP) and hemoglobin (Hb). CCP is a peroxide-utilizing enzyme (Yonetani, 1976), whose chemistry is determined by interactions between the heme prosthetic group and the protein residues lining the heme pocket. These interactions are many and subtle, and they can be varied systematically by site-directed mutagenesis of cloned protein (Mauro et al., 1989; Goodin et al., 1986). RR spec-

[†] This work was supported by NIH Grant GM33576.

[‡] Princeton University.

[§] Università di Firenze.

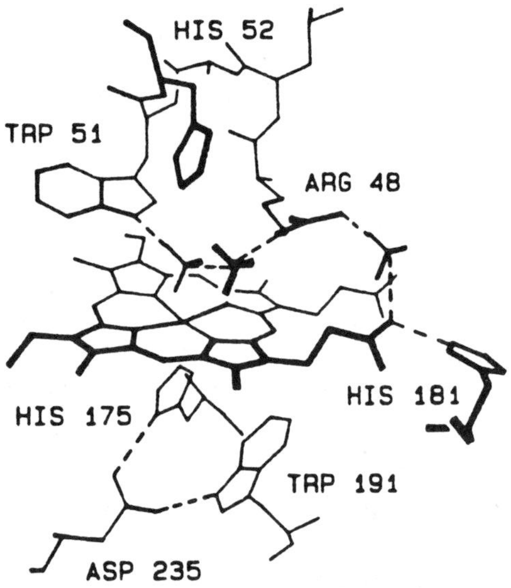

FIGURE 1: View of the heme group in bakers' yeast CCP and the surrounding residues that provide important interactions. H-Bonds are indicated by dotted lines. The tetrahedra represent water molecules localized on the distal side of the heme. From Finzel et al. (1986).

troscopy has proven to be a useful monitor of these interactions. Hb is limited to the simple function of binding O_2, but it does so cooperatively, via an allosteric transition. Because of intensive studies in many laboratories, Hb has become the paradigmatic molecule with which to understand allostery (Perutz, 1975; Perutz et al., 1987). A promising approach is to investigate the kinetic steps in the quaternary transition between the R (relaxed) and T (tense) state of the molecule. RR spectroscopy, applied in a time-resolved mode, has provided structural information about the intermediates in this process.

The reader is referred to recent reviews on the detailed interpretation of metalloporphyrin RR spectra (Spiro & Li, 1988) and of the RR signals associated with endogenous (Kitagawa, 1988; Champion, 1988) and exogenous (Kerr & Yu, 1988) heme protein ligands, as well as of protein aromatic residues (Hudson & Mayne, 1988). This information is relevant to the ensuing discussion, but much of it has been omitted in the interest of brevity and clarity.

CYTOCHROME *c* PEROXIDASE: A BALANCE OF PROTEIN–HEME FORCES

Proximal Imidazole/Imidazolate. Figure 1 shows the disposition of residues in the heme pocket of CCP, as determined from the crystal structure of protein from bakers' yeast (Finzel et al., 1986). A very similar structure is found for the engineered protein CCP(MI), expressed in *Escherichia coli* (Wang et al., 1989). In common with the globins, the heme Fe atom in CCP is bound to the protein through an N atom of an imidazole side chain (His-175), but the H-bond status of the other imidazole N atom is distinctive. In the globins this proton interacts with a (neutral) peptide carbonyl group of the backbone, whereas in CCP the interaction is with the anionic carboxylate side chain of Asp-235. The proximal H-bond is much stronger in CCP, and gives the ligand substantial imidazolate character, thereby strengthening its donor interaction with Fe. This interaction can be monitored directly via the frequency of the Fe–N(His) stretching vibration, which gives rise to a prominent band near 200 cm^{-1} in RR spectra of Fe(II) heme proteins, identifiable via its characteristic ^{54}Fe shift. In myoglobin (Mb) this band is at 220 cm^{-1} (Kitagawa et al., 1979; Argade et al., 1984), whereas in CCP (Hashimoto et al., 1986; Smulevich et al., 1986; Dasgupta et al., 1989) and also in horseradish peroxidase (HRP) and other plant peroxidases (Teraoka & Kitagawa, 1981; Teraoka et al., 1983) it is near 240 cm^{-1}.

In a model study with 2-methylimidazole (2-MeImH) bound to Fe(II) protoheme, the Fe–2-MeImH frequency was found to be 205 cm^{-1} in benzene but 220 cm^{-1} in water and 239 cm^{-1} when the imidazole proton was removed with a strong base in dimethylformamide (Stein et al., 1980). These are large shifts and reflect considerable sensitivity of the Fe–imidazole bond strength to the status of the proton on the bound imidazole. When this proton is removed, leaving an anionic imidazolate ligand, the bond strength increases by about one-third, judging from the 35% increase in force constant which is implied by the frequency shift from 205 cm^{-1} in benzene to 239 cm^{-1} in strong base/dimethylformamide. The 220-cm^{-1} frequency observed in water implies that H-bond donation to water molecules produces about half as much increase in the bond strength as does deprotonation. In comparing these values with the frequencies observed in protein, about 10 cm^{-1} should be added to the former in order to account for the steric encumbrance of the 2-methyl group in the model complexes (M. Mitchell and T. G. Spiro, unpublished results); [this steric encumbrance serves to inhibit bis-imidazole adduct formation (Collman & Reed, 1973)]. Thus, the 220-cm^{-1} frequency of Mb indicates that the proximal H-bond to the backbone carbonyl group is weaker than an H-bond between Fe-bound imidazole and water, while the 240-cm^{-1} frequency of CCP indicates that the H-bond to the Asp-235 carboxylate is stronger.

The proximal H-bond may play several roles in peroxidase activity. The increased imidazolate character certainly helps to lower the Fe(III/II) reduction potential, −194 vs +50 mV for Mb (Cassatt et al., 1975), by stabilizing the Fe(III) form, the form that reacts with peroxide. By the same token, imidazolate stabilizes the $Fe^{IV}{=}O$ intermediate (Edwards et al., 1987) resulting from heterolytic cleavage of the bound peroxide. Finally, the rate of release of the product, H_2O, of the electron-transfer reaction, is increased because of lowered H_2O affinity in the resting enzyme; both electronic [lowered Fe(III) charge density due to the partially anionic character of the ligand] and steric (anchoring of the Fe ion below the heme plane by the proximal H-bond) factors may be involved in the stabilization of 5-coordinate Fe(III) heme in CCP (see below).

Dramatic confirmation of the importance of the His-175–Asp-235 H-bond to the Fe–N(His) bond strength was obtained via the Asn-235 mutant, whose RR spectrum (Smulevich et al., 1988a) is compared with that of CCP(MI) in Figure 2. The broad ~240-cm^{-1} Fe–N(His) band of the latter is missing for the mutant and is replaced by a narrow band at 205 cm^{-1}, which is tentatively assigned to Fe–N(His) stretching. This is a much lower frequency than that of Mb and implies that the H-bond to the Asn-235 side chain is extremely weak. Interestingly, the crystal structure of the Asn-235 mutant shows the indole ring of Trp-191, which is also H-bonded to the Asp-235 carboxylate side chain in native CCP, to be flipped over and H-bonded to the carbonyl group of Leu-177 (Mauro et al., 1989). Thus, the carboxylate interactions with both of the H-bond donors, His-175 and Trp-191, are greatly attenuated upon replacement with the neutral carboxamide side chain. Trp-191 has been shown to be the residue giving rise to a unique axial EPR radical signal in the two-electron-oxidized ES intermediate arising from the reaction of CCP with hydroperoxides (Sivaraja et al., 1989). This signal is lost, and activity is greatly reduced, when Trp-191 is replaced by Phe. Thus another role for Asp-235 may be to anchor Trp-191 in

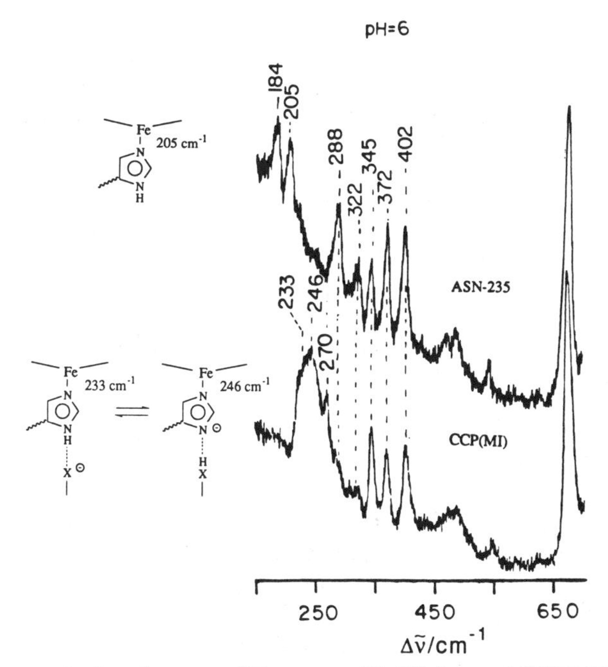

FIGURE 2: Low-frequency RR spectra of Fe(II) forms of CCP(MI) and of the Asn-235 mutant, obtained with 441.6-nm excitation [experimental details in Smulevich et al. (1988a)]. The broad $\sim$240-cm^{-1} Fe–N(His) stretching band of CCP(MI) has 233- and 246-cm^{-1} components, which are suggested to arise from a double-well potential in which the proton can reside on the proximal imidazole (233 cm^{-1}) or on the interacting carboxylate group of Asp-235 (246 cm^{-1}). Replacement of the carboxylate group by the carboxamide side chain of Asn eliminates the 240-cm^{-1} band and replaces it with one at 205 cm^{-1}, indicating that the proximal H-bond is very weak.

an orientation close to (and parallel with; see Figure 1) the proximal heme ligand, thereby facilitating electron transfer.

The breadth of the Fe–N(His) band is of interest. There are actually two components, at 233 and 246 cm^{-1}. Their relative intensities are altered at high pH (Hashimoto et al., 1986), leading to an apparent downshift in the entire band (Dasgupta et al., 1989), and also when various side chains are altered in the heme pocket (Smulevich et al., 1988). This behavior implies two forms of the proximal ligand, the populations being sensitive to the disposition of other side chains in the pocket. The 246-cm^{-1} frequency is so high that complete deprotonation is indicated, while the 233 cm^{-1} frequency is consistent with a fairly strong H-bond. It was therefore suggested (Smulevich et al., 1988) that the two forms arise from the proton being in a double-well potential. It resides on the imidazole in one minimum, corresponding to the 233-cm^{-1} frequency, and on the Asp-235 carboxylate in the other, corresponding to the 244-cm^{-1} frequency. The comparable intensities of the two components suggest that the two minima are nearly isoenergetic. At first sight it is surprising that the proton should be nearly equally shared by the imidazolate and carboxylate anions, given their vastly different pK_a's, $\sim$14 and $\sim$4.5, in aqueous solution. But the proximal linkage is isolated from solvating water molecules, and the imidazolate basicity is diminished by coordination to the Fe(II). The latter effect should be even more pronounced for higher Fe oxidation states, which should shift the balance further toward imidazolate. Indeed, NMR data on the CN$^-$ adduct of Fe(III) HRP show that the proximal imidazole proton is transferred to another protein residue (La Mar & de Ropp, 1979).

Distal H-Bonding to the CO Adduct. When RR and IR spectra are examined for CO adducts of CCP variants, three vibrational patterns emerge, as exemplified by the spectra of CCP(MI) and its Phe-51 mutant (Smulevich et al., 1988b),

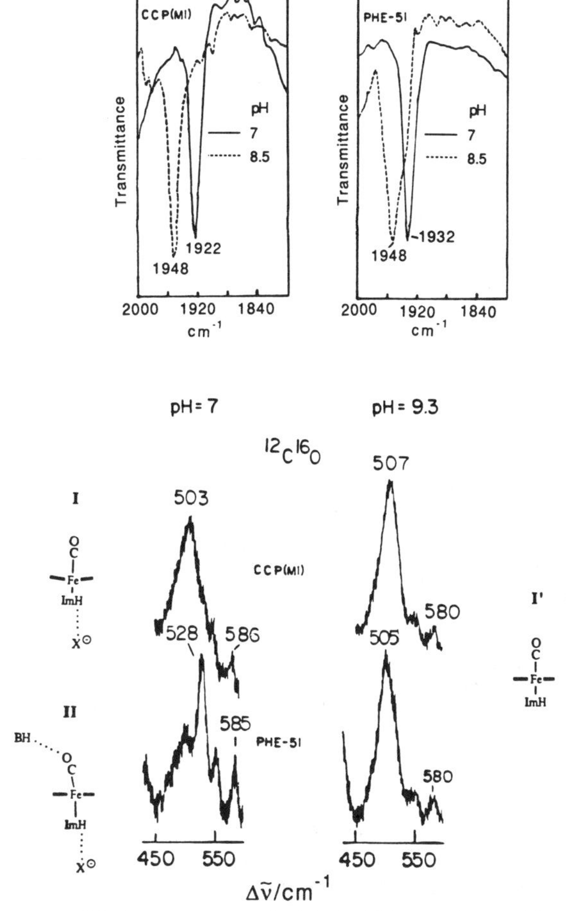

FIGURE 3: RR bands associated with Fe–CO stretching (500–530 cm^{-1}) and Fe–C–O bending (585 cm^{-1}) and the IR band due to C–O stretching ($\sim$1920–1960 cm^{-1}) for CO adducts of CCP(MI) and the Phe-51 mutant at neutral and high pH [experimental details in Smulevich et al. (1988b)]. These adducts illustrate the three forms, I and II (neutral and low pH) and I′ (high pH), encountered among CCP variants. The inferred bonding arrangements (see text) are shown in the diagram.

shown in Figure 3. These patterns are interpreted as arising from different structures, labeled form I, II, and I′, respectively. Form I′ is exhibited by all CCP variants at alkaline pH and is believed to reflect a standard CO adduct with an ordinary imidazole ligand. Forms I and II are formed at acid pH and are favored by different variants. Form I is attributed to a CO adduct with a ligand having substantial imidazolate character whereas in form II the CO adduct receives a distal H-bond to its O atom while the Fe–imidazole bond is concomitantly weakened.

These structural inferences are based on the correlation of the frequencies of the Fe–CO ($\sim$500 cm^{-1}) and C–O ($\sim$1950 cm^{-1}) stretching vibrations, ν_{Fe-CO} and ν_{C-O}, seen in Figure 4. For a large number of CO adducts of Fe(II) porphyrins having imidazole ligands, the data follow the straight line (Kerr & Yu, 1988; Uno et al., 1987; Li & Spiro, 1988b) shown in Figure 4. Its negative slope reflects the role in Fe–CO bonding of the back-donation of Fe electrons in the filled d_π orbitals to the empty π^* orbitals of the CO ligand. As back-donation increases, so does the Fe–CO bond strength, whereas the C–O bond strength decreases. Likewise, the stretching frequencies change in an inverse manner. Influences that increase back-donation move the data points upward along the line in Figure 4. The polarity of the CO environment is an important factor, since back-bonding increases the negative charge on the O atom and the developing charge is stabilized by interactions

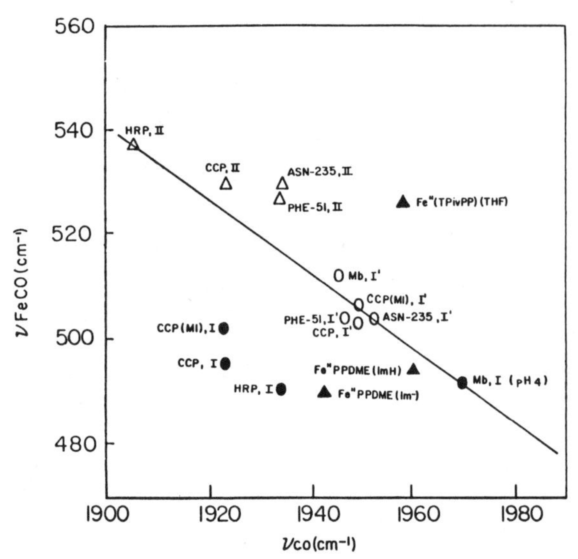

FIGURE 4: Correlation of ν_{Fe-CO} with ν_{C-O} for heme–CO adducts. The straight line (ν_{Fe-CO} = 1965–0.75ν_{C-O} cm^{-1}) fits those adducts having an axial imidazole ligand with a weak or absent H-bond (Li & Spiro, 1988b). Stronger σ donor ligands produce negative deviations from the lines while weaker donors produce positive deviations. Symbols: (▲) model compounds from protoporphyrin dimethyl ester (PPDME) or "picket fence" porphyrin (TPivPP) with imidazole (ImH), imidazolate (Im), and tetrahydrofuran (THF) ligands; (●) form I, (O) form I′, and (△) form II CO adducts of the indicated protein.

with polar solvent molecules or distal protein residues. A major influence of this kind is H-bonding from a distal residue, which enhances back-donation by compensating the negative charge built up on the O atom of the CO. Distal H-bonding has been demonstrated for form II CO adducts of HRP (Smith et al., 1983) and CCP (Satterlee & Erman, 1984) via ν_{C-O} shifts in D_2O. This interaction is the reason that the data points for form II adducts lie at the high end of the plot in Figure 4. H-Bonding is not present for the alkaline form I′ adducts, which lie much lower on the line, near the point for Mb; the CO adduct is known not to be H-bonded in Mb (Kuriyan et al., 1986).

Data points fall significantly above or below the line in Figure 4 when the axial ligand is a weaker or stronger σ donor than imidazole (Li & Spiro, 1988b). As the σ donation increases, so does the electron density on the Fe. Consequently, back-donation increases, and ν_{C-O} is lowered, but the expected increase in ν_{Fe-CO} is offset by competition between CO and the donor ligand for the Fe σ orbital (d_{z^2}). This competition weakens the Fe–CO bond. For example, negative deviations are shown by CO adducts of cytochrome P-450 enzymes (Uno et al., 1987), which have strongly donating thiolate ligands. Likewise, deprotonation of imidazole bound to a heme–CO adduct (Figure 4) produces a negative deviation, since imidazolate is a stronger σ donor. The negative deviations of the form I adducts are attributed to the imidazolate character of the proximal ligand, as demonstrated by the RR spectroscopy on the Fe(II) protein. Since, however, the data points for the form II and form I′ adducts fall close to the line in Figure 4, it may be inferred that imidazolate character is no longer pronounced in these forms. Either the proximal H-bond is significantly weakened, or the Fe–N(His) bond is actually stretched in the protein. In fact, this bond does appear to be stretched in the recently determined crystal structure of the CO adduct of bakers' yeast CCP (Edwards & Poulos, 1989).

The Fe–N(His) bond weakening in the form II adducts suggests that the steric requirements for distal and proximal H-bonding are not mutually compatible. This structural hypothesis is consistent with the observations that CCP(MI) favors form I, while the Asn-235 mutant, which lacks the proximal H-bond, favors form II (Smulevich et al., 1988). A mixture of forms I and II is obtained in the Phe-191 mutant; apparently, the loss of the Trp-191–Asp-235 H-bond, which anchors the His-175–Asp-235 H-bond, allows the distal CO H-bond to form in a fraction of the molecules. Likewise, a mixture is seen for the His-181 mutant (Smulevich et al., 1989b), probably reflecting the increased mobility of the six-residue loop connecting the proximal His-175 with His-181, which is H-bonded to a heme propionate group (Figure 1). Residue replacements on the distal side also alter the form I/II energetics. Thus, the Phe-51 mutant favors form II (Figure 3); this may result from a small displacement of the entire distal (B) helix seen in the crystal structure (Wang et al., 1989). On the other hand, replacement of Arg-48 with Leu or Lys does not favor form II. Crystal structures are not yet available for these mutants, but it seems likely that the substitutions influence the position of His-52.

Interestingly, a mixture of forms I and II is also seen for bakers' yeast CCP (Smulevich et al., 1986) but not for CCP(MI). The structures of these two proteins are very similar but not identical. The main difference is a slight (0.5 Å) shift of the His-52 residue relative to the heme, resulting from the altered helix interactions associated with the Thr-53 → Ile substitution in CCP(MI) (Wang et al., 1989). The crystal structure of the CO adduct of bakers' yeast CCP shows that His-52 does H-bond to the CO, so this shift could well account for the stability of form II. Actually, the form II stability was found to be marginal (Smulevich et al., 1986) since it was the dominant form in bakers' yeast CCP only when the protein was exposed to CO at 1-atm pressure. At lower pressure, form I was favored. This behavior was interpreted to mean that binding of a second molecule of CO somewhere in the protein is required to tip the energy balance toward form II. Similar behavior was found for HRP (Evangelista-Kirkup et al., 1986).

While the weakening of the proximal linkage in the form II adducts can be understood in terms of the distal H-bond requirements, a more delocalized protein rearrangement is required to explain the proximal weakening in the alkaline form I′ adducts, for which no significant distal interaction is evident. It may be significant that the acid/alkaline transition involves a two-proton titration (Iizuka et al., 1985; Miller et al., 1990). One proton is probably removed from His-52, accounting for the loss of the distal H-bond, and the other may come from His-181. The latter deprotonation might trigger a proximal alteration via the His-181–His-175 loop.

In addition to ν_{Fe-CO} and ν_{C-O}, the Fe–C–O bending mode, δ_{Fe-C-O}, which is absent in RR spectra of unconstrained CO adducts, in which the CO assumes its normal upright position, becomes observable in some heme proteins, including Mb (Tsubaki et al., 1982), as well as for Fe porphyrins having a synthetic "strap" (Yu et al., 1983) or "cap" (G. Ray, X.-Y. Li, J. L. Sessler, and T. G. Spiro, unpublished results) that hinders upright binding sterically. The MbCO crystal structure (Kuriyan et al., 1986) shows that the distal His residue provides steric hindrance to upright binding, and the CO is off axis. Whether the FeCO unit is bent or tilted is not readily determined because of disorder in the crystal, and this question has been a matter of some debate. On the basis of force constant considerations, a concerted distortion model has been proposed, in which the FeCO is mostly tilted, but slightly bent, and the off-axis displacement is also accommodated by a degree of porphyrin buckling (Li & Spiro, 1988b). This model was strikingly borne out by the subsequent high-resolution crystal structure of the CO adduct of a sterically hindered "capped" porphyrin, which showed just these features (Kim et al., 1989). The δ_{Fe-C-O} mode gains RR activity in these constrained systems because the loss of axial symmetry provides an enhancement mechanism which is otherwise absent (Li & Spiro, 1988b). Thus the RR intensity of this mode can be taken as an indication of off-axis binding, although it is possible that other symmetry-lowering perturbations, e.g., asymmetric electrostatic fields in the heme pocket, may play a role. The mode is fairly prominent in form II CCP adducts (Figure 3), consistent with the distal H-bond being off-axis and resulting in a tilted Fe–C–O unit (Edwards & Poulos, 1989). This potential for off-axis H-bonding to the diatomic ligand is undoubtedly relevant to the mechanism of O–O bond heterolysis in the peroxide reaction (Poulos, 1988).

The weakness of the δ_{Fe-C-O} mode in the form I and form I′ adducts suggests minimal off-axis influence in these forms. Thus the residues on the distal side of the heme offer little or no steric hindrance to upright binding of CO, consistent with the open pocket seen on the distal side in the crystal structure (Figure 1). CCP–CO offers an interesting structural contrast with MbCO. The data points of the alkaline forms (I′) of both proteins fall near one another on the line in Figure 4, indicating similar electronic states, but MbCO has a prominent δ_{Fe-C-O} RR band, consistent with the known off-axis geometry. When the distal His residue of Hb, His-64, is protonated, the δ_{FeCO} mode is deactivated, indicating upright binding of the CO (Ramsden & Spiro, 1989). Moreover, the low-pH Mb I data point in Figure 4 moves down the line. This indication of decreased back-bonding probably results from movement of His-64 out of the distal pocket, leaving only nonpolar residues near the CO. [The frequencies are close to those of an imidazole–heme–CO adduct in benzene (Ramsden & Spiro, 1989).] Protonation of the distal histidine does not lead to H-bond formation with the CO, as it does in the CCP form II adduct; if it did, the data point would move up the line in Figure 4. Moreover, the pK_a is strongly depressed, to ~4.2, in MbCO (Ramsden & Spiro, 1989), whereas it is quite normal, ~7.5, in CCP-CO. The difference must lie in the different distal stereochemistry in the two proteins. The distal His residue crowds the CO in Mb and is poorly disposed for H-bonding upon protonation. Consequently, protonation of the distal His is not stabilized by interaction with the CO. In CCP-CO the distal His-52 is further from the heme and can H-bond to the O atom of the CO (Edwards & Poulos, 1989).

Fe Ligation State in Native Protein: A Balancing Act. Subtle interactions between the heme and the residues lining the pocket also influence the coordination state of CCP in the absence of added ligands. The coordination state is conveniently monitored via the frequencies of porphyrin skeletal modes in the 1450–1650-cm^{-1} region of the RR spectrum (Li & Spiro, 1988a). These frequencies vary inversely with the porphyrin core size (Parthasarathi et al., 1987), since they involve stretching of the bonds to the methine bridges, where core size changes are largely accommodated. The core size depends on the ligation state because of the electronic properties of the Fe ions (Li & Spiro, 1988a). When two strong-field ligands are bound to the heme, the Fe valence electrons are maximally paired (low spin) in the nonbonding d_π orbitals, and the bonds to Fe are short, ~2.00 Å. When two weak-field ligands are bound, the Fe electrons occupy the antibonding d_σ orbitals, in order to minimize interelectronic repulsion, and are maximally unpaired (high spin). The metal–ligand bonds are thereby lengthened, and the core size increases, to a radius of ~2.04 Å for Fe^{III} and ~2.06 Å for Fe^{II}. Alternatively, the high-spin heme can be 5-coordinate, in which case the Fe atom moves out of the plane toward the fifth ligand and the core relaxes to an intermediate size. When imidazole is the fifth ligand, as in the globins and the peroxidases, the Fe(II) state is 5-coordinate (5-c) and high spin, unless there is a strong-field sixth ligand, either exogenous or endogenous, which produces a low-spin 6-coordinate (6-c) complex. In the Fe^{III} state the heme can also be 5-c, but it can bind weak-field (e.g., F^-) as well as strong-field ligands. When H_2O or OH^- is bound, the resulting complexes are typically mixed spin, and the spin population is temperature dependent, since these ligands, in combination with imidazole, produce Fe^{III} ligand fields that are close to the spin crossover.

The best core-size marker mode is ν_3, near 1500 cm^{-1}, a spectral region free of overlaps from other modes. ν_3 has characteristic frequencies for 5- and 6-c, high- and low-spin Fe(II) and Fe(III) hemes. For Fe(II) CCP its frequency is characteristic of 5-coordination (Smulevich et al., 1986), as expected since the heme is high spin. At alkaline pH the absorption spectrum shows a clear-cut transition to a low-spin product (Iizuka et al., 1985; Miller et al., 1990), but in several studies the RR spectrum was reported (Hashimoto et al., 1986; Dasgupta et al., 1989; Smulevich et al., 1988) to show a high-spin heme. This discrepancy has recently been traced to a reversible photolysis of the low-spin product by the Raman laser (Smulevich et al., 1989c). The alkaline conversion to the low-spin state requires binding of an endogenous ligand, which is evidently photolabile. The only plausible endogenous ligand is the distal His-52 side chain [an alkaline low-spin complex is still found if Trp-51 is replaced by Phe or Arg-48 is replaced by Leu (Smulevich et al., 1988)], but the resulting bis-imidazole complex would not be expected to be photolabile, on the precedent of model complexes or other heme proteins. Recent picosecond transient absorption measurements have shown, however, that the apparent photostability of the Fe–imidazole bond is due to rapid and efficient recombination of the imidazole following photolysis (Jogenward et al., 1988). Thus, the appearance of a high-spin heme during laser irradiation of alkaline Fe(II) CCP can be explained if the recombination of photodissociated His-52 is less efficient than a normal imidazole. The His-52 is 5.8 Å from the Fe in the crystal structure of Fe(III) protein (Finzel et al., 1986), and its binding to Fe(II) at alkaline pH must involve a substantial reorganization of the protein structure. If the energy required for this reorganization induces a strain in the Fe–His-52 bond, then the inefficient recombination could be readily explained. Such a strain would also provide an explanation for the finding that the rate of CO binding to Fe(II) CCP is unusually fast in the alkaline form, despite its being 6-c (Miller et al., 1990).

In the Fe(III) form of CCP the heme is likewise low spin at alkaline pH, as illustrated in Figure 5 (bottom). The sixth ligand may again be His-52, but it is more likely to be OH^-, which has a much higher affinity for Fe(III) than for Fe(II). OH^- is definitely bound to alkaline Fe(III) HRP as shown by the identification of an Fe–OH stretching band in the RR spectrum (Sitter et al., 1988). At acid pH the Fe(III) heme is high spin and 5-c, in both CCP and HRP. The CCP crystal structure shows a water molecule (Figure 1), W595, above the Fe but far enough away, 2.4 Å [2.7 Å in CCP(MI)], to preclude coordination. At low temperatures both EPR (Yonetani & Anni, 1987) and RR (Smulevich et al., 1989b) spectroscopies show conversion to a low-spin heme, suggesting that the factors inhibiting the binding of W595 to the Fe are overcome when thermal fluctuations are damped out.

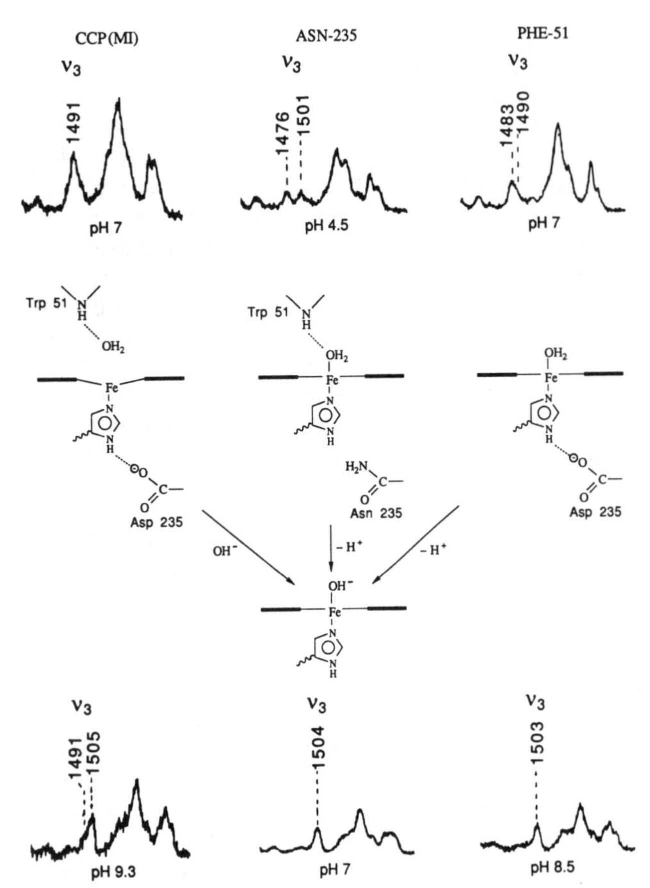

FIGURE 5: High-frequency region of the RR spectra (413.1-nm excitation) of Fe(III) forms of CCP(MI) and the Asn-235 and Phe-51 mutants at acid and alkaline pH (Smulevich et al., 1988a). The frequency of ν_3 is marked, from which the ligation state can be determined. The presumed interaction with a distal water molecule and with H-bonding residues is indicated in the structural diagram.

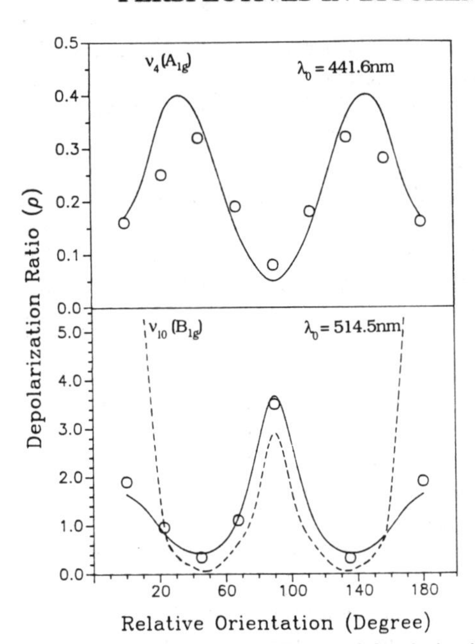

FIGURE 6: Orientation dependence of RR band depolarization ratios for single crystals of CCP mutant proteins. Experimental values (circles) for the ν_4 (A_{1g}) band (1374 cm^{-1}) of the Phe-191 mutant (top) and the ν_{10} (B_{1g}) band (1638 cm^{-1}) of the Asn-235 mutant (bottom) are compared with curves calculated on the basis of the heme orientations, assuming D_{4h} symmetry. Two directions for the x,y axes in the plane were calculated for the B_{1g} mode (the A_{1g} values are independent of the x,y directions), along the N–N directions (solid line) and rotated by 23° (dotted line). This rotation is attributed to the orienting effects of the porphyrin substituents and the protein environment. From Smulevich et al. (1990b).

Studies of CCP mutants indicate that these factors involve interactions on both the distal and proximal side of the heme. The proximal H-bond from His-175 to Asp-235 is an important interaction since it restrains movement on the Fe toward the heme plane and also reduces the affinity for a sixth ligand electronically, by increasing the anionic character of the imidazole. When Asp-235 is replaced by Asn, the heme is completely 6-c (Smulevich et al., 1988a); a high/low-spin mixture is seen (Figure 5). Also, the pK_a for the alkaline low-spin conversion is much lower than it is for CCP(MI) (Figure 5). The H-bond from Trp-191 to Asp-235 also plays a role, since the Phe-191 mutant shows coordination behavior intermediate between those of CCP and the Asn-235 mutant (Smulevich et al., 1988). The Gly-181 mutant also shows this (Smulevich et al., 1989b), indicating a similar role for the H-bond from His-181 to the heme propionate. Both of these H-bonds provide anchor points for the His-175–Asp-235 pair.

On the distal side, W595 is H-bonded by the indole NH proton on Trp-51. It also interacts weakly with His-52 [more strongly in CCP(MI), in which His-52 is slightly closer to the heme (Wang et al., 1989)] and with another water molecule, W596, which is in turn H-bonded by Arg-48. When Trp-51 is replaced by Phe, W595 is pulled slightly farther from the Fe (Wang et al., 1989), which remains 5-c in the crystal, as confirmed by the RR spectrum of the crystal obtained with a microprobe apparatus (Smulevich et al., 1990b). In solution, however, the RR spectrum is characteristic of a mainly high-spin 6-c heme (Figure 5), implying coordination of W595. This crystal-solution difference was traced to the presence of 30% methylpentanediol in the crystallizing medium. Apparently, loss of the restraining Trp-51 H-bond allows W595 to bind to the Fe, but the amphiphilic additive reverses this tendency, perhaps by strengthening the remaining distal H-bond network. At low temperature the effect of the additive is overcome, and high-spin 6-c RR (Smulevich et al., 1990b) and EPR (Yonetani & Anni, 1987) spectra are seen.

At pH values below 5, a high-spin 6-c state is also seen for bakers' yeast CCP (Hashimoto et al., 1986; Smulevich et al., 1986), but this effect has been shown (Yonetani & Anni, 1987; Dasgupta et al., 1989) to result from aging of the protein to a form that is much less active than fresh protein. This form retains its high-spin character at low temperature (10 K), as does the Phe-51 mutant. In both cases the Fe–water bond must be unusually weak, because all other CCP variants show a low-spin ground state. It may be that the aging process involves a conformation change and/or irreversible modification, one of whose effect is to alter the disposition of Trp-51.

Single-crystal RR spectroscopy has also permitted the investigation of electronic effects in the heme pocket. Figure 6 shows the dependence of the depolarization ratio as a function of crystal orientation for RR bands arising from a totally symmetric mode of the Phe-191 mutant and from a nontotally symmetric mode of the Asn-235 mutant (Smulevich et al., 1990b). The data are accurately reproduced (full curves) by calculations using the crystallographically determined orientation of the heme planes and assuming 4-fold symmetry of the porphyrin. From the nontotally symmetric mode data, the electronic axes in the porphyrin plane were found empirically to be rotated by 23° relative to the lines connecting opposite pyrrole N atoms. This effect is attributed to the electronic influences of the porphyrin substituents and of the axial ligands, as well as the electrostatic field of the surrounding residues. This angle shift was different from that found for bakers' yeast CCP, 44° (Smulevich et al., 1990a), reflecting electronic changes produced by the mutation.

Fe=O Bond in the ES Complex. Reaction of CCP with H_2O_2 produces the ES complex, in which the O–O bond has been broken, leaving a Fe^{IV}=O ferryl adduct. The low-temperature crystal structure (Edwards et al., 1987) shows the Arg-48 residue moving into position to form an H-bond with the ferryl O atom, just as it does in the Fe^{III} fluoride adduct (Edwards et al., 1984). This strong H-bond probably accounts for the low Fe=O stretching frequency, 767 cm^{-1} (Hashimoto et al., 1986) or 753 cm^{-1} (Reczek et al., 1989) reported from the RR spectrum. H-Bonding is known to decrease the frequency of oxo–metal bonds substantially (Su et al., 1988). Interestingly, the frequency increases to 782 cm^{-1}, close to that observed for the analogous HRP intermediate, compound II (Hashimoto et al., 1986; Sitter et al., 1985, 1986), when the distal residue Trp-51 is replaced by Phe (J. Terner, J. R. Schifflett, J. M. Mauro, L. A. Fishel, and J. Kraut, unpublished results). In HRP the residue homologous with Trp-51 is also Phe. Thus, it appears that the Trp-51/Phe substitution disrupts the H-bond from Arg-48. For HRP compound II there is a 12-cm^{-1} downshift, 787 → 775 cm^{-1}, between high pH and neutral pH, suggesting a rather weak H-bond from the distal His residue when it is protonated. All of these frequencies are substantially lower than those observed for protein-free ferryl–heme complexes with imidazole complexes, 807–820 cm^{-1} (Schappacher et al., 1986; Kean et al., 1987), an effect attributed (Su et al., 1988) to the imidazolate character of the ligand in HRP and CCP.

Summary. Used in conjunction with site-directed mutagenesis and X-ray crystallography, RR spectroscopy can provide important information about chemical interactions at an enzyme active site. In the case of CCP, the proximal His-175–Asp-235 H-bond has been shown to be very strong by its effect on the Fe–His RR band. Indeed, the frequency and contour of this band strongly suggest that the proton in the H-bond spends about equal time on the carboxylate and imidazolate anions. RR spectroscopy can also monitor the bonding of the heme Fe with water molecules or other sixth ligands, via porphyrin marker bands of the coordination and spin state. The strong influence of the proximal H-bond is again shown by the demonstration that the substitution of Asn for Asp-235 changes the Fe(III) coordination number from 5 to 6. This is consistent with the crystallographic finding that the W595 distal water molecule moves to within bonding distance of the heme, and the RR data show, in addition, that the Fe–water bond is quite strong since the adduct is partially low spin. In contrast, a purely high-spin 6-c heme is found when Phe replaces Trp-51, but addition of an amphiphile (30% methylpentanediol) reestablishes 5-c heme, perhaps by reorganizing the distal H-bond network to facilitate the interaction of W595 with His-52 and Arg-48. Thus, in the case of the Phe-51 mutant RR spectroscopy uncovered a clear-cut difference between crystal and solution structures, which was traceable to the crystallization conditions. The technique also uncovered unusual photolability in the alkaline form of Fe(II) CCP, probably resulting from formation of a strained distal bond between the Fe and His-52 in the acid–alkaline transition. Finally, RR in conjunction with IR spectroscopy has been informative with regard to the interplay of distal and proximal interactions when CO is used as a probe molecule. The status of the Fe–N(His) bond and of the distal H-bond to the bound CO can be monitored simultaneously. The data strongly imply that the distal and proximal H-bonds are not mutually compatible and that alternative structures are formed, forms I and II, depending on which H-bond is dominant. The RR spectra indicate that a distal H-bond is not formed when CO binds to CCP(MI) but it is formed in bakers' yeast CCP when a second CO molecule binds to the protein in saturated CO solution. This difference probably results from the slightly different disposition of the His-52 side chain seen in the crystal structures. The acid–alkaline transition eliminates the distal H-bond and weakens the proximal linkage, producing a common structure for all CCP variants, form I′, whose vibrational signature is much like that of MbCO, except that the inactivity of the δ_{FeCO} mode indicates no restraint to upright binding of CO. In MbCO this mode is activated by off-axis binding, due to steric crowding by the distal His side chain. The vibrational data indicate that when the distal His is protonated in MbCO, it does not H-bond to the CO, as it does in CCP, but swings out of the pocket leaving upright CO in a hydrophobic environment. RR spectroscopy also reveals the strength of the Fe^{II}=O bond in the ES complex via the Fe–O stretching frequency and again shows large effects of both distal and proximal H-bonding.

Hemoglobin: Dynamics of Allostery

Hemoglobin is arguably the most studied of proteins. Along with myoglobin it was the first protein to have its X-ray crystal structure solved, an accomplishment for which Perutz and Kendrew won the Nobel prize in 1962. Subsequently, Monod et al. (1965) published their famous two-state theory to explain the cooperativity of ligand binding to Hb, a subject that had long excited scientific interest, and continues to do so. In this theory one state of the tetrameric protein, T (for "tense"), has a low binding affinity and is the stable form when the protein is unligated, while the other state, R (for "relaxed"), has a high binding affinity and is the stable form when the protein is ligated. The T–R switch can account for cooperative binding. The finding from crystallography that deoxyHb has one arrangement of the four subunits while ligated forms have another (Fermi & Perutz, 1981) lent plausibility to the theory and gave the two thermodynamic states a physical identification with the two quaternary structures. Subsequent research has shown this picture to be incomplete; there are more than two states available to Hb (Viggiano & Ho, 1979; Blough & Hoffman, 1984; Simolo et al., 1985; Smith & Ackers, 1985; Marden et al., 1986). But the two-state model is a very good approximation under normal conditions of ligand binding to native Hb (Shulman et al., 1975), and there is little question that the T–R switch is the central event in the regulation of ligand binding.

The mechanism of this switch is the most interesting question about Hb from a molecular point of view. How is a large-scale rearrangement of the subunits initiated by the simple act of binding ligands to the heme groups? The most fruitful approach to this question is to elucidate the kinetic steps in the R–T transition consequent to photodissociation of heme-bound ligands. Following absorption of a photon by the heme group, the axial ligands dissociate efficiently and in most cases recombine on the picosecond time scale (Jongeward et al., 1988); CO however recombines slowly enough that most of it escapes from the protein (Hofrichter et al., 1983). A deoxy heme absorption spectrum develops immediately (0.35 ps) after photolysis (Martin et al., 1983) and remains essentially unchanged out to nanoseconds (Greene et al., 1978; Sawicki & Gibson, 1976). Small changes in this spectrum have been detected on a longer time scale, yielding relaxation times of ~0.1, ~1, and ~20 μs (Hofrichter et al., 1983). These changes no doubt reflect the influence of protein rearrangements on the heme electronic structure. The 20-μs process coincides with the transition between fast and slow second-order CO binding processes discovered by Gibson and

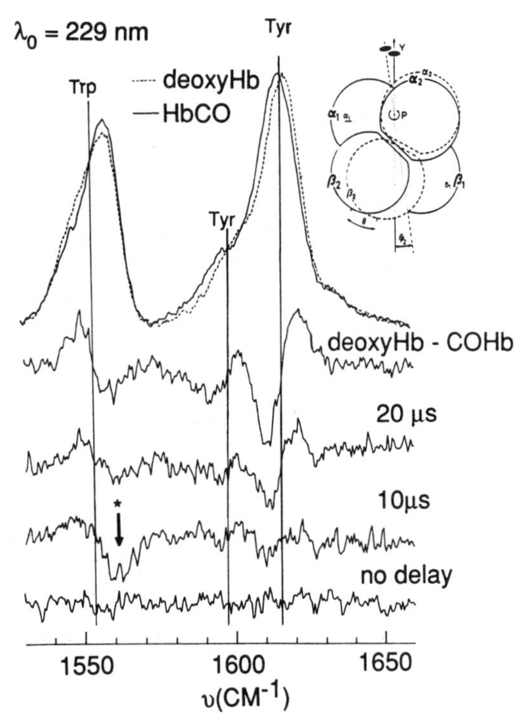

FIGURE 7: UVRR spectra of HbCO and deoxyHb (top) and their difference, compared with difference spectra of the HbCO photoproduct generated with a 532-nm photolysis pulse at the indicated delay prior to a 239-nm probe pulse. The difference signals are attributed to the Tyr-α_{42} and Trp-β_{37} residues at the $\alpha_1\beta_2$ interface, whose H-bonding is altered via the subunit rearrangement (inset) in the R–T transition. From Su et al. (1989).

co-workers (Gibson, 1959; Sawicki & Gibson, 1976) and attributed to the R–T quaternary rearrangement. The earlier processes are then plausibly attributed to tertiary changes on the pathway to the R–T switch. Transient RR spectroscopy has provided insight into the nature of these processes.

Aromatic Residue Changes Monitor the R–T Transition. When the RR spectrum with UV excitation at 229 nm is measured carefully, small shifts are seen between HbO_2 or HbCO, which give identical spectra, and deoxyHb (Su et al., 1989) (Figure 7). The difference spectrum clearly shows a downshift in the 1555-cm^{-1} band, arising from tryptophan residues, and upshifts in the 1615-cm^{-1} band and its 1600-cm^{-1} shoulder, arising from tyrosine residues (Rava & Spiro, 1985). When the UV excitation pulses are overlapped with 532-nm photolyzing pulses in a flowing solution of HbCO, the difference spectrum, relative to unphotolyzed HbCO, is featureless and remains so when the probe pulses are delayed by up to 1 μs (Su et al., 1989). Difference bands begin to appear at 5 μs, and at a 20-μs delay one sees a difference spectrum that closely resembles the static deoxyHb – HbCO difference spectrum (Figure 7). Thus, the aromatic residues are sensing a process with the same timing as the last optical transient of the deoxy heme group, and the transition from fast to slow CO rebinding rates. The perturbation of the aromatic groups is the same as that shown between Hb in the T vs the R quaternary structure.

The Hb tetramer has three inequivalent Trp and six inequivalent Tyr residues, occurring in equivalent pairs. Most of these do not show significant changes in their environments when the HbCO and deoxyHb crystal structures are compared (Fermi & Perutz, 1981). Two of the residues, however, Trp-β_{37} and Tyr-α_{42}, are at the $\alpha_1\beta_2$ subunit interface, where most of the displacements occur between the R and T structures (Baldwin & Chothea, 1979). The indole NH proton of Trp-β_{37} is H-bonded to the backbone carbonyl of Asn-β_{102} in the R state but to the carboxylate side chain of Asp-α_{94} in the T state. The Tyr-α_{42} OH group accepts an H-bond from the peptide NH of Asp-α_{94} in both states, but in the T state it also forms a donor interaction with the side chain of Asp-β_{99}. The UVRR band shifts are plausibly attributed to these H-bonding changes. Thus, the transient UVRR difference spectra provide direct structural evidence that the 20-μs transition is the R–T subunit rearrangement.

Fe–His Stretch Evolves over Several Time Scales. Although the proximal His side-chain is H-bonded to a backbone carbonyl group in both Mb and Hb (Fermi & Perutz, 1981; Baldwin & Chothea, 1979), the Fe–His stretching frequency is at a distinctly lower frequency, 215 vs 220 cm^{-1}, in Hb (Kincaid et al., 1979; Nagai et al., 1980) than in Mb (Kitagawa et al., 1979; Argade et al., 1984). When, however, chemically modified or mutant Hb's are examined, in which the quaternary constraints are relaxed and the R state is stable even in the deoxy form, then the Fe–His frequency is 222 cm^{-1} (Nagai et al., 1980), near that in Mb. Likewise, a 223-cm^{-1} frequency was observed when the R state was isolated kinetically, by photolyzing HbCO in a flowing stream of sample, with a 0.3-μs residence time in the laser beam (Stein et al., 1982), shorter than the R–T relaxation time. Thus the Fe–His frequency decreases in the T state and is linked to the quaternary structure. Since a decreased frequency implies a weakened bond, the RR data provide a definite indication that there really is molecular tension in the T state (Perutz, 1975).

The molecular mechanism responsible for the bond weakening is not entirely clear, however. Weakened H-bonding to the backbone carbonyl group has been considered (Stein et al., 1980), but the behavior of the imidazole proton NMR resonances seems to argue against this interpretation (La Mar & De Ropp, 1982). Alternatively, tilting of the imidazole ring relative to the heme normal, as seen in the deoxyHb crystal structure (Baldwin & Chothea, 1979), has been suggested to weaken the Fe–His bond due to the resulting nonbonded repulsion between the imidazole and the pyrrole ring toward which it bends (Gelin & Karplus, 1977; Friedman et al., 1982; Scott & Friedman, 1984). The observed tilt is greater in Mb [11° (Takano, 1977)] than in Hb [7° (Baldwin & Chothea, 1979)], however, and Fermi et al. (1984) have questioned whether the nonbonded contact is significant in the deoxyHb structure. It has also been argued that rotation of the imidazole relative to the Fe–pyrrole bonds, together with the tilt, can modulate the Fe–His bonding via a repulsive interaction between the Fe d_{z^2} and the porphyrin π orbitals (Bangcharoenparpong et al., 1984). Finally, it cannot be ruled out that a straightforward mechanical force stretches the Fe–His bond, inasmuch as the F-helix, which contains the proximal His residue, is displaced by about 1 Å relative to the heme plane in the R–T transition (Baldwin & Chothea, 1979). Any or all of these factors could, of course, operate simultaneously. Interestingly, the T-state Fe–His frequency shift is much greater for the α than for the β chains, in line with Pertuz' view that the lowering of ligand affinity in the T state is due to proximal constraints in the α chains but to steric hindrance on the distal side in the β chains (Perutz, 1975). Using chain-specific valency hybrids, Nagai and Kitagawa (1980) showed that the R- and T-state frequencies were 222 and 218 cm^{-1} for the β chains but 220 and 205 cm^{-1} for the α chains. Similar results were obtained with Co,Fe hybrids (Ondrias et al., 1982).

While the Fe–His frequency is higher in the R than in the T state, Friedman and co-workers have shown that it is still higher, 230 cm^{-1}, in the HbCO photoproduct generated with

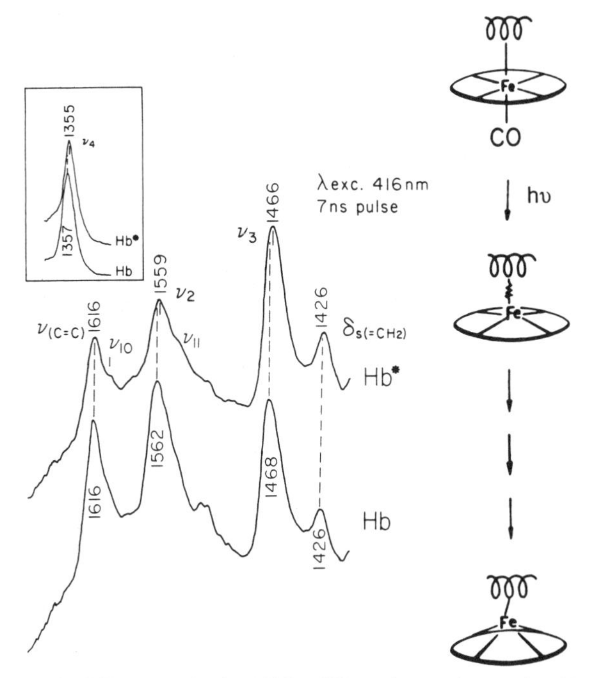

FIGURE 8: RR spectra in the 1400–1650-cm^{-1} core-size marker band region for the HbCO photoproduct (Hb*) generated with 7-ns 416-nm pulses, showing 2–3-cm^{-1} downshifts relative to deoxyHb, of ν_2 and ν_3 and also of ν_4 (inset). The $\nu_{C=C}$ and $\delta_{2(=CH2)}$ vinyl modes are unshifted, however. Schematic structures illustrate the proposed restraint on the Fe out-of-plane motion by the F-helix in Hb*. Adapted from Dasgupta and Spiro (1986).

10-ns (Friedman et al., 1982) or 25-ps (Findsen et al., 1985) laser pulses. Thus the frequency is elevated immediately after photolysis and remains constant for the first few nanoseconds. It evolves toward the R-state frequency on a time scale of about a microsecond, as shown by pulse–probe experiments (Scott & Friedman, 1984). Thus, during the ~1-μs tertiary transient seen in the absorption kinetics, the Fe–His bond relaxes to its R-state character. This behavior has been discussed extensively in terms of the evolution of the imidazole from an upright orientation in HbCO to a tilted one in deoxyHb (Scott & Friedman, 1984). But it seems likely that the frequency elevation immediately after photolysis is at least partly due to protein restraint on the out-of-plane displacement of the Fe atom, as discussed in the next section.

Early Restraint on the Fe Out-of-Plane Displacement. When the porphyrin RR bands of the HbCO photoproduct were examined with 30-ps (Terner et al., 1981) or 7-ns (Dasgupta & Spiro, 1986) laser pulses, the core-size marker band frequencies were found to be 2–3 cm^{-1} lower than those in deoxyHb. As shown in Figure 8, these differences do not arise from a generalized spectral shift, since no shift is seen for the vinyl modes at 1616 and 1426 cm^{-1} but only for the core-size markers ν_2 and ν_3 at 1562 and 1468 cm^{-1} [also ν_{10}, ν_{11}, and ν_{19}, which are brought out with 532-nm excitation (Dasgupta & Spiro, 1986)]. These results imply that the core size is slightly larger than that in the final deoxyHb. If as is generally accepted photolysis occurs via intersystem crossing to a high-spin state of the 6-c heme-CO adduct (Greene et al., 1978; Waleh & Lowe, 1982), then the porphyrin core must initially be substantially expanded. But it relaxes to a somewhat smaller value as the Fe moves out of the plane in the 5-c deoxy heme. The Fe motion may, however, be constrained by protein forces, and full relaxation of the core size can thereby be coupled to protein motions. Detailed consideration of the geometry and RR frequencies of 5-c and 6-c high-spin Fe(II) hemes led to the estimate that the restraint on the out-of-plane displacement in the prompt photoproduct is about 0.1 Å out of the 0.4-Å displacement in deoxyHb (Dasgupta & Spiro, 1986). Dynamics calculations (Henry et al., 1985) indicate very rapid (50–150 fs) motion of the Fe atom out of the plane, due to nonbonded imidazole–pyrrole forces which are uncompensated when the Fe–CO bond breaks (Friedman et al., 1982), but the amplitude of the motion was slightly smaller when the protein was included in the calculation, consistent with the proposed 0.1-Å pause.

The core-size frequencies are fully relaxed to the deoxyHb values when examined within 0.3 μs (Stein et al., 1982), suggesting that the 0.1-μs optical transient is associated with a protein motion allowing full out-of-plane displacement of the Fe. Interestingly, the marker bands of the MbCO photoproduct are also downshifted relative to those of deoxyMb when examined with 30-ps pulses but are fully relaxed when examined with 7-ns pulses; preliminary data suggest a 0.2-ns relaxation time (Dasgupta et al., 1985). Thus, the protein motion reponsible for the core-size relaxation is about 3 orders of magnitude faster for Mb than for Hb. This motion may be associated with the proximal F-helix, which is displaced by 1.0 Å relative to the heme group in deoxyHb vs ligated Hb (Baldwin & Chothea, 1979) but by only 0.1 Å in deoxyMb vs ligated Mb (Baldwin & Chothea, 1979).

If protein forces restrain the Fe out-of-plane displacement, then the Fe–His bond might be compressed at early times, perhaps accounting for at least part of the elevation of the Fe–His frequency in the prompt HbCO photoproduct. Part of the relaxation of this frequency does occur in the first 0.4 μs (Scott & Friedman, 1984), but part of it continues into the microsecond regime, after the core size has relaxed completely. Thus, the first protein motion (0.1 μs) allows full Fe displacement, while further relaxation of the Fe–His bond continues with the second protein motion (1 μs).

Mystery of ν_4. The porphyrin skeletal mode ν_4, a breathing vibration at ~1360 cm^{-1} which involves the pyrrole C–N bonds primarily (Li & Spiro, 1988a), gives rise to the strongest band in heme protein RR spectra when in resonance with the dominant Soret electronic transition near 400 nm. It has been studied extensively, but interpretation is somewhat clouded by its apparent sensitivity to multiple stereoelectronic factors. Thus, its frequency is only weakly sensitive to the core size (Parthasarathi et al., 1987) but strongly dependent on the occupancy of the porphyrin HOMO and LUMO, as seen in the spectra of radical cation (Czernuszewicz et al., 1989) and anion (Atamain et al., 1989) species, and therefore sensitive to the electronic effects of back-bonding (Spiro & Burke, 1976). It is also sensitive to the Fe oxidation state (Spiro & Strekas, 1974; Yamamoto et al., 1973), although this may be a fortuitous result of separate deviations from the frequencies expected on the basis of core size for low- and high-spin Fe(II), the former due to back-bonding and the latter due to doming of the porphyrin ring (Parthasarathi et al., 1987).

Like the core-size markers, ν_4 is 2 cm^{-1} lower in the HbCO photoproduct monitored with 7-ns laser pulses than it is in deoxyHb, but it does not relax within 0.3 μs, as the core size sensitive ν_2 does (Stein et al., 1982). Rather, the ν_4 downshift appears to relax in concert with the 1- and 20-μs optical transients (Friedman & Lyons, 1980), just as the Fe–His frequency does. Indeed, the Fe–His frequency correlates accurately with ν_4 for a variety of deoxyHb's, in both R and T states (Rousseau & Ondrias, 1983). Thus within the molecular architecture of Hb (the correlation does not extend to other heme proteins, such as the peroxidases), the factors that determine the Fe–His frequency also determine the ν_4 frequency. Whether this effect is due to the electron density on the Fe atom, as modulated by the strength of the Fe–His bond, or to some correlated parameter of the porphyrin ring, e.g., doming, is uncertain.

Recently, Petrich et al. (1987) have set a timing record for transient RR spectroscopy by monitoring ν_4 of the HbCO photoproduct with 0.2-ps pump and probe pulses. Although the pump pulse had a large frequency width, 35 cm^{-1}, they deconvoluted the RR band from the instrument response to obtain 2–11-cm^{-1} frequency differences relative to deoxyHb. The largest difference was seen at the earliest delay, 0.9 ps. The difference quickly fell to 2 cm^{-1}, at a 10-ps delay, and then rose again to 4 cm^{-1} at 95 ps. The authors attributed the first phase to vibrational cooling of the temperature rise associated with the excess energy of the photolysis pulse and the second phase to a structural relaxation, perhaps associated with the Fe–His bond. Dynamical calculations do indicate that local heating from the photolysis pulse is significant on the scale of 10 ps (Henry et al., 1985), but it is not clear that an 11-cm^{-1} shift can be produced by this effect since ν_4 is known to vary only 3 cm^{-1} in deoxyHb between 4 and 300 K (Rousseau & Friedman, 1988); the temperature dependence of the ν_{10} band of NiOEP (Asher & Murtaugh, 1983), which was used to rationalize the result, is not an appropriate reference, since it stems from porphyrin ruffling dynamics, which are specific to the Ni porphyrin (Czernuszewicz et al., 1989b). It is also hard to understand how the frequency upshift between 10 and 95 ps could be caused by Fe–His bond relaxation. The Fe–His frequency itself shows no change between 25 ps and 7 ns (Findsen et al., 1985). Thus, these intriguing results suggest the need for further experiments with very short pulses, monitoring other modes besides ν_4, including, if possible, the Fe–His mode itself.

Summary. Because vibrational frequencies are sensitive to structure, RR spectroscopy can provide structural information about kinetic steps in protein transformations when carried out in a time-resolved mode. UVRR spectroscopy has shown that the aromatic groups of the HbCO photoproduct respond with a delay of 20 μs and has provided direct structural evidence that the 20-μs kinetic step is the R–T quaternary rearrangement of the subunits. RR bands of the porphyrin ring show that the core relaxes via a 0.1-μs protein motion, which probably allows the Fe atom to attain its full out-of plane displacement. The Fe–His stretching frequency has an elevated value immediately after CO photolysis, in part, perhaps, because of the protein constraint on the Fe displacement. It relaxes on both the 0.1- and 1-μs time scales to its value in R-state Hb and then decreases further to its T-state value. These changes may be connected with reorientation of the proximal His side chain. At very early times after a photolysis pulse, heating effects may be an important aspect of the protein dynamics, but further experiments are needed to understand the RR response.

Registry No. CCP, 9029-53-2.

REFERENCES

Argade, P. V., Sassaroli, M., Rousseau, D. L., Inubushi, T., Ikeda-Saito, M., & Lapidot, A. (1984) *J. Am. Chem. Soc. 106*, 6593–6596.

Asher, S. A., & Murtaugh, J. (1983) *J. Am. Chem. Soc. 105*, 7244–7251.

Asher, S. A., Johnson, C. R., & Murtaugh, J. (1983) *Rev. Sci. Instrum. 54*, 1657.

Atamain, M., Donahoe, R. J., Lindsay, J. S., & Bocian, D. F. (1989) *J. Phys. Chem. 93*, 2236.

Baldwin, J. M., & Chothea, C. J. (1979) *J. Mol. Biol. 129*, 175–201.

Bancharoenpaurpong, O., Shoemacker, K. T., & Champion, P. M. (1984) *J. Am. Chem. Soc. 106*, 5688.

Beinert, H., & Thomson, A. J. (1983) *Arch. Biochem. Biophys. 222*, 333.

Blough, N. V., & Hoffman, B. M. (1984) *Biochemistry 23*, 2875–2882.

Carey, P. R. (1988) in *Biological Applications of Raman Spectroscopy* (Spiro, T. G., Ed.) Vol. 2, Chapter 6, Wiley, New York.

Cassatt, J. C., Marini, C. P., & Bender, J. (1975) *Biochemistry 14*, 5470.

Champion, P. M. (1988) in *Biological Applications of Raman Spectroscopy* (Spiro, T. G., Ed.) Vol. 3, pp 249–992, Wiley, New York.

Chen, D., Yue, K.-T., Martin, C., Rhee, K. W., Sloan, D., & Callender, R. (1987) *Biochemistry 26*, 4776.

Chernoff, D. A., Hochstrasser, R. M., & Steele, A. W. (1980) *Proc. Natl. Acad. Sci. U.S.A. 77*, 5606.

Collman, J. P., & Reed, C. A. (1973) *J. Am. Chem. Soc. 95*, 2068.

Czernuszewicz, R. S., Macor, K. A., Kincaid, J. R., & Spiro, T. G. (1989a) *J. Am. Chem. Soc. 111*, 3860.

Czernuszewicz, R. S., Li, X.-Y., & Spiro, T. G. (1989b) *J. Am. Chem. Soc. 111*, 7024–7031.

Dasgupta, S., & Spiro, T. G. (1986) *Biochemistry 25*, 5941–5948.

Dasgupta, S., Spiro, T. G., Johnson, C. K., Delickas, G. A., & Hochstrasser, R. M. (1985) *Biochemistry 24*, 5295–5297.

Dasgupta, S., Rousseau, D., Anni, H., & Yonetani, T. (1989) *J. Biol. Chem. 264*, 654.

Derguini, F., Dunn, D., Eisenstein, L., Nakanishi, K., Odashima, K., Rao, V. J., Sastry, L., & Termini, J. (1986) *Pure Appl. Chem. 58*, 719.

Edwards, S. L., & Poulos, T. L. (1990) *J. Biol. Chem. 265*, 2588.

Edwards, S. L., Poulos, T. L., & Kraut, J. (1984) *J. Biol. Chem. 259*, 12984.

Edwards, S. L., Yuong, Ng. H., Hamlin, R. C., & Kraut, J. (1987) *Biochemistry 26*, 1503.

Evangelista-Kirkup, R., Smulevich, G., & Spiro, T. G. (1986) *Biochemistry 25*, 4420.

Fermi, G., & Perutz, M. F. (1981) *Atlas of Molecular Structures in Biology, to Hemoglobin and Myoglobin*, Clarendon Press, Oxford.

Fermi, G., Perutz, M. F., Shanan, B., & Foerm, R. (1984) *J. Mol. Biol. 175*, 159.

Findsen, W., Friedman, J. M., Ondrias, M. R., & Simon, S. R. (1985) *Science 229*, 661–665.

Finzel, B. C., Poulos, T. L., & Kraut, J. (1984) *J. Biol. Chem. 259*, 13027.

Fodor, S. P. A., Rava, R. P., Copeland, R. A., & Spiro, T. G. (1986) *J. Raman Spectrosc. 17*, 471.

Friedman, J. M., & Lyons, K. B. (1980) *Nature 284*, 570.

Friedman, J. M., Rousseau, D. L., Ondrias, M. R., & Stepnoski, R. A. (1982) *Science 218*, 1244.

Gelin, B. R., & Karplus, M. (1977) *Proc. Natl. Acad. Sci. U.S.A. 74*, 801.

George, S. J., Richards, A. J. M., Thomson, A. J., & Yates, M. G. (1984) *Biochem. J. 224*, 247.

Ghosh, D., O'Donnell, S., Furey, W., Jr., Robbins, A. H., & Stout, C. D. (1982) *J. Mol. Biol. 158*, 73.

Gibson, Q. H. (1959) *Biochem. J. 71*, 293–303.

Goodin, D. B., Mauk, A. G., & Smith, H. (1986) *Proc. Natl. Acad. Sci. U.S.A. 83*, 1295.

Greene, B. I., Hochstrasser, R. M., Weisman, R. B., & Eaton, W. A. (1978) *Proc. Natl. Acad. Sci. U.S.A. 75*, 5255.

Hashimoto, S., Tatsuno, Y., & Kitagawa, T. (1986) *Proc. Natl. Acad. Sci. U.S.A. 83*, 2417.

Henry, E. R., Levitt, M., & Eaton, W. A. (1985) *Proc. Natl. Acad. Sci. U.S.A. 82*, 2034–2038.

Hofrichter, J., Sommer, J. H., Henry, E. R., & Eaton, W. A. (1983) *Proc. Natl. Acad. Sci. U.S.A. 80*, 2235.

Hudson, B. S., & Mayne, L. C. (1988) in *Biological Applications of Raman Spectroscopy* (Spiro, T. G., Ed.) Vol. 2, Chapter 4, Wiley, New York.

Iizuka, T., Makino, R., Ishimura, Y., & Yonetani, T. (1985) *J. Biol. Chem. 260*, 1407.

Jogenward, K. A., Magde, D., Taube, D. J., & Traylor, T. G. (1988) *J. Biol. Chem. 263*, 6027.

Johnson, M. K., Czernuszewicz, R. S., Spiro, T. G., Fee, J. A., & Sweeney, W. V. (1983) *J. Am. Chem. Soc. 105*, 6671.

Jones, C. M., Devito, V. L., Harmon, P. A., & Asher, S. A. (1987) *Appl. Spectrosc. 41*, 1268.

Jongeward, K. A., Magde, D., Taube, D. J., & Traylor, T. G. (1988) *J. Biol. Chem. 263*, 6027.

Kean, R. T., Oertling, W. A., & Babcock, G. T. (1987) *J. Am. Chem. Soc. 109*, 2185–2187.

Kerr, E. R., & Yu, N.-T. (1988) in *Biological Applications of Raman Spectroscopy* (Spiro, T. G., Ed.) Vol. 3, pp 39–96, Wiley, New York.

Kim, K., Fettinger, J., Sessler, J. L., Cyr, M., Hugdahl, J., Collman, J. P., & Ibers, J. A. (1989) *J. Am. Chem. Soc. 111*, 403.

Kincaid, J., Stein, P., & Spiro, T. G. (1979) *Proc. Natl. Acad. Sci. U.S.A. 76*, 4156.

Kitagawa, T. (1988) in *Biological Applications of Raman Spectroscopy* (Spiro, T. G., Ed.) Vol. 3, pp 97–132, Wiley, New York.

Kitagawa, T., Nagai, K., & Tsubaki, M. (1979) *FEBS Lett. 104*, 376.

Kuriyan, J., Wilz, S., Karplus, M., & Petsko, G. A. (1986) *J. Mol. Biol. 192*, 133.

La Mar, G. N., & de Ropp, J. S. (1979) *Biochem. Biophys. Res. Commun. 90*, 36.

La Mar, G. N., & de Ropp, J. S. (1982) *J. Am. Chem. Soc. 104*, 5203–5206.

Li, X.-Y., & Spiro, T. G. (1988a) in *Biological Applications of Raman Spectroscopy* (Spiro, T. G., Ed.) Vol. 3, Chapter 1, Wiley, New York.

Li, X.-Y., & Spiro, T. G. (1988b) *J. Am. Chem. Soc. 105*, 7781.

Marden, M. C., Hazard, E. S., & Gibson, Q. H. (1986) *Biochemistry 25*, 7591–7596.

Martin, J. L., Migus, A., Poyart, C., LeCarpentier, T., Astier, R., & Antosetti, A. (1983) *Proc. Natl. Acad. Sci. U.S.A. 80*, 173.

Mauro, J. M., Miller, M. A., Edwards, S. L., Wang, J., Fishel, L. A., & Kraut, J. (1989) *Met. Ions Biol. Syst. 25*, 477–503.

Miller, M. A., Coletta, M., Mauro, J. M., Putnam, L. D., Farnum, M. F., Kraut, J., & Traylor, T. G. (1990) *Biochemistry 29*, 1777.

Monod, J., Wyman, J., & Changeux, J. P. (1965) *J. Mol. Biol. 12*, 88.

Nagai, K., & Kitagawa, T. (1980) *Proc. Natl. Acad. Sci. U.S.A. 77*, 2033–2037.

Nagai, K., Kitagawa, T., & Morimoto, H. (1980) *J. Mol. Biol. 136*, 271–289.

Ondrias, M. R., Rousseau, D. L., Kitagawa, T., Ikeda-Saito, M., Inobushi, T., & Yonetani, T. (1982) *J. Biol. Chem. 257*, 8766–8770.

Parthasarathi, N., Hansen, C., Yamaguchi, S., & Spiro, T. G. (1987) *J. Am. Chem. Soc. 109*, 3865.

Perutz, M. F. (1975) *Br. Med. Bull. 32*, 195–207.

Perutz, M. F., Fermi, G., Luisi, B., Shaanan, B., & Liddington, R. C. (1987) *Acc. Chem. Res. 20*, 309.

Petrich, J. W., Martin, J. L., Houda, D., Poyart, C., & Orszag, A. (1987) *Biochemistry 26*, 7914–7923.

Poulos, T. L. (1988) *Adv. Inorg. Biochem. 7*, 1–36.

Poulos, T. L., & Kraut, J. (1980) *J. Biol. Chem. 255*, 8199.

Ramsden, J., & Spiro, T. G. (1989) *Biochemistry 28*, 3125.

Rava, R. P., & Spiro, T. G. (1985) *J. Phys. Chem. 89*, 1856.

Ray, G., Li, X.-Y., Sessler, J. L., & Spiro, T. G. (1989) (in preparation).

Reczek, C. M., Sitter, A. J., & Terner, J. (1989) *J. Mol. Struct.* (in press).

Rousseau, D. L., & Ondrias, M. R. (1983) *Annu. Rev. Biophys. Bioeng. 12*, 357–380.

Rousseau, D. L., & Friedman, J. M. (1988) in *Biological Applications of Raman Spectroscopy* (Spiro, T. G., Ed.) Vol. 3, pp 172–180, Wiley, New York.

Satterlee, I. D., & Erman, J. E. (1984) *J. Am. Chem. Soc, 106*, 1139.

Sawicki, C., & Gibson, Q. H. (1976) *J. Biol. Chem. 251*, 1533.

Schappacher, M., Chottard, G., & Weiss, R. (1986) *J. Chem. Soc., Chem. Commun.*, 93–94.

Scott, T. W., & Friedman, J. M. (1984) *J. Am. Chem. Soc. 106*, 5677.

Shulman, R. G., Hopfield, J. J., & Ogawa, S. (1975) *Q. Rev. Biophys. 8*, 325–420.

Simolo, K., Stucky, G., Chen, S., Bailey, M., Scholes, C., & McLendon, G. (1985) *J. Am. Chem. Soc. 107*, 2865–2872.

Sitter, A. J., Shiffett, J. R., & Terner, J. (1988) *International Conference on Raman Spectroscopy* (Clark, R. J. H., & Long, D. A., Eds.) pp 659–660, Wiley, London.

Sivaraja, M., Goodin, D. B., Mauk, A. G., Smith, M., & Hoffman, B. M. (1989) *Science 245*, 738.

Smith, F. R., & Ackers, G. K. (1985) *Proc. Natl. Acad. Sci. U.S.A. 82*, 5347–5351.

Smith, M. L., Ohlsson, P.-J., & Paul, K. G. (1983) *FEBS Lett. 163*, 303.

Smulevich, G., Evangelista-Kirkup, R., English, A. M., & Spiro, T. G. (1986) *Biochemistry 25*, 4426.

Smulevich, G., Mauro, J. M., Fishel, L. A., English, A. M., Kraut, J., & Spiro, T. G. (1988a) *Biochemistry 27*, 5477.

Smulevich, G., Mauro, J. M., Fishel, L. A., English, A. M. Kraut, J., & Spiro, T. G. (1988b) *Biochemistry 27*, 5486.

Smulevich, G., Mantini, A. R., English, A. M., & Mauro, J. M. (1989a) *Biochemistry 28*, 5058.

Smulevich, G., Miller, M. A., & Spiro, T. G. (1989b) (in preparation).

Smulevich, G., Miller, M. A., Gosztola, D., & Spiro, T. G. (1989c) *Biochemistry 28*, 9905.

Smulevich, G., Wang, Y., Edwards, S. L., Poulos, T. L., English, A. M., & Spiro, T. G. (1990a) *Biochemistry 29*, 2586.

Smulevich, G., Wang, Y., Mauro, J. M., Wang, J., Fishel, L. A., Kraut, J., & Spiro, T. G. (1990b) *Biochemistry* (in press).

Spiro, T. G., Ed. (1988) *Biological Applications of Raman Spectroscopy*, Vols. 1–3, Wiley, New York.

Spiro, T. G., & Strekas, T. C. (1974) *J. Am. Chem. Soc. 96*, 338.

Spiro, T. G., & Burke, J. M. (1976) *J. Am. Chem. Soc. 98*, 5482–5489.

Spiro, T. G., & Stein, P. (1977) *Annu. Rev. Phys. Chem. 28*, 501.

Spiro, T. G., & Li, X.-Y. (1988) in *Biological Applications of Raman Spectroscopy* (Spiro, T. G., Ed.) Vol. 3, pp 1–38, Wiley, New York.

Stein, P., Mitchell, M., & Spiro, T. G. (1980) *J. Am. Chem. Soc. 102*, 7795.

Stein, P., Terner, J., & Spiro, T. G. (1982) *J. Phys. Chem. 86*, 168–170.

Stout, C. D. (1988) *J. Biol. Chem. 263*, 9256.

Stout, G. H., Turley, S., Sieker, L. C., & Jensen, L. H. (1988) *Proc. Natl. Acad. Sci. U.S.A. 85*, 1020.

Su, C., Park, Y. D., Liu, G.-Y., & Spiro, T. G. (1989) *J. Am. Chem. Soc. 111*, 3457.

Takano, T. (1977) *J. Mol. Biol. 110*, 569–584.

Teraoka, J., & Kitagawa, T. (1981) *J. Biol. Chem. 256*, 3969.

Teraoka, J., Job, D., Morita, Y., & Kitagawa, T. (1983) *Biochim. Biophys. Acta 747*, 10.

Terner, J., Stong, J. D., Spiro, T. G., Nagumo, M., Nicol, M. S., & El-Sayed, M. A. (1981) *Proc. Natl. Acad. Sci. U.S.A. 78*, 1313–1317.

Terner, J., Schifflett, J. R., Mauro, J. M., Fishel, L. A., & Kraut, J. (1989) (in preparation).

Tsubaki, M., Srivastava, R. B., & Yu, N.-T. (1982) *Biochemistry 21*, 1132.

Tsuboi, M., Nishimura, Y., Hirakawa, A. Y., & Peticolas, W. (1988) in *Biological Applications of Raman Spectroscopy* (Spiro, T. G., Ed.) Vol. 2, Chapter 3, Wiley, New York.

Uno, T., Nishimura, Y., Tsuboi, M., Makino, R., Iizuka, T., & Ishimura, Y. (1987) *J. Biol. Chem. 262*, 4549.

Viggiano, G., & Ho, C. (1979) *Proc. Natl. Acad. Sci. U.S.A. 76*, 3673–3677.

Waleh, A., & Lowe, G. H. (1982) *J. Am. Chem. Soc. 104*, 2346.

Wang, J., Mauro, M. J., Fishel, L. A., Edwards, S. L., Oatley, S. J., Yuong, Ng. H., & Kraut, J. (1989) *Biochemistry* (submitted for publication).

Yamamoto, T., Palmer, G., Gill, D., Salmeen, I. T., & Remi, L. (1973) *J. Biol. Chem. 248*, 5211.

Yonetani, T. (1976) *Enzymes (3rd Ed.) 13*, 345.

Yonetani, T., & Anni, H. (1987) *J. Biol. Chem. 262*, 9547.

Yu, N.-T., Kerr, E. A., Ward, B., & Chang, C. K. (1983) *Biochemistry 22*, 4534.

Ziegler, L. D., & Hudson, B. S. (1983) *J. Chem. Phys. 79*, 1139.

ENZYME STRUCTURE AND MECHANISM

Chapter 9

Human Leukocyte and Porcine Pancreatic Elastase: X-ray Crystal Structures, Mechanism, Substrate Specificity, and Mechanism-Based Inhibitors[†,‡]

Wolfram Bode,[§] Edgar Meyer, Jr.,[‖] and James C. Powers*

Max-Planck-Institut für Biochemie, D-8033 Martinsried, West Germany, Department of Biochemistry, Texas A&M University, College Station, Texas 77843, and School of Chemistry, Georgia Institute of Technology, Atlanta, Georgia 30332

Received November 2, 1988; Revised Manuscript Received December 19, 1988

The serine protease family of enzymes is one of the most widely studied group of enzymes, as evidenced by the fact that more crystal structures are available for individuals of this superfamily than for any other homologous group of enzymes. These enzymes contain a conserved triad of catalytic residues including Ser-195, His-57, and Asp-102. The active-site serine is very nucleophilic, and serine proteases are inhibited by specific serine protease reagents such as diisopropyl phosphorofluoridate (DFP), phenylmethanesulfonyl fluoride, and 3,4-dichloroisocoumarin (Harper et al., 1985). The structure, chemistry, and biochemistry of serine proteases are discussed in recent reviews (Bieth, 1986; Bode & Huber, 1986; Kraut, 1977; Neurath, 1986; Powers & Harper, 1986).

Elastases are a group of proteases that possess the ability to cleave the important connective tissue protein elastin (Bieth, 1986; Werb et al., 1982). Elastin has the unique property of elastic recoil, is widely distributed in vertebrate tissue, and is particularly abundant in the lungs, arteries, skin, and ligaments. This flexible protein is highly cross-linked with unusual amino acid residues such as desmosine and isodesmosine, which contain pyridinium rings, it is rich in amino acids with small side chains (Ala, Ser, Val) and is poor in aromatic or basic amino acids. A wide variety of proteases possess the ability to cleave elastin including thiol proteases such as papain, metalloproteases such as *Pseudomonas aeruginosa* elastase and related enzymes secreted from a variety of virulent pathogenic organisms (e.g., *Schistosoma mansoni*), and many important serine proteases.

Human neutrophil elastase and pancreatic elastase are two major serine proteases that cleave elastin. Neutrophil elastase is found in the dense granules of polymorphonuclear leukocytes and is essential for phagocytosis and defense against infection by invading microorganisms. Pancreatic elastase is stored as an inactive zymogen in the pancreas and is secreted into the intestines where it becomes activated by trypsin and then participates in digestion. Both elastases cleave substrates at peptide bonds where the P_1 residue is an amino acid residue with a small alkyl side chain.[1] Although PP[2] and HL elastase cleave elastin, elastin is neither their only substrate nor necessarily their most important physiological substrate. In particular, the powerful proteolytic activity of neutrophil elastase is essential for migration of neutrophils through connective tissue and for the destruction of foreign bacterial invaders which do not contain elastin.

Elastases can be extremely destructive if not controlled because they can destroy many connective tissue proteins. Under normal physiological conditions these proteases are carefully regulated by compartmentalization or by natural circulating plasma protease inhibitors. Any elastase that reaches the circulation is quickly complexed by the natural inhibitors α_1-protease inhibitor (α_1-antitrypsin) and α_2-

† Supported by grants from the Deutsche Forschungsgemeinschaft to W.B. (SFB 207/H-1), from the Robert A. Welch Foundation (A328) to E.M., and from the National Institutes of Health to J.C.P. (HL29307 and HL34035).

‡ This perspective is dedicated to the memory of Aaron Janoff, who did so much to elucidate the biological function of human leukocyte elastase.

* Address correspondence to this author at Georgia Institute of Technology.

§ Max-Planck-Institut für Biochemie.

‖ Texas A&M University.

[1] The nomenclature of Schechter and Berger (1967) is used to designate the individual amino acid residues (P_2, P_1, P_1', P_2', etc.) of a peptide substrate and the corresponding subsites (S_2, S_1, S_1', S_2', etc.) of the enzyme. The scissile bond is the P_1–P_1' peptide bond. Amino acid residues of the turkey ovomucoid inhibitor third domain (TOM) residues are labeled with I.

[2] Abbreviations: Ahe, 2-aminohexanoic acid (Nle, norleucine); Ape, 2-aminopentanoic acid (Nva, norvaline); HL, human leukocyte; HLE, human leukocyte (neutrophil) elastase; MeLeu, *N*-methylleucine; MeOSuc, methoxysuccinyl; PP, porcine pancreatic; PPE, porcine pancreatic elastase; rms, root mean square; Boc, *tert*-butyloxycarbonyl; TOM, turkey ovomucoid inhibitor third domain; Tos, *p*-toluenesulfonyl.

Table I: Sequence Alignment of Human Leukocyte Elastase (HLE) and Porcine Pancreatic Elastase (PPE) Based on Topological Criteria[a]

	16	17	18	19	20	21	22	23	24	25	26	27	28	29	30
HLE	ILE	VAL	GLY	GLY	ARG	ARG	ALA	ARG	PRO	HIS	ALA	TRP	PRO	PHE	MET
PPE	VAL	VAL	GLY	GLY	THR	GLU	ALA	GLN	ARG	ASN	SER	TRP	PRO	SER	GLN
	31	32	33	34	35	36	36A	36B	36C	37	38	39	40	41	42
HLE	VAL	SER	LEU	GLN	LEU	ARG	-	-	-	-	GLY	GLY	HIS	PHE	CYS
PPE	ILE	SER	LEU	GLN	TYR	ARG	SER	GLY	SER	SER	TRP	ALA	HIS	THR	CYS
	43	44	45	46	47	48	49	50	51	52	53	54	55	56	57
HLE	GLY	ALA	THR	LEU	ILE	ALA	PRO	ASN	PHE	VAL	MET	SER	ALA	ALA	HIS
PPE	GLY	GLY	THR	LEU	ILE	ARG	GLN	ASN	TRP	VAL	MET	THR	ALA	ALA	HIS
	58	59	60	61	62	63	63A	63B	63C	64	65	65A	66	67	68
HLE	CYS	VAL	ALA	ASN	VAL	ASN	VAL	ARG	ALA	VAL	ARG	-	VAL	VAL	LEU
PPE	CYS	VAL	ASP	ARG	GLU	LEU	-	-	-	THR	PHE	ARG	VAL	VAL	VAL
	69	70	71	72	73	74	75	76	77	78	79	80	81	82	83
HLE	GLY	ALA	HIS	ASN	LEU	SER	ARG	ARG	GLU	PRO	THR	ARG	GLN	VAL	PHE
PPE	GLY	GLU	HIS	ASN	LEU	ASN	GLN	ASN	ASN	GLY	THR	GLU	GLN	TYR	VAL
	84	85	86	87	88	89	90	91	92	93	94	95	96	97	99
HLE	ALA	VAL	GLN	ARG	ILE	PHE	GLU	-	ASN	GLY	TYR	ASP	PRO	VAL	ASN
PPE	GLY	VAL	GLN	LYS	ILE	VAL	VAL	HIS	PRO	TYR	TRP	ASN	THR	ASP	ASP
	99A	99B	100	101	102	103	104	105	106	107	108	109	110	111	112
HLE	-	-	LEU	ASN	ASP	ILE	VAL	ILE	LEU	GLN	LEU	ASN	GLY	SER	ALA
PPE	ALA	ALA	GLY	TYR	ASP	ILE	ALA	LEU	LEU	ARG	LEU	ALA	GLN	SER	VAL
	113	114	115	116	117	118	119	120	121	122	123	124	125	126	127
HLE	THR	ILE	ASN	ALA	ASN	VAL	GLN	VAL	ALA	GLN	LEU	PRO	ALA	GLN	GLY
PPE	THR	LEU	ASN	SER	TYR	VAL	GLN	LEU	GLY	VAL	LEU	PRO	ARG	ALA	GLY
	128	129	130	131	132	133	134	135	136	137	138	139	140	141	142
HLE	ARG	ARG	LEU	GLY	ASN	GLY	VAL	GLN	CYS	LEU	ALA	MET	GLY	TRP	GLY
PPE	THR	ILE	LEU	ALA	ASN	ASN	SER	PRO	CYS	TYR	ILE	THR	GLY	TRP	GLY
	143	144	145	146	147	148	150	151	152	153	154	155	156	157	158
HLE	LEU	LEU	GLY	ARG	ASN	ARG	GLY	ILE	ALA	SER	VAL	LEU	GLN	GLU	LEU
PPE	LEU	THR	ARG	THR	ASN	GLY	GLN	LEU	ALA	GLN	THR	LEU	GLN	GLN	ALA
	159	160	161	162	163	164	165	166	167	168	169	170	170A	170B	171
HLE	ASN	VAL	THR	VAL	VAL	THR	SER	LEU	-	CYS	-	-	-	-	-
PPE	TYR	LEU	PRO	THR	VAL	ASP	TYR	ALA	ILE	CYS	SER	SER	SER	SER	TYR
	172	173	174	175	176	177	178	179	180	181	182	183	184	185	186
HLE	-	-	-	-	-	ARG	ARG	SER	ASN	VAL	CYS	THR	LEU	VAL	ARG
PPE	TRP	GLY	SER	THR	VAL	LYS	ASN	SER	MET	VAL	CYS	ALA	GLY	GLY	ASP
	186A	186B	187	188	188A	189	190	191	192	193	194	195	196	197	198
HLE	GLY	ARG	GLN	ALA	-	GLY	VAL	CYS	PHE	GLY	ASP	SER	GLY	SER	PRO
PPE	-	-	GLY	VAL	ARG	SER	GLY	CYS	GLN	GLY	ASP	SER	GLY	GLY	PRO
	199	200	201	202	203	204	205	206	207	208	209	210	211	212	213
HLE	LEU	VAL	CYS	ASN	-	-	-	-	GLY	LEU	ILE	HIS	GLY	ILE	ALA
PPE	LEU	HIS	CYS	LEU	VAL	ASN	GLY	GLN	TYR	ALA	VAL	HIS	GLY	VAL	THR
	214	215	216	217	217A	218	219	220	220A	221	222	223	224	225	226
HLE	SER	PHE	VAL	ARG	-	GLY	GLY	CYS	ALA	SER	GLY	LEU	TYR	PRO	ASP
PPE	SER	PHE	VAL	SER	ARG	LEU	GLY	CYS	ASN	VAL	THR	ARG	LYS	PRO	THR
	227	228	229	230	231	232	233	234	235	236	237	238	239	240	241
HLE	ALA	PHE	ALA	PRO	VAL	ALA	GLN	PHE	VAL	ASN	TRP	ILE	ASP	SER	ILE
PPE	VAL	PHE	THR	ARG	VAL	SER	ALA	TYR	ILE	SER	TRP	ILE	ASN	ASN	VAL
	242	243	244	245											
HLE	ILE	GLN	-	-											
PPE	ILE	ALA	SER	ASN											

[a] Chymotrypsinogen numbering.

macroglobulin. The complexes are cleared from the plasma by the liver and/or macrophages and are degraded. When an imbalance occurs due to a deficiency of effective α_1-protease inhibitor or abnormally high levels of elastases, severe permanent tissue damage may occur. Pancreatic elastase participates in the usually fatal disease pancreatitis, which occurs when pancreatic zymogens are activated and released into the circulation. Neutrophil elastase has been linked to pulmonary emphysema, acute respiratory distress syndrome, shocked lung, glomerulonephritis, rheumatoid arthritis, and other inflammatory disorders.

ELASTASE STRUCTURES

Sequences. The primary structures of porcine pancreatic elastase (PPE) and human leukocyte elastase (HLE) are shown in Table I. The alignment is based on tertiary structure similarities and follows the common practice of using the bovine chymotrypsinogen A numbering (Hartley, 1964). PPE is a single peptide chain of 240 amino acids beginning with Val-16 and terminating with Asn-245. It contains four disulfide bridges and no carbohydrate (Shotton & Hartley, 1970).

The sequence of HLE was established by a combination of peptide sequencing (Sinha et al., 1987) and crystallographic methods (Bode et al., 1986b). HLE is a glycoprotein with a single peptide chain of 218 amino acid residues and four disulfide bridges. Analyses of the cDNA sequence of HLE (Farley et al., 1988; Takahashi et al., 1988a) confirmed the sequence with the addition of a carboxy-terminal 20 amino acid extension. The extension is probably removed during posttranslational trimming and packaging in the lysosomal granules. Medullasin, an inflammatory serine proteinase derived from bone marrow cells, is similar to if not identical with HLE (Aoki, 1978; Okano et al., 1987).

HLE is homologous with other elastolytic serine proteases such as PPE (40%), rat pancreatic elastase II (40%; MacDonald et al., 1982), human pancreatic elastase I (37%; Tani et al., 1988), and human pancreatic elastase E (36%; Shen et al., 1987). The sequence identity with less related proteases such as rat mast cell protease II (Woodbury et al., 1978), porcine pancreatic kallikrein (Bode et al., 1983), rat tonin (Fujinaga & James, 1987), bovine chymotrypsin (Cohen et al., 1981; Tsukuda & Blow, 1985), human cathepsin G, human lymphocyte proteases, and human plasminogen is ca. 32–35%. PPE exhibits the same degree of sequence identity with bovine pancreatic chymotrypsin (39%) and trypsin (37%), but the identity is much higher with functionally related mammalian pancreatic elastases including human elastase I (90%), human protease E (57%), rat elastase I (85%; MacDonald et al., 1982), and rat elastase II.

HLE and PPE: Tertiary Structures. Crystal structures of 19 or more elastase derivatives have been determined to atomic resolution (Table II). PPE (as the tosyl derivative) was the second serine protease whose tertiary structure was elucidated (Watson et al., 1970; Shotton & Watson, 1970; Sawyer et al., 1978), and the structure of native elastase has recently been refined crystallographically to 1.65-Å resolution (Meyer et al., 1988a). Native HLE produces small crystals unsuitable for analysis (E. F. Meyer and W. Bode, unpublished results), but the structures of two separate HLE–inhibitor complexes have been determined (Bode et al., 1986b; Wei et al., 1988). The structure analyses of two more complexes of HLE with protein protease inhibitors are nearly complete, but attempts to cocrystallize HLE with other small ligands have been unsuccessful up to now. Fortunately, unligated PPE yields beautiful single crystals with a binding site that is relatively open and accessible to low molecular weight substrate analogues and inhibitors except at remote subsites. Small inhibitors can readily be soaked into the crystals, and the crystal structures of several PPE complexes formed with various peptidic and heterocyclic inhibitors have been analyzed (Table II). All the PPE structures have been based on the same isomorphous crystal form while the HLE complexes exhibit different crystal packing arrangements.

Along with other serine proteases, the polypeptide chains of PPE and HLE are organized as two structurally similar, interacting antiparallel β-barrel cylindrical domains. Only an intermediate segment and the carboxy-terminal segment are organized as helices (Figure 1). Most of the catalytic residues, especially those of the active-site triad Ser-195, His-57, and Asp-102, are localized in the crevice formed between both domains. Across this crevice is the substrate binding site, which includes parts of both domains.

Of all the serine proteases whose tertiary structures have been determined, PPE is topologically most similar to HLE,

Table II: Human Leukocyte Elastase and Porcine Pancreatic Elastase Crystal Structures

structure	resoln (Å)	*R*-factor (%)	reference
	Human Leukocyte Elastase Complexes		
turkey ovomucoid inhibitor third domain (TOM)	1.80	16.7	Bode et al., 1986b
MeO-Suc-Ala-Ala-Pro-Val-CH_2Cl	2.30	14.5	Wei et al., 1988
MeO-Suc-Ala-Ala-Pro-Ala-CH_2Cl	1.84	16.4	Navia et al., 1989
	Porcine Pancreatic Elastase Structures		
tosyl elastase	2.50	32.6	Sawyer et al., 1978
native (70% methanol buffer)	1.65	16.6	Meyer et al., 1988a
CF_3CO-Lys-Ala-NH-C_6H_4-CF_3	2.50	21	Hughes et al., 1982
CH_3CH_2CO-Ala-Pro-NH-Et (and -NH-C_5H_9)	3.50		Hassal et al., 1979
Ac-Ala-Pro-Ala (pH 5.0)	1.65	18.4	Meyer et al., 1986
Ac-Ala-Pro-Ala (pH 7.5)	1.65	18.6	unpublished results
Ac-Pro-Ala-Pro-Tyr	1.80	19.2	Clore et al., 1986
Thr-Pro-Ape*MeLeu-Tyr-Thr[a]	1.80	18.8	Meyer et al., 1988b
Ac-Ala-Pro-Val-CF_3	2.50	15.0	Takahashi et al., 1988b
Ac-Ala-Pro-Val-CF_2CO-NH-$CH_2CH_2C_6H_5$	1.78	16.0	Takahashi et al., 1989
5-Me-2-[Boc-NHCH(*i*-Pr)]benzoxazinone	2.10	15.3	Radhakrishnan et al., 1987
5-Cl-2-[Boc-NHCH(*i*-Pr)]benzoxazinone	1.74	17.2	Radhakrichnan et al., 1987
7-amino-4-chloro-3-methoxyisocoumarin	1.80	17.5	Meyer et al., 1985
4-chloro-3-ethoxy-7-guanidinoisocoumarin	1.70	21.0	E. F. Meyer and R. Radhakrishnan, unpublished results
cephalosporin derivative	1.84	16.8	Navia et al., 1987

[a] The asterisk indicates the peptide bond corresponding to the scissile bond of a good substrate.

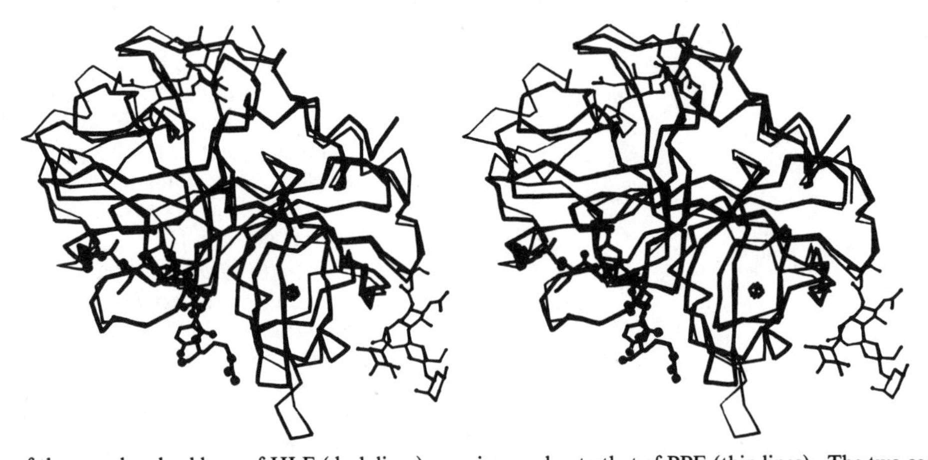

FIGURE 1: Stereoview of the α-carbon backbone of HLE (dark lines) superimposed onto that of PPE (thin lines). The two carbohydrate moieties of HLE are included together with their points of attachment (Asn-159 and Asn-109). The calcium atom bound to PPE is drawn as a ball; this site is occupied by Arg-80 in HLE. The primary binding segment of TOM is shown (with heteroatoms drawn as small spheres) in order to indicate the substrate binding site.

followed by rat mast cell protease II (Remington et al., 1988), porcine pancreatic kallikrein (Bode et al., 1983), rat tonin (Fujinaga & James, 1987), and bovine chymotrypsin (Cohen et al., 1988; Tsukuda & Blow, 1985). Most of the structural differences between the elastases and other serine proteases are located in surface loops. PPE and HLE have approximately 150 equivalent α-carbon atoms at a rms deviation of 0.6 Å (W. Bode, unpublished results). The topological alignment of PPE and HLE is shown in Table I; a similar alignment of HLE and α-chymotrypsin has been reported (Wei et al., 1988).

The two elastases are surprisingly similar especially in their active-site regions. The catalytic triad (Ser-195, His-57, and Asp-102) and the residues forming the central core of the binding site (i.e., peptide segments 189 to Ser-195, 213–216, 226–228, residue 41) are structurally similar, with their α-carbon atoms exhibiting a rms deviation of only 0.35 Å. When the backbone atoms of these residues are superimposed, 55 atoms agree to 0.26 Å, the greatest divergence occurring at amino acids 192 and 226.

There are significant structural differences between the two elastases that include 10 deletions (comprising 27 residues) and 2 insertions (comprising 5 residues). In addition, larger spatial deviations between the peptide chains occur at amino acid residues 25, 75–79, 88, 130–133, 147–150, 177–180, and 240–243. The largest structural differences are observed around the "methionine loop" (centered around Met-180 in PPE and α-chymotrypsin; Asn in HLE). In addition, one of the four disulfide bridges (Cys-168–Cys-182) is of quite different size and has a different conformation due to a long deletion in HLE. There is no indication in HLE of an intermediate helix which is normally observed around Cys-168 in other vertebrate proteases including PPE. The other three disulfide bridges (42–58, 136–201, and 191–220) are geometrically similar in both elastases.

PPE and the other digestive enzymes trypsin (Bode & Schwager, 1975) and α-chymotrypsin (Birktoft & Blow, 1972) possess a calcium binding loop composed of the peptide segment Glu-70–Glu-80 with the Glu residues acting as ligands. Binding of calcium to this loop stabilizes the molecule but does not affect its catalytic activity (Bode & Schwager, 1975). The equivalent peptide segment in HLE is spatially similar to the calcium loop in PPE and trypsin; however, Glu-80 (PPE and trypsin) is replaced by Arg-80 in HLE. The terminal guanidino group of Arg-80 in HLE occupies the site where calcium is found in PPE and trypsin. Thus HLE carries its own stabilizing cation and does not depend on calcium for stability.

Both PPE and HLE contain several internal or "buried"

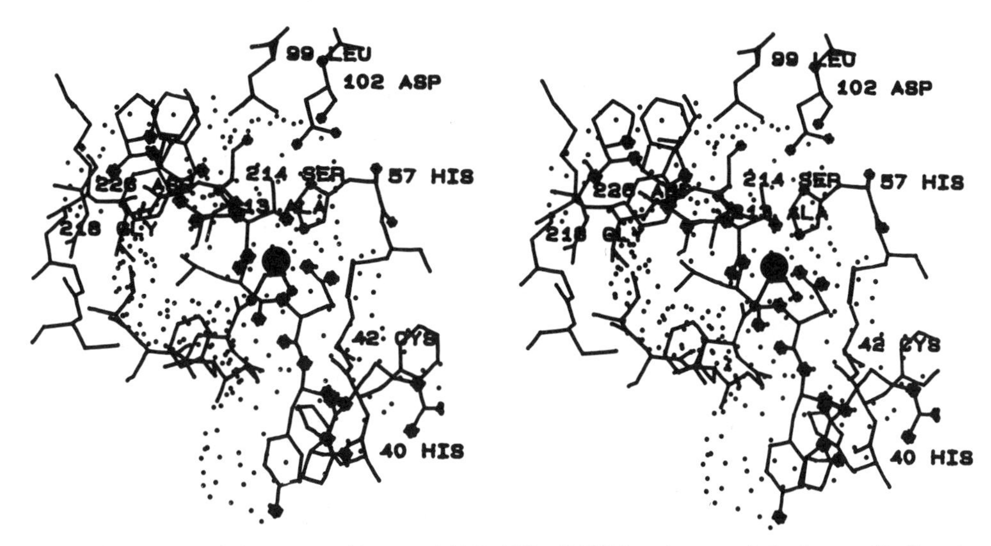

FIGURE 2: Stereoview of the binding of the octapeptide strand (14I–22I) of TOM to the extended substrate binding site of HLE. For the sake of discrimination, heteroatoms in the inhibitor segment are drawn as large spheres and those in the catalytic triad as small spheres. A double van der Waals' contact surface illustrates the structurally convoluted extended binding site of HLE. Ser-195 Oγ is drawn as an enlarged sphere for the purposes of orientation.

channels or domains of water that are common to other serine proteases. The seven water domains in PPE (Meyer et al., 1988a) are present in HLE, even though the enzymes have considerable sequence and structural differences in the immediate vicinity of the water domains. The presence of a channel of water linking the buried portion of the catalytic triad to the surface of the enzyme is especially noteworthy. The presence of similar water channels in other hydrolases suggests that this channel may have a functional role (E. F. Meyer, unpublished observations).

HLE is much more hydrophobic and at the same time more basic than PPE. In HLE approximately 90 hydrophobic residues (>40% of its total residues) are accessible to bulk water molecules, compared to only 70 such residues (30%) in PPE. HLE owes its high basicity to the presence of 19 arginines, which are balanced with only 9 acidic residues (three of which, Asp-102, Asp-194, and Asp-226, are buried). All of the arginine residues, with the exception of Arg-80, are arranged on the surface of the enzyme in a horseshoe-like manner around the active site, with several forming clusters of two to four arginines. The surface arrangement of arginines around the active site explains the preferred binding of linear sulfated polysaccharides to HLE (Baici et al., 1980).

HLE along with many other human serine proteases such as the human pancreatic protease E (I. A. Szigoleit, personal communication) is a glycoprotein, while PPE is devoid of prosthetic groups. Two of HLE's three Asn-X-Ser/Thr consensus glycosylation sequences (Asn-109 and Asn-159) are linked to different degrees with carbohydrate chains (Figure 1; Bode et al., 1986b; Sinha et al., 1987), giving rise to three isoenzyme forms of HLE (E1, E2, and E3) with almost identical enzymatic activity (Baugh & Travis, 1976). Only the first three to four sugar residues at both glycosylation sites are defined by proper electron density and have a rigid relationship to the polypeptide backbone in the crystalline complex. Most of the noncovalent contacts between the polysaccharide and the peptide backbone are formed with the fucose ring. Both sugar chains extend out into solution and are located away from the substrate binding site. Consequently, they should not interfere with binding of substrates and inhibitors at the active site. However, the carbohydrate chains affect crystallization. Different HLE isoenzyme forms, when crystallized with TOM, crystallize under different conditions (W. Bode, W. Watorek, and J. Travis, unpublished results). The carbohydrate structure in the HLE–TOM complex is probably the best defined of any glycoprotein now available at high resolution.

Peptide Inhibitors and the Extended Substrate Binding Site

HLE–TOM Complex. HLE is inhibited by several naturally occurring or engineered protein protease inhibitors including the following: α_1-protease inhibitor (Travis & Salvesen, 1983) and some genetically engineered variants (Rosenberg et al., 1984); eglin c (Baici & Seemueller, 1984; Bode et al., 1986a; Braun et al., 1987); bovine pancreatic trypsin inhibitor and some variants (Tschesche et al., 1988); ovomucoid inhibitors and inhibitors derived from the human pancreatic trypsin inhibitor; human seminal plasma inhibitor/human secretory leukocyte inhibitor/human mucous protease inhibitor (Seemueller et al., 1986; Thompson & Ohlsson, 1986; Gruetter et al., 1988). Most protein protease inhibitors bind to their cognate serine proteases in the manner of a good substrate and form extremely stable Michaelis complexes in which the inhibitor's reactive-site peptide bond remains intact or is only cleaved extremely slowly (Huber & Bode, 1978; Laskowski & Kato, 1980; Marquart et al., 1983). Reactive-site loops of the inhibitors are complementary to the substrate recognition site of the cognate serine proteases, and thus these complexes are ideal for probing substrate binding subsites due to extensive interactions and tight binding between the enzyme and inhibitor.

The first structure of HLE was determined as a complex with the third domain of the turkey ovomucoid inhibitor (TOM). TOM forms tight complexes with both HLE and PPE with association constants of 6.2×10^9 M^{-1} and 4.1×10^{10} M^{-1}, respectively (M. Laskowski, S. J. Parks, M. Tashiro, and R. Wynn, personal communication). This inhibitor, like other Kazal-type inhibitors (Laskowski & Kato, 1980), consists of a molecular scaffold made up of an α-helical segment, a three-stranded β-pleated sheet, and an extended protease binding loop centered around the reactive-site scissile peptide bond Leu-18I–Glu-19I (Papamokos et al., 1982). The HLE binding site makes direct contact with eight residues of the "primary binding segment" (P_5 Pro-14I to P_3' Arg-21I) of TOM (Figure 2) and with an additional three (to five) residues (Gly-32I, Asn-33I, and Asn-36I) of a "secondary binding segment". The majority of the total intermolecular contacts

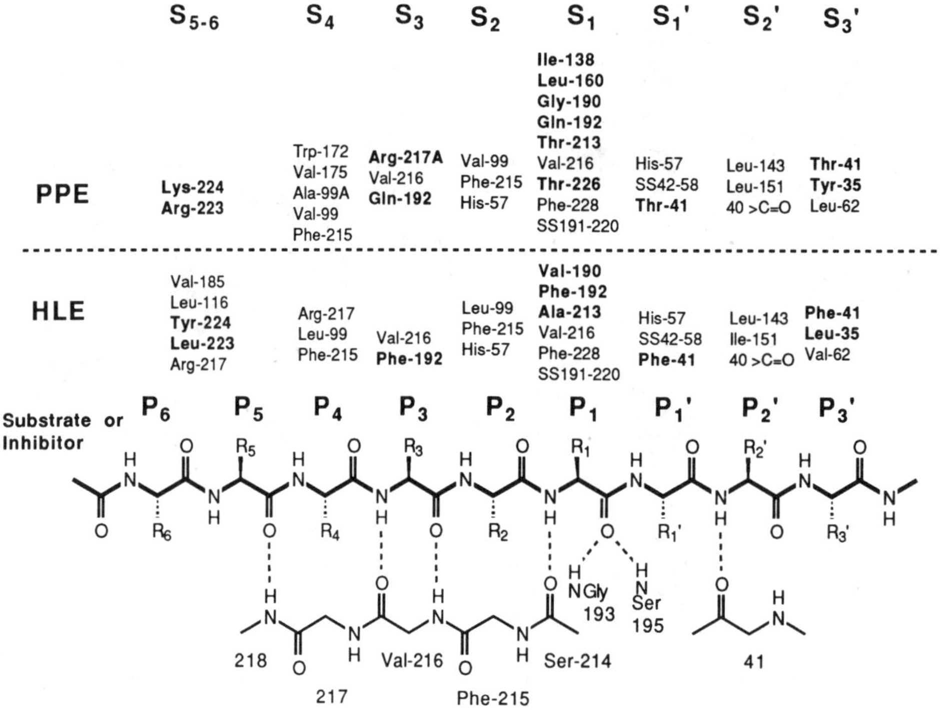

FIGURE 3: Schematic representation of the main-chain and side-chain interactions between the primary binding segment of TOM or a peptide substrate with HLE and PPE. Significant amino acid sequence changes that might influence specificity are shown in bold type. The subsites of the elastases are represented by S_1, S_1', etc. and the residues of the peptide by P_1, P_1', etc. The carbonyl of the scissile peptide bond is shown interacting with the oxyanion hole (hydrogen bonds to the backbone NHs of Gly-193 and Ser-195).

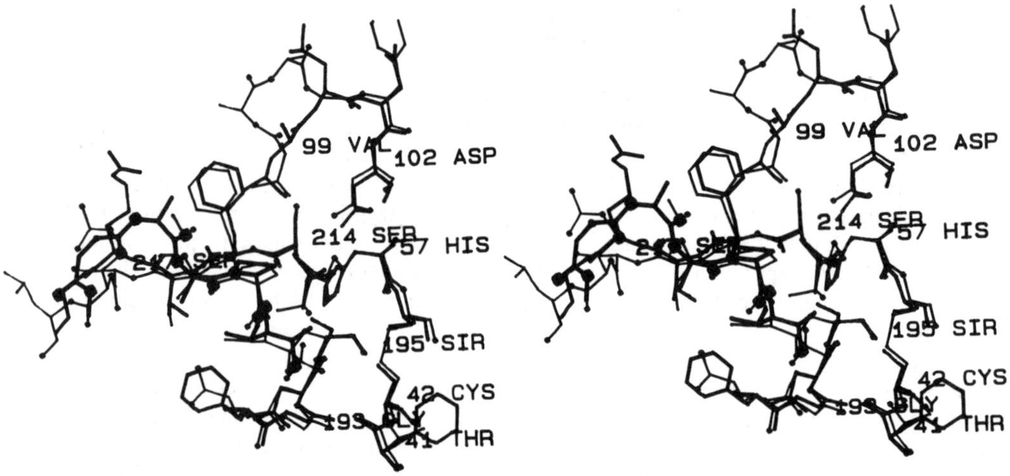

FIGURE 4: Comparison of the binding of peptide halo ketone inhibitors to HLE and PPE showing the similarity of binding modes. The complex of HLE with MeO-Suc-Ala-Ala-Pro-Val-CH_2Cl is drawn with heavy lines, while the PPE complex with Ac-Ala-Pro-Val-CF_3 is drawn with thin lines. Large and small spheres, respectively, indicate heteroatoms, including the fluorines of the fluoro ketone. The bonds between Ser-195 and His-57 of HLE and the chloro ketone are not shown. Significant differences in the two active sites (HLE vs PPE) are seen at residues 41 (Phe vs Thr) and 192 (Phe vs Gln). The 99A–99B insertion loop of PPE is shown at the top. Arg-217A, an insertion residue in PPE, is drawn on the left side of the figure in collision with the N-terminal methoxysuccinyl group of the chloromethyl ketone inhibitor of HLE.

(94 out of 106 contacts within 4 Å) involve the primary binding loop, which mimics a bound substrate. Seven hydrogen bonds are formed between the peptide backbones of the inhibitor segment and HLE (Figure 3). The P_3, P_2, and P_1 residues of the inhibitor form an antiparallel β-sheet structure with the peptide backbone of Ser-214–Val-216 of HLE, an interaction that is typical for the binding of peptides to serine proteases. The carbonyl group of Leu-18I is located in the oxyanion hole (Gly-193N, Ser-195N), and the P_2' residue makes an antiparallel arrangement with the backbone of Phe-41.

Complex of HLE and MeO-Suc-Ala-Ala-Pro-Val-CH_2Cl. The complex formed between HLE and the specific chloromethyl ketone irreversible inhibitor MeO-Suc-Ala-Ala-Pro-Val-CH_2Cl (Powers et al., 1977) was crystallized and its structure (Figure 4) determined by Patterson methods (Wei et al., 1988) (Table II). A similar structure of HLE inhibited by MeO-Suc-Ala-Ala-Pro-Ala-CH_2Cl will be available shortly (Navia et al., 1989; Williams et al., 1987). The inhibitor peptide chains of both chloromethyl ketones are bound in a conformation that is similar to the corresponding P_4 to P_1 residues of TOM. The MeO-Suc group (formally the P_5 group) of the valyl chloromethyl ketone is not rigidly fixed, but runs antiparallel to the backbone residues 216–218. In solution this group could be placed close to Arg-217, and a negatively charged succinyl group on an inhibitor could interact with the guanidino group of Arg-217.

Complexes of Peptides with PPE. The relatively open active site of crystalline PPE at S_3 to S_4' has facilitated isomorphous

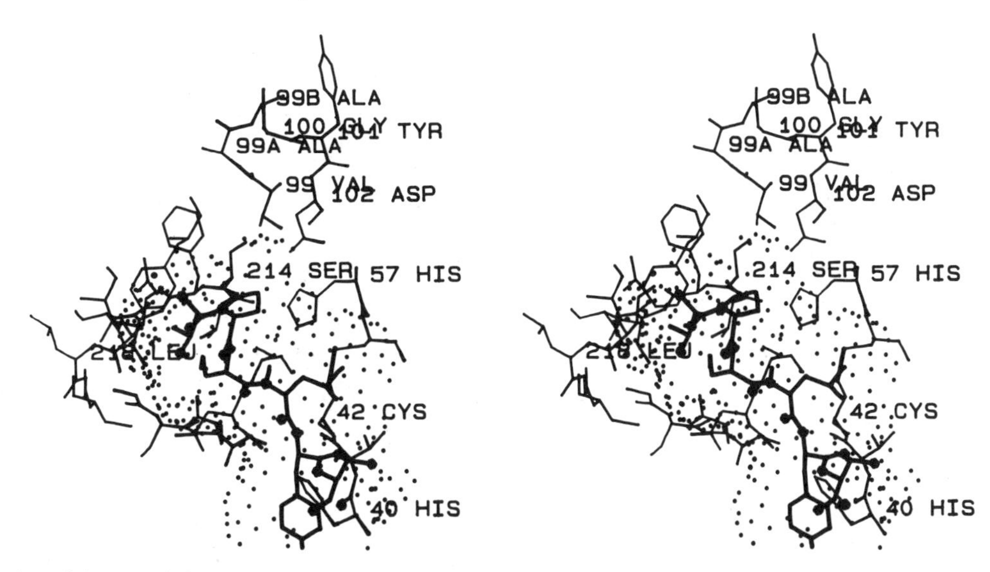

FIGURE 5: Stereoview of the extended binding site of PPE complexed with the hexapeptide Thr-Pro-Ape*MeLeu-Tyr-Thr (the asterisk indicates the peptide bond that corresponds to the scissile bond of a substrate). One of the insertion loops in PPE (99A–99B) is shown at the top of the figure.

crystallographic studies of the binding of peptides to this enzyme. However, remote subsites such as S_4 are blocked in crystals of PPE by Glu-62 from a symmetry-related neighbor. Low-resolution structures of PPE complexes with the dipeptides CH_3CH_2CO-Ala-Pro-NH-Et (or -NH-C_5H_9) first showed that peptides were binding to PPE in the vicinity of residues 214–216 (Hassal et al., 1979). Interestingly, the first peptide that was investigated at high resolution was found to bind backward in the active site of PPE, although this was not recognized at the time (Hughes et al., 1982). In this complex of PPE and CF_3CO-Lys-Ala-NH-C_6H_4-*p*-CF_3, the CF_3 group of the trifluoroacetyl group occupies the S_1 subsite and the peptide chain makes a parallel β-sheet structure with Ser-214–Phe-215–Val-216 instead of the more typical antiparallel β-sheet juxtaposition.

A number of peptides containing P_2 prolyl residues have been investigated at high resolution. While Pro at P_2 is not a prerequisite for binding, it eliminates (or reduces) the possibility of multiple binding in longer peptides and thus has been included in the design strategy of a number of ligands and substrates for PPE. This strategy does not always apply for short peptides; Ac-Pro-Ala-Pro-Tyr-NH_2 (Clore et al., 1986) binds on the S′ side of the active site, while two molecules of Ac-Ala-Pro-Ala (Meyer et al., 1986) bind backward on both the S and S′ sides of the binding site. Directed hydrogen bonds between the substrate peptide backbone atoms and PPE have a greater influence on binding architecture than do the side chains, which frequently point out into solution. This regular spacing of backbone interactions, especially on the acyl side of the scissile bond (S_1–S_3), permits the inverse binding that is observed with both peptides. Backward binding is also characterized by a translational half-step shift with respect to PPE's extended substrate binding site to enable formation of hydrogen bonds when a peptide is bound in the reverse orientation.

The structure of the complex of PPE with the hexapeptide Thr-Pro-Ape*MeLeu-Tyr-Thr (Figure 5) was the first PPE structure that showed the nature of subsites on both sides of the scissile bond and comes closest to mimicking the binding of good peptide substrates (Meyer et al., 1988b). The interaction pattern of this complex is similar to that observed in the HLE–TOM complex (Figures 2 and 5). The 24 backbone atoms superimpose with a rms deviation of 1.2 Å. As may be expected, the greatest variations are at the P_2 Pro and the terminal regions. Even though the residues fill the space quite well, the P_2'–P_3' Tyr-Thr terminus is relatively free to move within the van der Waals' surface of PPE's S′ binding region.

Primary Specificity Site (S_1). One-third of the total contacts between HLE and TOM are made by the P_1 residue (Leu-18I) of TOM (Figure 6). This pocket has its entrance between the flat sides of the peptide backbones of 214–216 and 191–192 and is constricted toward its bottom by residues Val-190, Phe-192, Ala-213, Val-216, and Phe-228 and the disulfide bridge Cys-191–Cys-220. The S_1 pocket has a hemispheric shape and is rather hydrophobic in character. Thus it is well adapted to accommodate medium-sized aliphatic side chains such as leucine and isoleucine. There is no specific anchoring point in the S_1 pocket, such as Asp-189 in trypsin, so that the aliphatic P_1 side chain is not fixed in a distinct orientation and can revolve like a ball and socket joint. Accommodation of larger side chains such as Phe would require a considerable expansion of the pocket, and as a consequence, substrates with P_1 phenylalanine are not normally cleaved by HLE.

In the structure of the valyl chloromethyl ketone–HLE complex (Wei et al., 1988), the β-branched bulky side chain of the P_1 valyl residue is accommodated with a slight tilting of its main chain. Simultaneously, the HLE S_1 pocket shrinks slightly and adapts to the reduced size of the smaller side chain. Serine proteases are generally characterized by the rigid "lock and key" model since no significant changes in conformation have been observed crystallographically upon binding of substrates or inhibitors. In the case of HLE, the lock and key model may have to be modified due to the small induced fit conformational changes observed upon binding of the chloromethyl ketone.

Leucine is slightly preferred over valine at P_1 among ovomucoid inhibitors (M. Laskowski, personal communication), in contrast to BPTI-derived HLE inhibitors, where valine seems to have the greater affinity (Tschesche et al., 1987). With nitroanilide substrates, a P_1 Val is favored over Leu (Zimmerman & Ashe, 1977); in the case of thiobenzyl ester substrates, the straight-chain Ape (Nva) has the highest k_{cat}/K_M value (relative values: Ape, 12.5; Ile, 5.3; Val, 4.1; Leu, 3.0; Ahe or Nle, 2.5; Met, 1.2; Ala, 1.0; Phe, 0; Harper

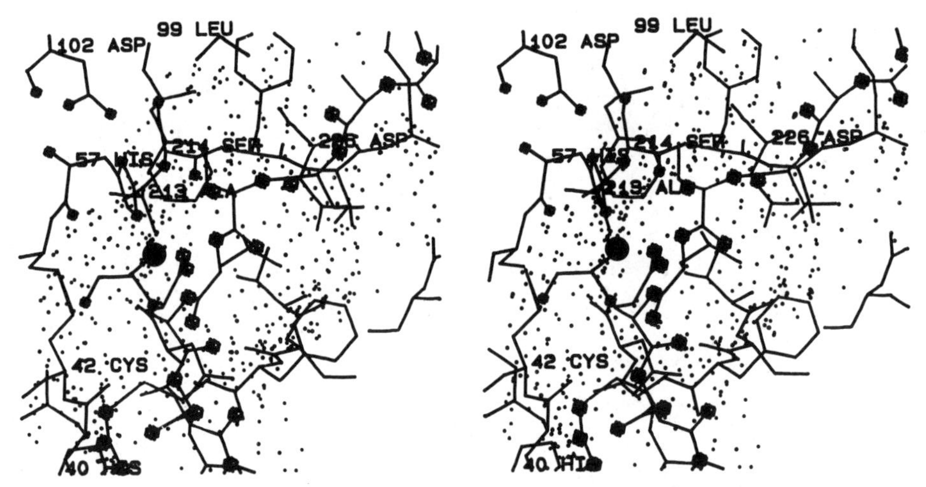

FIGURE 6: "Inside-out" view of the primary substrate specificity site (S_1) of HLE, as observed in the HLE–TOM complex. The heteroatoms of the inhibitor are represented by large spheres, and the catalytic tetrad of HLE is represented by small spheres. The double van der Waals' contact surface is represented as heavy dots. Superimposed on this is the contact surface of the PPE complex with the hexapeptide Thr-Pro-Ape-MeLeu-Tyr-Thr (small dots). This view depicts the marginally deeper S_1 pocket of HLE. Ser-195 Oγ is drawn as an enlarged sphere.

et al., 1984). Modeling predicts that an isoleucine side chain at P_1, with its Cδ atom trans to Cα, would fill the pocket of HLE even better than Leu, which is in accord with the solution experiments.

The P_1 residue of HLE's natural plasma inhibitor (α_1-protease inhibitor) is Met-358. Modeling experiments show that a methionine residue at P_1 should easily fit (with a typical bent conformation) into the S_1 pocket. Peptides with P_1 Met residues are reasonable HLE substrates, while oxidation of the Met to the sulfoxide results in decreased binding and absence of hydrolysis characterized by a significant drop in k_{cat}/K_M values (McRae et al., 1980; Nakajima et al., 1979). The extra oxygen of the methionine sulfoxide creates severe steric hindrance with the S_1 pocket, providing a structural basis for loss of the inhibitory capacity of α_1-protease inhibitor upon oxidation of Met-358 by cigarette smoke (Janoff et al., 1983; Matheson et al., 1982).

The bottom of the S_1 pocket in HLE contains an acidic residue, Asp-226, which is completely buried in the interior of the HLE molecule without any balancing counterion. The carboxylate group of Asp-226 is hydrogen-bonded with three interior water molecules which might serve to dissipate its negative charge. Asp-226 is somewhat shielded from the exterior of the pocket by Val-216 and Val-190. Model building experiments indicate that the gap between the side chains of Val-190 and Val-216 is just barely large enough to allow penetration of a basic side chain of a P_1 residue to form a salt link with Asp-226. Recent experiments have shown that HLE does not cleave either Lys or *S*-aminocysteine (an analogue of homolysine) substrates at significant rates, indicating that these substrates do not bind well in S_1 (T. Ueda and J. C. Powers, unpublished results).

The primary specificity pocket of PPE is slightly less hydrophobic and only marginally smaller in spite of considerable changes in the lining amino acids (Figures 3 and 6). This is consistent with PPE's perference for Ala and Nva (norvaline) rather than larger or branched amino acids in P_1 (Zimmerman & Ashe, 1977; Harper et al., 1984). In both elastases, residue 189 (Gly in HLE, Ser in PPE, Asp in trypsin) is not directly accessible to P_1 side chains. The bottom of the S_1 pocket in PPE is coated by the hydrophobic portion of Thr-226 (equivalent to Val-190 in HLE) and Ile-138 (Ser-138 of HLE is not accessible). The polar amino acid Gln-192 (Phe-192 in HLE) is located at the entrance to the pocket. Only Val-216 at the entrance, Phe-228, forming the "ceiling", and the disulfide bridge Cys-191–Cys-220 are common to both elastases (Figure 3).

Remaining Subsites. The second most important set of interactions in the HLE–TOM complex involves Thr-17I (P_2, Figure 2). The complementary subsite S_2, lined by Phe-215, Leu-99, and the flat side of the imidazole ring of His-57, is bowl-shaped and quite hydrophobic, but similar to that found in all other mammalian proteases, including PPE. Medium-sized hydrophobic side chains including proline are preferred.

The P_3 Cys-16I group is only in contact with HLE through main-chain atoms. More elongated side chains would, however, interact with hydrophobic surfaces of HLE, mainly with Phe-192 and Val-216. In common with interactions seen in other serine protease complexes with peptides, the P_3–S_3 contact is characterized by two intermolecular hydrogen bonds made with the backbone NH and carbonyl of Val-216. The side-chain Ala-15I (P_4) points along the side chains of residues Phe-215 and Arg-217 in HLE toward the bulk water. Interestingly, substrates with Lys at P_3 or P_4 are much less reactive than the corresponding Ala substrates (Yasutake & Powers, 1981). In contrast, the corresponding substrates with aromatic groups or large hydrophobic groups with a charge are more reactive than the corresponding Ala substrates. These substrates resemble desmosine, the charged hydrophobic cross-linking amino acid residue of elastin, and suggest that HLE may selectively bind to and cleave elastin near cross-linking regions through interaction at the S_3 and S_4 subsites.

The position and conformation of Pro-14I (P_5) are clearly dictated by intramolecular constraints in the ovomucoid inhibitor itself. The very weak hydrogen bond formed between the NH of Gly-218 in HLE with the carbonyl of P_5 may not be representative of binding of substrates with flexible peptide chains. Addition of a P_5 residue in most cases increases the reactivity of substrates (Wenzel & Tschesche, 1981; Lestienne & Bieth, 1980).

On the leaving group side of the scissile bond, Tyr-20I (P_2') of TOM makes by far the most contacts with HLE, mainly with Ile-151. Affinity measurements with ovomucoid variants lacking tyrosine emphasize the favorable interaction made by the phenolic side chain at this subsite (M. Laskowski, personal communication). Arg-21I (P_3') nestles with its (primary hydrophobic) side chain and the flat side of its guanidino group toward the quite hydrophobic surface of HLE's S_3' subsite. This shallow subsite is formed mainly by the side chains of Phe-41, Leu-35, and Val-63 and probably represents the hydrophobic subsite that binds peptide inhibitors [e.g., Boc-Val-Val-Val-NH$(CH_2)_{11}CH_3$, K_I = 0.21 μM] with long carboxyl-terminal aliphatic chains (Lentini et al., 1987).

The S_1' subsite consists of a relatively hydrophobic pocket lined by Cys-42–Cys-58 and Phe-41 and can accommodate all amino acid residues at P_1', with the exception of Pro, other N-methylated residues, and large hydrophobic residues (e.g., Trp). In the ovomucoid inhibitors, Glu-19I (P_1') forms an intramolecular hydrogen bond with Thr-17I (P_2). This hydrogen bond becomes stronger in the course of complex formation and makes a significant contribution to the overall binding of ovomucoid inhibitors to serine proteases (M. Laskowski, personal communication). Thus, a cyclic structure containing this tripeptide segment (P_2–P_1') might be an excellent template for the design of potent new HLE inhibitors.

In general, the PPE subsites are less hydrophobic than those in HLE (Figure 3). This is most apparent for subsites at extreme ends of the substrate binding site (S_5–S_6 and S_3'), but is also true for sites in the immediate vicinity of the specificity pocket (Phe-192 and Phe-41 in HLE compared to Gln-192 and Thr-41 in PPE). The substantial increase in hydrophobic character and the reduced capability to form hydrogen bonds may explain the unique property of HLE to be inhibited by cis-unsaturated oleic acids and other related fatty acids (Ashe & Zimmerman, 1977). In addition, many hydrophobic peptide and heterocyclic inhibitors are much more reactive with HLE than PPE.

Specificity and Mechanism

Interactions at remote subsites can profoundly influence the catalytic mechanism and specificity of both elastases. With nitroanilide substrates of HLE, extension of the peptide chain by the addition of a P_3 residue results in an 100-fold increase in the acylation rate and a change in the rate-limiting step from acylation to deacylation (Stein et al., 1987a). There also appears to be communication between remote subsites and the S_1 pocket because some amino acid residues (e.g., Phe) are tolerated in monomeric reactive substrates and not in extended substrates (Stein, 1985). In addition, proton inventory studies have shown that minimal substrates (monomeric or dipeptides) are hydrolyzed by simple general-base catalysis by His-57 involving transfer of one proton in the rate-limiting transition state (Stein et al., 1987b). However, tri- and tetrapeptides are hydrolyzed by a mechanism that involves full functioning of the catalytic triad and transfer of two protons in the rate-limiting step.

The P_3–S_3 contact (Figure 3) is characterized by the two intermolecular hydrogen bonds made with the backbone NH and carbonyl of Val-216. Clearly this interaction communicates directly with the S_1 subsite since the isopropyl side chain of Val-216 forms one side of the pocket and the pocket changes slightly in size upon binding the tetrapeptide valyl chloromethyl ketone inhibitor. Communication with the catalytic triad may involve the hydrogen-bonding network between Asp-102 and Ser-214, which in turn is part of the peptide backbone (including Val-216) that forms the β-sheet structure with a peptide substrate. Indeed, Ser-214 is conserved in serine proteases and forms a catalytic tetrad with Ser-195, His-57, and Asp-102.

Peptide inhibitors such as TOM possess a reactive-site sequence that matches that of an ideal substrate and yet is cleaved extremely slowly. Nature has devised a number of "tricks" to keep this bond intact: complementary surfaces of enzyme binding sites and inhibitor binding segments; restricted mobility of binding loops within the enzyme–inhibitor complex; and favorable hydrogen bonds. The rate of cleavage of the hexapeptide Thr-Pro-Ape*MeLeu-Tyr-Thr might be slowed down because stereochemical inversion at the P_1' nitrogen atom of the scissile peptide bond in N-methylated P_1' substrates is internally restricted (Bizzozero & Zweifel, 1975; Bizzozero & Dutler, 1981) due to stereoelectronic control (Deslongchamps, 1975). A similar rationale could be used to explain the stability of complexes of proteases with TOM and other protein protease inhibitors.

Synthetic Elastase Inhibitors

Peptide Chloromethyl Ketone and Fluoro Ketone Inhibitors. Peptide halo ketone inhibitors have been widely investigated with serine proteases, and several crystal structures of serine proteases complexed with peptide chloromethyl ketones have been reported previously [see Powers (1977) and Powers and Harper (1986) for reviews]. The existence of the extended substrate binding site composed of the backbone atoms of 214–216 was first demonstrated in chloromethyl ketone serine protease structures. With chloromethyl ketones, two bonds are formed between the enzyme and inhibitor: one between His-57 and the methylene group of the inhibitor and one between the Ser-195 and the ketonic carbonyl group. In the HLE complex with MeO-Suc-Ala-Ala-Pro-Val-CH_2Cl (Figure 4), similar covalent bonds are formed between the methylene group of the P_1 valine chloromethyl ketone residue and Nϵ of His-57 of HLE and between the Val carbonyl group and Oγ of Ser-195. The P_1 residue is pulled toward the S_1 pocket, with a simultaneous strengthening of the hydrogen bond between the P_1 amide nitrogen and the amide carbonyl of Ser-214 (Figure 3) and with pyramidalization of its carbonyl group with the addition of Oγ of Ser-195. This structure is similar to the presumed tetrahedral intermediate formed during serine protease peptide bond cleavage.

Peptide fluoro ketones are potent reversible inhibitors of HLE, PPE, and other serine proteases (Stein et al., 1987c; Imperali & Abeles, 1986; Trainor, 1987). These inhibitors, unlike chloromethyl ketones, are not potent alkylating agents and have greater therapeutic potential. The structures of two fluoro ketones bound to PPE have been determined (Figure 4; Takahashi et al., 1988b, 1989). Both form tetrahedral adducts with Ser-195 of PPE. In both the Ac-Ala-Pro-Val-CF_3 complex (K_I = 9.5 μM, pH 5.5) and the Ac-Ala-Pro-Val-CF_2CO-NH-$CH_2CH_2C_6H_5$ complex (K_I = 5.5 μM, pH 5.5), the oxyanion oxygen atom (derived from the fluoroketone carbonyl) is located in the oxyanion hole and forms hydrogen bonds with the NH of Ser-195 and Gly-193. The imidazole ring of His-57, occupying the "in" position also observed in native PPE, forms hydrogen bonds with a water molecule and with one fluorine atom of each inhibitor. The two inhibitors form antiparallel β-sheet structures with residues 214–216 of PPE. Comparison of the peptide backbone of the trifluoromethyl ketone bound to PPE with the HLE–chloromethyl ketone complex gives an excellent rms agreement of 0.38 Å when the 12 common backbone atoms are superimposed.

Heterocyclic Inhibitors. A wide variety of heterocyclic compounds have been described as inhibitors for serine proteases, including β-lactams, isocoumarins, benzoxazinones, pyrones, and saccharin derivatives (Powers & Harper, 1986). Such heterocyclic compounds are much more likely to yield practical therapeutic agents than peptide inhibitors, which are often poorly adsorbed and are often cleaved by other proteases. The crystal structures of four heterocyclic complexes have been determined to high resolution with PPE: the isostructural benzoxazinones (Radhakrishnan et al., 1987) and two isocoumarins (Meyer et al., 1985; R. Radhakrishnan and E. F. Meyer, unpublished results).

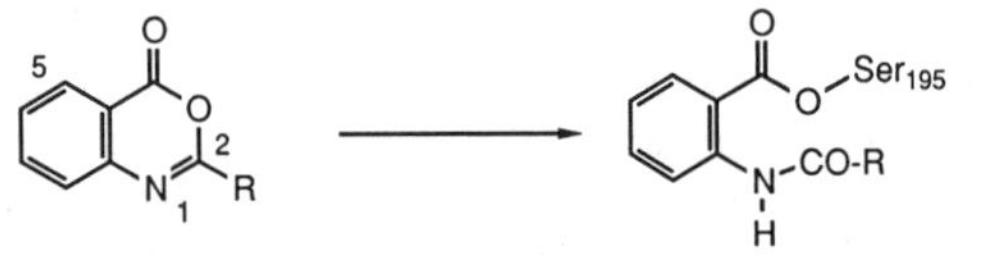

FIGURE 7: Mechanism of inhibition of elastase by benzoxazinone inhibitors.

Substituted benzoxazin-4-ones are potent inhibitors of HLE and PPE along with other serine proteases (Teshima et al., 1982). Benzoxazin-4-ones with 2-alkyl or fluoroalkyl substituents are particularly potent with K_I values as low as 92 nM, probably due to an interaction of the alkyl substituent with the S_1 pocket. The mechanism of benzoxazinone inhibition involves reaction with Ser-195 to form stable acyl enzyme derivatives (Stein et al., 1987d; Figure 7). The acylation rates, deacylation rates, and hydrolytic stability of the benzoxazin-4-ones can be altered by changing the nature of the substituents in both rings (Spencer et al., 1986; Krantz et al., 1987). Electron-withdrawing substituents on both rings increase the acylation and hydrolysis rates, electron-donating substituents increase stability, and bulky substitutes often slow the deacylation rates.

Two isomorphous benzoxazinone complexes, 5-Me-2-[Boc-NHCH(*i*-Pr)]benzoxazin-4-one and 5-Cl-2-[Boc-NHCH(*i*-Pr)]benzoxazin-4-one, have been studied crystallographically (Table II; Radhakrishnan et al., 1987). The resultant, isomorphous structures have an 8-Cl or 8-Me group pointing into the S_1 primary specificity site. The hydrophobic Boc-Val groups make van der Waals contacts with both the S_2' or S_1' sites of the enzymic receptor site (Figure 11).

In all PPE complexes with heterocyclic inhibitors thus far studied, the imidazole ring of His-57 is observed in the "out" position; in the place of the "in"-positioned imidazole ring, a water molecule is always found bridging the Asp-102 carboxylate with the benzoyl carbonyl group of the heterocycle or with Ser-195 Oγ. Surprisingly, the benzoyl ester group formed upon reaction with Ser-195 is not in the oxyanion hole but is rather directed to the new water molecule. Instead, the valyl carbonyl group, newly formed during the opening of the benzoxazinone ring, is located in the oxyanion hole. Thus deacylation seems in this case not to be facilitated by oxyanion hole activation, and this may be the reason for the slow hydrolysis of this complex. It is likely that the bulky Boc and *i*-Pr groups of the inhibitor, by leverage through the phenyl group of the opened benzoxazinone ring, contribute to this particular conformation. Benzoxazinone inhibitors without bulky groups might thus be bound in quite different conformations in the active site of PPE.

Isocoumarin Inhibitors. Isocoumarins are one class of heterocyclic structures that are rich in possible masked functional groups. Thus far, isocoumarins containing latent acid chloride (or ketenes) and quinone imine methide functional groups have been described as serine protease inhibitors. The initial reaction involves an acylation of Ser-195 of the serine protease by the isocoumarin ring carbonyl to form an acyl enzyme in which the isocoumarin ring is opened and new functional groups are unmasked. These acyl enzymes are often quite stable, but with the appropriately substituted isocoumarins, further reactions can occur between the newly released reactive groups and the side chains of neighboring amino acid residues of the enzyme.

HLE and PPE are both inhibited effectively by 3,4-dichloroisocoumarin with $k_{obsd}/[I]$ values of 8900 and 2500 M^{-1} s^{-1}, respectively (Harper et al., 1985). Although the elastases are most effectively inhibited, dichloroisocoumarin is a more general serine protease inhibitor and reacts with all serine proteases that have been observed thus far. Increased specificity and reactivity have been observed with 3-alkoxy-7-amino-4-chloroisocoumarins (Figure 8, **1**) containing varous substituents on the 7-amino group and in the 3-position (Harper & Powers, 1985). In the initial enzyme–inhibitor complex, the 3-alkoxy group is probably interacting with the S_1 pocket because derivatives with small alkoxy groups (3-methoxy, 3-ethoxy, 3-propoxy) inhibit HLE and PPE most effectively, while those with aromatic groups (3-benzyloxy) inhibit chymotrypsin and rat mast cell protease II most effectively.

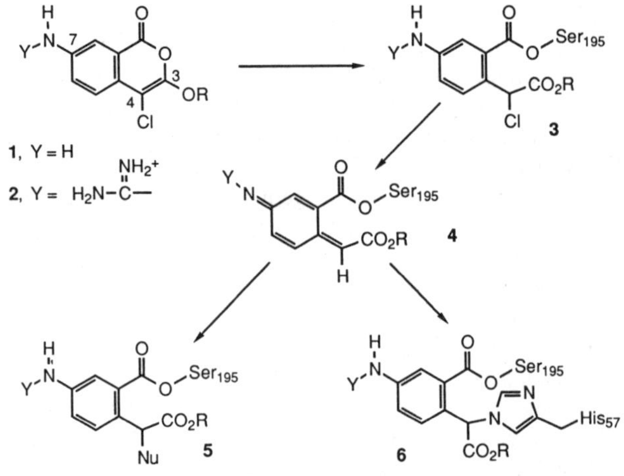

1, Y = H
2, Y = $H_2N-C(=NH_2^+)-$

FIGURE 8: Mechanism of inhibition of elastase by isocoumarin inhibitors.

The available evidence points to the inactivation mechanism shown in Figure 8. Enzyme acylation results in the formation of an acyl enzyme (**3**), which can undergo an elimination reaction to produce a 4-quinone imine methide (**4**) in the active site. This can react either with an enzyme nucleophile (probably His-57) to give an irreversibly inhibited enzyme structure (**6**) or with a solvent nucleophile to give a stable acyl enzyme (**5**). Partial reactivation by hydroxylamine (42% at pH 7.5 with PPE) suggests a partitioning between the two enzyme–inhibitor complexes in solution with the residual 58% probably representing the nonreactivatable complex **6** containing an alkylated histidine residue. Both the 7-amino and 4-chloro groups are required for formation of a stable inactivated enzyme; isocoumarins that lack these features inhibit serine proteases but deacylate fairly rapidly.

The crystal structures of two complexes of isocoumarins with PPE have been solved to atomic resolution (Table II). The structure analysis of PPE inactivated by 7-amino-4-chloro-3-methoxyisocoumarin at pH 5 in 0.1 M acetate buffer (Meyer et al., 1985) confirmed the postulated acylation mechanism concurrent with isocoumarin ring–ester cleavage since a single ester bond was formed with Ser-195 of PPE. It also contained an element of surprise: an acetate ion from the buffer had displaced the chlorine stereospecifically to give a stable acyl enzyme (Figure 8, **5**, Nu = acetoxy). The resulting structure has the acetoxy group in the S_1 pocket, the methoxy of the ester extends out into solution, and the benzoyl carbonyl points approximately into the oxyanion hole. The carbonyl group of the carbomethoxy moiety forms a hydrogen bond with His-57. The 7-amino group points toward the bulk water and makes no important intermolecular contacts.

Isocoumarins with basic substitutes such as 3-alkoxy-4-chloro-7-guanidinoisocoumarin (**2**) are potent inhibitors for trypsin, blood coagulation enzymes (Kam et al., 1988), complement enzymes, and natural killer cell tryptases (Hudig et

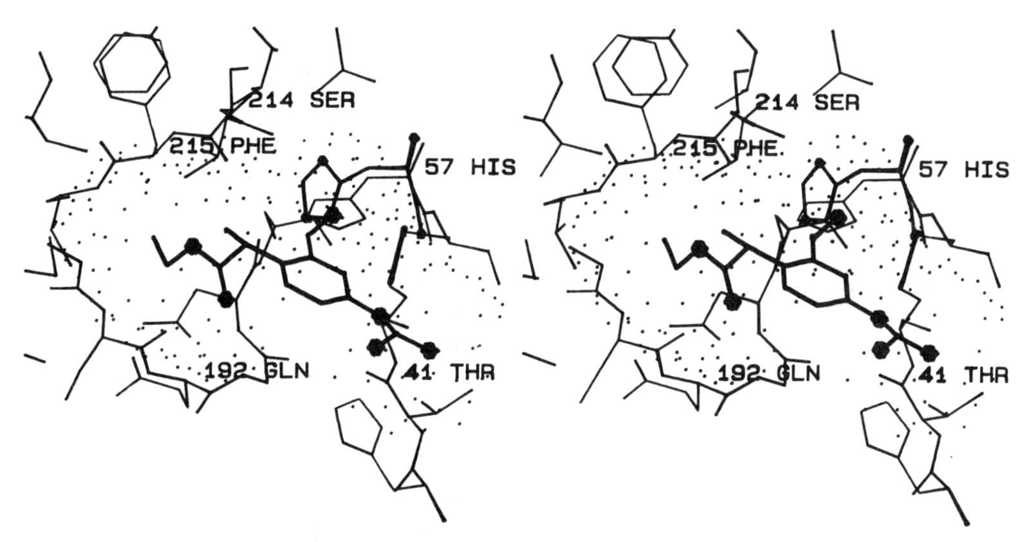

FIGURE 9: Stereo drawing of the complex formed upon reaction of PPE with 4-chloro-3-ethoxy-7-guanidinoisocoumarin. The heteroatoms of the inhibitor are drawn as spheres. His-57 in native PPE is drawn with heavier lines in the "in" orientation, as found in the native enzyme and peptide complexes. In this complex, the imidazole ring of His-57 (thin lines) is shifted to the "out" position, stacking on the phenyl group of the inhibitor. The ethoxy group of the inhibitor is in the S_1 pocket (center left), and the guanidino group is hydrogen-bonded to Thr-41. The covalent bond between the inhibitor and Ser-195 Oγ has not been drawn. The interface contact surface is depicted as small dots.

al., 1987). Unexpectedly, 4-chloro-3-ethoxy-7-guanidinoisocoumarin (**2**, R = Et) proved to be a potent inhibitor of HLE ($k_{obsd}/[I]$ = 81 000 M^{-1} s^{-1}) and a slow inhibitor of PPE ($k_{obsd}/[I]$ = 2300 M^{-1} s^{-1}). Due to its good solubility, it reacted readily with crystals of PPE (Figure 9; E. F. Meyer and R. Radhakrishnan, unpublished results). As in the cases of the other benzoxazinone and isocoumarin inhibitors, the reaction product is covalently bound to Ser-195 Oγ (Figure 9). The ethoxy group is found in the S_1 pocket, in agreement with the P_1 preference of PPE for small hydrophobic groups.

The phenyl group is approximately parallel to and in van der Waals contact with the outward rotated imidazole groups of His-57. The chlorine atom of the isocoumarin has not been displaced but is near His-57. The 7-guanidino group makes three good hydrogen bonds to Thr-41. The benzoyl carbonyl group produced by the isocoumarin ring opening is not located in the oxyanion hole, perhaps again explaining the stability of the complex toward deacylation. The enzyme contains no charge-compensating groups near the 7-guanidino moiety.

Molecular modeling of the 7-guanidinoisocoumarin–PPE complex suggested that the addition of small alkyl groups (*tert*-butyl) to the guanidino group might increase affinity due to the presence of a small hydrophobic pocket near the terminal nitrogen atom of the guanidino group. Indeed, replacement of the 7-H_2N–C(=NH_2^+)–NH group with a *tert*-butyl–NH–CO–NH– group led to a 3.5-fold increase in $k_{obsd}/[I]$ (both the 7-guanidino and the 7-H_2N–CO–NH variants are equally potent inhibitors, and the *tert*-butylurea derivative was synthetically more accessible than a *tert*-butylguanidine derivative).

Thr-41 is replaced in HLE by Phe-41, which cannot form the same side-chain hydrogen bonds with the guanidino group. Thus it is not clear why 4-chloro-3-ethoxy-7-guanidinoisocoumarin is such a good inhibitor of HLE and whether it indeed binds in the same manner as with PPE. However, replacement of the 7-guanidino group with large aromatic acyl groups such as 7-(Tos-Phe-NH-) yields even more potent HLE inhibitors ($k_{obsd}/[I]$ = 190 000 M^{-1} s^{-1}; Harper & Power, 1985), which could partially be explained by an interaction with Phe-41 in HLE.

β-Lactam Elastase Inhibitors. β-Lactam antibiotics are widely prescribed and well-tolerated inhibitors of D-Ala-D-Ala transpeptidases and other enzymes involved in bacterial cell wall biosynthesis. The active drugs typically have a free carboxyl group and are thought to mimic D-Ala-D-Ala. Medicinal chemists have tailored other β-lactams to inhibit some β-lactamases. A group at Merck recognized the similarity of these β-lactamases to serine proteases and tested β-lactams as inhibitors of HLE. Esters of cephalosporin and other β-lactams were thus discovered to be potent acylating inhibitors of HLE (Doherty et al., 1986). In contrast to other enzymes inhibited by β-lactams, HLE requires the carboxyl group of the antibiotic to be esterified for effective inhibition. The cephalosporin *tert*-butyl ester shown in Figure 10 irreversibly inhibits HLE (K_I = 0.18 μM; k_2/K_I = 161 000 M^{-1} s^{-1}). Most of the derivatives reported were also excellent inhibitors of PPE.

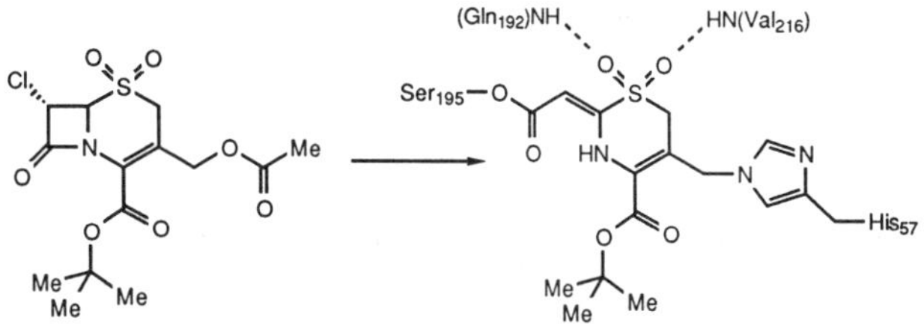

FIGURE 10: Structure of the acyl enzyme formed upon inhibition of PPE by a cephalosporin β-lactam inhibitor.

The structure of PPE inhibited by 3-(acetoxymethyl)-7α-chloro-3-cephem-4-carboxylate 1,1-dioxide (Figure 10) has been determined (Navia et al., 1987). Two bonds are formed between the inhibitor and the enzyme, an ester linkage with Ser-195 and a bond between the imidazole ring of His-57 and the 3′-methylene group. During the inhibition reaction, the β-lactam ring is opened and both the chloro and acetoxy groups are lost, obviously an excellent example of a mechanism-based or suicide inhibitor. In addition to the two covalent bonds, the sulfone oxygen atoms are hydrogen-bonded to the NH of Val-216 and the side chain of Gln-192. The 7′α-chloro group probably occupies the S_1 pocket during the initial acylation reaction, while the *tert*-butyl group fits at the edge of a large open area on the periphery of the enzyme.

Modeling and Inhibitor Design

Due to the absence of strong (i.e., Coulombic) interactions in elastase–ligand complexes, these studies have helped to

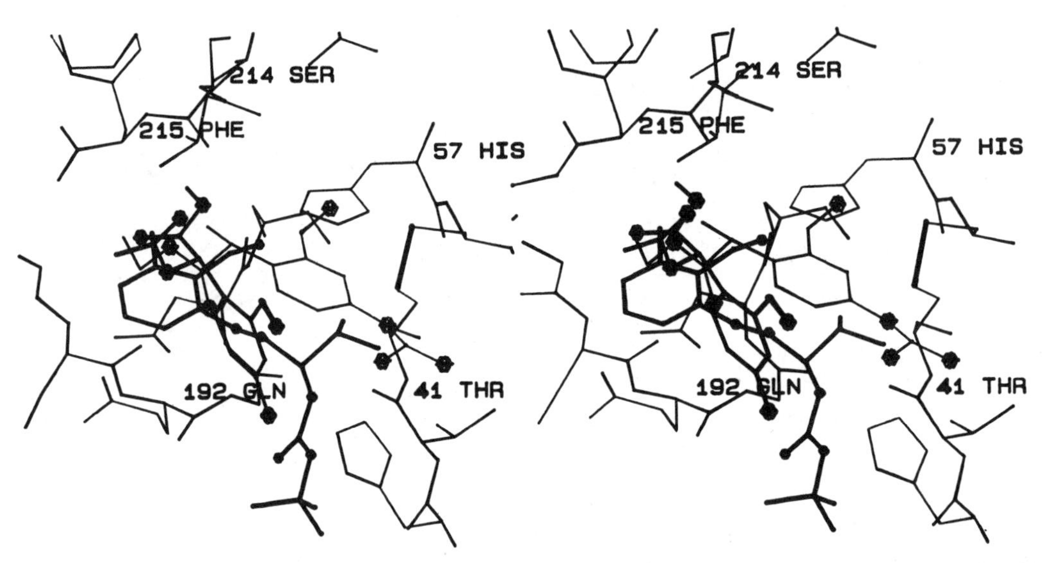

FIGURE 11: Stereo drawing of the strikingly different orientations of three heterocyclic inhibitors bound to the active site of PPE. The PPE complex with 5-Cl- or 5-Me-2-[Boc-NHCH(*i*-Pr)]benzoxazinone is drawn with heavy lines. The *tert*-butyl group of the inhibitor reaches down to the S_2' subsite of PPE, and the 5-chloro (or 5-methyl) substituent of the inhibitor is in the primary (S_1) specificity pocket (center left). Lines of intermediate thickness depict the 7-amino-4-chloro-3-methoxyisocoumarin complex. An acetate has displaced chloride from the inhibitor and occupies the S_1 pocket. The benzoyl carbonyl oxygen atom is in the oxyanion hole. Thin lines are used to depict the 4-chloro-3-ethoxy-7-guanidinoisocoumarin complex with PPE in the same orientation as in Figure 8. The PPE active-site conformation depicted here corresponds to the recently determined structure of the complex of PPE with 4-chloro-3-ethoxy-7-guanidinoisocoumarin (Figure 8).

define the decisive role played by weak (i.e., van der Waals) interactions in enzyme–ligand interactions. A common factor of the elastase–inhibitor complexes is the close fit of van der Waals interaction surfaces between the inhibitor structure and the convoluted surface of the enzyme. A remarkable feature of the PPE complexes formed with heterocyclic inhibitors is the variety of binding modes of chemically and structurally related compounds (Figure 11). This observation must be regarded as a "caveat" to chemists who wish to use an analogous crystal structure for molecular modeling and inhibitor design. We clearly do not yet know all the rules for predicting a preferred binding conformation of a small molecule inhibitor to the active site of these enzymes. Energy minimization is an obedient servant which will bring a proposed model structure to its local minimum, but molecular dynamics and energy minimization methods still must be improved considerably to generate the actual conformation. Thus, there is a compelling need for additional structural information on many complexes of small molecule inhibitors of varying structures with elastases and other serine proteases. Only with such information in hand will it be possible to model new inhibitor structures into the active sites of serine proteases with any degree of confidence.

Summary and Future Directions

The crystal structures of HLE and PPE with peptides and peptide derivatives are remarkably similar (a property shared with other homologous serine proteases). Both enzymes have similar primary substrate specificities by virtue of the similar geometric and hydrophobic character of their S_1 binding pockets. However, the enhanced hydrophobicity of the HLE active site compared with the PPE active site that corresponds to differences in primary sequence is probably responsible for its affinity for longer aliphatic chains at the P_1 residue of substrates and for the binding of long alkyl chain inhibitors (e.g., oleic acid).

In contrast to peptide derivatives, the crystal structures of heterocyclic inhibitors of PPE show dramatic differences. Again the increased reactivity of hydrophobic inhibitors, such as isocoumarins for HLE, may reflect the increased hydrophobicity of the HLE active site. Molecular modeling of the crystal structures, when used with caution, often provides new insights into the binding mode of inhibitors, suggests ways to exploit differences in homologous enzymes to improve specificity, and leads to the discovery of novel inhibitor structures.

A variety of elastase inhibitors have been shown to be effective in animal models of pulmonary emphysema, inflammation, and other related diseases (Powers & Bengali, 1986). These include peptide chloromethyl ketones (Stone et al., 1981), peptide aldehydes (Hassal et al., 1985), peptide boronic acids (Sosket et al., 1986), β-lactams (Doherty et al., 1986), and the protein protease inhibitor eglin c. Human α_1-protease inhibitor has already been used in therapy with PiZ (α_1-protease inhibitor deficiency) patients (Gadek et al., 1981). Several additional inhibitors will be tested in the near future for treatment of disease in humans. At present it is most likely that practical therapeutic drugs will originate from either the β-lactam, peptide boronic acid, or peptide fluoro ketone classes of elastase inhibitors since current classes of heterocyclic inhibitors suffer from low plasma stability. The structural information obtained with elastase–inhibitor complexes should be invaluable for future design work with all classes of elastase inhibitors and should improve the prospects for the treatment of chronic diseases such as pulmonary emphysema.

Acknowledgments

We thank Dr. James Travis and his research group at the University of Georgia for the neutrophil enzymes used in this research and for many hours of valuable and stimulating discussions.

References

Aoki, Y. (1978) *J. Biol. Chem. 253*, 2026–2032.

Ashe, B. M., & Zimmerman, M. (1977) *Biochem. Biophys. Res. Commun. 75*, 184–189.

Baici, A., & Seemueller, U. (1984) *Biochem. J. 218*, 829–833.

Baici, A., Salgam, P., Fehr, K., & Boni, A. (1980) *Biochem. Pharmacol. 29*, 1723–1727.

Baugh, R. J., & Travis, J. (1976) *Biochemistry 15*, 836–841.

Bieth, J. G. (1986) in *Regulation of Matrix Accumulation* (Mechan, R. D., Ed.) pp 217–320, Academic Press, New York.

Birktoft, J. J., & Blow, D. M. (1972) *J. Mol. Biol. 68*, 187–240.

Bizzozero, S. A., & Zweifel, B. O. (1975) *FEBS Lett. 59*, 105–108.

Bizzozero, S. A., & Dutler, H. (1981) *Bioorg. Chem. 10*, 46–62.

Bode, W., & Schwager, P. (1975) *J. Mol. Biol. 98*, 693–717.

Bode, W., & Huber, R. (1986) in *Molecular and Cellular Basis of Digestion* (Desnuelle, P., Sjostrom, H., & Noren, O., Eds.) pp 213–234, Elsevier, Amsterdam.

Bode, W., Chen, Z., Bartels, K., Kutzbach, C., Schmidt-Kastner, G., & Bartunik, H. (1983) *J. Mol. Biol. 164*, 237–282.

Bode, W., Papmokos, E., Musil, D., Seemueller, U., & Fritz, H. (1986a) *EMBO J. 5*, 813–818.

Bode, W., Wei, A.-Z., Huber, R., Meyer, E., Travis, J., & Neumann, S. (1986b) *EMBO J. 5*, 2453–2458.

Braun, N. J., Bodmer, J. L., Virca, G. D., Metz-Virca, G., Maschler, R., Bieth, J. G., & Schnebli, H.-P. (1987) *Biol. Chem. Hoppe-Seyler 368*, 299–308.

Clore, G. M., Gronenborn, A. M., Carlson, G., & Meyer, E. (1986) *J. Mol. Biol. 190*, 259–267.

Cohen, G. H., Silverton, E. W., & Davies, D. (1981) *J. Mol. Biol. 148*, 449–479.

Deslongchamps, P. (1977) *Heterocycles 7*, 2463–2490.

Doherty, J. B., Ashe, B. M., Argenbright, L. W., Barker, P. L., Bonney, R. J., Chandler, G. O., Dahlgren, M. E., Dorn, C. P., Finke, P. E., Firestone, R. A., Fletcher, D., Hagmann, W. K., Mumford, R., O'Grady, L., Maycock, A. L., Pisano, J. M., Shah, S. K., Thompson, K. R., & Zimmerman, M. (1986) *Nature 322*, 192–194.

Farley, D., Salvesen, G., & Travis, J. (1988) *Biol. Chem. Hoppe-Seyler 369 (Suppl.)*, 3–7.

Fujinaga, M., & James, M. N. G. (1987) *J. Mol. Biol. 195*, 373–396.

Gadek, J. E., Klein, H. G., Holland, P. V., & Crystal, R. G. (1981) *J. Clin. Invest. 68*, 1158–1165.

Gruetter, M. G., Fendrich, G., Huber, R., & Bode, W. (1988) *EMBO J. 7*, 345–351.

Harper, J. W., & Powers, J. C. (1985) *Biochemistry 24*, 7200–7213.

Harper, J. W., Cook, R. R., Roberts, C. J., McLaughlin, B. J., & Powers, J. C. (1984) *Biochemistry 23*, 2995–3002.

Harper, J. W., Hemmi, K., & Powers, J. C. (1985) *Biochemistry 24*, 1831–1841.

Hartley, B. S. (1964) *Nature 201*, 1284.

Hassall, C. H., Johnson, W. H., & Roberts, N. A. (1979) *Bioorg. Chem. 8*, 299–309.

Hassal, C. H., Johnson, W. H., Kennedy, A. J., & Roberts, N. A. (1985) *FEBS Lett. 183*, 201–205.

Huber, R., & Bode, W. (1978) *Acc. Chem. Res. 11*, 114–122.

Hudig, D., Gregg, N. J., Kam, C., & Powers, J. C. (1987) *Biochem. Biophys. Res. Commun. 149*, 882–888.

Hughes, D. L., Sieker, L. C., Bieth, J., & Dimicoli, J. L. (1982) *J. Mol. Biol. 162*, 645–658.

Imperiali, B., & Abeles, R. (1986) *Biochemistry 25*, 3760–3767.

Janoff, A., Carp, H., Laurent, P., & Raju, L. (1983) *Am. Rev. Respir. Dis. 127*, 31–38.

Kam, C., Fujikawa, K., & Powers, J. C. (1988) *Biochemistry 27*, 2547–2557.

Krantz, A., Spencer, R. W., Tam, T. F., Thomas, E., & Copp, L. J. (1987) *J. Med. Chem. 30*, 589–591.

Kraut, J. (1977) *Annu. Rev. Biochem. 46*, 331–358.

Laskowski, M., & Kato, I. (1980) *Annu. Rev. Biochem. 49*, 593–626.

Lentini, A., Farchione, F., Bernai, B., Kreula-Ongarjnukool, N., & Tovivich, P. (1987) *Biol. Chem. Hoppe-Seyler 368*, 369–378.

Lestienne, P., & Bieth, J. (1980) *J. Biol. Chem. 255*, 9289–9294.

MacDonald, R. J., Swift, G. H., Quinto, C., Swain, W., Pictet, R. L., Nikovits, W., & Rutter, W. J. (1982) *Biochemistry 21*, 1453–1463.

Marquart, M., Walter, J., Deisenhofer, J., Bode, W., & Huber, R. (1983) *Acta Crystallogr. B39*, 480–490.

Matheson, N. R., Janoff, A., & Travis, J. (1982) *Mol. Cell. Biochem. 45*, 65–71.

McRae, B., Nakajima, K., Travis, J., & Powers, J. C. (1980) *Biochemistry 19*, 3973–3978.

Meyer, E., Presta, L. G., & Radhakrishnan, R. (1985) *J. Am. Chem. Soc. 107*, 4091–4093.

Meyer, E. F., Radhakrishnan, R., Cole, G. M., & Presta, L. G. (1986) *J. Mol. Biol. 189*, 533–539.

Meyer, E. F., Cole, G., Radhakrishnan, R., & Epp, O. (1988a) *Acta Crystallogr. B44*, 26–38.

Meyer, E. F., Clore, G. M., Gronenborn, A. M., & Hansen, H. A. S. (1988b) *Biochemistry 27*, 725–730.

Nakajima, K., Powers, J. C., & Zimmerman, M. (1979) *J. Biol. Chem. 254*, 4027–4032.

Navia, M. A., Springer, J. P., Lin, T.-Y., Williams, H. R., Firestone, R. A., Pisano, J. M., Doherty, J. B., Finke, P. E., & Hoogsteen, K. (1987) *Nature 327*, 79–82.

Navia, M. A., McKeever, B. M., Springer, J. P., Lin, T.-Y., Williams, H. R., Fluder, E. M., Dorn, C. D., & Hoogsteen, K. (1989) *Proc. Natl. Acad. Sci. U.S.A. 86*, 7–11.

Neurath, H. (1986) *J. Cell. Biochem. 32*, 35–49.

Okano, K., Aoki, Y., Sakurai, T., Kajitani, M., Kanai, S., Shimazu, H., & Naruto, M. (1987) *J. Biochem. 102*, 13–16.

Papamokos, E., Weber, E., Bode, W., Huber, R., Empie, M. W., Kato, I., & Laskowski, M. (1982) *J. Mol. Biol. 158*, 515–537.

Powers, J. C. (1977) in *Chemistry and Biochemistry of Amino Acids, Peptides and Proteins* (Weinstein, B., Ed.) Vol. 4, pp 65–178, Marcel Dekker, New York.

Powers, J. C., & Bengali, Z. H. (1986) *Am. Rev. Respir. Dis. 134*, 1097–1100.

Powers, J. C., & Harper, J. W. (1986) in *Proteinase Inhibitors* (Barrett, A. J., & Salvensen, G., Eds.) pp 55–152, Elsevier, Amsterdam.

Powers, J. C., Gupton, B. F., Harley, A. D., Nishino, N., & Whitley, R. J. (1977) *Biochim. Biophys. Acta 485*, 156–166.

Radhakrishnan, R., Presta, L. G., Meyer, E. F., & Wildonger, R. (1987) *J. Mol. Biol. 198*, 417–424.

Remington, S. J., Woodbury, R. G., Reynolds, R. A., Matthews, B. W., & Neurath, H. (1988) *Biochemistry 27*, 8097–8105.

Rosenberg, S., Barr, P. S., Najarian, R., & Hallewell, R. A. (1984) *Nature 312*, 77–80.

Sawyer, L., Shotton, D. M., Campbell, J. W., Wendell, P. L., Muirhead, H., Watson, H. C., Diamond, R., & Ladner, R. C. (1978) *J. Mol. Biol. 118*, 137–208.

Schechter, I., & Berger, A. (1967) *Biochem. Biophys. Res. Commun. 27*, 157–162.

Seemueller, U., Arnhold, M., Fritz, H., Wiedenmann, K., Machleidt, W., Heinzel, R., Appelhans, H., Gassen, H. G.,

& Lottspeich, F. (1986) *FEBS Lett. 199*, 43–48.
Shen, W.-F., Fletcher, T. S., & Largman, C. (1987) *Biochemistry 26*, 3447–3452.
Shotton, D. M., & Hartley, B. S. (1970) *Nature 225*, 802–806.
Shotton, D. M., & Watson, H. C. (1970) *Nature 225*, 811–816.
Sinha, S., Watorek, W., Karr, S., Giles, J., Bode, W., & Travis, J. (1987) *Proc. Natl. Acad. Sci. U.S.A. 84*, 2228–2232.
Soskel, N. T., Watanabe, S., Hardie, R., Shenvi, A. B., Punt, J. A., & Kettner, C. (1986) *Am. Rev. Respir. Dis. 133*, 639–642.
Spencer, R. W., Copp, L. J., Bonaventura, B., Tam, T. F., Liak, T. J., Billedeau, R. J., & Kranz, A. (1986) *Biochem. Biophys. Res. Commun. 140*, 928–933.
Stein, R. L. (1985) *Arch. Biochem. Biophys. 236*, 677–680.
Stein, R. L., Strimpler, A. M., Hori, H., & Powers, J. C. (1987a) *Biochemistry 26*, 1301–1305.
Stein, R. L., Strimpler, A. M., Hori, H., & Powers, J. C. (1987b) *Biochemistry 26*, 1305–1314.
Stein, R. L., Strimpler, A. M., Edwards, P. E., Lewis, J. J., Mauger, R. C., Schwartz, J. A., Stein, M. M., Trainor, D. A., Wildonger, R. A., & Zottola, M. A. (1987c) *Biochemistry 26*, 2682–2689.
Stein, R. L., Strimpler, A. M., Viscarello, B. R., Wildonger, R. A., Mauger, R. C., & Trainor, D. A. (1987d) *Biochemistry 26*, 4126–4130.
Stone, P. J., Lucey, E. C., Calore, J. D., Snider, G. L., Franzblau, C., Castillo, M. J., & Powers, J. C. (1981) *Am. Rev. Respir. Dis. 124*, 56–59.
Takahashi, H., Nukiwa, T., Basset, T., & Crystal, R. G. (1988a) *J. Biol. Chem. 263*, 2543–2547.
Takahashi, L. H., Radhakrishnan, R., Rosenfield, R. E., Meyer, E. F., Trainor, D. A., & Stein, M. (1988b) *J. Mol. Biol. 201*, 423–428.
Takahashi, L. H., Radhakrishnan, R., Rosenfield, R. E., Meyer, E. F., & Trainor, D. A. (1989) *J. Am. Chem. Soc.* (submitted for publication).
Tani, T., Ohsumi, J., Mita, K., & Takiguchi, J. (1988) *J. Biol. Chem. 263*, 1231–1239.
Teshima, T., Griffin, J. C., & Powers, J. C. (1982) *J. Biol. Chem. 257*, 5085–5091.
Thompson, R. D., & Ohlsson, K. (1986) *Proc. Natl. Acad. Sci. U.S.A. 83*, 6692–6696.
Trainor, D. A. (1987) *Trends Pharmacol. Sci. 8*, 303–307.
Travis, J., & Salvesen, G. (1983) *Annu. Rev. Biochem. 52*, 655–709.
Tschesche, H., Beckmann, J., Mehlich, A., Schnabel, E., Truscheit, E., & Wenzel, H. (1987) *Biochim. Biophys. Acta 913*, 97–101.
Tsukuda, H., & Blow, D. M. (1985) *J. Mol. Biol. 184*, 703–711.
Watson, H. C., Shotton, D. M., Cox, J. C., & Muirhead, H. (1970) *Nature 225*, 806–811.
Wei, A.-Z., Mayr, I., & Bode, W. (1988) *FEBS Lett. 234*, 367–373.
Wenzel, H. R., & Tschesche, H. (1981) *Hoppe-Seyler's Z. Physiol. Chem. 362*, 829–831.
Werb, Z., Banda, M. J., McKerrow, J. H., & Sandhaus, R. A. (1982) *J. Invest. Dermatol. 79*, 154s–159s.
Williams, H. R., Lin, T.-Y., Navia, M. A., Springer, J. P., McKeever, B. M., Hoogsteen, K., & Dorn, C. P. (1987) *J. Biol. Chem. 262*, 17178–17181.
Woodbury, R. G., Katunuma, N., Kobayashi, K., Titani, K., & Neurath, H. (1978) *Biochemistry 17*, 811–819.
Yasutake, A., & Powers, J. C. (1981) *Biochemistry 20*, 3675–3679.
Zimmerman, M., & Ashe, B. M. (1977) *Biochim. Biophys. Acta 480*, 241–245.

Chapter 10

How Do Serine Proteases Really Work?†

A. Warshel,* G. Naray-Szabo,‡ F. Sussman, and J.-K. Hwang
Department of Chemistry, University of Southern California, Los Angeles, California 90089-0482
Received December 27, 1988; Revised Manuscript Received February 16, 1989

ABSTRACT: Recent advances in genetic engineering have led to a growing acceptance of the fact that enzymes work like other catalysts by reducing the activation barriers of the corresponding reactions. However, the key question about the action of enzymes is not related to the fact that they stabilize transition states but to the question to *how* they accomplish this task. This work considers the catalytic reaction of serine proteases and demonstrates how one can use a combination of calculations and experimental information to elucidate the key contributions to the catalytic free energy. Recent reports about genetic modifications of the buried aspartic group in serine proteases, which established the large effect of this group (but could not determine its origin), are analyzed. Two independent methods indicate that the buried aspartic group in serine proteases stabilizes the transition state by electrostatic interactions rather than by alternative mechanisms. Simple free energy considerations are used to eliminate the double proton-transfer mechanism (which is depicted in many textbooks as the key catalytic factor in serine proteases). The electrostatic stabilization of the oxyanion side of the transition state is also considered. It is argued that serine proteases and other enzymes work by providing electrostatic complementarity to the *changes* in charge distribution occurring during the reactions they catalyze.

One of the most basic problems in biochemistry is the understanding of the molecular origin of enzyme catalysis [for a recent review, see Kraut (1988)]. The enormous advances in site-directed mutagenesis of enzymes [for review, see Knowles (1987)] and the production of catalytic antibodies (Tramontano et al., 1986; Jacobs et al., 1987) have led to a renewed interest in this question. Unfortunately, however, it seems that the problem is sometimes being reduced to the statement that "enzymes work because they work". That is, it is now repeatedly stated that Pauling (1946) was right and that enzymes work because they stabilize the transition state of the reacting system. Although a great insight has been provided by the pioneering statement about transtion-state stabilization (Pauling, 1946), those who have accepted the main premises of transition-state theory knew clearly that enzymes (as well as other catalysts) work by reducing the activation free energies of the corresponding reactions. This is in fact the figure drawn in countless textbooks [e.g., Dickerson et al. (1984)]. Thus the real question and the real puzzle do not boil down to whether or not enzymes reduce activation barriers but to how do they accomplish it? This problem is far from being trivial and cannot be resolved quantitatively by saying that enzymes provide "templates" that bind the transition state. That is, an enzyme can reduce the free energy difference between the ground state and the transition state by 10 kcal/mol, relative to the same system in water. Apparently, it is hard to account for such an enormous effect by a consistent interpolation of related factors from chemical reactions in solutions to enzyme active sites. For example, an optimal complement between the van der Waals interactions of an active site and the substrate's geometry cannot account for more than 2 kcal/mol if the flexibility of the enzyme is taken into account (Warshel & Levitt, 1976). In fact, many proposals attempted to address this crucial question [for a review, see Page (1987)] but only now with the emergence of genetic engineering can one distinguish between different proposals and assess their merits.

This paper explores what is probably the most extensively studied enzyme family (the serine protease family) and demonstrates the feasibility of quantitative examination of different

†This work was supported by Grant GM-24492 from the National Institutes of Health and by Contract N00014-87-K0507 from the Office of Naval Research.

*To whom correspondence should be addressed.
‡Permanent address: CHINOIN Pharmaceutical and Chemical Works, P.O. Box 110, H-1325 Budapest, Hungary.

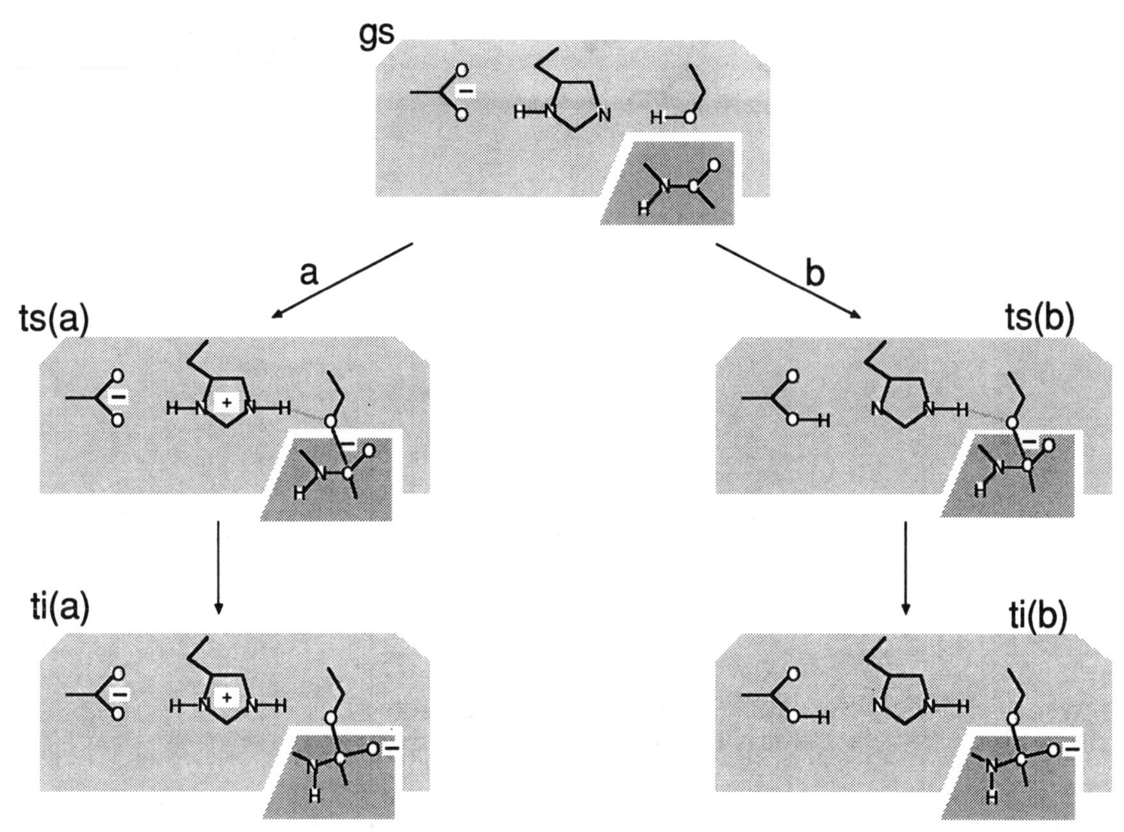

FIGURE 1: Schematic drawings displaying the possible reaction mechanisms for the formation of the tetrahedral intermediate (ti) in serine proteases. Route a, electrostatic catalysis; route b, double PT mechanism. gs, ts, and ti denote ground state, transition state, and tetrahedral intermediate state, respectively. The figure is merely a schematic diagram, and the actual structure of the transition state is given in Figure 3 (see caption of Table I for more details on the transition state). The shaded substrate region does not indicate planarity but only designates the substrate region.

proposals for the catalytic action of enzymes. The proposal that electrostatic free energy is the key factor in enzyme catalysis (Warshel, 1978, 1981; Naray-Szabo & Bleha, 1982) gains further support in this specific test case.

The central machinery of serine proteases is composed of the three invariant residues Asp_c–His_c–Ser_c (where c denotes a catalytic residue), which are referred to as the *catalytic triad*. These residues catalyze the hydrolysis of peptide bonds in proteins and peptides by a mechanism whose rate-limiting step can be formally described as

$$\mathrm{Im} + \mathrm{R'{-}OH} + \mathrm{X{-}C({=}O)R} \rightarrow \mathrm{Im^{+}{-}H} + \mathrm{R'OC(O^{-})(R)X} \quad (1)$$

where Im and R′—OH designate His_c and Ser_c, respectively. The origin of the great catalytic activity of this system and the specific role of Asp_c have been the subject of continuous debate since the early structural studies of serine proteases [for reviews, see Kraut (1977), Stroud et al. (1977), Markley (1979), and Schowen (1988)].

The pioneering work of Blow and co-workers (Matthews et al., 1967; Blow et al., 1969; Henderson et al., 1971), which was the first to reveal the structure of this catalytic system, suggested an electronic rearrangement of the form Asp_c^- His_c Ser_c → Asp_c His_c Ser_c^- as a factor that increases the nucleophilicity of Ser_c. This implied that the Asp_c His_c pair has a zero net charge at the transition state but did not specify explicitly its internal charge distribution and protonation state. However, the bonding diagram drawn in the original work could be interpreted as assuming the (Asp_c–H His_c) rather than the (Asp_c^- H–His_c^+) tautomeric form. In fact, subsequent workers [e.g., Hunkapillar et al. (1973)] interpreted the electronic rearrangement idea as an explicit mechanism that involves a concerted transfer of two protons (from Ser_c to His_c and from His_c to Asp_c, as described in Figure 1b). This mechanism [see discussion in Markley (1979)], which is now depicted in many textbooks, will be called here the double proton-transfer (double PT) mechanism. The acceptance of this mechanism in the chemical community might have been motivated by the recognition that ion pairs are not stable in nonpolar environments so that Asp_c^- His_c^+ must be less stable than Asp_c His_c in nonpolar active sites [see Dewar (1986)]. This, however, overlooks the fact that the active site around Asp_c is very polar (see below). Some workers [e.g., Polgar and Bender (1969), Rogers and Bruice (1974), and Brayer et al. (1978)] have objected to the double PT mechanism and more recent nuclear magnetic resonance experiments (Bachouchin & Roberts, 1978; Markley, 1979) as well as neutron diffraction experiments (Kossiakoff & Spencer, 1981) gave evidence against ground-state proton transfer but could not exclude a transfer at the transition state.

Most quantum mechanical calculations that consider only the catalytic triad without its surrounding protein [e.g., Scheiner et al. (1975), Umeyama et al. (1973), and Dewar and Storch (1985)] supported the double PT mechanism. One such calculation (Kollman & Hayes 1981) that considered the effect of Ser_c concluded that Asp_c^- His_c^+ Ser_c^- is more stable than Asp_c His_c Ser_c^- in the gas phase. Calculations that attempted to include the effect of the protein active site on the catalytic triad (Umeyama et al., 1981; Warshel et al., 1982; Naray-Szabo, 1982, 1983; Warshel & Russell, 1986) contradicted the double PT mechanism, but these latter calculations [with the exception of Warshel et al. (1982) and Warshel and Russell (1986)] did not include the key effect of the solvent around the protein and did not calibrate the intrinsic energy of the ion pair on reliable experimental information. This prevented a quantitative assessment of the

Table I: Electrostatic Energy Changes Calculated by the PDLD[a] Method for the Native (N) and Mutant (M) Enzymes

	native			mutant			
	ts	gs	$\Delta G^{\ddagger}_{N} - \lambda^{\ddagger}$	ts	gs	$\Delta G^{\ddagger}_{M} - \lambda^{\ddagger}$	$\Delta\Delta G^{\ddagger}_{M\to N}$
Trypsin							
V_{QQ}	−95.3	−36.5	−58.8	−47.5	−9.8	−37.7	21.1
$V_{Q\mu}$	−42.5	−28.2	−14.3	−25.2	−10.9	−14.3	0.0
$V_{Q\alpha}$	−4.4	0.0	−4.4	−13.2	0.0	−13.2	−8.8
G_{Qw}	−11.7	−14.8	3.1	−14.6	−10.8	−3.8	−6.9
sum	−153.9	−79.5	−74.4	−100.4	−31.5	−69.0	5.4[b] (6.1)[c]
Subtilisin							
V_{QQ}	−97.5	−35.7	−61.8	−46.4	−6.4	−40.0	21.8
$V_{Q\mu}$	−61.5	−53.5	−8.0	−12.0	−4.0	−8.0	0.0
$V_{Q\alpha}$	−0.7	0.0	−0.7	−8.7	0.0	−8.7	−8.0
G_{Qw}	−22.6	−17.9	−4.7	−32.3	−18.2	−14.1	−9.4
sum	−182.3	−107.1	−75.2	−99.4	−28.6	−70.8	4.4 (6.0)[a]

[a] Energies are in kilocalories per mole. Experimental values, $\Delta\Delta G^{\ddagger}_{exp} = -1.38[\log (k^{M}_{cat}/k^{N}_{cat})]$, are in parentheses. The experiments are pH dependent, and the results quoted for subtilisin (Carter & Wells, 1988) were obtained at pH = 8.6, while in the case of trypsin the experimental result was obtained from the hydroxide-independent rate constant [see Craik et al. (1987) for details]. ts and gs refer to transition and ground states, respectively. $\lambda^{\ddagger}$ is the nonelectrostatic contribution to the formation of the transition state (see eq 4). For other notations, see text. The three-dimensional coordinates were taken from X-ray diffraction studies of trypsin (Huber et al., 1974) and subtilisin (Robertus et al., 1972). Atomic charges of region 1 were taken from Warshel and Russell (1986) except for the ti, whose charges were modified to −0.5, 0.7, and −1.2, respectively, for the serine oxygen, the carbonyl carbon, and the carbonyl oxygen [the set of Warshel and Russell (1986) has zero charge on the serine oxygen of ti]. The transition-state charges (−0.70, 0.58, −0.88) and structure were obtained from the EVB calculations of Warshel et al. (1988), weighting the charges of the O^{-} C═O and O—C—O^{-} resonance forms by 0.4 and 0.6, respectively (the sensitivity of the calculations to the contributions of the two resonance structures can be judged by comparing the calculations for ts and ti in Figure 2). The weighting factors used for locating the transition state are obtained by the EVB procdure as described in Warshel et al. (1988). The O···C bond length at the transition state is around 2.2 Å, and the out-of-plane deformation of the carbonyl carbon is around 50°. The grid size and spacing for the Langevin dipoles, modeling the water molecules, were 12.0 Å and 3.0 Å, respectively. [b] The calculated value does not include the contribution of 1.8 kcal/mol (see text) associated with the rotation of His_c frosm its ground-state improper configuration (Sprang et al., 1987) to the configuration that is adequate for accepting a proton from Ser_c. The later configuration was used in the PDLD calculations. [c] This value is taken by using k_{enz} of Table II in Craik et al. (1987).

feasibility of the double PT mechanism.

It should also be mentioned that the search for catalysis by the double PT mechanism is a subject of intensive effort in biomimetic chemistry, and various model compounds that were meant to mimic the catalytic triad have been designed in recent years [e.g., Mallick et al. (1984) and Cram et al. (1986)].

The emergence of genetic engineering allows one to determine the contribution of individual residues to the catalytic activity of an enzyme. Recent site-directed mutagenesis of trypsin (Sprang et al., 1987; Craik et al., 1987) and subtilisin (Carter & Wells, 1988) has indicated clearly that Asp_c is important since its removal leads to a reduction of 4 orders of magnitude in k_{cat}. However, while these experimental findings establish that Asp_c contributes around 6 kcal/mol to the reduction of the activation free energy of amide hydrolysis, they cannot tell us how this effect is being obtained. To examine this point, one should determine the free energy contribution associated with Asp_c for different possible mechanisms. Such an examination is presented in this work.

Microscopic Calculations of the Electrostatic Contribution of Asp_c

In order to clarify the role of Asp_c, one has to consider the following alternatives (cf. Figure 1): (i) Asp_c is left in its ionized form and serves to stabilize the ionic transition state (route a), (ii) Asp_c is used to accept a proton from His_c in the transition state, thus creating by a double PT-type mechanism a neutral Asp_c His_c pair (route b), and (iii) Asp_c helps in orienting His_c in a proper position to interact with the substrate contributing entropically to the rate acceleration.

The examination of the possible electrostatic role of Asp_c is far from being trivial. Attempts to calculate the electrostatic interaction between Asp_c^{-} and the His_c^{+} t^{-} pair [where t^{-} designates the $(OCO)^{-}$ fragment] by conventional macroscopic models may amount to *assuming* the effect of Asp_c rather than calculating it. That is, one can use a Coulomb's law type expression for the interaction

$$\Delta G = -332[1/(R_1\epsilon) - 1/(R_2\epsilon)] \quad (2)$$

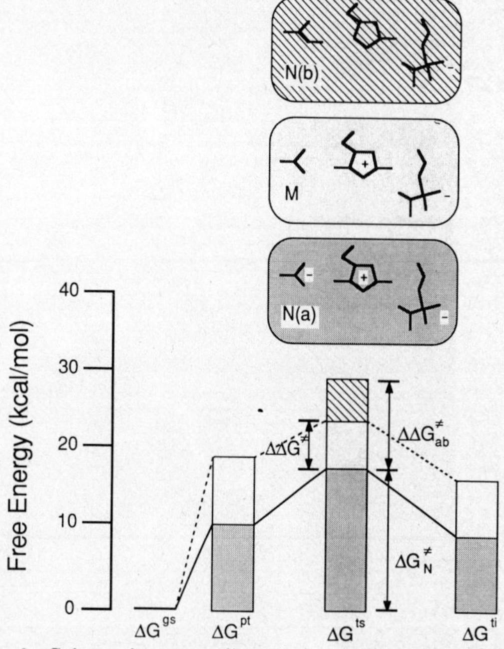

FIGURE 2: Schematic energy diagram for the formation of the tetrahedral intermediate in trypsin. For notations, see eq 5. Numerical values for ΔG^{i}_{N} (i = pt, ts, ti) were taken from earlier calculations (Warshel & Russell, 1986). $\Delta G^{ts}_{N(b)}$ denotes the activation energy in the double PT mechanism, $\Delta\Delta G^{\ddagger}_{ab} = \Delta G^{ts}_{N(b)} - \Delta G^{ts}_{N(a)}$, with $\Delta G^{ts}_{N(a)} = \Delta G^{ts}_{N}$. Subscripts a and b refer to routes a and b in Figure 1.

where R_1 and R_2 are respectively the distances (in angstroms) between the charge center of Asp_c^{-} and that of His_c^{+} and t^{-}, respectively, ϵ is the assumed dielectric constant, and the free energy ΔG is given in kilocalories per mole. R_1 and R_2 of around 3.4 and 6.3 Å reproduce the actual charge–charge interaction (as given by the V_{QQ} of Table I), but more sophisticated calculations should also consider the electrostatic energy in the ground state (where His_c is uncharged). A dielectric constant of about 40 [which is the typical value for

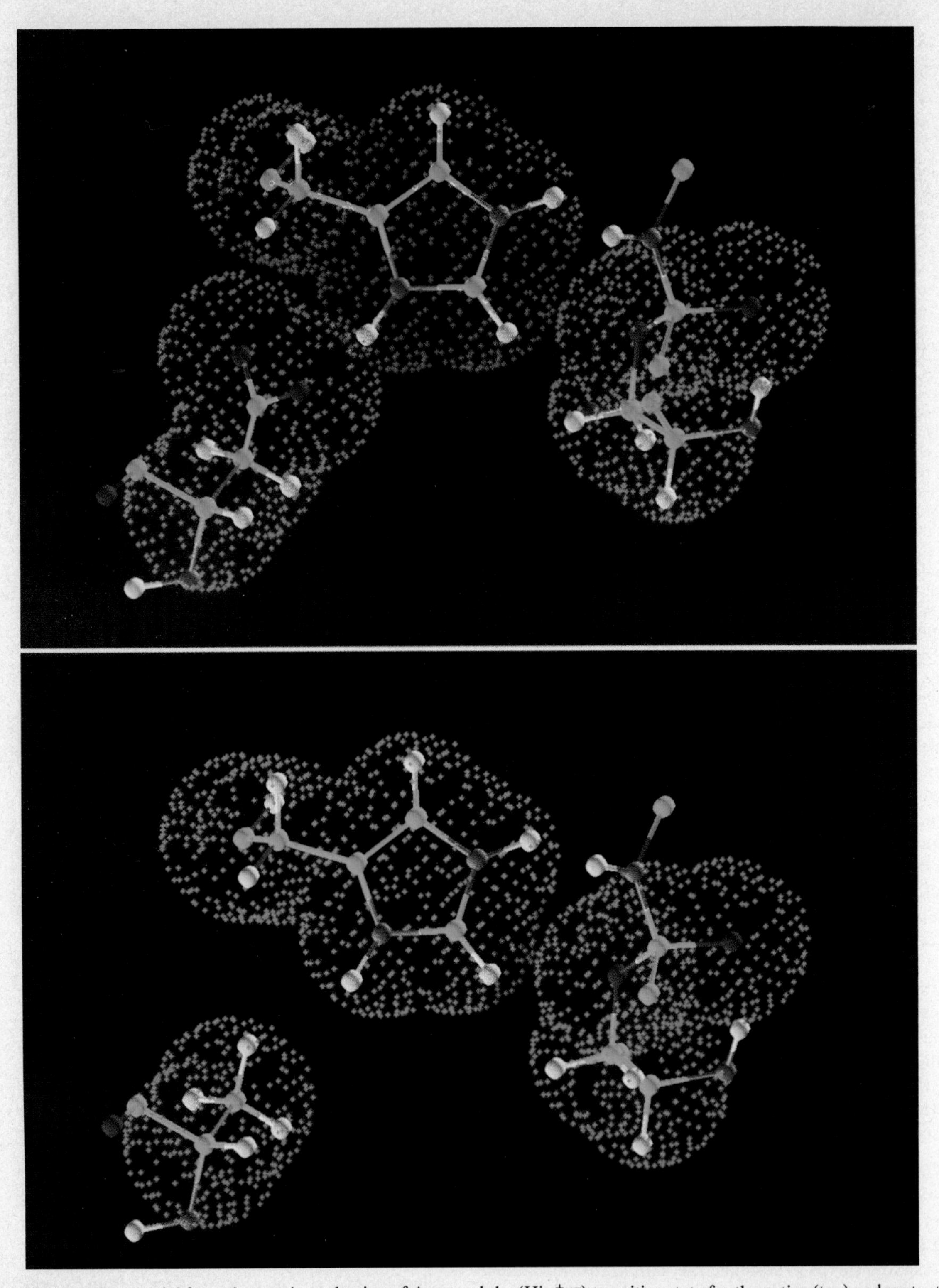

FIGURE 3: Electrostatic potential from the protein at the sites of Asp_c^- and the (His_c^+ t^-) transition state for the native (top) and mutant (bottom) subtilisin. The Asp_c, His_c, and t^- are located respectively at the left, center, and right of the figure. The electrostatic potential surface at 1.5 Å from the atoms of a given residue (for example, the His_c^+) is calculated by summing all the contributions from *other* residues that are within 7 Å of the given residue. Note that the contribution to the electrostatic potential on each residue is from its surrounding environment and not from the residue itself. Positive and negative potentials are given in blue and red, respectively. The figure shows that the large negative potential at the site of His_c, in the native enzyme, disappears upon substitution of Asp_c by Ala_c. This demonstrates that Asp_c plays a major role in stabilizing the (His^+ t^-) transition state.

charge–charge interactions between surface groups in proteins; see Warshel and Russell (1984)] will give $\Delta G \simeq 1$ kcal/mol, which is negligible compared to the ~6 kcal/mol observed effect of Asp_c. In fact, experiments in model compounds in solutions that gave such small effect (Rogers & Bruice, 1974) could serve as evidence that electrostatic stabilization by Asp_c cannot be important. Similar results would be obtained by the modified Tanford–Kirkwood method that places the ionized groups on the surface of the protein (Matthew et al., 1979). On the other hand, assuming a dielectric constant of around 2 for the entire system will give $\Delta G \simeq -22$ kcal/mol (or $\Delta G = -12$ kcal/mol if the energy of the neutral ground state is taken into account), which is consistent with the observed effect. The effect of Asp_c will be overestimated even if the solvent around the protein is modeled consistently with a large dielectric constant and a discretized continuum approach, while leaving the protein interior with $\epsilon = 2$. Furthermore, Asp_c will not be ionized in such a model. Only semimacroscopic models that explicitly include the protein polar groups (see eq 9 at the end of this paper) can reproduce the correct energetics, but the value of the relevant ϵ cannot be obtained from macroscopic considerations. Similar problems will arise with quantum mechanical studies that neglect the effect of the protein-induced dipoles and the solvent around the protein [e.g., Umeyama et al. (1981) and Naray-Szabo (1983)].

From the above discussion it appears that using a macroscopic model with an assumed dielectric constant cannot tell us what the actual role of Asp_c is (unless we known from experience with related systems what ϵ to use). It appears that the most effective way to determine the actual electrostatic interaction between Asp_c^- and His_c^+ is to use a microscopic model, which avoids the need of a dielectric constant altogether (in the microscopic world there is no need for a dielectric constant since all interactions are taken into account explicitly). The simplest and the first model used for microscopic calculations of electrostatic energies in proteins (Warshel & Levitt, 1976) is the protein dipoles langevin dipoles (PDLD). This model [see Russell and Warshel (1985)], which is now implemented in the convenient program POLARIS, is based on the assumption that the electrostatic energy of the average structure of the system is a good approximation for the electrostatic free energy of the system. The crystallographic coordinates are taken as the average protein structure, and the electrostatic contributions associated with the permanent residual charges and induced dipoles of the protein atoms are taken into account explicitly (the polarization of each induced dipole is evaluated self-consistently by considering the field from the charges of the system and other induced dipoles). The average polarization of the solvent around the protein is simulated by using a grid of polarizable point dipoles whose response to the local field is evaluated by a Langevin-type relationship (which is calibrated by fitting the model to microscopic simulations of solvated ions). The actual calculations are implemented by dividing the reacting system into the following three regions: (i) The catalytic triad (region 1), whose atoms are represented by residual charges ($\mathbf{Q}^i$) that change during the course of the reaction, proceeding from the ground state (gs) via a proton-transfer stage (pt) and the transition state (ts) to the tetrahedral intermediate (ti); (ii) the rest of the protein environment (region 2), which is represented by a standard set of permanent residual charges and atomic induced dipoles; (iii) the water molecules surrounding the active site (region 3), which are modeled by Langevin-type dipoles. More details on the definition and characteristics of these regions are given in Russell and Warshel (1985).

The total electrostatic free energy of a given catalytic triad charge distribution, $G(\mathbf{Q}^i)$, is estimated as the sum of the Coulombic interaction energy within region 1 (V_{QQ}), the interaction between the charges of region 1 and the permanent dipoles of region 2 ($V_{Q\mu}$), the energy of polarization of the protein-induced dipoles by the charges of region 1 ($V_{Q\alpha}$), and the free energy of polarization of the surrounding water molecules (G_{Qw})

$$G(\mathbf{Q}^i) = V^i_{QQ} + V^i_{Q\mu} + V^i_{Q\alpha} + G^i_{Qw} \quad (3)$$

Using the charge distributions of the ground state and state i, evaluated in earlier works by the empirical valence bond (EVB) method (Warshel & Russell, 1986), we can approximate the corresponding free energy difference by

$$\Delta G^i = G(\mathbf{Q}^i) - G(\mathbf{Q}^{gs}) + \lambda^i \quad (4)$$

where λ^i is the nonelectrostatic contribution associated with the formation of state i from the ground state. The change in activation energy as a result of a given mutation is given by

$$\Delta\Delta G^{\ddagger}_{M\to N} = \Delta G^{\ddagger}_M - \Delta G^{\ddagger}_N \sim [G_M(\mathbf{Q}^{ts}) - G_N(\mathbf{Q}^{ts})] - [G_M(\mathbf{Q}^{gs}) - G_N(\mathbf{Q}^{gs})] \quad (5)$$

where N and M designate the native and mutant enzymes, respectively.

The results of the calculations are summarized in Table I and Figure 2. The calculated values of $\Delta\Delta G^{\ddagger}$ are in a semiquantitative agreement with the corresponding observed values. According to these calculations (which consider only electrostatic effects) the replacement of Asp_c by a neutral residue results in destabilization of the transition state by more than 4 kcal/mol. Apparently, as illustrated in Figure 2, the transition state involves an ion pair between His_c^+ and t^-. Asp_c^- is located in an optimal position to stabilize the positive charge on His_c. This electrostatic effect can be examined in a very qualitative way by evaluating the electrostatic potential on the (His_c^- t^-) pair from the rest of the protein. As shown in Figure 3, the replacement of Asp_c by a noncharged residue leads to a major reduction in the negative potential on His_c, destabilizing the transition state.

The results of the PDLD calculations might still be considered by the reader as an accidental success, since nonelectrostatic contributions were not included. To establish the validity of these results, we repeated the calculations using the more sophisticated (and much more expensive) combination of the empirical valence bond (EVB) and the free energy perturbation (FEP) methods (Warshel & Sussman, 1986; Warshel et al., 1988). This approach uses the detailed all-atom model for the solvent and evaluates (at least in principle) the rigorous free energy of the ground state of a model Hamiltonian [for details, see Warshel et al. (1988)], allowing one to evaluate both the total $\Delta\Delta G^{\ddagger}$ and its electrostatic component. The results of EVB/FEP calculations, summarized in Table II, are quite similar to those obtained by the PDLD and to the corresponding experimental results.

ENERGETICS OF THE DOUBLE PT MECHANISM

The agreement between the calculated and observed results strongly supports the role of Asp_c as a key element in the electrostatic stabilization of the transition state. It is, however, important to examine other options. Let us first consider option 2, namely, that Asp_c is involved in the double PT mechanism (route b in Figure 1). This important option could not be explored quantitatively by early quantum mechanical calculations [e.g., Umeyama et al. (1983) and Naray-Szabo (1983)] since (as will be shown below) the actual difference between the two options is smaller than the error associated with the neglect of the surrounding water molecules and with the errors associated with calculations of the intrinsic gas-phase energy of the reacting fragments. However, one can explore the *difference* between route b and route a without the uncertainty associated with quantum mechanical calculations of large systems. That is, the energetics of the double PT mechanism can be obtained from the difference between the activation barriers of route b and route a by $\Delta G^{\ddagger}_b = \Delta G^{\ddagger}_a + \Delta\Delta G^{\ddagger}_{ab}$. If $\Delta\Delta G^{\ddagger}_{ab}$ is positive, then the double PT mechanism is not important. The free energy $\Delta\Delta G^{\ddagger}_{ab}$ is basically the free energy associated with a proton transfer from His_c to Asp_c at the transition state. This free energy can be evaluated in two

Table II: Summary of Results of the EVB–FEP Calculations[a]

enzyme	energy	$\Delta\Delta G^{\ddagger}$	$\Delta\Delta G^{\ddagger}_{exp}$
trypsin	electrostatic	5.4 ± 1.6	
	total	6.2[b] ± 1.6	6.1[c]
subtilisin	electrostatic	3.6 ± 1.2	
	total	4.1 ± 1.2	6.0[d]

[a] $\Delta\Delta G^{\ddagger}$ denotes the differences between the activation free energies of the native and the mutant enzymes [the transition-state structure is the same as obtained in Warshel et al. (1988)]. All energy values are in kilocalories per mole. Thes EVB–FEP calculations were performed by our molecular simulation package MOLARIS. The parameters used are given in Warshel et al. (1988). The reacting systems were represented by the EVB formalism with resonance structures that describe the ground state, proton transfer state, oxyanion state, and other relevant states [for more details, see Warshel and Russell (1986) and Warshel et al. (1988)]. The calculations involved in initial equilibration of 3 ps at 300 K, followed by free energy calculations in which the systems were driven by the EVB mapping potential in 20 discrete steps from the ground state, via the proton-transfer step to the oxyanion state. Each of the mapping steps involved 0.5-ps equilibration and 0.5-ps data collection. The differences between the forward and backward integration was less than 3 kcal/mol, and the average value was taken ass the calculated $\Delta\Delta G$. The calculated $\Delta\Delta G$ values changes by less than 0.2 kcal/mol with longer integration time. [b] The calculated value includes a contribution of 1.8 kcal/mol from the free energy associated with rotating His_c from its improper ground-state orientation to the configuration that is adequate for accepting a proton from Ser_c (see text). [c] Using k_{enz} of Craik et al. (1987). [d] Carter & Wells, 1988.

steps. First we estimate the free energy difference for the reference reaction in water. This is done by evaluating the free energy of proton transfer from His to Asp in water at the transition-state configuration $r^{\ddagger}$. The corresponding thermodynamic cycle can be expressed as the free energy of taking the fragments of [Asp^-–His^+–t^-] from $r^{\ddagger}$ to infinity, transferring the proton from His to Asp at $r = \infty$, and then bringing the fragments back from infinity to $r^{\ddagger}$. The free energy for the proton-transfer process is evaluated by using the expression (Warshel & Russell, 1984)

$$\Delta G^{w}_{PT}(\infty) = 1.38[pK_a(\text{His}) - pK_a(\text{Asp})] \simeq 4 \text{ kcal/mol} \quad (6)$$

The sum of the electrostatic free energies of taking the fragments of [Asp^-···His^+···t^-] from $r^{\ddagger}$ to ∞ and bringing them back to $r^{\ddagger}$ with the charge distribution of [Asp···His···t^-] can be evaluated by

$$\Delta\Delta G^{w}_{el}(a{\rightarrow}b,r{=}r^{\ddagger}) \simeq \Delta(332\sum_{ij} Q_iQ_j/r_{ij}\epsilon) = \Delta V^{\ddagger}_{QQ}(a{\rightarrow}b,r{=}r^{\ddagger})/\epsilon \simeq 2 \pm 1 \text{ kcal/mol} \quad (7)$$

where V_{QQ} designates charge–charge interaction in the vacuum and r_{ij}'s are the distances (in angstroms) between the atoms of the different fragments and where we use a uniform dielectric constant ($\epsilon = 40$). The approximation associated with using a large ϵ for charge–charge interaction in water is completely justified on the basis of countless experimental facts (Warshel & Russell, 1984), and the error associated with changing ϵ from 40 to 80 is around 1 kcal/mol. The same result obtained from eq 7 can be obtained by the reader from eq 2 using just one effective charge center for each fragment. Combining ΔG^{w}_{PT} and $\Delta\Delta G^{w}_{el}$ completes our thermodynamic cycle and gives an overall estimate of 6 kcal/mol for $\Delta\Delta G^{\ddagger}_{ab}$ in water (Figure 4a). This indicates that the double PT mechanism is quite unfavorable in an aqueous medium [see also Rogers and Bruice (1974)]. Replacing the solvent cage by the protein environment *increases* the energy loss by an additional 6 kcal/mol [as estimated by the PDLD calculations of Warshel and Russell (1986)], yielding a value of 12 kcal/mol for $\Delta\Delta G^{\ddagger}_{ab}$ (Figure 4b). This estimate indicates that the double PT mechanism is strongly unfavorable as compared

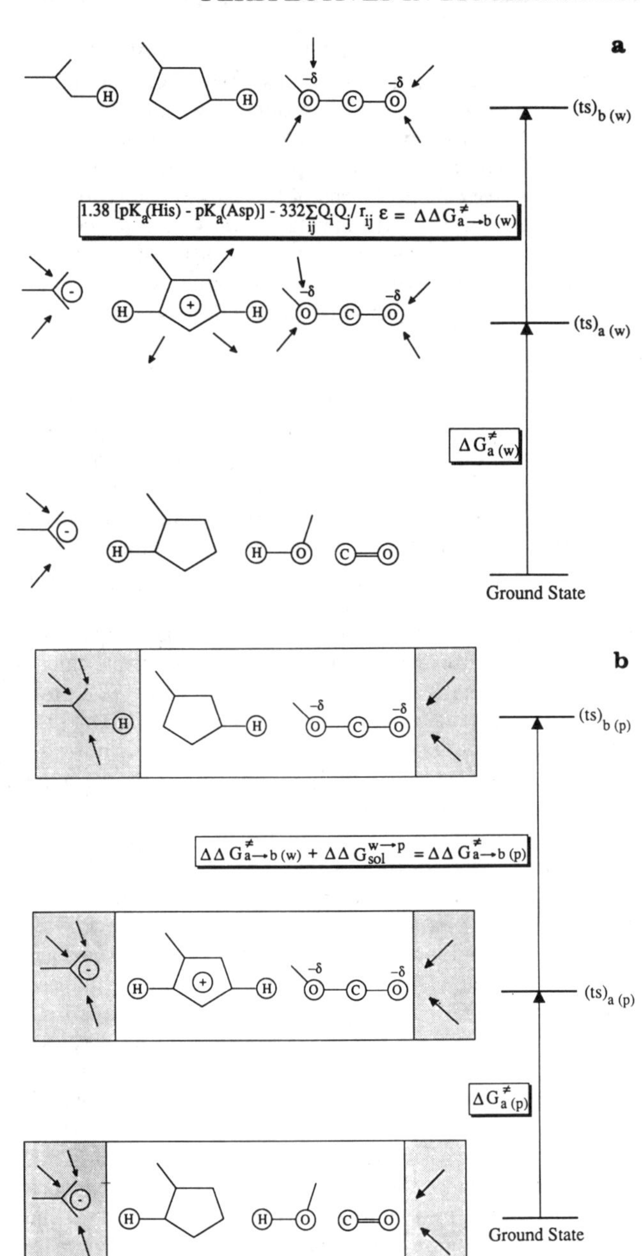

FIGURE 4: (a) Energetics of the double PT mechanism in water. The figure gives the energy difference between the transition states of routes a and b in terms of the free energy of forming the ionized fragments at infinite separation in water (the pK_a term) and the electrostatic energy associated with bringing the fragments together. Note that the dielectric constant ϵ is larger than 40. (b) Energetics of the double PT mechanism in the protein active site. This energetic is given by taking the corresponding free energy in water (a) and adding the free energy associated with moving the relevant fragments from water to the protein active site, $\Delta\Delta G^{w\rightarrow p}_{sol}$, where w and p designate water and protein, respectively.

to the electrostatic one. To realize this from a simple intuitive point of view, it is important to recognize that the ionized form of Asp_c is even more stable in the protein active site than in water, as is apparent from its observed pK_a being equal to 3 in chymotrypsin and the fact that this group is stabilized by three hydrogen bonds [cf. Brayer et al. (1978) and Warshel and Russell (1986) for discussion]. The more stable the negative charge on Asp_c, the less advantageous a proton transfer from His_c to Asp_c would be.

Let us now consider option iii for the catalytic rate acceleration, namely, that Asp_c contributes to catalysis by fixing His_c in an exact orientation for accepting a proton from Ser_c. The X-ray study of the $Asp_c \rightarrow Asn_c$ mutant of trypsin reveals two configurations of His_c, where in one of them the histidine is rotated out of the catalytic site. The population ratio of this rotated conformation to the one where His_c is inside the active site is 1:2. The corresponding increase in ground-state entropy [$R \ln (3/2)$] increases $\Delta\Delta G^{\ddagger}_{M \to N}$ by less than 0.3 kcal/mol. However, the configuration with His_c in the correct position appears to stabilize a tautomer that is unable to accept a proton from Ser_c [see Sprang et al. (1987)]. Thus one has to consider the free energy associated with the 180° inversion of His_c in the active site of the Asn_c mutant as a part of the free energy contribution to $\Delta G^{\ddagger}_M$. The calculated value of this contribution is around 2 kcal/mol (see Table II). This is a significant contribution that is, however, not entropic in nature but simply the free energy required to move an incorrectly oriented group to the correct orientation. Also note that the incorrect orientation of the Asn_c mutant of trypsin is not expected in the Ala_c mutant of subtilisin. Nevertheless, one can still argue that the actual role of Asp_c is to fix His_c in a very exact catalytic orientation. This possibility cannot be completely excluded by the present calculations, which do not give converging results for $\Delta\Delta S$. Yet the fact that the electrostatic contribution to $\Delta\Delta G^{\ddagger}$ (Table II) accounts for almost all the calculated effect of the mutation can be used to argue that other catalytic effects should be small.

This paper argues that Asp_c is ionized at the ti state, and the reader might wonder how this is related to the original charge relay system (Blow et al., 1969). This original proposal considered two resonance structures at pH 8 [they are **1** and **4** in Figure 2 of Warshel and Russell (1986)], and one more [**2** in Warshel and Russell (1986)] could also be added. The relative contribution to each resonance structure depends strongly on the nuclear coordinates. For example, the Asp_c^- His_c^+ is much less stable than the Asp_c His_c when the proton is closer to Asp_c than to His_c, and the reverse occurs when the proton is closer to His_c (Figure 5). Both resonance structures contribute to the energy of the transition state, but the calculations of Warshel and Russell (1986) indicate that the Asp^- His^+ resonance structure gives a much larger contribution (for the lower energy position of the proton) and that Asp_c is almost fully charged at the transition state.

The present work argues that the removal of Asp_c leads to large electrostatic destabilization of the transition state. Thus one might expect a large effect of this removal on the pK_a of His_c (in the absence of substrate). However, the situation in the two cases is entirely different: In the transition state the water molecules are largely removed from the active site, and the "dielectric constant" for the $Asp_c^- \cdots (His_c^+\ t^-)$ interaction is much lower than that of the Asp_c^- His_c^+ interaction in the absence of the substrate. That is, in the latter case the water molecules on the Ser_c side of His_c^+ can compensate for the removal of Asp_c^- by increasing their solvation effects. Furthermore, if the ionized His_c^+ is unstable, it will be displaced out to water, as is indeed the case in one of the conformers of the mutated trypsin (Craik et al., 1987).

It is important to mention that the effect of Asp 102 was estimated experimentally before the emergence of genetic engineering. This pioneering study (Henderson, 1971; Henderson et al., 1971) involved methylation of His_c, which reduced k_{cat} by more than 3 orders of magnitudes. The resultsant rate reduction was attributed to the incorrect orientation of His_c and to the absence of the polarizing effect of Asp_c.

Some readers might feel that the arguments presented here are valid only for a mechanism that involves a full proton transfer from Ser_c to His_c, while the actual mechanism in serine proteases might involve a concerted proton transfer from Ser_c to the amide nitrogen. Although our calculations do not support such a mechanism, they give a linear correlation between the electrostatic stabilization of the fully ionic resonance structures used here and the corresponding stabilization of the transition state of the concerted mechanism. Thus, the ionized Asp_c would also provide a major electrostatic stabilization effect to the concerted mechanism.

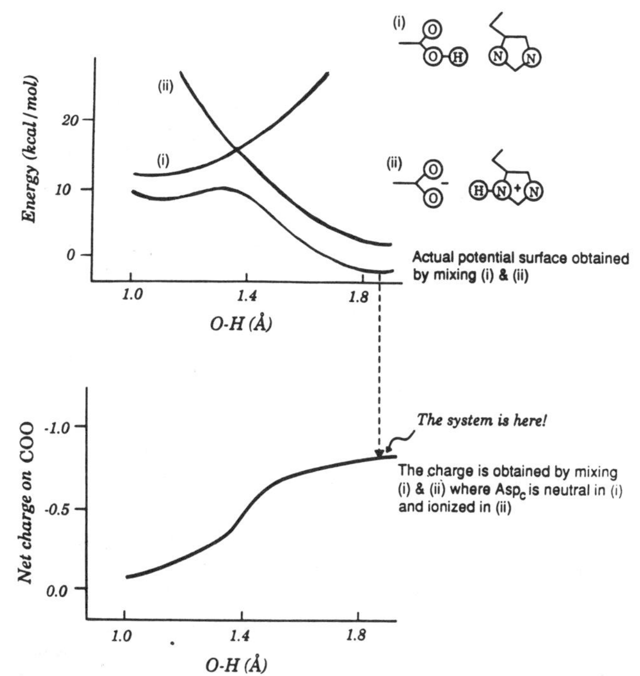

FIGURE 5: Plots showing the relationship between electronic distribution and nuclear coordinates for the Asp_c His_c system at the transition state of the catalytic reaction of serine protease. The upper figure demonstrates that the Asp_c^- His_c^+ configuration is less stable than the Asp_c His_c configuration when the proton is closer to Asp_c than to His_c, while it becomes more stable when the proton is closer to His_c. The lower figure gives the charge on Asp_c as a function of the proton position [this charge is obtained by mixing (i) and (ii)]. As seen from the figure, both resonance structures can contribute to the actual potential surface and charge distribution, but at the transition state the minimum of (ii) is lower than that of (i) [as found by Warshel and Russell (1986) and the present considerations] and the Asp_c^- His_c^+ is the most important configuration, so that Asp_c is almost fully charged.

What Stabilizes the Tetrahedral Intermediate?

The previous sections concentrated on the electrostatic role of Asp_c and did not address the stabilization of t^-. This issue, which was addressed in our earlier works [e.g., Hwang and Warshel (1987)], is quite instructive. That is, as was noted in early structural studies of serine proteases (Henderson, 1970; Robertus et al., 1972), the oxyanion intermediate, t^-, is stabilized by hydrogen bonding from the so-called "oxyanion hole" of the protein. However, this important discovery has not been emphasized as a major catalytic factor since the X-ray structure by itself could not be used to determine the actual energy contribution associated with the oxyanion hole. As much as anyone knew at that time, the energy of interaction between hydrogen bonds to a charge inside a protein could be the same as between hydrogen bonds to a charge in water (which would mean no catalytic advantage of the oxyanion hole). In fact, only microscopic theoretical calculations [e.g., Warshel et al. 1982)] could estimate the effect of the oxyanion hole until the emergence of genetic engineering experiments on this system (Wells et al., 1986; Bryan et al., 1986). These experiments and FEP calculations (Hwang & Warshel, 1987; Rao et al., 1987) indicated that one hydrogen bond from the oxyanion hole contributes about 5 kcal/mol to the stabilization of t^-, and similar results have been found in related cases [e.g., Wilkinson et al. (1984)]. This electrostatic stabilization effect is not a univeral effect for hydrogen bonds in proteins; it only occurs in sites designed to stabilize a charge for a specific purpose. The electrostatic stabilization of the oxyanion is provided by the folding energy of the protein that is invested in aligning the dipoles of the oxyanion hole (i.e., the hydrogen bonds) toward the carbonyl oxygen. Using prealigned dipoles can minimize the "reorganization energy", usually associated with the polarization of solvent molecules around charges. That is, in water we invest approximately half of the gain in solvent–charge interaction, $V_{Q\mu}$, on bringing the solvent molecules to a configuration where the solvent–solvent interaction, $V_{\mu\mu}$, is unfavorable. A "solvent" that is already polarized in the right direction does not have to pay the reorganization energy and can stabilize charges more than water does [see Warshel (1978)].

It has been frequently implied that the tetrahedral geometry of the t^- system is somehow very important. In fact, some books do not draw negative charge on t^- and only emphasize its tetrahedral geometry, perhaps assuming that the protein stabilizes the transition state by being a template to the tetrahedral geometry. This overlooks the fact that the change of the carbonyl carbon of the substrates from the sp^2 to the sp^3 geometry is mainly the result of change in intrinsic bonding structure and not the effect of the surrounding active site. In fact, the change in the geometry around the carbonyl carbon is associated with less than 0.3-Å displacement of each atom, and the flexible enzyme can accommodate these changes for less than 1 kcal/mol strain energy. For related discussion on strain in lysozyme, see Warshel and Levitt (1976).

Can We Use Semimacroscopic Electrostatic Models To Study Enzyme Catalysis?

Macroscopic models with assumed dielectric constants could not be used as a predictive tool before the emergence of both genetic engineering and microscopic electrostatic calculations (see Microscopic Calculations of the Electrostatic Contribution of Asp_c). However, at the current stage when we can evaluate the actual electrostatic energy of a given transition state, it is tempting to look for a macroscopic model that reproduces this energy. Accomplishing such a task is not so simple, as the macroscopic nature of the enzyme–water system is not defined uniquely by any dielectric model [see Warshel and Russell (1984)]. Yet, a useful and consistent macroscopic approximation can be obtained by "scaling" the PDLD model. That is, we can take the following two-step procedure: First, assign a uniform internal dielectric constant (ϵ_p) to the protein and then surround the protein by a very large sphere of the same dielectric, which is then surrounded by water ($\epsilon_w = 80$). The free energy of the relevant charges (Q) in this "extended protein" is given by the sum of the energy (ΔG_1) of transferring the charges from water to ϵ_p and the energy of bringing these charges to their actual positions in the protein (ΔG_2). This sum is given by

$$\Delta G_1 + \Delta G_2 = 166\sum_i (Q_i^2/a_i)/(1/\epsilon_p - 1/\epsilon_w) + 332\sum_{ij} Q_i q_j / r_{ij}\epsilon_p \quad (8)$$

where a_i is the cavity radius of ith charge and the q_j's are all the protein residual charges, as well as other charges not included in region 1. The first term in this equation can be obtained in a more quantitative way by defining the cavity radius a_i, using the solvation free energy of the given charge in water [$\Delta G^i_{sol,w} = -166(Q_i^2/a_i)$]. The second term in eq 8 is given by the PDLD model as $(V_{QQ} + V_{Q\mu})/\epsilon_p$, where the contribution of $V_{Q\alpha}$ is incorporated in ϵ_p [see Figure 13 of Warshel and Russell (1984)]. In the second step we evaluate the energy (ΔG_3) associated with changing back the surrounding of the protein from ϵ_p to ϵ_w. This quantity is also given by the PDLD model as $\Delta G_3 = \Delta G_{Qw}(1/\epsilon_p - 1/\epsilon_w)$. Thus we obtain the estimate

$$\Delta G_1 + \Delta G_2 + \Delta G_3 = [(\sum_i -\Delta G^i_{sol}) + \Delta G_{Qw}](1/\epsilon_p - 1/\epsilon_w) + (V_{QQ} + V_{Q\mu})/\epsilon_p \quad (9)$$

This macroscopic expression is more precise than the corresponding microscopic expression (although its results might be less accurate). That is, in the microscopic approaches one evaluates the electrostatic free energy as the sum of two very large numbers (ΔV_{QQ} and ΔG_{sol}) that compensate each other as shown in Figure 16 of Warshel and Russell (1984). On the other hand, in the macroscopic world one *assumes* the compensation by scaling the energy associated with moving the charges by $1/\epsilon_p$ and obtains small numbers for large ϵ_p. Adopting eq 9, one can find that $\epsilon_p = 4$ reproduces the microscopic results of Table I. Unfortunately, the parameter ϵ_p is not a universal parameter. For example, in studying the catalytic effect of lysozyme, one has to use $\epsilon_p = 7$ in eq 9 to reproduce the results of microscopic calculations (Warshel, 1978). It is also important to note that the dielectric constant that should be used in eq 9 is entirely different from the one that *should* have been used in traditional macroscopic methods [e.g., Tanford and Kirkwood (1957)], which describe the protein as a uniform medium with a single dielectric constant, that should also represent the effect of the protein polar groups. Such models, which are much more consistent with the macroscopic philosophy than the formulation of eq 9, require the use of $\epsilon_p \simeq 80$ to reproduce the actual energetics of the Asp_c^- His_c^+ ion pair, since they do not take into account the protein polarity in an explicit way.

It seems to us, in view of the above discussion, that the semimacroscopic formulation of eq 9 (which has not been reported before) can provide a useful estimate of electrostatic effects in enzyme active sites, if ϵ_p is evaluated by microscopic simulations. It is instructive to note that the ΔG_{Qw} contribution in eq 9 can be evaluated by discretized continuum approaches [e.g., Gilson and Honig (1987) and Sternberg et al. (1987)] but the $V_{Q\mu}$ term must be evaluated explicitly.

Concluding Remarks

In summary, we have concluded on the basis of both sophisticated and simple free energy calculations that the buried aspartate contributes to catalysis by stabilizing the charges of the His^+ t^- transition state. The overall rate acceleration appears to be associated with two elements: the ionized Asp_c that stabilizes the ionic transition state from the His_c side and the oxyanion hole that stabilizes the t^- side. The stabilization by the oxyanion hole is manifested by hydrogen bonding to t^-, and since hydrogen bonding is mostly electrostatic in nature (Umeyama & Morokuma, 1977), we can state that the rate acceleration in serine proteases is largely due to electrostatic stabilization of the ($His^+ \cdots t^-$) transition state. This point can be best realized by considering the fact that the formation of the transition state involves a major change in charge distribution that can be described as a formation of an ion pair from a neutral ground state (see Figure 1). The easiest way to reduce the relevant free energy difference is by stabilizing the charge distribution of the ion pair. This fact led to the evolutionary constraint that results in the design of the negatively charged Asp_c and the dipolar oxyanion hole. It is important to realize in this respect that Asp_c^- would not be kept ionized without its stabilizing hydrogen bonds [see Warshel and Russell (1986)], which can be called the "aspartate hole". The enzyme, in fact, stabilizes the ($Asp^- \cdots His^+ \cdots t^-$) transition state by a preorganized cryptate-like network of oriented hydrogen bonds (Warshel, 1981). Thus, it would be interesting in this respect to destroy the aspartate hole by mutating Ser 214 in trypsin.

ACKNOWLEDGMENTS

We are grateful to Dr. J. Wells and Dr. P. Carter for making their results available before publication.

REFERENCES

Bachouchin, W. W., & Roberts, J. D. (1978) *J. Am. Chem. Soc. 100*, 8041–8047.
Brayer, G. D., Delbaere, J. T. J., & James, M. N. G. (1978) *J. Mol. Biol. 124*, 261–283.
Bryan, P., Pantoliano, M. W., Quill, S. G., Hsiao, H. Y., & Poulos, T. (1986) *Proc. Natl. Acad. Sci. U.S.A. 83*, 3743–3745.
Blow, D. M., Birktoft, J. J., & Hartley, B. S. (1969) *Nature 21*, 337–340.
Carter, P., & Wells, J. A. (1988) *Nature 332*, 564–568.
Craik, C. S., Roczniak, S., Largeman, C., & Rutter, W. J. (1987) *Science 237*, 909–913.
Cram, D. J., Lam, P. Y.-Y., & Ho, S. P. (1986) *J. Am. Chem. Soc. 108*, 839.
Dewar, M. J. S. (1986) *Enzymes 36*, 8–20.
Dewar, M. J. S., & Storch, D. M. (1985) *Proc. Natl. Acad. Sci. U.S.A. 82*, 2225–2229.
Dickerson, R. E., Gray, H. B., Darensbourg, M. Y., & Darensbourg, D. J. (1984) *Chemical Principles*, p 873, Benjamin/Cummings, Menlo Park, CA.
Gilson, M. K., & Honig, B. H. (1987) *Nature 330*, 84–86.
Henderson, R. (1970) *J. Mol. Biol. 54*, 341–354.
Henderson, R. (1971) *Biochem. J. 124*, 13–18.
Henderson, R., Wright, C. S., Hess, G. P., & Blow, D. M. (1971) *Cold Spring Harbor Symp. Quant. Biol. 36*, 63–70.
Huber, R., Kukla, D., Bode, W., Schwager, P., Bartels, K., Deisenhofer, J., & Steigemann, W. (1974) *J. Mol. Biol. 89*, 73–101.
Hunkapillar, M. W., Smallcombe, S. H., Whitaker, D. R., & Richards, J. H. (1973) *Biochemistry 12*, 4732–4743.
Hwang, J.-K., & Warshel, A. (1987) *Biochemistry 26*, 2669–2673.
Jacobs, J., Schultz, P. G., Sugasawara, R., & Powell, M. (1987) *J. Am. Chem. Soc. 109*, 2174–2176.
Knowles, J. R. (1987) *Science 236*, 1252–1258.
Kollman, P. A., & Hayes, D. M. (1981) *J. Am. Chem. Soc. 103*, 2955–2961.
Kossiakoff, A. A., & Spencer, S. A. (1981) *Biochemistry 20*, 6462–6473.
Kraut, J. (1977) *Annu. Rev. Biochem 46*, 331–358.
Kraut, J. (1988) *Science 28*, 533–540.
Mallick, I. M., D'Souza, V. T., Yamaguchi, M., Lee, J., Chalabi, P., Gadwood, R. C., & Bender, M. L. (1984) *J. Am. Chem. Soc. 106*, 7252.
Markley, J. L. (1979) in *Biological Applications of Magnetic Resonance* (Shulman, R. G., Ed.) pp 397–461, Academic Press, New York.
Mattew, J. B., Hanania, G. I. H., & Gurd, F. R. N. (1979) *Biochemistry 18*, 1919–1928.
Matthews, B. W., Sigler, P. B., Henderson, R., & Blow, D. M. (1967) *Nature 214*, 652.
Naray-Szabo, G. (1982) *Int. J. Quantum Chem. 22*, 575–582.
Naray-Szabo, G. (1983) *Int. J. Quantum Chem. 23*, 723–728.
Naray-Szabo, G., & Bleha, M. (1982) in *Progress in Theoretical Organic Chemistry* (Csizmadia, I. G., Ed.) Vol. 3, pp 267–336, Elsevier, Amsterdam.
Page, M. I. (1987) in *Enzyme Mechanisms* (Page, M. I., & Williams, A., Ed.) pp 1–13, Royal Society of Chemistry, London.
Pauling, L. (1946) *Chem. Eng. News 263*, 294–297.
Polgar, L., & Bender, M. L. (1969) *Proc. Natl. Acad. Sci. U.S.A. 64*, 1335–1342.
Rao, S. N., Singh, U. C., Bash, P. A., & Kollman, P. A. (1987) *Nature 328*, 551–554.
Robertus, J. D., Kraut, J., Richard, A. A., & Birktoft, J. J. (1982) *Biochemistry 11*, 4293–4303.
Rogers, G. A., & Bruice, T. C. (1974) *J. Am. Chem. Soc. 96*, 2473–2481.
Russell, S., & Warshel, A. (1985) *J. Mol. Biol. 185*, 389–404.
Rutter, W. J., & Craik, C. S. (1987) *Science 237*, 905–909.
Showen, L. R. (1988) in *Principles of Enzyme Activity* (Liebman, J. F., & Greenberg, A., Eds.) Vol. 9, Molecular Structure and Energetics, VCH Publishers, Weinheim, FRG.
Sprang, S., Standing, T., Fletterick, R. J., Finer-Moore, J., Stroud, R. M., Xuong, N.-H., Hamlin, R., Rutter, W. J., & Craik, C. S. (1987) *Science 237*, 905–909.
Sternberg, M. J. E., Hayes, F. R. F., Russell, A. J., Thomas, P. J., & Fersht, A. R. (1987) *Nature 330*, 86–88.
Stroud, R. M. (1974) *Sci. Am. 231*, 74–88.
Stroud, R. M., Kossiakoff, A. A., & Chambers, J. L. (1977) *Annu. Rev. Biophys. Bioeng. 6*, 177–193.
Tanford, C., & Kirkwood, J. G. (1957) *J. Am. Chem. Soc. 79*, 5333–5339.
Tramontano, A., Janda, K. D., & Lerner, R. A. (1986) *Science 234*, 1566–1570.
Umeyama, H., & Morokuma, K. (1977) *J. Am. Chem. Soc. 99*, 1316–1322.
Umeyama, H., Imamura, A., Nagato, C., & Hanano, M. (1973) *J. Theor. Biol. 41*, 485–502.
Umeyama, H., Nakagawa, S., & Kudo, T. (1981) *J. Mol. Biol. 150*, 409–421.
Warshel, A. (1978) *Proc. Natl. Acad. Sci. U.S.A. 75*, 5250–5254.
Warshel, A. (1981) *Biochemistry 20*, 3167–3177.
Warshel, A., & Levitt, M. (1976) *J. Mol. Biol. 103*, 227–249.
Warshel, A., & Russell, S. T. (1984) *Q. Rev. Biophys. 17*, 283–422.
Warshel, A., & Sussman, F. (1986) *Proc. Natl. Acad. Sci. U.S.A. 83*, 3806–3810.
Warshel, A., & Russell, S. T. (1986) *J. Am. Chem. Soc. 108*, 6569–6579.
Warshel, A., Russell, S., & Weiss, R. M. (1982) in *Biomimetic Chemistry and Transition-State Analogs* (Green, B. S., Ashani, Y., & Chipman, D., Eds.) pp 267–273, Elsevier, Amsterdam.
Warshel, A., Sussman, F., & Hwang, J.-K. (1988) *J. Mol. Biol. 201*, 139–159.
Wells, J. A., Cunningham, B. C., Graycar, T. P., & Estell, D. A. (1986) *Philos. Trans. R. Soc. London, A No. 317*, 415–423.
Wilkinson, A. J., Fersht, A. R., Blow, D. M., Carter, C., & Winter, G. (1984) *Nature 307*, 187–188.

Chapter 11

Angiotensin-Converting Enzyme: New Concepts Concerning Its Biological Role†

Mario R. W. Ehlers‡ and James F. Riordan*

Center for Biochemical and Biophysical Sciences and Medicine, Harvard Medical School, Boston, Massachusetts 02115

Received February 1, 1989; Revised Manuscript Received March 6, 1989

Angiotensin-converting enzyme (ACE,[1] EC 3.4.15.1) is a zinc metalloprotease catalyzing the hydrolysis of carboxy-terminal dipeptides from oligopeptide substrates, most notably the decapeptide angiotensin I (AI) and the nonapeptide bradykinin (BK) (Soffer, 1981). ACE is a component of the renin–angiotensin system (RAS), and in view of recent advances in our understanding of the physiology and pathophysiology of the RAS (Campbell, 1987; Dzau, 1988), it is important that the implications of these new concepts are explored in terms of what is known about the biochemistry of ACE. Further, there has been a dramatic increase in the use of potent orally effective, active-site-directed ACE inhibitors (Johnston, 1988) in the absence of a complete understanding of the molecular basis for their activity or of the full physiological significance of their use. The biochemical and enzymatic properties of ACE have recently been reviewed in extensive detail (Soffer, 1976, 1981; Ondetti & Cushman, 1982; Erdös, 1987; Ehlers & Riordan, 1989) and need not be reiterated here. However, a perspective in terms of the role of ACE, both within and independent of the framework of the RAS, would appear to be both timely and useful.

Although ACE was discovered in 1954 (Skeggs et al., 1954), more than a decade elapsed before it became clear that ACE plays a critical role in the RAS, after the finding that the most important site for the generation of circulating AII was the lungs (Ng & Vane, 1967, 1968). In this organ ACE is both abundant (Cushman & Cheung, 1971; Lieberman & Sastre, 1983) and uniquely positioned to act on circulating substrates by its localization on the luminal surface of endothelial cell plasma membranes in the lung's rich vasculature (Ryan et al., 1975; Caldwell et al., 1976; Wigger & Stalcup, 1978). ACE was therefore incorporated into the central dogma of the circulating RAS: the acid protease renin is elaborated into plasma from the juxtaglomerular apparatus in the kidney in response to stimuli such as hypotension and diminished delivery of sodium to the distal tubular macula densa sites. Renin acts on liver-derived angiotensinogen (a 50-kDa α_2-globulin) to release the largely inactive AI, which is then converted to the active peptide hormone AII by ACE in the pulmonary vascular bed (Oparil & Haber, 1974a,b). AII, the most important effector of the RAS, is a potent vasoconstrictor and also stimulates the release of the sodium-retaining steroid hormone aldosterone from the adrenal cortex (Sancho et al., 1976). The circulating RAS was thus considered to play a central role in the maintenance of blood pressure and in fluid and electrolyte homeostasis, as well as in the etiology of at least some forms of hypertension (Skeggs et al., 1981).

This classical endocrine view of the RAS has recently been challenged by newer concepts that have emerged as a result of the cumulative weight of data acquired over the past 10–15 years from such sources as the purification and characterization of all the components of the RAS, the production of antibodies which allow the use of sensitive radioimmunoassay and immunohistochemical methods for the detection and quantitation of every component of the system, the development of multiple inhibitors such as the orally active ACE inhibitors and the AII antagonist saralasin, and the introduction of molecular biological techniques that provided clearer insights at the molecular level into some of the components of the system. Together, these studies have expanded the traditional systemic (endocrine) mode of action of the RAS to include effects on local tissues (paracrine), on their cells of synthesis (autocrine), and perhaps even on intracellular

†This work was supported in part by NIH Grant HL 34704.

*Address correspondence to this author at the Center for Biochemical and Biophysical Sciences and Medicine, Harvard Medical School, 250 Longwood Ave., Boston, MA 02115.

‡Associated with the MRC Liver Research Centre, Department of Medicine, University of Cape Town Medical School, Cape Town, South Africa, and supported in part by the South African Medical Research Council.

[1] Abbreviations: ACE, angiotensin-converting enzyme; AI, angiotensin I (Asp-Arg-Val-Tyr-Ile-His-Pro-Phe-His-Leu); AII, angiotensin II (Asp-Arg-Val-Tyr-Ile-His-Pro-Phe); AIII, angiotensin III (Arg-Val-Tyr-Ile-His-Pro-Phe); BK, bradykinin (Arg-Pro-Pro-Gly-Phe-Ser-Pro-Phe-Arg); LH-RH, luteinizing hormone-releasing hormone (<Glu-His-Trp-Ser-Tyr-Gly-Leu-Arg-Pro-Gly-NH_2); RAS, renin–angiotensin system; SP, substance P (Arg-Pro-Lys-Pro-Gln-Gln-Phe-Phe-Gly-Leu-Met-NH_2).

events (intracrine) (Re, 1984; Re & Rovigatti, 1988). Further, while the importance of the RAS for blood pressure and fluid and electrolyte homeostasis remains undisputed, the RAS and, consequently, ACE have been implicated in at least two other areas, unrelated to the former or each other, both of major physiological and pathophysiological interest: reproduction and immunity (Weinstock, 1986). In addition to its expanding roles as a member of the RAS, ACE has also been localized to at least one tissue containing no other components of the RAS, notably the basal ganglia in the brain, where ACE may hydrolyze non-angiotensin peptides such as substance P (SP) (Strittmatter et al., 1985b) or opioid peptides (Stewart et al., 1981; Kase et al., 1986).

Biochemistry of ACE

ACE is an acidic glycoprotein composed of a single, large polypeptide chain and 1 mol/mol zinc [for reviews, see Soffer (1976, 1981), Ondetti and Cushman (1982), Erdös (1987), Ehlers and Riordan (1989)]. In most mammalian species the predominant form appears to have a molecular weight of 140000–160000. It occurs in most tissues, although in humans primarily in kidney and lung. A lower molecular weight form has been shown to be present in testis (El-Dorry et al., 1982a,b; Lanzillo et al., 1985; Soffer et al., 1987). An mRNA encoding ACE was first isolated from bovine lung tissue (Deluca-Flaherty et al., 1987), and the corresponding cDNA has been partially sequenced. An ACE cDNA has also been obtained by using mRNA isolates from mouse kidney (Bernstein et al., 1988) and from rabbit testes (Roy et al., 1988). The complete amino acid sequence of human ACE has recently been determined on the basis of the DNA complementary to mRNA obtained from human umbilical vein endothelial cells (Soubrier et al., 1988). The initial translation product contains 1306 amino acids including an N-terminal 29-residue signal peptide. There is a hydrophobic region near the C-terminus, which is thought to be involved in binding to the plasma membrane. There is a high degree of internal homology indicative of gene duplication, and within each half of the molecule there is a region corresponding to the putative metal binding sites of several metalloproteins. This raises the intriguing question of why ACE appears to contain only a single catalytically active zinc atom. Genomic DNA analysis by Southern blotting indicates a single ACE gene is present in both the human and mouse genome (Soubrier et al., 1988; Bernstein et al., 1988). Differential splicing of the gene transcript has been proposed to account for the generation of the lung/kidney and testicular forms of the enzyme (Soubrier et al., 1988).

ACE and the RAS

By enzymatic activity, immunoreactivity, and/or the presence of mRNA, all the components of the RAS (renin, angiotensinogen, AI and AII, AII receptors, and ACE) have been found to coexist locally in various tissues, including blood vessels, kidney, adrenal, heart, and brain [reviewed by Dzau (1988)]. These intrinsic tissue RASs form autocrine or paracrine systems that provide tonic control of vascular resistance and influence local tissue function, while the circulating or endocrine RAS serves to provide short-term cardiorenal homeostasis (Dzau, 1988).

It is becoming increasingly accepted that many of the most important effects of the RAS are exerted by local tissue systems, rather than by the circulating RAS (Dzau, 1986; Campbell, 1987; Dzau, 1988). However, the implications of these revised concepts for our understanding of ACE have not been fully explored. Most radical is the notion that in some tissues the entire RAS, including ACE, may be localized and perform their functions intracellularly (Inagami et al., 1983); the evidence for and implications of this view are discussed below.

ACE and an Intracellular RAS

Consequent to the development of the concept of multiple local tissue RASs is the hypothesis of the complete intracellular generation of AI and AII, implying the intracellular localization of some or all of the components of the RAS, including ACE (Inagami et al., 1983; Re, 1984). By immunoreactivity and/or enzymatic activity, the various components of the RAS have been colocalized in cell cultures of neuroblastoma × glioma cells (Fishman et al., 1981), neuroblastoma cells (Okamura et al., 1981), and murine Leydig tumor cells (Pandey & Inagami, 1986) and in tissue sections of rat kidney juxtaglomerular cells (Celio & Inagami, 1981; Naruse et al., 1982). While the intracellular generation of angiotensins is one explanation for these data, no conclusive evidence has been provided that the entire renin–angiotensin cascade occurs intracellularly, since the uptake of extracellularly produced angiotensin is equally likely (Campbell, 1987). Indeed, one recent study (Urata et al., 1988) provided strong evidence that in the rat adrenal zona glomerulosa AII generation occurred extracellularly rather than intracellularly and that intracellular AII probably represents receptor endocytosed hormone. However, the possibility of intracellular ACE is an intriguing one which deserves to be explored.

ACE is generally thought, in keeping with its extensive glycosylation, to be a transmembrane peptidase, bound to the external surface of plasma membrane of varius cell types by a hydrophobic anchor (Erdös, 1987). This is consistent with a stretch of hydrophobic amino acids at the C-terminal region of ACE (Soubrier et al., 1988). Numerous immunohistochemical studies have confirmed this extracellular membrane-bound position on endothelial, epithelial, neuroepithelial, and miscellaneous other cell types (Ryan et al., 1975; Caldwell et al., 1976; Wigger & Stalcup, 1978; Defendini et al., 1983; Danilov et al., 1987). However, two studies have demonstrated the existence of large amounts of ACE in the cytoplasm of swine and rabbit spermatids and spermatozoa and in their discarded cytoplasmic droplets (Yotsumoto et al., 1984; Brentjens et al., 1986). Further, ultrastructural immunohistochemical studies have localized the enzyme to intracellular positions in the endocytoplasmic reticulum, nuclear envelope, and endocytotic vesicles of human renal vascular and proximal tubular epithelial cells (Bruneval et al., 1986). While this may merely reflect synthesis and cellular processing of ACE, it is equally conceivable that some of the ACE, following processing in the Golgi apparatus, remains sequestered in intracellular vesicles and is copackaged with other components of the RAS, namely, renin and angiotensinogen, with subsequent generation of AII (Re, 1984). Alternatively, endocytosis of extracellular membrane-bound ACE, with or without the other components of the RAS, can be envisaged. Intracellularly generated AII may be exported to serve local paracrine or autocrine functions, or it may even remain and exert its effects intracellularly. The latter possibility is supported by evidence that internalized AII localizes to mitochondria and nuclei (Robertson & Khairallah, 1971), by preliminary evidence suggesting that nuclear chromatin contains AII receptors (Re et al., 1984), and by the finding of a cytosolic AII-binding protein in rabbit liver (Rosenberg et al., 1988).

The evidence presented above suggests that ACE may exist intracellularly in microsomes or endocytotic vesicles, that is, in an environment which is essentially extracellular in its properties, since it lies outside the cytosol. It certainly would be highly unlikely that a heavily glycosylated protein such as ACE is active in both an extracellular, oxidizing environment, where it is known to act on various peptide substrates in flux, and in an intracytoplasmic, reducing environment, which would likely profoundly affect the enzyme's folded conformation required for activity (Creighton, 1984).

Lastly, any hypothesis postulating an intracellular location for ACE must take into account the anion requirement of the enzyme. This is a complex phenomenon that involves both essential and nonessential mechanisms of activation, dependent on substrate, pH, and probably enzyme source (species) (Cheung et al., 1980; Bünning & Riordan, 1983; Shapiro et al., 1983; Ehlers & Kirsch, 1988). Although a great deal of kinetic data has been collected, the molecular basis for this activation remains obscure and its physiological relevance, if any, unclear. Studies with rabbit ACE led to the suggestion that chloride activation represents a potential regulatory mechanism (Bünning & Riordan, 1983; Shapiro et al., 1983), but recent data obtained for the conversion of AI to AII by human ACE indicate that maximal activation for this substrate is achieved already at 30 mM Cl^- at physiological pH (Ehlers & Kirsch, 1988). Thus, anion activation is unlikely to play a regulatory role in the extracellular sites where ACE is known to occur, with the possible exception of the intestinal microvilli. On the other hand, it is conceivable that chloride and hydrogen ion concentration fluxes may regulate ACE intracellularly, where the ionic composition differs significantly from that of the extracellular milieu: $[Cl^-]$ is approximately 4 mM, and there are marked intercompartmental pH differences (Darnell et al., 1986). Endosomes are acidic with the pH as low as 5.0 (Helenius et al., 1983), while a pH of 5.5 has been noted for some secretory vesicles (Russell, 1984). Such environments would lead to near-total inhibition of ACE activity (Bünning et al., 1983; Ehlers & Kirsch, 1988). However, other vesicles, such as the small vesicles and tubules associated with the Golgi complex, have a pH of 6.4 (Yamashiro & Maxfield, 1984), and it is tempting to speculate that if ACE is indeed active intracellularly, it is regulated by transient intravesicular hydrogen and chloride ion concentration changes.

ACE and the RAS: Functions Unrelated to Blood Pressure and Fluid and Electrolyte Homeostasis

The circulating RAS and the local tissue RASs described above, i.e., in the kidney, brain, adrenal, heart, and vessel walls, are concerned with the maintenance of blood pressure and fluid and electrolyte homeostasis by such mechanisms as vasoconstriction, aldosterone release resulting in sodium and water retention, regulation of intrarenal hemodynamics, stimulation of thirst, and release of vasopressin and catecholamines (Dzau, 1988; Unger et al., 1988). Further, the ACE found in high concentrations on epithelial surfaces such as the intestinal, choroid plexus, and placental brush borders, probably coexisting with other components of the RAS, is, at least in part, also involved in fluid and electrolyte balance at these fluid/membrane interfaces (Defendini et al., 1983; Erdös, 1987). On the other hand, ACE and the RAS have recently been linked to two functions unrelated to fluid and pressure balance: reproduction and immunity.

Reproduction

Testicular ACE is of great interest: it is 30% smaller than its pulmonary counterpart, has different N- and C-terminal sequences and mRNA, yet arises from a common gene (Soubrier et al., 1988). It probably represents an internal portion of pulmonary ACE and has similar enzymatic properties (El-Dorry et al., 1982a,b; Iwata et al., 1982). Further, testicular ACE is under hormonal control: it is absent in immature rats and develops with puberty (Cushman & Cheung, 1971; Strittmatter et al., 1985a; Velletri et al., 1985). Testicular ACE has been localized to the cytoplasmic droplets of sperm (Yotsumoto et al., 1984; Brentjens et al., 1986) and to Leydig cells (Brentjens et al., 1986; Pandey et al., 1984). A complicating factor in the male genital tract is that ACE is also very abundant in the prostate and epididymis, from which it is secreted or sloughed off into the seminal plasma (Yokoyama et al., 1982; Strittmatter et al., 1985a), and this ACE is almost certainly of the lung, rather than testicular, type (El-Dorry et al., 1983; Strittmatter & Snyder, 1984). Recently other components of the RAS have been localized to the testis: renin mRNA has been detected by Northern blot analysis in testis (Pandey et al., 1984) and by in situ hybridization in Leydig cells (Deschepper et al., 1986); AI, AII, and AIII immunoreactivity was found in Leydig cell homogenates (Pandey et al., 1984); and AII receptors were identified in rat and primate Leydig cells (Khanum & Dufau, 1988; Millan & Aguilera, 1988). While prostatic and epididymal ACE may primarily be involved in the regulation of fluid and electrolyte balance of the seminal plasma, testicular ACE, in the germinal and Leydig cells, may play a more direct role in testicular function. It has been thought that ACE may contribute to male fertility by affecting sperm motility and capacitation through BK inactivation (Hohlbrugger & Dahlheim, 1983). It can also play a role in the modulation of steroidogenesis and/or regulation of cell growth and differentiation (Millan & Aguilera, 1988). In this latter regard its actions may be analogous to those of ovarian ACE, discussed below.

Ovaries appear to contain all the components of the RAS, as evidenced by the detection, in rat and human ovaries, of angiotensinogen mRNA (Ohkubo et al., 1986); prorenin and renin activity, immunoreactivity, and mRNA (Glorioso et al., 1986; Do et al., 1988; Kim et al., 1987); AII and AII receptors (Husain et al., 1987; Pucell et al., 1987; Speth & Husain, 1988); and ACE by immunohistochemistry (Brentjens et al., 1986) and in vitro autoradiographic localization by ^{125}I-labeled inhibitor binding (Speth & Husain, 1988). The RAS may have important paracrine functions in the ovary, since it has been shown that AII increases steroidogenesis (Pucell et al., 1988), and thus AII may be involved in follicular development (Speth & Husain, 1988). Indeed, the AII antagonist saralasin blocks ovulation in the rat (Pellicer et al., 1988). The wider implications of these findings for the use of ACE inhibitors and their effects on both the male and female reproductive systems will be discussed below. Lastly, it is worth noting that estrogens stimulate an increased rate of angiotensinogen synthesis in the liver and elevate plasma angiotensinogen levels (Tewksbury, 1981), and thus there appears to be a general link between the RAS and steroidogenesis.

Immunity

Since Lieberman (1975) reported the association between an elevated level of serum ACE and the granulomatous disease sarcoidosis, numerous studies have localized ACE to cells of the monocyte lineage, including monocytes, macrophages, and their derivatives, and ACE is thought to be important in mediating and/or modulating inflammation [reviewed by Weinstock (1986)]. One mechanism is through the conversion of AI to AII: angiotensinogen, renin-like activity, ACE, and AI, AII, and AIII are all present within granulomas of murine schistosomiasis (Weinstock & Blum, 1983); AII is synthesized by granuloma macrophages (Weinstock & Blum, 1987) and enhances phagocytotic activity of such macrophages (Foris et al., 1983); and AII is chemotactic for a T-lymphocyte subset

(Weinstock et al., 1987). Alternatively, the effect of ACE on inflammation may be due either to the inactivation of BK, a proinflammatory vasoactive peptide, or to the metabolism of neuropeptides such as SP and neurotensin (Weinstock, 1986).

Non-RAS ACE Functions

Although ACE is increasingly being described as a ubiquitous mammalian dipeptidyl carboxypeptidase, present in most tissues and body fluids (Erdös, 1987), it is also becoming increasingly evident that in the great majority of instances it is colocalized with some or all of the components of the RAS, and it is likely that its most important function remains its critical role in that system. Nevertheless, a search for roles outside the RAS was stimulated not only by its widespread tissue distribution but also by the apparent lack of substrate specificity. Thus, in addition to AI and BK, ACE also hydrolyzes, in vitro, other biologically active peptides such as the enkephalins (Stewart et al., 1981) and neurotensin (Skidgel et al., 1984), as well as the B chain of insulin (Igic et al., 1972). Further, although classically regarded as a dipeptidyl carboxypeptidase with a requirement for a free C-terminal carboxyl, ACE can cleave amidated peptides such as SP (Yokosawa et al., 1983; Cascieri et al., 1984; Skidgel et al., 1984; Thiele et al., 1985) and substance K (Thiele et al., 1985). Also, tripeptidyl carboxypeptidase activity is observed with SP (Cascieri et al., 1984; Thiele et al., 1985), des-Arg^9-BK (Inokuchi & Nagamatsu, 1981), and LH-RH (Skidgel & Erdös, 1985); ACE even exhibits tripeptidyl aminopeptidase activity, as with LH-RH which has a blocked N-terminus (Skidgel & Erdös, 1985). However, while these activities have been observed in vitro, there is no convincing evidence that most of these are physiologically relevant, and only BK and SP can seriously be considered, in addition to the well-established AI, as biological substrates of mammalian ACE.

BK is a nonapeptide that is inactivated by ACE by sequential cleavage of two C-terminal dipeptides (Cushman & Ondetti, 1980). It is an interesting substrate because, in common with other "class II" substrates (Shapiro et al., 1983), it is tight binding (K_m ~10^{-6} M) with a favorable k_{cat}/K_m (~10^8 M^{-1} min^{-1}) (Bünning et al., 1983) and its hydrolysis is not absolutely dependent on chloride, occurring maximally at a low concentration of ~10 mM (Dorer et al., 1974; Bünning et al., 1983). Twenty years ago it was suggested that the pulmonary inactivation of BK and the conversion of AI to AII were due to the same enzyme (Ng & Vane, 1967, 1968) and that this enzyme was ACE (Yang et al., 1972). However, while it is an attractive idea that ACE performs its blood pressure regulating role by both generating the vasoconstrictor AII and inactivating the vasodilator BK, the physiological importance of ACE in the inactivation of BK remains debatable. Thus, inhibition of ACE in isolated rat lungs has little effect on the rate of BK inactivation (Ryan et al., 1970), studies on plasma BK levels during administration of ACE inhibitors have yielded conflicting results [reviewed by Filep et al. (1987)], and urinary kinins are not increased following ACE inhibition in the spontaneously hypertensive rat (Filep et al., 1987). It should be noted, of course, that BK is susceptible to the action of proteases and peptidases other than ACE.

SP, an amidated undecapeptide, is a potential neuropeptide transmitter in the central and peripheral nervous system (Nicoll et al., 1980). The k_{cat}/K_m for the ACE-catalyzed hydrolysis of SP is 2×10^5 M^{-1} min^{-1} (Cascieri et al., 1984), which is 50 times less than the ~10^7 M^{-1} min^{-1} determined for AI (Bünning et al., 1983; Ehlers et al., 1986). Evidence that ACE plays a role in its degradation in vivo is 3-fold. First, ACE colocalizes with SP to a striatonigral pathway (Strittmatter et al., 1984; Strittmatter & Snyder, 1987) in which there is no evidence for AII (Brownfield et al., 1982), AII receptors (Mendelsohn et al., 1984), or BK (Correa et al., 1979). Although the rat striatal enzyme is smaller than its lung counterpart and has been termed an ACE isozyme analogous to the testicular form (Strittmatter et al., 1985b; Strittmatter & Snyder, 1987), this difference is likely the result of differential glycosylation rather than separate isozymes (Hooper & Turner, 1987). Second, ACE inhibitors increase SP-induced salivation in rats (Cascieri et al., 1984) and also lead to an increase in SP-like immunoreactivity in the substantia nigra and trigeminal nucleus when administered intraventricularly (Hanson & Lovenberg, 1980). Third, cough, an increasingly recognized side effect of ACE inhibitor therapy (Morice et al., 1987), may be due to SP accumulation (Morice et al., 1987), since both the bronchoconstrictor response to intravenously administered SP and its plasma level were increased by the ACE inhibitor captopril (Shore et al., 1988). In addition, recent studies with rodents have shown that ACE inhibitors enhance short-term memory and alleviate the amnesia induced by the muscarinic receptor antagonist scopolamine (Usinger et al., 1988; Costall et al., 1988). Although the basis for these effects is unknown, it could involve the metabolism of acetylcholine, which is thought to be regulated by both SP and Met-enkephalin (Sastry & Tayeb, 1982). Taken together, these various lines of evidence suggest that ACE may, at least in part, be involved in the metabolism of SP, and therefore ACE may play a role in neurotransmission.

ACE Inhibition

The development of captopril (D-3-mercapto-2-methylpropanoyl-L-proline) (Cushman et al., 1977) and the two carboxyalkyl dipeptides enalaprilat {*N*-[(*S*)-1-carboxy-3-phenylpropyl]-L-alanyl-L-proline} and lisinopril {*N*-[(*S*)-1-carboxy-3-phenylpropyl]-L-lysyl-L-proline} (Patchett et al., 1980), among others, has revolutionized the treatment of hypertension and congestive cardiac failure (Edwards & Padfield, 1985; Johnston, 1988). These compounds are active-site-directed, zinc-coordinating, slow- and tight-binding competitive inhibitors with subnanomolar K_i values (K_i ~ 10^{-10}–10^{-11} M) (Shapiro & Riordan, 1984a,b; Bull et al., 1985). A detailed kinetic analysis of their mechanism of action revealed it to be complex and substrate dependent (Shapiro & Riordan, 1984a,b). Inhibitor binding appears to follow a two-step mechanism with the initial EI complex undergoing slow isomerization to a more tightly bound EI* complex (Shapiro & Riordan, 1984a,b; Bull et al., 1985), and this slow step is chloride dependent (Shapiro & Riordan, 1984b). It is worth noting that, in the absence of structural information, interpretation of the complex kinetic data remains difficult, as the inhibition is both substrate and anion dependent, and may even be species dependent, since it was recently shown that the hydrolysis of AI by human ACE shows some differences from that by the rabbit enzyme (Ehlers & Kirsch, 1988) (all kinetic data on ACE inhibition have, to date, been obtained with the rabbit enzyme).

In view of the widespread use of these inhibitors in clinical practice, it is becoming increasingly important to understand the catalytic mechanism of the enzyme in molecular terms, permitting the rational design and synthesis of newer agents with specific properties, as well as to understand the diverse physiological roles of the enzyme, facilitating the anticipation and evaluation of specific inhibitor-related adverse reactions. In the context of the RAS, ACE, as discussed earlier, is involved in blood pressure and fluid and electrolyte homeostasis in most tissues. Thus, inhibition of endothelial ACE, in both the lungs and peripheral vasculature, and inhibition of ACE in the kidneys, heart, adrenal, and the circumventricular organs in the brain is, in all these various tissues, likely to result in a hypotensive effect (Dzau, 1988; Unger et al., 1988). Yet what of ACE in those loci where, although a component of the RAS, it participates in functions other than fluid and pressure balance and/or where there is no evidence of an independent RAS? In a recent study (Sakaguchi et al., 1988) tissue ACE was assessed by quantitative in vitro autoradiography using the ACE inhibitor [^{125}I]351A as a ligand: following oral administration of lisinopril, ACE activity was markedly inhibited in kidney, adrenal, duodenum, and lung but was not altered in testis, an organ considered to be protected by a blood–testis barrier (Waites & Gladwell, 1982). Similarly, ACE in the choroid plexus and the basal ganglia of the brain was not inhibited, while that in the circumventricular organs, which are considered to lie outside the blood–brain barrier, was inhibited (Sakaguchi et al., 1988). Captopril has also been shown not to cross the blood–brain barrier readily when administered peripherally (Vollmer & Boccagno, 1977), but it does seem to have an effect on cognitive processes in laboratory animals (Costall et al., 1988). Nevertheless, the ACE present in at least two sites where it may have important functions unrelated to blood pressure control, a striatonigral pathway and the testis, may be largely unaffected by the present generation of ACE inhibitors under ordinary circumstances.

On the other hand, intestinal ACE, which is present in high concentration on the microvilli (Defendini et al., 1983; Ward et al., 1980) and may, by the generation of AII, be involved in fluid and sodium absorption (Crocker & Munday, 1970), or function as a digestive peptidase (Yoshioka et al., 1987), is vulnerable to inhibition by orally active agents (Sakaguchi et al., 1988; Stevens et al., 1988), although this does not appear to have any pathophysiological consequences (Gavras & Gavras, 1988). Where ACE and the RAS are associated with the immune system, the administration of ACE inhibitors results in suppression of granulomatous inflammation, which was beneficial when associated with a nonreplicating antigen such as schistosoma eggs (Weinstock & Boros, 1981) or BCG inoculum (Schrier et al., 1982), but worsened the severity of acute infection with the yeast *Histoplasma capsulatum* (Deepe et al., 1985). Lastly, ovarian ACE may be inhibited by orally active ACE inhibitors, since in the ovary there is no barrier comparable to the blood–testis barrier (Richards, 1980). This may have a significant effect on ovarian function since it was recently shown that ovulation is blocked by an AII antagonist (Pellicer et al., 1988).

From these considerations, and from the paucity of side effects specifically related to the inhibition of ACE (rather than to nonspecific effects resulting from the chemical structure of the inhibitor) (Gavras & Gavras, 1988), it seems likely that the bulk of mammalian ACE forms part of the circulating and tissue RAS and is primarily concerned with blood pressure and fluid and electrolyte homeostasis. In those loci where it performs a different, more specialized function, either the enzyme is inaccessible to current inhibitors or its function is not rate limiting and thus its inhibition is not pathophysiologically important. Despite the apparent effectiveness and specificity of current ACE inhibitors, it nevertheless remains desirable to develop new generations of compounds, for two reasons. First, drugs with similar mechanisms of action as the current agents but with different physical properties allowing their penetration across the blood–brain and blood–testis barriers would be of interest; compounds with much higher lipophilicity than captopril, enalapril, or lisinopril have already been synthesized (Ondetti, 1988). Second, once the substrate-binding and catalytic mechanisms of ACE are understood in molecular terms, it may be possible to design inhibitors that preferentially inhibit hydrolysis of some substrates more than others. Thus inhibitors that do not disable the catalytic zinc or other moieties indispensable for the catalysis of all substrates, but instead bind only to substrate-binding residues in the active site, may displace some substrates but not others. This is feasible since the ACE active site has been hypothesized to form an extended linear trench with both obligatory and auxilliary binding sites (Cushman et al., 1981). Binding to the latter may explain the good inhibitory activity of compounds like teprotide, which have a C-terminal tripeptide sequence with poor affinity for the obligatory binding sites (Ondetti, 1988), as well as the finding that the binding of substrates is significantly influenced by the N-terminal residues distant from the site of hydrolysis (Gaynes et al., 1978). Further, binding of different components of substrates to the various binding sites may be synergistic, and the ACE active site may be dependent upon an induced-fit mechanism for optimal catalytic activity (Pfuetzner & Chan, 1988). The synthesis of inhibitors that, for example, minimally retard the hydrolysis of AI but significantly interfere with the breakdown of either BK or SP could therefore be envisaged.

Conclusion

Great strides have been made in our understanding of the biological role of ACE. Considerable insights have been provided by the use of ACE inhibitors, and these have confirmed the importance of this enzyme in the RAS and of the latter's role in blood pressure regulation and hypertension. Further, recent advances in our understanding of the RAS have led to newly evolving concepts, including an expansion of its role as a purely endocrine system to one having important additional paracrine/autocrine effects; the probability of an entirely intracellular RAS; and the participation of the RAS in functions unrelated to blood pressure regulation, namely, reproduction and immunity. The introduction of molecular biologic techniques has provided important additional information, such as the primary structure of this enzyme, and should lay the groundwork for an eventual elucidation of its catalytic mechanism in molecular terms. In addition, the use of cDNA probes that are now available will no doubt provide an insight into the expression and regulation of ACE at the tissue and cellular level.

Registry No. ACE, 9015-82-1; angiotensin, 1407-47-2; renin, 9015-94-5.

References

Bernstein, K. E., Martin, B. M., Bernstein, B. A., Linton, J., Striker, L., & Striker, G. (1988) *J. Biol. Chem. 263*, 11021–11024.

Brentjens, J. R., Matsuo, S., Andres, G. A., Caldwell, P. R. B., & Zamboni, L. (1986) *Experientia 42*, 399–402.

Brownfield, M. S., Field, I. A., Ganten, D., & Ganong, W. F. (1982) *Neuroscience 7*, 1759–1769.

Bruneval, P., Hinglais, N., Alhenc-Gelas, F., Tricottet, V., Corval, P., Menard, J., Camilleri, J. P., & Bariety, J. (1986) *Histochemistry 85*, 73–80.

Bull, H. G., Thornberry, N. A., Cordes, M. H. J., Patchett, A. A., & Cordes, E. H. (1985) *J. Biol. Chem. 260*, 2952–2962.

Bünning, P., & Riordan, J. F. (1983) *Biochemistry 22*, 110–116.

Bünning, P., Holmquist, B., & Riordan, J. F. (1983) *Biochemistry 22*, 103–110.

Caldwell, P. R. B., Seegal, B. C., Hsu, K. C., Das, M., & Soffer, R. L. (1976) *Science 191*, 1050–1051.

Campbell, D. J. (1987) *J. Clin. Invest. 79*, 1–6.

Cascieri, M. A., Bull, H. G., Mumford, R. A., Patchett, A. A., Thornberry, N. A., & Liang, T. (1984) *Mol. Pharmacol. 25*, 287–293.

Celio, M. R., & Inagami, T. (1981) *Proc. Natl. Acad. Sci. U.S.A. 78*, 3897–3900.

Cheung, H. S., Wang, F. L., Ondetti, M. A., Sabo, E. F., & Cushman, D. W. (1980) *J. Biol. Chem. 255*, 401–407.

Correa, F. M. A., Innis, R. B., Uhl, G. R., & Snyder, S. H. (1979) *Proc. Natl. Acad. Sci. U.S.A. 76*, 1489–1493.

Costall, B., Horovitz, Z. P., Kelly, M. E., Naylor, R. J., & Tomkins, D. M. (1988) *Br. J. Pharmacol. 96*, 882P.

Creighton, T. E. (1984) *Proteins: Structures and Molecular Properties*, pp 61–91, 265–333, W. H. Freeman, New York.

Crocker, A. D., & Munday, K. A. (1970) *J. Physiol. 206*, 323–333.

Cushman, D. W., & Cheung, H. S. (1971) *Biochim. Biophys. Acta 250*, 261–265.

Cushman, D. W., & Ondetti, M. A. (1980) *Prog. Med. Chem. 17*, 41–104.

Cushman, D. W., Cheung, H. S., Sabo, E. F., & Ondetti, M. A. (1977) *Biochemistry 16*, 5484–5491.

Cushman, D. W., Cheung, H. S., Sabo, E. F., & Ondetti, M. A. (1981) in *Angiotensin Converting Enzyme Inhibitors* (Horovitz, Z. P., Ed.) pp 3–25, Urban & Schwarzenberg, Baltimore and Munich.

Danilov, S. M., Faerman, A. I., Printseva, O. Y., Martynov, A. V., Sakharov, I. Y., & Trakht, I. N. (1987) *Histochemistry 87*, 487–490.

Darnell, J., Lodish, H., & Baltimore, D. (1986) *Molecular Cell Biology*, pp 617–666, Scientific American Books, New York.

Deepe, G. S., Taylor, C. L., Srivastava, L., & Bullock, W. E. (1985) *Infect. Immun. 48*, 395–401.

Defendini, R., Zimmerman, E. A., Weare, J. A., Alhenc-Gelas, F., & Erdös, E. G. (1983) *Neuroendocrinology 37*, 32–40.

Deluca-Flaherty, C., Schullek, J. R., Wilson, I. B., & Harris, R. B. (1987) *Int. J. Pept. Protein Res. 29*, 678–684.

Deschepper, C. F., Mellon, S. H., Cumin, F., Baxter, J. D., & Ganong, W. F. (1986) *Proc. Natl. Acad. Sci. U.S.A. 83*, 7552–7556.

Do, Y. S., Sherrod, A., Lobo, R. A., Paulson, R. J., Shinagawa, T., Chen, S., Kjos, S., & Hsueh, W. A. (1988) *Proc. Natl. Acad. Sci. U.S.A. 85*, 1957–1961.

Dorer, F. E., Kahn, J.R., Lentz, K. E., & Skeggs, L. T. (1974) *Circ. Res. 34*, 824–827.

Dzau, V. J. (1986) *Hypertension 8*, 553–559.

Dzau, V. J. (1988) *Circulation 77* (*Suppl. I*), I-4–I-13.

Edwards, C. R. W., & Padfield, P. L. (1985) *Lancet i*, 30–34.

Ehlers, M. R. W., & Kirsch, R. E. (1988) *Biochemistry 27*, 5538–5544.

Ehlers, M. R. W., & Riordan, J. F. (1989) in *Hypertension: Pathophysiology, Diagnosis and Management* (Laragh, J. H., & Brenner, B. M., Eds.) Raven Press, New York (in press).

Ehlers, M. R. W., Maeder, D. L., & Kirsch, R. E. (1986) *Biochim. Biophys. Acta 883*, 361–372.

El-Dorry, H. A., Bull, H. G., Iwata, K., Thornberry, N. A., Cordes, E. H., & Soffer, R. L. (1982a) *J. Biol. Chem. 257*, 14128–14133.

El-Dorry, H. A., Pickett, C. B., MacGregor, J. S., & Soffer, R. L. (1982b) *Proc. Natl. Acad. Sci. U.S.A. 79*, 4295–4297.

El-Dorry, H. A., MacGregor, J. S., & Soffer, R. L. (1983) *Biochem. Biophys. Res. Commun. 115*, 1096–1100.

Erdös, E. G. (1987) *Lab. Invest. 56*, 345–348.

Filep, J., Rigter, B., & Földes-Filep, E. (1987) *J. Cardiovasc. Pharmacol. 10*, 222–227.

Fishman, M. C., Zimmerman, E. A., & Slater, E. E. (1981) *Science 214*, 921–923.

Foris, G., Dezso, B., Medgyesi, G. A., & Füst, G. (1983) *Immunology 48*, 529–535.

Gavras, H., & Gavras, I. (1988) *Hypertension 11* (*Suppl. II*), II-37–II-41.

Gaynes, R. P., Szidon, J. P., & Oparil, S. (1978) *Biochem. Pharmacol. 27*, 2871–2877.

Glorioso, N., Atlas, S. A., Laragh, J. H., Jewelewicz, R., & Sealy, J. E. (1986) *Science 233*, 1422–1424.

Hanson, G. R., & Lovenberg, W. (1980) *J. Neurochem. 35*, 1370–1374.

Helenius, A., Mellman, I., Wall, D., & Hubbard, A. (1983) *Trends Biochem. Sci. 8*, 245–250.

Hohlbrugger, G., & Dahlheim, H. (1983) *Adv. Exp. Med. Biol. 156B*, 845–853.

Hooper, N. M., & Turner, A. J. (1987) *Biochem. J. 241*, 625–633.

Husain, A., Bumpus, F. M., De Silva, P., & Speth, R. C. (1987) *Proc. Natl. Acad. Sci. U.S.A. 84*, 2489–2493.

Igic, R., Erdös, E. G., Yeh, H. S. J., Sorrells, K., & Nakajima, T. (1972) *Circ. Res. 30 and 31* (*Suppl. II*), II-51–II-61.

Inagami, T., Okamura, T., Clemens, D., Celio, M. R., Naruse, K., & Naruse, M. (1983) *Clin. Exp. Hypertens., Part A A5* (*7 and 8*), 1137–1149.

Inokuchi, J., & Nagamatsu, A. (1981) *Biochim. Biophys. Acta 662*, 300–307.

Iwata, K., Lai, C. Y., El-Dorry, H. A., & Soffer, R. L. (1982) *Biochem. Biophys. Res. Commun. 107*, 1097–1103.

Johnston, C. I. (1988) *Med. J. Aust. 148*, 488–489.

Kase, R., Sekine, R., Katayama, T., Takayi, H., & Hazato, T. (1986) *Biochem. Pharmacol. 35*, 4499–4503.

Khanum, A., & Dufau, M. L. (1988) *J. Biol. Chem. 263*, 5070–5074.

Kim, S. J., Shinjo, M., Fukamizu, A., Miyazaki, H., Usuki, S., & Murakami, K. (1987) *Biochem. Biophys. Res. Commun. 142*, 169–175.

Lanzillo, J. J., Stevens, J., Dasarathy, Y., Yotsumoto, H., & Fanburg, B. L. (1985) *J. Biol. Chem. 260*, 14938–14944.

Lieberman, J. (1975) *Am. J. Med. 59*, 365–372.

Lieberman, J., & Sastre, A. (1983) *Lab. Invest. 48*, 711–717.

Mendelsohn, F. A. O., Quirion, R., Saavedra, J. M., Aguilera, G., & Catt, K. J. (1984) *Proc. Natl. Acad. Sci. U.S.A. 81*, 1575–1579.

Millan, M. A., & Aguilera, G. (1988) *Endocrinology 122*, 1984–1990.

Morice, A. H., Lowry, R., Brown, M. J., & Higenbottam, T. (1987) *Lancet ii*, 1116–1118.

Naruse, K., Inagami, T., Celio, M. R., Workman, R. J., & Takii, Y. (1982) *Hypertension 4* (*Suppl. II*), II-70–II-74.

Ng, K. K. F., & Vane, J. R. (1967) *Nature 216*, 762–766.

Ng, K. K. F., & Vane, J. R. (1968) *Nature 218*, 144–150.

Nicoll, R. A., Schenker, C., & Leeman, S. E. (1980) *Annu. Rev. Neurosci. 3*, 227–268.

Ohkubo, H., Nakayama, K., Tanaka, T., & Nakanishi, S. (1986) *J. Biol. Chem. 261*, 319–323.

Okamura, T., Clemens, D. L., & Inagami, T. (1981) *Proc. Natl. Acad. Sci. U.S.A. 78*, 6940–6943.

Ondetti, M. A. (1988) *Circulation 77* (*Suppl. I*), I-74–I-78.

Ondetti, M. A., & Cushman, D. W. (1982) *Annu. Rev. Biochem. 51*, 283–308.

Oparil, S., & Haber, E. (1974a) *N. Engl. J. Med. 291*, 389–401.

Oparil, S., & Haber, E. (1974b) *N. Engl. J. Med. 291*, 446–457.

Pandey, K. N., & Inagami, T. (1986) *J. Biol. Chem. 261*, 3934–3938.

Pandey, K. N., Misorno, K. S., & Inagami, T. (1984) *Biochem. Biophys. Res. Commun. 122*, 1337–1343.

Patchett, A. A., Harris, E., Tristram, E. W., Wyvratt, M. J., Wu, M. T., Taub, D., Peterson, E. R., Ikeler, T. J., ten Broeke, J., Payne, L. G., Ondeyka, D. L., Thorsett, E. D., Greenlee, W. J., Lohr, N. S., Hoffsommer, R. D., Joshua, H., Ruyle, W. V., Rothrock, J. W., Aster, S. D., Maycock, A. L., Robinson, F. M., Hirschmann, R., Sweet, C. S., Ulm, E. H., Gross, D. M., Vassil, T. C., & Stone, C. A. (1980) *Nature 288*, 280–283.

Pellicer, A., Palumbo, A., DeCherney, A. H., & Naftolin, F. (1988) *Science 240*, 1660–1661.

Pfuetzner, R. A., & Chan, W. W. C. (1988) *J. Biol. Chem. 263*, 4056–4058.

Pucell, A. G., Bumpus, F. M., & Husain, A. (1987) *J. Biol. Chem. 262*, 7076–7080.

Pucell, A. G., Bumpus, F. M., & Husain, A. (1988) *J. Biol. Chem. 263*, 11954–11961.

Re, R. N. (1984) *Arch. Intern. Med. 144*, 2037–2041.

Re, R. N., & Rovigatti, U. (1988) *Circulation 77* (*Suppl. I*), I-14–I-17.

Re, R. N., Vizard, D. L., Brown, J., & Bryan, S. E. (1984) *Biochem. Biophys. Res. Commun. 119*, 220–227.

Richards, J. S. (1980) *Physiol. Rev. 60*, 51–89.

Robertson, L., & Khairallah, P. A. (1971) *Science 172*, 1138–1139.

Rosenberg, E., Kiron, M. A. R., & Soffer, R. L. (1988) *Biochem. Biophys. Res. Commun. 151*, 466–472.

Roy, S. N., Kusari, J., Soffer, R. L., Lai, C. Y., & Sen, G. C. (1988) *Biochem. Biophys. Res. Commun. 155*, 678–684.

Russell, J. T. (1984) *J. Biol. Chem. 259*, 9496–9507.

Ryan, J. W., Roblero, J., & Stewart, J. M. (1970) *Adv. Exp. Med. Biol. 8*, 263–272.

Ryan, J. W., Ryan, U. S., Schultz, D. R., Whitaker, C., Chung, A., & Dorer, F. E. (1975) *Biochem. J. 146*, 497–499.

Sakaguchi, K., Chai, S. Y., Jackson, B., Johnston, C. I., & Mendelsohn, F. A. O. (1988) *Hypertension 11*, 230–238.

Sancho, J., Re, R., Burton, J., Barger, A. C., & Haber, E. (1976) *Circulation 53*, 400–405.

Sastry, B. V. R., & Tayeb, O. S. (1982) *Adv. Biosci. 38*, 165–172.

Schrier, D. J., Ripani, L. M., Katzenstein, A., & Moore, V. L. (1982) *J. Clin. Invest. 69*, 651–657.

Shapiro, R., & Riordan, J. F. (1984a) *Biochemistry 23*, 5225–5233.

Shapiro, R., & Riordan, J. F. (1984b) *Biochemistry 23*, 5234–5240.

Shapiro, R., Holmquist, B., & Riordan, J. F. (1983) *Biochemistry 22*, 3850–3857.

Shore, S. A., Stimler-Gerard, N. P., Coats, S. R., & Drazen, J. M. (1988) *Am. Rev. Respir. Dis. 137*, 331–336.

Skeggs, L. T., Marsh, W. H., Kahn, J. R., & Shumway, N. P. (1954) *J. Exp. Med. 99*, 275–282.

Skeggs, L. T., Dorer, F. E., Kahn, J. R., Lentz, K. E., & Levine, M. (1981) in *Biochemical Regulation of Blood Pressure* (Soffer, R. L., Ed.) pp 4–38, Wiley, New York.

Skidgel, R. A., & Erdös, E. G. (1985) *Proc. Natl. Acad. Sci. U.S.A. 82*, 1025–1029.

Skidgel, R. A., Engelbrecht, S., Johnson, A. R., & Erdös, E. G. (1984) *Peptides 5*, 769–776.

Soffer, R. L. (1976) *Annu. Rev. Biochem. 45*, 73–94.

Soffer, R. L. (1981) in *Biochemical Regulation of Blood Pressure* (Soffer, R. L., Ed.) pp 123–164, Wiley, New York.

Soffer, R. L., Berg, T., Sulner, J., & Lai, C. Y. (1987) *Clin. Exp. Hypertens., Part A A9* (*2 and 3*), 229–234.

Soubrier, F., Alhenc-Gelas, F., Hubert, C., Allegrini, J., John, M., Tregear, G., & Corvol, P. (1988) *Proc. Natl. Acad. Sci. U.S.A. 85*, 9386–9390.

Speth, R. C., & Husain, A. (1988) *Biol. Reprod. 38*, 695–702.

Stevens, B. R., Fernandez, A., Kneer, C., Cerda, J. J., Phillip, M. I., & Woodward, R. R. (1988) *Gastroenterology 94*, 942–947.

Stewart, T. A., Weare, J. A., & Erdös, E. G. (1981) *Peptides 2*, 145–152.

Strittmatter, S. M., & Snyder, S. H. (1984) *Endocrinology 115*, 2332–2341.

Strittmatter, S. M., & Snyder, S. H. (1987) *Neuroscience 21*, 407–420.

Strittmatter, S. M., Lo, M. M. S., Javitch, J. A., & Snyder, S. H. (1984) *Proc. Natl. Acad. Sci. U.S.A. 81*, 1599–1603.

Strittmatter, S. M., Thiele, E. A., De Souza, E. B., & Snyder, S. H. (1985a) *Endocrinology 117*, 1374–1379.

Strittmatter, S. M., Thiele, E. A., Kapiloff, M. S., & Snyder, S. H. (1985b) *J. Biol. Chem. 260*, 9825–9832.

Tewksbury, D. A. (1981) in *Biochemical Regulation of Blood Pressure* (Soffer, R. L., Ed.) pp 95–121, Wiley, New York.

Thiele, E. A., Strittmatter, S. M., & Snyder, S. H. (1985) *Biochem. Biophys. Res. Commun. 128*, 317–324.

Unger, T., Badoer, E., Ganten, D., Lang, R. E., & Rettig, R. (1988) *Circulation 77* (*Suppl. I*), I-40–I-54.

Urata, H., Khosla, M. C., Bumpus, F. M., & Husain, A. (1988) *Proc. Natl. Acad. Sci. U.S.A. 85*, 8251–8255.

Usinger, P., Hock, F. J., Wiemer, G., Gerhards, H. J., Henning, R., & Urbach, H. (1988) *Drug Dev. Res. 14*, 315–324.

Velletri, P. A., Aquilano, D. R., Bruckwick, E., Tsai-Morris, C. H., Dufau, M. L., & Lovenberg, W. (1985) *Endocrinology 116*, 2516–2522.

Vollmer, R. R., & Boccagno, J. A. (1977) *Eur. J. Pharmacol. 45*, 117–125.

Waites, G. M., & Gladwell, R. T. (1982) *Physiol. Rev. 62*, 624–671.

Ward, P. E., Sheridan, M. A., Hammon, K. J., & Erdös, E. G. (1980) *Biochem. Pharmacol. 29*, 1525–1529.

Weinstock, J. V. (1986) *Sarcoidosis 3*, 19–26.

Weinstock, J. V., & Boros, D. L. (1981) *Gastroenterology 81*, 953–958.

Weinstock, J. V., & Blum, A. M. (1983) *J. Immunol. 131*, 2529–2532.

Weinstock, J. V., & Blum, A. M. (1987) *Cell. Immunol. 107*, 273–280.

Weinstock, J. V., Blum, A. M., & Kassab, J. T. (1987) *Cell. Immunol. 107*, 180–187.

Wigger, H. J., & Stalcup, S. A. (1978) *Lab. Invest. 38*, 581–585.

Yamashiro, D. J., & Maxfield, F. R. (1984) *J. Cell. Biochem. 26*, 231–246.

Yang, H. Y. T., Erdös, E. G., & Levin, Y. (1972) *J. Pharmacol. Exp. Ther. 177*, 291–300.

Yokosawa, H., Endo, S., Ogura, Y., & Ishii, S. (1983) *Biochem. Biophys. Res. Commun. 116*, 735–742.

Yokoyama, M., Takada, Y., Iwata, H., Ochi, K., Takeuchi, M., Hiwada, K., & Kokubu, T. (1982) *J. Urol. 127*, 368–370.

Yoshioka, M., Erickson, R. H., Woodley, J. F., Gulli, R., Guan, D., & Kim, Y. S. (1987) *Am. J. Physiol. 253*, G781–G786.

Yotsumoto, H., Shoichiro, S., & Shibuya, M. (1984) *Life Sci. 35*, 1257–1261.

Chapter 12

The Role of Pyridoxal 5′-Phosphate in Glycogen Phosphorylase Catalysis[†]

Dieter Palm, Helmut W. Klein,[‡] Reinhard Schinzel, Manfred Buehner, and Ernst J. M. Helmreich*

Department of Physiological Chemistry, The University of Würzburg School of Medicine, Koellikerstrasse 2, D-8700 Würzburg, Federal Republic of Germany

Received May 25, 1989; Revised Manuscript Received August 2, 1989

Pyridoxal 5′-Phosphate in Phosphorylase

Glycogen phosphorylase is found in every species from unicellular organisms and bacteria to the complex tissues of higher plants and mammals where it plays a key role in carbohydrate metabolism. The best studied example of the highly regulated mammalian glycogen phosphorylases is the rabbit muscle phosphorylase, which was isolated and characterized by G. T. Cori and C. F. Cori [cf. Brown and Cori (1961)], who also described two forms (Cori & Green, 1943): one of which, phosphorylase *a*, was active without 5′-AMP and the other, phosphorylase *b*, was inactive unless 5′-AMP was added. Shortly after Baranowski and colleagues (Baranowski et al., 1957) in Cori's laboratory discovered pyridoxal 5′-phosphate in rabbit skeletal muscle phosphorylase, Fischer et al. (1958) reported that the bond linking the 4′-aldehyde group of pyridoxal 5′-phosphate to the ϵ-amino group of a lysine of the enzyme could be reduced with sodium borohydride without substantial loss of activity. Since all pyridoxal 5′-phosphate dependent enzymes in amino acid metabolism require a functional aldehyde group of the cofactor for activity and become inactive on reduction, it became clear at once that if pyridoxal 5′-phosphate in phosphorylase should also participate in catalysis, it must do so in a manner different from that operative in all the other known vitamin B_6 dependent enzymes (Snell & DiMari, 1970). This is also borne out by the different disposition and geometry of pyridoxal 5′-phosphate bound to aspartate aminotransferase, the only other pyridoxal 5′-phosphate containing enzyme whose 3-D structure is known (Kirsch et al., 1984).

Ever since pyridoxal 5′-phosphate was shown to be indispensable for activity of rabbit skeletal muscle phosphorylase, the role of this cofactor in phosphorylase has puzzled enzymologists. Shaltiel et al. (1966) found that the cofactor could be removed quantitatively once the protein was deformed with imidazole citrate. Removal of pyridoxal 5′-phosphate was accompanied by complete loss of enzymatic activity, but activity was restored on reconstitution with pyridoxal 5′-phosphate (Hedrick et al., 1966). This made it possible to test more than 30 pyridoxal 5′-phosphate analogues modified in every single position of the substituted pyridine ring for their ability to reactivate apophosphorylase [see Graves and Wang (1972)]. These systematic studies gave the first indication that the 5′-phosphate group of the cofactor was the most likely candidate for participation in catalysis, since only pyridoxal 5′-phosphate analogues that contribute a reversibly protonatable dianion, like $-OPO_3^{2-}$ or $-CH_2PO_3^{2-}$, in the 5′-position were able to reconstitute an active enzyme. This generalization was supported by NMR spectroscopy (Feldmann & Hull, 1977), which revealed that the 5′-phosphate of pyridoxal 5′-phosphate in the nonactivated form of phosphorylase *b* is a monoprotonated phosphate species (form I), whereas in phosphorylase activated allosterically (or by phosphorylation) the phosphorus NMR signal could be assigned to the dianionic form (form III) [see also Hörl et al. (1979)]. This significant finding was corroborated by phosphorus NMR studies with potato (Klein & Helmreich, 1979) and *Escherichia coli* maltodextrin phosphorylase (Palm et al., 1979) in both of which the activity is neither allosterically nor covalently controlled. Hence, as was to be expected, the 5′-phosphate group of the cofactor was in the dianionic form (form III) in these phosphorylases (see Figure 1). More recently it was shown that replacement of Glu 672[1] by Asp or Gln by site-directed mutagenesis blocks the transition between protonation states I and III and results in a 1000-fold decrease of k_{cat} (Schinzel and Palm, unpublished experiments). On the basis of these and additional NMR data in the presence of maltoheptaose or glucose (see Figure 1; Klein & Helmreich, 1979; Helmreich & Klein, 1980), it was suggested that the 5′-

[†] This work was supported by grants from the DFG (He 22/36-4; Pa 92-19) and the Fonds der Chemischen Industrie e.V.

* Address correspondence to this author.

[‡] Present address: Institute for Diabetes Research, The University of Düsseldorf, Düsseldorf, Federal Republic of Germany.

[1] Sequence numbers for rabbit muscle phosphorylase are used (see Table III).

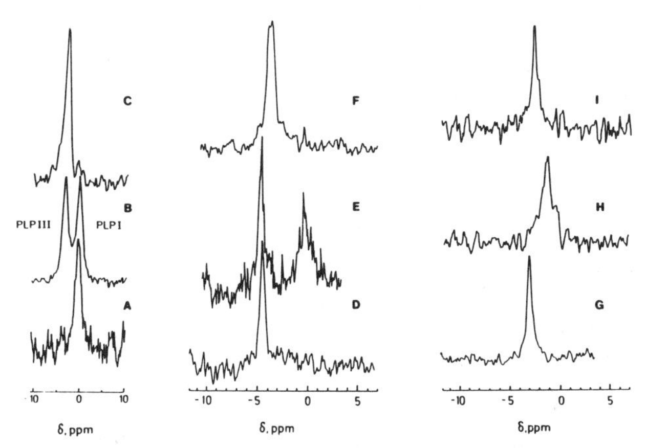

FIGURE 1: ^{31}P NMR spectra of pyridoxal 5'-phosphate in phosphorylases. The dianionic state ($PLOPO_3^{2-}$) and monoprotonated state ($PLOPO_3H^-$) are assigned to the low- and high-field forms, respectively. (Left panel) Rabbit muscle phosphorylase *b*: (A) no ligands, pH 7.8; (B) 0.4 mM AMP-S (adenosine 5'-thiophosphate) and 50 mM arsenate, pH 8.0; (C) 1.1 mM AMP-S, pH 8.1. (Middle panel) Potato phosphorylase: (D) no ligands, pH 6.5; (E) 1 mM maltoheptaose, pH 6.5; (F) 15 mM maltoheptaose and 100 mM arsenate, pH 6.5. (Right panel) *E. coli* maltodextrin phosphorylase: (G) no ligands, pH 6.9; (H) 100 mM maltoheptaose, pH 6.4; (I) 50 mM maltoheptaose and 100 mM arsenate, pH 7.0. The spectra are reproduced from Feldmann and Hull (1977), Hörl et al. (1979), Klein and Helmreich (1979), Palm et al. (1979), Helmreich and Klein (1980), and Feldmann et al. (1978).

phosphate group of pyridoxal 5'-phosphate functions in phosphorylase as a proton donor–acceptor shuttle in a general-acid–base catalysis (Feldmann et al., 1978; Helmreich & Klein, 1980; Klein et al., 1981). However, this proposal was not accepted by Graves (Parrish et al., 1977; Chang et al., 1983) and by Madsen and their colleagues (Withers et al., 1981a, 1982). Parrish et al. (1977) had shown that phosphorylase reconstituted separately with pyridoxal and fluorophosphate shows significant activity, e.g., about one-fifth of that of the native enzyme. On the basis of these experiments, they argued that this would be incompatible with a proton donor–acceptor function for the 5'-phosphate of the cofactor, since fluorophosphate has a p$K \approx 4.8$ and cannot become protonated at pH 6.8 where phosphorylase is active. However, in phosphorylase catalysis fluorophosphate cannot replace a protonatable "substrate phosphate anion" (see later). Moreover, phosphate or fluorophosphate does not bind in the pyridoxal-reconstituted phosphorylase to the same position to which these anions bind when linked covalently to the 5'-OH of pyridoxal (Oikonomakos et al., 1987). It was therefore not surprising that phosphorylase reconstituted with pyridoxal 5'-fluorophosphate was inactive (Klein et al., 1982), since the latter, in contrast to pyridoxal 5'-phosphate, has only one ionizable site (Klein et al., 1984a). Graves and colleagues (Chang et al., 1983; Soman et al., 1983) still maintain that the 5'-phosphate of the cofactor is mainly a structural determinant holding other reactive groups at the active site in the right orientation for catalysis. On the other hand, Madsen, Fukui, and colleagues (Withers et al., 1981a, 1982; Takagi et al., 1982) proposed a different role for the dianionic 5'-phosphate of the cofactor in which a "constrained" dianion functions as an electrophilic catalyst. But aside from that crucial difference of opinion, Madsen, Fukui, and their colleagues (Withers et al., 1981a,b, 1982; Takagi et al., 1982; Madsen, 1986) did clearly recognize the important role of noncovalent interactions between the 5'-phosphate group of the cofactor and the substrate phosphates.

The concept, which we shall now describe in detail, was profoundly influenced by the direct experimental proof of protonation of glycosylic substrates in the presence of phosphate by phosphorylase (Klein et al., 1982). This led us to postulate that the 5'-phosphate of pyridoxal 5'-phosphate transfers its proton to another "substrate" phosphate which in turn protonates the glycosidic bond. But a precise understanding of the role of pyridoxal 5'-phosphate in phosphorylase catalysis did actually emerge only recently from studies employing a variety of approaches reaching from classical enzymology to NMR spectroscopy and X-ray crystallography. Thus, it was the use of glycosylic substrate analogues that not only made possible the conception of a catalytic mechanism but also allowed for the visualization of enzyme–product complexes in the crystal by high-resolution X-ray diffraction (McLaughlin et al., 1984; Hajdu et al., 1987).

Reaction of Glycogen Phosphorylase

α-Glucan phosphorylases catalyze the sequential phosphorolytic degradation of oligo- or polysaccharides and the formation of α-1,4-glycosidic bonds in poly- or oligosaccharides from glucose 1-phosphate:

$$(\text{glucose})_n + P_i \rightleftharpoons (\text{glucose})_{n-1} + \alpha\text{-D-glucose 1-phosphate}$$

The reaction is readily reversible in vitro ($K_{eq} = P_i$/glucose 1-phosphate = 3.6, pH 6.8).

Early proposals of a catalytic mechanism centered around the idea that the reaction proceeds with retention of configuration of the α-glycosidic bond involving sequential double inversion, whereby the glucose residue passes through a hypothetical intermediate (Brown & Cori, 1961). Such an intermediate was shown to be formed in sucrose phosphorylase, which does not contain pyridoxal 5'-phosphate but likewise catalyzes reversible transfer of glucosyl units with retention of configuration between glucose 1-phosphate and sucrose, but not oligo- or polysaccharides (Doudoroff, 1961; Voet & Abeles, 1970). However, in glycogen phosphorylase, lack of phosphate exchange between glucose 1-phosphate and [^{32}P]P_i (Cohn & Cori, 1948) and random bi–bi kinetics (Chao et al., 1969; Engers et al., 1969) were not in favor of a covalent glucosyl intermediate. The finding by Kokesh and Kakuda (1977) of an exchange of the sugar ester oxygens and the phosphoryl oxygens of α-D-glucose 1-phosphate catalyzed by potato phosphorylase in the presence of cyclodextrin could be interpreted in favor of a covalent glucosyl intermediate, but does not exclude a mechanism where a stabilized glucosyl cation is formed.

D-Glucal, which yields a 2-deoxyglucosyl moiety on protonation, would seem to be especially suited to form covalent bonds with the enzyme in the absence of an acceptor (Legler et al., 1979), but D-glucal did not give rise to a stable derivative of glycogen phosphorylase (Helmreich & Klein, 1980; Klein et al., 1981, 1984a). Although no evidence for an enzyme-bound intermediate was obtained, the study of the D-glucal reaction revealed an obligatory requirement for phosphate ions for substrate activation, and this was the starting point of an extended study with glycosylic substrates.

Studies with Glycosylic Substrates

α-Glucan phosphorylases have in common with carbohydrases the fact that they are glycosyl transferases. This could explain why phosphorylases accept glycals as substrates and catalyze the reaction with them at significant rates (Klein et al., 1982, 1984a,b). Glycosylic substrates are compounds of nonglycosidic structure with the potential anomeric carbon linked via an electron-rich bond (Table I). They were introduced by Hehre et al. (1973) to the study of glycosylases.

Table I: Glycosylic Substrate Analogues Reactive with Phosphorylase[a]

malto-heptaose	D-glucal	α-D-glucosyl-fluoride	heptenitol

[a] Cf. Klein et al. (1982, 1986) and Palm et al. (1983).

Table II: Reaction Pathways Catalyzed by Phosphorylase with Natural Substrates and Substrate Analogues

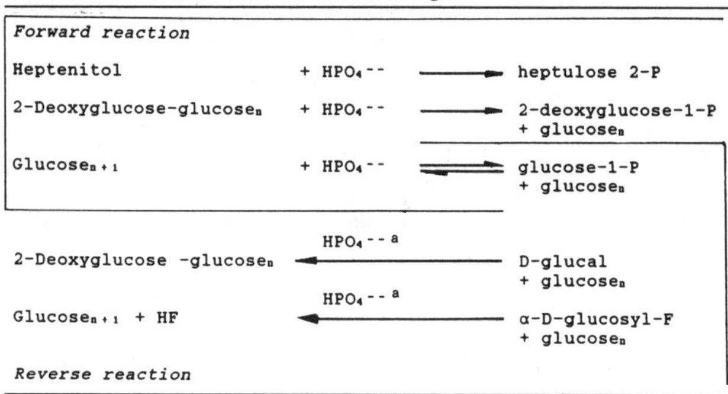

Forward reaction			
Heptenitol	+ HPO_4^{--}	→	heptulose 2-P
2-Deoxyglucose-glucose$_n$	+ HPO_4^{--}	→	2-deoxyglucose-1-P + glucose$_n$
Glucose$_{n+1}$	+ HPO_4^{--}	⇌	glucose-1-P + glucose$_n$
2-Deoxyglucose -glucose$_n$		← (HPO_4^{--} [a])	D-glucal + glucose$_n$
Glucose$_{n+1}$ + HF		← (HPO_4^{--} [a])	α-D-glucosyl-F + glucose$_n$
Reverse reaction			

[a] HPO_4^{2-} is required for substrate activation but does not participate as a reactant.

Formation of a transferable glycosyl residue from a glycosylic substrate requires a stereoselective protonation which, in the case of endocyclic or exocyclic enitols such as D-glucal or heptenitol, leads to products retaining the activating proton. Glycosylic substrates are therefore ideally suited to the study of a protonation-type reaction mechanism with participation of the 5′-phosphate of the cofactor.

Reaction with D-Glucal and α-D-Glucosyl Fluoride

A major conclusion drawn by Lehmann and Schröter (1973), Hehre et al. (1977), and Chiba et al. (1988) from their studies with carbohydrases and glycals such as D-galactal and D-glucal was that one carboxyl group in the enzyme is acting as a general acid which protonates the glycosylic (or the glycosidic) bond while another carboxyl group acts as a nucleophile or a base stabilizing the charge and determining the stereospecific course of the reaction. Our results with phosphorylases and glycosylic substrates are in agreement with these conclusions, as far as a requirement for a general acid and a nucleophile is concerned. However, an important difference between phosphorylases and carbohydrases is that the former require, in addition to the first substrate (which can be the same as that for carbohydrases), a primer and phosphate anions, whereby arsenate is the only anion that can replace phosphate (see Table II). In order to distinguish these anions from the 5′-phosphate dianion of pyridoxal 5′-phosphate, we have named the former the "substrate" or "mobile" phosphate anion (Klein et al., 1986). Besides D-glucal, α-D-glucosyl fluoride was also accepted as a phosphorylase substrate in the presence of phosphate, but it reacted at a slow rate (Palm et al., 1983). The dependence of the phosphorylase reaction with glycosylic substrate analogues on both pyridoxal 5′-phosphate and mobile phosphate anions raised two questions: first, whether the mobile phosphate is directly involved in the protonation of glycosylic and glycosidic substrates, and second, how pyridoxal 5′-phosphate might assist the mobile phosphate in its action.

To answer these questions, the steric and functional consequences of a substitution of glucose 1-phosphate by D-glucal, heptenitol, or α-D-glucosyl fluoride and phosphate were traced by "model building" (not shown) of substrate complexes with phosphorylase *b*. In accordance with the structural information of McLaughlin et al. (1984) and Barford et al. (1988), analysis of the catalytic site showed that there is scope for the substrate phosphate in closely similar positions to act as a proton donor either to heptenitol or to glucal as proposed in Figure 2. The equally variable orientation of the phosphate supports its role as a general acid in protonating the electron-rich glycosylic bond of the substrate, whereby pyridoxal 5′-phosphate assists the mobile phosphate because the latter can form a proton shuttle with the cofactor phosphate.

Reaction with D-Glucoheptenitol

The reaction of phosphorylase with an exocyclic glycal, D-glucoheptenitol (heptenitol), which Hehre et al. (1980) and Schlesselmann et al. (1982) had introduced as a prochiral enzymic substrate and as a glycosyl donor for α-glucosidase, seemed at first sight to proceed like the reaction with the endocyclic D-glucal. However, a closer look at the product revealed that the primer was not elongated in the course of the reaction (Klein et al., 1986). Thus, heptenitol is a substrate for only one direction, e.g., glucosyl transfer to phosphate yielding heptulose 2-phosphate (see Table II). It now becomes clear, whatever the direction may be, that in each case the presence of phosphate is obligatory. The fact that heptulose 2-phosphate formed from heptenitol is a "suicide inhibitor" that binds with high affinity (K_i = 2–14 μM) and the absence of a requirement for primer made heptenitol a welcome tool for the X-ray crystallographers. It enabled them to determine for the first time the structure of the catalytic site in an active (R-state) conformation (McLaughlin et al., 1984). Formation of heptulose 2-phosphate in the crystal demonstrated unequivocally direct interactions of the phosphate of the cofactor with the substrate phosphate as required by the "general-acid" mechanism (Hajdu et al., 1987). Such a direct interaction was also shown to occur by ^{31}P NMR (Klein et al., 1984a) while the proximity of the phosphates was apparent from work of Madsen, Fukui, and colleagues (Withers et al., 1981a; Takagi et al., 1982). X-ray crystallographic structure determination of the phosphorylase–heptulose 2-phosphate complex (McLaughlin et al., 1984; Hajdu et al., 1987) also clarified two other important points: first, it showed that there is no nucleophile or carboxylate near the 1β-position of the glycosylic substrate that could stabilize a transient oxocarbonium ion (or a covalent intermediate) in the case of a double inversion; second, there are no positive charges positioned in such a manner that they could constrain the 5′-phosphate dianion of pyridoxal 5′-phosphate as one might have expected on the basis of the electrophilic mechanism postulated by Madsen and his colleagues [see Withers et al. (1982)].

Figure 2 summarizes the reactions for glycosylic substrates and phosphorylase. Figure 2, mechanism I, shows the D-glucal reaction: It is proposed that orthophosphate (or arsenate) protonates the endocyclic double bond from below C-2 of the pyranose ring and, subsequently, participates in charge stabilization of the incipient 2-deoxyglucosyl carbonium ion. Nucleophilic attack by an incoming primer glucosyl on the carbonium ion terminates the reaction. Before the next turnover, the mobile phosphate must be recharged (reprotonated) by the protonated phosphate anion of pyridoxal 5′-phosphate. The reaction with heptenitol in the direction of phosphorolysis is shown in Figure 2, mechanism II. In this sequence of reactions the mobile phosphate anion stays in place and functions subsequently as a nucleophile. Accordingly, the reaction heptenitol + P_i → heptulose 2-phosphate is analogous

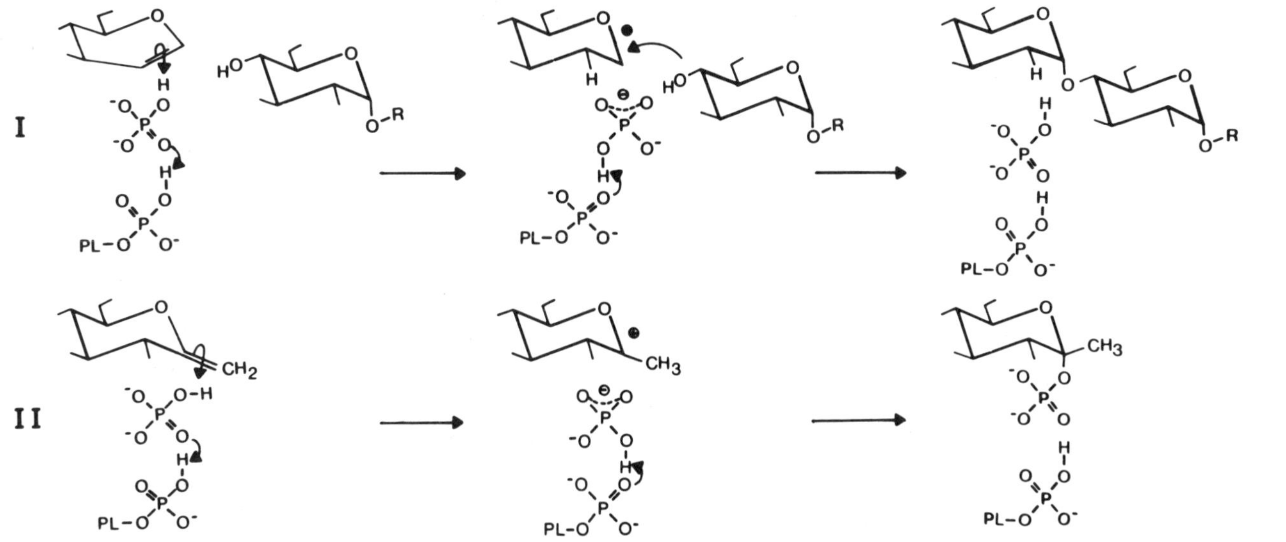

FIGURE 2: Activation of phosphorylase substrates by protonation: (I) D-glucal; (II) heptenitol. PL = pyridoxal; R = α-1,4-oligosaccharides.

Table III: Evolutionary Conserved Sites in Phosphorylases[a]

```
         ▼▼■ ▼          ■       ○○       ■             ▼▼  ■
RabP  131-LGNGGLGRLA-283-DN-309-RR-338-NDTHP-373-AYTNHTVLPEALE-
YeaP     -LGNGGLGRLA-   -DN-   -RR-   -NDTHP-   -AYTNHTVLPEALE-
PotP     -LGNGGLGRLA-   -DE-   -SR-   -NDTHP-   -AYTNHTVLPEALE-
EcoM     -LGNGGLGRLA-   -DN-   -RR-   -NDTHP-   -AYTNHTVLPEALE-

            ■       ■ ▼■   ▼       ▼▼■ ■ ■       ▼ ▼  ▼
RabP  452-VNGV-483-TNGITPRRW-566-QVKRIHEYKRQ-648-YRVSLAE-
YeaP     -VNGV-   -TNGITPRRW-   -QVKRIHEYKRQ-   -YNVSKAE-
PotP     -VNGV-   -TNGVTPRRW-   -QVKRIHEYKRQ-   -YNVSVAE-
EcoM     -VNGV-   -TNGITPRRW-   -QIKRLHEYKRQ-   -YCVSAAE-

                   ▼▼      ■■   ■■ ▼*   ▼▼
RabP          661-DLSEQISTAGTEASGTGNMKFMLNG-685
SsuP                              -MKFMLNG-
YeaP             -DLSEHISTAGTEASGTSNMKFVMNG-
PotP             -DLSEHISTAGMEASGTSNMKFAMNG-
EcoM             -DISEQISTAGKEASGTGNMKLALNG-
```

[a] Conserved amino acid residues are related to proposed substrate (■), pyridoxal 5'-phosphate (▼), or effector (○) sites in glycogen phosphorylase *b* (*Rab*P), yeast phosphorylase (*Yea*P), potato tuber phosphorylase (*Pot*P), maltodextrin phosphorylase from *E. coli* (*Eco*M), and dogfish phosphorylase (*Ssu*P). K680 contains the phosphopyridoxyl residue and is marked by an asterisk and bold type. Underlined sequences correspond to fluorescent phosphopyridoxyl peptides isolated from $NaBH_4$-reduced phosphorylases (see text). The one-letter amino acid code is used. Differences from the *Rab*P sequences are italicized.

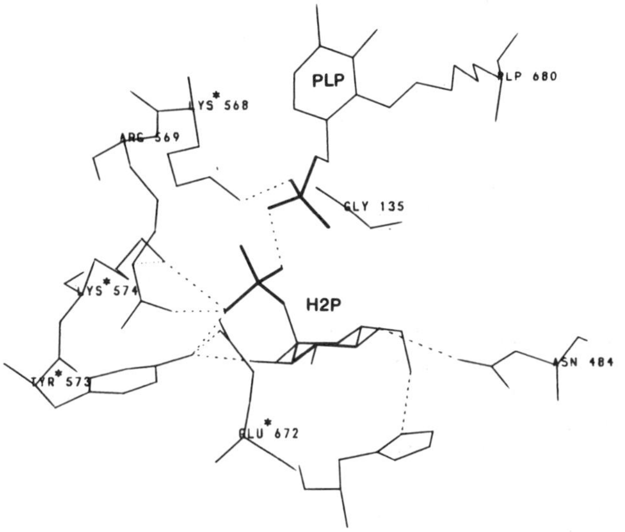

FIGURE 3: The active site in phosphorylase *b* [adapted from Hajdu et al. (1987) for the unrefined heptulose 2-phosphate–phosphorylase *b* complex]. The picture shows polar contacts and hydrogen bonds in the environment of heptulose 2-phosphate (H2P) and pyridoxal 5'-phosphate (PLP) bound to Lys 680. The phosphates are set off by dark bonds. Amino acids marked by asterisks indicate positions that were altered by site-directed mutagenesis as is discussed in the text.

to an addition without discharge of the product, a fact of which Johnson and colleagues took advantage when they studied the heptenitol reaction in the crystal (McLaughlin et al., 1984; Hajdu et al., 1987).

SITE-DIRECTED MUTAGENESIS WITH *E. coli* MALTODEXTRIN PHOSPHORYLASE

Work in Fischer's laboratory revealed an impressive similarity of sequences at the pyridoxal 5'-phosphate site in rabbit muscle (Forrey et al., 1971), dogfish (Cohen et al., 1973), and yeast (Lerch & Fischer, 1975) phosphorylases. Although this argued for a catalytic role of pyridoxal 5'-phosphate, the evidence based on the complete sequences of rabbit muscle glycogen phosphorylase (Titani et al., 1977; Nakano et al., 1986) and *E. coli* maltodextrin phosphorylase (Schächtele et al., 1978; Palm et al., 1985, 1987a) showed that more than 30 "conserved" residues are spread over the entire sequence with the exception of the N-terminal or "regulatory" tail (Table III). But taking the crystallographic evidence on substrate and effector binding sites into consideration, a footprint of conserved amino acid residues, which are functionally relevant, can be traced from bacteria to yeast (Hwang & Fletterick, 1986), higher plants (Nakano & Fukui, 1986; Nakano et al., 1978), and mammals (Table III). On the basis of this information and the structure of the phosphorylase–heptulose 2-phosphate complex shown in Figure 3, the amino acids most likely to be involved in catalysis were selected and replaced by site-directed mutagenesis.

The cloned *mal*P gene from *E. coli* (Raibaud et al., 1983; Bloch & Raibaud, 1986) provided an opportunity to generate site-directed mutants of maltodextrin phosphorylase (Palm et al., 1987a,b; Palm & Schinzel, 1989; Schinzel and Palm, unpublished experiments). Since a 3-D structure of *E. coli* maltodextrin phosphorylase is not available, its structure was inferred from the crystal structure of rabbit muscle phosphorylase, and the sequences were aligned accordingly. All mutated amino acids were near the pyridoxal 5'-phosphate and sugar phosphate binding sites in a region corresponding to the active site in rabbit skeletal muscle phosphorylase *b*.[1]

A prominent arrangement of basic residues, Lys 568, Arg 569, and Lys 574, is located close enough for interactions with

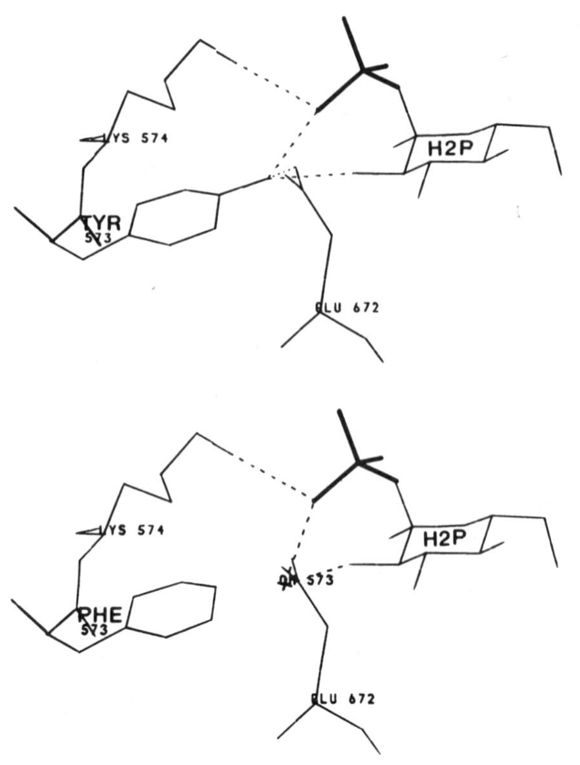

FIGURE 4: View of the substrate binding site of phosphorylase showing the network of polar contacts before and after a substitution by mutagenesis of Tyr 573 to Phe. OH 573 is the position of a water molecule derived from model building (see legend to Figure 3).

the phosphates of pyridoxal 5′-phosphate and the substrate phosphate (Figure 3). Lys 568, which is in hydrogen-bond distance to the cofactor phosphate (Oikonomakos et al., 1987), was substituted for by uncharged amino acids of smaller or comparable size, e.g., Ser or Gln. A 200–1000-fold reduction of k_{cat} in the Lys 568 mutant enzyme may be rationalized by assuming that Lys 568 helps to position the phosphates of pyridoxal 5′-phosphate and the substrate in an orientation to allow for catalysis (see Figure 3). The removal of a neutralizing charge from one of the two phosphate groups could explain the 20–40-fold increase in K_m for P_i in the mutant enzyme. In the case of Arg 569 an interaction with the substrate phosphate is indicated by a reorientation of the side chain of Arg 569, which swings in the phosphorylase *b*–heptulose 2-phosphate complex releasing the whole loop Pro 281 → Gly 288 in order to make contact with the substrate phosphate (Hajdu et al., 1987; Johnson et al., 1989; see also Figure 3). Concerning Lys 574, we have not been able to recover a viable phosphorylase mutant.

We have also considered that the binding interactions between phosphates and positively charged groups could be altered by negatively charged amino acid residues in the vicinity. Therefore, Glu 672, the only acidic amino acid residue within this region, was replaced by the isoionic Asp or the isosteric Gln. The carboxyl group of Glu 672 is close to the 2-OH and 3-OH of the glucopyranose ring of the substrate, and Glu 672 is moreover within H-bonding distance of the OH or Tyr 573 (see Figure 4) and is part of the hydrogen-bond network connecting the substrate phosphates. All mutations of Glu 672 lowered the k_{cat} value by several orders of magnitude without changes of the K_m values for maltoheptaose and P_i but with small but significant changes in the case of glucose 1-phosphate. Hence, interactions with the substrate in the ground state are not changed much when Glu 672 is replaced, whereas the much more pronounced decrease in k_{cat} in the mutant enzyme signals the large effect of the charged carboxyl group of Glu 672 on the binding of the substrates in the transition state and points to a role in the protonation of the 5′-phosphate of pyridoxal 5′-phosphate (see above).

An interaction of Tyr 573 with the glucopyranosyl moiety of various substrates is reflected in the K_m changes resulting from the Tyr 573 → Phe replacement. The more impressive differences in affinity between D-glucal and glucose 1-phosphate both in the wild-type and in the Tyr 573 → Phe mutant enzyme are indicative of the role of additional hydrogen bonds to the 2-OH of the glucose in the case of glucose 1-phosphate (Klein et al., 1984a; Street et al., 1986).

Model building with Phe 573 and water replacing Tyr 573 shows that H_2O can substitute for the Tyr-OH for most of its hydrogen bonds in connection with the substrate (see Figure 4). The H_2O must, however, be properly placed next to the hydrophobic Phe and exactly at the place normally occupied by the OH of tyrosine [cf. Oxender and Fox (1987) and Benkovic et al. (1988)]. The formation of a water site in the mutant enzyme could be responsible for a significant increase in "error" frequency as expressed in terms of the ratio of phosphorolysis/hydrolysis: The ratio of the formation of glucose 1-phosphate versus glucose is 9000:1 for the wild-type maltodextrin phosphorylase but 500:1 for the Tyr 573 → Phe mutant. Thus, in this respect the mutant phosphorylase behaved more like sucrose phosphorylase than glycogen phosphorylase (Doudoroff, 1961; Silverstein et al., 1967; Klein & Helmreich, 1985).

A Mechanism for the Phosphorylase Reaction

Stereospecific protonation of D-glucal provided the first direct proof of a general- (Brønsted-) acid-type activation mechanism for the phosphorylase reaction in direction of synthesis and gave considerable support for our earlier suggestions and proposal (Kastenschmidt et al., 1968; Feldmann et al., 1978; Helmreich & Klein, 1980) of a catalytic role for the phosphate group of pyridoxal 5′-phosphate in a proton shuttle. Subsequently, heptenitol was found to react in a similar fashion, but exclusively in the direction of phosphorolysis (Table II). Hence, it is now established that the reactions in both directions are initiated by activation of the substrate by protonation. This information allowed the presentation of a plausible reaction mechanism (see Figure 5): In the forward direction, e.g., phosphorolysis of α-1,4-glycosidic bonds in oligo- or polysaccharides, the reaction is started by the protonation of the glycosidic oxygen with orthophosphate functioning as the general acid, e.g., $H_2PO_4^-$. Subsequently, the glycosidic bond is cleaved and the incipient oxocarbonium ion is stabilized on the front side by the phosphate dianion. The reaction proceeds by nucleophilic attack of the phosphate on the carbonium ion, which might force the orthophosphate to readjust its position prior to bond formation. The product, glucose 1-phosphate, dissociates and is replaced by a new incoming phosphate. In the reverse direction, α-1,4-glycosidic bond formation and primer elongation, the reaction starts with the protonation of the "acid-labile" phosphate of glucose 1-phosphate, which weakens the glycosidic bond and promotes stabilization of the complex of the carbonium ion and the phosphate anion. A sequential double inversion of configuration is not required, nor is it possible, since the structural prerequisites are missing (see Figure 3). In this case the stabilization by phosphate assures retention of configuration. Once the phosphate is separated, it might also serve as a hydrogen-bond acceptor for the 4-OH of the terminal nonreducing glucose residue of the oligosaccharide primer. This would facilitate a nucleophilic attack of the primer at C_1. The elongated primer must dissociate

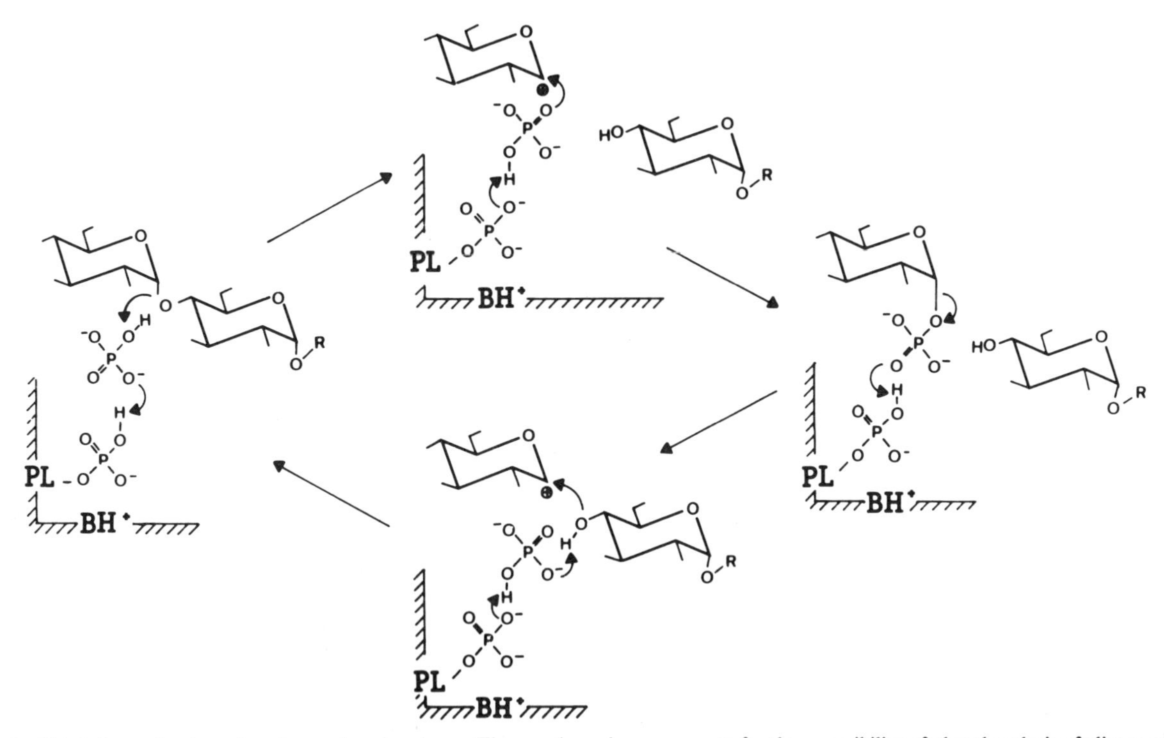

FIGURE 5: Catalytic mechanism of α-glucan phosphorylases. The reaction scheme accounts for the reversibility of phosphorolysis of oligosaccharides (R) in the presence of orthophosphate (upper half) and primer-dependent synthesis in the presence of glucose 1-phosphate (lower half). PL = enzyme-bound pyridoxal; BH^+ = a protonatable general base contributed by the enzyme protein.

before a new round of synthesis can begin. The action pattern of glucose addition to the primer is in agreement with this mechanism (Palm et al., 1973).

The mobile phosphate anion acquires its proton from the phosphate of pyridoxal 5′-phosphate [cf. Klein et al. (1986)]. To accomplish this, the phosphates of the cofactor and the substrate must approach each other within a hydrogen-bond distance (2.8–3.0 Å) corresponding to a distance of the phosphorus atoms of 4.7–5.3 Å. This was actually shown to be the case, since the phosphorus atoms in the phosphorylase–heptulose 2-phosphate complex are only 5.1 Å apart. On the other hand, the mechanism proposed by Madsen and Fukui (Withers et al., 1981a, 1982; Takagi et al., 1982) is based on a pyrophosphate-like alignment of phosphates in analogy with a phosphoryl-transfer reaction as in the case of adenylate kinase [cf. Pai et al. (1977)]. Accordingly, this type of electrophilic mechanism would require an even closer approach of the cofactor phosphate and the substrate phosphate and a much tighter coordination which in turn would make it difficult, if not impossible, to explain the ready reversibility of the phosphorylase reaction.

A mechanism of the kind proposed in Figure 5, although chemically plausible, would not be expected to proceed without assistance from the enzyme protein. This is exemplified by stereospecific protonation giving rise to products, all with glycosidic bonds having α-configurations. Moreover, the role in catalysis of certain amino acid residues of the enzyme, localized by X-ray crystallography at the active site, is impressively demonstrated by site-directed mutagenesis. The basic amino acids Lys 574 and Arg 569 have a decisive role to play in productive binding of the substrate phosphates, indicated by our inability to isolate an active mutant phosphorylase in which Lys 574 was replaced by Ser or Gln and by evidence of others that the interaction between the phosphates and Arg 569 triggers the T → R transition in the course of activation (Dreyfus et al., 1980; Hajdu et al., 1987; Barford et al., 1988). On the other hand, the consequences of substitution of Lys 568 and Glu 672 support the proposed role of the phosphate of the cofactor and the substrate phosphate in catalysis.

The mechanism proposed requires stabilization of the activated glucosyl moiety. X-ray crystallography has shown that there are no charged amino acid side chains at the back side (*re* side; above the plane) and no negatively charged groups in the vicinity of O-5 and C-1 of the sugar. This leaves the anionic substrate phosphate at the front side as the best candidate for the stabilization and the subsequent nucleophilic attack on C_1 of the glucosyl residue (Klein et al., 1984a, 1986; McLaughlin et al., 1984; Hajdu et al., 1987; Palm et al., 1987b). Stabilization of the intermediate transition-state-like complex from the *si* side by the substrate phosphate would at the same time prevent the phosphate from diffusing away and would be compatible with the exchange of the ester and phosphoryl oxygens of α-D-glucose 1-phosphate (Kokesh & Kakuda, 1977).

The proposed mechanism is a concerted mechanism with front-side attack (see Figure 5). This mechanism is a special case of a more general treatment of phosphorylases as glucosyl transferases. It was already anticipated by Mildred Cohn in 1961 as she wrote: "The most likely mechanism consistent with the experimental data for polysaccharide phosphorylase is a single displacement reaction with a front-side attack by the phosphate anion on carbon 1 of the terminal glucose". In a mechanism of this kind the functioning of the 5′-phosphate group of pyridoxal 5′-phosphate as a general acid (PL–OPO_3H^- or PL–OPO_3^{2-}) that transfers its proton to a substrate phosphate (or arsenate) can readily be imagined.

X-ray crystallographic and kinetic data and stereochemical considerations all suggest that the phosphorylase mechanism does not require distortion of the terminal sugar of the saccharide or of the glucosyl residue of glucose 1-phosphate prior to bond breaking. Barford et al. (1988) have pointed out that C–OR cleavage is unfavorable in the ground-state conformation of β-D-glycosides. In this case the sugar must be forced into a half-chair or sofa conformation in order to make the reaction possible. This is the case with lysozyme which handles β-D-glycosides and where the sofa conformation of the glucosyl residue is preferred (Phillips, 1967; Imoto et al., 1972; Ford et al., 1974). On the other hand, in the case of α-D-glycosides, C–OR bond cleavage is stereoelectronically favorable in the ground-state conformation. This is the case for the phosphorylase reaction. Moreover, it should be recalled that in the phosphorylase reaction the phosphate stabilizes the glucosyl oxocarbonium ion intermediate. Accordingly, there is no need for the glucopyranose ring to be distorted. The role of the phosphate is emphasized by the consequences of mutations of charged amino acids (e.g., Lys 568, Lys 574, and Glu 672).

The mechanism proposed is compatible with all the X-ray structural evidence available at present. Recently, additional support came from the analysis of crystals of rabbit muscle phosphorylase *b* tetramers in the R state in the presence of sulfate and phosphate ions (Barford & Johnson, 1989). These new data reveal a close relationship of the active-site conformation of the R-state phosphorylase with that of the heptulose 2-phosphate–phosphorylase complex [cf. Hajdu et al. (1987)] on which our mechanistic interpretation rests.

A Rationale for a Catalytic Role of Pyridoxal 5′-Phosphate in Glycogen Phosphorylase

It was previously pointed out (Klein & Helmreich, 1985; Palm & Schinzel, 1989) that the important evolutionary advantage gained on the emergence of phosphorylases was associated with their capability of cleaving "energy-rich" glucose polymers phosphorolytically rather than hydrolytically like the majority of polymer-degrading enzymes. Phosphorolysis preserves the energy of the glycosidic bond in sugar phosphate esters which then can generate energy in subsequent glycolytic reactions. This requires an efficient way of excluding water from the catalytic site. That this is realized in the case of phosphorylase has now been shown by site-directed mutagenesis which indicated that Tyr 573, which is hydrogen-bonded to the 2-OH of the glucosyl residue, assists in the exclusion of water because its substitution by Phe increased the ratio of hydrolysis to phosphorolysis. We would like to speculate that the preference of α-glucan phosphorylases for phosphorolysis rather than hydrolysis is related to a novel catalytic feature that these enzymes have acquired and that has enabled them to take advantage of a "mobile phosphate anion" receiving protons from the phosphate of the cofactor. Thus, in phosphorylases, a pair of phosphates, which directly interact, has replaced the pair of amino acid carboxyl groups in carbohydrases. One of the reasons why nature has made use of a vitamin B_6 analogue in phosphorylase catalysis is probably that no amino acid side chain besides histidine could accomplish an equally well balanced proton transfer. In addition, pyridoxal 5′-phosphate has a different positional mobility than do amino acid side chains constrained by protein structure. An intriguing question is whether these features of the catalytic mechanism facilitated the selection of the extremely versatile means of activity control by promoting substrate and effector binding (Sprang et al., 1988; Barford & Johnson, 1989). But all what one can say at present is that in a general way it appears that the multiplicity of phosphate binding sites both in the catalytic and in the regulatory domain is a characteristic feature in the evolution of regulatory enzymes. This applies to muscle phosphorylase as well as to phosphofructokinase (Poorman et al., 1984).

Conclusions

A mechanism for the phosphorylase reaction is proposed: In the forward direction, e.g., phosphorolysis of α-1,4-glycosidic bonds in oligo- or polysaccharides, the reaction is started by protonation of the glycosidic oxygen by orthophosphate, followed by stabilization of the incipient oxocarbonium ion by the phosphate anion and subsequent covalent binding of the phosphate to form α-glucose 1-phosphate. In the reverse direction, protonation of the phosphate of glucose 1-phosphate destabilizes the glycosidic bond and promotes formation of a glucosyl oxocarbonium ion–phosphate anion pair. In the subsequent step the phosphate anion becomes essential for promotion of the nucleophilic attack of a terminal glucosyl residue on the carbonium ion. This sequence of reactions brings about α-1,4-glycosidic bond formation and primer elongation. This mechanism accounts for retention of configuration in both directions without requiring sequential double inversion of configuration. It also provides for a plausible explanation of the essential role of pyridoxal 5′-phosphate in glycogen phosphorylase catalysis: The phosphate of the cofactor, pyridoxal 5′-phosphate, and the substrate phosphates approach each other within a hydrogen-bond distance allowing proton transfer and making the phosphate of pyridoxal 5′-phosphate into a proton shuttle which recharges the substrate phosphate anion.

Acknowledgments

Our efforts to interpret the catalytic mechanism of α-glucan phosphorylases and the role of pyridoxal 5′-phosphate in catalysis profited greatly from productive and harmonious cooperation and exchange of ideas between the Oxford group and our group. With respect to this paper, we are especially indebted to Dr. Louise N. Johnson and Dr. David Barford and their colleagues from the Laboratory of Molecular Biophysics, The University of Oxford, Oxford, U.K., for making available unpublished data on the 3-D structure of phosphorylase *b* and modeling of the active site (Figures 3 and 4). We are grateful to C. Möller for secreterial help.

Registry No. Pyridoxal 5′-phosphate, 54-47-7; glycogen phosphorylase, 9035-74-9.

References

Baranowski, T., Illingworth, B., Brown, D. H., & Cori, C. F. (1957) *Biochim. Biophys. Acta 23*, 16.

Barford, D., & Johnson, L. N. (1989) *Nature 340*, 609.

Barford, D., Schwabe, J. W. R., Oikonomakos, N. G., Acharya, K. R., Hajdu, J., Papageorgiou, A. C., Martin, J. L., Knott, J. C. A., Vasella, A., & Johnson, L. N. (1988) *Biochemistry 27*, 6733.

Benkovic, S. J., Fierke, C. A., & Naylor, A. M. (1988) *Science 239*, 1105.

Bloch, M.-A., & Raibaud, O. (1986) *J. Bacteriol. 168*, 1220.

Brown, D. H., & Cori, C. F. (1961) *Enzymes, 2nd Ed. 5*, 207.

Chang, Y. C., McCalmont, T., & Graves, D. J. (1983) *Biochemistry 22*, 4987.

Chao, J., Johnson, G. F., & Graves, D. J. (1969) *Biochemistry 8*, 1459.

Chiba, S., Brewer, C. F., Okada, G., Matsui, H., & Hehre, E. J. (1988) *Biochemistry 27*, 1564.

Cohen, P., Saari, J. C., & Fischer, E. H. (1973) *Biochemistry 12*, 5233.

Cohn, M. (1961) *Enzymes, 2nd Ed. 5*, 179.

Cohn, M., & Cori, G. T. (1948) *J. Biol. Chem. 175*, 89.

Cori, G. T., & Green, A. A. (1943) *J. Biol. Chem. 151*, 31.

Doudoroff, M. (1961) *Enzymes, 2nd Ed. 5*, 229.

Dreyfus, M., Vanderbunder, B., & Buc, H. (1980) *Biochemistry 19*, 3634.

Engers, H. D., Bridger, W. A., & Madsen, N. B. (1969) *J. Biol. Chem. 244*, 5936.

Feldmann, K., & Hull, W. E. (1977) *Proc. Natl. Acad. Sci. U.S.A. 74*, 856.

Feldmann, K., Hörl, M., Klein, H. W., & Helmreich, E. J. M. (1978) *Proc. FEBS Meet. 42*, 205.

Fischer, E. H., Kent, A. B., Snyder, E. R., & Krebs, E. G. (1958) *J. Am. Chem. Soc. 80*, 2906.

Ford, L. O., Johnson, L. N., Machin, P. A., Phillips, D. C., & Tjian, T. (1974) *J. Mol. Biol. 88*, 349.

Forrey, A. W., Sevilla, C. L., Saari, J. C., & Fischer, E. H. (1971) *Biochemistry 10*, 3132.

Graves, D. J., & Wang, H. J. (1972) *Enzymes (3rd Ed.) 7*, 435.

Hajdu, J., Acharya, K. R., Stuart, D. I., McLaughlin, P. J., Barford, D., Oikonomakos, N. G., Klein, H., & Johnson, L. N. (1987) *EMBO J. 6*, 539.

Hedrick, J. L., Shaltiel, S., & Fischer, E. H. (1966) *Biochemistry 5*, 2117.

Hehre, E. J., Okada, G., & Genghof, D. (1973) *Adv. Chem. Ser. No. 117*, 309.

Hehre, E. J., Genghof, D. S., Sternlicht, H., & Brewer, C. F. (1977) *Biochemistry 16*, 1780.

Hehre, E. J., Brewer, C. F., Uchiyama, T., Schlesselmann, P., & Lehmann, J. (1980) *Biochemistry 19*, 3557.

Helmreich, E. J. M., & Klein, H. W. (1980) *Angew. Chem. 92*, 429; *Angew. Chem., Int. Ed. Engl. 19*, 441.

Hörl, M., Feldmann, K., Schnackerz, K. D., & Helmreich, E. J. M. (1979) *Biochemistry 18*, 2457.

Hwang, P. K., & Fletterick, R. J. (1986) *Nature 324*, 80.

Illingworth, B., Jansz, H. S., Brown, D. H., & Cori, C. F. (1958) *Proc. Natl. Acad. Sci. U.S.A. 44*, 1180.

Imoto, T., Johnson, L. N., North, A. C. T., Phillips, D. C., & Rupley, J. A. (1972) *Enzymes (3rd Ed.) 7*, 665.

Johnson, L. N., Hajdu, J., Acharya, K. R., Stuart, D. I., McLaughlin, P. J., Oikonomakos, N. G., & Barford, D. (1989) in *Allosteric Enzymes* (Hervé, G., Ed.) p 81, CRC Press, Boca Raton, FL.

Kastenschmidt, L. L., Kastenschmidt, J., & Helmreich, E. (1968) *Biochemistry 7*, 3590.

Kirsch, J. F., Eichele, G., Ford, G. C., Vincent, M. G., Jansonius, J. N., Gehring, H., & Christen, P. (1984) *J. Mol. Biol. 174*, 497.

Klein, H. W., & Helmreich, E. J. M. (1979) *FEBS Lett. 108*, 209.

Klein, H. W., & Helmreich, E. J. M. (1985) *Curr. Top. Cell. Regul. 26*, 281.

Klein, H. W., Schiltz, E., & Helmreich, E. J. M. (1981) *Cold Spring Harbor Conf. Cell Proliferation 8*, 305.

Klein, H. W., Palm, D., & Helmreich, E. J. M. (1982) *Biochemistry 21*, 6675.

Klein, H. W., Im, M. J., Palm, D., & Helmreich, E. J. M. (1984a) *Biochemistry 23*, 5853.

Klein, H. W., Im, M. J., & Helmreich, E. J. M. (1984b) in *Chemical and Biological Aspects of Vitamin B_6 Catalysis* (Evangelopoulos, A. E., Ed.) Part A, p 147, Liss, New York.

Klein, H. W., Im, M. J., & Palm, D. (1986) *Eur. J. Biochem. 157*, 107.

Kokesh, F. C., & Kakuda, Y. (1977) *Biochemistry 16*, 2467.

Legler, G., Roeser, K. R., & Illig, H. K. (1979) *Eur. J. Biochem. 101*, 85.

Lehmann, J., & Schröter, E. (1973) *Carbohydr. Res. 23*, 359.

Lerch, K., & Fischer, E. H. (1975) *Biochemistry 14*, 2009.

Madsen, N. B. (1986) *Enzymes (3rd Ed.) 17*, 365.

McLaughlin, P. J., Stuart, D. I., Klein, H. W., Oikonomakos, N. G., & Johnson, L. N. (1984) *Biochemistry 23*, 5862.

Nakano, K., & Fukui, T. (1986) *J. Biol. Chem. 261*, 8230.

Nakano, K., Wakabayashi, S., Hase, T., Matsubara, H., & Fukui, T. (1978) *J. Biochem. (Tokyo) 83*, 1085.

Nakano, K., Hwang, P. K., & Fletterick, R. J. (1986) *FEBS Lett. 204*, 283.

Oikonomakos, N. G., Johnson, L. N., Acharya, K. R., Stuart, D. I., Barford, D., Hajdu, J., Varvill, K. M., Melpidou, A. E., Papageorgiou, T., Graves, D. J., & Palm, D. (1987) *Biochemistry 26*, 8381.

Oxender, D. L., & Fox, C. F., Eds. (1987) *Protein Engineering*, A. R. Liss, New York.

Pai, E. F., Sachsenheimer, W., Schirmer, R. H., & Schultz, G. E. (1977) *J. Mol. Biol. 114*, 37.

Palm, D., & Schinzel, R. (1989) in Functional and Regulatory Aspects of Enzyme Action, Proceedings of the Leopoldina Symposium, *Nova Acta Leopold. No. 61*, 143.

Palm, D., Starke, A., & Helmreich, E. (1973) *FEBS Lett. 33*, 213.

Palm, D., Schaechtele, K. H., Feldmann, K., & Helmreich, E. J. M. (1979) *FEBS Lett. 101*, 403.

Palm D., Blumenauer, G., Klein, H. W., & Blanc-Muesser, M. (1983) *Biochem. Biophys. Res. Commun. 111*, 530.

Palm, D., Goerl, R., & Burger, K. J. (1985) *Nature 313*, 500.

Palm, D., Goerl, R., Weidinger, G., Zeier, R., Fischer, B., & Schinzel, R. (1987a) *Z. Naturforsch. 42C*, 394.

Palm, D., Schinzel, R., Zeier, R., & Klein, H. W. (1987b) in *Biochemistry of Vitamin B_6* (Korpela, T., & Christen, P., Eds.) p 83, Birkhäuser, Basel.

Parrish, T., Uhing, R. J., & Graves, D. J. (1977) *Biochemistry 16*, 4824.

Phillips, D. C. (1967) *Proc. Natl. Acad. Sci. U.S.A. 57*, 484.

Poorman, R. A., Randolph, A., Kemp, R. G., & Heinrikson, R. L. (1984) *Nature 309*, 467.

Raibaud, O., Débarbouillé, M., & Schwartz, M. (1983) *J. Mol. Biol. 163*, 395.

Schächtele, K. H., Schiltz, E., & Palm, D. (1978) *Eur. J. Biochem. 92*, 427.

Schlesselmann, P., Fritz, H., Lehmann, J., Uchiyama, T., Brewer, C. F., & Hehre, E. J. (1982) *Biochemistry 21*, 6606.

Shaltiel, S., Hedrick, J. L., & Fischer, E. H. (1966) *Biochemistry 5*, 2108.

Silverstein, R., Voet, J., Reed, D., & Abeles, R. H. (1967) *J. Biol. Chem. 242*, 1338.

Snell, E. E., & DiMari, S. J. (1970) *Enzymes (3rd Ed.) 2*, 335.

Soman, G. M., Chang, Y. C., & Graves, D. J. (1983) *Biochemistry 22*, 4994.

Sprang, S. R., Acharya, K. R., Goldsmith, E. J., Stuart, D. I., Varvill, K., Fletterick, R. J., Madsen, N. B., & Johnson, L. N. (1988) *Nature 336*, 215.

Street, I. P., Armstrong, C. R., & Withers, S. G. (1986) *Biochemistry 25*, 6021.

Takagi, M., Fukui, T., & Shimomura, S. (1982) *Proc. Natl. Acad. Sci. U.S.A. 79*, 3716.

Titani, K., Koide, A., Hermann, J., Ericsson, L. H., Kumar, S., Wade, R. D., Neurath, H., & Fischer, E. H. (1977) *Proc. Natl. Acad. Sci. U.S.A. 74*, 4762.

Voet, J. G., & Abeles, R. H. (1970) *J. Biol. Chem. 245*, 1020.
Withers, S. G., Madsen, N. B., Sykes, B. D., Takagi, M., Shimomura, S., & Fukui, T. (1981a) *J. Biol. Chem. 256*, 10759.
Withers, S. G., Madsen, N. B., & Sykes, B. D. (1981b) *Biochemistry 20*, 1748.
Withers, S. G., Madsen, N. B., Sprang, S. R., & Fletterick, R. J. (1982) *Biochemistry 21*, 5372.

Chapter 13

Fatty Acid Synthase, A Proficient Multifunctional Enzyme†

Salih J. Wakil

The Verna and Marrs McLean Department of Biochemistry, Baylor College of Medicine, One Baylor Plaza, Houston, Texas 77030

Received January 4, 1989; Revised Manuscript Received February 28, 1989

The animal fatty acid synthase is the most sophisticated entry in the newly recognized class of multifunctional enzymes. The subunit protein of this elegant multienzyme has a molecular weight of 260 000 and contains, in separate domains, seven different catalytic activities and a site for the prosthetic group, 4′-phosphopantetheine, of the acyl carrier protein. Investigations of fatty acid biosynthesis not only yielded information about this multienzyme system but uncovered significant concepts basic to biochemistry at large. For instance, the notion that synthetic pathways are the reversal of degradative reactions was refuted in part by studies on fatty acid biosynthesis (Wakil et al., 1957). It is now well accepted that anabolic pathways may not be the same as the degradative pathways. Also, the role of the vitamin biotin in biological reactions was first recognized in studies of fatty acid synthesis by the discovery of biotin as a prosthetic group of acetyl-CoA carboxylase (Wakil et al., 1958). Several such biotin-containing enzymes have since been recognized and studied. Moreover, the activation of acetyl-CoA carboxylase by citrate (Waite & Wakil, 1962; Martin & Vagelos, 1962) provided one of the early examples that led to the formulation of the model for allosteric modification of proteins by Monod, Wyman, and Changeux (1965). The finding that a protein, acyl carrier protein (ACP), acts as a coenzyme was first demonstrated in enzymatic reactions involved in fatty acid synthesis (Majerus et al., 1964; Wakil et al., 1964). This protein binds substrates and all acyl intermediates as thioesters and channels them into the synthetic pathway. Finally, studies of this system (Knobling et al., 1975; Stoops et al., 1975) led to the recognition of multifunctional proteins as a new class of enzymes with two or more catalytic domains associated with a single polypeptide (Kirschner & Bisswanger, 1976). This multifunctional protein is encoded by a single gene which may have evolved by fusion of component genes. This review presents our current knowledge of this multifunctional enzyme and provides a summary of the organization of component activities on the protein and their function in the synthesis of long-chain fatty acids.

Long-chain fatty acids are essential constituents of membrane lipids and are important substrates for energy metabolism of the cell. Palmitate, the most abundant acid, is synthesized de novo from acetyl-CoA, malonyl-CoA, and NADPH by the fatty acid synthase according to the following reaction:

$$CH_3COS\text{–}CoA + 7HOOCCH_2COS\text{–}CoA + 14NADPH + 14H^+ \rightarrow CH_3CH_2(CH_2CH_2)_6CH_2COOH + 14NADP^+ + 6H_2O + 8CoA\text{–}SH + 7CO_2 \quad (1)$$

Basically, this reaction consists of elongating the acetyl group by C_2 units derived from malonyl-CoA in a stepwise and sequential manner. For instance, in the synthesis of palmitate there are over 40 steps with at least 30 acyl intermediates.

The nature of these reactions and the intermediates involved became known primarily from studies of fatty acid synthesis in cell-free extracts of *Escherichia coli* (Wakil, 1970; Volpe & Vagelos, 1977; Bloch & Vance, 1977). A protein known as acyl carrier protein (ACP), with its 4′-phosphopantetheine prosthetic group (M_r 8847), was identified as the coenzyme that binds all acyl intermediates as thioester derivatives. The individual enzymes were then isolated and utilized in the reconstitution of the synthesis pathway. The following enzymes and reactions are involved in the synthesis of palmitate:

acetyl transacylase

$$CH_3COS\text{–}CoA + ACP\text{–}SH \rightleftharpoons CH_3COS\text{–}ACP + CoA\text{–}SH \quad (2)$$

malonyl transacylase

$$HOOCCH_2COS\text{–}CoA + ACP\text{–}SH \rightleftharpoons HOOCCH_2COS\text{–}ACP + CoA\text{–}SH \quad (3)$$

† This work was supported in part by grants from the National Institutes of Health (GM191091), The Robert A. Welch Foundation, and the Clayton Foundation.

β-ketoacyl synthase (condensing enzyme)

$$CH_3COS\text{–}ACP + synthase\text{–}SH \rightleftharpoons CH_3COS\text{–}synthase + ACP\text{–}SH \quad (4)$$

$$CH_3COS\text{–}synthase + HOOCCH_2COS\text{–}ACP \rightarrow CH_3COCH_2COS\text{–}ACP + CO_2 + synthase\text{–}SH \quad (5)$$

β-ketoacyl reductase

$$CH_3COCH_2COS\text{–}ACP + NADPH + H^+ \rightleftharpoons \text{D-}CH_3CHOHCH_2COS\text{–}ACP + NADP^+ \quad (6)$$

β-hydroxyacyl dehydratase

$$\text{D-}CH_3CHOHCH_2COS\text{–}ACP \rightleftharpoons trans\text{-}CH_3CH{=}CHCOS\text{—}ACP + H_2O \quad (7)$$

enoyl reductase

$$trans\text{-}CH_3CH{=}CHCOS\text{—}ACP + NADPH + H^+ \rightarrow CH_3CH_2CH_2COS\text{–}ACP + NADP^+ \quad (8)$$

thioesterase

$$CH_3CH_2(CH_2CH_2)_6CH_2COS\text{–}ACP + H_2O \rightarrow CH_3CH_2(CH_2CH_2)_6CH_2COOH + CoA\text{–}SH \quad (9)$$

These reactions are essentially the same in all organisms. A malonyl group derived from malonyl-CoA is condensed with an acetyl group, as a primer, with loss of carbon dioxide (reactions 4 and 5). The β-ketoacyl derivative is reduced in three consecutive steps (reactions 6, 7, and 8) to the saturated acyl derivative, which then acts as a primer for further elongation and reduction cycles to yield ultimately a palmitoyl derivative. The latter is either hydrolyzed to free palmitate as in bacteria and animal tissue (reaction 9), transferred to CoA–SH to form palmitoyl-CoA as in yeast (Schreckenbach et al., 1977), or utilized directly in the synthesis of phosphatidic acid as in *E. coli* (Rock et al., 1981).

In prokaryotes and plants, the enzymes of fatty acid synthesis can be readily separated by conventional procedures (Wakil, 1970; Volpe & Vagelos, 1977; Stumpf, 1984). In yeast and animal cells the enzyme activities purify together and were initially thought to be tightly associated complexes. However, later investigations clearly showed that yeast (Schweizer et al., 1975; Stoops et al., 1978) and animal (Stoops et al., 1975) fatty acid synthases are multifunctional in nature. As multifunctional enzymes, the catalytic sites are arranged as a series of connected globular domains. In yeast and other fungi, the enzyme activities are distributed on two separate subunits: subunit α (M_r 207 863), which contains an attachment site of the prosthetic group of ACP, 4′-phosphopantetheine, and domains for two enzyme activities, β-ketoacyl reductase and β-ketoacyl synthase (Mohamed et al., 1988); subunit β (M_r 220 077), which contains domains for the remaining five enzymes, the acetyl transacylase, the enoyl reductase, the dehydratase, and the malonyl/palmitoyl transacylases (Chirala et al., 1987). The active enzyme is an $\alpha_6\beta_6$ complex with a molecular weight of 2.4×10^6 (Schweizer et al., 1975; Stoops et al., 1975, 1978; Lynen, 1980).

In animal cells the component enzymes of fatty acid synthesis are covalently linked on a single polypeptide chain. The native enzyme (M_r ~500 000) isolated from animal cells consists of two such multifunctional polypeptides (Arslanian et al., 1976; Buckner & Kolattukudy, 1976; Stoops et al., 1979; Smith et al., 1985). The mRNA coding for the avian (Zehner et al., 1980; Morris, 1982) and rat mammary gland (Mattick et al., 1981) synthases was shown to be a single contiguous mRNA which can be translated to a protein of M_r 260 000 that specifically binds to the anti-synthase antibodies and can be competed for by native synthase.

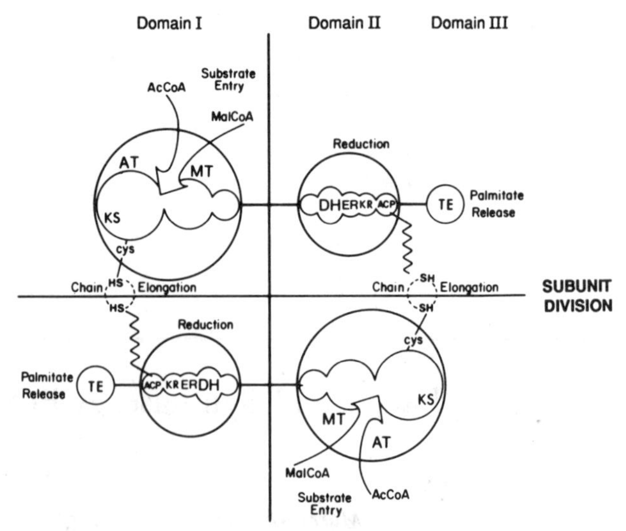

FIGURE 1: Functional map and organization of the subunit proteins of the chicken liver fatty acid synthase. The two subunits with their domains I, II, and III are drawn in an antiparallel arrangement (subunit division) so that two sites of palmitate synthesis are constructed (functional division). The abbreviations for the partial activities used are as follows: KS, β-ketoacyl synthase; AT, acetyl transacylase; MT, malonyl transacylase; DH, β-hydroxyacyl dehydratase; ER, enoyl reductase; KR, β-ketoacyl reductase; ACP, acyl carrier protein; TE, thioesterase. The wavy line represents the 4′-phosphopantetheine prosthetic group of ACP.

The organization of the various component activities on the subunit protein was determined after detailed proteolytic mapping of the synthase protein and subsequently confirmed by localization of their active sites in the primary sequence. Proteolytic cleavage of chicken liver fatty acid synthase by different proteases (trypsin, chymotrypsin, subtilisin, kallikrein, and *Myxobacter* protease) used either individually or in combination was performed, and the results were analyzed with respect to the kinetics and the size of fragments generated (Mattick et al., 1983a,b; Wong et al., 1983; Tsukamoto et al., 1983; Tsukamoto & Wakil, 1988). The results show that the fatty acid synthase subunit protein consists primarily of three major domains (I, II, and III) of M_r 127 000, 107 000, and 33 000, respectively. Each of these domains can be further subdivided by various proteases into a number of distinct regions that can be aligned and mapped. A proteolytic map was generated and served as a reference for the controlled cleavage of the fatty acid synthase, thereby allowing the isolation, determination, and placement of the functional activities within the subunit protein (Figure 1).

Chymotrypsin makes the most restricted cut of the synthase by hydrolyzing its subunits into two fragments of M_r 230 000 and 33 000. These fragments can be readily separated into pure fractions. The large fragment contains all the core activities of fatty acid synthesis, that is, acetyl and malonyl transacylases, β-ketoacyl synthase, β-ketoacyl and enoyl reductases, dehydratase, and ACP. The smaller fragment comprises the thioesterase activity and contains the COOH-terminal end of the synthase subunit (Mattick et al., 1983a). The properties of the isolated thioesterase were studied with long-chain acyl-CoA derivatives as substrates. The enzyme shows the highest rate of hydrolysis with palmitoyl-CoA but is also active with stearoyl-CoA and to a lesser extent with myristoyl-CoA (Lin & Smith, 1978; Libertini & Smith, 1979; Crisp & Wakil, 1982), reflecting the production of these acids by the fatty acid synthase (Bressler & Wakil, 1961).

Cleavage of the synthase by other proteases made it possible to assign locations to the other component activities on the subunit protein. Domain I contains the NH_2 terminus of the polypeptide, the active cysteine–SH of the β-ketoacyl synthase, and the common serine–OH of the acetyl and malonyl transacylases (Joshi et al., 1970; Stoops & Wakil, 1982; Tsukamoto et al., 1983; McCarthy et al., 1983; Mikkelsen et al., 1985b). Hence, this domain functions as a site for entry of the substrates, the acetyl and malonyl groups, and their subsequent condensation to form carbon–carbon bonds (Figure 1). The β-ketoacyl derivative formed by this chain elongation is then reduced by the component enzymes associated with domain II, the reduction domain. In this domain the dehydratase and enoyl and β-ketoacyl reductases are located and ordered as shown in Figure 1 (Wong et al., 1983; Tsukamoto et al., 1983; Tsukamoto & Wakil, 1988). The ACP and its serine–OH for attachment of the 4′-phosphopantetheine cofactor are also located in domain II and connect the β-ketoacyl reductase (domain II) to the thioesterase (domain III).

This arrangement of the partial activities on the synthase subunit protein was recently confirmed by cloning and sequencing the cDNA of the component enzymes of chicken synthase (Chirala et al., 1989). In the predicted amino acid sequences, the active sites of the component activities were identified, and their location on the polypeptide confirmed the proteolytic mapping of the activities on the synthase subunit protein (Tsukamoto et al., 1983; Tsukamoto & Wakil, 1988). The complete amino acid sequence of the thioesterase domain of chicken liver synthase was determined by sequencing the protein (Yang et al., 1988) and confirmed by nucleotide sequencing of the cDNA (Kasturi et al., 1988; Yuan et al., 1988; Chirala et al., 1989). The amino acid sequence of the thioesterase of rat mammary gland synthase has also been predicted from nucleotide sequencing of the cDNA (Naggert et al., 1988) and shows about 70% homology to that of the thioesterase of chicken synthase. However, a much higher homology (~90%) was noted around the active serine sites and at the COOH-terminal ends of the two thioesterases, suggesting that these sequences are highly conserved and may be essential to the function of the thioesterase.

The complete amino acid sequence of the ACP domain prepared after tryptic digestion of the chicken liver fatty acid synthase has been determined (Huang et al., 1989) and confirmed by nucleotide sequencing of the cDNA (Chirala et al., 1989). Comparison of amino acid sequences of chicken synthase ACP to those of *E. coli* ACP and rat synthase ACP shows 23% and 65% homology, respectively. However, the sequence similarities are particularly remarkable around the seryl residue where the phosphopantetheine is attached. The high degree of homology in this region among the ACP domains of animal, plant, and bacterial synthases suggests the conserved nature of these sequences and reinforces the central functional role of ACP in fatty acid synthesis.

The nucleotide sequence of the chicken synthase cDNA shows that the ACP domain connects directly to the thioesterase domain through a stretch of 15 amino acid residues which does not appear to be conserved when compared to the rat synthase. The seryl residue to which the phosphopantetheine is attached is only 141 amino acids upstream from the active serine–OH of the thioesterase (Chirala et al., 1989), indicating that the two domains are highly compacted. The close proximity of the two active sites may be essential in facilitating the scanning of the growing fatty acyl chain by the thioesterase prior to its hydrolysis. Also, it may explain why the regeneration of synthase activity after chymotryptic cleavage of the thioesterase could not be reconstituted by addition of the thioesterase.

Recently, we have expressed the cDNA coding for the ACP and thioesterase domains in *E. coli* using a PL-promoter-based expression vector (unpublished results). The recombinant protein purified from *E. coli* extracts has thioesterase activity, and the ACP domain has 4′-phosphopantetheine attached to it by the *E. coli* ACP-synthetase (Elovson & Vagelos, 1968). The recognition, therefore, of the apo-ACP by the ACP–synthetase suggests that the ACP domain of the recombinant protein has a similar secondary structure as that of *E. coli* ACP and is a substrate for the synthetase, even though its amino acid sequence homology is only 23% (Huang et al., 1989). Similarly, the ACP–thioesterase recombinant protein appears to behave structurally and functionally equivalently to that of the native synthase domains. Thus, the folding of each of the domains appears to be an inherent property of the domain's primary sequence and is independent of the rest of the domains associated with the subunit polypeptide. This observation suggests that the functional domains of the multifunctional fatty acid synthase are folded independently and probably have arisen from fusion of genes coding for separate activities.

The native fatty acid synthase isolated from animal tissues is a dimer and is active in palmitate synthesis. Moreover, active enzyme centrifugation studies showed that the dimer is the only active species involved in fatty acid synthesis (Stoops et al., 1979). Dissociation of the native enzyme to monomers resulted in the loss of palmitate synthesis. However, six of the component enzymes remain active, including the acetyl and malonyl transacylases, β-ketoacyl and enoyl reductases, β-hydroxyacyl dehydratase, and thioesterase (Butterworth et al., 1967; Yung & Hsu, 1972; Muesing et al., 1975; Stoops et al., 1979; Stoops & Wakil, 1981). The one activity lost from the monomer is the β-ketoacyl synthase (condensing enzyme), yet its active cysteine–SH is present on the subunit (Stoops & Wakil, 1981). This result was puzzling since each subunit contains the domains of all component activities.

Earlier studies suggested that the acetyl and malonyl groups are bound to the synthase via thioester linkages (Joshi et al., 1970; Philips et al., 1970; Stoops & Wakil, 1982; Mikkelsen et al., 1985b). The acetyl group is bound to the cysteine–SH of the β-ketoacyl synthase and to the cysteamine–SH of the 4′-phosphopantetheine. The malonyl group, on the other hand, is bound only to the cysteamine–SH (Joshi et al., 1970; Plate et al., 1970; Philips et al., 1970; Stoops & Wakil, 1981, 1982). The cysteine–SH is unusually reactive at neutral pH or lower (Stoops & Wakil, 1981), indicating that this thiol most likely exists as a thiolate ion and therefore is highly susceptible to alkylation. The pantetheine–SH is also susceptible to alkylation and is specifically alkylated by chloroacetyl-CoA (Kumar et al., 1980; McCarthy & Hardie, 1982). The latter compound and reagents such as iodoacetamide, 1,3-dibromopropanone, or Ellman's reagent [5,5′-dithiobis(2-nitrobenzoic acid) (DTNB)] alkylate the fatty acid synthase, resulting in the inhibition of palmitate synthesis. For instance, chicken liver synthase in the presence of 0.5 mM iodoacetamide is completely inhibited with a rate constant of 0.033 min^{-1} and $t_{1/2}$ of 21 min. The reaction follows first-order kinetics for over 90% inhibition of palmitate synthesis and is therefore consistent with the assumption that the inhibition resulted from the reaction of iodoacetamide with one catalytic site on the enzyme. When [^{14}C]iodoacetamide is used, over 80% of the ^{14}C label is recovered as *S*-([^{14}C]carboxymethyl)cysteine after HCl hydrolysis, indicating that the inhibition of the enzyme is due

Scheme I: Proposed Reactions of DTNB with Chicken Liver Fatty Acid Synthase Homodimer[a]

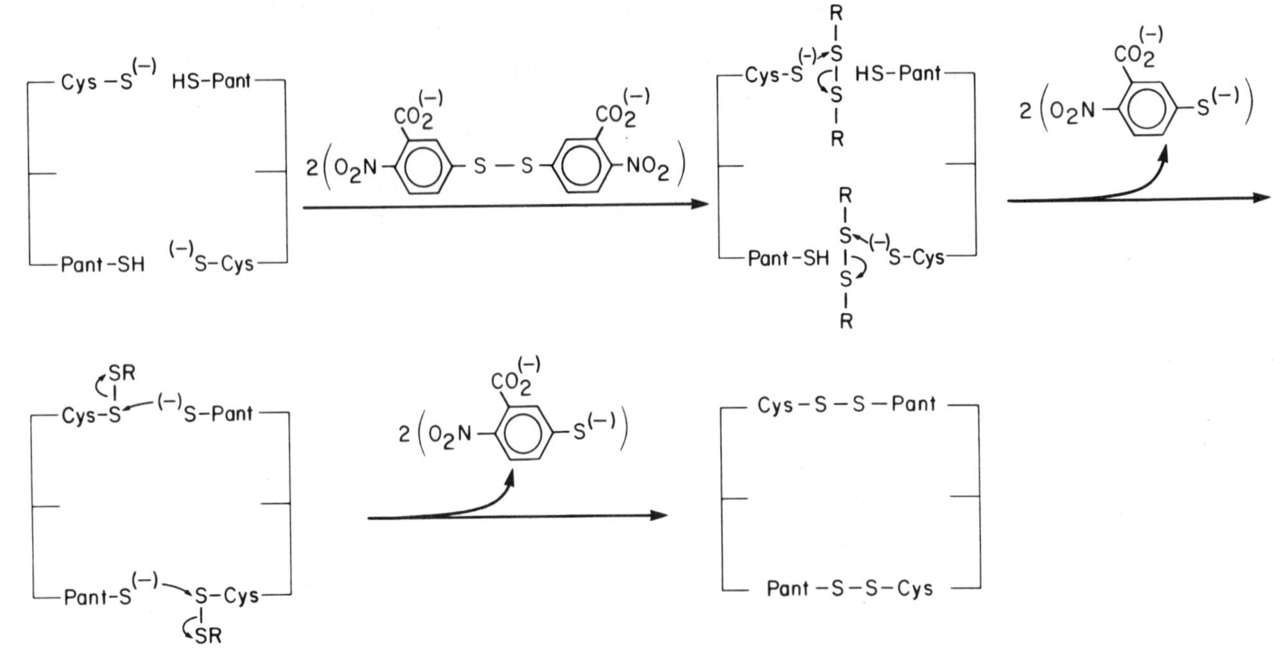

[a] Cys–S⁻ is cysteinyl–S⁻ and Pant–SH is pantetheine–SH.

to alkylation of an active cysteine–SH. This thiol is the active cysteine–SH of the β-ketoacyl synthase component of the fatty acid synthase since it is the only partial activity lost among the seven partial activities associated with the synthase (Stoops & Wakil, 1981). Preincubation of synthase with acetyl-CoA prior to its reaction with iodoacetamide protects the enzyme against inhibition, while preincubation with malonyl-CoA does not render such protection. These observations strongly indicated that alkylating reagents compete with the acetyl group for the same cysteine–SH at the β-ketoacyl synthase site.

In contrast, the reagents 1,3-dibromopropanone and DTNB react very rapidly (within 30 s) with the fatty acid synthase and inhibit palmitate synthesis almost completely (Stoops & Wakil, 1981, 1982). Survey of the partial activities associated with the inhibited enzyme showed that the loss of palmitate synthesis was due to loss of the β-ketoacyl synthase activity. All other partial activities remain intact after reaction of the fatty acid synthase with the alkylating reagents. Scatchard analysis of the inhibition of synthase by dibromopropanone or DTNB indicated that the binding of about 2 mol of the inhibitor/mol of dimer was required for the complete loss of palmitate synthesis.

When the dibromopropanone-inhibited synthase is analyzed by sodium dodecyl sulfate–polyacrylamide gel electrophoresis, the synthase subunit of 260 kDa is nearly absent and new protein bands of doubled molecular weight (500 000–550 000) are found. If [^{14}C]dibromopropanone is used, nearly all the radioactivity is associated with the oligomers of higher molecular weight, indicating that synthase subunits are cross-linked by the bifunctional reagent, dibromopropanone. Preincubation of synthase with acetyl-CoA or malonyl-CoA prior to its reaction with dibromopropanone prevents the cross-linking of the subunits. Similarly, pretreatment of the enzyme with iodoacetamide prior to its reaction with dibromopropanone prevents cross-linking of the synthase subunits. These observations suggest that dibromopropanone cross-linking of the subunit takes place between the cysteine–SH (the site of acetyl binding or iodoacetamide reaction) and the cysteamine–SH of the prosthetic group, 4′-phosphopantetheine (the site of malonyl binding).

Further support for this organization is based on studies of the residues involved in cross-linking synthase subunits by dibromopropanone. When the fatty acid synthase was inhibited with [^{14}C]dibromopropanone and the cross-linked enzyme was subjected to performic acid oxidation [Baeyer–Villeger reaction (Hassal, 1957)] followed by HCl hydrolysis (Crestfield et al., 1963), equal amounts of ^{14}C-labeled sulfones of (carboxymethyl)cysteine and (carboxymethyl)cysteamine were obtained, indicating that the residues involved in the cross-linking are cysteine–SH and pantetheine–SH of the prosthetic group. This conclusion was supported further by the isolation of secondary propyl thioether derivatives of cysteine and cysteamine as the major products after reduction of the [^{14}C]dibromopropanone-cross-linked synthase with borohydride followed by HCl hydrolysis (Stoops & Wakil, 1983).

The subunits of DTNB-inhibited synthase are also cross-linked as shown by analysis with sodium dodecyl sulfate–polyacrylamide gel electrophoresis in an inert atmosphere (under Ar or N_2) and in the absence of 2-mercaptoethanol (Stoops & Wakil, 1982; Tian et al., 1985). When the DTNB-inhibited synthase is treated with 2-mercaptoethanol, the cross-linking is reversed with a concomitant regeneration of palmitate synthesis activity (Stoops & Wakil, 1982). Altogether, these observations indicate that the DTNB inhibition of the synthase results in formation of disulfide bridges between the active cysteine–SH and the cysteamine–SH of the two subunits as shown in Scheme I. Detailed kinetics studies of the reactivity of the synthase with DTNB and the effect of pH, salt concentration, and $NADP^+$ and NADPH on this reaction were reported by Tian et al. (1985).

Scheme I presupposes a head-to-tail organization of the two synthase subunits and the juxtaposition of their thiols within 2 Å of each other. This is a remarkable arrangement of the two subunit proteins considering their unusually large sizes.

Further support for this arrangement was obtained by hybridization experiments in which the phosphopantetheine–SH group was chemically modified by chloroacetyl-CoA and the cysteine–SH group by iodoacetamide (Wang et al., 1984). The resulting enzyme variants were dissociated and hybridized.

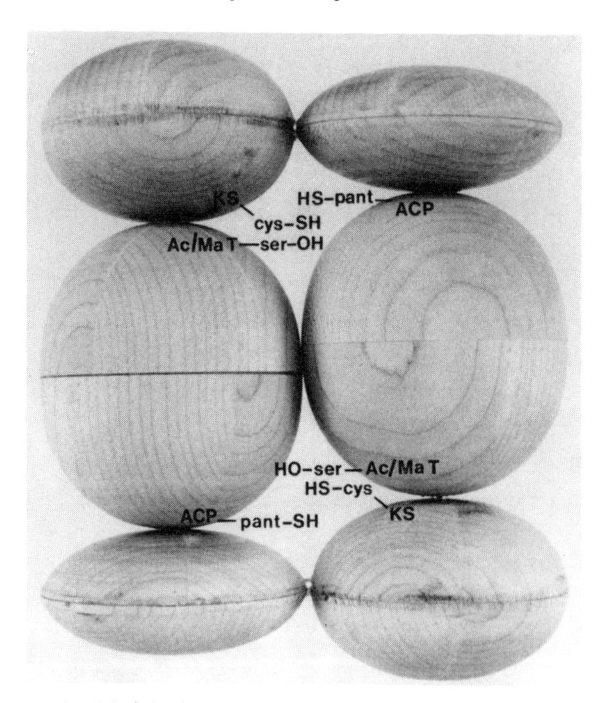

FIGURE 2: Model of chicken liver fatty acid synthase based on small-angle neutron-scattering studies. The model is two side-by-side cylinders with dimensions of 160 × 146 × 73 Å. Each cylinder has three domains measuring, respectively, 32, 82, and 62 Å in length and 36.5 Å in radius, and the domains are related by the dyad axis. Probable locations of the β-ketoacyl synthase and its active cysteine–SH (KS–cys–SH), the acetyl and malonyl transacylases and their common serine–OH (Ac/MaT–ser–OH), and the acyl carrier protein and its 4′-phosphopantetheine prosthetic group (ACP–pant–SH) are shown in the two crevices generated.

The hybrid containing an intact phosphopantetheine–SH group on one subunit and a cysteine–SH group on the other subunit was active in palmitate synthesis. Altogether, the aforementioned results strongly support the conclusion that the two identical subunits of animal synthases are organized in a head-to-tail arrangement so that the active cysteine–SH of one subunit is juxtapositioned opposite a pantetheine–SH of the adjacent subunit, generating two identical centers (Figure 1). Since the cysteine–SH is located at the active site of the β-ketoacyl synthase, the two centers obtained by this arrangement constitute novel β-ketoacyl synthase (condensing) sites. The acetyl group is attached to the cysteine–SH site, and the malonyl group is bound to the pantetheine–SH and therefore makes it possible for the condensation to occur yielding the β-ketoacyl-*S*-pantethenyl derivative.

This conclusion is supported further by physical studies of the synthase using electron microscopy and small-angle neutron-scattering techniques (Stoops et al., 1987). On the basis of these studies a structural model was proposed with an overall appearance of two side-by-side cylinders with dimensions of 160 × 146 × 73 Å (Figure 2). Each cylinder (160 Å in length and 36.5 Å in diameter) is divided into three domains having lengths of 32, 82, and 46 Å, respectively. In the antiparallel arrangement of the two cylinders, two crevices are generated on the major axis of the model at opposite ends of the molecular dyad (Figure 2). It is probable that the β-ketoacyl synthase is located in these crevices with its cysteine and pantetheine residues derived from adjacent subunits.

Since the dimeric complex possesses two complements of each catalytic activity involved in fatty acid synthesis, the model presented above predicts the presence of two active centers for fatty acid synthesis (Figures 1 and 2). Evidence in support of this prediction was recently obtained from chicken liver fatty acid synthase with its thioesterase activity either cleaved by chymotrypsin or inhibited by phenylmethanesulfonyl fluoride or diisopropyl fluorophosphate, thus ensuring that the fatty acyl products remain bound to the protein (Singh et al., 1984). By use of stoichiometric amounts of these enzyme preparations, it was possible to determine the kinetics and the amounts of NADPH oxidized and long-chain fatty acids synthesized. The results show that the amounts of NADPH consumed were sufficient to account for all the fatty acid synthesized and that 1.0 mol of fatty acyl enzyme is synthesized per mole of phosphopantetheine available, indicating that two sites for fatty acid synthesis are active and function simultaneously (Figure 1). The fatty acid synthase has, therefore, full-site reactivity and not half-site reactivity as was claimed earlier (Clements & Barden, 1979; Libertini & Smith, 1979).

Analysis of the fatty acyl products obtained with modified synthase that does not have thioesterase activity showed that $C_{20:0}$ and $C_{22:0}$ constituted over 85% of the fatty acids produced (Libertini & Smith, 1979; Singh et al., 1984). This is in contrast to the native enzyme, which yields mainly palmitate (70%) with stearate (20%) and myristate (10%) (Bressler & Wakil, 1961). These results indicate that the thioesterase is responsible for the chain termination of the process of fatty acid synthesis, since synthase is capable of chain elongating the bound fatty acyl group beyond the $C_{16:0}$ acid. However, in the presence of active thioesterase, the acyl group is preferentially hydrolyzed as it attains the $C_{16:0}$ chain length. This action reflects the specificity of the thioesterase, which has most of its activity on the palmitoyl-CoA and to a lesser extent on the stearoyl- and myristoyl-CoA derivatives (Lin & Smith, 1978; Crisp & Wakil, 1982).

The two centers of palmitate synthesis resulting from the antiparallel arrangement of the two identical subunits of synthase derive the required component activities from complementing halves of the two subunits. As depicted in Figure 1, one subunit contributes domain I while the second subunit contributes domains II and III. This novel organization of these proteins is not surprising since the reactions leading to palmitate synthesis are complex and the component activities are distributed all along the lengthy polypeptide chain, thereby making their interaction difficult if the active synthase is a monomer.

Domain I catalyzes the entry of the acetyl and malonyl substrates into the process of palmitate synthesis. These entries are catalyzed by the acetyl and malonyl transacylases. The acetyl transacylase is specific for acetyl-CoA, but can to a lesser degree utilize propionyl- and butyryl-CoAs as primers in fatty acid synthesis. Malonyl transacylase, on the other hand, is specific for malonyl-CoA. Both transacylases transfer the acyl groups from CoA to an active serine–OH residue on the protein. In the animal synthases the acetyl and malonyl groups competitively inhibit each other's binding to the protein, suggesting that they bind to a common active site (Joshi et al., 1970; Plate et al., 1970; Stern et al., 1982; McCarthy & Hardie, 1983). This was confirmed by the isolation and sequencing of the peptides derived from acetyl and malonyl synthases (McCarthy et al., 1983; Mikkelsen et al., 1985a). Both the acetyl and malonyl groups are bound as an O-ester prior to their transfer to appropriate thiols. The sequence of the amino acids of the peptide containing this serine is highly conserved and retains the sequence motif of -Gly-X-Ser-Y-Gly- that characterizes enzymes with a reactive serine, such as yeast acetyl and malonyl/palmitoyl transacylases, synthase thioesterases, human plasmin, and bovine trypsin and carboxyl

Scheme II: Diagrammatic Scheme of Fatty Acid Synthesis by Full-Sites-Active Chicken Liver Fatty Acid Synthase

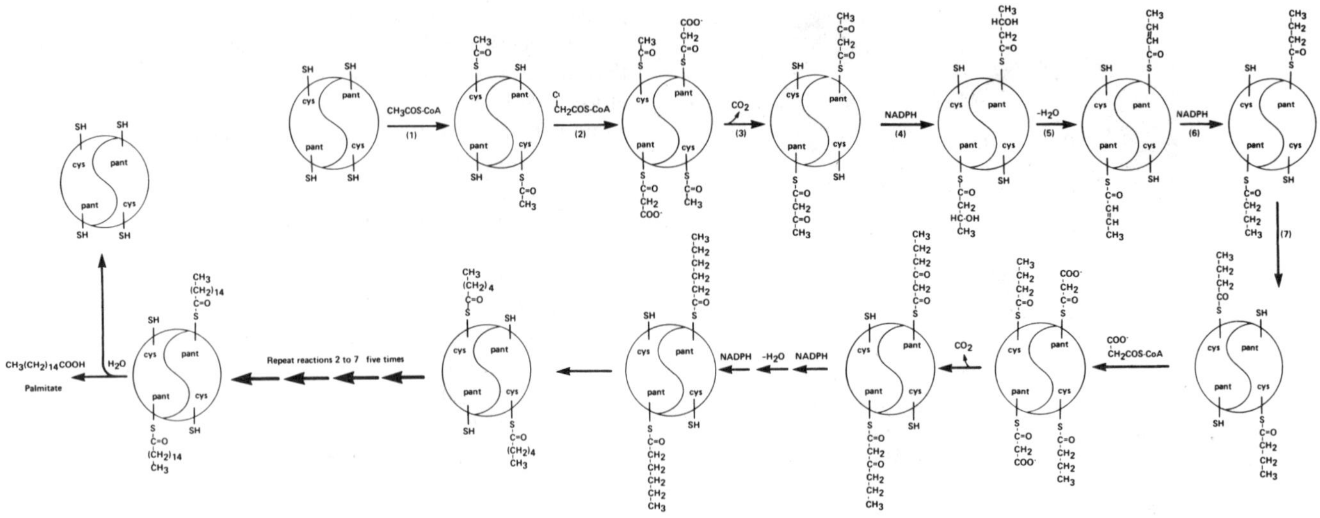

esterase (Barker & Dayhoff, 1972; Hardie & McCarthy, 1986; Yang et al., 1988).

The binding of the proper substrate (acetyl or malonyl) to the enzyme serine–OH is critical in ensuring proper entry of substrates into fatty acid synthesis. To facilitate this process, free CoA–SH plays an important role in the binding of the correct substrate, an acetyl or malonyl group (Lin et al., 1980; Stern et al., 1982; Soulie et al., 1984). Removal of CoA–SH by a scavanging reaction such as that of citrate lyase stops fatty acid synthesis. If the wrong substrate is bound, CoA–SH acts as an acceptor and removes the acyl group, thereby regenerating the active serine–OH for binding of the proper acyl group needed at that particular moment in the repetitive process (Figure 3). If a primer is needed, an acetyl group is bound, and if a source of C_2 units for chain elongation is needed, a malonyl group is bound.

The next step in the entry of the substrates is the channeling of the acetyl and malonyl groups to their proper thiols. The malonyl group binds only to the cysteamine–SH of the prosthetic group, 4′-phosphopantetheine; hence, it is presumed to be channeled directly from the active serine O-ester to the ACP–pantetheine–SH (Figure 3) (Joshi et al., 1970; Phillips et al., 1970). The acetyl group, on the other hand, binds to both the active cysteine–SH of the β-ketoacyl synthase and to the cysteamine–SH of ACP (Joshi et al., 1970; Mikkelsen et al., 1985b). It is not clear, therefore, whether the acetyl group is transferred directly from the serine O-ester to the cysteine–SH or whether it is first transferred to the cysteamine–SH and then to the cysteine–SH of the condensing domain. In the fatty acid synthesizing system of bacteria, the answer is clear because both acetyl-S-ACP and malonyl–S–ACP are required for the β-ketoacyl synthase reaction. The acetyl group is transferred from the ACP to the active cysteine–SH of the condensing enzyme prior to its condensation with the malonyl group to form acetoacetyl–S–ACP (reactions 4 and 5). In the animal fatty acid synthase it has not been possible as yet to determine which of the two pathways the acetyl group follows.

Recently, the acetylation of the yeast fatty acid synthase was studied in detail with specific alkylating reagents for the cysteine–SH and cysteamine–SH (N. Singh, J. K. Stoops, and S. J. Wakil, unpublished data). For instance, iodoacetamide specifically alkylates the cysteine–SH, chloroacetyl-CoA alkylates the cysteamine–SH, and dibromopropanone alkylates both thiols. The kinetics and stoichiometry of acylation of the alkylated yeast synthase were then studied with p-nitrophenyl

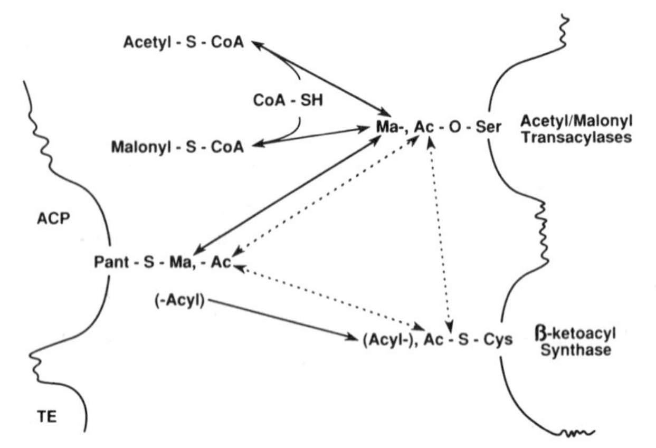

FIGURE 3: Role of CoA–SH in the self-editing of the loading of acetyl (Ac–) and malonyl (Ma–) groups to the common active Ser–OH of the acetyl and malonyl transacylases of animal fatty acid synthase. During each of the eight loading steps the enzyme site is partitioned between competently and incompetently bound substrate molecules. The transfer of the malonyl group from the Ser–OH to the pantetheine–SH (Pant–S–) and of the acyl product of each elongation step (Acyl–) from Pant–S– to the ketoacyl synthase–Cys–SH is shown by solid arrows denoting their more likely route. The transfer of the Ac from Ser–O to Pant–S to Cys–S is represented by dotted arrows denoting the uncertainty in the involvement of the Pant–SH in the Ac loading of Cys–SH of the ketoacyl synthase.

thioacetate. The results show that acetylations of the seryl–OH and cysteine–SH are fast, while acetylation of the cysteamine–SH is very slow and could be discounted. In the yeast synthase at least, the pathway of acetyl binding to the enzyme proceeds from CoA to the serine–OH of the acetyl transacetylase to the cysteine–SH of the β-ketoacyl synthase. No such studies have been carried out with the animal synthase because of technical difficulties. However, Yuan and Hammes (1985) have investigated this problem in chicken liver synthase and have concluded that the acetyl moiety is transferred to the cysteine–SH from the cysteamine–SH, a proposal parallel to reaction 5 (i.e., all sites are acetylated from CoA). Their kinetic data, however, could be explained equally well by direct transfer of the acetyl group from serine–OH to the cysteine–SH.

On the basis of these observations the following mechanism for palmitate synthesis has been proposed (Scheme II). The active synthase is charged with an acetyl group at the cysteine–SH of the β-ketoacyl synthase and with the malonyl group at the cysteamine–SH of the adjacent subunit. Condensation takes place by coupling the carbonyl group of the acetyl moiety to the β-carbon of the malonyl group with a simultaneous release of CO_2 and formation of an acetoacetyl product. The cysteine–SH of the β-ketoacyl synthase is reset to the free thiol state. The acetoacetyl-*S*-pantethenyl derivative is then reduced with NADPH (β-ketoacyl reductase; reaction 6) to the β-hydroxybutyryl derivative, which is then dehydrated by the dehydratase (reaction 7) and reduced by NADPH (enoyl reductase; reaction 8) to the butyryl derivative. All these enzymes are present within domain II and accessible to the pantetheine prosthetic group at the ACP domain. The butyryl product is then transferred to the cysteine–SH of the condensing enzyme in domain I of the adjacent subunit, thereby freeing the cysteamine–SH to accept a malonyl group from malonyl-CoA. The butyryl and malonyl groups condense to form a β-ketohexanoyl-*S*-pantethenyl intermediate which is then reduced to a hexanoyl derivative. The process is then repeated five more times; with each cycle the acyl group is elongated by a C_2 unit, ultimately yielding a palmitoyl derivative. The latter is then hydrolyzed by the neighboring thioesterase to form palmitic acid and the cysteamine–SH. The essence of this mechanism is (1) the organization of the two homodimers of the synthase so that the binding sites of the primer and the elongating malonyl groups are within bond distances, (2) the involvement of complementary domains derived from the two subunits in the condensation reaction and the transfer of the acyl group between the cysteine–SH and pantetheine–SH of the two subunits with each cycle adding C_2 units, and (3) the presence of two active centers for the synthesis of palmitate within the synthase dimer, with each independent center having its own complement of enzymes. The multifunctional nature of the subunits and their head-to-tail organization produce a highly efficient enzyme complex capable of carrying out multiple reactions leading to the synthesis of a palmitate molecule from one acetyl and seven malonyl moieties. Moreover, the entire system is designed so that intermediate acyl derivatives remain efficiently anchored to the enzyme during conversion to the product acid. With the exception of some highly specialized tissues (e.g., lactating mammary gland), the useless shorter chain byproducts (C_6 through C_{12}) are not produced. This remarkable multifunctional enzyme system has evolved into a proficient machine that carries out the synthesis of long-chain fatty acids from simple building blocks of C_2 units.

ACKNOWLEDGMENTS

I thank Drs. Subrahmanyam S. Chirala, A. Habib Mohamed, and James K. Stoops for their helpful discussions and Pamela Powell for her editorial assistance.

Registry No. Fatty acid synthase, 9045-77-6.

REFERENCES

Arslanian, M. J., Stoops, J. K., Oh, Y. H., & Wakil, S. J. (1976) *J. Biol. Chem. 251*, 3194–3196.

Banker, W. C., & Dayhoff, M. O. (1972) in *Atlas of Protein Sequences and Structure*, Vol. 5, pp 53–66, National Medical Research Foundation, Washington, DC.

Bloch, K., & Vance, D. (1977) *Annu. Rev. Biochem. 46*, 263–298.

Bressler, R., & Wakil, S. J. (1961) *J. Biol. Chem. 236*, 1643–1651.

Buckner, J. S., & Kolattukudy, P. E. (1976) *Biochemistry 15*, 1948–1957.

Butterworth, P. H. W., Yang, P. C., Bock, R. M., & Parker, J. W. (1967) *J. Biol. Chem. 250*, 1814–1823.

Chirala, S. S., Kuziora, M. A., Spector, D. M., & Wakil, S. J. (1987) *J. Biol. Chem. 262*, 4231–4240.

Chirala, S. S., Kasturi, R., Pazirandeh, M., Stolow, D. T., Huang, W.-Y., & Wakil, S. J. (1989) *J. Biol. Chem. 264*, 3750–3757.

Clements, P. R., & Barden, R. E. (1979) *Biochem. Biophys. Res. Commun. 86*, 278–284.

Crestfield, A. M., Stein, M. H., & Moore, S. (1963) *J. Biol. Chem. 238*, 2413–2420.

Crisp, D., & Wakil, S. J. (1982) *J. Protein Chem. 1*, 241–255.

Elovson, J., & Vagelos, P. R. (1968) *J. Biol. Chem. 243*, 3603–3611.

Hardie, D. G., & McCarthy, A. D. (1986) in *Multidomain Proteins Structure and Evolution* (Hardie, D. G., & Coggins, J. R., Eds.) pp 229–256, Elsevier Science, Amsterdam.

Hassal, C. H. (1957) *Org. React. 9*, 73–106.

Huang, W.-Y., Stoops, J. K., & Wakil, S. J. (1989) *Arch. Biochem. Biophys.* (in press).

Joshi, V. C., Plate, C. A., & Wakil, S. J. (1970) *J. Biol. Chem. 245*, 2857–2867.

Kasturi, R., Chirala, S. S., Pazirandeh, M., & Wakil, S. J. (1988) *Biochemistry 27*, 7778–7785.

Kirschner, K., & Bisswanger, H. (1976) *Annu. Rev. Biochem. 45*, 143–166.

Knobling, A., Schiffman, D., Sickinger, H.-D., & Schweizer, E. (1975) *Eur. J. Biochem. 56*, 359–364.

Kumar, S., Opas, E., & Alli, P. (1980) *Biochem. Biophys. Res. Commun. 95*, 1642–1649.

Libertini, L. J., & Smith, S. (1979) *Arch. Biochem. Biophys. 192*, 47–60.

Lin, C. Y., & Smith, S. (1978) *J. Biol. Chem. 253*, 1954–1962.

Lynen, F. (1980) *Eur. J. Biochem. 114*, 431–442.

Majerus, P. W., Alberts, A. W., & Vagelos, P. R. (1964) *Proc. Natl. Acad. Sci. U.S.A. 51*, 1231–1238.

Martin, D. B., & Vagelos, P. R. (1962) *J. Biol. Chem. 237*, 1787–1792.

Mattick, J. S., Zehner, Z. E., Calabro, M. A., & Wakil, S. J. (1981) *Eur. J. Biochem. 114*, 643–651.

Mattick, J. S., Nickless, J., Mizugaki, M., Yang, C. Y., Uchiyama, S., & Wakil, S. J. (1983a) *J. Biol. Chem. 258*, 15300–15304.

Mattick, J. S., Tsukamoto, Y., Nickless, J., & Wakil, S. J. (1983b) *J. Biol. Chem. 258*, 15291–15299.

McCarthy, A. D., & Hardie, D. G. (1982) *FEBS Lett. 147*, 256–259.

McCarthy, A. D., & Hardie, D. G. (1983) *Eur. J. Biochem. 130*, 185–193.

McCarthy, A. D., Aitken, A., Hardie, D. G., Santikarn, S., & Williams, D. H. (1983) *FEBS Lett. 160*, 296–300.

Mikkelsen, J., Hoyrup, P., Rasmussen, M. M., Roepstorff, P., & Knudsen, J. (1985a) *Biochem. J. 227*, 21–27.

Mikkelsen, J., Smith, S., Stern, A., & Knudsen, J. (1985b) *Biochem. J. 230*, 435–440.

Mohamed, A. H., Chirala, S. S., Mody, N. H., Huang, W. Y., & Wakil, S. J. (1988) *J. Biol. Chem. 263*, 12315–12325.

Monod, J., Wyman, J., & Changeux, J. P. (1965) *J. Mol. Biol. 133*, 517–532.

Morris, S. M., Jr., Nilson, J. H., Jenik, R. A., Winberry, L. K., McDevitt, M. A., & Goodridge, A. G. (1982) *J. Biol. Chem. 257*, 3225–3229.

Muesing, R. A., Lornitzo, F. A., Kumar, S., & Porter, J. W. (1975) *J. Biol. Chem. 250*, 1814–1823.

Naggert, J., Witkowski, A., Mikkelsen, J., & Smith, S. (1988) *J. Biol. Chem. 263*, 1146–1150.

Phillips, G. T., Nixon, J. E., Abramovitz, A. S., & Porter, J. W. (1970) *Arch. Biochem. Biophys. 138*, 357–371.

Plate, C. A., Joshi, V. C., & Wakil, S. J. (1971) *J. Biol. Chem. 245*, 2868–2875.

Rock, C. D., Goelz, S. E., & Cronan, J. E., Jr. (1981) *J. Biol. Chem. 256*, 736–742.

Schreckenbach, T., Wobser, H., & Lynen, F. (1977) *Eur. J. Biochem. 80*, 13–23.

Schweizer, E., Dietlein, G., Knobling, A., Tahedl, H. W., Schweitz, H., & Schweizer, M. (1975) *FEBS Proc. 40*, 85–97.

Singh, N., Wakil, S. J., & Stoops, J. K. (1984) *J. Biol. Chem. 259*, 3605–3611.

Smith, S., Stern, A., Randhawa, Z. I., & Knudsen, J. (1985) *Eur. J. Biochem. 85*, 547–555.

Soulie, J.-M., Sheplocks, G. J., Tian, W.-X., & Hsu, R. Y. (1984) *J. Biol. Chem. 259*, 134–140.

Stoops, J. K., & Wakil, S. J. (1981) *J. Biol. Chem. 256*, 5128–5133.

Stoops, J. K., & Wakil, S. J. (1982) *Biochem. Biophys. Res. Commun. 104*, 1018–1024.

Stoops, J. K., & Wakil, S. J. (1983) *J. Biol Chem. 258*, 12482–12486.

Stoops, J. K., Arslanian, M. J., Oh, Y. H., Aune, K. C., Vanaman, T. C., & Wakil, S. J. (1975) *Proc. Natl. Acad. Sci. U.S.A. 72*, 1940–1944.

Stoops, J. K., Awad, E. S., Arslanian, M. J., Gunsberg, S., Wakil, S. J., & Oliver, R. M. (1978) *J. Biol. Chem. 253*, 4464–4475.

Stoops, J. K., Ross, P. R., Arslanian, M. J., Aune, K. C., Wakil, S. J., & Oliver, R. M. (1979) *J. Biol. Chem. 254*, 7418–7426.

Stoops, J. K., Wakil, S. J., Uberbacher, E. C., & Bunick, E. J. (1987) *J. Biol. Chem. 262*, 10246–10251.

Stumpf, P. K. (1984) in *Fatty Acid Metabolism and Its Regulation* (Numa, S., Ed.) pp 155–180, Elsevier Science, Amsterdam.

Tian, W.-X., Hsu, R. Y., & Wang, Y.-S. (1985) *J. Biol. Chem. 260*, 11375–11387.

Tsukamoto, Y., & Wakil, S. J. (1988) *J. Biol. Chem. 263*, 16225–16229.

Tsukamoto, Y., Wong, H., Mattick, J. S., & Wakil, S. J. (1983) *J. Biol. Chem. 258*, 15312–15322.

Volpe, J. J., & Vagelos, P. R. (1977) *Physiol. Rev. 56*, 339–417.

Waite, M., & Wakil, S. J. (1962) *J. Biol. Chem. 237*, 2750–2757.

Wakil, S. J. (1970) in *Lipid Metabolism* (Wakil, S. J., Ed.) pp 1–48, Academic Press, New York.

Wakil, S. J., Porter, J. W., & Gibson, D. M. (1957) *Biochim. Biophys. Acta 24*, 453–461.

Wakil, S. J., Titchener, E. B., & Gibson, D. M. (1958) *Biochim. Biophys. Acta 29*, 225–226.

Wakil, S. J., Pugh, E. L., & Sauer, F. (1969) *Proc. Natl. Acad. Sci. U.S.A. 52*, 106–114.

Wang, Y.-S., Tian, W.-X., & Hsu, R. Y. (1984) *J. Biol. Chem. 259*, 13644–13647.

Wong, H., Mattick, J. S., & Wakil, S. J. (1983) *J. Biol. Chem. 258*, 15305–15311.

Yang, C. H., Huang, W. Y., Chirala, S. S., & Wakil, S. J. (1988) *Biochemistry 27*, 7773–7777.

Yuan, Z., & Hammes, G. G. (1985) *J. Biol. Chem. 260*, 13532–13538.

Yung, S., & Hsu, R. (1972) *J. Biol. Chem. 247*, 2689–2698.

Zehner, Z. E., Mattick, J. S., Stuart, R., & Wakil, S. J. (1980) *J. Biol. Chem. 255*, 9519–9522.

MEMBRANE PROTEINS AND TRANSMEMBRANE PHENOMENA

Chapter 14

Signal Sequences[†,‡]

Lila M. Gierasch

Departments of Pharmacology and Biochemistry, University of Texas Southwestern Medical Center, 5323 Harry Hines Boulevard, Dallas, Texas 75235-9041

Received October 7, 1988; Revised Manuscript Received November 7, 1988

While considerable progress has been made in the last 15 years in elucidating the mechanism of protein secretion [for reviews, see Verner and Schatz (1988), Randall et al. (1987), Briggs and Gierasch (1986), Rapoport (1986), Wickner and Lodish (1986), Walter and Lingappa (1986), and Walter et al. (1984)], the roles of the signal sequence are still poorly understood. Ironically, these 15–30 residue long, highly hydrophobic sequences constitute the most general requirement for export of a protein whether from yeast, higher eukaryotes, or bacteria. Several lines of evidence argue that signal sequences from these various organisms work in much the same way. Many features of the export pathway appear to be shared by all species, since most exported proteins can be translocated and processed correctly by the export machinery from several organisms [for an example, see Mueller et al. (1982); for an exception, see Bird et al. (1987)]. Recombinant proteins composed of a signal sequence from one organism and a mature secretory protein from another organism are frequently export competent (Yost et al., 1983; Jabbar & Nayak, 1987). Yet, despite this striking conservation of a critical cellular function, signal sequences display a remarkable lack of primary sequence homology, even among closely related proteins. This perspective first briefly reviews present understanding of signal sequence functions and then discusses results of several approaches that may enhance our understanding of the way these intriguing sequences perform their functions.

Interest in signal sequences is high. In addition to the practical motivation of finding more effective vehicles for production of proteins in recombinant systems, a better understanding of how signal sequences work will shed light on several pressing biological, biophysical, and biochemical questions. Signal sequences are essential for the efficient and selective targeting of nascent protein chains either to the endoplasmic reticulum, in eukaryotes, or to the cytoplasmic membrane, in prokaryotes. As such, they are representative of a much broader class of targeting sequences that serve as organizers and zip codes for cellular traffic of macromolecules (Warren, 1987). Furthermore, signal sequences play a central, although poorly understood, role in the translocation of polypeptide chains across membranes.

The ability of signal sequences to facilitate these complex processes despite their high degree of sequence variability (Perlman & Halvorson, 1983; Watson, 1984; von Heijne, 1985) pointedly raises the issue of the relationship between amino acid sequence and the conformations and interactions of a polypeptide chain (the so-called second half of the genetic code). Furthermore, while the importance of amino acid sequence in determining the three-dimensional structure of a mature protein has been recognized and actively investigated for the last two decades, much less attention has been devoted to the *process* of protein folding in vivo (Tsou, 1988). The sequences of existing proteins have been selected through evolution not only to adopt a functional three-dimensional structure after folding but also to optimize the protein folding process both *temporally* and *spatially*, given the constraints of the cellular context. Clearly, presence of the signal sequence (or other transient sequences) may influence the folding of the nascent chain (Park et al., 1988), and many recent results emphasize the coupling of folding and targeting (Randall & Hardy, 1986; Eilers & Schatz, 1988).

Roles and Interactions of Signal Sequences

In both prokaryotes and eukaryotes, considerable progress has been made in the last few years in the identification of components of the export or secretion machinery. However, current understanding stops abruptly at perhaps the most interesting stage of protein export: translocation across the membrane, be it cytoplasmic or ER.[1] The components and

† Supported by grants from the NIH (GM34962), the NSF (DCB-8896144), and the Robert A. Welch Foundation.

‡ Dedicated to the late E. Thomas Kaiser.

[1] Abbreviations: ER, endoplasmic reticulum; SRP, signal recognition particle; SSR, signal sequence receptor; SDS, sodium dodecyl sulfate; MBP, maltose-binding protein; LPP, lipoprotein; PhoA, alkaline phosphatase; PhoE, phosphate limitation protein; PTH, parathyroid hormone; FT-IR, Fourier transform infrared; BIP, heavy chain binding protein.

mechanistic steps worked out so far are involved in *targeting* the nascent chain to the membrane and then in *cleaving* the transient signal sequence from the mature chain. Any discussion of the intermediate stages falls necessarily in the realm of speculation. Despite these gaps in our understanding, there are several points at which the signal sequence clearly must play direct or indirect roles. We consider now what is known about prokaryotic and eukaryotic protein export in light of the involvement of the signal sequence.

In higher eukaryotes, where the components of the secretory apparatus have been more fully characterized (Rapoport, 1986; Walter & Lingappa, 1986), the first interaction of the signal sequence appears to be with the signal recognition particle (SRP). This interaction is probably the first committed step in protein secretion; it ensures, by virtue of the subsequent specific binding between SRP and its receptor in the ER membrane (SRP receptor or docking protein), that the nascent chain will be correctly targeted. Under some experimental conditions, SRP binding leads to an arrest or a pause in translation (Walter & Blobel, 1981b), which is relieved by release of SRP upon its binding to SRP receptor (Meyer et al., 1982; Gilmore et al., 1982; Gilmore & Blobel, 1982). This arrest or pause may or may not be a feature of the in vivo process (Meyer, 1985); if so, it would couple synthesis to translocation by preventing translation unless delivery to the membrane had taken place.

Interaction of the signal sequence with SRP has been probed by cross-linking experiments with a photoactivatable probe. These experiments indicated that the 54-kDa subunit of the SRP ribonucleoprotein complex was the site of signal sequence binding (Kurzchalia et al., 1986; Krieg et al., 1986). This binding step occurs after the signal sequence emerges from the ribosome, viz., when a chain of about 70–80 residues has been synthesized (Wiedmann et al., 1987a). All evidence points to the existence of *only one* SRP in a particular organism; hence, several different signal sequences must be recognized by the same SRP. The binding site for signal sequences may include the ribosome, since a ternary complex of SRP, nascent chain, and ribosome forms prior to membrane targeting (Walter et al., 1981). Furthermore, signal sequence binding may cause conformational changes within SRP, since the SRP/ribosome affinity increases by 4 orders of magnitude in the presence of nascent chain (Walter & Blobel, 1981a).

The involvement of the signal sequence in the next steps, viz., association with the membrane and translocation, is unclear. Cross-linking studies analogous to those used to identify the SRP-signal sequence interaction have revealed the presence of a 35-kDa species in the ER membrane that is proposed to serve as a "signal sequence receptor" (SSR) (Wiedmann et al., 1987b). Cross-linking of isolated signal peptides identified a 45-kDa species in the microsomal membrane (Robinson et al., 1987). There may in fact be multiple interactions of the nascent chain at the level of the membrane, including the possibility of binding to phospholipids, which has often been suggested on the basis of the hydrophobicity of signal sequences (see below; von Heijne & Blomberg, 1979; Engelman & Steitz, 1983; Briggs et al., 1986). Initial binding (either to a proteinaceous receptor or to the bilayer) may be followed by interaction with a protein or complex that facilitates translocation. Since nothing is known about this process, one can speculate freely. Signal peptidase recognition and cleavage constitute the final steps in eukaryotic protein secretion that involve the signal sequence; these processes occur on the lumenal side of the ER membrane. Evans et al. (1986) isolated signal peptidase from canine pancreas as a complex of six polypeptide chains. Signal peptidase has more recently been purified from hen oviduct in a solubilized form requiring only two polypeptide chains (Baker & Lively, 1987). The multiple components of the canine microsomal peptidase complex are potential candidates for an apparatus to translocate the polypeptide chain.

The steps in prokaryotic protein export are less well defined (Randall et al., 1987), but recent findings promise clarification of the mechanism in the very near future. As is true presently for the eukaryotic systems, virtually nothing is known about translocation; the components identified thus far are nearly all involved in target or cleavage. Much of current knowledge came originally from genetic evidence [for reviews, see Bankaitis et al. (1985), Benson et al. (1985), and Oliver (1985)], which implicated the products of several genes in bacterial protein export: SecA,[2] PrlA (also known as SecY), SecB, PrlD, and the two signal peptidases, leader peptidase (or signal peptidase I) and signal peptidase II, which processes lipoproteins. Biochemical evidence has led to the identification of other species, including soluble factors that are required for in vitro translocation (Mueller & Blobel, 1984; Weng et al., 1988), and a protein, called trigger factor, which forms a complex with the precursor to OmpA and stabilizes this precursor in a translocation-competent form in vitro (Crooke & Wickner, 1987; Crooke et al., 1988a,b).

Putting together all available data at the present time, using a eukaryotic paradigm, suggests the following steps: Upon translation of the nascent protein, part or all of the precursor protein binds to cytoplasmic factors that may include SecA, trigger factor, and/or SecB, depending on the protein to be exported and the kinetic relationship of translation and translocation.[3] SecB (Collier et al., 1988; Kumamoto & Gannon, 1988) and trigger factor (Crooke & Wickner, 1987; Crooke et al., 1988a,b; Lill et al., 1988) may be most critical to export in cases where the synthesis of the precursor is complete or nearly complete prior to its entry into the export pathway. These proteins seem to be important in maintaining an export-competent conformation in the precursor; also, only a subset of proteins depends on SecB for export. The binding of precursors to SecB does not seem to require interaction with the signal sequence (Collier et al., 1988); it is not known whether binding to trigger factor does.

SecA is known to be an essential player in bacterial protein export and likely serves a role similar to that of eukaryotic SRP. Defects in SecA cause pleiotropic effects on protein export (Oliver & Beckwith, 1981). This protein has recently been purified and its gene sequenced; it has 901 amino acids and no apparent homology with any known protein (Schmidt et al., 1988). A complex of nascent chain and SecA (possibly plus trigger factor) may, by analogy with SRP/docking protein, bind to the cytoplasmic membrane and facilitate targeting of the nascent chain to export sites. Purified SecA can be added back to membranes depleted or defective in SecA

[2] The genes associated with protein export were named for this putative function, hence *sec* for secretion or *prl* for protein localization. Then different genes were named A, B, etc. The products of these various genes are designated SecA, PrlA, etc.

[3] Results of different experiments [recently reviewed by Lee and Beckwith (1986)] have been interpreted to indicate cotranslational translocation [for example, Smith et al. (1977)], posttranslational translocation (Koshland & Botstein, 1982), or domain-by-domain cotranslational translocation mechanisms (Randall, 1983). It seems increasingly clear that the relative rates of translation and of export vary as a function of the nature of the protein (for example, its size and rapidity of folding) and of the cellular conditions (i.e., whether there is high export activity and consequent saturation of export sites).

product in vitro and will reconstitute protein translocation (Cabelli et al., 1988).

PrlA (SecY) is a membrane protein (Akiyama & Ito, 1985); mutations in the *prlA* gene cause pleiotropic effects on protein export (Ito et al., 1983). PrlA has been shown to be essential in protein export both in vivo (Ito et al., 1984) and in vitro (Fendl & Tai, 1987). PrlA may be a receptor for the signal sequence (analogous to SSR), may play a direct role in translocation (as a pore or tunnel), or may serve as a receptor for the SecA/export complex (like docking protein). The first suggestion is supported by the finding that several signal sequence mutations normally associated with severe export defects are suppressed very effectively by mutations within PrlA (Emr et al., 1981). As pointed out by Randall et al. (1987), this argument is not unequivocal, since indirect effects cannot be ruled out. For example, the PrlA mutations may alter binding to another species such that its interaction with signal sequences becomes less restrictive. Nonetheless, inspection of the types of signal sequence mutations that can be tolerated in different PrlA backgrounds, and of the changes in the PrlA sequence itself, is of interest in efforts to relate the required sequence characteristics of signal peptides to their ability to function (see below). The suggestion that SecY (PrlA) is a receptor for a SecA/export complex is supported by the recent finding that purified SecA can suppress a temperature-sensitive SecY defect in translocation activity in membrane vesicles (Fandl et al., 1988).

Possible mechanisms for the translocation steps in prokaryotic protein export are, as in the eukaryotic case, speculative. The possibility of direct interaction between the nascent chain and membrane lipids has been discussed frequently (von Heijne & Blomberg, 1979; Engelman & Steitz, 1983; Briggs et al., 1986) but again lacks direct evidence. The last step involving the signal sequence is recognition and cleavage by the leader or signal peptidase. The active site of the transmembrane leader peptidase (signal peptidase I) is situated on the periplasmic side of the membrane (Zimmermann et al., 1982), requiring that the signal sequence cleavage site be oriented appropriately.

While the specific components of the eukaryotic and prokaryotic export machinery are not the same, one conclusion applies to both: The signal sequence is required to perform several roles which probably involve interactions with a variety of species. From the above discussion, we can extract a list of possible roles and interactions of signal sequences:

(*1*) *Binding to SRP or Prokaryotic Equivalent.* In the eukaryotic system, identification of a nascent chain as a secretory protein is mediated by the signal sequence/SRP interaction. Delivery of the nascent chain to the ER membrane is catalyzed by the SRP/SRP receptor (docking protein) binding step. In prokaryotes, these steps may involve SecA, trigger factor, SecB, and/or other cytoplasmic factors. SecA is a good candidate for facilitating membrane targeting of the export complex, since evidence supports its association with the membrane (Oliver & Beckwith, 1982), possibly via PrlA (SecY) (Ryan & Bassford, 1985;[4] Fandl et al., 1988). The identification of SecA mutations that suppress signal sequence defects (Ryan & Bassford, 1985;[4] Fikes & Bassford, 1989) suggests that SecA may interact directly with the signal sequence, but alternative explanations cannot be excluded. Trigger factor, on the other hand, has been found to associate with ribosomes and to interact in a saturable way with membrane vesicles in in vitro translocation assays (Lill et al., 1988). It is possible that the multiple functions of SRP, which in eukaryotes are carried out by different polypeptide chains in one ribonucleoprotein complex (Siegel & Walter, 1988a,b), are associated with separate species in prokaryotes.

(*2*) *Binding to the Membrane To Be Translocated.* This role may be mediated by a proteinaceous receptor molecule (SSR or PrlA) or by direct association with membrane lipids, or possibly both.

(*3*) *Facilitation of Translocation.* The greatest mystery of protein secretion at present is the mechanism of translocation across the membrane. The signal sequence is present at the time of initiation of translocation but may be cleaved during the transfer of the mature portion of the nascent chain. Hence, its potential role in this process might be to facilitate initiation of translocation.

(*4*) *Recognition by Signal Peptidase.* One of the most clear-cut requirements of all cleaved signal sequences is that they be recognized and productively bound by the processing enzyme. This step may involve "traditional" enzyme/substrate interactions but also is likely to be influenced by the topology of the translocating chain in the membrane. The signal sequence must be compatible with the arrangement of the peptidase and nascent chain that enables cleavage to take place (e.g., depth in the membrane, specific conformational features, interaction with the mature segment).

What Makes a Sequence Function as a Signal Sequence?

The traditional approaches to determining sequence/function correlations are quickly stymied by signal sequences. Comparison of all known signal sequences reveals no regions of strict homology; the cleavage site shows the strongest conservation, as might be expected since it must be recognized by signal peptidase. Although other portions of signal sequences lack homology, they do display common distributions of residue type. Von Heijne (1985) has shown by detailed analyses of known signal sequences that their variability is limited: Three recognizable regions with specific characteristics emerge from his comparisons. These characteristics are shared by signal sequences from both eukaryotes and prokaryotes. Counting from the cleavage site, there are usually five to seven residues [including the "−1, −3 rule" residues (von Heijne, 1983; Perlman & Halvorson, 1983)] that comprise the so-called c-region. Although not generally charged, these residues are of higher polarity on average than those in the "h-region" immediately N-terminal to the c-region. The h-region is rich in Leu, Ala, Met, Val, Ile, Phe, and Trp but may contain an occasional Pro, Gly, Ser, or Thr residue. This hydrophobic core (h-region) is the true hallmark of signal sequences. Its length (10 ± 3) distinguishes it from membrane-spanning sequences (24 ± 2 residues long) and from hydrophobic segments of globular proteins (6–8 residues in length) (G. von Heijne, personal communication). Statistical results suggest that overall hydrophobicity is the major requirement in the h-region (von Heijne, 1985). The n-region is of highly variable length and composition, but always carries a net positive charge (on average +1.7). In eukaryotes, this charge is contributed by the N-terminus and any charged residues; in prokaryotes, the N-terminus retains a formyl-Met, and the charge comes exclusively from basic residues.

This sort of analysis of signal sequences convinces one that they indeed have defining characteristics. However, relating

[4] In the paper by Ryan and Bassford (1985), the SecA mutation was referred to as PrlD2. Subsequent to sequencing, it was found to be in the SecA gene (Fikes & Bassford, 1989). This allele has an effect on export of MBP with a defective signal sequence that is synergistic with mutations in PrlA (SecY), arguing for an interaction between these PrlA and SecA.

these characteristics to the functional roles listed above is difficult. As pointed out by von Heijne (1985), the lack of specific patterns in signal sequences does not seem consistent with specific protein/protein interactions: "Both regions (n and h) seem well suited for binding in a rather unspecific way to the surface (n-region) and to the interior (h-region) of membranes."

On the other hand, there have been many reports of alterations in signal sequences, including point mutations, that lead to loss of function. In fact, examination of these sequences and the specific nature of their export defects is a promising route to determining sequence/function correlations. An early example of this strategy was the incorporation of β-hydroxyleucine in the preprolactin signal sequence, which led to a cytoplasmic protein that escaped SRP binding (Hortin & Boime, 1980; Walter et al., 1981; Walter & Blobel, 1981a). Substitution of this polar Leu analogue in a nascent protein whose signal sequence has no or few Leu residues did not impair export. This result suggests that the hydrophobic core of eukaryotic signal sequences mediates their recognition by SRP. There is a plethora of data on mutations in bacterial signal sequences that impair export to varying degrees, and in some cases in quite distinct ways (Benson et al., 1985). Examples drawn from four *Escherichia coli* proteins are gathered in Table I. Two of these proteins (LamB, the λ phage receptor, and LPP, the major lipoprotein) are in the outer membrane, and two (MBP, maltose-binding protein, and PhoA, alkaline phosphatase) are periplasmic. Most of these mutations lead to accumulation of precursor in the cytoplasm. Examples of point mutations in the n-region that lead to decreased synthesis (translation) of the exported protein have been found in both lipoprotein (Inouye et al., 1982; Vlasuk et al., 1983) and LamB (Hall et al., 1983). In the case of the LamB Arg6 → Ser mutation, evidence has been presented that the synthesis-down phenotype was not a consequence of mRNA structure or stability (Benson et al., 1987) and therefore argues for a coupling of export and synthesis, as has been shown in eukaryotes under certain conditions. Suppressors of the translation-down phenotype were found and arose from incorporation of a hydrophilic residue in the h-region or deletion of a large segment of the signal sequence. One interpretation of this result is that the first mutation prevents release of an SRP-like block of translation and the second mutation bypasses this block altogether by disrupting the binding site for the SRP-like species.

The bulk of the export-defective mutants have suffered alterations in the hydrophobic core, usually introduction of a charged amino acid or sometimes a deletion. Also shown in this table are some pseudorevertants which indicate the nature of compensating changes that can again yield a functional signal sequence. In most cases, the changes leading to a reversion phenotype restore the hydrophobic core. Generally, the introduction of a charge in the h-region has a major effect regardless of the specific position; an exception was found in LamB where a charge at position 17 is only modestly deleterious, but a charge at position 19 nearly abolishes export.

On the one hand, all of these mutations demonstrate the sensitivity of signal sequences to quite modest changes. On the other hand, it seems as though debilitating mutations are very rare: Even fairly substantial alterations in the signal sequence usually show somewhat "leaky" phenotypes (Ferenci & Silhavy, 1987). It may be that the numerous cytoplasmic factors (SecB, trigger, SecA, etc.) can rescue the cell from accumulation of precursors in the case of weak signal sequences. Furthermore, a point mutation in PrlA (the PrlA4

Table I: Examples of Signal Sequence Mutations Causing Export Defects in *E. coli*[a]

Protein	Signal Sequence	Relative Function[b]
LamB	1 M 2 M 3 I 4 T 5 L 6 R 7 K 8 L 9 P 10 L 11 A 12 V 13 A 14 V 15 A 16 A 17 G 18 V 19 M 20 S 21 A 22 Q 23 A 24 M 25 A / V	++++
	6: S	Synth↓
	12: D	0
	13: D	+
	14: D	+
	15: E	0
	16: E	0
	17: R	+++
	17: R → D	+++
	19: R	0
	19: R → K	+
	▨ (10–13)	0[c]
	9: L; ▨ (10–13)	++++[c]
	▨ (10–13); 17: R → D → C	++[c]
LPP	1 M 2 K 3 A 4 T 5 K 6 L 7 V 8 L 9 G 10 A 11 V 12 I 13 L 14 G 15 S 16 T 17 L 18 L 19 A 20 G / C	++++
	3: D	Synth↓
	2: ▨; 3: D → A	Synth↓
	2: ▨; 3: D → A → D	Synth↓
	2: ▨ → E; 3: D → A → D → D	Synth↓
	9: V	++++
	9: ▨	++++
	14: D	np,++++
	14: D → V	++++
	14: ▨	++
	15: A	++[d]
	16: A	++++
MBP[e]	1 M 2 K 3 I 4 K 5 T 6 G 7 A 8 R 9 I 10 L 11 A 12 L 13 S 14 A 15 L 16 T 17 T 18 M 19 M 20 S 21 A 22 S 23 A 24 L 25 A / K	++++
	10: P	+
	10: P → R	+++
	11: E	+
	11: E → P	++++
	11: E → P → S	++++
	14: E	+
	14: E → P	++++
	16: K	+
	18: R	+
	19: R	+
	▨ (12–17)	0
	8: L; ▨ (12–17)	++++
	8: L → C; ▨ (12–17)	++++
PhoA	1 M 2 K 3 Q 4 S 5 T 6 I 7 A 8 L 9 A 10 L 11 L 12 P 13 L 14 L 15 F 16 Y 17 P 18 V 19 T 20 K 21 A / R	++++
	8: Q	+
	9: E	0
	10: H	0
	14: R	0
	14: R → Q	+

[a] Sequences shown were compiled by Benson et al. (1985); original references can be found there except as noted. Hatched boxes represent deletions; vertical arrows indicate point mutations. Mutant sequences are otherwise unchanged from wild type. [b] Wild-type levels of export are designated and those severely defective by 0. Intermediate levels of export are then qualitatively indicated by +++, ++, or +. Reduced levels of synthesis are noted (Synth ↓). np indicates not processed. [c] Emr and Silhavy (1983). [d] This mutant shows slow processing to mature form. [e] Bankaitis et al. (1985).

allele) suppresses several severe mutations of signal sequences (Stader et al., 1986). While the mechanism of this suppression is a puzzle,[5] it is tempting to speculate that PrlA forms a pore or channel that becomes more permissive in the PrlA4 strains.

In light of the variability of wild-type signal sequences and the apparent tolerance to many mutational variations, one might well ask the question "Are there sequences that will *not* work as signal sequences?" This question was addressed by Kaiser et al. (1987), who substituted random sequences for the signal sequence of yeast invertase and asked for secretion. Their assay for function was relative growth on sucrose, and

[5] The PrlA4 allele arises from a Leu → Asn substitution (Stader et al., 1986) in the last predicted transmembrane-spanning helix of the protein (Akiyama & Ito, 1987). The predicted structure has 10 membrane-spanning segments, which are consistent with alkaline phosphatase topology mapping (Akiyama & Ito, 1987).

Table II: Signal Sequences Studied as Isolated Peptides

peptide	sequence	conformation[a] aqueous	nonpolar[b]	reference
pre-pro-PTH[c]	SAKDMVKVMIVMLAICFLARSDGKSVKKR(Y)	β	α	Rosenblatt et al. (1980)
M13 coat protein	MKKSLVLKASVAVATLVPMLSFA-NH_2	rc	α	Shinnar and Kaiser (1984)
modified pretrypsinogen	Ac-NPKKAKLFLFLALLLAYVA	rc	α	Austen and Ridd (1982)
PhoE	MKKSTLALVVMGIVASASVQA	β	α	Batenburg et al. (1988a,b)
LamB (and several mutants)	MMITLRKLPLAVAVAAGVMSAQAMA	rc	α	Briggs (1986)
OmpA	MKKTAIAIAVALAGFATVAQA/APKD	rc/β[d]	α	D. W. Hoyt, unpublished results

[a] Predominant conformation from CD analysis; rc designates random or unordered structure. [b] Nonpolar environments include trifluoroethanol, SDS micelles, or hexafluoroisopropyl alcohol. [c] The underlined residues are from the pro region of the hormone; the C-terminal Tyr residue was D. [d] As noted in the text, this peptide undergoes a time-dependent conformational change from a random ensemble of states to β-structure.

they found remarkably that 20% of random sequences from a human DNA library would work. Not surprisingly, it is difficult to score *functional* versus *nonfunctional* in a clear-cut way. The measure used by these authors did not always correlate with near wild-type levels of secreted invertase. Nonetheless, those sequences that facilitated invertase export at reasonable levels had the characteristics of signal sequences as described above. Hence, a result that initially seemed to point a pronounced lack of constraints on signal sequences actually reconfirms that we have an idea of what defines a signal sequence. Furthermore, as pointed out by Ferenci and Silhavy (1987), known signal sequences have been optimized for the particular passenger protein and the needs of the cell. Their ability to function in vivo in most cases will go well beyond meeting some minimal level of export. The multiple roles played by signal sequences and the likelihood of additional mechanisms for facilitating or "rescuing" export (see above) also confound the interpretation of the invertase/random sequence results.

Another approach to determining the limits on signal sequences is to idealize them and ask whether the assumptions used in the idealization were justified. Kendall et al. (1986) applied this approach to the alkaline phosphatase signal sequence and have been able to replace the entire hydrophobic core by Leu, or more recently by Ile (Kendall & Kaiser, 1988), while retaining function. In a similar study, the h-region of the hen lysozyme signal sequence was replaced by $(Leu)_n$, and the amount of mature lysozyme secreted to the medium by *Saccharomyces cerevisiae* was determined. Best export occurred with a core length of 8–10 residues (Yamamoto et al., 1987).

Emerging from all of these approaches is the generalization that primary structure is not critical to signal sequence functions. Clearly, disruption of the hydrophobic core leads to a less effective signal sequence. In all cases, it is difficult to deduce the point in the export pathway where a defect is manifesting itself. In vitro translocation assays may help to sort out the steps, as may genetic tests for suppression.

The Study of Isolated Signal Peptides

We and others have sought a better understanding of how signal sequences work by studying them as isolated peptides. We can then analyze their conformations, their interactions, and their responses to changes in environment. This dissection strategy can be risky, but the characteristics of signal sequences reviewed in the previous section seem to invite such an approach. The fact that they can in many instances be transferred from one protein to another and still function implies that they act quite independently of their context (the sequences adjacent). Signal sequences perform their multiple roles while they are attached as N-terminal extensions on their cognate mature proteins; yet they are probably relatively free of interactions with the rest of the nascent chain.[6] Signal sequences seem likely to interact with many cellular components, some of which have been characterized (signal recognition particle, signal peptidase) and others postulated ["translocon" (Walter & Lingappa, 1986), membrane lipids, signal sequence receptor], but they apparently do so by virtue of their overall properties (residue type and patterns of residues) as opposed to specific sequence. Characterization of isolated signal sequences has the potential to reveal what the critical properties are, particularly if *functional* signal sequences are compared to variants that are *nonfunctional*. Results to date using this strategy have been enlightening.

A potential limitation in the study of isolated signal sequences is that they are not likely to have strongly preferred conformations. Linear peptides of fewer than 30 residues generally sample several conformations in aqueous solution (Wright et al., 1988). Even in structure-promoting environments, most short polypeptides are likely to be interconverting among different structures, with at best a bias toward one. Characterization of such a dynamic state is extremely difficult. On the other hand, the biological roles of signal sequences may require them to be conformationally dynamic and to respond to different environments by conformational changes. Studies of isolated signal peptides suggest this to be the case.

Signal sequences as isolated peptides have generally demonstrated similar conformational preferences (Table II), which had been predicted from secondary structure analysis (Austen, 1979; Rosenblatt et al., 1980; Emr & Silhavy, 1983).[7] For the most part they are unordered in aqueous solution, and interactions with nonpolar solvents or with micellar solutions induce adoption of α-helix. Preproparathyroid hormone (signal sequence plus six residues of the pro region plus one residue of the mature hormone) was found to exist predominantly in β-structure in aqueous solution (Rosenblatt et al., 1980), and other examples have been reported of a conformational equilibrium that includes β forms (Batenburg et al., 1988). It is clear that conformational interconvertibility is a feature of signal sequences, and it has been suggested to be of functional importance (Austen & Ridd, 1981; Bedouelle & Hofnung, 1981; Rosenblatt et al., 1980; Batenburg et al., 1988a). As a complicating factor, isolated signal peptides are sparingly soluble, and one must interpret with caution the presence of β-structure. We have followed the circular dichroism spectra

[6] Evidence in favor of this idea includes the observation that signal sequences are accessible to antibodies (Baty & Lazdunski, 1979) and that they can be proteolytically clipped from the precursor species and will bind nonionic detergents while they are still linked to the precursor (Dierstein & Wickner, 1985). Evidence against this image includes the recent report that the presence of the signal sequence on maltose-binding protein modulates its rate of folding in vitro (Park et al., 1988), which suggests direct interaction between the signal sequence and the mature region of the protein.

[7] It is quite surprising that these predictions would apply to signal sequences, since they have been derived from the behavior of sequences within globular proteins.

of aqueous solutions of the OmpA signal sequence from *E. coli* as a function of time; this peptide begins in an unordered conformational ensemble and gradually changes to nearly 100% β (David W. Hoyt, unpublished results). The rate of this conformational transition is increased by higher concentration and is decreased at low pH. Intermolecular association is the apparent driving force for the conformational change. Nonetheless, the fact that these sequences visit both α-helical and β-structures argues that these states are of very similar energy.

Assessing the importance of these preferred conformations of isolated signal peptides in terms of their function in vivo is not straightforward. It is difficult to mimic the microenvironments likely to be encountered in the export process, and it is not clear whether a particular conformational propensity is required for function. To address these problems, we have made use of the families of export-impaired mutant signal sequences from *E. coli* to draw correlations between physical properties and ability to facilitate export in vivo. The LamB system was chosen and offers several particularly interesting comparisons: For example, as shown in Table I, a deletion of four residues in the h-region of the LamB sequence causes a severe export defect. This is not surprising given the generality of the requirement for a 10–12 residue hydrophobic core. What is surprising is that two pseudorevertant strains with restored ability to facilitate export were isolated from the deletion mutant strain; the pseudorevertants had secondary point mutations that apparently compensate for the loss of four residues (Emr & Silhavy, 1983). When Emr and Silhavy found these strains, they argued that α-helicity is required for signal sequence function, since the deletion mutant would be predicted (Chou & Fasman, 1974a,b) to have a much reduced tendency to adopt helix (relative to wild type) because of the proximity of a Pro and a Gly in its sequence. The two pseudorevertants replace either the Pro or the Gly with a helix-favoring residue and hence restore predicted helix formation. Conformational analysis of these sequences as isolated peptides confirms this interpretation (Briggs & Gierasch, 1984; Briggs, 1986). We find that the wild-type LamB signal sequence adopts a largely α-helical conformation in SDS micellar environments, in lipid vesicles, or in water/trifluoroethanol mixtures. The deletion mutant has much less helix under the same conditions, and the pseudorevertants show increased helicity.

The ability to take up an α-helix in nonpolar or interfacial environments thus seems to be a property of functional signal sequences, but it is clearly not sufficient for a given sequence to function as a signal sequence. For example, we have also examined the two LamB signal sequence mutants that harbor a charge (A13D and G17R,[8] Table I) as isolated peptides (C. J. McKnight, M. S. Briggs, and L. M. Gierasch, unpublished results). Although the extent to which they cause an export-defective phenotype in vivo is quite different, their tendency to adopt α-helix is not; both behave similarly to wild type.

As noted above, the possibility that signal sequences interact with the membrane has been suggested on many occasions. Isolated signal peptides provide a means of exploring the likelihood and mechanism of such an interaction. Furthermore, comparison of the various mutant signal sequences confirms that a high affinity for a phospholipid membrane is also characteristic of functional signal sequences. We have compared the abilities of the various LamB mutant signal sequences to insert either into a lipid monolayer or into a lipid bilayer in a vesicle (Briggs et al., 1985; C. J. McKnight, M. S. Briggs, and L. M. Gierasch, unpublished results). We found the wild type, the G17R, and the Pro → Leu pseudorevertant to have the highest affinities. The A13D mutant, which is severely export defective in vivo yet folds into helix equally as well as the G17R, has a reduced affinity for a membrane. Others have reported high-affinity lipid interactions for signal sequences from M13 (Shinnar & Kaiser, 1984), from PhoE (Batenburg et al., 1988b), and from ovalbumin (Fidelio et al., 1987).

In order to describe more fully the conformational states of the LamB wild-type signal sequence upon its interaction with a membrane, we have carried out spectroscopy on peptide/lipid monolayers transferred onto solid supports [either quartz plates for CD or germanium crystals for Fourier transform infrared (FT-IR) spectroscopy] (Briggs et al., 1986). The transfer was done under two conditions: either at a high packing density (surface pressure) of the lipid, such that the signal peptide did not insert but instead associated with the surface, or at a lower lipid packing density (surface pressure resembling that of a biological membrane), such that the signal peptide inserted into the lipid acyl chain region. We found that the peptide adopted a β-structure when associated with the surface but was predominantly α-helical when inserted. From differential scanning calorimetry (M. Kodama, M. S. Briggs, C. J. McKnight, L. M. Gierasch, and E. Freire, unpublished results), fluorescence studies of Trp-containing signal peptides (C. J. McKnight and M. Rafalski, unpublished results), and polarized FT-IR (D. G. Cornell, R. A. Dluhy, C. J. McKnight, and L. M. Gierasch, unpublished results), we have concluded that α-helical, inserted form of the LamB wild-type signal peptide is oriented parallel to the acyl chains. Assuming that the N-terminus does not traverse the membrane, this mode of interaction suggests that an isolated signal peptide can facilitate the insertion and translocation of its C-terminus to the opposite side of the membrane. We have incorporated this idea and the associated conformational interconversions into a model for the initial interactions of the signal sequence with a membrane in protein export (Briggs et al., 1986). We have now synthesized the LamB wild-type signal sequence plus a segment of the mature protein in order to ask whether the signal sequence can cause the C-terminal segment to be translocated in the absence of any other components of the export apparatus.

These observations on isolated signal sequences serve to point out just what a functional signal sequence will do, by virtue of its inherent properties that arise from its amino acid sequence. Yet, there is no question that protein export in vivo involves additional components and that the signal sequence interacts with proteins that target and possibly translocate the bulk of the nascent chain. In fact, isolated signal peptides can be used as probes of the export machinery. As noted above, Robinson et al. (1987) used this approach with a photolabile cross-linker on the signal peptide to find a possible component of the ER translocation apparatus. Addition of synthetic signal peptides at approximately micromolar concentration to in vitro translocation systems causes inhibition of translocation both in prokaryotes (Chen et al., 1987) and in eukaryotes (Majzoub et al., 1980; Austen & Ridd, 1983; Austen et al., 1984). The LamB mutant signal sequences inhibited the translocation of pre-alkaline phosphatase and pre-OmpA to an extent that paralleled their in vivo function (Chen et al., 1987). This result supports the interpretation that the inhibition arises from an

[8] Substitution mutations are designated by the single-letter code for the original residue, the position (numbered from the N-terminus), and then the single-letter code for the substituted residue: hence, A13D, etc.

intervention of the added signal peptide at a normal step in export, despite the relatively high concentrations required. However, we could not distinguish a mode of inhibition based on competition between the signal peptide and the precursor for a proteinaceous receptor (cytoplasmic or membrane associated) from one based on membrane insertion and an indirect effect on translocation. Recently, we found that an all-D LamB wild-type signal peptide inhibits translocation of the pre-OmpA less than does the all-L peptide, arguing that there is a recognition by protein, which would distinguish the opposite handedness of the all-D peptide (A. R. Sgrignoli, L. L. Chen, P. C. Tai, and L. M. Gierasch, unpublished results).

CONCLUSIONS: IMPLICATIONS FOR SIGNAL SEQUENCE FUNCTION

Signal sequences mediate a critical cellular function: correct and efficient localization of nascent secretory proteins. Yet, paradoxically their amino acid sequences are not highly constrained. As discussed in this perspective, they must interact with several components of the export pathway, whether in prokaryotes or in eukaryotes. These interactions are intriguing in their lack of a requirement for specific sequences. Similar binding mechanisms may be operative in other systems: for example, in presentation of antigens by the major histocompatibility complexes (Bjorkman et al., 1987), in binding to BIP in the ER lumen (Gething et al., 1986), and in degradative proteolysis as mediated by protease La in *E. coli* (Waxman & Goldberg, 1986). In all of these examples, as in the case of signal sequences, the overall properties of sequences are the key recognition features. In addition, the way signal sequences are presented probably contributes to their ability to facilitate the correct targeting of a nascent chain despite their lack of sequence specificity. Since they are on or near the N-terminus and accessible (not sequestered by folding), it is likely that SRP or its prokaryotic equivalent binds to the growing polypeptide chain and specifies targeting to the ER or cytoplasmic membrane whenever a signal-sequence-like pattern of residues emerges early in translation in a largely unfolded form. As demonstrated by Kaiser et al. (1987), many sequences within a mature polypeptide *could* function as signal sequences; that they do not is probably a consequence of their mode of presentation and their relationship to the three-dimensional structure of the protein. It could be said that cytoplasmic proteins have to be selected *not* to reveal any targeting sequences so as not to be incorrectly localized. Perhaps more rapid folding is required of nascent chains destined to remain in the cytoplasm.

Because of these characteristics of signal sequences—that they function by virtue of their overall properties and quite independently of their context, work on isolated signal peptides has been particularly fruitful. Signal sequences are clearly conformationally flexible, responding to their environment by pronounced conformational changes. They also have a strong tendency to insert into phospholipid membranes. This biophysical attribute may have a direct functional significance, implying interactions with lipids in vivo. Alternatively, the binding sites for signal sequences on the various proteinaceous components of the export pathway may require the same linear amphiphilicity that favors lipid interactions, perhaps because at any earlier stage in evolution there were direct lipid interactions. Further understanding of these questions awaits dissection of the components required for export and analysis of their interactions with signal sequences.

ACKNOWLEDGMENTS

The work from my laboratory has been carried out through the efforts of a number of people: Marty Briggs, Jamie McKnight, Dave Hoyt, Maria Rafalski, Anita Sgrignoli, Robin Huff, and Jennifer Johnston. Fruitful collaborations with Don Cornell, Rich Dluhy, Tom Silhavy, P. C. Tai, Masayori Inouye, and Ernesto Freire have also made possible several aspects of the work. I thank the several colleagues who provided results prior to publication.

REFERENCES

Akiyama, Y., & Ito, K. (1985) *EMBO J. 4*, 3351–3356.

Akiyama, Y., & Ito, K. (1987) *EMBO J. 6*, 3465–3470.

Austen, B. A. (1979) *FEBS Lett. 103*, 308–313.

Austen, B. A., & Ridd, D. H. (1981) *Biochem. Soc. Symp. 46*, 235–258.

Austen, B. M., & Ridd, D. H. (1983) *Biochem. Soc. Trans. 11*, 160–161.

Austen, B. M., Hermon-Taylor, J., Kaderbhai, M. A., & Ridd, D. H. (1984) *Biochem. J. 224*, 317–325.

Baker, R. K., & Lively, M. O. (1987) *Biochemistry 26*, 8561–8567.

Bankaitis, V. A., Ryan, J. P., Rasmussen, B. A., & Bassford, P. J., Jr. (1985) *Curr. Top. Membr. Transp. 54*, 105–150.

Batenburg, A. M., Brasseur, R., Ruysschaer, J.-M., van Scharrenburg, G. J. M., Slotboom, A. J., Demel, R. A., & de Kruijff, B. (1988a) *J. Biol. Chem. 263*, 4202–4207.

Batenburg, A. M., Demel, R. A., Verkleij, A. J., & de Kruijff, B. (1988b) *Biochemistry 27*, 5678–5685.

Baty, D., & Lazdunski, C. (1979) *Eur. J. Biochem. 102*, 503–507.

Bedouelle, H., & Hofnung, M. (1981) in *Membrane Transport and Neuroreceptors*, pp 399–403, Alan R. Liss, New York.

Benson, S. A., Hall, M. N., & Silhavy, T. J. (1985) *Annu. Rev. Biochem. 54*, 101–134.

Benson, S. A., Hall, M. N., & Rasmussen, B. A. (1987) *J. Bacteriol. 169*, 4686–4691.

Bird, P., Gething, M.-J., & Sambrook, J. (1987) *J. Cell Biol. 105*, 2905–2914.

Bjorkman, P., Saper, M. A., Samroui, B., Bennett, W. S., Strominger, J. L., & Wiley, D. C. (1987) *Nature 329*, 512–517.

Briggs, M. S. (1986) Ph.D. Dissertation, Yale University.

Briggs, M. S., & Gierasch, L. M. (1984) *Biochemistry 23*, 3111–3114.

Briggs, M. S., & Gierasch, L. M. (1986) *Adv. Protein Chem. 38*, 109–180.

Briggs, M. S., Gierasch, L. M., Zlotnick, A., Lear, J. D., & DeGrado, W. F. (1985) *Science 228*, 1096–1099.

Briggs, M. S., Cornell, D. G., Dluhy, R. A., & Gierasch, L. M. (1986) *Science 233*, 206–208.

Cabelli, R. J., Chen, L., Tai, P. C., & Oliver, D. B. (1988) *Cell 55*, 683–692.

Cerretti, D. P., Dean, D., Davis, G. R., Bedwell, D. M., & Nomura, M. (1983) *Nucleic Acids Res. 11*, 2599–2616.

Chen, L., Tai, P. C., Briggs, M. S., & Gierasch, L. M. (1987) *J. Biol. Chem. 262*, 1427–1429.

Chou, P. Y., & Fasman, G. D. (1974a) *Biochemistry 13*, 211–222.

Chou, P. Y., & Fasman, G. D. (1974b) *Biochemistry 13*, 222–245.

Collier, D. N., Bankaitis, V. A., Weiss, J. B., & Bassford, P. J., Jr. (1988) *Cell 53*, 273–283.

Crooke, E., & Wickner, W. (1987) *Proc. Natl. Acad. Sci. U.S.A. 84*, 5216–5220.

Crooke, E., Brundage, L., Rice, M., & Wickner, W. (1988a) *EMBO J. 7*, 1831–1835.

Crooke, E., Guthrie, B., Lecker, S., Lill, R., & Wickner, W. (1988b) *Cell 54*, 1003–1011.

Dierstein, R., & Wickner, W. (1985) *J. Biol. Chem. 260*, 15919–15924.
Eilers, M., & Schatz, G. (1988) *Cell 52*, 481–483.
Emr, S. D., & Silhavy, T. J. (1983) *Proc. Natl. Acad. Sci. U.S.A. 80*, 4599–4603.
Emr, S. D., Hanley, S., & Silhavy, T. J. (1981) *Cell 23*, 79–88.
Engelman, D. M., & Steitz, T. A. (1981) *Cell 23*, 411–422.
Evans, E. A., Gilmore, R., & Blobel, G. (1986) *Proc. Natl. Acad. Sci. U.S.A. 83*, 581–585.
Fandl, J. P., & Tai, P. C. (1987) *Proc. Natl. Acad. Sci. U.S.A. 84*, 7048–7052.
Fandl, J. P., Cabelli, R., Oliver, D., & Tai, P. C. (1988) *Proc. Natl. Acad. Sci. U.S.A. 85*, 8953–8957.
Ferenci, T., & Silhavy, T. J. (1987) *J. Bacteriol. 169*, 5339–5340.
Fidelio, G. D., Austen, B. M., Chapman, D., & Lucy, J. A. (1987) *Biochem. J. 244*, 295–301.
Fikes, J. D., & Bassford, P. J., Jr. (1989) *J. Bacteriol.* (in press).
Gething, M.-J., McCammon, K., & Sambrook, J. (1986) *Cell 46*, 939–950.
Gilmore, R., & Blobel, G. (1983) *Cell 35*, 677–685.
Gilmore, R., Walter, P., & Blobel, G. (1982) *J. Cell Biol. 95*, 453–469.
Hall, M. N., Gabay, J., & Schwartz, M. (1983) *EMBO J. 2*, 15–19.
Hortin, G., & Boime, I. (1980) *Proc. Natl. Acad. Sci. U.S.A. 77*, 1356–1360.
Inouye, S., Soberon, X., Franceschini, T., Nakamura, K., Itakura, K., & Inouye, M. (1982) *Proc. Natl. Acad. Sci. U.S.A. 79*, 3438–3441.
Ito, K., Wittekind, M., Nomura, M., Shiba, K., Yura, T., Miura, A., & Nashimoto, H. (1983) *Cell 32*, 789–797.
Ito, K., Yura, T., & Cerretti, D. (1984) *EMBO J. 3*, 631–635.
Jabbar, M. A., & Nayak, D. P. (1987) *Mol. Cell. Biol. 7*, 1476–1485.
Kaiser, C. A., Preuss, D., Grisafi, P., & Botstein, D. (1987) *Science 235*, 312–317.
Kendall, D. A., & Kaiser, E. T. (1988), *J. Biol. Chem. 263*, 7261–7265.
Kendall, D. A., Bock, S. C., & Kaiser, E. T. (1986) *Nature 321*, 706–708.
Koshland, D., & Botstein, D. (1982) *Cell 30*, 893–902.
Kumamoto, C. A., & Gannon, P. M. (1988) *J. Biol. Chem.* (in press).
Lee, C., & Beckwith, J. (1986) *Annu. Rev. Cell Biol. 2*, 315–336.
Lill, R., Crooke, E., Guthrie, B., & Wickner, W. (1988) *Cell 54*, 1013–1018.
Majzoub, J. A., Rosenblatt, M., Fennick, B., Maunus, R., Kronenberg, H. M., Potts, J. T., Jr., & Habener, J. F. (1980) *J. Biol. Chem. 255*, 11478–11483.
Meyer, D. I. (1985) *EMBO J. 4*, 2031–2033.
Meyer, D. I., Krause, E., & Dobberstein, B. (1982) *Nature 297*, 503–508.
Mueller, M., & Blobel, G. (1984) *Proc. Natl. Acad. Sci. U.S.A. 81*, 7737–7741.
Mueller, M., Ibrahimi, I., Chang, C. N., Walter, P., & Blobel, G. (1982) *J. Biol. Chem. 257*, 11860–11863.
Oliver, D. B., & Beckwith, J. (1981) *Cell 25*, 765–772.
Oliver, D. B., & Beckwith, J. (1982) *Cell 30*, 311–319.
Park, S., Liu, G., Topping, T. B., Cover, W. H., & Randall, L. L. (1988) *Science 239*, 1033–1035.
Perlman, D., & Halvorson, H. O. (1983) *J. Mol. Biol. 167*, 391–409.
Randall, L. L. (1983) *Cell 33*, 231–240.
Randall, L. L., & Hardy, S. J. S. (1986) *Cell 46*, 921–928.
Randall, L. L., Hardy, S. J. S., & Thom, J. R. (1987) *Annu. Rev. Microbiol. 41*, 507–541.
Rapoport, T. A. (1986) *CRC Crit. Rev. Biochem. 20*, 73–137.
Robinson, A., Kaderbhai, M. A., & Austen, B. M. (1987) *Biochem. J. 242*, 767–777.
Rosenblatt, M., Beaudette, N. V., & Fasman, G. D. (1980) *Proc. Natl. Acad. Sci. U.S.A. 77*, 3983–3987.
Ryan, J. P., & Bassford, P. J., Jr. (1985) *J. Biol. Chem. 260*, 14832–14837.
Schmidt, M. G., Rollo, E. E., Grodberg, J., & Oliver, D. B. (1988) *J. Bacteriol. 170*, 3404–3414.
Shinnar, A. E., & Kaiser, E. T. (1984) *J. Am. Chem. Soc. 106*, 5006–5007.
Siegel, V., & Walter, P. (1988a) *Trends Biochem. Sci. 13*, 314–315.
Siegel, V., & Walter, P. (1988b) *Cell 52*, 39–49.
Smith, W. P., Tai, P. C., Thompson, R. C., & Davis, B. D. (1977) *Proc. Natl. Acad. Sci. U.S.A. 74*, 2830–2834.
Stader, J., Benson, S. A., & Silhavy, T. J. (1986) *J. Biol. Chem. 261*, 15075–15080.
Tsou, C. L. (1988) *Biochemistry 27*, 1809–1812.
Verner, K., & Schatz, G. (1988) *Science 241*, 1307–1313.
Vlasuk, G., Inouye, S., Ito, H., Itakura, K., & Inouye, M. (1983) *J. Biol. Chem. 258*, 7141–7148.
von Heijne, G. (1983) *Eur. J. Biochem. 133*, 17–21.
von Heijne, G. (1985) *J. Mol. Biol. 184*, 99–105.
von Heijne, G. (1986) *J. Mol. Biol. 192*, 287–290.
von Heijne, G., & Blomberg, C. (1979) *Eur. J. Biochem. 97*, 175–181.
Walter, P., & Blobel, G. (1981a) *J. Cell Biol. 91*, 551–557.
Walter, P., & Blobel, G. (1981b) *J. Cell Biol. 91*, 557–561.
Walter, P., & Lingappa, V. (1986) *Annu. Rev. Cell Biol. 2*, 499–516.
Walter, P., Gilmore, R., & Blobel, G. (1984) *Cell 38*, 5–8.
Warren, G. (1987) *Nature 327*, 17–18.
Watson, M. E. E. (1984) *Nucleic Acids Res. 12*, 5145–5164.
Waxman, L., & Goldberg, A. L. (1986) *Science 232*, 500–503.
Weng, Q. P., Chen, L. L., & Tai, P. C. (1988) *J. Bacteriol. 170*, 126–131.
Wickner, W., & Lodish, H. (1985) *Science 230*, 400–407.
Wiedmann, M., Kurzchalia, T. V. Bielka, H., & Rapoport, T. A. (1987a) *J. Cell Biol. 104*, 201–209.
Wiedmann, M., Kurzchalia, T. V., Hartmann, E., & Rapoport, T. A. (1987b) *Nature 328*, 830–833.
Wright, P. E., Dyson, H. J., & Lerner, R. A. (1988) *Biochemistry 27*, 7167–7175.
Yamamoto, Y., Taniyama, Y., Kikuchi, M., & Ikehara, M. (1987) *Biochem. Biophys. Res. Commun. 149*, 431–436.
Yost, C. S., Hedgepeth, J., & Lingappa, V. R. (1983) *Cell 34*, 759.
Zimmermann, R., Watts, C., & Wickner, W. (1982) *J. Biol. Chem. 257*, 6529–6536.

Chapter 15

Membrane Protein Folding and Oligomerization: The Two-Stage Model

J.-L. Popot*,‡ and D. M. Engelman*,§

Institut de Biologie Physico-Chimique and Collège de France, 13 rue Pierre et Marie Curie, F-75005 Paris, France, and Department of Molecular Biophysics and Biochemistry, Yale University, 260 Whitney Avenue, New Haven, Connecticut 06511

Received October 30, 1989

ABSTRACT: We discuss the view that the folding of many, perhaps most, integral membrane proteins can be considered as a two-stage process. In stage I, hydrophobic α-helices are established across the lipid bilayer. In stage II, they interact to form functional transmembrane structures. This model is suggested by the nature of transmembrane segments in known structures, refolding experiments, the assembly of integral membrane protein from fragments, and the existence of very small integral membrane protein subunits. It may extend to proteins with a variety of functions, including the formation of transmembrane aqueous channels. The model is discussed in the context of the forces involved in membrane protein folding and the interpretation of sequence data.

In the following, we examine the suggestion that the folding of many integral membrane proteins can be understood in terms of two energetically distinct stages. In the first stage, independently stable helices are formed across the hydrophobic region of the membrane lipid bilayer. In the second, the helices interact with one another to give a functional, globular membrane protein. Similar helix–helix interactions participate in the stabilization of membrane protein oligomers.

Integral membrane proteins are partially buried in the nonpolar environment of the lipid bilayer, where the hydrophobic effect is absent and intrachain hydrogen bonds take on a much greater significance than in water, since the lipid solvent is unable to form them. This energy balance led to the expectation that the transmembrane region of membrane proteins would consist of predominantly hydrophobic segments with regular secondary structure and, more specifically, of bundles of hydrophobic α-helices (Henderson, 1975, 1977). This expectation has been largely fulfilled, although some exceptions do exist (see below).

Methods for predicting transmembrane helical regions from amino acid sequence data rely on scanning the sequence for stretches of residues long enough to span the nonpolar region of a lipid bilayer as α-helices and hydrophobic enough to be expected to be at lower free energy across a membrane than in an aqueous environment [reviewed by Engelman et al. (1986)]. Only two structures are known with enough certainty to provide a test of these predictions: bacterial photosynthetic reaction centers and bacteriorhodopsin. The existence and approximate position in the sequence of the 18 transmembrane helices in these two structures are well predicted by a simple hydrophobicity analysis (Michel et al., 1986a; Engelman et al., 1982; Ovchinnikov et al., 1985). This success is all the more striking given that the helices involved, predicted to be stable as independent entities in the bilayer, actually have more contact with other helices and pigments than they do with lipids. Therefore, we propose a conceptual division of the process of folding of these molecules into factors giving rise to the transmembrane helices and those determining their assembly into the final, folded structure. In this two-stage model (Figure 1), transmembrane helices are regarded as autonomous folding domains.

Similarly, the formation of oligomeric complexes of proteins in membranes will frequently involve the side to side interaction of transmembrane helices, as is seen in the oligomer of the photosynthetic reaction center and in the trimeric association of bacteriorhodopsin molecules. Association of single transmembrane helices is encountered in complexes of the photosynthetic and respiratory chains [see Popot and de Vitry (1990)] and may play a role in the dimerization of some anchored proteins (Bormann et al., 1989). In these instances, the packing of helices at subunit interfaces is like that within the subunits themselves. Thus, understanding helix–helix interactions in bilayers may clarify both folding of individual

‡ Institut de Biologie Physico-Chimique.
§ Yale University.

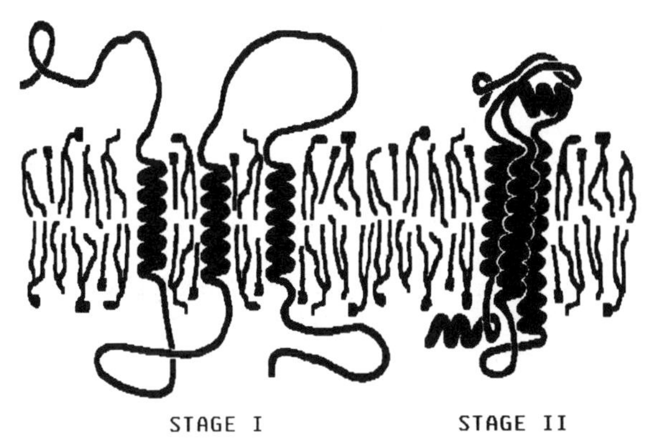

FIGURE 1: Two-stage model for the folding of α-helical integral membrane protein. The first stage is the formation of independently stable trans-bilayer helices, principally in response to the hydrophobic effect and the formation of main-chain hydrogen bonds in the nonaqueous environment. The second stage is the interaction of the helices to form the tertiary fold of the polypeptide. Factors that could contribute to the energetics of stage II are the links between helices, packing of helices and lipid molecules, polar interactions between helices, and, when applicable, association with prosthetic groups or with other proteins.

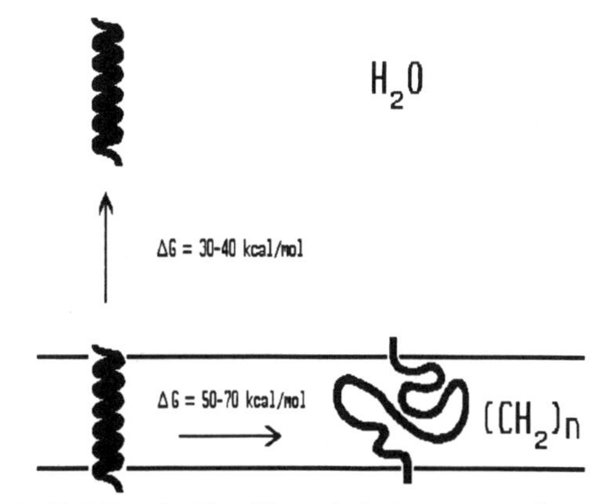

FIGURE 2: Stability of a 20-residue polyalanine transmembrane α-helix. Estimates are indicated of the free energy cost of removing the helix from the lipid phase (top) or breaking all of the hydrogen bonds (right). Modified from Engelman and Steitz (1981).

molecules and factors involved in oligomerization.

The two-stage model posits that the final structure in the transmembrane region results from the accretion of smaller elements (helices), each of which has reached thermodynamic equilibrium with the lipid and aqueous phases before packing. Some rearrangements are to be expected upon assembly (see below), but some others, like flipping helices through the membrane or inserting new segments, are considered as kinetically forbidden. It is therefore implicit in the model that the final structure may be the one with the lowest free energy according to these constraints, but not necessarily that with the lowest overall free energy. This interplay of thermodynamic and kinetic factors has been recently discussed by Finkelstein and Ptitsyn (1987) in the context of a model in which soluble globular proteins fold by packing preformed secondary structure elements. The major difference between soluble and membrane proteins, in this respect, is that transmembrane α-helices in lipids are much more stable than secondary structure elements of the same length in water and more akin to autonomous folding domains. In the same way as folding domains do not unfold and refold when they associate, the two-stage model excludes structures incorporating transmembrane segments that would not be individually stable. While a number of other circumstances can be imagined, such as the stabilization of an otherwise unstable helix or extended segment through a hairpin link to a stable helix (Engelman & Steitz, 1981), it seems that the simpler view may suffice in most of the known cases.

In the following paragraphs, we discuss helix formation in the bilayer and helix–helix associations. We have deliberately restricted ourselves to a consideration of the energetics of polypeptides inserted into a lipid bilayer, leaving aside the mechanism of insertion itself. One of the leading contentions of this discussion, indeed, is that many aspects of integral membrane protein folding can be understood without reference to the actual way insertion is achieved.

HYDROPHOBIC HELICES IN LIPID BILAYERS ARE HIGHLY STABLE STRUCTURES

Helical structure is known to be induced in polypeptides in nonaqueous environments (Singer, 1962, 1971). The large free energy cost of transferring an unsatisfied hydrogen-bond donor or acceptor from an aqueous to a nonpolar environment or of breaking such a bond in a nonpolar environment suggests that most hydrogen bonds must be satisfied when peptides are inserted into a membrane environment. α-Helices accomplish this in a systematic fashion that links nearby parts of the chain. If we consider a polypeptide with nonpolar side chains traversing a membrane bilayer, its removal from the bilayer is opposed by the hydrophobic effect and amounts to tens of kilocalories. If the peptide is unfolded within the nonpolar environment, breaking all of the hydrogen bonds, even more energy is required (Figure 2).

Thus, the nonpolar transbilayer helix is a very stable structure—so stable, in fact, that it can accommodate several polar groups without becoming unstable. Inserting polar groups into the bilayer may be required for function, as when histidine residues serve as ligands for chlorophylls or hemes. The term "hydrophobic" therefore can be ambiguous, inasmuch as many hydrophobic α-helices may contain a number of polar residues that may also make them "amphipathic". Several different scales and approaches for identifying transmembrane helices on the basis of their hydrophobicity have been developed and appear to be largely successful. In many cases, indeed, the local hydrophobicity is so high that putative transbilayer helices can be identified by inspection. In other cases, it may be impossible to decide on the basis of hydrophobicity alone whether a given segment is more likely to span the membrane or to belong to the hydrophobic core of an extramembrane domain [cf. Eisenberg (1984), Klein et al. (1985), Engelman et al. (1986), and Popot and de Vitry (1990)]. Throughout the present paper, we mean by "hydrophobic α-helices" helices that are hydrophobic enough to partition into lipid bilayers with a strongly favorable free energy change.

If transbilayer helices are the stable building blocks of many membrane proteins and complexes, at least two interesting issues arise. If hydrogen bonding of the main chain and the hydrophobic effect are not important factors in helix–helix association (as they are already taken into account in the formation of the helices themselves), how are the folded structures of the membrane proteins maintained? Further, ion channels contain aqueous transmembrane paths—how can such paths be provided by largely nonpolar helices?

FACTORS GOVERNING HELIX–HELIX ASSOCIATION

A membrane protein formed from independently stable helices must be held together by interactions that permit the detailed close packing of the helices, overcoming the entropy that favors helix separation (in the range of 1–10 kcal/mol for a pair of helices). Possibilities for such interactions include hydrogen bonds, ion pairs, the interaction of helix dipoles, packing differentials involving lipid/protein interactions, interactions with prosthetic groups, and external constraints such as links between the helices or interactions with other molecules.

It has been pointed out that external structures of the subunits may stabilize the transmembrane helix bundle in the photosynthetic reaction center (Yeates et al., 1987). Extramembrane regions may play an important role in helix assembly in proteins such as the Na^+/K^+-ATPase α-subunit or the subunits of the nicotinic acetylcholine receptor, where large extramembrane regions appear to separate putative transmembrane helices (see below). It is of interest to note in this context that the size of a stable extramembrane protein domain may be somewhat less than for globular soluble proteins, since constraint of the polypeptide ends will favor folding. Conversely, links between helices will limit their possible diffusion in the plane of the lipid bilayer so that the unfavorable entropic term in the free energy of association will be reduced. These two entropic effects are analogous to those of disulfide bridges in soluble proteins (Flory, 1956; Pace et al., 1988).

Hydrogen bonds and ion pairs can be used to drive the association of local regions of a pair of helices. Van der Waals interactions would then promote detailed close packing and further stabilize the association. As has been noted, the strength of a hydrogen bond in a nonaqueous milieu is in the range of 4–6 kcal/mol (Allen, 1975). Thus, a single hydrogen bond could, in principle, match the entropic term. Ion pairing or strong hydrogen bonding could provide even larger association energies; these would depend very much on the local environment of the ion pair [see Honig et al. (1986)]. While the detailed structure of bacteriorhodopsin is not yet known, early indications that the interior of the protein might be more polar than its lipid-exposed surface (Engelman & Zaccai, 1980) are consistent with results from hydrophobic labeling (Brunner et al., 1985) and site-directed mutagenesis (Khorana, 1988) and with the orientation of the two sequence segments whose arrangement into the structure has been established with some certainty (Popot et al., 1989). The photosynthetic reaction center transmembrane region features no interhelical salt bridges and few hydrogen bonds (Yeates et al., 1987), but the helices are oriented with their more polar surfaces toward the interior of the complex (Rees et al., 1989).

Interactions with prosthetic groups also provide packing constraints. These will be particularly strong when liganding involves residues belonging to several helices. Such is the case in the photoreaction centers (Michel et al., 1986b) and probably in most of the complexes of the respiratory and photosynthetic chains.

Because α-helices have polar ends arising from charge separation in their peptide linkages ("dipoles"), it has been proposed that antiparallel association between nearest-neighbor helices is energetically favored compared with a parallel association [see Hol (1985) and references cited therein]. As noted by Yeates et al. (1987), the photosynthetic reaction center contains mainly associations of antiparallel helices; the most likely folding model for bacteriorhodopsin also involves mostly antiparallel interactions [see Popot et al. (1989)]. The weakening influences of solvent dielectric ions and counterions at the helix ends, however, can be important (Rogers & Sternberg, 1984; Honig et al., 1986; Gilson & Honig, 1989); parallel helices are actually found in several instances. Stabilization by helix–dipole interactions may be weak in bacteriorhodopsin, which has only short extramembrane loops and therefore fairly exposed helix ends. In the photoreaction center from *Rhodopseudomonas viridis*, helix–dipole interactions might be reinforced when the cytochrome and the H subunit bind to the core and screen the two ends of the α-helix bundle from the solvent.

The contribution of packing effects deserves close examination. It may be difficult for lipid chains to pack well against the surface of an α-helix given the irregular contour presented by protruding side chains. As helices are known to engage in detailed close packing (Yeates et al., 1987; Richards, 1977), a less favorable packing at the helix–lipid interface should act to favor association of helices and association of lipids in separate regions. Further, the restriction of lipid-chain conformations in the vicinity of a comparatively rigid helix may give rise to some entropic preference of lipid molecules to be next to other lipids. The magnitudes of such contributions are difficult to calculate, but a rough estimate can be made for the effect of close packing. The free energy cost of creating cavities in proteins has been estimated by Rashin et al. (1986) to be 60 $cal \cdot mol^{-1} \cdot Å^{-3}$ (within a factor of 2 or so). On this basis, filling a 30–80-$Å^3$ cavity (the size of a small side chain) could provide a large portion of the free energy required to associate two helices. Destabilizing free energy changes of comparable magnitude have been recently determined after voids were introduced genetically in the interior of a soluble enzyme (Kellis et al., 1988).

Thus, it is seen that a variety of possible interactions may drive helix association. It is likely that each of the factors considered above will prove to play a role but that a different balance will exist in different cases, e.g., depending on the protein's function [cf. Michel et al. (1986a)].

An Experimental Study of Factors in Folding: Bacteriorhodopsin

Strong experimental tests of the two-stage folding model have come from studies of the refolding of proteolytically cleaved and denatured bacteriorhodopsin. Bacteriorhodopsin (BR) consists of a bundle of seven transmembrane helices that surround the retinal prosthetic group (Henderson & Unwin, 1975; Kouyama et al., 1981; Jubb et al., 1984). It is the only membrane protein to have been wholly renatured starting from the completely unfolded polypeptide as well as from two denatured proteolytic fragments (Huang et al., 1981; Liao et al., 1983). The fragments are produced by a single chymotryptic cleavage between the second and third of the seven transmembrane segments. A covalent link between them is not required for the molecule to refold properly. Further, it has proven possible to refold the BR fragments in lipid bilayers under conditions such that the formation of a two-dimensional lattice can subsequently be induced (Popot et al., 1987). The structure of the renatured molecules can be studied crystallographically and is shown to be indistinguishable from that of native BR (Popot et al., 1986). These observations bring further strong support to the contention that the native structure of BR lies at a free energy minimum (Huang et al., 1981). This is an important conclusion, as the anisotropic environment of membrane proteins and the intrinsic asymmetry of their mode of insertion during synthesis make it possible that a native structure might be biosynthetically trapped at a state of higher energy that could not be reached during refolding in vitro.

When the two BR fragments are separately refolded into lipid vesicles, they regain helical structure. Upon fusion of such vesicles, the two fragments interact and bind retinal. The resulting complex has the visible absorption spectrum of native BR (Popot et al., 1987). The kinetics of chromophore regeneration are the same whether retinal is added to native (chymotryptically cleaved) apoprotein or to the complex formed by reassociating prerefolded fragments. All steps occur in bilayers of *Halobacterium* lipids, in an environment that must be similar to that of the natural plasma membrane.

These observations indicate that, while the isolated fragments have been refolded under conditions vastly different from those prevailing during natural folding in vivo, their structure as isolated entities is close to that which they adopt in native BR. This permits recognition and correct reassociation to occur under conditions (absence of detergent) where major reorientations, e.g., flipping of helices through the bilayer, must be kinetically forbidden. It seems probable that, despite the absence of the rest of the molecule, the correct number of α-helices per fragment must form, that they must span the bilayer of the reconstituted vesicles, and that they must be close to their correct position in the sequence. The first point is directly substantiated by ultraviolet–circular dichroism spectra, which indicate each fragment to have recovered a highly α-helical structure. The other two points are currently under investigation. The refolded fragments in lipid vesicles are highly stable and their folded state must correspond to a deep free energy minimum.

The outcome of these experiments is such as would be expected from the two-stage model, where each transmembrane helix behaves as an independent folding domain. In this case, the link between the second and third helices and the presence of the prosthetic group are not required for fragment assembly to occur. The implication is that the remaining sources of helix–helix interactions, polar forces and packing effects, must dominate.

As mentioned before, at the level of the bilayer there is not much difference between packing helices within a subunit and packing subunits into an oligomer. This is illustrated in vitro by the above experiments, in which refolded BR fragments behave as the two subunits of a heterodimer. Equivalent cases are observed in vivo, either naturally or as the result of genetic experiments, and produce split or "microassembled" proteins that are functional; integral protein complexes in the inner membranes of chloroplasts and mitochondria contain very small subunits apparently comprised of a single transmembrane α-helix and little more [reviewed by de Vitry and Popot (1989) and Popot and de Vitry (1990)].

The formation of separately stable domains by individual transmembrane helices may be related to the fact that most introns in genes coding for polytopic proteins are located in the loops between helices (Jennings, 1989; P. Slonimski, personal communication). Transmembrane helices may have been recombined by exon shuffling in the course of evolution, in response to the need for new or modified functions.

Structures Involved in Transmembrane Aqueous Channels

Aqueous channels present a challenge to the idea that bundles of individually stable helices may form most transmembrane protein structures. It has been thought that channels must have strongly polar linings and that such linings cannot be formed from the weakly polar surfaces found in stable transmembrane helices. The nicotinic acetylcholine receptor (nAChR) provides an instructive example.

Several folding models of the nAChR subunits have been advocated. Most feature either four mainly hydrophobic transmembrane helices ("four-helix" model) or five transmembrane segments, one or two of them carrying numerous charges [reviewed by Popot and Changeux (1984), Hucho (1986), Numa (1989), and Changeux (1990)]. Immunological evidence in favor of models with a charged channel lining and an odd number of transmembrane segments has been presented [see Ratnam et al. (1986b) and references cited therein]. Recent experiments, however, have tended not to support these models. Namely, (i) biochemical data favor an even number of transmembrane segments (McCrea et al., 1986, 1987; DiPaola et al., 1989); (ii) the various charged segments proposed as candidates for the channel lining are probably external to the membrane (Dennis et al., 1988; Dwyer, 1988; Atassi et al., 1987); (iii) electrophysiological measurements indicate that, in its narrowest part, the nAChR channel is uncharged (Dani & Eisenman, 1987); (iv) biochemical (Giraudat et al., 1986, 1987, 1989; Hucho et al., 1986; Oberthür et al., 1986; Revah et al., 1990) and genetic evidence (Imoto et al., 1986, 1988; Leonard et al., 1988) suggests that each subunit contributes the second of its four hydrophobic helices toward lining the channel.

While no folding model of the nAChR subunits can be considered as established, it seems therefore probable that their transmembrane region (and, presumably, that of all chemically gated channels sequenced to date; Grenningloh et al., 1987a,b; Schofield et al., 1987; Hollmann et al., 1989; Wada et al., 1989; Gregor et al., 1989) comprises only hydrophobic α-helices. A simple hypothesis is that the two-stage model applies, but with an additional level of assembly: following folding of the subunits, they would pack without a drastic change in their transmembrane region to form the oligomeric channel. It has often been argued that in large membrane proteins or in channel-forming proteins some sequence segments are not in contact with lipids, and therefore need be neither α-helical, nor hydrophobic, nor long enough to span the bilayer, and that restricting topological models to long hydrophobic transmembrane α-helices would be misleading. In the case of the nAChR, such hypotheses do not seem warranted.

Two differences between BR and the nAChR subunits deserve comment. First, the transmembrane region of the nAChR represents only about one-fourth of its residues, rather than three-fourths in bacteriorhodopsin. From the amino terminus on, the four-helix model of the subunits features a large extracellular region, followed by three closely spaced hydrophobic helices, a cytoplasmic region, and a carboxy-terminal hydrophobic helix. Unlike BR, the extramembrane regions are large enough to contain folding domains of their own (they comprise about 200 and 100 residues, respectively).

A second and important difference is the formation of the channel. When BR monomers pack into trimers, they trap between them half a dozen lipid molecules (Glaeser et al., 1985). Would not we expect the same thing to occur upon assembly of the nAChR subunits, resulting in a "channel" filled with lipids? An answer might lie in the channel's dimensions. While it is known to be blocked by many amphipathic compounds [see Popot and Changeux (1984)], its narrowest width is estimated to be 6–7 Å [Maeno et al., 1977; Dwyer et al., 1980; see also Furois-Corbin & Pullman (1989) and references cited therein]. This is too small to accommodate a phospholipid molecule (but close to the dimension of cholesterol). The overall profile of the channel is not known (although the narrow section has been shown to be short; Dani & Eisenman, 1987; Dani, 1989), but narrowness or poor steric adaptation might exclude membrane lipids. It is worth recalling that ca. 22-residue-long hydrophobic peptides inserted in lipid bilayers do manage to open aqueous channels, which are thought to result from aggregation of independently stable transmembrane α-helices [see, e.g., Molle et al. (1988), Lear et al. (1988), and Oiki et al. (1988)].

To form a stable water-filled channel from fully folded nAChR subunits in a membrane requires that the unfavorable energy of creating a largely hydrophobic surface in contact with water be compensated by the energy of other interactions. Simple calculations show that this can easily be the case. While the actual geometry of the nAChR channel is not known, reasonable assumptions about the area and composition of its lining give an upper bound of 40 kcal/mol as the likely penalty for creating it. Dissociation constants for oligomeric proteins are usually less than 10^{-8}, corresponding to a free energy of dissociation of more than 10 kcal per interface [see, e.g., Chothia and Janin (1975)]. As the receptor has large extramembrane domains, the association of the five subunits could more than compensate for the energy cost of forming the channel, even if the additional contribution from subunit–subunit interactions in the lipid phase is disregarded. The structure of the tryptophan synthase complex (Hyde et al., 1988) shows the presence of a largely hydrophobic tunnel 25 Å long in the oligomeric enzyme complex. The tunnel seems filled with solvent and is large enough for an indole molecule (ca. 7 Å wide) to pass. This striking structural finding provides a clear basis for the contention that oligomeric association energies can create an aqueous channel through a hydrophobic region.

Other Structures

Space limitations prevent a detailed consideration of other structures. It should be noted, however, that, while the hydrophobic helix bundle might be the most common sort of transmembrane region, it is not the only one. Porins, which open aqueous channels in the outer membranes of Gram-negative bacteria, are neither α-helical nor markedly hydrophobic. Rather than being a cluster of individually stable helices, they are thought to be made up of an assembly of β-sheets stabilized by a dense network of hydrogen bonds [see, e.g., Kleffel et al. (1985), Nabedryk et al. (1988), and references cited therein]. Porins are the only integral membrane proteins for which the presence of nonhelical transmembrane secondary structure is well established. Their peculiar structure might be related to difficulties in exporting proteins containing hydrophobic sequence segments to the outer membrane [cf. Davis and Model (1985), MacIntyre et al. (1988), and Popot and de Vitry (1990)]. As individual β-strands, or pairs of β-strands, leave many hydrogen bonds unsatisfied, some concerted mechanism of insertion must be expected.

Voltage-gated channels such as the Na^+, K^+, and Ca^{2+} channels present a special problem, as their structure must include a charged sensor of transmembrane voltage [for a discussion on structural requirements for sensors, see Honig et al. (1986)]. Comparative examination of their sequences suggests that the sensor function could be accomplished by transmembrane α-helices including one positively charged residue every turn in an otherwise hydrophobic segment (Noda et al., 1984; Tempel et al., 1987; Tanabe et al., 1987; Numa, 1989). The rest of the transmembrane region would be made up of numerous hydrophobic helices. If these structural premises are correct, these proteins contain helices that would not be expected to spontaneously partition into the lipid phase. While the protein as a whole might lie at its free energy minimum, some of the transmembrane segments would not. This again would require a concerted mechanism of insertion. For instance, the insertion of a polar helix could be stabilized by interaction with a very hydrophobic helix, as has been proposed in the helical hairpin hypothesis (Engelman & Steitz, 1981).

Model Building

When one attempts to build structural models from a sequence, the safest approach is probably to restrict postulates to hydrophobic transmembrane segments, unless structural and functional data impose a more elaborate hypothesis. The two-stage model, whether or not it describes exactly the folding pathway that is followed in vivo, provides an interesting approach to structural predictions and justifies attempts at building up transmembrane regions by assembling computationally preformed helices [see, e.g., Holm et al. (1987) or Furois-Corbin and Pullman (1989)]. It should be realized, however, that the difficulties of this task are considerable. Even if one assumes that the identification of the transmembrane helices is error free, the exact limits of each cannot be precisely defined, and the dynamics of tilting, bobbing, twisting, and bending [cf. Vogel et al. (1988)] permit a range of helix–helix contacts at the time of assembly. In BR, there is indirect evidence that reassociation of the two chymotryptic fragments involves proline–imide bond cis–trans isomerization (Popot et al., 1987). The presence of prolyl residues in transmembrane segments will result in various types of deviations from perfect α-helices, as in helix C of the reaction center L subunit (Deisenhofer et al., 1985; Michel et al., 1986a) and in helix B of BR (Popot et al., 1989). Given that the C^α positions in real helices deviate from the ideal and that side chains can adopt many different conformations, packing helices is not a trivial computational problem. It can probably best be approached when additional constraints can be resorted to, such as the geometrical constraints introduced by multiple liganding of prosthetic groups, as in cytochrome *c* oxidase [see, e.g., Holm et al. (1987)].

Conclusion

The observations and experiments summarized here can be simply understood by assuming that hydrophobic helices in transmembrane regions of proteins behave as autonomous folding domains, analogous to the larger domains characterized in soluble proteins. This type of structure is probably not particular to electron or proton pumps and may extend to proteins that delimit transmembrane aqueous channels. Membrane proteins that include strongly hydrophilic transmembrane segments do exist, but they may turn out to be the exception.

While undoubtedly an oversimplification, the two-stage model represents a useful way of rationalizing the structure and behavior of integral proteins in terms of assemblies of hydrophobic transmembrane helices. It provides a reasonable basis for model building that emphasizes what can be included in models over what cannot be excluded.

References

Allen, L. C. (1975) *Proc. Natl. Acad. Sci. U.S.A. 72*, 4701–4705.

Atassi, M. Z., Mulac-Jericevic, B., Yokoi, T., & Manshouri, T. (1987) *Fed. Proc., Fed. Am. Soc. Exp. Biol. 46*, 2538–2547.

Bormann, B. J., Knowles, W. J., & Marchesi, B. T. (1989) *J. Biol. Chem. 264*, 4033–4037.

Brunner, J., Franzusoff, A. J., Luscher, B., Zugliani, C., & Semenza, G. (1985) *Biochemistry 24*, 5422–5430.

Changeux, J.-P. (1990) *Fidia Res. Found. Neurosci. Award Lect.* (in press).

Chothia, C., & Janin, J. (1975) *Nature (London) 256*, 705–708.

Dani, J. A. (1989) *J. Neurosci. 9*, 884–892.

Dani, J. A., & Eisenman, G. (1987) *J. Gen. Physiol. 89*, 959–983.
Davis, N. G., & Model, P. (1985) *Cell 41*, 607–614.
Deisenhofer, J., Epp, O., Miki, K., Huber, R., & Michel, H. (1985) *Nature (London) 318*, 618–624.
Dennis, M., Giraudat, J., Kotzyba-Hibert, F., Goeldner, M., Hirth, C., Chang, J.-Y., Lazure, C., Chrêtien, M., & Changeux, J.-P. (1988) *Biochemistry 27*, 2346–2357.
de Vitry, C., & Popot, J.-L. (1989) *C. R. Acad. Sci., Ser. 3 309*, 709–714.
DiPaola, M., Czajkôwski, C., & Karlin, A. (1989) *J. Biol. Chem. 264*, 15457–15463.
Dwyer, B. P. (1988) *Biochemistry 27*, 5586–5592.
Dwyer, T. M., Adams, D. J., & Hille, B. (1980) *J. Gen. Physiol. 75*, 469–492.
Eisenberg, D. (1984) *Annu. Rev. Biochem. 53*, 595–623.
Engelman, D. M., & Zaccai, G. (1980) *Proc. Natl. Acad. Sci. U.S.A. 77*, 5894–5898.
Engelman, D. M., & Steitz, T. A. (1981) *Cell 23*, 411–422.
Engelman, D. M., Goldman, A., & Steitz, T. A. (1982) *Methods Enzymol. 88*, 81–88.
Engelman, D. M., Steitz, T. A., & Goldman, A. (1986) *Annu. Rev. Biophys. Biophys. Chem. 15*, 321–353.
Finkelstein, A., & Ptitsyn, O. B. (1987) *Prog. Biophys. Mol. Biol. 50*, 171–190.
Flory, P. J. (1956) *J. Am. Chem. Soc. 78*, 5222–5235.
Furois-Corbin, S., & Pullman, A. (1989) *Biochim. Biophys. Acta 984*, 339–350.
Gilson, M. K., & Honig, B. (1989) *Proc. Natl. Acad. Sci. U.S.A. 86*, 1524–1528.
Giraudat, J., Dennis, M., Heidmann, T., Chang, J.-Y., & Changeux, J.-P. (1986) *Proc. Natl. Acad. Sci. U.S.A. 83*, 2719–2723.
Giraudat, J., Dennis, M., Heidmann, T., Haumont, P. T., Lederer, F., & Changeux, J.-P. (1987) *Biochemistry 26*, 2410–2418.
Giraudat, J., Galzi, J.-L., Revah, F., Changeux, J.-P., Haumont, P.-Y., & Lederer, F. (1989) *FEBS Lett. 253*, 190–198.
Glaeser, R. M., Jubb, J. S., & Henderson, R. (1985) *Biophys. J. 48*, 775–780.
Gregor, P., Mano, I., Maoz, I., McKeown, M., & Teichberg, V. (1989) *Nature (London) 342*, 689–692.
Grenningloh, G., Rienitz, A., Schmitt, B., Methfessel, C., Zensen, M., Beyreuther, K., Gundelfinger, E. D., & Betz, H. (1987a) *Nature (London) 328*, 215–220.
Grenningloh, G., Gundelfinger, E. D., Schmitt, B., Betz, H., Darlison, M. G., Barnard, E. A., Schofield, P. R., & Seeburg, P. H. (1987b) *Nature (London) 330*, 25–26.
Henderson, R. (1975) *J. Mol. Biol. 93*, 123–138.
Henderson, R. (1977) *Annu. Rev. Biophys. Bioeng. 6*, 87–109.
Henderson, R., & Unwin, P. N. T. (1975) *Nature (London) 257*, 28–32.
Hol, W. G. J. (1985) *Prog. Biophys. Mol. Biol. 45*, 149–195.
Hollmann, M., O'Shea-Greenfield, A., Rogers, S. W., & Heinemann, S. (1989) *Nature (London) 342*, 643–648.
Holm, L., Saraste, M., & Wikström, M. (1987) *EMBO J. 6*, 2819–2823.
Honig, B. H., Hubbell, W. L., & Flewelling, R. F. (1986) *Annu. Rev. Biophys. Biophys. Chem. 15*, 163–193.
Huang, K.-S., Bayley, H., Liao, M.-J., London, E., & Khorana, H. G. (1981) *J. Biol. Chem. 256*, 3802–3809.
Hucho, F. (1986) *Eur. J. Biochem. 158*, 211–226.
Hucho, F., Oberthür, W., & Lottspeich, F. (1986) *FEBS Lett. 205*, 137–142.
Hyde, C. C., Ahmed, S. A., Padlan, E. A., Miles, E. W., & Davies, D. R. (1988) *J. Biol. Chem. 263*, 17857–17871.
Imoto, K., Methfessel, C., Sakmann, B., Mishina, M., Mori, Y., Konno, T., Fukuda, K., Kurasaki, M., Bujo, H., Fujita, Y., & Numa, S. (1986) *Nature (London) 324*, 670–674.
Imoto, K., Busch, C., Sakmann, B., Mishina, M., Konno, T., Nakai, J., Bujo, H., Mori, Y., Fukuda, K., & Numa, S. (1988) *Nature (London) 335*, 645–648.
Jennings, M. L. (1989) *Annu. Rev. Biochem. 58*, 999–1027.
Jubb, J. S., Worcester, D. L., Crespi, H. L., & Zaccai, G. (1984) *EMBO J. 3*, 1455–1461.
Kellis, J. T., Jr., Nyberg, K., Sali, D., & Fersht, A. R. (1988) *Nature (London) 333*, 784–786.
Khorana, H. G. (1988) *J. Biol. Chem. 263*, 7439–7442.
Kleffel, B., Garavito, R. M., Baumeister, W., & Rosenbusch, J. P. (1985) *EMBO J. 4*, 1589–1592.
Klein, P., Kanehisa, M., & DeLisi, C. (1985) *Biochim. Biophys. Acta 815*, 468–476.
Kouyama, T., Kimura, Y., Kinosita, K., Jr., & Ikegami, A. (1981) *J. Mol. Biol. 153*, 337–359.
Lear, J. B., Wasserman, Z. R., & DeGrado, W. F. (1988) *Science 240*, 1177–1181.
Leonard, R. J., Labarca, C. G., Charnet, P., Davidson, N., & Lester, H. A. (1988) *Science 242*, 1578–1581.
Liao, M.-J., London, E., & Khorana, H. G. (1983) *J. Biol. Chem. 258*, 9949–9955.
MacIntyre, S., Freudl, R., Eschbach, M.-L., & Henning, U. (1988) *J. Biol. Chem. 263*, 19053–19059.
Maeno, T., Edwards, C., & Anraku, M. (1977) *J. Neurobiol. 8*, 173–184.
McCrea, P. D., Popot, J.-L., & Engelman, D. M. (1986) *Biophys. J. 49*, 355a.
McCrea, P. D., Popot, J.-L., & Engelman, D. M. (1987) *EMBO J. 6*, 3619–3626.
Michel, H., Weyer, K. A., Gruenberg, H., Dunger, I., Oesterhelt, D., & Lottspeich, F. (1986a) *EMBO J. 5*, 1149–1158.
Michel, H., Epp, O., & Deisenhofer, J. (1986b) *EMBO J. 5*, 2445–2451.
Molle, G., Dugast, J. Y., Duclohier, H., Daumas, P., Heitz, F., & Spach, G. (1988) *Biophys. J. 53*, 193–203.
Nabedryk, E., Garavito, R. M., & Breton, J. (1988) *Biophys. J. 53*, 671–676.
Noda, M., Shimizu, S., Tanabe, T., Takai, T., Kayano, T., Ikeda, T., Takahashi, H., Nakayama, H., Kanaoka, Y., Minamino, N., Kangawa, K., Matsuo, H., Raftery, M. A., Hirose, T., Inayama, S., Hayashida, H., Miyata, T., & Numa, S. (1984) *Nature (London) 312*, 121–127.
Numa, S. (1989) *Harvey Lect. 83*, 121–165.
Oberthür, W., Muhn, P., Baumann, H., Lottspeich, F., Wittmann-Liebold, B., & Hucho, F. (1986) *EMBO J. 5*, 1815–1819.
Oiki, F., Danho, W., Madison, V., & Montal, M. (1988) *Proc. Natl. Acad. Sci. U.S.A. 85*, 8703–8707.
Ovchinnikov, Y. A., Abdulaev, N. G., Vasilov, R. G., Vturine, I. Y., Kuryatov, A. B., & Kiselev, A. V. (1985) *FEBS Lett. 179*, 343–350.
Pace, C. N., Grimsley, G. R., Thompson, J. A., & Barnett, B. J. (1988) *J. Biol. Chem. 263*, 11820–11825.
Popot, J.-L., & Changeux, J.-P. (1984) *Physiol. Rev. 64*, 1162–1239.
Popot, J.-L., & de Vitry, C. (1990) *Annu. Rev. Biophys. Biophys. Chem. 19*, 369–403.
Popot, J.-L., Trewhella, J., & Engelman, D. M. (1986) *EMBO J. 5*, 3039–3044.

Popot, J.-L., Gerchman, S.-E., & Engelman, D. M. (1987) *J. Mol. Biol. 198*, 655–676.

Popot, J.-L., Engelman, D. M., Gurel, O., & Zaccai, G. (1989) *J. Mol. Biol. 210*, 829–847.

Rashin, A. A., Iofin, M., & Honig, B. H. (1986) *Biochemistry 25*, 3619–3625.

Ratnam, M., Le Nguyen, D., Rivier, J., Sargent, P. B., & Lindstrøm, J. (1986) *Biochemistry 25*, 2633–2643.

Rees, D. C., Komiya, H., Yeates, T. O., Allen, J. P., & Feher, G. (1989) *Annu. Rev. Biochem. 58*, 607–633.

Revah, F., Galzi, J.-L., Giraudat, J., Haumont, P.-Y., Lederer, F., & Changeux, J.-P. (1990) *Proc. Natl. Acad. Sci. U.S.A.* (in press).

Richards, F. (1977) *Annu. Rev. Biophys. Bioeng. 6*, 151–176.

Rogers, N. K., & Sternberg, M. J. E. (1984) *J. Mol. Biol. 174*, 527–542.

Schofield, P. R., Darlison, M. G., Fujita, N., Burt, D. R., Stephenson, F. A., Rodriguez, H., Rhee, L. M., Ramachandran, J., Reale, V., Glencorse, T. A., Seeburg, P. H., & Barnard, E. A. (1987) *Nature (London)* 328, 221–227.

Singer, S. J. (1962) *Adv. Protein Chem. 17*, 1–68.

Singer, S. J. (1971) in *Structure and Function of Biological Membranes*, pp 145–222, Academic Press, New York.

Tanabe, T., Takeshima, H., Mikami, A., Flockerzi, V., Takahashi, H., Kangawa, K., Kojima, M., Matsuo, H., Hirose, T., & Numa, S. (1987) *Nature (London) 328*, 313–318.

Tempel, B. A., Papazian, D. M., Schwartz, T. L., Jan, Y. N., & Jan, L. Y. (1987) *Science 237*, 770–775.

Vogel, H., Nillson, L., Rigler, R., Voges, K.-P., & Jung, G. (1988) *Proc. Natl. Acad. Sci. U.S.A. 85*, 5067–5071.

Wada, K., Dechesne, C. J., Shimasaki, S., King, R. G., Kusano, K., Buonanno, A., Hampson, D. R., Banner, C., Wenthold, R. J., & Nakatani, Y. (1989) *Nature (London) 342*, 684–689.

Yeates, T. O., Komiya, H., Rees, D. C., Allen, J. P., & Feher, G. (1987) *Proc. Natl. Acad. Sci. U.S.A. 84*, 6438–6442.

Chapter 16

Homology and Analogy in Transmembrane Channel Design: Lessons from Synaptic Membrane Proteins†

Heinrich Betz

ZMBH, Universität Heidelberg, Im Neuenheimer Feld 282, D-6900 Heidelberg, FRG

Received September 26, 1989; Revised Manuscript Received November 20, 1989

Transport of ions and small molecules across biological membranes is an essential process in living cells. Specific transport systems—pumps and channels—have evolved to allow translocation against and along existing solute concentration gradients. The highest rates of transport are achieved by channel proteins that in their activated ("open") state allow flux rates approaching that of free diffusion (Hille, 1984).

In the nervous system, channel proteins provide the molecular basis for electrical signaling and selective information transfer between excitable cells and thus are particularly abundant and diversified. Many neuronal channel proteins are enriched at synapses, i.e., the cell–cell contacts specialized for interneuronal communication. Voltage-sensitive ion channels control the propagation of action potentials along the neuronal plasma membrane and regulate the release of neurotransmitter from presynaptic nerve terminals. Neurotransmitter-activated channel proteins serve as receptors for rapid transmembrane signaling at the postsynaptic membrane of classical "chemical" synapses. Gap junction channels connect the cytoplasm of neighboring cells and provide direct electric coupling via "electrical" synapses.

During the past 5 years, a wealth of sequence information has been obtained by cDNA cloning on various members of different classes of synaptic channel proteins [reviewed in Miller (1989) and Jan and Jan (1989)]. Comparison of the primary structures and their hydropathy analysis disclosed common structural designs characteristic of functionally related protein superfamilies and facilitated identification of domains implicated in transport and/or activation and inactivation ("gating"). Here, the prominent structural features recognized thus far for three major classes of synaptic channels are illustrated by discussing selected examples of the different protein families. Attempts will be made to correlate structural motifs with functional properties and to delineate the evolutionary diversification creating the present set of channel gene products.

Potassium Channel Subunits: Evolutionary Ancestors of the Voltage-Gated Channel Protein Family

Conduction of action potentials along excitable membranes depends on the sequential activation and inactivation of voltage-sensitive Na^+ and K^+ channels (Hodgkin & Huxley, 1952). Due to the availability of selective toxins from scorpions and marine organisms, Na^+ channels have been purified by several laboratories and shown to represent large transmembrane protein complexes containing a predominant high molecular weight α-subunit (apparent M_r 260K) associated with a variable set of smaller polypeptides, the β-subunits [(mammalian brain) $\beta 1$, 36 kDa; $\beta 2$, 33 kDa; (mammalian muscle) $\beta 1$, 38 kDa; for review, see Catterall (1988)]. The primary structure of the α-subunit of different vertebrate and *Drosophila* Na^+ channels has been deduced from cDNA sequences and found to contain four repeated domains of 300–400 amino acids with about 50% overall amino acid identity (Noda et al., 1984, 1986a; Salkhoff et al., 1987; Kayano et al., 1988; Auld et al., 1988; Trimmer et al., 1989). Each domain contains six largely hydrophobic segments (S1 to S6) long enough to form a membrane-spanning α-helix. Most of these segments possess predominantly hydrophobic residues with few negative charges. The S4 segment, however, is unique in being highly positively charged by having an arginine or lysine residue at every fourth position of its sequence (Figure 2). This segment is thought to play a crucial role in channel gating by serving as a voltage sensor (see below).

On the basis of hydrophobicity analysis and secondary structure prediction, Noda et al. (1986a) and others [see Catterall (1988)] have proposed a transmembrane topology of the Na^+ channel α-subunit as shown in Figure 1A. Accordingly, the four homologous domains are assumed to follow the same general transmembrane folding scheme with six membrane-spanning segments, each. Both the C- and N-

†Work in the author's laboratory was funded by the Bundesministerium für Forschung und Technologie (BCT 365/1), Deutsche Forschungsgemeinschaft (SFB 317 and Leibniz Program), Fonds der Chemischen Industrie, and German–Israeli Foundation.

A. VOLTAGE - GATED

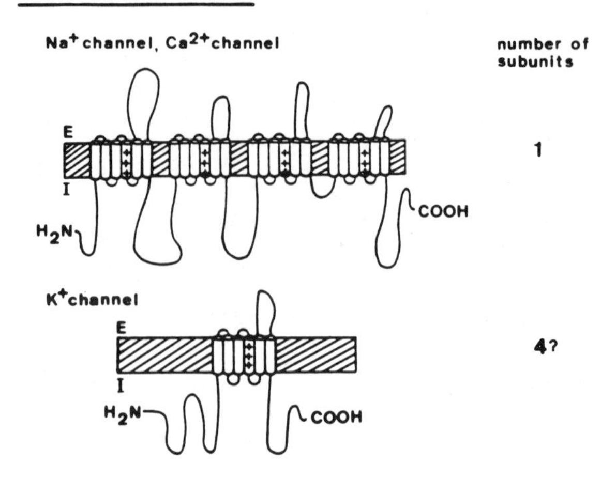

B. LIGAND - GATED

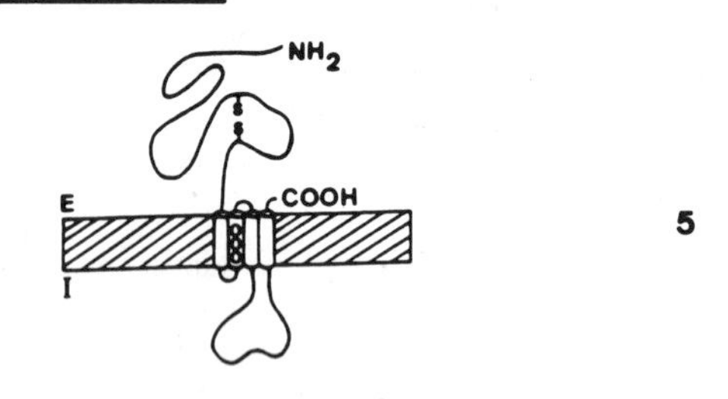

C. GAP JUNCTION/SYNAPTOPHYSIN

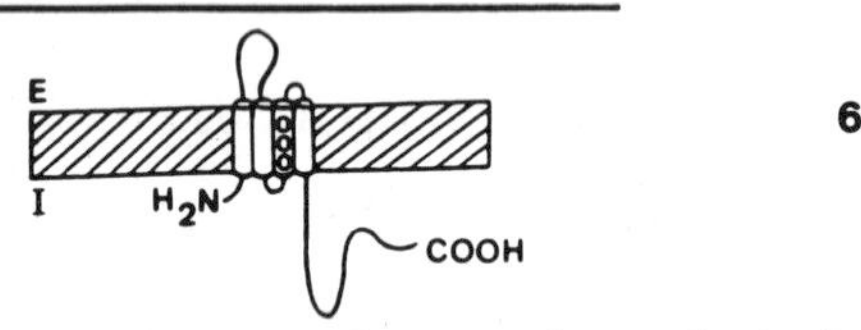

FIGURE 1: Proposed transmembrane topology and subunit number of different types of channel proteins. The protein-folding models shown represent idealized drawings deduced from various Na^+ channel, Ca^{2+} channel, K^+ channel, nicotinic acetylcholine receptor, glycine receptor, $GABA_A$ receptor, synaptophysin, and gap junction protein sequences (see text). The length of individual extramembrane stretches may differ considerably between individual members of the channel protein families shown and do not correspond to a particular protein. E denotes the extracellular (or vesicular) and I the cytoplasmic sides of the membrane. The S4 segment of voltage-gated channels is indicated by (+) and the putative channel-forming M2 and M3 segments of ligand-gated channels and connexins and synaptophysin, respectively, by (O).

termini of the polypeptide then are located at the cytoplasmic side of the plasma membrane. An ion channel may be formed by symmetrically arranging the four repeat domains around a central axis perpendicular to the plane of the membrane, thus creating a pseudotetrameric transmembrane protein. Mapping of antibody epitopes and phosphorylation sites as well as protease digestion experiments are consistent with this model (Gordon et al., 1988; Catterall, 1988). Presently, however, it is unclear which transmembrane segments contribute to the channel proper by lining the ion path. Segments S2, S3, and S4 together with an additional hypothetical intramembrane stretch, S7, all have been considered in theoretical models (Greenblatt et al., 1985; Noda et al., 1986a; Guy & Seetharamulu, 1986). Also, a synthetic peptide corresponding to segment S3 of domain I of a rat Na^+ channel α-subunit has been shown to form cation-selective channels upon incorporation into lipid bilayers (Oiki et al., 1988).

Functional expression in *Xenopus* oocytes confirmed that the α-subunit sequence of the Na^+ channel indeed contains the elements required for ion conduction and voltage-dependent gating (Noda et al., 1986b; Auld et al., 1988; Stühmer et al., 1989; Trimmer et al., 1989). Injection of in vitro transcribed α-subunit cRNA directs the synthesis and incorporation into the oocyte membrane of depolarization-activated Na^+-selective channels which display the pharmacological characteristics of their counterparts in brain. Noteworthy, kinetic characteristics of channel inactivation are altered by coexpression of other brain mRNAs (Auld et al., 1988). This may indicate a regulatory role of associated β-subunits or modifying enzymes like protein kinases.

The delineation of Ca^{2+} channel subunit sequences and, recently, their functional expression disclosed similar structural principles. Dihydropyridines antagonize Ca^{2+} currents in skeletal and heart muscles and find wide therapeutic applications in various human diseases. The dihydropyridine receptor has been purified from skeletal muscle [reviewed in Catterall (1988)], and cDNAs of its major antagonist binding subunit have been isolated from both muscle and heart (Tanabe et al., 1987; Ellis et al., 1988; Mikami et al., 1989). The deduced primary structures show four homologous repeats as found in the Na^+ channel α-subunit and display considerable sequence similarity to the latter, in particular within the highly charged S4 segment (Figure 2); major variations concern the N- and C-terminal intracellular regions. Thus, a common transmembrane topology and probably a similar tertiary structure of the channel-forming domains seem to be shared by these voltage-gated cation channels.

The striking conservation of the S4 segment between different members of the voltage-gated channel protein family has raised speculations on its possible function. From elec-

Electrophorus Na^+ channel	DLRNVSALRTFRVLRALKTITIFP	Domain I
" " "	NMQGMSVLRSLRLLRIFKLAKSWP	Domain II
" " "	ELGAIKNLRTIRALRPLRALSRFE	Domain III
" " "	LFRVIRLARIARVLRLIRAAKGIR	Domain IV
Rat brain Na^+ channel, type I	LFRVIRLARIGRILRLIKGAKGIR	Domain IV
Rabbit muscle Ca^{2+} channel	SSAFFRLFRVMRLIKLLSRAEGVR	Domain IV
Drosophila K^+ channel	ILRVIRLVRVFRIFKLSRHSK	-
Rat brain K^+ channel	ILRVIRLVRVFRIFRLSRHSK	-

FIGURE 2: S4 segment of voltage-gated ion channels. Amino acid sequences shown are taken from the following: *Electrophorus* Na^+ channel, amino acids 202–225, 649–672, 1090–1113, and 1415–1438 (Noda et al., 1984); rat brain Na^+ channel, type I, amino acids 1634–1657 (Noda et al., 1986); rabbit muscle Ca^{2+} channel, amino acids 1231–1254 (Tanabe et al., 1987); *Drosophila* K^+ channel, amino acids 432–452 (Tempel et al., 1987); rat brain K^+ channel, amino acids 290–310 (Baumann et al., 1988). Positively charged residues are indicated in boldface letters.

trophysiological measurements it was known that channel activation involves translocation of charges through the protein which are detectable as "gating current" (Armstrong, 1981). Different investigators have proposed that alterations of the transmembrane electrical field might induce movements of the charged S4 segments, which could quantitatively account for this gating current, thus producing a conformational change resulting in channel opening (Noda et al., 1984, 1986a; Guy & Seetharamulu, 1986; Catterall, 1988). Recently, this hypothesis has been verified by mutagenesis of up to three of the basic residues within the S4 segments of a brain Na^+ channel α-subunit (Stühmer et al., 1989). Upon expression of the mutant mRNAs in the oocyte system, a direct correlation was found between the net positive charge in the first S4 segment and the steepness of the potential dependence of activation. The same study also confirmed that, as previously deduced from intracellular protease and anti-peptide antibody treatments (Armstrong et al., 1973; Vassilev et al., 1988), the cytoplasmic domain linking the homology repeats III and IV is involved in channel inactivation.

K^+ channels comprise the most diverse family of voltage-sensitive channel proteins (Hille, 1984). They are found in most cells of the body and control many cellular functions including duration of action potentials, neurotransmitter release from presynaptic nerve terminals, and cardiac pacemaking. Cross-linking and binding studies using "facilitatory" polypeptide toxins from snake and bee venom indicate that putative voltage-sensitive K^+ channel proteins in vertebrate brain are large complexes (M_r about 400000) which contain at least two types of polypeptides (Rehm & Lazdunski, 1988; Schmidt & Betz, 1989). A major subunit species (M_r's about 70K–90K) carries binding sites for different polypeptide toxins (Rehm & Betz, 1983; Black & Dolly, 1986; Schmidt et al., 1988; Rehm et al., 1988); the role of smaller accessory polypeptides (28–38 kDa) seen upon purification (Rehm & Lazdunski, 1988) or cross-linking (Schmidt & Betz, 1989) is not clear. However, reconstitution experiments show that this set of polypeptides can form functional K^+ channels upon incorporation into a planar bilayer (Rehm et al., 1989b). Thus, vertebrate K^+ channels apparently do not contain the high molecular weight subunit characteristic of voltage-gated Na^+ and Ca^{2+} channel proteins.

A breakthrough in the elucidation of K^+ channel structure came from molecular genetic analysis of a neurological mutant of *Drosophila*, the "*shaker*" fly. Electrophysiological analysis of this mutant had indicated that its phenotype, continuous trembling of appendages, resulted from alterations in a voltage-sensitive K^+ current (Salkhoff & Wyman, 1981). Mutational breakpoint localization and chromosome walking techniques led to the identification of a large transcription unit that covers >100 kb of genomic DNA (Papazian et al., 1987; Kamb et al., 1987; Baumann et al., 1987). Analysis of the corresponding transcripts revealed a set of alternatively spliced gene products which displayed striking homology to Na^+ and Ca^{2+} channel sequences (Tempel et al., 1987; Schwarz et al., 1988; Kamb et al., 1988; Pongs et al., 1988). The deduced protein sequences are characterized by six hydrophobic putative transmembrane regions including a typical S4 segment (Figure 2). Although diverging in their 5′ and 3′ coding regions, all fully processed mRNAs contained a central stretch encoding these predicted transmembrane domains. In other words, the *shaker* proteins apparently correspond to individual repeat domains of Na^+ and Ca^{2+} channel α-subunits but vary strikingly in their putative cytoplasmic N- and C-terminal regions (Figure 1). K^+ channels therefore are thought to consist of tetrameric assemblies of subunits, a view supported by genetic evidence indicating that *shaker* gene products interact with one another (Salkhoff & Wyman, 1983; Timpe & Jan, 1987). Na^+ and Ca^{2+} channel proteins then may have evolved from *shaker*-related sequences by gene duplication and fusion events. This is consistent with the phylogeny of K^+ channels (Hille, 1984). Accordingly, K^+ channels belong to the evolutionarily oldest ion channels known.

Upon expression in *Xenopus* oocytes, *shaker* mRNAs generate functional K^+ channels, presumably by assembly of homooligomeric protein complexes. Their activation and inactivation kinetics differ for the various *shaker* transcripts, indicating a role of the variable C- and N-terminal regions in the determination of channel characteristics (Timpe et al., 1988a,b; Iverson et al., 1988). Up to now, five or more functionally different K^+ channel subunits have been shown to result from regulated splicing of the primary transcript of the *shaker* gene. Whether mixing of different subunits at variable stoichiometries further amplifies channel diversity in vivo is presently unclear.

In *Drosophila* and vertebrates, K^+ channel diversity may indeed be much larger than previously anticipated. Recently, a set of cDNAs encoding *shaker*-related proteins (*shab*, *shaw*, and *shal*) has been described, which all may correspond to functional K^+ channel subunits (Butler et al., 1989). These gene products exhibit high (>50%) homology to *shaker* transcripts within the central domain containing the putative six membrane-spanning segments. Interestingly, the number of positively charged residues within different S4 segments of *shaker*, *shab*, *shaw*, and *shal* is variable, suggesting that the corresponding channels may differ in activation kinetics (Butler et al., 1989).

Availability of *Drosophila* K^+ channel cDNA probes has allowed cloning of their homologues in the vertebrate central nervous system. From both mouse and rat brain libraries, *shaker*-related recombinants have been isolated which correspond to transcripts of different mammalian genes (Tempel et al., 1988; Baumann et al., 1988; McDonald et al., 1989). Also, a rat brain cDNA resembling *shab* has been recently obtained by expression cloning (Frech et al., 1989). Polypeptides of 55–95 kDa are predicted from these sequences. Rat proteins have been expressed in *Xenopus* oocytes and shown to functionally differ from the *shaker* proteins in that they form slowly inactivating K^+ channels of the delayed rectifier type prevalent in mammalian neurons and other excitable cells (Stühmer et al., 1988; Frech et al., 1989). The expressed channels were sensitive to K^+ channel blockers and, in one case, to facilitatory neurotoxins, and thus probably correspond to putative K^+ channel proteins identified in toxin binding studies (see above). Indeed, antibodies raised against sequences of a mouse *shaker* homologue recognize the purified mammalian dendrotoxin receptor (Rehm et al., 1989a). In other words, the structural features deduced from *shaker* transcripts appear to be generally valid for voltage-gated K^+ channel proteins.

Inhibitory Glycine Receptor, an Archetypic Ligand-Gated Chloride Channel

Signal transmission at chemical synapses requires specific receptors that transduce neurotransmitter binding into electrical signals, e.g., alterations of membrane potential. Receptors containing integral ion channels mediate rapid (in the less than or equal to millisecond range) transduction events, whereas receptors activating G-protein-coupled channels operate at slower time scales (in the millisecond to second range). At resting membrane potential, excitation is generated by

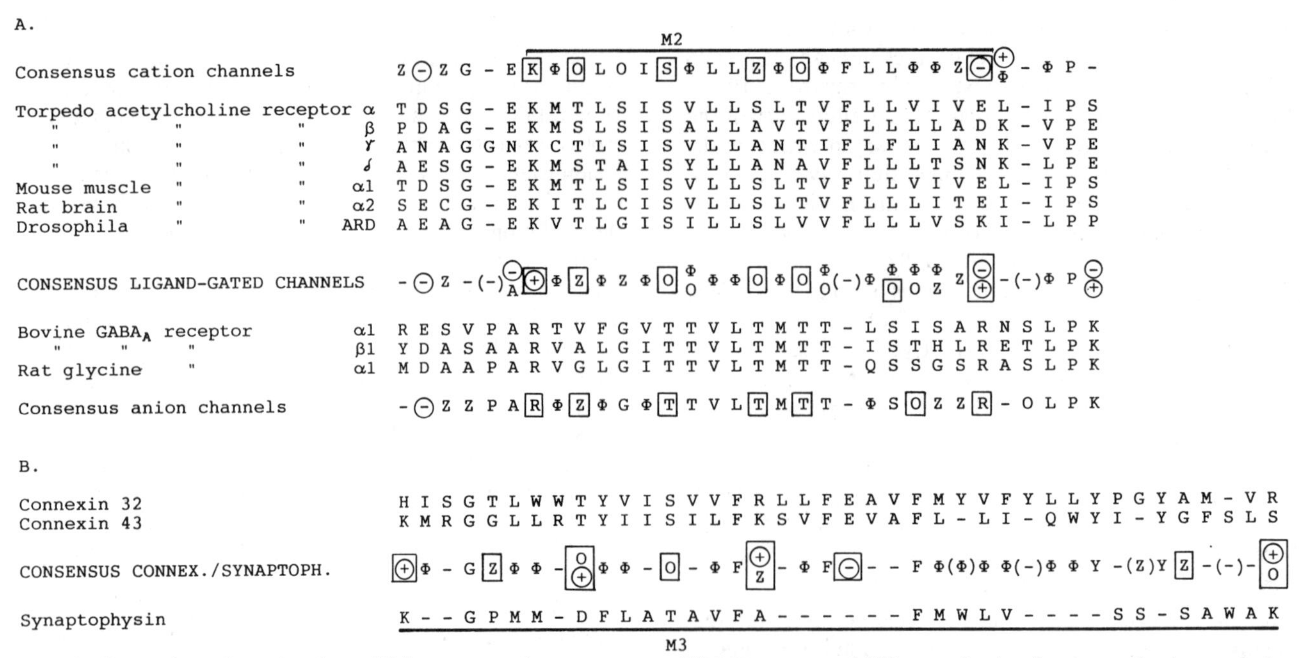

FIGURE 3: Comparison of putative channel-lining transmembrane segments. (A) M2 segments of different subunits of cation- and anion-conducing ligand-gated channel proteins. The sequences are taken from the following: *Torpedo* nicotinic acetylcholine receptor (Noda et al., 1983); mouse muscle nicotinic acetylcholine receptor (Boulter et al., 1985); rat brain nicotinic acetylcholine receptor (Wada et al., 1988); *Drosophila* nicotinic acetylcholine receptor (Hermans-Borgmeyer et al., 1986); bovine $GABA_A$ receptor (Schofield et al., 1987); rat glycine receptor (Grenningloh et al., 1987). (B) M3 segments of synaptophysin and different connexins. Sequences are from the following: connexin 32 (Paul, 1986); connexin 43 (Beyer et al., 1987); synaptophysin (Leube et al., 1987). The following symbols are used in addition to the IUPAC single-letter amino acid code: (O) hydroxylated side chain; (Φ) large hydrophobic residue; (Z) small or hydroxylated side chain; (+ and −) positively and negatively charged amino acids. Consensus sequences are aligned above or between the corresponding channel protein families. Charged, small, and hydroxylated residues at about every fourth position of the consensus sequences are boxed.

cation influx, but inhibition of neuronal firing results from increased Cl^- permeability.

The nicotinic acetylcholine receptor at the neuromuscular junction initiates muscle contraction; due to its abundance in fish electric organ, it is the best characterized cation-conducing channel protein known [reviewed in Changeux et al. (1984)]. The primary structures of its subunits have been determined in different species (Noda et al., 1983; Claudio et al., 1983; Devilliers-Thiery et al., 1984), and homologous cDNAs have been isolated from vertebrate and *Drosophila* brain (Boulter et al., 1986; Goldman et al., 1987; Deneris et al., 1988; Nef et al., 1988; Hermanns-Borgmeyer et al., 1986; Bossy et al., 1988). The major inhibitory neurotransmitters at central synapses, glycine and γ-aminobutyric acid (GABA), gate chloride channel-forming receptors of similar conductance properties (Bormann et al., 1987), but distinct pharmacology. For example, the convulsive alkaloid strychnine antagonizes postsynaptic inhibition by glycine, the predominant inhibitory neurotransmitter in the brain stem and spinal cord, whereas benzodiazepines and barbiturates modify inhibitory $GABA_A$ receptor responses in many regions of the central nervous system.

The glycine receptor was the first ligand-gated channel protein to be isolated from mammalian nervous tissue (Pfeiffer et al., 1982). Affinity-purified preparations of this receptor contain two glycosylated integral membrane proteins of M_r = 48K (α) and 58K (β), respectively (Pfeiffer et al., 1982; Graham et al., 1985; Becker et al., 1986). These polypeptides are thought to form the chloride channel of the receptor (Pfeiffer et al., 1984; Betz & Becker, 1988). A copurifying peripheral membrane protein of 93 kDa associated with cytoplasmic domains of the postsynaptic glycine receptor (Schmitt et al., 1987; Becker et al., 1989) has been implicated in its synaptic localization and/or anchoring to cytoskeletal elements.

Peptide mapping (Pfeiffer et al., 1984) and cDNA sequencing (Grenningloh et al., 1987a; Grenningloh et al., unpublished results) of the α and β glycine receptor subunits revealed a high homology between these proteins. Both possess a cleavable signal sequence and, in the C-terminal half of the polypeptide, four hydrophobic segments (M1 to M4) long enough to form transmembrane α-helices. This arrangement resembles that of nicotinic acetylcholine receptor (Noda et al., 1983; Changeux et al., 1984) and $GABA_A$ receptor proteins (Schofield et al., 1987; Levitan et al., 1988), suggesting that all channel-forming receptors are composed of subunits sharing a common transmembrane topology (Figure 1). Furthermore, significant amino acid sequence homology exists between the subunits of different ligand-gated ion channels (Grenningloh et al., 1987a,b; Schofield et al., 1987). Thus, these receptors constitute a protein superfamily that evolved by gene duplication from a common ancestor early in phylogeny.

The membrane-spanning segments M1 to M3 are highly conserved between glycine and $GABA_A$ receptor subunits, pointing to their potential importance in chloride channel function (Grenningloh et al., 1987b). Segment M2 contains many uncharged polar amino acid residues and therefore is thought to provide the hydrophilic inner lining of the chloride channel. Here, eight consecutive amino acid residues are identical in $GABA_A$ and glycine receptor subunits (Figure 3A). Interestingly, transmembrane segment M2 of nicotinic acetylcholine receptor proteins is known to be involved in cation transport and channel blocker binding (Giraudat et al., 1986; Hucho et al., 1986; Imoto et al., 1986). Segment M2 thus seems to be a common structural determinant of ligand-gated ion channel function. Indeed, comparison of both "cationic" and "anionic" M2 sequences unravels some common periodicity where a bulky hydrophobic residue at roughly every fourth position, i.e., on the same side of a potential transmembrane α-helix, is followed by a small or polar side chain (Figure 3A).

Having the latters arranged toward the channel lumen, a pore sufficiently large and polar for passage of permeating ions may be created in the center of only five symmetrically arranged α-helices (Bormann et al., 1987). In support of a structural model assuming that homologous M2 segments from each receptor subunit associate to form the lining of the channel (Hucho, 1986), exchange of up to three serine residues within different M2 transmembrane segments of nicotinic acetylcholine receptor subunits was found to decrease single channel outward currents and to accelerate channel blocker dissociation (Leonard et al., 1988).

The M2 segments of anion-selective glycine and $GABA_A$ receptor proteins terminate with positively charged residues at both ends (Figure 3A). Furthermore, positively charged residues surround the M1 to M3 sequences both intra- and extracellularly. Cation-conducing nicotinic acetylcholine receptors in contrast have negatively charged side chains in addition to positive charges bordering transmembrane segment M2 (Figure 3A). Patch clamp data indicate two sequentially occupied anion binding sites in both glycine and $GABA_A$ receptor channels (Bormann et al., 1987). The terminal charged residues of the M2 segments may be the structural correlates of these sites at the presumptive inner and outer mouths of receptor ion channels, and thus provide their ion selectivity filter. Indeed, mutation of the negatively charged residues bordering the M2 regions of nicotinic acetylcholine subunits has been found to modulate the conductivity of this cation channel (Imoto et al., 1988). From single-channel analysis of the expressed mutant receptors, three rings of negatively charged and glutamine residues neighboring the M2 segments have been identified (see Figure 3A), which serve as major determinants of the rate of cation transport through the nicotinic receptor. Also, a synthetic peptide corresponding to segment M2 of the glycine receptor α-subunit (Figure 3A) has been shown to produce randomly gated "channels" upon incorporation into planar lipid bilayers (D. Langosch, K. Hartung, E. Grell, E. Bamberg, and H. Betz, unpublished results). Interestingly, the ion selectivity of these channels was changed upon inversing the terminal charges of the peptide. As outlined above, an accumulation of hydroxylated residues is highly conserved within the M2 regions of the anion channel forming glycine and $GABA_A$ receptor subunits (Figure 3A). The high positive ion potential of the hydroxyl terminus may stabilize permeating anions and thus additionally contribute to ion selection in these channels.

A proline residue in segment M1 of glycine and $GABA_A$ receptor polypeptides is also present in nicotinic acetylcholine receptor proteins. Proline residues causing a bend in transmembrane-spanning α-helices may provide the structural flexibility required for reversible conformational transitions of intramembrane regions of transport proteins (Brandl & Deber, 1986). Their conservation thus may point to a conserved machinery of different ligand-gated channels for transforming external ligand binding into activation of a presumptive intramembrane gating mechanism.

Besides considerable conservation of transmembrane sequences, homology also exists in the putative extracellular N-terminal domain. Remarkable are two precisely conserved cysteine residues, which also are present in nicotinic acetylcholine receptor polypeptides. For the acetylcholine receptor, these cysteines have been proposed to form a disulfide bridge essential for receptor tertiary structure (Mishina et al., 1985; Stroud & Finer-Moore, 1985). Similar folding patterns thus may exist in the extracellular portion of channel-forming receptor proteins.

Photoaffinity-labeling experiments using [^{3}H]strychnine have localized the ligand binding site of the glycine receptor on its α-subunit (Graham et al., 1983). From theoretical considerations, a stretch of charged residues preceding the first transmembrane segment has been proposed to be part of the binding pocket (Grenningloh et al., 1987a). Interestingly, a corresponding region containing two neighboring cysteine residues is known to be important for acetylcholine binding to the α-subunits of the nicotinic acetylcholine receptor (Kao et al., 1984). Here, however, many other extracellular residues are, in addition, labeled by covalent acetylcholine analogues and derivatives (Dennis et al., 1989). This suggests that agonist binding to ligand-gated channels generally may involve multiple interactions with an extended extracellular domain of the respective ligand binding subunits.

Analysis of the subunit composition of the glycine receptor indicates a pentameric channel core probably composed of three α-subunits and two β-subunits, respectively (Langosch et al., 1988). This subunit stoichiometry resembles that of the nicotinic acetylcholine receptor, which also contains five membrane-spanning subunits (Changeux et al., 1984; Stroud & Finer-Moore, 1985). In view of the sequence homology and similar predicted transmembrane topology of the different receptor proteins discussed above, a quasisymmetrical pentameric complex of transmembrane polypeptides around a central ion channel is proposed as the common quaternary structure of all members of the ligand-gated ion channel superfamily (Langosch et al., 1988).

Expression of individual agonist binding subunits of glycine and $GABA_A$ receptors in *Xenopus* oocytes and mammalian cell lines generates functional ligand-gated channels that display most of the typical pharmacology of their corresponding receptors (Pritchett et al., 1988; Schmieden et al., 1989; Sontheimer et al., 1989). Although the efficiency of assembly of such homooligomeric receptors is low (Sontheimer et al., 1989), their formation clearly indicates that individual receptor subunits must have similar exchangeable oligomerization sites. This may be exploited in vivo for generating functional diversity from a limited set of subunit subtypes.

Recent biochemical and cDNA sequence data have established subtype diversity as a general phenomenon for brain nicotinic acetylcholine (Goldman et al., 1987; Deneris et al., 1988; Wada et al., 1988), $GABA_A$ (Levitan et al., 1988; Pritchett et al., 1989; Ymer et al., 1989), and glycine (Becker et al., 1988; Hoch et al., 1989) receptor subunits. In the case of $GABA_A$ and neuronal nicotinic acetylcholine receptors, expression of different subtype combinations has been shown to produce functionally and/or pharmacologically distinct receptor entities (Levitan et al., 1988a; Pritchett et al., 1989; Deneris et al., 1988). A particularly well-investigated example of variable subunit composition is the nicotinic acetylcholine receptor in vertebrate skeletal muscle. There, developmentally regulated embryonic and adult isoforms differing in channel properties are generated by exchange of a single subunit within the receptor complex (Mishina et al., 1986). A similar mechanism may underly glycine receptor heterogeneity in neonatal and adult spinal cord (Becker et al., 1988); here, however, neonatal receptors may be homooligomers of a single developmentally regulated subunit (Hoch et al., 1989).

Synaptophysin, an Intracellular Analogue of Gap Junction Proteins

Neurotransmitter release from presynaptic nerve terminals involves rapid exocytosis of low molecular weight transmitters and neuropeptides from their storage compartments, synaptic vesicles. Fusion of secretory vesicles with specialized plasma membrane domains is commonly assumed to underly exocytosis of hormones and neurotransmitters (Cecarelli & Hurlbut, 1980). To identify the molecules involved in vesicle–plasma membrane interaction, peripheral and integral membrane proteins of both synaptic vesicle and presynaptic plasma membrane fractions have lately been studied with great intensity [reviewed in Kelly (1988)].

Synaptophysin has orginally been identified as a major transmembrane glycoprotein of apparent M_r = 38K in synaptic vesicle preparations isolated from rat brain (Wiedenmann & Franke, 1985; Jahn et al., 1985). Its detergent binding properties (Rehm et al., 1986) and amino acid composition deduced from the cDNA sequence (Buckley et al., 1987; Leube et al., 1987; Südhoff et al., 1987) classify synaptophysin as a hydrophobic integral membrane protein. On the basis of hydropathy analysis, synaptophysin is thought to span the vesicle membrane four times, with both its N- and C-termini on the cytoplasmic side of the vesicular membrane (Figure 1). This transmembrane topology is supported by protease digestion and epitope mapping experiments (Leube et al., 1987; Johnston et al., 1989) and resembles that of gap junction proteins (or "connexins"), in particular the well-characterized junction protein connexin 32 from liver (Paul, 1986; Kumar & Gilula, 1986) (Figure 1). Although synaptophysin exhibits no significant overall sequence homology to any other known protein, there is a common motif of hydrophobic amino acids in all predicted transmembrane regions of synaptophysin that reappears in one of the hydrophobic segments of junction proteins (Thomas et al., 1989). Furthermore, synaptophysin and connexins share extended C-terminal cytoplasmic tail regions containing many charged amino acid residues. These tail regions may mediate interactions with cytoskeletal elements; in the case of synaptophysin, the cytoplasmic C-terminal domain has been shown to contain a Ca^{2+} binding site (Rehm et al., 1986).

Biochemical and reconstitution studies indicate that the similarity in presumed transmembrane topology of both synaptophysin and the connexins reflects common structural and functional properties. Liver gap junctions consist of paired connexons (Unwin & Zampighi, 1980). Each connexon forms a transmembrane channel composed of six symmetrically arranged copies of connexin 32. Synaptophysin also is a homooligomeric membrane protein containing six identical subunits (Thomas et al., 1988). Ultrastructural analysis of negatively stained synaptophysin preparations revealed rosette-like 7.8-nm particles that closely resemble other transmembrane channels, i.e., the nicotinic acetylcholine receptor, the voltage-gated Na^+ channel, and hepatic gap junction preparations. Upon incorporation into planar lipid bilayers, synaptophysin displays voltage-sensitive channel activity of average conductance $\approx$ 150 pS (Thomas et al., 1988). This value is similar to the conductance of electrically coupled cells and gap junction protein containing membranes. Thus, synaptophysin not only resembles gap junction connexons in predicted transmembrane topology and hexameric structure but also forms channels of comparable size and possibly corresponding function. The occurrence of such channels in intracellular membranes may appear surprising. However, many channel activities have recently been disclosed in cell organelles by the patch clamp method. Among others, a channel resembling reconstituted synaptophysin in conductance and opening time has been described in synaptic vesicles isolated from *Torpedo* electric organ (Rahamimoff et al., 1988).

The channel-forming domains of synaptophysin and the connexins are not identified. From helical modeling of connexin 32, transmembrane segment M3 has been postulated to line the channel proper (Milks et al., 1988). A large portion of this segment of different gap junction proteins is characterized by having charged, polar, or small side chains repeated at every fourth position adjacent to large hydrophobic residues (Figure 3B). This organization is shared by transmembrane segment M3 of synaptophysin and reminiscent of the consensus sequence of the M2 segment of ligand-gated channel proteins (Figure 3A). Thus, common structural features of pore-forming transmembrane α-helices appear to be present in different channel protein families.

On the basis of high-resolution electron microscopic data, rotational sliding of entire subunits has been proposed in opening and closing of gap junction connexons (Unwin & Ennis, 1984). Gating of synaptophysin channels may involve similar subunit rearrangements within the hexameric membrane protein. Presently, the physiological trigger for synaptophysin channel activation is unknown; in reconstituted planar bilayers, positive transmembrane voltages are required. These may be sensed by the charged cytoplasmic tail regions characterizing synaptophysin and most members of the connexin family and, in analogy to the variable cytoplasmic N- and C-terminal domains of voltage-gated Na^+ and K^+ channels, modify the kinetics of channel opening. Alternatively, ion binding to these domains may be important; as outlined above, synaptophysin has a cytoplasmic Ca^{2+} binding site. Also, gap junction conductances are known to be regulated by changes in internal pH, Ca^{2+}, and cAMP levels.

The physiological role of synaptophysin in the synaptic vesicle membrane is presently unknown. It may act as a transmembrane channel that allows exchange of low molecular weight components between the cytoplasm and vesicle interior. Such a function has been proposed for vacuolar Ca^{2+}-regulated channels in plants which are thought to serve for metabolite transport (Hedrich & Neher, 1987). Alternatively, synaptophysin might be involved in transport processes across two apposed lipid bilayers, in analogy to gap junction proteins. A tempting speculation is that, during exocytosis, synaptophysin may couple its associated vesicle to a related channel ("vesicle docking protein") in the presynaptic membrane, thus mediating the formation of a transient two-membrane channel or "fusion pore" (Thomas et al., 1988, 1989). The occurrence of such gap junction-like structures in the initial step of exocytosis in mast cells has indeed been indicated by electrophysiological experiments (Breckenridge & Almers, 1987). There, these pores are assumed to represent transitory structures preceding vesicle plasma membrane fusion. In the case of synaptic vesicles containing low molecular weight neurotransmitters, a channel mechanism of vesicular release may not only account for initiation of exocytosis but explain all features of the secretory event by assuming reversible pore formation via docking and undocking of vesicular and plasma membrane channels.

Conclusions and Perspectives

The present structural data on different synaptic channel protein superfamilies indicate that their functional diversity largely arose by divergent evolution of common ancestral building blocks or subunits. These originally may have assembled into homooligomeric transmembrane structures being the precursors of modern K^+ channels, or of glycine and $GABA_A$ receptor homooligomers, which still are detected upon expression of a single subunit of these proteins. For creating stably assembled entities, the corresponding gene segments then were duplicated and eventually fused to generate single-subunit proteins of the voltage-gated Na^+ or Ca^{2+} channel prototype. On the other hand, with functional specialization being the primary goal, divergence of the duplicated genes allowed generation of heterooligomeric complexes of variable subunit composition. This seems to be realized to a large extent in the ligand-gated channel protein family where exchange of single subunits was shown to drastically modify receptor properties.

Although divergent evolution certainly accounts for the different voltage- and ligand-gated channel proteins of synaptic membranes, related channel structures probably also arose via convergence of unrelated polypeptide sequences. In case of the synaptophysin/connexin hexameric channel protein superfamily, similar transmembrane organization and functional properties are found for proteins barely related in amino acid sequence. Thus, as in the case of various soluble proteins, common folding designs may originate from distinct primary structures.

Despite great differences in sequence and transmembrane organization displayed by the various synaptic channel protein superfamilies, some common principles of transmembrane channel architecture emerge from their comparison: (a) All channel proteins are assembled from homologous "blocks" or subunits of transmembrane-spanning sequences. (b) Each subunit or homology domain contributes a membrane-spanning α-helix to the lining of the channel. (c) The organization of these channel-lining α-helices appears to follow some general rules. Commonly, small or polar (or even charged) residues adjacent to bulky hydrophobic side chains are accumulated on that side of the α-helix which is assumed to be exposed to the channel lumen. (d) The pore size of the channel is largely determined by the number of homology blocks or subunits, i.e., of available lining segments [see Unwin (1986)]. (e) Rings of charged residues bordering the channel-lining α-helices are major determinants of ion selectivity and current flow rates, probably by acting as transiently occupied binding sites for the transported ion species.

Apart from these rather general deductions, most of the pertinent structure–function relationships of channel proteins are not understood. In particular, the domains involved in channel activation are still poorly defined. Although extended cytoplasmic and extracellular domains certainly are important determinants of gating in voltage- and ligand-gated ion channels, the transduction of conformational changes in these regions into movement of transmembrane segments resulting in opening and closing remains enigmatic. Determination of the crystal structure of representative members of the different channel protein superfamilies hopefully will lead to a more detailed picture of this important class of membrane proteins.

ACKNOWLEDGMENTS

I thank my colleagues for critical reading of the manuscript and I. Baro and I. Veit-Schirmer for help during its preparation.

REFERENCES

Armstrong, C. (1981) *Physiol. Rev. 61*, 644–682.

Armstrong, C. M., Benzanilla, F., & Rojas, E. (1973) *J. Gen. Physiol. 62*, 375–391.

Auld, V. J., Goldin, A. L., Krafte, D. S., Marshall, J., Dunn, J. M., Catterall, W. A., Lester, H. A., Davidson, N., & Dunn, R. J. (1988) *Neuron 1*, 449–461.

Baumann, A., Krah-Jentgens, I., Mueller, R., Mueller-Holtkamp, F., Seidel, R., Kecskemethy, N., Casal, J., Ferrus, A., & Pongs, O. (1987) *EMBO J. 6*, 3419–3429.

Baumann, A., Grupe, A., Ackermann, A., & Pongs, O. (1988) *EMBO J. 7*, 2457–2463.

Becker, C.-M., Hermans-Borgmeyer, I., Schmitt, B., & Betz, H. (1986) *J. Neurosci. 6*, 1358–1364.

Becker, C.-M., Hoch, W., & Betz, H. (1988) *EMBO J. 7*, 3717–3726.

Becker, C.-M., Hoch, W., & Betz, H. (1989) *J. Neurochem. 53*, 125–131.

Betz, H., & Becker, C.-M. (1988) *Neurochem. Int. 13*, 137–146.

Beyer, E. C., Paul, D. L., & Goodenough, D. A. (1987) *J. Cell Biol. 105*, 2621–2629.

Black, A. R., & Dolly, J. O. (1986) *Eur. J. Biochem. 156*, 609–617.

Bormann, J., Hamill, O. P., & Sakmann, B. (1987) *J. Physiol. (London) 385*, 243–286.

Bossy, B., Ballivet, M., & Spierer, P. (1988) *EMBO J. 7*, 611–618.

Boulter, J., Evans, K., Goldman, D., Martin, G., Treco, D., Heinemann, S., & Patrick, J. (1986) *Nature (London) 319*, 368–374.

Brandl, C. J., & Deber, C. M. (1986) *Proc. Natl. Acad. Sci. U.S.A. 83*, 917–921.

Breckenridge, L. J., & Almers, W. (1987) *Nature (London) 328*, 814–817.

Buckley, K. M., Floor, E., & Kelly, R. B. (1987) *J. Cell Biol. 105*, 2447–2456.

Butler, A., Wei, A., Baker, K., & Salkoff, L. (1989) *Science 243*, 943–947.

Catterall, W. A. (1988) *Science 242*, 50–61.

Cecarelli, B., & Hurlbut, W. P. (1980) *Physiol. Rev. 60*, 369–441.

Changeux, J.-P., Devilliers-Thiéry, A., & Chenouilli, P. (1984) *Science 225*, 1335–1345.

Claudio, T., Ballivet, M., Patrick, J., & Heinemann, S. (1983) *Proc. Natl. Acad. Sci. U.S.A. 80*, 1111–1115.

Deneris, E. S., Conolly, J., Boulter, J., Wada, E., Swanson, L. W., Patrick, J., & Heinemann, S. (1988) *Neuron 1*, 45–54.

Dennis, M., Giraudat, J., Kotzyba-Hibert, F., Goeldner, M., Hirth, C., Chang, J.-Y., Lazure, C., Chrétien, M., & Changeux, J.-P. (1988) *Biochemistry 27*, 2345–2351.

Devillers-Thiéry, A., Giraudat, J., Bentaboulet, M., & Changeux, J.-P. (1983) *Proc. Natl. Acad. Sci. U.S.A. 80*, 2067–2071.

Ebihara, L., Beyer, E. C., Swenson, K. I., Paul, D. L., & Goodenough, D. A. (1989) *Science 243*, 1194–1195.

Ellis, S. B., Williams, M. E., Ways, N. R., Brenner, R., Sharp, A. H., Leung, A. T., Campbell, K. P., Mc Kenna, E., Koch, W. J., Hui, A., Schwartz, A., & Harpold, M. M. (1988) *Science 241*, 1661.

Frech, G. C., VanDongen, A. M. J., Schuster, G., Brown, A. M., & Joho, R. H. (1989) *Nature (London) 340*, 642–645.

Giraudat, J., Dennis, M., Heidmann, T., Chang, J.-Y., & Changeux, J.-P. (1986) *Proc. Natl. Acad. Sci. U.S.A. 83*, 2719–2723.

Goldman, D., Deneris, E., Luyten, E., Kochhar, A., Patrick, J., & Heinemann, S. (1987) *Cell 48*, 965–973.

Gordon, D., Merrick, D., Wollner, D. A., & Catterall, W. A. (1988) *Biochemistry 27*, 7032–7037.

Graham, D., Pfeiffer, F., & Betz, H. (1983) *Eur. J. Biochem. 131*, 519–525.

Graham, D., Pfeiffer, F., Simler, R., & Betz, H. (1985) *Biochemistry 24*, 990–994.

Greenblatt, R., Blatt, Y., & Montal, M. (1985) *FEBS Lett. 193*, 125–130.

Grenningloh, G., Rienitz, A., Schmitt, B., Methfessel, C., Zensen, M., Beyreuther, K., Gundelfinger, E. D., & Betz, H. (1987a) *Nature* (*London*) *328*, 215–220.

Grenningloh, G., Gundelfinger, E., Schmitt, B., Betz, H., Darlison, M. G., Barnard, E. A., Schofield, P. R., & Seeburg, P. H. (1987b) *Nature* (*London*) *330*, 25–26.

Guy, H. R., & Seetharamulu, P. (1986) *Proc. Natl. Acad. Sci. U.S.A. 83*, 508–512.

Hedrich, R., & Neher, E. (1987) *Nature* (*London*) *329*, 833–835.

Hermanns-Borgmeyer, I., Zopf, D., Ryseck, R.-P., Hovemann, B., Betz, H., & Gundelfinger, E. D. (1986) *EMBO J. 5*, 1503–1508.

Hille, B. (1984) *Ionic Channels of Excitable Membranes*, Sinauer Associates, Sunderland, MA.

Hoch, W., Betz, H., & Becker, C.-M. (1989) *Neuron 3*, 339–348.

Hodgkin, A. L., & Huxley, A. F. (1952) *J. Physiol.* (*London*) *117*, 500–544.

Hucho, F. (1986) *Eur. J. Biochem. 158*, 211–226.

Hucho, F., Oberthür, W., & Lottspeich, F. (1986) *FEBS Lett. 205*, 137–142.

Imoto, K., Methfessel, C., Sakmann, B., Mishina, M., Mori, Y., Konno, T., Fukuda, K., Kurasaki, M., Bujot, H., Fujita, Y., & Numa, S. (1986) *Nature* (*London*) *324*, 670–674.

Imoto, K., Busch, C., Sakmann, B., Mishina, M., Konno, T., Nakai, J., Bujo, H., Mori, Y., Fukuda, K., & Numa, S. (1988) *Nature* (*London*) *335*, 645–648.

Iverson, L. E., Tanouye, M. A., Lester, H. A., Davidson, N., & Rudy, B. (1988) *Proc. Natl. Acad. Sci. U.S.A. 85*, 5723–5727.

Jahn, R., Schiebler, W., Ouimet, C., & Greengard, P. (1985) *Proc. Natl. Acad. Sci. U.S.A. 82*, 4137–4141.

Jan, L. Y., & Jan, Y. N. (1989) *Cell 56*, 13–25.

Johnston, P. A., Jahn, R., & Südhoff, T. (1989) *J. Biol. Chem. 264*, 1268–1273.

Kamb, A., Iverson, L. E., & Tanouye, M. A. (1987) *Cell 50*, 405–413.

Kamb, A., Tseng-Crank, J., & Tanouye, M. A. (1988) *Neuron 1*, 421–430.

Kao, P., Dwork, A., Kaldany, R., Silver, M., Wideman, J., Stein, S., & Karlin, A. (1984) *J. Biol. Chem. 259*, 11662–11665.

Kayano, T., Noda, M., Flockerzi, V., Takahashi, H., & Numa, S. (1988) *FEBS Lett. 228*, 187–194.

Kelly, R. B. (1988) *Neuron 1*, 431–438.

Kumar, N. M., & Gilula, N. B. (1986) *J. Cell Biol. 103*, 767–776.

Langosch, D., Thomas, L., & Betz, H. (1988) *Proc. Natl. Acad. Sci. U.S.A. 85*, 7394–7398.

Leonard, R. J., Labarca, C. G., Charnet, P., Davidson, N., & Lester, H. A. (1988) *Science 242*, 1578–1581.

Leube, R. E., Kaiser, P., Seiter, A., Zimbelmann, R., Franke, W. W., Rehm, H., Knaus, P., Prior, P., Betz, H., Reinke, H., Beyreuther, K., & Wiedenmann, B. (1987) *EMBO J. 6*, 3261–3268.

Levitan, E. S., Schofield, P. R., Burt, D. R., Rhee, L. M., Wisden, W., Köhler, M., Fujita, N., Rodriguez, H. F., Stephenson, A., Darlison, M. G., Barnard, E. A., & Seeburg, P. H. (1988) *Nature* (*London*) *335*, 76–79.

MacDonald, J. C., Adelman, J. P., Douglas, J., & North, R. A. (1989) *Science 244*, 221–224.

Mikami, A., Imoto, K., Tanabe, T., Niidome, T., Mori, Y., Takesshima, H., Narumiya, S., & Numa, S. (1989) *Nature* (*London*) *340*, 230–233.

Milks, C., Kumar, N. M., Houghten, R., Unwin, N., & Gilula, N. B. (1988) *EMBO J. 7*, 2967–2975.

Miller, C. (1989) *Neuron 2*, 1195–1205.

Mishina, M., Takai, T., Imoto, K., Noda, M., Takahashi, T., Numa, S., Methfessel, C., & Sakmann, B. (1986) *Nature* (*London*) *321*, 406–410.

Nef, P., Oneyser, C., Alliod, C., Couturier, S., & Ballivet, M. (1988) *EMBO J. 7*, 595–601.

Noda, M., Furutani, Y., Takahashi, H., Toyosato, M., Tanabe, T., Shimizu, S., Kikyotani, S., Kayano, T., Inayama, S., & Numa, S. (1983) *Nature* (*London*) *305*, 818–823.

Noda, M., Shimizu, S., Tanabe, T., Takai, T., Kayano, T., Ikeda, T., Takahashi, H., Nakayama, H., Kanaoka, Y., Minamino, N., Kangawa, K., Matsuo, H., Raftery, M. A., Hirose, T., Inayama, S., Hayashida, H., Miyata, T., & Numa, S. (1984) *Nature* (*London*) *312*, 121–127.

Noda, M., Ikeda, T., Kayano, T., Suzuki, H., Takeshima, H., Kurasaki, M., Takahashi, H., & Numa, S. (1986a) *Nature* (*London*) *320*, 188–192.

Noda, M., Ikeda, T., Suzuki, H., Takeshima, H., Takahashi, I., Kuno, M., & Numa, S. (1986b) *Nature* (*London*) *322*, 826–828.

Oiki, S., Danho, W., & Montal, M. (1988) *Proc. Natl. Acad. Sci. U.S.A. 85*, 2393–2397.

Papazian, D. M., Schwarz, T. L., Tempel, B. L., Jan, Y. N., & Jan, L. Y. (1987) *Science 237*, 749–753.

Paul, D. L. (1986) *J. Cell Biol. 103*, 123–134.

Pfeiffer, F., Graham, D., & Betz, H. (1982) *J. Biol. Chem. 257*, 9389–9393.

Pfeiffer, F., Simler, R., Grenningloh, G., & Betz, H. (1984) *Proc. Natl. Acad. Sci. U.S.A. 81*, 7224–7227.

Pongs, O., Kecskemethy, N., Müller, R., Krah-Jentgens, I., Baumann, A., Kiltz, H. H., Canal, I., Llamazares, S., & Ferrus, A. (1988) *EMBO J. 7*, 1087–1096.

Pritchett, D. B., Sontheimer, H., Gorman, C. M., Kettenmann, H., Seeburg, P. H., & Schofield, P. R. (1988) *Science 242*, 1306–1309.

Pritchett, D. B., Sontheimer, H., Shivers, B. D., Ymer, S., Kettenmann, H., Schofield, P. R., & Seeburg, P. H. (1989) *Nature* (*London*) *338*, 582–585.

Rahaminoff, R., De Riemer, S. A., Sakmann, B., Stadler, H., & Yakir, N. (1988) *Proc. Natl. Acad. Sci. U.S.A. 85*, 5310–5314.

Rehm, H., & Betz, H. (1983) *EMBO J. 2*, 1119–1122.

Rehm, H., & Lazdunski, M. (1988) *Proc. Natl. Acad. Sci. U.S.A. 85*, 4919–4923.

Rehm, H., Wiedenmann, B., & Betz, H. (1986) *EMBO J. 5*, 535–541.

Rehm, H., Bidard, J.-N., Schweitz, H., & Lazdunski, M. (1988) *Biochemistry 27*, 1827–1832.

Rehm, H., Newitt, R. A., & Tempel, B. L. (1989a) *FEBS Lett. 249*, 224–228.

Rehm, H., Pelzer, S., Cochet, C., Chambaz, E., Tempel, B. L., Trautwein, W., Pelzer, D., & Lazdunski, M. (1989b) *Biochemistry 28*, 6455–6460.

Salkhoff, L., & Wyman, R. J. (1981) *Nature* (*London*) *293*, 228–230.

Salkhoff, L., & Wyman, R. J. (1983) *J. Physiol.* (*London*) *337*, 687–708.

Salkhoff, L., Butler, A., Wei, A., Scavarda, N., Giffen, K., Ifune, C., Goodman, R., & Mandel, G. (1987) *Science 237*, 744–749.

Schmidt, R. R., & Betz, H. (1989) *Biochemistry 28*, 8346–8350.

Schmidt, R. R., Betz, H., & Rehm, H. (1988) *Biochemistry 27*, 963–967.

Schmieden, V., Grenningloh, G., Schofield, P. R., & Betz, H. (1989) *EMBO J. 8*, 695–700.

Schmitt, B., Knaus, P., Becker, C-M., & Betz, H. (1987) *Biochemistry 26*, 805–811.

Schofield, P. R., Darlison, M. G., Fujita, N., Rodriguez, H., Burt, D. R., Stephenson, F. A., Rhee, L. M., Ramachandran, J., Glencorse, T. A., Reale, V., Seeburg, P. H., & Barnard, E. A. (1987) *Nature (London) 328*, 221–227.

Schwarz, T. L., Tempel, B. L., Papazian, D. M., Jan, Y. M., & Jan, L. Y. (1988) *Nature (London) 331*, 137–145.

Sontheimer, H., Becker, C.-M., Pritchett, D. B., Schofield, P. R., Grenningloh, G., Kettenmann, H., Betz, H., & Seeburg, P. H. (1989) *Neuron 2*, 1491–1497.

Stroud, R. M., & Finer-Moore, J. (1985) *Annu. Rev. Cell. Biol. 1*, 317–351.

Stühmer, W., Stocker, M., Sakmann, B., Seeburg, P. H., Baumann, A., Grupe, A., & Pongs, O. (1988) *FEBS Lett. 242*, 199–206.

Stühmer, W., Conti, F., Suzuki, H., Wang, X., Noda, M., Yahagi, N., Kubo, H., & Numa, S. (1989) *Nature (London) 339*, 597–603.

Südhoff, T. C., Lottspeich, F., Greengard, P., Mehl, E., & Jahn, R. (1987) *Science 238*, 1142–1144.

Tanabe, T., Takeshima, H., Mikami, A., Flockerzi, V., Takahashi, H., Kangawa, K., Kojima, M., Matsuo, H., Hirose, T., & Numa, S. (1987) *Nature (London) 328*, 313–318.

Tempel, B. L., Papazian, D. M., Schwarz, T. L., & Jan, L. N. (1987) *Science 237*, 770–775.

Tempel, B. L., Jan, Y. N., & Jan, L. Y. (1988) *Nature (London) 332*, 836–843.

Thomas, L., Hartung, K., Langosch, D., Rehm, H., Bamberg, E., Franke, W. W., & Betz, H. (1988) *Science 242*, 1050–1053.

Thomas, L., Knaus, P., & Betz, H. (1989) in *Molecular Biology of Neuroreceptors and Ion Channels* (Maelicke, A., Ed.) pp 283–289, Springer, Berlin, Heidelberg, and New York.

Timpe, L. C., & Jan, L. Y. (1987) *J. Neurosci. 7*, 1307–1317.

Timpe, L. C., Jan, J. N., & Jan, L. Y. (1988a) *Neuron 1*, 659–667.

Timpe, L. C., Schwarz, T. L., Tempel, B. L., Papazian, D. M., Jan, Y. N., & Jan, L. Y. (1988b) *Nature (London) 331*, 143–145.

Trimmer, J. S., Cooperman, S. S., Tomiko, S. A., Zhou, J., Crean, S. M., Boyle, M. B., Kallen, R. G., Sheng, Z., Barchi, R. L., Sigworth, F. J., Goodman, R. H., Agnew, W. S., & Mandel, G. (1989) *Neuron 3*, 33–49.

Unwin, N. (1986) *Nature (London) 323*, 12–13.

Unwin, P. N. T., & Zampighi, G. (1980) *Nature (London) 283*, 545–549.

Unwin, P. N. T., & Ennis, P. D. (1984) *Nature (London) 307*, 609–613.

Vassilev, P. M., Scheuer, T., & Catterall, W. A. (1988) *Science 241*, 1658–1661.

Wada, K., Ballivet, M., Boulter, J., Connolly, J., Wada, E., Deneris, E., Swanson, L. W., Heinemann, S., & Patrick, S. (1988) *Science 240*, 330–334.

Wiedenmann, B., & Franke, W. W. (1985) *Cell 41*, 1017–1028.

Ymer, S., Schofield, P. R., Draguhn, A., Werner, P., Köhler, M., & Seeburg, P. H. (1989) *EMBO J. 8*, 1665–1670.

Chapter 17

Fatty Acylated Proteins as Components of Intracellular Signaling Pathways†

Guy James and Eric N. Olson*

Department of Biochemistry and Molecular Biology, The University of Texas M. D. Anderson Cancer Center, 1515 Holcombe Boulevard, Box 117, Houston, Texas 77030

Received October 10, 1989

An increasing number of proteins of diverse origin are being found to undergo covalent modification by the addition of long-chain, saturated fatty acids. The two most common fatty acids found associated with proteins are myristate, linked through an amide bond to N-terminal glycine, and palmitate, which is usually attached to cysteine residues via a thioester bond. These two fatty acids differ in chain length by only two carbon atoms, but the enzymes catalyzing their transfer to proteins, as well as the subcellular distributions of the proteins to which they are attached, exhibit striking differences. Although clues as to the function of covalent fatty acids on proteins are available in only a few cases, the specificity with which these modifications occur suggests that amide- and ester-linked fatty acids confer distinct properties on the proteins to which they are attached. An intriguing observation has been that many fatty acylated proteins participate in intracellular growth factor signaling pathways in which numerous intermolecular interactions must occur. In addition, a number of such proteins are associated with the cytoplasmic face of the plasma membrane, where the increased hydrophobicity provided by a long-chain fatty acid might be expected to facilitate interaction with the lipid bilayer and/or other membrane proteins. Indeed, studies to date indicate that covalent fatty acids, in most cases, are necessary for optimum biological activity of the parent protein. In this paper, we will briefly review the enzymology of cellular protein fatty acylation and will focus on current knowledge concerning those fatty acylated proteins believed to function in growth factor signaling pathways.

Enzymology of Protein Fatty Acylation

The 16-carbon saturated fatty acid palmitate is attached to proteins posttranslationally, usually through labile thioester bonds that can be readily cleaved by hydroxylamine or potassium hydroxide (Olson et al., 1985; Magee & Courtneidge, 1985; McIlhinney et al., 1985). Two distinct classes of cellular proteins contain covalent palmitate, suggesting the existence of multiple palmitoyltransferases capable of catalyzing the esterification of palmitate to proteins.

The majority of palmitoylated proteins appear to be synthesized on free polysomes and transported to the plasma membrane posttranslationally. Identification of fatty acylated proteins by metabolic labeling of tissue culture cells with ^{3}H-labeled fatty acids revealed that the majority of these proteins are localized to the inner face of the plasma membrane, where they are resistant to extraction by agents known to release classical peripheral membrane proteins (Wilcox & Olson, 1987). The absence of palmitoylated proteins in the soluble fraction indicates that the enzymes responsible for ester-linked fatty acylation of nonsecretory proteins reside in or near the plasma membrane. Many palmitoylated proteins can be acylated several hours after inhibition of protein synthesis (Olson & Spizz, 1986). This may reflect relatively slow transport of the proteins to the intracellular site of palmitoylation, reversible fatty acylation, or both.

A minor subset of cellular palmitoylated proteins are transmembrane glycoproteins and, in general, appear to acquire covalent palmitate a short time after synthesis, probably in the endoplasmic reticulum or Golgi apparatus. Included in this class are the insulin and β_2-adrenergic receptors (see below), as well as the α subunit of the voltage-sensitive sodium channel (Schmidt & Catterall, 1987). Palmitoylation of the transferrin receptor has also been demonstrated, but this apparently takes place at the plasma membrane and is therefore likely to be catalyzed by an enzyme distinct from that responsible for acylation of glycoproteins early in the secretory pathway (Omary & Trowbridge, 1981).

In a few cases, ester-linked fatty acids have been found to undergo turnover, indicating that reversible fatty acylation may provide an additional mechanism by which certain activities of a given protein may be modulated (Magee et al., 1987; Jing & Trowbridge, 1987; Staufenbiel, 1987). We have recently identified a palmitoylated protein in BC_3H1 cells that is deacylated following serum or growth factor stimulation, pro-

† Work supported by grants from the American Cancer Society and NIH. E.N.O. is an Established Investigator of the American Heart Association.

* Author to whom correspondence should be addressed.

1887–0/91/0174$06.00/0

viding direct evidence for modulation of fatty acylation in response to external stimuli (James & Olson, 1989a). Thus, it will be important to identify the enzymes involved in protein deacylation and to define their mode of regulation. In this regard, a fatty acyl esterase has been described in microsomal membrane preparations that removes palmitate from proteins (Berger & Schmidt, 1986).

The enzymes catalyzing ester-linked fatty acylation are somewhat relaxed with regard to fatty acyl CoA substrates, in that they transfer palmitoyl-CoA preferentially but recognize fatty acyl CoA's of shorter and longer chain lengths as well (Berger & Schmidt, 1984; Olson et al., 1985). A question that arises is whether this apparent preference for palmitate reflects true enzyme specificity or rather is a result of the high abundance of palmitate in cells compared to fatty acids of different chain lengths. The relaxed specificity of these enzymes for fatty acid substrates is in contrast to the strict specificity of *N*-myristoyltransferase (NMT) for myristoyl-CoA (see below) and may reflect in ester-linked acyl proteins the need for a general hydrophobic domain, as opposed to the apparent absolute requirement for myristate in proteins acylated by NMT. Thus far, no palmitoyltransferases have been purified or cloned.

The rare 14-carbon fatty acid myristate is covalently attached via a stable amide linkage to the α-amino group of N-terminal glycine residues, constituting the other major type of protein fatty acylation [for reviews, see Olson (1988) and Schultz et al. (1988)]. Although the somewhat leaky nature of enzymes involved in ester-linked fatty acylation of proteins allows a low level of posttranslational incorporation of myristate by certain proteins, acylation of classical myristoylated proteins occurs cotranslationally on nascent polypeptide chains and is thus completely abolished by inhibitors of protein synthesis (Wilcox et al., 1987). In contrast to the near exclusive localization of palmitoylated proteins to the plasma membrane, myristoylated proteins are found in a number of subcellular compartments, including the cytosol, plasma membrane, endoplasmic reticulum, and nucleus [see Olson and James (1989)]. The existence of soluble myristoylated proteins, together with the specific targeting of myristoylated proteins to distinct locations within the cell, suggests that the myristoyl moiety serves as more than just a hydrophobic membrane anchor. Rather, it appears that this relatively rare modification is necessary for correct orientation of the protein within appropriate membrane structures and/or interaction with critical substrates.

N-Myristoyltransferase (NMT), the enzyme responsible for myristoylation of proteins, has been purified and cloned from *Saccharomyces cerevisiae* and its substrate specificity thoroughly characterized (Towler & Glaser, 1986; Towler et al., 1987a,b, 1988a). The properties of this enzyme have been reviewed elsewhere and will not described in detail here (Towler et al., 1988b; Olson, 1988). The enzyme exhibits remarkable substrate specificity for fatty acyl donor and protein acceptor: only N-terminal glycine can be acylated, and myristoyl-CoA is highly favored as fatty acyl donor. Using myristoylation of synthetic peptides as an assay, Towler and co-workers have defined additional amino acids near the N-terminus which influence the ability of a protein to serve as a substrate for NMT. These studies have resulted in the formation of a loose consensus sequence for myristoylation that has been found in all proteins known to be myristoylated in vivo.

In all known cases, myristoylation of cellular protein occurs on penultimate glycine residues, thus requiring prior removal of the initiating methionine. Although NMT does not po an intrinsic aminopeptidase activity (Towler et al., 1987b), an aminopeptidase is known to catalyze removal of amino-terminal methionine residues after elongation of the first 30–40 amino acids (Palmiter, 1977). NMT activity is found in both the crude membrane and soluble fractions (Towler & Glaser, 1986), suggesting that it may be a peripheral membrane protein. Alternatively, it may associate with ribosomes as part of an amino-terminal processing complex, as is the case with some *N*-acetyltransferases (Pestana & Pitot, 1975), thus facilitating stoichiometric myristoylation of substrate proteins. Indeed, the fact that purified myristoylated proteins appear to be quantitatively blocked at their N-termini suggests that myristoylation occurs stoichiometrically. The cotranslational nature of myristoylation also indicates that NMT must interact with the elongating polypeptide soon after initiation of protein synthesis (Wilcox et al., 1987).

The extreme specificity exhibited by NMT for fatty acyl CoA donor, coupled with the fact that myristate comprises only 1–3% of total fatty acids in eukaryotic cells (Khandwala & Casper, 1971), suggests that the myristoyl moiety provides a structural feature that is critical for the normal function of the parent protein and that fatty acids other than myristate cannot duplicate this feature. Indeed, incorporation of a myristate analogue (11-oxymyristic acid) of reduced hydrophobicity into cellular proteins resulted in altered membrane association of $p60^{v\text{-}src}$ and a related protein from BC_3H1 cells (Heuckeroth & Gordon, 1989). This finding is of particular interest in light of the recent identification of a $p60^{src}$ receptor in the plasma membrane which binds only the myristoylated form of the protein (Resh, 1989). Taken together, these data strongly suggest a specific role for myristoylation in the sorting of myristoyl proteins within cells and demonstrate that the precise structural features of myristate may be required for this function.

Fatty Acylated Proteins in Growth Factor Signaling Pathways

Studies involving metabolic labeling of tissue culture cells with ^{3}H-labeled fatty acids have revealed numerous proteins acylated by either palmitate or myristate (Olson et al., 1985; Magee & Courtneidge, 1985; McIlhinney et al., 1985). Of the small number that have been identified, many (but not all) are known to be critical components of growth factor signaling pathways, and in at least some cases, this modification has been found to be essential for efficient functioning of the relevant pathway. To date, proteins that function as cell surface receptors, tyrosine and serine/threonine kinases, their substrates, a phosphatase, G-proteins, and Ca^{2+}-binding proteins are known to be fatty acylated (for references, see below). It thus seems likely that additional fatty acylated proteins will be identified which perform key regulatory functions in growth control. In the following sections, we will discuss evidence, much of it very recent, concerning the functions of fatty acylation in the establishment of intracellular signaling pathways.

RAS and RAS-Related Proteins. The mammalian ras gene products, as well as their yeast homologues, bind guanine nucleotides and exhibit intrinsic GTPase activity. Point mutations in mammalian RAS proteins that abolish GTPase activity induce cellular transformation, presumably due to persistent activation of one or more growth factor signaling pathways [see Barbacid (1987) for review]. RAS proteins incur a number of posttranslational modifications (now known to include C-terminal proteolysis of three amino acids, carboxyl methylation, polyisoprenylation, and palmitoylation), which

. in their binding to the inner face of the plasma ... amanoi et al., 1988; Clarke et al., 1988; Guti- ... 989; Hancock et al., 1989). Proteolytic mapping ... have shown that RAS is not deeply embedded in ..., despite its tight association with the membrane ... al., 1987).

Pa... oylation of RAS was first demonstrated by Buss and Sefton (1986), and it was postulated to provide a mechanism for targeting the protein to the membrane. A C-terminal sequence of four amino acids (C-A-A-X, where C = cysteine, A = aliphatic R group, and X = any amino acid) is conserved among virtually every member of the RAS and RAS-related family of proteins. Initial mutational analyses revealed that these four C-terminal amino acids are required for membrane localization, palmitoylation, and transformation (Willumsen et al., 1984a,b). These studies suggested that the cysteine residue in this conserved sequence served as the acceptor amino acid for acylation. However, more recent work has demonstrated that this C-terminal cysteine actually acquires a polyisoprenoid moiety in the cytosol (Hancock et al., 1989). Palmitoylation of RAS proteins occurs at a cysteine residue located two to six amino acids upstream of the C-terminal cysteine (the exact position varies between different members of the family). The earlier finding, that deletion of the C-terminal cysteine residue resulted in a nonacylated protein, is now explained by the fact that this residue is polyisoprenylated in the cytosol, which apparently facilitates interaction of RAS proteins with the membrane. Palmitoylation is thus dependent upon prior isoprenylation. It is not known for certain whether the dependence of palmitoylation on isoprenylation reflects a recognition event by the palmitoyltransferase or is due to the specific localization of the palmitoyltransferase to the plasma membrane. Isoprenylated but nonpalmitoylated mutants of RAS interact weakly with the plasma membrane and exhibit reduced biological activity. Palmitoylation increases both membrane affinity and biological activity, indicating that the palmitate moiety is required for optimal functioning of RAS proteins, perhaps by mediating appropriate interactions between RAS and other proteins in the membrane.

Additional data that suggest a critical functional role for palmitoylation in regulating the activity of RAS proteins come from a study that demonstrated rapid turnover of the fatty acid moiety associated with $p21^{N\text{-}ras}$ (Magee et al., 1987). In T15 cells overexpressing the human N-ras gene, fatty acid turnover on $p21^{N\text{-}ras}$ occurred with a half-life of $\sim$20 min, whereas the half-life of the protein itself was $\sim$1 day. This implies that a rapid cycle of acylation/deacylation takes place during normal functioning of RAS proteins. Whether fatty acid turnover occurs with similar kinetics, if at all, on transforming RAS mutants has not been reported.

Further attempts to define the role of fatty acylation in RAS function have employed chimeric proteins possessing a myristoylation signal at their N-termini. Membrane binding and transforming potential of activated H-ras proteins, lacking the C-terminal cysteine residue required for proper processing and membrane localization, was restored by fusion of either 15 amino-terminal amino acids of $p60^{src}$ or 11 amino acids of the Gag protein of Rasheed leukemia virus, both of which direct myristoylation of the fusion protein (Buss et al., 1989). These results indicated that activated RAS proteins are capable of transmitting an oncogenic signal regardless of the lipid moiety that anchors them to the plasma membrane and suggest that palmitoylation and myristoylation may be interchangeable for RAS function. However, it was found that normal cellular RAS proteins also acquired transforming potential upon addition of a myristoylation signal. These data can be interpreted to indicate that reversible fatty acylation (i.e., palmitoylation) serves to modulate the avidity of membrane binding or coupling between RAS proteins and key effector proteins, whereas myristoylated forms of these proteins remain tightly associated with the membrane and thus are unable to be uncoupled from other regulatory proteins. It will be important to determine if cellular RAS proteins possessing a myristoylation signal still become palmitoylated at their C-termini and, if so, whether their ability to become deacylated has been altered.

Much knowledge regarding the processing and function of proteins in the RAS family has come from studies in yeast. RAS1 and RAS2 proteins from *S. cerevisiae* are structurally similar to mammalian RAS proteins (Powers et al., 1984). These proteins bind GTP, possess intrinsic GTPase activity, and have been shown to modulate the activity of yeast adenylate cyclase (Broek et al., 1985; Toda et al., 1985). At least one RAS protein is required in yeast, as mutants defective in both *ras* genes are incapable of vegetative growth, whereas haploid cells lacking only one grow normally (Kataoka et al., 1984; Tatchell et al., 1984). Consistent with the situation in mammalian systems, RAS1 and RAS2 proteins are palmitoylated posttranslationally (Fujiyama & Tamanoi, 1986). Mutational analysis of RAS2 has also shown that fatty acylation is required for membrane association and complementation of $RAS1^-$ mutants (Deschenes & Broach, 1987). Three allelic yeast mutants (supH, ste16, and dpr1) have been identified which are required for RAS function as well as for production of the mating hormone α-factor (Powers et al., 1986; Fujiyama et al., 1987). It was originally proposed that this allele, termed RAM (RAS protein and α-factor maturation function), might encode an acyltransferase (Powers et al., 1986). At that time it was believed that yeast α-factor precursors were also palmitoylated. The predicted C-terminal sequence of both α-factor polypeptide precursors consists of Cys-Val-Ile-Ala (Betz et al., 1987), matching the sequence previously believed to be a consensus for palmitoylation of ras proteins. However, the recent finding by Hancock et al. that the C-A-A-X sequence directs polyisoprenylation rather than palmitoylation, combined with another recent report that yeast α-factor precursors are indeed isoprenylated at their C-termini (Anderegg et al., 1988), strongly suggests that the RAM gene actually encodes an enzyme involved in isoprenylation. Although the precise order of events involved in C-terminal modification of RAS proteins is not known, it has been proposed that polyisoprenylation occurs first and creates a recognition signal for a carboxypeptidase which removes the three C-terminal amino acids (Hancock et al., 1989). The newly generated α-carboxyl group of the C-terminal cysteine would then be available for methylation. The fact that isoprenoid chains could not be detected on newly synthesized pro-$p21^{ras}$ suggests that the conversion from pro-p21 to processed c-p21 (involving isoprenylation, C-terminal proteolysis, and carboxyl methylation) occurs very rapidly and may be mediated by a single enzyme or multienzyme complex. Reconstitution experiments utilizing in vitro translated RAS and RAM gene products should aid in defining the catalytic activity encoded by the RAM gene. If the enzymes responsible for polyisoprenylation and C-terminal proteolysis/carboxyl methylation are encoded by distinct loci, it should be possible to isolate yeast mutants that are polyisoprenylation positive but defective in one or both of the latter.

Additional yeast and mammalian proteins have been identified which share significant homology with mammalian RAS

Table I: RAS and Related Proteins

protein	C-terminal sequence	polyisoprenylated[a]	palmitoylated	refs
H-ras	ESGPGCMSCKCVLS	+	+	Sefton et al., 1982
N-ras	DGTQGCMGLPCVVM	+	+	Magee et al., 1987
K(A)-ras	KTPGCVKIKKCVIM	+	+	Shimizu et al., 1983; Capon et al., 1983; Buss & Sefton, 1986
K(B)-ras	KKKKKKSKTKCVIM	+	−	Shimizu et al., 1983; Capon et al., 1983; Buss & Sefton, 1986
Krev-1	VEKKKPKKKSCLLL	+	−	Kitayama et al., 1989
Ras1	NARKEYSGGCCIIC	+	+	Powers et al., 1984; Fujiyama & Tamano, 1986
Ras2	EASKSGSGGCCIIS	+	+	Powers et al., 1984; Fujiyama & Tamano, 1986
α-factor	IIKGVFWAPACVIA	+	+	Betz et al., 1987; Anderegg et al., 1988
YPT1	KGNVNLKGGCC	−	+	Molenaar et al., 1988; Gallwitz et al., 1983
rhoA	QARRGKKKSGCLVL	+	−	Hall, 1989
rhoB	YGSQNGCINCCKVL	+	+	Hall, 1989
rap-1a	VEKKKPKKKSCLLL	+	−	Hall, 1989
rap-2	PDKDDPCCSACNIQ	+	+	Hall, 1989

[a] Either demonstrated or predicted, on the basis of studies by Hancock et al. (1989).

Table II: GTP-Binding Proteins

protein	N-terminal sequence	C-terminal sequence	predicted substrate for NMT	myristoylated	pertussis toxin sensitive[a]	refs
$G_{i\alpha}$	MGCTLSA	CGLF	+	+	+	Buss et al., 1987; Nukada et al., 1986; Bray et al., 1987
$G_{o\alpha}$	MGCTLSA	CGLY	+	+	+	Van Meurs et al., 1987; Itoh et al., 1986
$G_{s\alpha}$	MGCLGNS	YELL	−	−	−	Robishaw et al., 1986; Yatsunami & Khorana, 1985
$G_{t\alpha}$[b]	MGAGASA	GCLF	+	−	−	Lochrie et al., 1985; Tanabe et al., 1985
$G_{x\alpha}$	MGCRQSS	IGLC	o[c]	unknown	−	Matsuoka et al., 1988
G_{olf}	MGCLGNS	YELL	−	unknown	−	Jones & Reed, 1989
$G_{t\gamma}$	MPVINIE	CVIS	−	−	+	Hurley et al., 1984
G_{γ}	MASNNIA	CAIL	−	−	+	Gautam et al., 1989

[a] Either known or predicted, on the basis of the presence of a cysteine near the carboxy terminus. [b] Two $G_{t\alpha}$ proteins have been characterized. However, their amino acid sequences are identical within the first six residues. [c] o indicates "consensus" sequence for myristoylation.

proteins. These include yeast and mammalian YPT1 (Gallwitz et al., 1983) and Krev-1 (a suppressor of ras transformation) (Kitayama et al., 1989) and members of the rho, rab, and rap gene families [for review, see Hall (1989)]. Yeast YPT1, which is required for cell viability, lacks a C-terminal consensus for polyisoprenylation but does possess two C-terminal cysteine residues and has been shown to be palmitoylated (Molenaar et al., 1988). Although the protein is found in both the membrane and soluble fractions, palmitoylated YPT1 is localized to the membrane, indicating that palmitoylation occurs immediately prior to, or concomitant with, membrane association. Mutants lacking either one of the C-terminal cysteines were functionally indistinguishable from wild-type proteins. Deletion of both C-terminal cysteine residues, however, yielded a nonpalmitoylated mutant that was biologically inactive and found exclusively in the soluble fraction, suggesting that palmitoylation does indeed regulate the subcellular distribution of YPT1 (Molenaar et al., 1988).

Hancock et al. demonstrated that a C-terminal C-A-A-X sequence is sufficient to direct polyisoprenylation and weak membrane binding of a heterologous protein. However, only proteins that also contain a cysteine residue a short distance upstream of the C-terminus are predicted to be palmitoylated. Those proteins lacking a site for palmitoylation, however, contain a polybasic domain, consisting of at least five positively charged basic amino acids immediately upstream of the C-A-A-X motif, which might be expected to facilitate interaction with negatively charged polar head groups on the surface of a lipid bilayer. As noted in this study, it is interesting that the ras, rho, and rap families are each predicted to encode both palmitoylated and nonpalmitoylated proteins (see Table I). Thus, the confinement of these proteins to the plasma membrane indicates that mechanisms in addition to fatty acylation participate in determining their subcellular localization and suggests that specific protein–protein interactions, perhaps in the form of RAS "receptors", are also involved. It has not been reported whether binding of fully processed RAS proteins to the membrane can be duplicated in artificial liposomes or if protein components are also required. Determination of the precise functional role for palmitoylation of ras and ras-related proteins will likely await the identification of effector proteins whose interaction with ras is required for normal signaling.

Additional GTP-Binding Proteins. Heterotrimeric G proteins, composed of α, β, and γ subunits, participate in transmembrane signaling by coupling the intracellular domains of cell surface receptors with appropriate effector proteins. G_s and G_i are involved in stimulation and inhibition, respectively, of adenylate cyclase [see Gilman (1987) for review]. The function of G_o, another member of the G protein family (Sternwies & Robishaw, 1984; Neer et al., 1984), is unknown, but a role for this complex in regulation of calcium channels has been proposed (Hescheler et al., 1987). Transducin (G_t) is localized to disc membranes of retinal rod outer segments and activates cGMP phosphodiesterase in response to photosignal transduction [see Stryer et al. (1981)]. The GTP-binding and GTPase activities of G proteins reside in the α subunit, whereas the β/γ subunits are thought to help anchor the complex in the plasma membrane.

cDNA clones encoding at least nine different α subunits have now been isolated from various sources (see Table II). The deduced amino acid sequence of each of these contains glycine as the penultimate residue, and additional residues within the first six amino acids of each protein are compatible with the minimal sequence requirements established by Towler and co-workers for recognition by NMT in vitro. However, Buss et al. have reported that only the α subunits of G_i and G_o are myristoylated, whereas covalent myristate was not detected on $G_{s\alpha}$ or $G_{t\alpha}$ (Buss et al., 1987). Closer inspection of the N-terminal amino acid sequence of $G_{s\alpha}$ revealed glycine and asparagine at positions 4 and 5, respectively. In their in vitro studies of NMT, Towler and co-workers found that placing asparagine in position 5 of a known myristoylated peptide (corresponding to the N-terminus of the catalytic subunit of cAMP-dependent protein kinase, see below) yielded a substrate with an ~10-fold higher K_m compared to that of the parental peptide. Similarly, glycine in position 4 of the same parental peptide resulted in a 9-fold increase in K_m (Towler et al., 1988). While these data may explain the absence of myristate on $G_{s\alpha}$, similar peculiarities within the N-terminal sequence of $G_{t\alpha}$ were not found, and it is currently unclear why myristoylation of this protein was not detected. Perhaps the in vitro substrate specificities of NMT are subject to additional constraints in an in vivo setting, which might result in myristoylation of only a subset of proteins possessing an N-terminal myristoylation "consensus" sequence. Alternatively, the absence of myristate on $G_{t\alpha}$ may reflect failure to remove the initiating methionine. An additional G protein α subunit, designated $G_{x\alpha}$, has recently been identified which has significant amino acid similarity with $G_{i\alpha}$ and $G_{s\alpha}$ (including a glycine residue at position 2), but the possibility that this protein might be myristoylated was not examined (Matsuoka et al., 1988). Given the apparent exception, in the case of $G_{t\alpha}$, to the consensus sequence for N-terminal myristoylation, it is not possible to predict with certainty whether $G_{x\alpha}$ might be myristoylated. Finally, an olfactory neuron specific G protein α subunit, termed G_{olf} has been characterized which can activate adenylate cyclase in a heterologous system (Jones & Reed, 1989). The first six amino acids of the deduced sequence of G_{olf} are identical with those found in $G_{s\alpha}$, and therefore, by inference, one would not expect this protein to be myristoylated. This remains to be demonstrated, however, as sequences beyond the first six amino acids may contribute to a protein's ability to be recognized by NMT. Thus, as for the ras family, the G protein gene family codes for both acylated and nonacylated proteins.

It is interesting, in light of the recent findings concerning targeting of ras proteins to the plasma membrane, that the predicted amino acid sequences of several members of the G protein family terminate with the C-A-A-X motif, which was shown to be sufficient to direct the isoprenylation of a heterologous protein (Hancock et al., 1989). This C-terminal sequence is found in all pertussis toxin sensitive G protein α subunits (see Table II), as well as in both G protein γ subunits characterized to date. Indeed, the cysteine residue of this sequence in $G_{t\alpha}$ has been identified as the ADP-ribose acceptor site (West et al., 1985). Pertussis toxin insensitive α subunits lack this cysteine residue. It will be interesting to determine if any members of this family are isoprenylated and, if so, how this might affect their availability as a substrate for pertussis toxin.

The functional consequences of myristoylation of $G_{i\alpha}$ and $G_{o\alpha}$ are currently unknown. It has been suggested that interaction of G proteins with the plasma membrane might involve formation of a β–γ subunit "anchor" to which α subunits bind (Gilman, 1987). The possibility of γ subunits being isoprenylated, as mentioned above, would support their proposed role as part of an anchor. Myristoylation of certain α subunits might be required for their interactions with β–γ complexes, or it might facilitate their coupling with receptors and effector proteins, thus allowing efficient utilization of the relevant signaling pathway. Mutational analyses of $G_{i\alpha}$ and $G_{o\alpha}$ should result in the definition of specific, myristoylation-dependent protein–protein interactions in which these two proteins are involved.

The α subunits of G_s and G_t are also subject to cholera toxin catalyzed ADP-ribosylation at an internal arginine residue, resulting in the irreversible activation of adenylate cyclase (Cassel & Selinger, 1977) and inhibition of the light-stimulated GTPase activity of rod outer segments (Abood et al., 1982), respectively. Numerous ADP-ribosylation factors (ARFs), 21-kDa GTP-binding proteins that serve as required cofactors in this reaction, have recently been described. Their ability to function as cofactors for α subunit ADP-ribosylation is dependent upon GTP binding, but they possess no detectable GTPase activity (Schleifer et al., 1982; Kahn & Gilman, 1984, 1986). Although ARF was originally purified from a membrane preparation, immunological studies demonstrated that the majority of at least one ARF is present in the cytosol (Kahn et al., 1988). cDNA clones encoding ARFs have been isolated from bovine, yeast, and chicken (Sewell & Kahn, 1988; Price et al., 1988; Alsip & Konkel, 1986). Kahn et al. first demonstrated the presence of covalent myristate on the N-terminal glycine residue from purified bovine brain ARF, and subsequent cDNA cloning revealed a predicted N-terminal amino acid sequence that is consistent with the protein serving as a substrate for NMT. Vaughan and co-workers have described two soluble ARF-like proteins, sARFI and sARFII, from bovine brain. They isolated a cDNA clone, using oligonucleotides based on the amino acid sequences of peptides derived from sARFII, that is nearly identical with that described by Sewell and Kahn (Price et al., 1988). Although the deduced amino acid sequence of the sARF cDNA does not perfectly match the peptide sequences from purified sARFII, it nevertheless appears to encode an ARF-like protein. Both cDNAs are predicted to encode proteins of 181 amino acids. Inspection of these amino acid sequences revealed that 173/181 positions are identical. However, one of the eight substitutions occurred at position 5: alanine in one versus glutamate in the other. The protein with alanine at position 5 has been shown to be myristoylated. Although possible myristoylation of the other clone (glutamate at position 5) was not examined, peptides with a charged residue at position 5 were not substrates for NMT in vitro (Towler et al., 1987b). Whether these sARFs might also associate with the membrane has not been reported. It is noteworthy that the sARFII peptide sequences used for designing oligonucleotide probes match perfectly with sequences found within the ARF described by Sewell and Kahn. This, combined with the observation that a majority of the ARF described by the latter group is found in the cytosol, suggests that these proteins may be identical. However, the isolation of a closely related but distinct ARF-like cDNA demonstrates that a family of genes encoding these proteins is likely to exist. It will be interesting to determine whether ARF-like proteins are universally myristoylated or, as is the case with the ras and G protein α subunit families, if only a subset are subject to fatty acylation. Selective myristoylation represents a potential mechanism whereby the interactions of various members of a given family of proteins with key components of a signaling pathway might be restricted. Identification of the normal physiological role of ARF proteins in eukaryotic cells will undoubtedly contribute to the understanding of how myristoylation participates in their function. However, with cDNA clones in hand, mutational analyses can now be performed to define the requirements, if any, for myristoylation in the interaction between ARFs and G protein α subunits.

Tyrosine Kinases and Substrates. Nonreceptor tyrosine kinases have been implicated as major participants in a number of cellular processes regulating growth and differentiation, and it is now apparent that a large family of these enzymes exists in mammalian cells (see Table III). Since the discovery that the transforming potential of Rous sarcoma virus is due to a

Table III: Nonreceptor Tyrosine Kinase

protein	N-terminal sequence	predicted substrate for NMT[a]	myristoylated	refs
SRC	MGSSKSK	+	+	Takeya & Hanafusa, 1983; Buss & Sefton, 1985
YES	MGCIKSK	+	+	Sukegawa et al., 1987; Sudel et al., 1988
SYN/FYN	MGCVQCK	+	+	Semba et al., 1986; Kypta et al., 1988; Cheng et al., 1988
FGR	MGCVPCK	+	unknown	Inoue et al., 1987
HCK	MGCMKSK	+	unknown	Ziegler et al., 1987; Quintrell et al., 1987
LCK	MGCVCSN	+	+	Voronova et al., 1984; Marchildon et al., 1984; Marth et al., 1985
LYN	MGCIKSK	+	unknown	Yamanashi et al., 1987
FPS/FES	MGFSSEL	–	–	Groffen et al., 1983
FER	MGFGSDL	–	–	Hao et al., 1989

[a] Sequences of the cellular-derived proteins are given.

protein tyrosine kinase encoded by the *src* gene, several other transforming retroviruses have been shown to encode tyrosine-specific protein kinases. These include *v-fgr* of Gardner–Rasheed feline sarcoma virus, *v-yes* of Y73 avian sarcoma virus, and *v-fps* of Fujinami sarcoma virus [for review, see Hunter and Cooper (1985)]. In addition, normal cellular homologues for each of these, as well as several tyrosine kinase encoding protooncogenes not found associated with retroviruses, have recently been identified (see Table III). Members of this family are characterized by a highly conserved C-terminal catalytic domain, whereas the N-terminal regions diverge significantly. Despite this overall divergence in their N-terminal domains, several members of this family are very similar within the first six amino acids following the initiating methonine. Consistent with this observation, SRC, YES, SYN/FYN, and LCK have been shown to be myristoylated, and additional members are predicted to undergo this modification, on the basis of sequence similarities with the three mentioned above (Table III).

Certain members of this family, such as HCK (Ziegler et al., 1987; Quintrell et al., 1987) and LCK (Marth et al., 1985), exhibit tissue or cell type specific expression, whereas others are found in a wide variety of cells. The simultaneous expression of two or more nonreceptor tyrosine kinases in a single cell type suggests that the functions of these enzymes may not be entirely overlapping. Indeed, the variable N-terminal domains of these proteins have been proposed to confer substrate specificity. However, key substrates involved in transformation may be accessible to multiple SRC-related kinases, as an activated form of $p59^{hck}$, which is expressed primarily in hematopoietic cells, has been shown to transform NIH 3T3 fibroblasts (Ziegler et al., 1989). Inspection of the first six N-terminal amino acids of several members of this family reveals another interesting distinction between them. The N-terminal sequence of the first seven proteins listed in Table III indicates that they are likely to serve as substrates for NMT, and myristoylation of SRC, YES, SYN/FYN, and LCK has indeed been demonstrated. In contrast, the N-terminal sequences of FPS/FES and FER, which are more closely related to each other than to other members of the family, are incompatible with recognition by NMR. Specifically, each contains phenylalanine at position 2 (with glycine at position 1) and glutamate (FPS/FES) or aspartate (FER) at position 5. In vitro studies demonstrated that peptides with aromatic side chains at position 2 or charged residues at position 5 were not myristoylated (Towler et al., 1987a,b). The fact that these two proteins fall into both categories, combined with the observation that 60–90% of $p92^{c\text{-}fps/fes}$ is found in the soluble fraction (Young & Martin, 1984), strongly suggests that they are not myristoylated in vivo. Interestingly, $p140^{gag\text{-}fps}$, the viral transforming counterpart of c-fps/fes, is normally membrane associated, while in a temperature-sensitive mutant defective for transformation it is soluble (Moss et al., 1984). The molecular basis for this change in subcellular location has not been reported. Thus, as is the case for the RAS and G protein families, the nonreceptor tyrosine kinase family is apparently composed of both acylated and nonacylated members.

Although the precise functions of nonreceptor tyrosine kinases remain unknown, it is now obvious in the case of SRC that the myristoyl moiety it acquires during its synthesis is essential for targeting to the plasma membrane and for specific protein–protein interactions. Initial studies established that SRC is translated and myristoylated in the cytosol and binds to the plasma membrane within 15 min after synthesis (Buss et al., 1984). A number of subsequent mutational analyses demonstrated that nonmyristoylated mutants of SRC, lacking glycine at the penultimate position, are defective in membrane association and transformation, despite having normal tyrosine kinase activities (Cross et al., 1984; Pellman et al., 1985; Kamps et al., 1985; Buss et al., 1986). In addition, the initiating methionine was removed from the N-terminus of these mutant proteins, indicating that lack of myristoylation was not due to nonrecognition by aminopeptidase. Consistent with their lack of membrane association, nonmyristoylated mutants were no longer substrates for protein kinase C (Buss et al., 1986), although their association with the soluble "carrier" proteins p59 and p80 was not affected. Together, these data demonstrated that the covalent modification of SRC with fatty acid plays a crucial role in determining its subcellular location and transforming potential. The basis for the transformation deficiency of these mutants was not immediately clear, as many known substrates of $p60^{v\text{-}src}$ were still phosphorylated by nonmyristoylated mutants (Kamps et al., 1986). However, studies using anti-phosphotyrosine antibodies have resulted in the identification of membrane-associated proteins that are substrates for myristoylated but not nonmyristoylated SRC, suggesting that these proteins participate in the process of transformation (Linder & Burr, 1988; Hamaguchi et al., 1988; Reynolds et al., 1989). Also, it was found that nonmyristoylated mutants of $p60^{v\text{-}src}$ retained their ability to stimulate cell proliferation, suggesting that the SRC substrates regulating proliferation are soluble and distinct from those that participate in transformation (Calothy et al., 1987). Recently, however, overexpression of $p60^{c\text{-}src}$ has been shown to confer enhanced mitogenic responsiveness on 10T1/2 murine fibroblasts in a myristoylation-dependent manner (Luttrell et al., 1988; Wilson et al., 1989). Therefore, while oncogenically activated SRC can induce cell proliferation in the absence of myristoylation, the role of normal $p60^{c\text{-}src}$ in this process is dependent upon membrane association. Interestingly, the hyperresponsiveness of these cells is expressed only in response to epidermal growth factor, as growth in 10% fetal calf serum yielded no increased mitogenic response. These findings not only provide important clues as to the function of normal SRC in controlling cell growth but also pinpoint a specific growth factor signaling pathway which may be involved in SRC-mediated transformation.

The mutational analyses discussed above demonstrated clearly that membrane association of SRC is dependent upon myristoylation. However, the fact that membrane-bound SRC is localized to the plasma membrane, rather than spread evenly among many intracellular membranes, indicated that the myristoyl moiety must provide a structural feature that confers properties more unique than a nonspecific hydrophobic anchor which might partition into any lipid bilayer. Futhermore, the existence of other myristoylated proteins that reside mainly in the cytosol suggested that myristoylation per se is not sufficient to direct a protein to the plasma membrane and that specific protein–protein interactions are involved as well. Also, the identification of transformation-defective SRC proteins that remain soluble despite being myristoylated provides further evidence that structural properties in addition to myristoylation are required for binding of SRC to the plasma membrane (Buss & Sefton, 1985; Garber et al., 1985). Thus, recent studies that demonstrated the presence of a proteinaceous, plasma membrane localized SRC-specific binding site, which recognizes only the myristoylated form of the protein, generated excitement but little surprise (Resh, 1989; Goddard et al., 1989).

Chimeric proteins, constructed by Hanafusa and co-workers, provided the first hint that the myristoylated N-terminus of SRC contains a recognition signal for targeting to the plasma membrane. They fused the first 14 amino acids of SRC to chimpanzee α-globin, which is normally soluble. The resulting fusion protein was myristoylated, and it fractionated in the crude membrane pellet, demonstrating the ability of the N-terminal region of SRC to target an otherwise soluble protein to the membrane (Pellman et al., 1985). The first evidence of a requirement for membrane components other than the phospholipid bilayer in the binding of SRC was obtained by studying the association of $p60^{v\text{-}src}$ with phospholipid vesicles (Resh, 1988). $p60^{v\text{-}src}$ obtained by detergent extraction of cell membranes was efficiently reconstituted into phospholipid vesicles in a manner dependent upon a myristoylated, N-terminal 10-kDa domain. In contrast, myristoylated SRC molecules obtained from a high-speed supernatant fraction were reconstituted only in the presence of added membrane proteins. Even more convincing evidence for an "SRC receptor" was provided by binding studies using in vitro translated SRC protein and a cellular plasma membrane enriched fraction (Resh, 1989). Binding of newly synthesized $p60^{v\text{-}src}$ occurred on the inner face of the plasma membrane and was saturable, myristoylation dependent, and sensitive to heat and trypsin. Furthermore, binding could be competed by a myristoylated peptide corresponding to the first 11 amino acids of SRC but not by the nonmyristoylated peptide or myristoylated peptides derived from the N-termini of other known myristoylated proteins. Similar results were obtained with an iodinated, 15 amino acid peptide from the N-terminus of SRC and red cell membrane vesicles (Goddard et al., 1989). Again, binding was restricted to the myristoylated SRC peptide and was destroyed by prior protease treatment of vesicles. These results demonstrate the existence of one or more membrane proteins that exhibit the characteristics of a high-affinity receptor, with specificity for a myristoylated SRC "ligand".

The identity of this putative receptor and the nature of its association with the plasma membrane are currently unknown. However, in light of the biological activities of both normal and oncogenic SRC discussed above and their dependence on myristoylation and membrane binding, it seems likely that this receptor represents a key regulatory element in SRC-mediated signaling pathways. The fact that membrane-bound SRC is resistant to salt extraction indicates that the receptor itself also interacts tightly with the membrane, possibly as an integral membrane protein. An intriguing possibility is that the SRC receptor represents the cytoplasmic domain of a transmembrane growth factor receptor, thus providing a direct link between an extracellular stimulus and an intracellular signaling protein, analogous to the coupling of G proteins with β-adrenergic receptors.

An additional question that arises is whether binding of SRC to its receptor can be competed by a SRC peptide with an N-terminal palmitate moiety. Given the extreme specificity for myristoyl-CoA with which NMT has evolved, one would not expect fatty acids of other chain lengths to substitute efficiently for myristate. The in vitro binding assays that have now been developed should allow this question to be answered. An alternative approach to investigating the importance of chain length and hydrophobicity in protein–membrane interactions of myristoylated proteins has recently been reported. An oxygen-substituted analogue of myristate, 11-oxymyristic acid, was synthesized and found to function as a fatty acyl donor for NMT in vitro, albeit at a reduced efficiency that varied among several peptide substrates tested (Heuckeroth & Gordon, 1989). Consistent with the in vitro data, this analogue was incorporated by only a subset of myristoylated proteins from yeast and BC_3H1 cells, a murine myocyte cell line. Incorporation of 11-oxymyristic acid by $p60^{v\text{-}src}$ and a 63-kDa protein from the BC_3H1 cell line resulted in their redistribution from the membrane to the soluble fraction. It appears, therefore, that incorporation of a myristate analogue of similar length but reduced hydrophobicity results in less efficient binding of SRC to its receptor. The ability of specific fatty acid analogues to be selectively incorporated into certain oncogenic proteins and to alter their subcellular distribution raises the possibility that certain analogues might be designed that would interfere with transformation.

We have also observed the 63-kDa protein during our studies of fatty acylated proteins in BC_3H1 cells. Its size, p*I*, and tyrosine phosphorylation suggested that it might represent SRC or a closely related protein. Indeed, cDNA cloning has revealed that this protein is encoded by the *syn/fyn* gene (G. James and E. Olson, unpublished data). The parallel altered distribution of this protein and SRC following incorporation of 11-oxymyristic acid would suggest that additional members of the nonreceptor tyrosine kinase family also possess membrane "receptors" with characteristics similar to those of the SRC receptor. Alternatively, two or more cellular tyrosine kinases may interact with the same receptor. Isolating and characterizing these molecules will be an important area for future studies.

The 36-kDa calpactin I heavy chain (also known as p36 and lipocortin II) is phosphorylated by $p60^{src}$ and exhibits calcium-dependent association with the plasma membrane [for recent review, see Klee (1988)]. This protein has been reported to be myristoylated in a transformation-sensitive manner (Soric & Gordon, 1985). In this study, transformation of chick embryo fibroblasts by $p60^{v\text{-}src}$ resulted in a reduction in the amount of [^{3}H]myristate incorporated by p36 compared to that in nontransformed cells. The radioactivity associated with this protein was identified as authentic myristate, and its resistance to hydroxylamine treatment suggested an amide linkage. However, subsequent molecular cloning of a cDNA encoding p36 revealed a predicted N-terminal amino acid sequence that cannot serve as a substrate for NMT (Saris et al., 1986). Also, the decreased myristoylation of p36 upon *src* transformation occurred without any apparent reduction in synthesis of the protein, which suggests that p36 is not subject to classical, cotranslational myristoylation. Thus, myristoylation of this protein may be mediated by a unique pathway. It is tempting to speculate that the fatty acyl moiety associated with p36 may participate in its acquisition of a hydrophobic domain during calcium-induced conformational changes that it is known to undergo, thereby facilitating interactions with the plasma membrane.

Serine/Threonine Kinases, Phosphatases, and Substrates. The catalytic subunit of cAMP-dependent protein kinase was among the first proteins demonstrated to contain covalent myristate (Carr et al., 1982), but no information is available concerning any possible role this modification may play in substrate recognition or subcellular localization of the enzyme. The inactive enzyme is a membrane-bound tetrameric complex composed of two regulatory (R) and two catalytic (C) subunits. Elevation of intracellular cAMP levels, provoked by a variety of stimuli, results in the binding of cAMP by an R_2 dimer and the release of two monomeric, active subunits into the soluble fraction [reviewed by Edelman et al. (1987)]. Given the requirement for covalent fatty acids in mediating protein–membrane and protein–protein interactions of other known fatty acylated proteins, it appears likely that myristoylation of the cAMP-dependent protein kinase catalytic subunit plays a significant role in its association with either the regulatory subunits or important substrates. Mutational analyses, similar to those described for SRC, should provide clues as to the role of covalent myristate in the functioning of cAMP-dependent protein kinase.

Another myristoylated protein that has been implicated in serine/threonine kinase signaling pathways in the B subunit of calcineurin (Aitken et al., 1982). Calcineurin is a Ca^{2+}/calmodulin-dependent protein phosphatase composed of two subunits. Subunit A interacts with calmodulin and contains the catalytic site, whereas subunit B binds calcium (Manalan & Klee, 1983; Tonks & Cohen, 1983). Catalytic activity of the A subunit is stimulated by either the B subunit or calmodulin, but only in the presence of calcium (Stewart et al., 1983). As is the case with cAMP-dependent kinase, however, requirements for myristoylation of the B subunit in regulating calcineurin activity have not been reported. The highly selective nature of this modification, combined with its propensity for proteins involved in cellular regulatory pathways, indeed suggests that it provides a necessary feature of calcineurin as well.

A myristoylated protein that has received considerable attention of late is the 67–87-kDa protein kinase C substrate ("80K protein"), which is found in a wide variety of cell types. Although no function has yet been assigned to the 80K protein, its phosphorylation in response to polypeptide growth factors, phorbol esters, and *ras* transformation suggests that it occupies a central position in one or more pathways mediated by protein kinase C (Rozengurt et al., 1983; Rodriguez-Pena & Rozengurt, 1986; Blackshear et al., 1985, 1986; Wolfman & Macara, 1987; Wang et al., 1989). Myristoylation of the 80K protein through an amide linkage was discovered during studies on the response of macrophages to bacterial lipopolysaccharide (LPS) (Aderem et al., 1986, 1988). It was reported that myristoylation of the 80K protein was induced following exposure of macrophages to LPS. The majority of ^{32}P-labeled 80K protein was in the cytosol, whereas the myristoylated protein was almost exclusively associated with the membrane fraction. The discrete subcellular locations of the myristoylated versus the ^{32}P-labeled forms of the protein prompted speculation that LPS-induced myristoylation of the 80K protein may target it to the membrane, where it would be more accessible to active protein kinase C.

We have recently studied myristoylation of the 80K protein in BC_3H1 myocytes in order to determine if it may become fatty acylated through a pathway that is distinct from that utilized by other myristoylated proteins. We found that the 80K protein exhibits the characteristics of a classical myristoylated protein. Its myristoylation occurred cotranslationally via an amide linkage, and there was no evidence for post-translational, stimulus-dependent fatty acylation of a preexisting pool of the protein (James & Olson, 1989b). Furthermore, pulse–chase experiments demonstrated that the 80K protein is not demyristoylated following phorbol dibutyrate stimulation of quiescent cells, despite a 6-fold increase in its level of phosphorylation. The majority of myristate-labeled 80K protein was indeed found in the membrane fraction, but a significant amount (~20%) was soluble, both before and after its phosphorylation by protein kinase C. This finding demonstrated that the overall distribution of this protein between the plasma membrane and cytosol is not influenced by its degree of phosphorylation and that myristoylation is not the sole mechanism by which it associates with the membrane.

Previous studies have reported that immunoreactive 80K is more abundant in the membrane fraction (Albert et al., 1986), whereas the ^{32}P-labeled protein has been found mostly in high-speed supernatants (Patel & Kligman, 1987). Our studies with BC_3H1 myocytes revealed that the intensity of ^{32}P labeling of the 80K protein is approximately equal in the membrane and cytosol, and stimulation with phorbol dibutyrate induced a parallel 6-fold increase in phosphorylation of the protein in each fraction. When compared to the subcellular distribution of the protein under identical conditions, as determined by myristate labeling, it was apparent that cytosolic 80K protein is phosphorylated to a 4-fold higher stoichiometry than the membrane-bound form.

Recent in vitro experiments have shown that phosphorylation of the 80K protein results in its release from membranes (Wang et al., 1989), suggesting that phosphorylation regulates the subcellular distribution of this protein. We found, however, that the in vivo subcellular distribution of the protein is unchanged after a 6-fold increase in its level of phosphorylation. Preliminary experiments indicate that in vitro phosphorylation of the 80K protein by exogenous protein kinase C in isolated membranes from BC_3H1 cells does not result in its solubilization. However, the possibility that soluble factors are also involved in regulating the subcellular distribution of this protein has not been thoroughly investigated in our system. Consistent with this notion, a soluble protein has been identified in bovine brain which stimulates the calcium-dependent phosphorylation of the 80K protein and its release from membranes in vitro (Kligman & Patel, 1986). Taken together, these data suggest that increased phosphorylation of the 80K protein may mediate its cycling between the plasma membrane and cytosol. Thus, phosphorylation at the membrane might result in translocation of the protein to the cytosol, whereas subsequent dephosphorylation in the cytosol would allow it to reassociate with the plasma membrane. If the overall distribution of the 80K protein is maintained in an equilibrium, as our studies suggest, this type of mechanism would not be revealed with intact cells. Indeed, active protein kinase C is known to be associated with the plasma membrane (Halsey et al., 1987; Spach et al., 1986), suggesting that phosphorylation of the 80K protein could occur there. Alternatively, cytosolic 80K protein may be phosphorylated by the soluble proteolytic fragment of protein kinase C, which is known to be generated following its translocation to the membrane (Tapley & Murray, 1985; Melloni et al., 1986).

In vitro phosphorylation of the 80K protein in synaptosomal cytosol has been shown to be inhibited by calmodulin in the presence of calcium (Wu et al., 1982). It should be reiterated here that calmodulin also associates with the protein phosphatase calcineurin A. Thus, the apparent calmodulin-mediated inhibition of 80K phosphorylation may be due to dephosphorylation of the protein. The amino acid sequence of the bovine form of the 80K protein has recently been reported (Stumpo et al., 1989), and the sites for phosphorylation by protein kinase C were found clustered within a 25 amino acid basic domain (Graff et al., 1989). Interestingly, this domain is absolutely conserved between the bovine and chicken proteins, and it was found to have similarity to calmodulin-binding domains of other proteins, suggesting that calmodulin may interact with this region of the protein.

The subcellular distribution of the 80K protein is similar to that observed for SRC ($\sim$10% soluble), and in both cases it is known that myristoylation alone is insufficient to confer membrane binding. The fact that membrane-bound 80K protein is resistant to high-salt extraction (Albert et al., 1986) suggests that its association with the membrane is mediated by specific protein–protein interactions. In light of the recent identification of a plasma membrane receptor for SRC, it is tempting to speculate that such receptors might exist for other myristoylated proteins, such as 80K, that partition between the membrane and cytosol. In any event, it is clear that the 80K protein interacts with regulatory proteins in both the plasma membrane and cytosol. It will be important to determine the requirements for myristoylation in these interactions.

Cell Surface Receptors. A recent study has shown that the human β_2-adrenergic receptor is palmitoylated and that the fatty acyl moiety is essential for the functional integrity of the G protein mediated pathway through which it acts (O'Dowd et al., 1989). Mutant receptor molecules in which cysteine 341 was replaced by glycine were no longer palmitoylated and exhibited a dramatic decrease in their ability to activate adenylate cyclase in response to isoproterenol. The palmitoylation site, cysteine 341, lies in the carboxyl-portion, cytoplasmic domain of the molecule, and it was pointed out in this study that this cysteine residue is conserved in every G protein coupled receptor examined, suggesting that palmitoylation may be a universal modification of this receptor family. The β_2-adrenergic receptor has seven putative membrane-spanning domains, thus creating three intracellular and three extracellular loops. Addition of palmitate to cysteine 341, located 12 amino acids from the cytoplasmic surface of the plasma membrane, might be expected to promote binding of this region to the membrane, thus creating a fourth intracellular loop. Reversible fatty acylation of this residue may modulate the formation of a functional G protein binding site. The nicotinic acetylcholine (Olson et al., 1984), insulin, and IGF-1 (Magee & Siddle, 1988) receptors are also palmitoylated, but neither the site for palmitoylation nor the functional consequences of these modification are known.

Other Regulatory Proteins. Transforming growth factor α is a secreted polypeptide growth factor with mitogenic properties, and has been implicated in initiation or maintenance of transformation via an autocrine pathway. TGF-α has been shown to be synthesized as part of a glycosylated and palmitoylated precursor that is transported to the cell surface via the secretory pathway (Bringman et al., 1987). Alternate proteolytic cleavage of the extracellular domain gives rise to multiple TGF-α species, leaving a membrane-associated 12.5-kDa C-terminal fragment containing covalent palmitate, presumably in the cysteine-rich cytoplasmic domain. Tunicamycin had no significant effect on expression or secretion of TGF-α, suggesting that N-glycosylation is not required for transport to the cell surface or proteolytic cleavage. Although the function of covalent palmitate near the C-terminus is unknown, the amino acid sequences of the transmembrane and cytoplasmic segments of the human and rat TGF-α precursors are nearly identical over approximately 60 residues (one conservative substitution), suggesting that these regions play an important biological role. It has been proposed that palmitoylation may serve to slow the movement of the precursor through the endoplasmic reticulum and Golgi, thus allowing for more efficient proteolytic processing. These ideas, however, remain to be tested.

SV40 large T-antigen (T-ag) is a polypeptide that provides various functions required for viral infection and transformation by Simian virus 40 (Butel & Jarvis, 1986). This protein exhibits an unusual subcellular distribution, such that $\sim$95% of intracellular T-ag is found in the nucleus, whereas a small fraction is localized to the plasma membrane. T-ag at the plasma membrane has been subdivided into two classes on the basis of its solubility in the nonionic detergent NP-40 (Klockmann & Deppert, 1983). Approximately one-third of T-ag associated with the plasma membrane was soluble in NP-40, with the remainder tightly bound to the detergent-resistant lamina of the plasma membrane. This detergent-resistant subclass of T-ag was found to contain covalent palmitate, while detergent-soluble and nuclear T-ag were nonacylated. Although it is possible that detergent-soluble T-ag at the plasma membrane may be an artifact of fractionation, selective fatty acylation of only the lamina-associated T-ag suggests that the presence of this subclass is due to specific interactions and that palmitoylation may facilitate tight binding.

Summary

From the studies presented above, it is obvious that fatty acylation is a common modification among proteins involved in cellular regulatory pathways, and in certain cases mutational analyses have demonstrated the importance of covalent fatty acids in the functioning of these proteins. Indeed, certain properties provided by fatty acylation make it an attractive modification for regulatory proteins that might interact with many different substrates, particularly those found at or near the plasma membrane/cytosol interface. In the case of intracellular fatty acylated proteins, the fatty acyl moiety allows tight binding to the plasma membrane without the need for cotranslational insertion through the bilayer. For example, consider the tight, salt-resistant interaction of myristoylated SRC with the membrane, whereas its nonmyristoylated counterpart is completely soluble. Likewise for the RAS proteins, which associate weakly with the membrane in the absence of fatty acylation, while palmitoylation increases their affinity for the plasma membrane and their biological activity. Fatty acylation also permits reversible membrane association in some cases, particularly for several myristoylated proteins, thus conferring plasticity on their interactions with various signaling pathway components. Finally, although this has not been demonstrated, it is conceivable that covalent fatty acid may allow for rapid mobility of proteins within the membrane.

Several questions remain to be answered concerning requirements for fatty acylation by regulatory proteins. The identity of the putative SRC "receptor" will provide important clues as to the pathways in which normal SRC functions, as well as into the process of transformation by oncogenic tyrosine kinases. The possibility that other fatty acylated proteins associate with the plasma membrane in an analogous manner also needs to be investigated. An intriguing observation that can be made from the information presented here is that at least three different families of proteins involved in growth factor signaling pathways encode both acylated and non-acylated members, suggesting that selective fatty acylation may provide a means of determining the specificity of their interactions with other regulatory molecules. Further studies of fatty acylated proteins should yield important information concerning the regulation of intracellular signaling pathways utilized during growth and differentiation.

References

Abood, M. E., Hurley, J. B., Pappone, M. C., Bourne, H. R., & Stryer, L. (1982) *J. Biol. Chem. 257*, 10540–10543.

Aderem, A. A., Keum, M. W., Pure, E., & Cohn, Z. A. (1986) *Proc. Natl. Acad. Sci. U.S.A. 83*, 5817–5821.

Aderem, A. A., Albert, K. A., Keum, M. M., Wang, J. K. T., Greengard, P., & Cohn, Z. A. (1988) *Nature 332*, 362–364.

Aitken, A., Cohen, P., Santikarn, S., Williams, D. H., Calder, G., Smith, A., & Klee, C. B. (1982) *FEBS Lett. 150*, 314–318.

Albert, K. A., Walaas, S. I., Wang, J. K. T., & Greengard, P. (1986) *Proc. Natl. Acad. Sci. U.S.A. 83*, 2822–2826.

Alsip, G. R., & Konkel, D. A. (1986) *Nucleic Acids Res. 14*, 2123–2130.

Anderegg, R. J., Betz, R., Carr, S. A., Crabb, J. W., & Duntze, W. (1988) *J. Biol. Chem. 263*, 18236–18240.

Barbacid, M. (1987) *Annu. Rev. Biochem. 56*, 779–827.

Berger, M., & Schmidt, M. F. (1984) *EMBO J. 3*, 713–719.

Berger, M., & Schmidt, M. F. (1986) *J. Biol. Chem. 261*, 14912–14918.

Betz, R., Crabb, J. W., Meyer, H. E., Wittig, R., & Duntze, W. (1987) *J. Biol. Chem. 262*, 546–548.

Blackshear, P. J., Witters, L. A., Girard, P. R., Kuo, J. F., & Quamo, S. N. (1985) *J. Biol. Chem. 260*, 13304–13315.

Blackshear, P. J., Wen, L., Glynn, B. P., & Witters, L. A. (1986) *J. Biol. Chem. 261*, 1459–1469.

Bray, P., Carter, A., Guo, V., Puckett, C., Kamholz, J., Spiegel, A., & Nirenberg, M. (1987) *Proc. Natl. Acad. Sci. U.S.A. 84*, 5115–5119.

Bringman, T. S., Lindquist, P. B., & Derynck, R. (1987) *Cell 48*, 429–440.

Broek, D., Samiy, N., Fasano, O., Fujiyama, A., Tamanoi, F., Northup, J., & Wigler, M. (1985) *Cell 41*, 763–769.

Buss, J. E., & Sefton, B. M. (1985) *J. Virol. 53*, 7–12.

Buss, J. E., & Sefton, B. M. (1986) *Mol. Cell. Biol. 6*, 116–122.

Buss, J. E., Kamps, M. P., & Sefton, B. M. (1984) *Mol. Cell. Biol. 4*, 2697–2704.

Buss, J. E., Kamps, M. P., Gould, K., & Sefton, B. M. (1986) *J. Virol. 58*, 468–474.

Buss, J. E., Mumby, S. M., Casey, P. J., Gilman, A. G., & Sefton, B. M. (1987) *Proc. Natl. Acad. Sci. U.S.A. 84*, 7493–7497.

Buss, J. E., Solski, P. A., Schaeffer, J. P., MacDonald, M. J., & Der, C. (1989) *Science 243*, 1600–1603.

Butel, J. S., & Jarvis, D. L. (1986) *Biochim. Biophys. Acta 865*, 171–195.

Calothy, G., Laugier, D., Cross, F. R., Jove, R., Hanafusa, T., & Hanafusa, H. (1987) *J. Virol. 61*, 1678–1681.

Capon, D. J., Seeberg, P. H., McGrath, J. P., Hayflick, J. S., Edman, A. D., Levinson, A. D., & Goeddel, D. V. (1983) *Nature 304*, 507–513.

Carr, S. A., Biemann, K., Shoji, S., Parmelee, D., & Titani, K. (1982) *Proc. Natl. Acad. Sci. U.S.A. 79*, 6128–6131.

Cassel, D., & Selinger, Z. (1977) *Proc. Natl. Acad. Sci. U.S.A. 74*, 3307–3311.

Cheng, S. H., Harvey, R., Espino, P. C., Semba, K., Yamamoto, T., Toyoshima, K., & Smith, A. E. (1988) *EMBO J. 7*, 3845–3855.

Clarke, S., Vogel, J. P., Deschenes, R. J., & Stock, J. (1988) *Proc. Natl. Acad. Sci. U.S.A. 85*, 4643–4647.

Cross, F. R., Garber, E. A., Pellman, D., & Hanafusa, H. (1984) *Mol. Cell. Biol. 4*, 1834–1842.

Deschenes, R. J., & Broach, J. R. (1987) *Mol. Cell. Biol. 7*, 2344–2351.

Edelman, A. M., βlumenthal, D. K., & Krebs, E. G. (1987) *Annu. Rev. Biochem. 56*, 567–613.

Fujiyama, A., & Tamanoi, F. (1986) *Proc. Natl. Acad. Sci. U.S.A. 83*, 1266–1270.

Fujiyama, A., Kunihiro, M., & Fuyuhiko, T. (1987) *EMBO J. 6*, 223–228.

Gallwitz, D., Donath, C., & Sander, C. (1983) *Nature 306*, 704–707.

Garber, E. A., Cross, F. R., & Hanafusa, H. (1985) *Mol. Cell. Biol. 5*, 2781–2788.

Gautam, N., Baetscher, M., Aebersold, R., & Simon, M. I. (1989) *Science 244*, 971–974.

Gilman, A. G. (1987) *Annu. Rev. Biochem. 56*, 615–649.

Goddard, C., Arnold, S. T., & Felsted, R. L. (1989) *J. Biol. Chem. 264*, 15173–15176.

Graff, J. M., Stumpo, D. J., & Blackshear, P. J. (1989) *J. Biol. Chem. 264*, 11912–11919.

Grand, R. J. A., Smith, K. J. U., & Gallimore, P. H. (1987) *Oncogene 1*, 305–314.

Groffen, J., Heisterkamp, N., Shibuya, M., Hanafusa, H., & Stephenson, J. R. (1983) *Virology 125*, 480–486.

Gutierrez, L., Magee, A. I., Marshall, C. J., & Hancock, J. F. (1989) *EMBO J. 8*, 1093–1098.

Hall, A. (1989) in *G Proteins as Mediators of Cellular Signaling Processes* (Houslay, M. D., & Milligan, G., Eds.) Wiley, London.

Halsey, D. H., Girard, P. R., Kuo, J. F., & Blackshear, P. J. (1987) *J. Biol. Chem. 262*, 2234–2243.

Hamaguchi, M., Grandori, C., & Hanafusa, H. (1988) *Mol. Cell. Biol. 8*, 3035–3042.

Hancock, J. F., Magee, A. I., Childs, J. E., & Marshall, C. J. (1989) *Cell 57*, 1167–1177.

Hao, Q.-L., Heisterkamp, N., & Groffen, J. (1989) *Mol. Cell. Biol. 9*, 1587–1593.

Hescheler, J., Rosenthal, W., Trautwein, W., & Schultz, G. (1987) *Nature 325*, 445–447.

Heuckeroth, R. O., & Gordon, J. I. (1989) *Proc. Natl. Acad. Sci. U.S.A. 86*, 5262–5266.

Hunter, T., & Copper, J. A. (1985) *Annu. Rev. Biochem. 54*, 897–931.

Hurley, J. B., Fong, H. K. W., Teplow, D. B., Dreyer, W. J., & Simon, M. I. (1984) *Proc. Natl. Acad. Sci. U.S.A. 81*, 6948–6952.

Inoue, K., Ikawa, S., Semba, K., Sukegawa, J., Yamamoto, T., & Toyoshima, K. (1987) *Oncogene 1*, 301–304.

Itoh, H., Kozasa, T., Nagata, S., Nakamura, S., Katada, T., Ui, M., Iwai, S., Ohtsuka, E., Kawasaki, H., Suzuki, K., & Kaziro, Y. (1986) *Proc. Natl. Acad. Sci. U.S.A. 83*, 3776–3780.

James, G., & Olson, E. N. (1989a) *J. Biol. Chem. 264*, 20998–21006.

James, G., & Olson, E. N. (1989b) *J. Biol. Chem. 264*, 20928–20933.

Jing, S. Q., & Trowbridge, I. S. (1987) *EMBO J. 6*, 327–331.

Jones, D. T., & Reed, R. R. (1989) *Science 244*, 790–795.

Kahn, R. A., & Gilman, A. G. (1984) *J. Biol. Chem. 259*, 6228–6234.

Kahn, R. A., & Gilman, A. G. (1986) *J. Biol. Chem. 261*, 7906–7911.

Kahn, R. A., Goddard, C., & Newkirk, M. (1988) *J. Biol. Chem. 263*, 8282–8287.

Kamps, M. P., Buss, J. E., & Sefton, B. M. (1985) *Proc. Natl. Acad. Sci. U.S.A. 82*, 4625–4628.

Kamps, M. P., Buss, J. E., & Sefton, B. M. (1986) *Cell 45*, 105–112.

Kataoka, T., Powers, S., McGill, C., Fasano, O., Strathern, J., Broach, J., & Wigler, M. (1984) *Cell 37*, 437–445.

Khandwala, A. S., & Casper, C. B. (1971) *J. Biol. Chem. 246*, 6242–6246.

Kitayama, H., Sugimoto, Y., Matsuzaki, T., Ikawa, Y., & Noda, M. (1989) *Cell 56*, 77–84.

Klee, C. B. (1988) *Biochemistry 27*, 6645–6653.

Kligman, D., & Patel, J. (1986) *J. Neurochem. 47*, 298–303.

Klockman, U., & Deppert, W. (1983) *EMBO J. 2*, 1151–1157.

Kypta, R. M., Hemming, A., & Courtneidge, S. A. (1988) *EMBO J. 7*, 3837–3844.

Linder, M. E., & Burr, J. G. (1988) *Proc. Natl. Acad. Sci. U.S.A. 85*, 2608–2612.

Lochrie, M. A., Hurley, J. B., & Simon, M. I. (1985) *Science 228*, 96–99.

Luttrell, D. K., Lutrell, L. M., & Parsons, S. J. (1988) *Mol. Cell. Biol. 8*, 497–501.

Magee, A. I., & Courtneidge, S. A. (1985) *EMBO J. 4*, 1137–1144.

Magee, A. I., Gutierrez, L., McKay, I. A., Marshall, C. J., & Hall, A. (1987) *EMBO J. 6*, 3353–3357.

Magee, A. I., & Siddle, K. (1988) *J. Cell. Biochem. 37*, 347–357.

Manalan, A. S., & Klee, C. B. (1983) *Proc. Natl. Acad. Sci. U.S.A. 80*, 4291–4295.

Marchildon, G. A., Casnellie, J. E., Walsh, K. A., & Krebs, E. G. (1984) *Proc. Natl. Acad. Sci. U.S.A. 81*, 7679–7682.

Marth, J. D., Peet, R., Krebs, E. G., & Perlmutter, R. M. (1985) *Cell 43*, 393–404.

Matsuoka, M., Itoh, H., Kozasa, T., & Kaziro, Y. (1988) *Proc. Natl. Acad. Sci. U.S.A. 85*, 5384–5388.

McIlhinney, R. A., Pelly, S. J., Chadwick, J. K., & Cowley, G. P. (1985) *EMBO J. 4*, 1145–1152.

Melloni, E., Pontremoli, S., Michetti, M., Sacco, O., Sparatore, B., & Horecker, B. L. (1986) *J. Biol. Chem. 261*, 4101–4105.

Molenaar, C. M. T., Prange, R., & Gallwitz, D. (1988) *EMBO J. 7*, 971–976.

Moss, P., Radke, K., Carter, V. C., Young, J., Gillmore, T., & Martin, G. S. (1984) *J. Virol. 52*, 557–565.

Neer, E. J., Lok, J. M., & Wolf, L. G. (1984) *J. Biol. Chem. 259*, 14222–14229.

Nukada, T., Tanabe, T., Takahashi, H., Noda, M., Haga, K., Haga, Y., Ichiyama, A., Kangawa, K., Hiranaga, M., Matsuo, H., & Numa, S. (1986) *FEBS Lett. 197*, 305–310.

O'Dowd, B. F., Hnatowich, M., Caron, M. G., Lefkowitz, R. J., & Bouvier, M. (1989) *J. Biol. Chem. 264*, 7564–7569.

Olson, E. N. (1988) *Prog. Lipid Res. 27*, 177–197.

Olson, E. N., & Spizz, G. (1986) *J. Biol. Chem. 261*, 2458–2466.

Olson, E. N., & James, G. (1989) *Dynamics and Biogenesis of Membranes*, NATO ASI Series, Springer-Verlag, New York (in press).

Olson, E. N., Merlie, J. P., & Glaser, L. (1984) *J. Biol. Chem. 259*, 5364–5367.

Olson, E. N., Towler, D. A., & Glaser, L. (1985) *J. Biol. Chem. 260*, 3784–3790.

Omary, M. B., & Trowbridge, I. S. (1981) *J. Biol. Chem. 256*, 4715–4718.

Palmiter, R. D. (1977) *J. Biol. Chem. 252*, 8781–8783.

Patel, J., & Kligman, D. (1987) *J. Biol. Chem. 262*, 16686–16691.

Pellman, D., Garber, E. A., Cross, F. R., & Hanafusa, H. (1985) *Nature 314*, 374–377.

Pestana, A., & Pitot, H. C. (1975) *Biochemistry 14*, 1404–1412.

Powers, S., Kataoka, T., Fasano, O., Goldfarb, M., Strathern, J., Broach, J., & Wigler, M. (1984) *Cell 36*, 607–612.

Powers, S., Michaelis, S., Broek, D., Santa Anna, A. S., Field, J., Herskowitz, I., & Wigler, M. (1986) *Cell 47*, 413–422.

Price, R. S., Nightingale, M., Tsai, S. C., Williamson, K. C., Adamik, R., Chen, H.-C., Moss, J., & Vaughan, M. (1988) *Proc. Natl. Acad. Sci. U.S.A. 85*, 5488–5491.

Quintrell, N., Lebo, R., Varmus, H., Bishop, J. M., Pettenati, M. J., Le Beau, M. M., Diaz, M. O., & Rowley, J. D. (1987) *Mol. Cell. Biol. 7*, 2267–2275.

Resh, M. D. (1988) *Mol. Cell. Biol. 8*, 1896–1905.

Resh, M. D. (1989) *Cell 58*, 281–286.

Reynolds, A. B., Roesel, D. J., Kanner, S. B., & Parsons, J. T. (1989) *Mol. Cell. Biol. 9*, 629–638.

Robishaw, J. D., Russell, D. W., Harris, B. A., Smigel, M. D., & Gilman, A. G. (1986) *Proc. Natl. Acad. Sci. U.S.A. 83*, 1251–1255.

Rodriguez-Pena, A., & Rozengurt, E. (1986) *EMBO J. 5*, 77–83.

Rozengurt, E., Rodriguez-Pena, M., & Smith, K. A. (1983) *Proc. Natl. Acad. Sci. U.S.A. 80*, 7244–7248.

Saris, C. J. M., Tack, B. F., Kristensen, T., Glenney, J. R., & Hunter, T. (1986) *Cell 46*, 201–212.

Schleifer, L. S., Kahn, R. A., Hanski, E., Northup, J. K., Sternweis, P. C., & Gilman, A. G. (1982) *J. Biol. Chem. 257*, 20–23.

Schmidt, J. W., & Catterall, W. A. (1987) *J. Biol. Chem. 262*, 13713–13723.

Schultz, A. M., Henderson, L. E., & Oroszlan, S. (1988) *Annu. Rev. Cell. Biol. 4*, 611–647.

Sefton, B. M., Trowbridge, I. S., Cooper, J. A., & Scolnick, E. M. (1982) *Cell 31*, 465–474.

Semba, K., Nishizawa, M., Miyajima, N., Yoshida, M. C., Sukegawa, J., Yamanashi, Y., Sasaki, M., Yamamoto, T., & Toyoshima, K. (1986) *Proc. Natl. Acad. Sci. U.S.A. 83*, 5459–5463.

Sewell, J. L., & Kahn, R. A. (1988) *Proc. Natl. Acad. Sci. U.S.A. 85*, 4620–4624.

Shimuzu, K., Birnbaum, D., Ruley, M. A., Fasano, O., Suard, Y., Edlund, L., Taparowsky, E., Goldfarb, M., & Wigler, M. (1983) *Nature 304*, 497–500.

Soric, J., & Gordon, J. A. (1985) *Science 230*, 563–566.
Spach, D. H., Nemenoff, R. A., & Blackshear, P. J. (1987) *J. Biol. Chem. 261*, 12750–12753.
Staufenbiel, M. (1987) *Mol. Cell. Biol. 7*, 2981–2984.
Sternweis, P. C., & Robishaw, J. D. (1984) *J. Biol. Chem. 259*, 13806–13813.
Stewart, A. A., Ingebritsen, T. S., & Cohen, P. (1983) *Eur. J. Biochem. 132*, 289–295.
Stryer, L., Hurley, J. B., & Fung, B. K.-K. (1981) *Curr. Top. Membr. Transp. 15*, 93–108.
Stumpo, D. J., Graff, J. M., Albert, K. A., Greengard, P., & Blackshear, P. J. (1989) *Proc. Natl. Acad. Sci. U.S.A. 86*, 4012–4016.
Sudol, M., Alvarez-Buylla, A., & Hanafusa, H. (1988) *Onc. Res. 2*, 345–355.
Sukegawa, J., Semba, K., Yamanashi, Y., Nishizawa, M., Miyajima, N., Yamamoto, T., & Toyoshima, K. (1987) *Mol. Cell. Biol. 7*, 41–47.
Takeya, T., & Hanafusa, H. (1983) *Cell 32*, 881–890.
Tamanoi, F., Hsueh, E. C., Goodman, L. E., Cobitz, A. R., Detrick, R. J., Brown, W. R., & Fujiyama, A. (1988) *J. Cell. Biochem. 36*, 261–273.
Tanabe, T., Nukada, T., Nishikawa, Y., Sugimoto, K., Suzuki, H., Takahashi, H., Noda, M., Haga, T., Ichiyama, A., Kangawa, K., Minamino, N., Matsuo, H., & Numa, S. (1985) *Nature 315*, 242–245.
Tapley, P. M., & Murray, A. W. (1985) *Eur. J. Biochem. 151*, 419–423.
Tatchell, K., Chaleff, D., DeFeo-Jones, D., & Scolnick, E. M. (1984) *Nature 309*, 523–527.
Toda, T., Uno, I., Ishikawa, T., Powers, S., Kataoka, T., Broek, D., Cameron, S., Broach, J., Matsumoto, K., & Wigler, M. (1985) *Cell. 40*, 27–36.
Tonks, N. K., & Cohen, P. (1983) *Biochim. Biophys. Acta 474*, 191–193.
Towler, D., & Glaser, L. (1986) *Proc. Natl. Acad. Sci. U.S.A. 83*, 2812–2816.
Towler, D., Eubanks, S. R., Towery, D. S., Adams, S. P., & Glaser, L. (1987a) *J. Biol. Chem. 262*, 1030–1036.
Towler, D., Adams, S. P., Eubanks, S. R., Towery, D. S., Jackson-Machelski, E., Glaser, L., & Gordon, J. I. (1987b) *Proc. Natl. Acad. Sci. U.S.A. 84*, 2708–2712.
Towler, D., Adams, S. P., Eubanks, S. R., Towery, D. S., Jackson-Machelski, E., Glaser, L., & Gordon, J. I. (1988a) *J. Biol. Chem. 263*, 1784–1790.
Towler, D., Gordon, J. I., Adams, S. P., & Glaser, L. (1988b) *Annu. Rev. Biochem. 57*, 69–99.
Van Meurs, K. P., Angus, C. W., Lavu, S., Kung, H.-F., Czarnecki, S. K., Moss, J., & Vaughan, M. (1987) *Proc. Natl. Acad. Sci. U.S.A. 84*, 3107–3111.
Voronova, A. F., Buss, J. E., Patschinsky, T., Hunter, T., & Sefton, B. M. (1984) *Mol. Cell. Biol. 4*, 2705–2713.
Wang, J. K. T., Walaas, S. I., Sihra, T. S., Aderem, A., & Greengard, P. (1989) *Proc. Natl. Acad. Sci. U.S.A. 86*, 2253–2256.
West, R. E., Moss, J., Vaughan, M., Liu, T., & Liu, T.-Y. (1985) *J. Biol. Chem. 260*, 14428–14430.
Wilcox, C. A., & Olson, E. N. (1987) *Biochemistry 26*, 1029–1036.
Wilcox, C., Hu, J.-S., & Olson, E. N. (1987) *Science 238*, 1275–1278.
Willumsen, B. M., Kjeld, N., Papageorge, A. G., Hubbert, N. L., & Lowy, D. R. (1984a) *EMBO J. 3*, 2581–2585.
Willumsen, B. M., Christensen, A., Hubbert, N. L., Papageorge, A. G., & Lowy, D. R. (1984b) *Nature 310*, 583–586.
Wilson, L. K., Luttrell, D. K., Parsons, J. T., & Parsons, S. J. (1989) *Mol. Cell. Biol. 9*, 1536–1544.
Wolfman, A., & Macara, I. G. (1987) *Nature 325*, 359–361.
Wu, W. C.-S., Walaas, S. I., Nairn, A. C., & Greengard, P. (1982) *Proc. Natl. Acad. Sci. U.S.A. 79*, 5249–5253.
Yamanashi, Y., Fukushige, S.-I., Semba, K., Sukegawa, J., Miyajima, N., Matsubara, K.-I., Yatsunami, K., & Khorana, H. G. (1987) *Mol. Cell. Biol. 7*, 237–243.
Young, J. C., & Martin, G. S. (1984) *J. Virol. 52*, 913–918.
Ziegler, S. F., Marth, J. D., Lewis, D. B., & Perlmutter, R. M. (1987) *Mol. Cell. Biol. 7*, 2276–2285.
Ziegler, S. F., Levin, S. D., & Perlmutter, R. M. (1989) *Mol. Cell. Biol. 9*, 2724–2727.

Chapter 18

Structure, Biosynthesis, and Function of Glycosylphosphatidylinositols

J. R. Thomas,[‡] R. A. Dwek, and T. W. Rademacher*

Glycobiology Unit, Department of Biochemistry, University of Oxford, South Parks Road, Oxford OX1 3QU, U.K.

Received November 29, 1989; Revised Manuscript Received February 26, 1990

Glycosylphosphatidylinositols (GPIs) are a recently discovered class of glycolipids that have been proposed to anchor either protein, polysaccharide, or small oligosaccharides to cellular membranes through covalent linkage. Only recently has the precise chemical nature of these molecules become apparent, and so far the complete structures of only a few have been elucidated. Nevertheless, there is a large body of relevant literature, and several comprehensive reviews of the field have been published (Cross, 1987; Ferguson & Williams, 1988; Low & Saltiel, 1988; Low, 1987, 1989).

Proteins from a diversity of eukaryotic organisms have been found to have GPI membrane anchors, and the list is surely to grow. They have been found in a wide variety of mammalian cells and tissues, squid brain, the slime mold *Dictyostelium*, the yeast (fungus) *Saccharomyces*, and protozoa. There have been no reports of GPIs in prokaryotes, algae, or plants, but neither have there been descriptions of attempts to look at these potential sources. GPIs, either unsubstituted or with polysaccharide rather than protein attached, have been described thus far only in the parasitic protozoan *Leishmania*. Despite the widespread occurrence of GPIs, and the large number of proteins that have GPI membrane anchors, few generalizations or conclusions can be made about their biosynthesis or function. At present only the chemical structures of GPIs can be described with any certainty.

Structures of GPIs and Related Molecules

The complete structures of only two protein-attached GPI membrane anchors [*Trypanosoma brucei* variable surface glycoprotein (VSG) and rat brain Thy-1] and the partial structure of a third, human erythrocyte AChE, have thus far been determined. The VSG and Thy-1 structures were determined by a combination of glycosyl-linkage and compositional analyses by GC–MS, specific chemical cleavages, sequential digestion with purified exoglycosidases combined with Bio-Gel P-4 column chromatography, and one- and two-dimensional ^{1}H NMR spectroscopy (Ferguson et al., 1988; Homans et al., 1988). Comparison of the VSG and Thy-1 anchor structures (Table I) reveals a conserved, linear 6-*O*-(ethanolamine-PO_4)-α-Man*p*-(1→2)-α-Man*p*-(1→6)-α-Man*p*-(1→4)-α-GlcNH_2*p*-(1→6)-*myo*-inositol-1-PO_4 core region. A structure of the VSG anchor from the *T. brucei* variant MITat.1.6 has been published in which the three mannosyl residues form a branch (Schmitz et al., 1987). This structure differs from that proposed for the anchor of VSG from variant MITat.1.4 (Ferguson et al., 1988), in which the mannosyl residues form a linear sequence (Table I). However, recent analysis of the anchor of VSG from the MITat.1.6 variant using one- and two-dimensional ^{1}H NMR has shown that its structure is in fact the same as that of VSG MITat.1.4 (Strang & van Halbeek, 1989).

Analysis by FAB-MS suggests that the human erythrocyte AChE anchor has the same core structure as the VSG and Thy-1 anchors (Roberts et al., 1988b) (Table I). Compositional analyses have indicated the presence of mannose as the only neutral hexose [see Roberts et al. (1988b)], *myo*-inositol (Roberts et al., 1987), and glucosamine (Haas et al., 1986). However, since the identities of potential phosphate-substituted monosaccharides, as well as the anomeric configurations and positions of substitution of the component glycosyl residues in the AChE anchor, are not yet known, it can only be presumed that the core of the human AChE anchor is identical with those of the VSG and Thy-1 anchors.

Heterogeneity in carbohydrate of the VSG anchor structure arises from a variable number of α-linked galactosyl side-chain residues linked to the core region. About 70% of the VSG anchor glycans are accounted for by structures bearing from two to four galactosyl residues, while another 15% most likely consist of structures bearing none, one, or five (or more) galactosyl residues (Ferguson et al., 1988). The two mammalian anchors that have been characterized (Thy-1 and AChE) both lack side chains composed of α-linked galactosyl residues (Homans et al., 1988; Roberts et al., 1988b). Instead, the rat brain Thy-1 anchor contains an additional mannose residue and an *N*-acetylgalactosamine (GalNAc) residue. About 30% of the rat brain Thy-1 structures do not contain the extra

* Correspondence should be addressed to this author.

[‡] Present address: Department of Biochemistry, University of Dundee, Dundee DD1 4HN, U.K.

Table I: Structure of GPIs for Which Data Have Been Published[h]

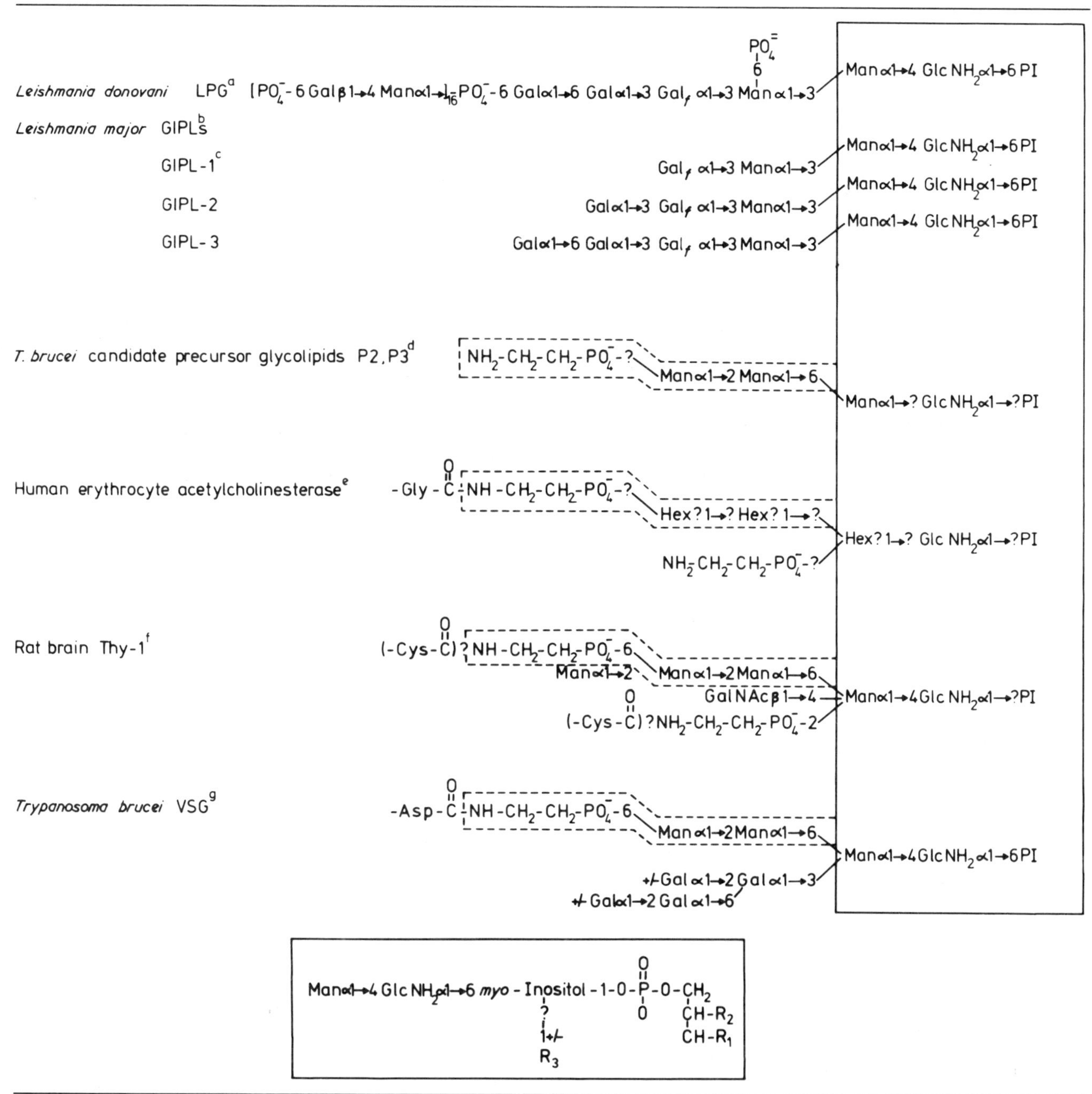

[a]See Turco et al. (1987, 1989). [b]*L. major* GPIs [see Rosen et al. (1989)] have also been reported to bear zero to two galactosyl residues attached to a Man*p* residue rather than the Gal*f* residue found in GIPLs by McConville et al. (1990) (see text for a discussion of these structures). [c]See McConville and Bacic (1989) and McConville et al. (1990b). [d]See Mayor et al. (1990a,b). [e]See Roberts et al. (1989a,b). Mannose is the only hexose detected by compositional analysis. An actylcholinesterose anchor bearing three ethanolamine residues has been observed, but the location of the third residue could not be determined (Roberts et al., 1989b). [f]See Homans et al. (1988). The ethanolamine through which the Thy-1 protein is linked to its anchor has not been determined. Compositional analysis of the rat thymocyte Thy-1 anchors (Tse et al., 1985) suggests that some have one less mannose and one less GalNAc residue than the brain structure. [g]See Ferguson et al. (1988). [h]All of the structures shown have in common a mannoglucosaminyl-PI (areas within solid lines) but can possess different aliphatic side chains R_1 and R_2 (see Table II). R_3 = palmitate in the acetylcholinesterase anchor (Roberts et al., 1988a,b) and probably in the *T. brucei* glycolipid P3 (Mayor et al., 1990b). Glycosyl residues are in the pyranose ring form, unless otherwise indicated, and in the D configuration on the basis of digestion with exoglycosidases. Unknown substitution positions are denoted by (?), glycosidic linkages by (→), and heterogeneity resulting from the presence or absence of residues by (+/−).

mannose residue, giving rise to heterogeneity within structures isolated from the same tissue. Tissue-specific differences in side-chain structure can also occur, as suggested by compositional analysis of rat thymocyte Thy-1 (Tse et al., 1985), which indicates that the extra mannose residue (in brain) is unlikely to be present at all and that only about one third of the anchors contain GalNAc.

Covalent linkage of the protein to the GPI-anchor core region via an ethanolamine–phosphate bridge has been demonstrated for *T. brucei* VSG (Holder, 1983), human AChE (Haas et al., 1986), and Thy-1 (Williams & Tse, 1985). Analysis of the human AChE anchor by FAB-MS revealed that the C-terminal glycine of the polypeptide is linked to the ethanolamine that is located on the nonreducing-terminal hexosyl residue (Roberts et al., 1988b). It has not been determined which, or indeed whether both of the two ethanolamines, bridges the Thy-1 anchor to the protein (Homans et al., 1988). Assuming the core regions of the VSG and human

AChE anchors are the same (see Table I), the proteins would then be linked via a bridging ethanolamine to their respective anchors at the same mannosyl residues in the glycan sequence. It is not known, however, whether linkage of the bridging ethanolamine–phosphate is through O-6 of a mannose residue in the AChE anchor.

In addition to at least one putative bridging ethanolamine, all GPI anchors for which data exist, except that from VSG, contain additional ethanolamine residues. Human red cell decay accelerating factor (DAF) (Medof et al., 1986; Walter et al., 1987) and scrapie prion (Stahl et al., 1987) GPI anchors contain about three ethanolamine residues per mole, while anchors from human and bovine AChE (Roberts et al., 1985; Haas et al., 1986), human placental alkaline phosphatase (Low et al., 1987; Takami et al., 1988), squid brain Sgp2 (Williams et al., 1988), and rat brain and thymus Thy-1 (Tse et al., 1985; Fatemi et al., 1987) contain about two ethanolamine residues per mole. These values may well be averages, since evidence obtained by FAB-MS analysis of the human AChE anchor indicates that it exists in two forms, one with two ethanolamine phosphate residues per molecule and one with three residues per molecule (Roberts et al., 1988b). No information about the location of the third ethanolamine phosphate residue in the AChE anchor was obtained, but the susceptibility of both the second and third ethanolamine residues to reductive methylation prior to Pronase treatment of the purified protein suggests that neither is substituted (Roberts et al., 1988b). The second ethanolamine phosphate residues of the Thy-1 and AChE anchors appear to be linked to analagous hexosyl residues in their respective core glycans (Table I), but it is not known whether the linkage is through O-2 of a mannosyl residue in the AChE anchor as it is in the Thy-1 anchor.

An interesting feature of the human erythrocyte AChE anchor is palmitoylation of the *myo*-inositol residue at a yet to be identified ring position (Roberts et al., 1988b). It was also demonstrated (Roberts et al., 1988a) that this unusual acylation of inositol is responsible for the resistance to PIPLC of a palmitoylated alkylacyl-PI derived from the intact protein by nitrous acid deamination. This result and the observation that the purified protein is resistant to PIPLC (Roberts et al., 1987) strongly suggest that it is indeed palmitoylation of inositol that confers PIPLC resistance to human erythrocyte AChE. Various acylated forms of dimannosyl-PI that contain extra fatty acids (Brennan & Ballou, 1967, Ballou 1972) have been reported in mycobacteria.

Palmitoylation of inositol might also explain the observed resistance to conversion by PIPLC from membrane-bound to soluble forms of other GPI-anchored proteins, which include human erythrocyte DAF (Davitz et al., 1986; Medof et al., 1986) and *Dictyostelium* antigen 117 (Sadeghi et al., 1988). However, no detailed information on the structures of the GPI anchors of these PIPLC-resistant proteins is available. Nevertheless, the TLC mobility of the nitrous acid deamination product of DAF is identical with that of the corresponding product derived from human erythrocyte AChE (Medof et al., 1986; Walter et al., 1987), suggesting that inositol acylation is responsible for PIPLC resistance in both cases (Roberts et al., 1988a). Interestingly, mouse and human erythrocyte AChE are resistant to PIPLC, but bovine, pig, ox, and rat erythrocyte AChE are not resistant (Low & Finean, 1977; Futerman et al., 1985a,b; Roberts et al., 1987; Haas et al., 1986). The functional significance of differences in the sensitivity of GPI-anchored proteins to enzymes that convert membrane-bound forms to soluble forms is unknown.

Table II: Summary of the Aliphatic Side Chains Found in Selected GPI Membrane Anchors, Related Glycolipids, and *Leishmania* Glycolipids[a]

GPI	R_1		R_2		R_3
	acyl	alkyl	acyl	alkyl	acyl
T. brucei VSG anchor[b]	14:0	–[h]	14:0	–	–
T. brucei P3 and glycolipid C[c]	14:0	–	14:0	–	16:0
T. brucei P2 and glycolipid A[c]	14:0	–	14:0	–	–
human erythrocyte AChE anchor[d]	–	18:0, 18:1	22:4, 22:5, 22:6	–	16:0
Leishmania GPIs A-C[e] and GIPLs **1–4**[f]	–	18:0, 24:0	14:0, 16:0, 18:0	–	–
Leishmania LPG[g] and GIPLs[f] **5** and **6**	–	24:0, 26:0	–	–	–

[a] See Table I for the structure of phosphatidylinositol showing the positions of R_1, R_2, and R_3. More abundant residues are underlined. [b] See Ferguson and Cross (1984) and Schmitz et al. (1986). [c] See Menon et al. (1990), Krakow et al. (1986, 1989), Mayor et al. (1990a,b), and the text. The presence of palmitate as R_3 is based solely on biosynthesis labeling (Mayor et al., 1990b). [d] See Roberts et al. (1988a,b). [e] See Rosen et al. (1989). [f] See McConville and Bacic (1989). Lyso forms of GIPLs 2 and 3 have also been found [McConville et al., (1990) and McConville and Bacic (1990)]. [g] See Orlandi and Turco (1987). [h] Indicates not present.

Treatment of GPI-anchored proteins with *T. brucei* GPI-specific PLC or bacterial PIPLC exposes a carbohydrate epitope known as the cross-reacting determinant (CRD). This epitope has been detected by Western blotting on a number of proteins following PIPLC digestion, including *T. brucei* VSG, *Leishmania major* gp63 surface protease, *Torpedo* and human erythrocyte AChE, the scrapie prion protein, Thy-1, and squid brain Sgp2 glycoprotein [see Zamze et al. (1988) and references cited therein], and cross-reaction is a criterion used for GPI anchor identification. Results of selective chemical and enzymatic modification, combined with a competitive ELISA assay, indicate that three overlapping epitopes are involved: (1) the inositol 1,2-cyclic phosphate (produced by PLC cleavage), (2) the non-N-acetylated glucosamine, and (3) the α-galactose branch (Zamze et al., 1988). Western blotting with anti-CRD antibody gives a positive reaction with anchors lacking α-galactose residues (Zamze et al., 1988), but immunoprecipitation may (Krakow et al., 1986) or may not (Zamze et al., 1988) be successful.

It is obvious from the foregoing discussion that failure to observe solubilization, changes in detergent binding, and/or exposure of the CRD upon PIPLC treatment does not necessarily rule out the presence of a GPI membrane anchor. In this event, where inositol acylation may be inhibiting the enzyme, deacylation with mild base may increase PIPLC susceptibility (Roberts et al., 1988a; Clayton & Mowat, 1989). Biosynthetic labeling with *myo*-inositol and ethanolamine or detection of these diagnostic components of GPIs by chemical analysis also provides compelling evidence of their presence. An additional and very useful criterion is the release of a fatty acid label upon deamination with nitrous acid; non-N-acetylated glucosamine is extremely rare and therefore diagnostic of GPIs. The various fragments that can be generated by chemical methods and have been used by various investigators for the structural analysis of GPI anchors are summarized in Figure 1.

GPI anchors display a great deal of variation in the structures of their glycerol-linked lipid components (Table II). Only myristic acid is found in the PI of the *T. brucei* VSG anchor (Ferguson & Cross, 1984; Gurnett et al., 1986), and the glycerol is in the *sn*-1,2-diacyl 3-phosphate configuration

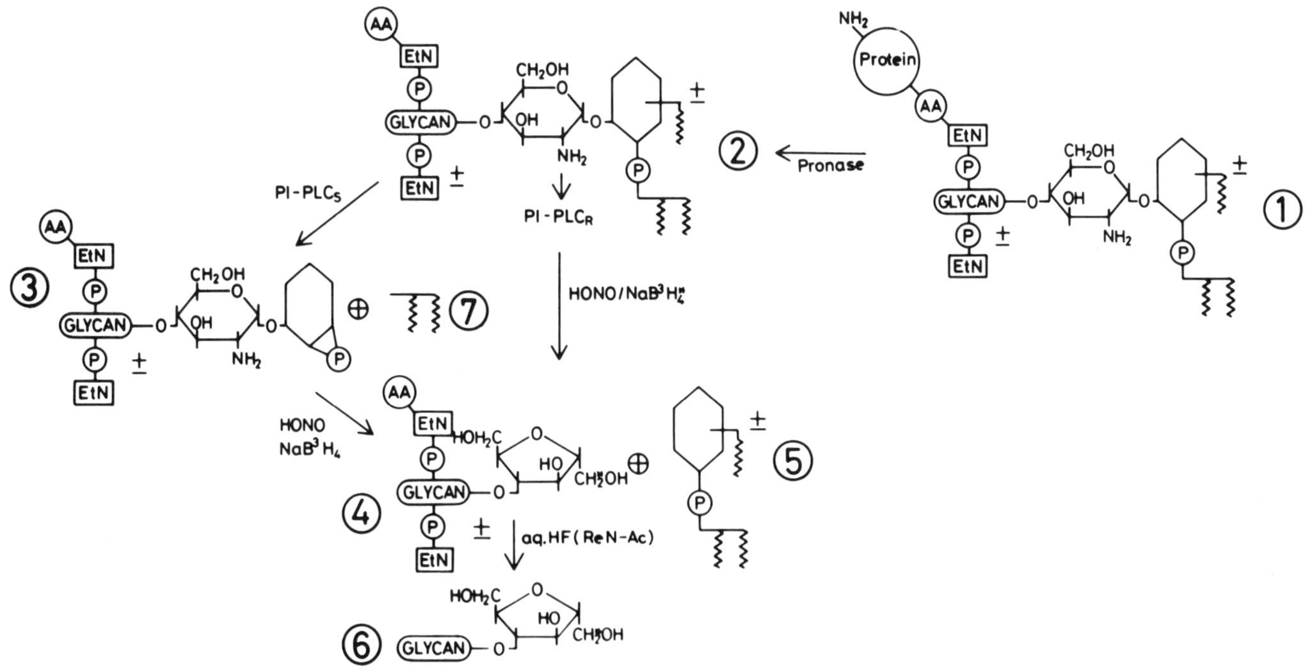

FIGURE 1: Summary of various fragments that have been used for structural determination of GPI. Compound 1 can be metabolically labeled at either the ethanolamine, glucosamine, inositol, or acyl moieties (Brodbeck & Bordier, 1988). Roberts et al. (1988b) used reductive methylation prior to Pronase treatment to label the amine groups of the glucosamine and the extra ethanolamine [see also Brodbeck and Bordier (1988)]. The lipid moieties can also be photolabeled with [^{125}I]TID [3-(trifluoromethyl)-3-(*m*-[^{125}I]iodophenyl)diazirine] (Roberts et al., 1988a). The reductive methylation however makes the GPI resistant to nitrous acid fragmentation. Compound **2** can be treated with nitrous acid to cleave off the inositol phospholipid moiety (compound **5**). The presence of an extra acyl group on the inositol ring makes the GPI anchor resistant to PI-PLC (Roberts et al., 1988b). If the inositol is not acyl substituted, PI-PLC treatment will give compound **3** plus the diacylglycerol moiety **7**. Radiolabeling of compound **2** and **3** can be achieved by treatment with nitrous acid and subsequent Na^3BH_4 reduction to convert the glucosamine to radiolabeled 2,5-dianhydromannitol **4** (Ferguson et al., 1988). The glycan portion **6** can be generated by treatment of compound **4** with HF. If acid-sensitive residues are present such as galactofuranose, then the HF treatment will also cleave at these points (Turco et al., 1989). Compound **3** contains the anti-CRD reactive epitopes for Western blotting of proteins containing a PI-PLC-sensitive GPI (Zamze et al., 1988). PI-PLC-resistant GPI anchors can be blotted after base treatment to cleave the acyl group, making the protein PI-PLC sensitive (Roberts et al., 1988a). Other chemical treatments useful in the structure analysis of GPI anchors include acetolysis, base hydrolysis, deacylation, and reacylation [see Brodbeck and Bordier (1988)]. A variety of analytical techniques have been used in the structural characterization of GPIs. Sequential exoglycosidase digestion combined with Bio-Gel P-4 chromatography is perhaps the most sensitive method (when radiolabeled oligosaccharides are used) and can, by virtue of the exquisite specificity of the exoglycosidases, define the identities (mannose, galactose, glucose, etc.), ring forms (furanose or pyranose), absolute configurations (D or L), anomeric configurations (α or β), in many cases the sequence, and in some cases the position(s) of substitution of glycosyl residues. The appropriate, highly purified enzymes are required for unambiguous results. Combined gas chromatography–mass spectrometry can yield the identity, molar ratios, absolute configurations, ring forms, and position(s) of substitution of most monosaccharide components, as well as the identities and molar ratios of lipid moieties, but provides only very limited information on the arrangement of the components with respect to each other. Fast atom bombardment mass spectrometry (FAB-MS) analysis provides the molecular weight and information on the sequence of glycosyl residues and both the nature and position of non-carbohydrate substituents; however, the identities of glycosyl residues and stereochemical information cannot be obtained by this method alone. NMR spectroscopy can also provide structural information, from composition to solution conformation, but has the drawback of a relative lack of sensitivity (about $^1/_{100}$ that of MS techniques). Combinations, or even all, of these techniques are necessary for a complete and unambiguous structural assignment of GPIs, due to their structural complexity and the different types of components that may be present (carbohydrate, ethanolamine, phosphate, protein, and lipid). Finally, structural analysis on heterogeneous mixtures is the greatest cause of incorrect structural assignments, necessitating separation of the various forms prior to analysis.

(Ferguson et al., 1985). VSGs of *Trypanosoma equiperdum* and *Trypanosoma congolense* can be biosynthetically labeled with [^{3}H]myristic acid but not [^{3}H]palmitic acid (Lamont et al., 1987), which suggests that they have the same fatty acid composition as their *T. brucei* counterpart. On the other hand, both the gp63 surface acid protease of *L. major* and the 195-kDa merozoite antigen of *Plasmodium falciparum* can be labeled with both fatty acids (Haldar et al., 1985). A variety of saturated and unsaturated fatty acids are found [see Ferguson and Williams (1988) and references cited therein]. 1-Alkyl-2-acylglycerol, rather than 1,2-diacylglycerol, is found in human and bovine erythrocyte AChE (Roberts et al., 1987, 1988b) and probably human erythrocyte DAF (Medof et al., 1986). The alkyl substituents are predominantly 18:0 and the acyl substituents 22:4 and 22:5 (Roberts et al., 1988a). Interestingly, an alkylacylglycerol structure has also been proposed for the insulin-sensitive GPI from H35 hepatoma cells (Mato et al., 1987), whereas the insulin-sensitive GPI from BC_3H1 monocytes contains diacylglycerol (Saltiel et al., 1986; Saltiel & Cuatrecasas, 1986). The LPGs and GPIs/GIPLs of *Leishmania* parasites (see below) also contain long-chain alcohols in ether linkage to glycerol. Finally, there is a report that the contact site A glycoprotein of *Dictyostelium* contains a ceramide-based phospholipid anchor (Stadler et al., 1989).

Some species of the parasitic protozoan *Leishmania* synthesize glycolipids with core structures resembling those of the GPIs that anchor membrane proteins. The GPI core of the *Leishmania donovani* LPG (Turco et al., 1989) resembles the GPI membrane protein anchors by virtue of possessing the sequence α-Man*p*-(1→4)-α-GlcNH_2*p*-(1→6)-*myo*-inositol-PO_4 (Table I). Beyond this conserved monosaccharide sequence, however, the carbohydrate portions of this protozoan glycolipid and the protein-linked GPI membrane anchors diverge completely, with the next mannosyl residue linked α(1→3) and α(1→6), respectively. The LPG of *L. donovani* also contains a polysaccharide made up of a repeating phosphorylated disaccharide unit (average of 16 repeats), 6-*O*-PO_4-β-Gal*p*-(1→4)-α-Man*p*-(1→ (Turco et al., 1987), that

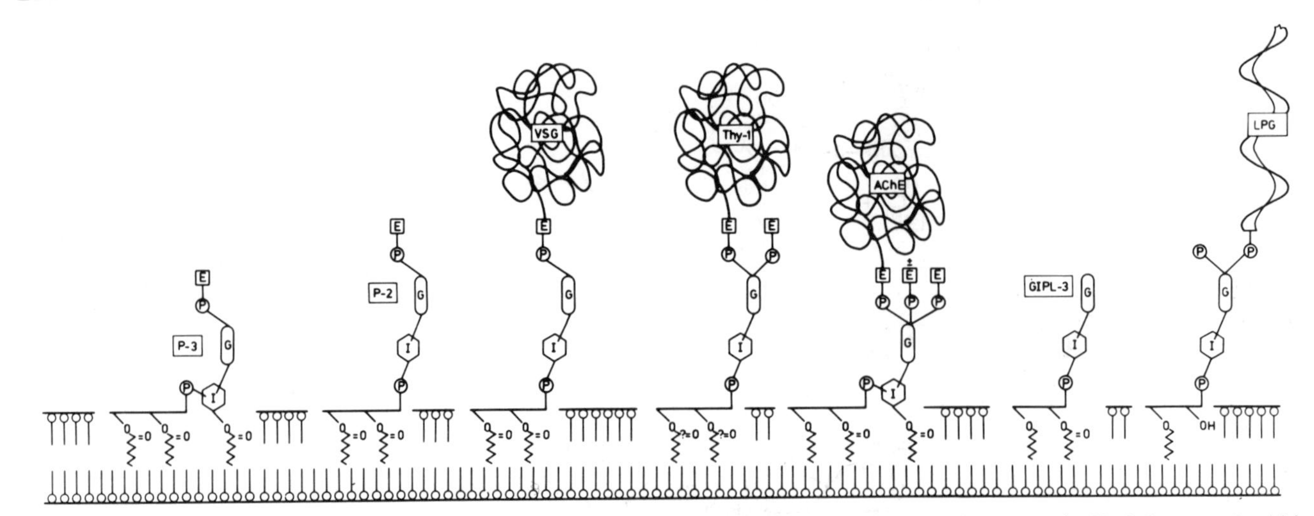

FIGURE 2: Schematic representation of the various forms of GPI anchors and related glycolipids so far reported. No inferences should be made as to the exact orientation of the GPI anchor in the membrane nor the conformational relationship between the protein (or polysaccharide) and the GPI anchor. The glycans (G) differ between the various structures and are tabulated in Table I. The lipids present on each of the molecules are presented in Table II. For Thy-1 it is not known whether the aliphatic side chains are alkyl or acyl. AChE exists in two forms, one with three ethanolamine phosphates and the other with two. The points of attachment of the phosphates (P) are represented schematically, and it should not be inferred that they are linked to the glycans (G) at a single residue (see Table I).

is linked to the glycolipid consisting of 6-*O*-PO_4-α-Gal*p*-(1→6)-α-Gal*p*-(1→3)-α-Gal*f*-(1→3)-α-Man*p*-(1→3)-α-Man*p*-(1→4)-α-GlcNH_2*p*-(1→6)-lysoalkyl-PI (Turco et al., 1989). The mannosyl residue to which the galactofuranosyl residue is linked in the LPG core bears a phosphate attached to O-6 (S. J. Turco, J. R. Thomas, J. Thomas-Oates, R. A. Dwek, and T. W. Rademacher, unpublished results). *L. major* also produces an LPG that contains mannose, galactose, arabinose, and glucose as phosphorylated tri- and tetrasaccharides (McConville et al., 1987) rather than the phosphorylated β-Gal-(1→4)-α-Man disaccharide repeats of the *L. donovani* LPG (Table I); however, the *L. major* LPG contains an internal galactofuranosyl residue in its glycan core (McConville & Bacic, 1990) and the same lipid moieties (see below) as its *L. donovani* counterpart.

L. major also produces glycolipids, termed GPIs (Rosen et al., 1989) or GIPLs (glycoinositol phospholipids) (McConville & Bacic, 1989, 1990), that contain an oligosaccharide of variable size linked to PI (Table I). The published structures of the GPIs from *L. major* strain LRC-L137 (Rosen et al., 1989) differ most notably from the *L. donovani* LPG structure in two respects: the absence and presence, respectively, of a 3-substituted galactofuranosyl residue and the replacement of the 6-substituted α-galactosyl residue in the LPG with a 3-substituted α-galactosyl residue at the homologous position in the GPI. Structures of the GIPLs from *L. major* strains V121 and LRC-L119 (McConville & Bacic, 1989, 1990; McConville et al., 1990) are shown in Table I. All of the GIPLs contain a 3-substituted galactofuranosyl residue, and the terminal α-galactosyl residue of GIPL-3 is linked α(1→6), which makes it structure homologous to that of the *L. donovani* LPG core. It remains to be determined whether the differences in the structures of *L. major* GPIs and GIPLs result from the use of different subclones of the same variant or from the interpretation of results.

Further divergence between the protozoan glycoconjugates (LPGs and GIPs/GPILs) and GPI anchors of membrane proteins is seen in their lipid portions (Table II). The *L. major* GPIs (Rosen et al., 1989) and some of the *L. major* GIPLs contain 1-alkyl-2-acylglycerol (Rosen et al., 1989; McConville & Bacic, 1989). As noted above, some protein GPIs have been found to contain 1-alkyl-2-acylglycerol. The lipid portions of both *L. donovani* LPG and some of the *L. major* GIPLs are even more unusual, containing a lyso-1-*O*-alkyl-PI (Orlandi & Turco, 1987; McConville et al., 1987; McConville & Bacic, 1989). These novel lipids contain only one ether-linked alkyl chain and no ester-linked fatty acid (Table I). Di-*O*-alkyl-PIs have been isolated from *Leishmania mexicana mexicana* (Singh et al., 1988); however, no such lipids bearing either oligosaccharide, polysaccharide, or protein have been isolated.

In summary, a variety of glycolipid structures based on PI have been characterized (Table I and Figure 2). The GPI membrane protein anchors contain a conserved core glycan to which are attached various carbohydrate side chains and ethanolamine phosphate(s). One of the ethanolamine phosphates links the protein to the glycolipid via amide and phosphodiester linkages. Acylation of the inositol in GPI anchors confers resistance to digestion by bacterial PIPLC. *Leishmania* glycolipids (LPGs, GPIs/GIPLs) appear to share a portion of the conserved core glycan of reported GPI membrane protein anchors, but beyond this α-Man*p*-(1→4)-α-Glc*Np*-(1→6)-*myo*-inositol sequence, the structures diverge. *L. donovani* LPG contains a small, highly phosphorylated polysaccharide attached to its glycan core, while *L. major* GPIs/GIPLs contain only a few glycosyl residues. Further investigation is required before any conclusion concerning the roles of the *L. major* GPIs/GIPLs as biosynthetic intermediates in either LPG or GPI membrane protein anchors can be inferred. It has been suggested (McConville & Bacic, 1990; McConville et al., 1990) that GIPL-3 may function as the biosynthetic precursor of LPG after undergoing selective deacylation to give the lyso-1-*O*-alkyl-PI analogue present in LPG. Since the structure of a GPI anchoring a membrane protein from *L. major* or *L. donovani* has yet to be reported, no conclusions can be made concerning the relationship between GPI anchors, GPIs, GIPLs, and LPGs. Preliminary data, however, suggest that the GIPLs are not precursors for membrane anchors of cell surface proteins (McConville et al., 1990).

BIOSYNTHESIS OF GPIs

Fully processed proteins containing GPI anchors lack a C-terminal sequence that is present in the immature, anchorless form [see Ferguson and Williams (1988) and references cited therein]. For example, nascent VSG contains a 20 amino acid C-terminal peptide, which is replaced with the

GPI anchor (Boothroyd et al., 1981). It is not known whether removal of the C-terminal sequence occurs before or simultaneously with addition of the anchor, but presumably this sequence functions as the signal for GPI addition. Fusion of the C-terminal domains of GPI-anchored Qa-2 antigen (Stroynowski et al., 1987; Waneck et al., 1988) or DAF (Caras et al., 1987) to proteins that are normally secreted or transmembrane results in the addition of GPI to the fusion proteins. All known GPI-anchored proteins are synthesized with a 10–20-residue hydrophobic domain at the C-terminus, which appears to be a requisite component of the addition signal. The hydrophobic domain must be a minimum length (Berger et al., 1988), but there seem to be no stringent sequence requirements, as replacement with other hydrophobic domains results in correct processing (Caras & Weddell, 1989). On the other hand, the degree of hydrophobicity may be important, since a single base change resulting in substitution of Asp with Val changes GPI-anchored Qa-2 to an apprently integral membrane protein (Waneck et al., 1988). In addition to a hydrophobic domain, information within the adjacent 20 residues is also important. Caras et al. (1989) found that either residues 291–310 or 311–330 of DAF could direct GPI addition but that other, nonspecific sequences could not. The cleavage/attachment site of DAF is not known; so far, Ala, Asn, Asp, Cys, Gly, and Ser have been identified as sites in other proteins [see Low (1989) and references cited therein]. Thus, the exact nature of the required signal within the 20 amino acid sequence adjacent to the hydrophobic domain is not known.

The rapidity with which GPI is added to proteins (1–5 min) (Bangs et al., 1985; Krakow et al., 1986; Ferguson et al., 1986; Conzelman et al., 1987; Takami et al., 1988) suggests that at least a portion of it exists in a preassembled form that is added en bloc. Two putative precursor glycolipids, termed glycolipid A (Krakow et al., 1986) and P2 (Menon et al., 1988), have been isolated from trypanosomes. Glycolipid A can be labeled by [^{3}H]myristrate but not by [^{3}H]palmitate and liberates PI when treated with nitrous acid and dimyristylglycerol when treated with PIPLC, and [^{3}H] mannose label can be immunoprecipitated by anti-CRD antibodies after PIPLC treatment (Krakow et al., 1986). P2 contains mannose, glucosamine with a free amino group, phosphate, and myristic acid, releases dimyristyl-PI upon nitrous acid deamination, and is susceptible to PIPLC (Menon et al., 1988). The structure of the deaminated, reduced, and phosphorylated glycan from P2 has been found to be α-Man-(1→2)-α-Man-(1→6)-α-Man-(1→?)-2,5-anhydromannitol (Mayor et al., 1990a), which suggests that its structure is the same as that of the conserved core region of the VSG and Thy-1 anchors (Table I). It is highly likely that glycolipids A and P2 are the same molecule, but detailed structural information on the glycan of glycolipid A is needed to confirm the identity of the two molecules. Although the two glycolipids possess structural features consistent with a precursor role and the kinetics of labeling of glycolipid A suggest that it is a metabolic intermediate (Krakow et al., 1986), there is no direct evidence that they are involved in GPI anchor formation.

A pathway for the biosynthesis of the putative glycolipid precursor of the membrane anchor of VSG has been proposed on the basis of results obtained with trypanosome cell free systems. The putative precursor is built up from PI by the sequential addition of a GlcNAc residue, three mannosyl residues, and finally ethanolamine phosphate to give ethanolamine-P-Man$_3$GlcNH$_2$-PI (Masterson et al., 1989; Menon et al., 1990). Doering et al. (1989) have obtained evidence suggesting that the GlcNH$_2$ residue is derived from GlcNAc, which is donated to PI by UDP-GlcNAc and subsequently de-N-acetylated. Addition of one or more mannosyl residues may involve dolichol-P-Man as the donor species. Treatment of trypanosomes with 2-fluoro-2-deoxy-D-glucose, an inhibitor of dolichol-P-Man synthesis in chick fibroblasts, inhibits incorporation of radiolabeled precursors into the putative precursor (Schwarz et al., 1989). Furthermore, a mutant lymphoma cell line that is deficient in dolichol-P-Man synthesis is also deficient in cell-surface expression of GPI-anchored Thy-1 (Conzelmann et al., 1986; Fatemi et al., 1987). Preliminary evidence (Conzelmann et al., 1990) indicates that the only available secretion and glycosylation mutant of *Saccharomyces cerevisiae* that does not produce GPI-anchored proteins is *sec* 53, which does not produce phosphomannomutase; although this suggests that GDP-mannose is also required for the biosynthesis of GPI anchors, it may only be needed for dolichol-P-Man synthesis.

Preliminary results (Masterson et al., 1989; Doering et al., 1990; Menon et al., 1990) suggest that the putative glycolipid precursors of GPI anchors undergo fatty acid remodeling. Myristate present in the mature glycolipid species is thought to replace more hydrophobic fatty acids at a late stage in the biosynthesis. Consistent with this scheme, the PI substrate for GlcNAc addition appears to be a heterogeneously acylated lipid rather than dimyristyl-PI (Menon et al., 1990), and [^{3}H]myristrate label is incorporated only into the mature glycolipids (Masterson et al., 1989; Menon et al., 1990). Myristoyl coenzyme A appears to be the donor for replacement at the *sn*-2 position (Doering et al., 1990), but little else is known of the remodeling system.

In addition to the PIPLC-sensitive glycolipids (glycolipids A and P2) described above, trypanosomes synthesize similar glycolipids that are resistant to the enzyme, termed either glycolipid C (Krakow et al., 1986) or glycolipid P3 (Menon et al., 1988). P3 and probably glycolipid C have the same glycan linked to PI as their enzyme-sensitive counterparts, and acylation of the inositol moiety appears to be the sole difference between the sensitive and resistant molecules (Mayor et al., 1990a,b; Krakow et al., 1989). Biosynthesis of the inositol-acylated, PIPLC-resistant glycolipids in trypanosomes has been investigated by Cross and co-workers using a cell-free system (Menon et al., 1990). The donor of the inositol-linked fatty acid does not appear to be a fatty acyl coenzyme A, and inositol-acylated GlcNH$_2$-PIs were observed. If inositol-acylated glycolipids are important in the assembly of GPI anchors, they could (1) make PI-containing glycolipids better substrates for glycosyltransferases, (2) protect mannosyl-PIs from digestion by GPI-specific PLC, and/or (3) effect the transbilayer distribution of mannosyl-PIs (Menon et al., 1990). In the case of P3, biosynthetic labeling suggests that the inositol is palmitoylated (Mayor et al., 1990b). Inositol palmitoylation is the same modification that renders the GPI anchor of human erythrocyte AChE resistant to PIPLC (Roberts et al., 1988a,b; see above). A palmitoylated, PIPLC-resistant glycolipid could serve as the precursor of a PIPLC-resistant, GPI-anchored cell-surface glycoprotein synthesized by procyclic (insect stage) trypanosomes (Clayton & Mowatt, 1989). It remains to be determined whether PIPLC-resistant glycolipids that could serve as precursors to proteins with resistant GPI anchors are synthesized by other eukaryotes.

Bangs et al. (1988) have obtained evidence suggesting that the galactosyl residues in the VSG anchor are added subsequent to GPI as an addition to the protein. They showed that the C-terminal glycopeptide from a mature 59-kDa form of

VSG can be converted by treatment with α-galactosidase to a form that coelutes from Bio-Gel P-6 with the glycopeptide isolated from an immature 58-kDa form. Although VSG appears to travel to the cell surface via the classical eukaryotic transport route through the Golgi apparatus (Duszenko et al., 1988), the intracellular site at which galactosylation occurs is not known. Since biosynthetic studies have been carried out only with trypanosomes, the timing and site of addition of the GalNAc and fourth mannosyl residues to the Thy-1 anchor are not known. Likewise, no conclusions can be made about the addition of further ethanolamine phosphates to those authors with more than one such residue (see above), since trypanosome VSG contains only one ethanolamine phosphate.

Fasel et al. (1989) have recently demonstrated selective incorporation of phospholipase-sensitive anchoring moieties into Thy-1 and DAF during in vitro translocation experiments. This indicates that rough microsomes are able to support and regulate GPI anchor incorporation and that all components for GPI-anchor processing are contained in this cell fraction. The availability of this cell-free translation system containing all the ingredients necessary for GPI addition will be useful as a experimental model to define the precursors and enzymes involved in the mammalian biosynthetic pathway.

Function of GPIs

No generalization can be made so far about the roles or properties of proteins that have GPI anchors. Most seem to be located at the cell surface, but one (the major protein of the pancreatic zymogen granule membrane; LeBel & Beattie, 1988) has been detected elsewhere in the cell. They vary in size from 10 to 300 kDa (Low, 1989) and fall into diverse functional groups, including hydrolytic enzymes (e.g., alkaline phosphatase, AChE, and the *Leishmania* 63-kDa surface protease), cell adhesion molecules (e.g., N-CAM, LFA-3, and *Dictyostelium* contact site A), and protective coat proteins (e.g., *Trypanosoma* VSG). The physiological roles of many GPI-anchored proteins are not yet known (e.g., scrapie prion protein, *Saccharomyces* 125-kDa glycoprotein, and mammalian antigens such as Thy-1, Qa-2, and carcinoembryonic antigen).

Phospholipid-containing membrane proteins from *S. cerevisiae* are >90% susceptible to the action of PIPLC (Conzelmann et al., 1990). However, there is no loss of the detergent-binding moiety upon PIPLC treatment for a large number of the proteins. These authors have proposed that many of the GPI-containing membrane proteins may have other hydrophobic, detergent- (membrane-) binding domains. These observations may be fundamental to our current perception of the role of GPI anchors as the sole membrane anchor of the proteins to which they are attached. The lack of hydrophobic amino acids near the C-terminal in proteins containing GPI anchors has been put forward as evidence that membrane attachment of these proteins is due entirely to the acyl chains of the phosphatidylinositol (Low & Kincade, 1985). Obviously other hydrophobic protein domains may still be present to interact with the lipid bilayer. Flow-cytometry reduction of intensity after PIPLC treatment of intact cells cannot be interpreted as release of a protein from the cell surface. Both internalization following PIPLC cleavage and a change in reactivity to the fluorescently labeled antibody to the protein under study would give a similar experimental result. Further, unless absolute calibration of fluorescent intensity and copy number is performed, these flow-cytometry experiments cannot be interpreted. Such multipoint attachment to membrane has been previously proposed for acylated (palmitoylated) proteins such as rhodopsin and the β_2-adrenergic receptor (Ovchinnikov et al., 1988; O'Dowd et al., 1989). In both of these molecules detachment from the acyl group leads to conformational and functional changes in the protein without release from the membrane. Figure 3 illustrates how such an arrangement may be useful in maintaining either an active or inactive conformation. Phospholipase action would be an on/off switch in this model. A further mechanism of multipoint anchoring can be envisaged for proteins that have their hydrophobic amino acid sequence (membrane spanning) near the N-terminus rather than near the C-terminus (i.e., the C-terminus is extracellular).

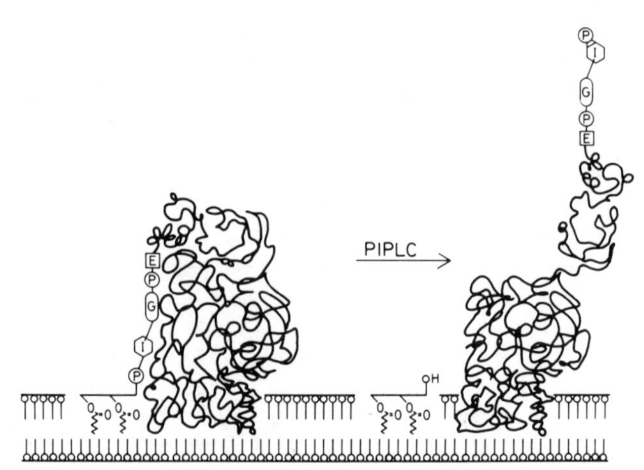

FIGURE 3: Hypothetical alternative function of GPI anchors. The inability to remove the detergent-binding (hydrophobic) domains of some GPI-containing proteins (see text) suggests that they may not be their sole membrane anchor and that the GPI moiety may function to tie down domains that can conformationally be modulated by the cleavage of the GPI anchor by PIPLC.

The physical properties imparted to proteins via GPIs may be crucial to their functioning. For example, it is easy to envisage that the mediation of cell adhesion by surface molecules such as N-CAM, LFA-3, and *Dictyostelium* contact site A is dependent upon, or at least affected by, their mode of membrane attachment. Proteins with GPI anchors do have higher lateral diffusion coefficients and different solubilities in detergents than do proteins that span the membrane [see Low (1987, 1989b) and Ferguson and Williams (1988) and references cited therein]. The GPI glycan of VSG has been proposed to serve a space-filling role, on the basis of the determination of its average solution conformation (Homans et al., 1989). The VSG glycan is proposed to lie along the plane of the membrane and span an area of 600 Å^2, which is comparable to the cross-sectional area of the protein's N-terminal domain (Figure 4), and could, therefore, be important in dimerization and/or the maintenance of the VSG coat as a diffusion barrier. Several proteins [e.g., the IgG receptor (CD16), LFA-3, N-CAM, AChE, and Leu 8/TQI] have been found to exist as GPI anchored, as well as transmembraneous [Camerini et al., (1989), Scallon et al. (1989), and Ferguson and Williams (1988) and references cited therein]. It remains to be determined whether functional differences result from the expression of alternatively anchored forms of the same protein.

The existence of phospholipases that can cleave GPI membrane anchors has led to speculation that release of membrane-bound proteins via this mechanism is of physiological significance. For example, degradation of the GPI anchor of GP-2 in the pancreatic zymogen granule has been proposed to lead to its release from the granule membrane (LeBel & Beattie, 1988); however, the enzyme responsible has yet to be identified. A GPI-PLC (so-called because of its specificity for GPI anchors and related molecules) has been isolated from

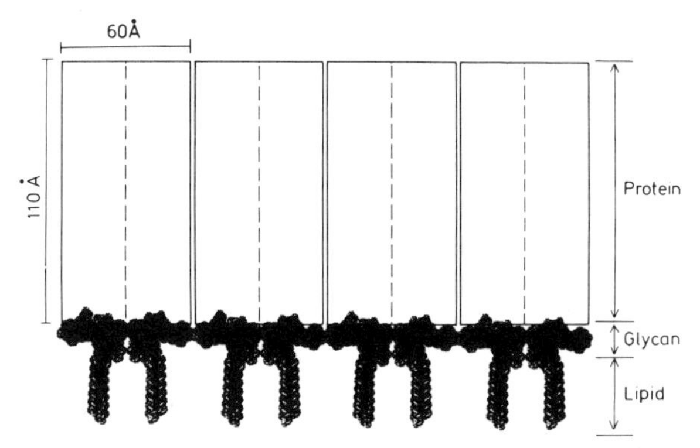

FIGURE 4: Two VSG glycan anchors can be accommodated by a rectangle of 30 Å × 50 Å (diagonal of 56 Å). The N-terminal two-thirds of a VSG protein (dimer) can be approximated as a 110-Å columnar structure (max diameter 60 Å) including a cap region which can be accommodated with a rectangle of 30 Å × 50 Å. No structural data are available concerning the C-terminal portion of the molecule. The figure was drawn with the dimensions given above to illustrate the relative widths of the VSG glycoprotein (monomer and dimer) relative to the GPI anchors. It should be noted, however, that a considerable amount of surface of the protein may still interact directly with the membrane. Preliminary data on the interaction between the Thy-1 peptide and its glycan anchor suggest that the glycan part of the GPI anchor may fit up into a pocket in the protein with only the terminal part of the lipids extending, thus allowing for extensive direct interactions between the Thy-1 peptide and the membrane surface (C. Edge, R. A. Dwek, and T. W. Rademacher, unpublished results).

T. brucei and rat liver, and a GPI-PLD is found in serum [see Ferguson and Williams (1988) and Low (1989) and references cited therein]. The *T. brucei* enzyme could function in the shedding of VSG during conversion of the bloodstream form to the coatless, insect-dwelling form of the organism (Bulow & Overath, 1985), and a model involving GPI-PLC-containing vesicles and both endo- and exocytosis has been proposed (Ferguson & Williams, 1988).

While VSG is being shed from the trypanosome surface, another surface protein (termed either procyclin or PARP) is expressed, and both VSG and PARP are for a time present simultaneously on the cell surface (Roditi et al., 1989). PARP is PIPLC resistant, and the differential susceptibility of PARP and VSG anchors could allow the selective release of VSG and retention of PARP during the life-cycle transformation (Clayton & Mowatt, 1989). Many GPI anchors are resistant to bacterial PIPLC (discussed above). Resistance to specific phospholipases could be a general mechanism whereby the release of GPI-anchored proteins is controlled. No unequivocal data, however, have been put forward that demonstrate that these proteins are in fact released by the action of a phospholipase.

The successful invasion and infection of hosts by *Leishmania* may depend on the presence of LPG and GIPLs on its surface. These parasites must survive in the harsh environments of the alimentary tract within the sandfly vector and of the phagolysosomes of infected mammalian host macrophages. LPG is the major cell-surface glycoconjugate of *Leishmania*, and protection probably is afforded by the presence or large amounts of LPG serving as a protective barrier against hydrolytic enzymes. In addition, both LPG and GIPLs from *Leishmania* have been shown to be efficient inhibitors of protein kinase C (McNeely et al., 1989) and may therefore prevent induction of the microbicidal oxidative burst within macrophage phagolysosomes (Turco, 1988a). LPG may also inhibit the oxidative burst by chelating intracellular calcium or other divalent cations and/or by scavenging oxygen free radicals [Chan et al. (1989) and Turco (1988b) and references cited therein]. It has been suggested that, besides having a protective role, LPG is also involved in attachment to and penetration of the host macrophage (Handman & Goding, 1985; Puentes et al., 1988). It is not known at present whether GPI anchors also have roles similar to those of LPGs and GIPLs.

Results from a large number of studies suggest that both GPI-anchored proteins and glycolipids that are structurally related to GPI anchors are involved in cell activation. Two of the more interesting cases where GPIs may be involved in transduction of extracellular signals are (1) the mitogenic effect of antibodies directed against GPI-anchored proteins on T-lymphocytes [see Robinson and Spencer (1988) and references cited therein; Robinson et al., 1989] and (2) the possible role of a GPI anchor or related glycolipids as a mediator of insulin action (Low & Saltiel, 1988). Unfortunately, more information is needed before any meaningful conclusions about the importance of GPIs in these processes can be made.

GPI membrane anchors appear to act as a targeting signal in Madin–Darby canine kidney (MDCK) cells, which form a polarized monolayer at confluency with an apical and a basolateral surface. Digestion with PIPLC released six proteins from the apical surface but none from the basolateral surface (Lisanti et al., 1988). Results of experiments in which fusion proteins are expressed in transfected cells show that a GPI anchor can direct proteins to the apical surface. Thus, GPI anchoring of a normally basolateral protein led to its apical expression, and replacement of the GPI anchor of an apical protein with the membrane-spanning domain of a basolateral protein led to its basolateral expression (Brown et al., 1989). Likewise, transfer of the GPI-addition signal of an apical membrane protein to either a basolateral or secreted protein resulted in apical expression of the fusion proteins (Lisanti et al., 1989). The mechanism by which GPIs influence intracellular trafficking of proteins in MDCK cells is unknown, but it has been suggested (Brown et al., 1989; Lisanti et al., 1989) that the sorting of glycosphingolipids and GPI-anchored proteins may be linked.

SUMMARY

The last few years have witnessed an explosion in our knowledge of GPI membrane anchors and related glycolipids and molecules where structure details are available, as illustrated in Figure 2. There is now sufficient information on a handful of these molecules to allow a detailed comparison of their chemical structures (Table I). Despite a common structural theme, i.e., the presence of mannoglucosaminyl-PI, a great deal of diversity exists in both the glycan structures and the glycerol-linked aliphatic substituents. The complexities of the structures clearly show that a multitechnique approach is required in the elucidation of their structures. The anticipated publication of more structures from a wider range of organisms may reveal even greater diversity, as well as suggesting possible biosynthetic pathways. The details of a potential biosynthetic pathway in *T. brucei* are becoming apparent, but confirmation of its importance awaits the isolation and characterization of the enzymes involved. *Leishmania*, in which LPG, GPIs, GIPLs, and GPI membrane anchors are produced, may also provide an interesting system for biosynthetic studies. The recent description of a GPI biosynthetic system in yeast may provide the crucial breakthroughs necessary in unraveling the enzymes and sugar donors involved in the biosynthetic pathway and possibly the role of the GPI membrane anchor in the functions of proteins containing these moieties.

Knowledge of the solution structure (conformation), in addition to the complete chemical structure, of the *T. brucei*

VSG anchor has led to speculation that the glycan fulfills a space-filling role in the VSG coat. Many other possible roles of GPI membrane anchors have been suggested, including the shedding and turnover of membrane proteins, signal transduction, and intracellular targeting. Nevertheless, the only function of GPIs that we can so far be certain of is that they anchor proteins or polysaccharide to a membrane. Regardless of the roles GPIs may or may not ultimately be shown to play, the fact that such a widely occurring structure has only recently been characterized serves as a reminder of the incompleteness of our knowledge of biological phenomena and the constant possibility of finding novel molecules in obvious places.

ACKNOWLEDGMENTS

The Oxford Glycobiology Unit is supported by Monsanto Co. We thank Professor G. Cross, Rockefeller University, Dr. Andreas Conzlemann, Lausanne University, and Dr. Malcolm McConville, Dundee University, for providing us with preprints of their work.

REFERENCES

Ballou, C. E. (1972) *Methods Enzymol. 28*, 493–500.

Bangs, J. D., Herald, D., Krakow, J. L., Hart, G. W., & Englund, P. T. (1985) *Proc. Natl. Acad. Sci. U.S.A. 82*, 3207–3211.

Bangs, J. D., Doering, T. L., Englund, P. T., & Hart, G. W. (1988) *J. Biol. Chem. 263*, 17697–17705.

Berger, J., Howard, A. D., Brink, L., Gerber, L., Hauber, J., Cullen, B. R., & Udenfriend, S. (1988) *J. Biol. Chem. 263*, 10014–10021.

Boothyroyd, J. C., Paytner, C. A., Cross, G. A. M., Bernards, A., & Borst, P. (1981) *Nucleic Acids Res. 9*, 4735–4743.

Brennan, P., & Ballou, C. E. (1967) *J. Biol. Chem. 242*, 3046–3056.

Brodbeck, U., & Bordier, C., Eds. (1988) *Post-Translational Modifications of Proteins by Lipids*, Springer-Verlag, Berlin.

Brown, D. A., Crise, B., & Rose, J. K. (1989) *Science 245*, 1499–1501.

Bulow, R., & Overath, P. (1985) *FEBS Lett. 187*, 105–110.

Camerini, D., James, S. P., Stamenkovic, I., & Seed, B. (1989) *Nature 342*, 78–82.

Caras, I. W., & Weddell, G. N. (1989) *Science 243*, 1196–1198.

Caras, I. W., Weddell, G. N., Devitz, M. A., Nussenzweig, V., & Martin, D. W., Hr. (1987) *Science 238*, 1280–1283.

Caras, I. W., Weddell, G. N., & Williams, S. R. (1989) *J. Cell Biol. 108*, 1387–1396.

Chan, J., Fujiwara, T., Brennon, P., McNeil, M., Turco, S. J., Sibille, J.-C., Snapper, M., Aisen, P., & Bloom, B. R. (1989) *Proc. Natl. Acad. Sci. U.S.A. 86*, 2453–2457.

Clayton, C. E., & Mowatt, M. R. (1989) *J. Biol. Chem. 264*, 15088–15093.

Conzelmann, A., Spiazzi, A., Hyman, R., & Bron, C. (1986) *EMBO J. 5*, 3291–3296.

Conzelmann, A., Spiazzi, A., & Bron, C. (1987) *Biochem. J. 246*, 605–610.

Conzelmann, A., Fankhauser, C., & Desponds, C. (1990) *EMBO J. 9*, 653–661.

Cross, G. A. M. (1987) *Cell 48*, 179–181.

Davitz, M. A., Low, M. G., & Nussenzweig, V. (1986) *J. Exp. Med. 163*, 1150–1161.

Doering, T. L., Masterson, W. J., Englund, P. T., & Hart, G. W. (1989) *J. Biol. Chem. 264*, 11168–11173.

Doering, T. L., Masterson, W. J., Englund, P. T., & Hart, G. W. (1990) *J. Biol. Chem. 265*, 611–614.

Duszenko, M., Ivanov, I. E., Ferguson, M. A. J., Plesken, H., & Cross, G. A. M. (1988) *J. Cell Biol. 106*, 77–86.

Fasel, P., Rousseaux, M., Schaerer, E., Medof, M. E., Tykocinski, M. L., & Bron, C. (1989) *Proc. Natl. Acad. Sci. U.S.A. 86*, 6858–6862.

Fatemi, S. H., Haas, R., Jentoft, N., Rossenberry, T. L., & Tartakoff, A. M. (1987) *J. Biol. Chem. 262*, 4728–4732.

Ferguson, M. A. J., & Cross, G. A. M. (1984) *J. Biol. Chem. 259*, 3011–3015.

Ferguson, M. A. J., & Williams, A. F. (1988) *Annu. Rev. Biochem. 57*, 285–320.

Ferguson, M. A. J., Haldar, K., & Cross, G. A. M. (1985) *J. Biol. Chem. 260*, 4963–4968.

Ferguson, M. A. J., Duszenko, M., Lamont, G. S., Overath, P., & Cross, G. A. M. (1986) *J. Biol. Chem. 261*, 356–362.

Ferguson, M. A. J., Homans, S. W., Dwek, R. A., & Rademacher, T. W. (1988) *Science 239*, 753–759.

Futerman, A. H., Low, M. G., Ackermann, K. E., Sherman, W. R., & Silman, I. (1985a) *Biochem. Biophys. Res. Commun. 129*, 312–317.

Futerman, A. H., Low, M. G., Michaelson, D. M., & Silman, I. (1985b) *Neurochemistry 45*, 1487–1494.

Gurnett, A. M., Ward, J., Raper, J., & Turner, M. J. (1986) *Mol. Biochem. Parasitol. 20*, 1–13.

Haas, R., Brandt, P. T., Knight, J., & Rosenberry, T. L. (1986) *Biochemistry 25*, 3098–3105.

Haldar, K., Ferguson, M. A. J., & Cross, G. A. M. (1985) *J. Biol. Chem. 260*, 4969–4974.

Handman, E., & Goding, J. W. (1985) *EMBO J. 4*, 329–336.

Holder, A. A. (1983) *Biochem. J. 209*, 261–262.

Homans, S. W., Ferguson, M. A. J., Dwek, R. A., Rademacher, T. W., Anand, R., & Williams, A. F. (1988) *Nature 333*, 269–272.

Homans, S. W., Edge, C. J., Ferguson, M. A. J., Dwek, R. A., & Rademacher, T. W. (1989) *Biochemistry 28*, 2881–2887.

Krakow, J. L., Herald, D., Bangs, J. D., Hart, G. W., & Englund, P. T. (1986) *J. Biol. Chem. 261*, 12147–12153.

Krakow, J. L., Doering, T. L., Masterson, W. J., Hart, G. W., & Englund, P. T. (1989) *Mol. Biochem. Parasitol. 36*, 263–270.

Lamont, G. S., Fox, J. A., & Cross, G. A. M. (1987) *Mol. Biochem. Parasitol. 24*, 131–136.

LeBel, D., & Beattie, M. (1988) *Biochem. Biophys. Res. Commun. 154*, 818–823.

Lisanti, M. P., Sargiacomo, M., Graeve, L., Saltiel, A. R., & Rodriguez-Boulan, E. (1988) *Proc. Natl. Acad. Sci. U.S.A. 85*, 9557–9561.

Lisanti, M. P., Caras, I. W., Davitz, M. A., & Rodriguez-Boulan, E. (1989) *J. Cell Biol. 109*, 2145–2156.

Low, M. G. (1987) *Biochem. J. 244*, 1–13.

Low, M. G. (1989) *FASEB J. 3*, 1600–1608.

Low, M. G., & Finean, J. B. (1977) *FEBS Lett. 82*, 143–146.

Low, M. G., & Kincade, P. W. (1985) *Nature 318*, 62–64.

Low, M. G., & Saltiel, A. R. (1988) *Science 239*, 268–275.

Masterson, W. J., Doering, T. L., Englund, P. T., & Hart, G. W. (1989) *Cell 56*, 793–800.

Mato, J. M., Kelly, K. L., Abler, A., & Jarett, L. (1987) *J. Biol. Chem. 262*, 2131–2137.

Mayor, S., Menon, A. K., Cross, G. A. M., Ferguson, M. A. J., Dwek, R. A., & Rademacher, T. W. (1990a) *J. Biol. Chem.* (in press).

Mayor, S., Menon, A. K., & Cross, G. A. M. (1990b) *J. Biol. Chem.* (in press).

McConville, M. J., & Bacic, A. (1989) *J. Biol. Chem. 264*, 757–766.
McConville, M. J., & Bacic, A. (1990) *Mol. Biochem. Parasitol. 38*, 57–68.
McConville, M. J., Bacic, A., Mitchell, G. F., & Handman, E. (1987) *Proc. Natl. Acad. Sci. U.S.A. 84*, 8941–8945.
McConville, M. J., Homans, S. W., Thomas-Oates, J. E., Dell, A., & Bacic, T. (1990) *J. Biol. Chem.* (in press).
McNeely, T. B., Rosen, G., Londner, M. V., & Turco, S. J. (1989) *Biochem. J. 259*, 601–604.
Medof, M. E., Walter, E. I., Roberts, W. F., Haas, R., & Rosenberry, T. L. (1986) *Biochemistry 25*, 6740–6747.
Menon, A. K., Mayor, S., Ferguson, M. A. J., Duszenko, M., & Cross, G. A. M. (1988) *J. Biol. Chem. 263*, 1970–1977.
Menon, A. K., Schwarz, R. T., Mayor, S., & Cross, G. A. M. (1990) *J. Biol. Chem.* (in press).
Metcalf, P., Blum, M., Freymann, D., Turner, M., & Wiley, D. C. (1987) *Nature 325*, 84–86.
Orlandi, P. A., Jr., & Turco, S. J. (1987) *J. Biol. Chem. 262*, 10384–10391.
Ovchinnikov, Y. A., Abdulaev, N. G., & Bogachuk, A. S. (1988) *FEBS Lett 230*, 1–5.
Puentes, S. M., Sacks, D. L., da-Silva, R. P., & Joiner, K. A. (1988) *J. Exp. Med. 169*, 887–902.
Roberts, W. L., & Rosenberry, T. L. (1985) *Biochem. Biophys. Res. Commun. 133*, 621–617.
Roberts, W. L., Kim, B. H., & Rosenberry, T. L. (1987) *Proc. Natl. Acad. Sci. U.S.A. 84*, 7817–7821.
Roberts, W. L., Myher, J. J., Kuksis, A., Low, M. G., & Rosenberry, T. L. (1988a) *J. Biol. Chem. 263*, 18766–18775.
Roberts, W. L., Santikarn, S., Reinhold, V. N., & Rosenberry, T. L. (1988b) *J. Biol. Chem. 263*, 18776–18784.
Robinson, P. J., & Spencer, S. C. (1988) *Immunol. Lett. 19*, 85–94.
Robinson, P. J., Millrain, M., Antoniou, J., Simpson, E., & Mellor, A. L. (1989) *Nature 342*, 85–87.
Roditi, I., Schwartz, H., Pearson, T. W., Beecoft, R. P., Liu, M. K., Richardson, J. P., Buhring, H.-J., Pliess, J., Bulow, R., & Overoth, P. (1989) *J. Cell. Biol. 108*, 737–746.
Rosen, G. P., Pahlsson, P., Londner, M. V., Wersterman, M. E., & Nilsson, B. (1989) *J. Biol. Chem. 264*, 10457–10463.
Sadeghi, H., Da Silva, A. M., & Klein, C. (1988) *Proc. Natl. Acad. Sci. U.S.A. 85*, 5512–5515.
Saltiel, A. R., & Cuatrecasas, P. (1986) *Proc. Natl. Acad. Sci. U.S.A. 83*, 5793–5797.
Saltiel, A. R., Fox, J. A., Sherline, P., & Cuatrecasas, P. (1986) *Science 233*, 967–972.
Scallon, B. J., Scigliano, E., Freedman, V. H., Miedel, M. C., Pan, Y.-C. E., Unkeless, J. C., & Kochan, J. P. (1989) *Proc. Natl. Acad. Sci. U.S.A. 86*, 5079–5083.
Schmitz, B., Klein, R. A., Egge, H., & Peter-Katalinic, J. (1986) *Mol. Biochem. Parasitol. 20*, 191–197.
Schmitz, B., Klein, R. A., Duncan, I. A., Egge, H., Gunawan, J., Peter-Katalinic, J., Dabrowski, U., & Dabrowski, J. (1987) *Biochem. Biophys. Res. Commun. 146*, 1055–1063.
Schwartz, R. T., Mayor, S., Menon, A. K., & Cross, G. A. M. (1989) *Biochem. Soc. Trans. 17*, 746–748.
Singh, B. N., Costello, C. E., Beach, D. H., & Holz, G. G., Jr. (1988) *Biochem. Biophys. Res. Commun. 157*, 1239–1246.
Stadler, J., Keenan, T. W., Bauer, G., & Gerisch, G. (1989) *EMBO J. 8*, 317–377.
Stahl, N., Borchelt, D. R., Hsiao, K., & Prusiner, S. B. (1987) *Cell 51*, 229–240.
Strang, A.-M., & van Halbeek, H. (1989) *Glycoconjugate J. 6*, 426.
Strang, A.-M., Williams, J. M., Ferguson, M. A. J., Holder, A. A., & Allen, A. K. (1986) *Biochem. J. 234*, 481–484.
Stroynowski, I., Soloski, M., Low, M. G., & Hood, L. (1987) *Cell 50*, 759–768.
Takami, N., Ogata, S., Oda, K., Misumi, Y., & Ikehara, Y. (1988) *J. Biol. Chem. 263*, 3016–3021.
Tse, A. G. D., Barclay, A. N., Watts, A., & Williams, A. F. (1985) *Science 230*, 1003–1008.
Turco, S. J. (1988a) *Biochem. Biophys. Trans. 16*, 259–261.
Turco, S. J. (1988b) *Parasitol. Today 4*, 255–257.
Turco, S. J., Hull, S. R., Orlandi, P. A., Jr., Shepherd, S. D., Homans, S. W., Dwek, R. A., & Rademacher, T. W. (1987) *Biochemistry 26*, 6233–6238.
Turco, S. J., Orlandi, P. A., Jr., Homans, S. W., Ferguson, M. A. J., Dwek, R. A., & Rademacher, T. W. (1989) *J. Biol. Chem. 264*, 6711–6715.
Walter, E. I., Roberts, W. F., Rosenberry, T. L., & Medof, M. E. (1987) *Fed. Proc., Fed. Am. Soc. Exp. Biol. 46*, 772.
Waneck, G. L., Sherman, D. H., Kincade, P. W., Low, M. G., & Flavell, R. A. (1968) *Proc. Natl. Acad. Sci. U.S.A. 85*, 577–581.
Williams, A. F., & Tse, A. G. D. (1985) *Biosci. Rep. 5*, 999–1005.
Williams, A. F., Tse, A. G. D., & Gagnon, J. (1988) *Immunogenetics 27*, 265–272.
Zamze, S. E., Ferguson, M. A. J., Collins, R., Dwek, R. A., & Rademacher, T. W. (1988) *Eur. J. Biochem. 176*, 527–534.

NUCLEIC ACIDS AND PROTEIN BIOSYNTHESIS

Chapter 19

Parameters for the Molecular Recognition of Transfer RNAs†

Paul Schimmel

Department of Biology, Massachusetts Institute of Technology, Cambridge, Massachusetts 02139

Received January 9, 1989; Revised Manuscript Received February 3, 1989

Transfer RNAs are highly differentiated nucleic acids comprised of 74–93 nucleotides that are folded into a compact three-dimensional pattern which is believed to accommodate most of the known tRNA sequences (Sprinzl, 1987). The molecules are differentiated from each other according to their amino acid acceptance. This is determined in the two-step aminoacylation reaction whereby an amino acid is activated by its cognate aminoacyl-tRNA synthetase, which catalyzes formation of a tightly bound aminoacyl adenylate; this complex than reacts with the 2′- or 3′-hydroxyl of a tRNA to form the aminoacylated species. After this, aminoacyl-tRNAs react with components of the translation apparatus that recognize features common to all tRNAs and enable amino acids to be inserted into growing polypeptide chains through the precise base-pairing interaction of trinucleotide anticodons (within tRNAs) with codons in messenger RNAs. Thus, the translation of mRNAs into polypeptides of defined sequences is a manifestation of the genetic code, but the code itself is based on the molecular recognition of transfer RNAs by aminoacyl-tRNA synthetases. This system of protein–nucleic acid recognition connects each amino acid with a trinucleotide sequence (anticodon) within the tRNA.

Because of degeneracy in the genetic code, whereby up to six trinucleotide sequences may code for a specific amino acid, there can be several tRNAs that are specific for a given amino acid. An example is serine, which in *Escherichia coli* has at least five distinct tRNA species that collectively recognize the six different codons for that amino acid (Sprinzl et al., 1987). These tRNAs differ not only in their anticodons but also in other parts of their respective sequences. These tRNA isoacceptors are generally recognized by the one aminoacyl-tRNA synthetase which is specific for that amino acid, however (Schimmel & Soll, 1979). This implies that, for at least some of the enzymes, the anticodon is not the primary determinant for recognition.

Unlike tRNAs that have a common structural motif and typically vary in length by not more than 20%, the aminoacyl-tRNA synthetases are diverse proteins with four different types of quaternary structures (α, α_2, $\alpha_2\beta_2$, and α_4) and subunit sizes that range from 334 to over 1000 amino acids (Schimmel, 1987). Although leucine, valine, and isoleucine tRNA synthetases have some sequence similarities (Heck & Hatfield, 1988), several of the enzymes have sequences which are not similar to that of any other synthetase. The enzymes arose early in evolution, probably appearing with the earliest life forms, and may never have been constrained to interact with elements of the protein synthesis apparatus other than tRNAs. In some cases, they have acquired additional biological functions. These considerations may explain in part the apparent uniqueness of the sequences and quaternary structures of many of the enzymes. However, similar structural motifs can in principle be assembled from diverse sequences, and there is evidence for a structural relatedness of portions of methionine and tyrosine tRNA synthetases, which have little sequence similarity (Blow et al., 1983).

The basis for the recognition of tRNAs by aminoacyl-tRNA synthetases has been difficult to solve [recent commentaries and summaries include de Duve (1988), Schulman and Abelson (1988), RajBhandary (1988), and Yarus (1988)]. There are significant differences from the system of recognition of DNA sequence elements by gene regulatory proteins. For protein–DNA complexes, the dissociation constant can be as small as 10^{-12} M, so that the complexes have long lifetimes (Pabo & Sauer, 1984). For synthetase–tRNA complexes, the dissociation constants at pH 7.5 are on the order of 10^{-6} M, which is comparable to the dissociation constant or Michaelis constant for an enzyme–small molecule complex (Schimmel & Soll, 1979). This means that the enzyme–tRNA complexes have short lifetimes, which facilites rapid turnover. The relatively high enzyme–tRNA dissociation constant limits the degree to which discrimination can be achieved at the binding step. There is a second step in which specificity can be manifested, however. Because tRNA is a substrate for the enzyme, there can be discrimination during the transition state of aminoacylation, which is expressed through the catalytic parameter k_{cat}. Early work suggested that this played a role

†Support of research on aminoacyl-tRNA synthetases and recognition of transfer RNAs through National Institutes of Health Grants GM15539 and GM23562 is acknowledged.

in the overall recognition process (Schimmel & Soll, 1979).

The tRNA molecule is folded, like many proteins, into a globular structure where some elements which are dispersed in the sequence are brought into close proximity (Kim et al., 1974; Robertus et al., 1974). Dissection of tRNAs into smaller pieces, as has been done in successfully identifying recognition sites in linear DNA molecules, sacrifices many of the structural features that only are present in the intact tRNA. The relatively short lifetimes of the synthetase–tRNA complexes make it difficult to obtain, from a nuclease digestion, a nuclease-resistant protein–nucleic acid complex—an approach that has also been successfully applied with the more stable DNA–protein complexes. Early work defined sites on tRNAs that are in close contact with bound synthetases [e.g., by using cross-linking among other approaches (Schimmel, 1977)] and investigated the basis for the mischarging of certain tRNAs when reacted with synthetases from another, heterologous organism [summarized in Schimmel and Soll (1979)]. The early application of an in vivo amber suppression assay also afforded an opportunity to explore determinants important for the recognition of tRNAs, although definitive conclusions were not reached at that time and the complexity of the recognition problem was highlighted (see below).

The recent availability of procedures for mutagenesis and expression of tRNAs in vivo, and for synthesis of tRNAs and variants in vitro, has stimulated a new generation of experiments. While much of the recent work has utilized in vivo amber suppression assays to study molecular recognition, the interpretation of some of these experiments is limited when there are no in vitro data. At the same time, in vitro experiments have their own limitations which can only be overcome by investigation of variant tRNAs in the context of all of the synthetases and tRNAs in vivo. Summarized below are recent experiments on the molecular recognition of transfer RNAs, with some consideration of binding and catalytic parameters that enable tRNAs to be distinguished from each other.

Locating Determinants for Identity

Amber Suppression Assay. Early experiments utilized the *E. coli supF* amber suppressor, which is based on the sequence of a tyrosine tRNA with the anticodon altered from GUA to CUA so that the UAG amber codon is recognized [summarized by Ozeki et al. (1980)]. This suppressor inserts tyrosine at amber codons, which demonstrates that the first position of the anticodon is not an essential determinant for the identity of a tyrosine tRNA. Mutants of *supF* that insert glutamine were isolated. The interpretation was not straightforward because some of the mutations that resulted in aminoacylation with glutamine did not create nucleotides that are in the analogous location of any *E. coli* $tRNA^{Gln}$ species (Figure 1a). Subsequently, it was demonstrated that simple alteration of the anticodon of $tRNA^{Trp}$ from CCA to CUA created an amber suppressor that inserts glutamine (Figure 1b). These results demonstrated that amino acid acceptance could be manipulated by simple mutations. One interpretation was that the mutations disturbed the interaction with the cognate enzyme and that glutamine tRNA synthetase had a relaxed specificity which was manifested on tRNA substrates that were less competitively aminoacylated by their cognate enzyme.

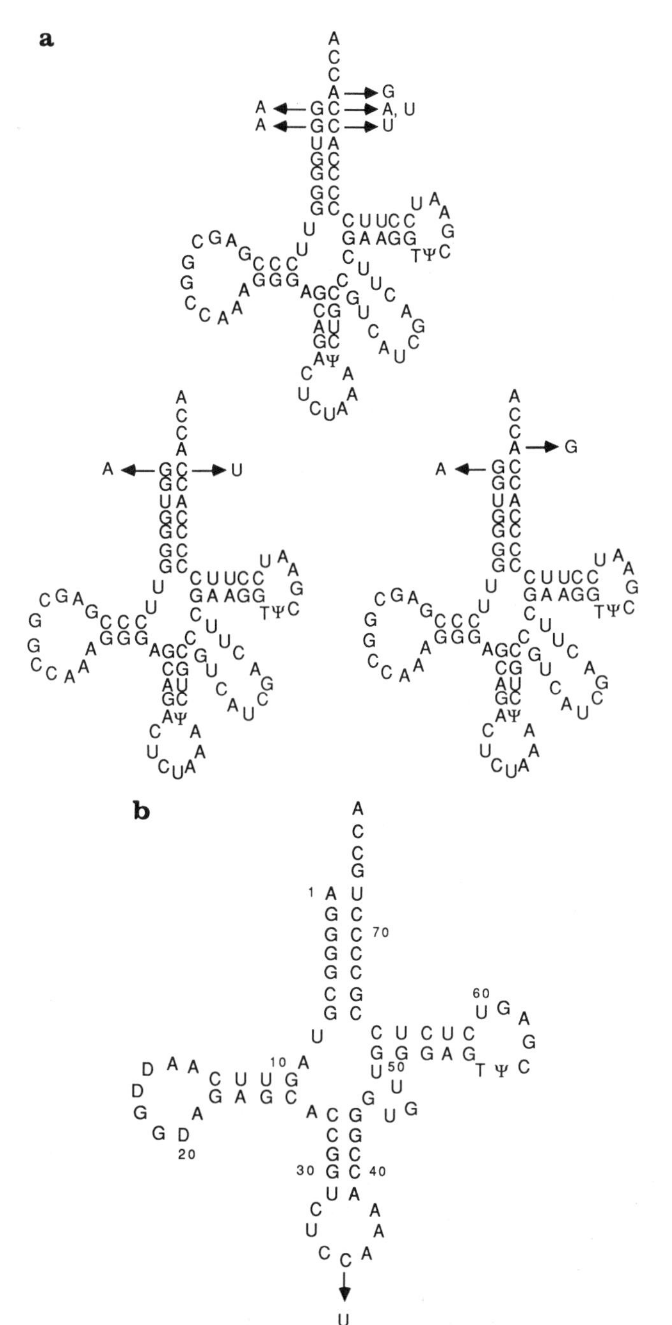

FIGURE 1: (a) Mutations in *E. coli* $tRNA^{Tyr/CUA}$ that cause insertion of glutamine at amber codons. The nucleotide changes are concentrated in the acceptor helix and in some cases do not recreate sequences found in any *E. coli* $tRNA^{Gln}$. The upper diagram shows the locations of single nucleotide substitutions that enable $tRNA^{Tyr/CUA}$ to be aminoacylated with glutamine, while the lower two diagrams show, separately, individual mutants that have two base changes. Data are summarized from Smith et al. (1970), Hooper et al. (1972), Shimura et al. (1972), Smith and Celis (1973), Celis et al. (1973), Ghysen and Celis (1974), Inokuchi et al. (1974), and Ozeki et al. (1980). (b) A single nucleotide substitution in *E. coli* $tRNA^{Trp}$ enables it to insert glutamine at amber codons. The substitution creates an amber CUA anticodon (Yaniv et al., 1974). See also Soll (1974), Celis et al. (1976), and Yarus et al. (1977).

Conversion of a Leucine into a Serine Transfer RNA. Although difficulties of interpretation raised concerns about the early work on amber suppressors with altered amino acid acceptance capability, the approach has clear advantages. By restriction of amino acid substitutions of mutant suppressor tRNAs to amber codons, the possibility is removed for deleterious substitutions in vivo in the translated regions of mRNAs for cellular proteins. The approach is limited to those tRNAs whose amino acid acceptance is not altered by change of the anticodon to CUA. One example is serine tRNA, where there are six known isoacceptors (five encoded by the *E. coli* genome and one by phage T_4) that can be aminoacylated by the *E. coli* serine tRNA synthetase and that have anticodons which include UGA, CGA, GGA, and GCU. These anticodons collectively vary all three positions, and furthermore, a change to CUA does not change the identity of this tRNA.

Normanly et al. (1986a) examined the 6 serine tRNAs and concluded that, apart from the bases that are conserved in all tRNAs, only 12 were held in common. These 12 nucleotides, which are located in the acceptor stem and dihydrouridine stem and loop, were transferred into a leucine tRNA isoacceptor (with a CUA amber anticodon). The resulting tRNA suppresses a serine-requiring amber allele at codon 68 of β-lactamase. The suppression efficiency of the strong parent leucine amber suppressor is 60%, and the transformed species with 12 substitutions has an efficiency of 0.5–1%. Direct information on the nature of the amino acid(s) inserted by the transformed $tRNA^{Leu/CUA}$ was obtained by protein sequence analysis of a suppressed dihydrofolate reductase gene which has an amber mutation at codon 10. This showed that serine was inserted, with a possibility of, but little evidence for, minor amounts of valine and/or leucine. No glutamine was detected at the position corresponding to the amber codon. Thus, the transformation of a strong leucine-inserting suppressor into a weak serine-inserting suppressor was achieved. Further work has focused on defining the minimal subset of nucleotides from among the 12 defined in these experiments, which are sufficient to confer serine acceptance, and determining whether that subset will function in the context of other tRNA sequences (Schulman & Abelson, 1988).

A Single Base Pair Is a Major Determinant of the Identity of an Alanine tRNA. Alanine tRNA synthetase has been extensively studied by biochemical and genetic approaches, and the segment important for tRNA binding has been delineated (Schimmel, 1987, 1989). In order to define further the amino acids that are critical for tRNA recognition, a large population of $tRNA^{Ala}$ variants were created by site-directed mutagenesis and tested for recognition by alanine tRNA synthetase (Hou & Schimmel, 1988). Those that are defective in recognition were to be used to isolate second-site revertants in the enzyme which compensate for the defect in the tRNA. In the course of this work it was discovered that a single base pair is a major determinant of the identity of $tRNA^{Ala}$.

A population of 28 mutant tRNAs were created that collectively varied over half of the nonconserved nucleotides. A $tRNA^{Ala/CUA}$ amber suppressor was used for these experiments. The sequence of this tRNA is based on that of $tRNA^{Ala/GGC}$ [Mims et al., 1985; cf. Normanly et al. (1986b) and Masson and Miller (1986)]. This suppressor inserts alanine, even though all three anticodon nucleotides have been changed. The mutant tRNAs, which in some cases had as many as five substitutions within a single species, were first introduced on a multicopy plasmid and checked for suppression of a *trpA*(*UAG234*) amber allele. There is evidence that this allele is suppressed by insertion of glycine or alanine but not by other amino acids (Murgola & Hijazi, 1983). Any mutant that did not suppress the *trpA*(*UAG234*) allele was then checked to determine whether a stable tRNA was made. Those examples of a sup^- phenotype for which the overproduced tRNA was not evident were not considered further. Of particular interest are those sup^- species that are not defective in tRNA biosynthesis.

This screen and analysis yielded only one site in the entire molecule where mutations resulted in a tRNA which has a sup^- phenotype on the *trpA*(*UAG234*) amber allele and which is clearly synthesized as a stable tRNA. This involved replacements of the G3·U70 base pair in the amino acid acceptor helix with A3·U70 or G3·C70 (Figure 2). In vitro aminoacylation measurements confirmed that the mutant tRNAs were not aminoacylated with catalytic amounts of purified alanine tRNA synthetase (Hou & Schimmel, 1988).

The G3·U70 base pair was introduced into $tRNA^{Cys/CUA}$ and $tRNA^{Phe/CUA}$. Each of these tRNAs has a C3·G70 base pair and differs by 38 ($tRNA^{Cys/CUA}$) and 31 ($tRNA^{Phe/CUA}$) nucleotides from $tRNA^{Ala/CUA}$. Introduction of the G3·U70 base pair into each of these tRNAs confers the ability to accept alanine in vivo. In the case of G3·U70 $tRNA^{Cys/CUA}$, only alanine was detected at the position of the suppressed amber codon in dihydrofolate reductase (Hou & Schimmel, 1988). This raises the possibility that the substitution at position 3·70 has perturbed a determinant for the cysteine tRNA synthetase in addition to conferring a determinant for alanine tRNA synthetase. Aminoacylation of G3·U70 $tRNA^{Cys/CUA}$ with purified alanine tRNA synthetase was demonstrated. The G3·U70 $tRNA^{Phe/CUA}$ species is aminoacylated with both alanine and phenylalanine in vivo. This suggests that the determinants for the identity of phenylalanine tRNA are located, at least in part, elsewhere.

The results indicate that the G3·U70 base pair is a major determinant for the identity of an alanine tRNA. The result of introducing a G3·U70 base pair into $tRNA^{Phe}$ was confirmed by McClain and Foss (1988a). These authors also did experiments with a variant of a glycine tRNA (which largely is aminoacylated with glutamine in vivo). Introduction of the G3·U70 base pair into this tRNA does not confer acceptance of alanine. This may mean that, at least in the context of the variant sequence, the glutamine enzyme competes more effectively than the alanine enzyme and, additionally or alternatively, there are negative determinants in the variant tRNA which block interaction with the alanine enzyme (see Concluding Remarks).

A mutant lysine *missense* suppressor that inserts glycine and/or alanine at lysine codons has been reported (Prather et al., 1984). This suppressor also is believed to insert lysine. The mutation creates a G3·U70 base pair, and while the insertion of alanine was not established, Prather et al. (1984) suggested that this was likely. The aforementioned results and recent experiments with a lysine tRNA amber suppressor (McClain et al., 1988) support this possibility.

Among the published *E. coli* tRNA sequences the G3·U70 base pair is unique to alanine. Further mutational analysis may uncover additional nucleotides that affect recognition by the alanine enzyme. If such nucleotides are uncovered, then it will be important to determine whether transfer of any of them into another tRNA framework will confer identity for alanine.

Role of the Anticodon in the Recognition of Methionine and Valine tRNAs. In vitro aminoacylation experiments have indicated that, for several tRNAs, single base changes in the anticodon affect the rate of aminoacylation. The availability of T7 RNA polymerase and synthetic DNA templates for transcription has made possible the preparation of synthetic tRNAs of any sequence whatsoever (Sampson & Uhlenbeck, 1988). This has afforded an opportunity to make substrates with varied anticodon sequences and to evaluate quantitatively the effect on aminoacylation.

The CAU anticodon of *E. coli* methionine tRNA is known to be important for in vitro aminoacylation with *E. coli* methionine tRNA synthetase (Schulman & Pelka, 1983, 1988). (This enzyme also aminoacylates initiator and elongator methionine tRNAs from other prokaryotes and from eukaryote organelles, all of which have the CAU anticodon.) At concentrations of tRNA that are below the Michaelis constant K_m, the initial rate of aminoacylation is given by $(k_{cat}/K_m)(E)_0(tRNA)_0$, where $(E)_0$ and $(tRNA)_0$ are total enzyme

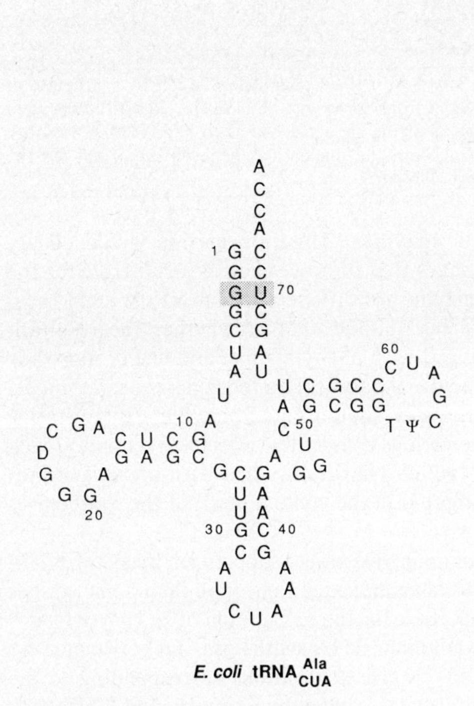

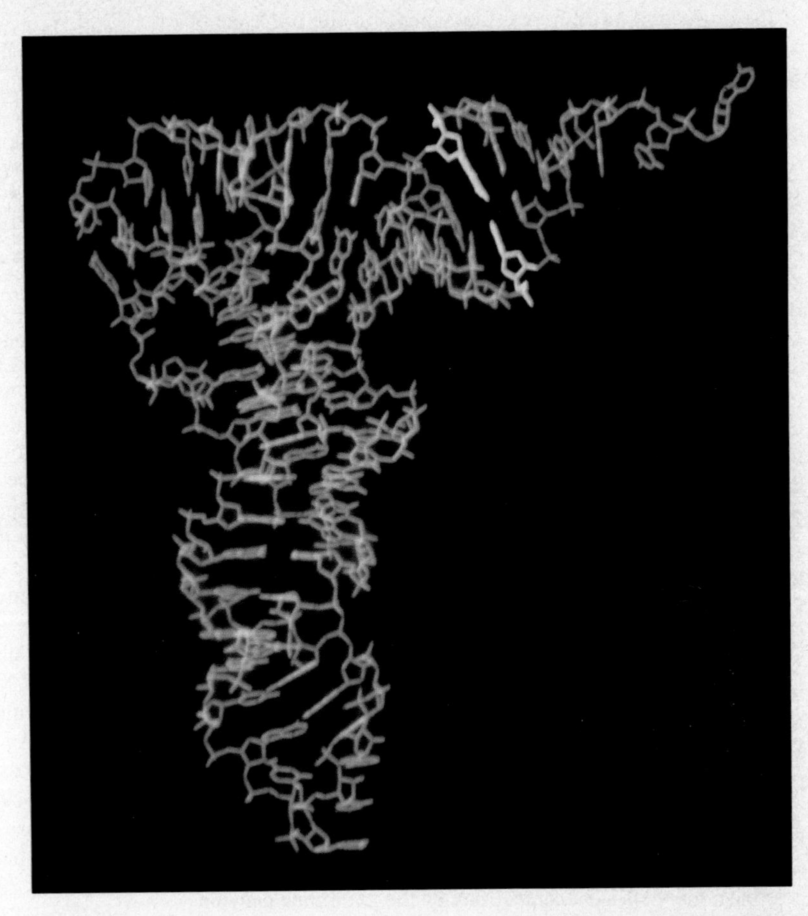

FIGURE 2: (a, left) Nucleotide sequence and cloverleaf structure of tRNA$^{Ala/CUA}$. The sequence is based on tRNA$^{Ala/GGC}$ (Mims et al., 1985) but where the GGC anticodon has been changed to CUA and a U38 → A substitution has been introduced to improve the efficiency of amber suppression (Raftery & Yarus, 1987). The G3·U70 base pair is highlighted. This base pair is a major determinant of the identity of an alanine tRNA (Hou & Schimmel, 1988). (b, right) Depiction of the three-dimensional structure of tRNA$^{Ala/CUA}$ with the G3·U70 base pair highlighted in white. The sequence of tRNA$^{Ala/CUA}$ was built into the known coordinates of yeast tRNA^{Phe} (obtained from Brookhaven Data Bank) utilizing the PS300 FRODO program (Bush et al., 1987).

and tRNA concentrations, respectively, and k_{cat} is the turnover number (unimolecular rate constant). The parameter k_{cat}/K_m has units of a second-order rate constant. Table Ia shows that the relative k_{cat}/K_m is reduced 4–5 orders of magnitude by substitutions at the first position of the anticodon (Schulman & Pelka, 1988). The large effect of substitutions at the first position demonstrates its importance, but it alone is not sufficient for recognition by the methionine enzyme, however. None of the amber suppressors (CUA anticodon) and natural tRNAs with a first position C are known to insert methionine in vivo.

An even larger effect is found when the anticodon is reversed from CAU to UAC. The UAC anticodon corresponds to that for a valine tRNA. Replacement of that anticodon in tRNA$_1^{Val}$ with CAU confers methionine acceptance on tRNA$_1^{Val/CAU}$ (Table Ia). The efficiency of aminoacylation is quantitatively close to that of tRNA^{Met}. In a reciprocal experiment, the CAU anticodon of elongator tRNA$_m^{Met}$ was replaced by UAC, and aminoacylation with valine tRNA synthetase was attempted. In this case the anticodon replacement converts the hybrid tRNA$^{Met/UAC}$ into a substrate for valine tRNA synthetase, with a relative k_{cat}/K_m that is 10-fold less than that of tRNA^{Val} (Table Ib). Thus the CUA → UAC substitution simultaneously eliminates aminoacylation with the methionine enzyme while conferring acceptance of valine. The tRNA$^{Met/UAC}$ and tRNA$^{Val/CAU}$ species have been checked for aminoacylation in vitro with glutamic, glutamine, lysine, isoleucine, and phenylalanine tRNA synthetases. These enzymes aminoacylate neither tRNA$^{Met/UAC}$ nor tRNA$^{Val/CAU}$ (Schulman & Pelka, 1988).

The data presented in Table I suggest that the anticodon has an important role in defining the identity of valine and methionine tRNAs and that the quantitative effects of simple nucleotide substitutions are large. In order to evaluate the role of the anticodon more fully, and to evaluate the significance

Table I: Relative Values of k_{cat}/K_m for Aminoacylation with Methionine and Valine of Anticodon Sequence Variants of tRNA$_m^{Met}$ and tRNA^{Val}

tRNA	rel k_{cat}/K_m
(a) Aminoacylation with Methionine[a]	
natural tRNA^{Met}	1.0
synthetic tRNA^{Met} (CAU)	0.5
synthetic tRNA^{Met} (UAU)	0.0001
synthetic tRNA^{Met} (GAU)	0.00001
synthetic tRNA^{Met} (UAC)	0.0000001
synthetic tRNA^{Val} (CAU)	0.8
(b) Aminoacylation with Valine[b]	
natural tRNA^{Val}	1.0
synthetic tRNA^{Val} (UAC)	0.4
synthetic tRNA^{Met} (CAU)	0.000002
synthetic tRNA^{Met} (UAU)	0.00001
synthetic tRNA^{Met} (UAC)	0.04

[a] Data were obtained with purified *E. coli* methionine tRNA synthetase at pH 7.5, 37 °C, and are taken from Shulman and Pelka (1988). [b] Data were obtained with purified *E. coli* valine tRNA synthetase at pH 7.5, 37 °C, and are taken from Schulman and Pelka (1988).

of the anticodon nucleotide replacements in the presence of all of the enzymes and tRNAs, experiments are being attempted in vivo. This necessitates special techniques because the amber suppression system (which requires a CAU anticodon) obviously cannot be used in these circumstances.

Recognition of E. coli Glutamine tRNA. The anticodon (CUG and UUG) has been implicated as a recognition site for glutamine tRNA synthetase. This apparently explains why certain amber suppressor tRNAs (CUA anticodon) mischarge at least partially with glutamine. The central U of the anticodon is believed to have an important role. Rogers and Soll (1988) have manipulated a serine-inserting $tRNA^{Ser/CUA}$ amber suppressor to examine one way to achieve discrimination of glutamine from serine tRNA synthetase. Because of the CUA anticodon, this suppressor potentially could be converted to a glutamine tRNA. A similar example had been previously demonstrated by isolation of mutations in the acceptor helix of a *supF* $tRNA^{Tyr/CUA}$ (Figure 1a).

The three base pairs at the end of the acceptor helix are implicated in recognition by the serine tRNA synthetase (Normanly et al., 1986a). The sequence at the beginning of the acceptor helix of $tRNA^{Ser}_1$ is G1·C72:G2·C71:A3·U70. This was changed to **U1·A72**:G2·C71:**G3·C70**, where the altered bases are indicated in bold type. These substitutions recreate the sequence of the first three base pairs of $tRNA^{Gln}_2$. While $tRNA^{Ser/CUA}$ inserts serine, the U1·A72:G3·C70 $tRNA^{Ser/CUA}$ inserts over 90% glutamine and about 5% serine in vivo (Rogers & Soll, 1988). One interpretation of this result is that substitution of four nucleotides has disrupted the interaction with the serine tRNA synthetase, thus making possible a more efficient competition by the glutamine enzyme. Additionally or alternatively, the substitutions have improved the interaction with glutamine tRNA synthetase. Analytical aminoacylation measurements would clarify this question, and further substitution and analysis will have to be done in order to define the determinants for the identity of a glutamine tRNA.

Determinants for Recognition of Yeast Phenylalanine tRNA. An in vitro analysis has been used to evaluate nucleotides important for recognition of *Saccharomyces cerevisiae* $tRNA^{Phe}$ by the homologous phenylalanine tRNA synthetase. Initial experiments established that replacements of any of the three anticodon nucleotides decreased k_{cat}/K_m by a factor of 3–10-fold (Bruce & Uhlenbeck, 1982). Additional experiments showed that replacement of G20 (in the dihydrouridine loop) or of A73 (in the single-stranded $ACCA_{OH}$ 3′ terminus) each resulted in a 12-fold reduction in catalytic efficiency [see Sampson and Uhlenbeck (1988, 1989)]. Because G20 is present in $tRNA^{Phe}$ but not in any of the other reported yeast tRNA sequences, it could act as an important discriminatory nucleotide.

E. coli $tRNA^{Phe}$ encodes four of the five aforementioned nucleotides, with a U replacing G20. Substitution of G for U20 improves k_{cat}/K_m for the yeast enzyme by 12-fold so that the yeast and *E. coli* substrates are almost equivalent (Sampson & Uhlenbeck, 1989) (Table II). This corresponds to a relatively small amount (1.5 kcal mol^{-1}) of free energy.

Yeast $tRNA^{Arg}$, $tRNA^{Met}$, and $tRNA^{Tyr}$ were reconstructed so that each contained a complete set of the five important nucleotides (Sampson & Uhlenbeck, 1989). Each is converted to a substrate for the homologous yeast phenylalanine tRNA synthetase (Table II). The parameter k_{cat}/K_m for each of the substituted tRNAs is within 50% of that for the cognate yeast $tRNA^{Phe}$. It is not known whether the nucleotide substitutions that were introduced in each case have an effect on the respective cognate enzyme. This information would clarify whether the changes that improve each as a substrate for the phenylalanine enzyme are sufficient to convert the amino acid specificity to phenylalanine only or whether the resultant tRNAs are charged with phenylalanine and one or more additional amino acids. Analytical aminoacylation experiments with other synthetases should make possible a quantitative evaluation of the various competitive effects. Eventually, each mutant tRNA (which charges in vitro with phenylalanine) should be investigated in the context of all of the synthetases and tRNAs in vivo.

Table II: Relative Values of k_{cat}/K_m for Aminoacylation with Phenylalanine of Synthetic tRNAs with Different Subtitutions

starting sequence	substitutions	rel k_{cat}/K_m[a]
yeast $tRNA^{Phe}$	none	1.0
yeast $tRNA^{Phe}$	U20	0.08
E. coli $tRNA^{Phe}$	none	0.04
E. coli $tRNA^{Phe}$	G20	0.5
yeast $tRNA^{Met}$	several[b]	0.7
yeast $tRNA^{Arg}$	several[b]	0.6
yeast $tRNA^{Tyr}$	several[b]	1.5

[a] Data were obtained with yeast phenylalanine tRNA synthetase at pH 7.45 and are taken from Samson et al. (1989). [b] Substitutions were introduced into the "starting sequence" so as to have G20, G34, A35, A36, and A73, which are the nucleotides believed important for the recognition of yeast $tRNA^{Phe}$.

On the basis of an in vivo amber suppression assay, McClain and Foss (1988b) have indicated that more than 5 nucleotides (i.e., 10) are important for the recognition of *E. coli* $tRNA^{Phe}$ by *E. coli* phenylalanine tRNA synthetase. Only two of these (positions 20 and 73) are at positions corresponding to the locations of the five sites studied by Sampson and Uhlenbeck (1989). The relative significance of each of these is unknown. It will be necessary to perform in vitro aminoacylation measurements similar to those done with the yeast enzyme in order to understand more fully the similarities and differences in the recognition by these two enzymes which are specific for the same amino acid.

Modified Bases. Examples Where They Do and Do Not Have an Important Role. Transfer RNAs isolated from natural sources contain several bases at specific locations that are posttranscriptionally modified. Some of these are common to most tRNAs (such as 7-methylguanosine, dihydrouridine, and pseudouridine), and others are unique to a particular amino acid specific tRNA [such as the wybutosine at position 37 of certain eukaryote phenylalanine tRNAs (Sprinzl et al., 1987)]. In the aforementioned studies of *E. coli* $tRNA^{Met}$ and $tRNA^{Val}$ and of yeast $tRNA^{Phe}$, the tRNA substrates were enzymatically synthesized in vitro. As a consequence, these substrates contain no modified bases. When these synthetic substrates were compared with their counterparts as isolated from natural sources (which thus contain the full complement of modified bases), only small differences in aminoacylation kinetics were observed (Sampson & Uhlenbeck, 1988; Schulman & Pelka, 1988). Thus, these are among the examples [see also Samuelsson et al. (1988) and Francklyn and Schimmel (1989)] where the modified bases do not play a major role in recognition by the cognate enzyme in vitro, although they could act to interfere with the interactions of noncognate enzymes.

In contrast, a modification of an isoacceptor of $tRNA^{Ile}$ is essential for recognition by isoleucine tRNA synthetase. In addition to promoting aminoacylation with isoleucine, the modification blocks misacylation by methionine tRNA synthetase. The $tRNA^{Ile}_2$ isoacceptor reads AUA codons. The gene for this isoacceptor encodes a CAT anticodon. Thus,

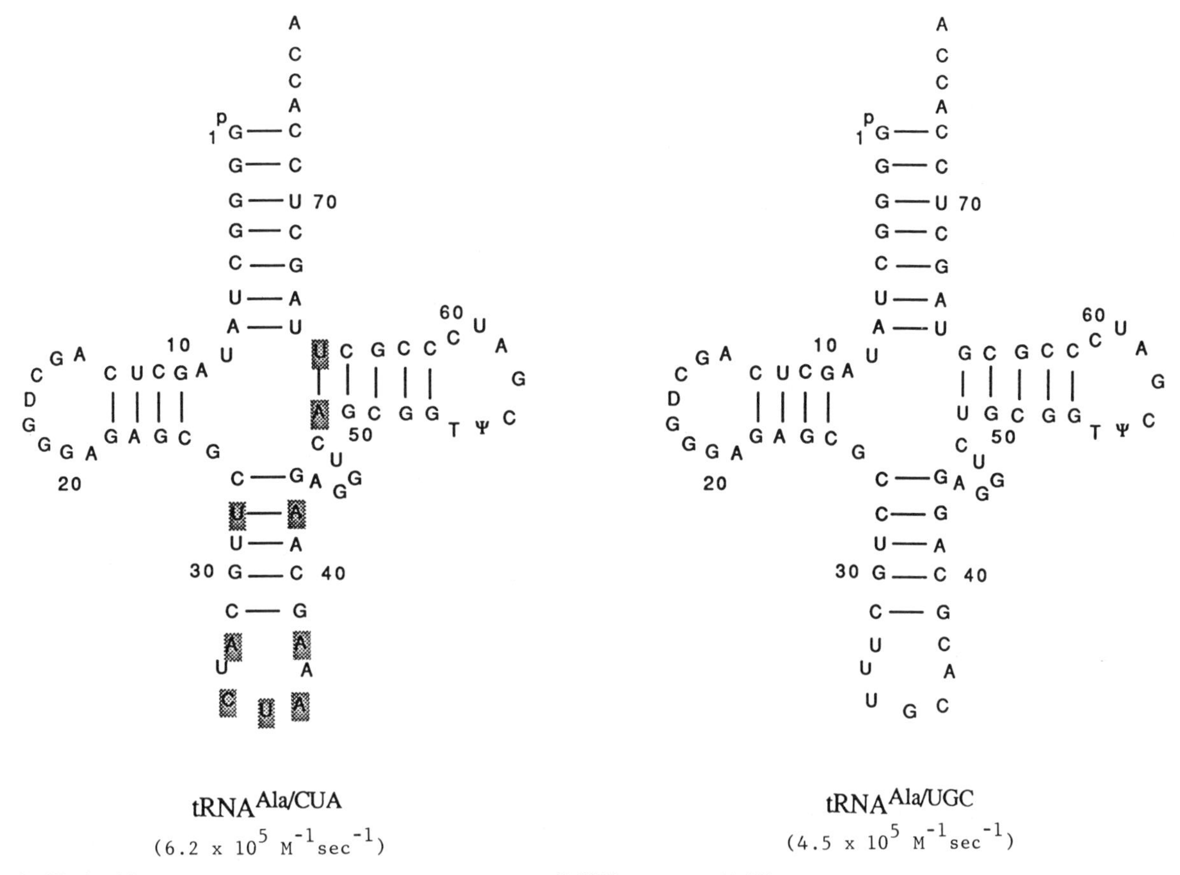

FIGURE 3: Nucleotide sequences and cloverleaf structures of tRNA$^{Ala/UGC}$ and tRNA$^{Ala/CUA}$. The nine nucleotide sequence differences between the natural species (tRNA$^{Ala/UGC}$) and the amber suppressor (tRNA$^{Ala/CUA}$) used in studies of Hou and Schimmel (1988) are shown by shading. Analytical aminoacylation measurements with purified alanine tRNA synthetase have shown that at pH 7.5, 37 °C, there is little difference in the apparent rate (k_{cat}/K_m) of aminoacylation (given in parentheses) of these two tRNA species. Adapted from Park et al. (1989).

without posttranscriptional modification the anticodon is CAU, which is the same as the anticodon for tRNA^{Met}. As shown by Schulman and co-workers, this triplet is recognized by the methionine tRNA synthetase (see Table Ia). Muramatsu et al. (1988a) have found that tRNA$_2^{Ile/CAU}$ is a substrate in vitro for methionine tRNA synthetase. By comparison, however, it is a poor substrate for isoleucine tRNA synthetase. In the mature tRNA, C34 is posttranscriptionally modified to lysidine L (Muramatsu et al., 1988b). This modification consists of a lysine substituted for O-2 of the cytidine ring and attached through the ϵ-amino group directly to C-2 of the pyrimidine base. With this modification, tRNA$_2^{Ile/LAU}$ is efficiently aminoacylated with isoleucine and cannot be aminoacylated with methionine. Thus, the amino acid acceptance in vitro switches according to the state of modification.

The same *E. coli* isoleucyl-tRNA synthetase aminoacylates the major isoacceptor, which has a GAU anticodon. Thus, if the anticodon is the primary site of recognition, then the enzyme recognizes a structural feature common to GAU and LAU.

Effect of a Single Base Pair on the Binding and Catalytic Parameters for the Molecular Recognition of Alanine Transfer RNA

Comparison of Kinetic Behavior of an Alanine-Inserting Amber Suppressor with a Naturally Occurring Isoacceptor. The alanine system has been used to explore the effects on tRNA binding and catalytic parameters of single nucleotide substitutions, which alter the G·U base pair, and thereby obtain greater insight into the molecular basis for discrimination of a simple structural feature. The tRNA$^{Ala/CUA}$ amber suppressor described above was based on the sequence of the tRNA$^{Ala/GGC}$ isoacceptor. The second known naturally occurring isoacceptor is tRNA$^{Ala/UGC}$. There are nine nucleotide differences between the synthetic tRNA$^{Ala/CUA}$ amber suppressor and tRNA$^{Ala/UGC}$. Eight of these are located in the anticodon stem and loop, and the ninth is a position 49·65 base pair in the TΨC stem (Figure 3). The kinetic parameters for these two tRNAs at pH 7.5, 37 °C, are k_{cat} = 1.0 s^{-1} and K_m = 2.2 μM for tRNA$^{Ala/UGC}$ and k_{cat} = 1.8 s^{-1} and K_m = 2.9 μM for tRNA$^{Ala/CUA}$ (Park et al., 1989). The small differences between the respective parameters may be experimentally insignificant, and the apparent second-order rate constants k_{cat}/K_m are close in value (Figure 3).

The binding of the two tRNA species has been measured at pH 5.5, where the nitrocellulose filter assay has a high efficiency for retention of synthetase–tRNA complexes. [Above pH 6.0 the efficiency drops so that the method cannot be used to measure binding (Yarus & Berg, 1970). In general, association of tRNAs with aminoacyl-tRNA synthetases increases at lower pH values (Schimmel & Soll, 1979).] At pH 5.5, 23 °C, the dissociation constants for the enzyme–tRNA complexes are within experimental error of each other (Park et al., 1989). Thus, by kinetic and equilibrium measurements, the nine nucleotide differences between tRNA$^{Ala/CUA}$ and tRNA$^{Ala/UGC}$ have little or no effect on binding and catalytic parameters. This is consistent with genetic studies which show that substitutions in the anticodon loop and stem, and of the position 49·65 base pair in the TΨC stem, do not interfere with acceptance of alanine (Hou & Schimmel, 1988). It is of interest to note that the anticodon of yeast tRNA^{Ala} is also not required for recognition by the yeast enzyme (Jin et al., 1987).

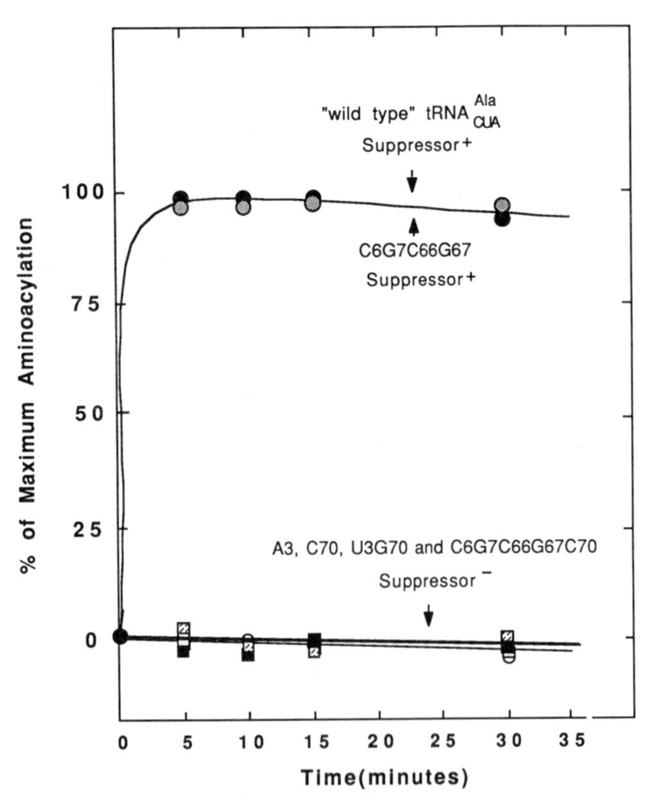

FIGURE 4: Substrate levels of alanine tRNA synthetase do not significantly aminoacylate A3·U70, G3·C70, or U3·G70 $tRNA^{Ala}$ at pH 7.5, 37 °C. For these experiments, the enzyme and tRNA concentrations were 20 and 4 μM, respectively. The suppressor⁺ or suppressor⁻ designation indicates which tRNAs are able to suppress the *trpA(UAG234)* amber allele in vivo. Note also that the C6G7C66G67 $tRNA^{Ala/CUA}$ has four substitutions in the acceptor helix and these do not prevent aminoacylation with alanine. Adapted from Park et al. (1989).

Effect of Sequence Variants of the G3·U70 Base Pair on Kinetic Parameters. The initial characterization of sequence variants of the G3·U70 base pair showed that they were defective for aminoacylation in vitro, but it was not determined whether the defect is in k_{cat}, K_m, or both (Hou & Schimmel, 1988). In further experiments, it was shown that addition of excess A3·U70 $tRNA^{Ala/CUA}$ does not inhibit aminoacylation of $tRNA^{Ala/CUA}$ at pH 7.5, 37 °C. The experiments were done under conditions where an inhibition constant of less than 95 μM would have been detected. Because the K_m for $tRNA^{Ala/CUA}$ is 2.2 μM, and because of evidence that the K_m can be regarded as an enzyme–tRNA dissociation constant, the result implies that at pH 7.5 (37 °C) the binding constant for the A3·U70 variant is at least 40-fold weaker than that for the natural $tRNA^{Ala/UGC}$ isoacceptor (Park et al., 1989).

This is not the only effect of the A3·U70 substitution, however. An excess of enzyme over tRNA substrate has been used to attempt aminoacylation of A3·U70, G3·C70, U3·G70, and C6C7G66G67C70 $tRNA^{Ala/CUA}$. (The latter species has five substitutions in the acceptor stem, which collectively change three base pairs, including the one at position 3·70.) Because of the excess of enzyme, it does not have to turn over in order to achieve complete aminoacylation. With 20 μM enzyme and 4 μM tRNA substrate, the "wild-type" $tRNA^{Ala/CUA}$ and C6C7G66G67 $tRNA^{Ala/CUA}$ species are immediately aminoacylated, as expected. Even after a 30-min incubation, however, the position 3·70 variants are not aminoacylated (Figure 4). If the position 3·70 variants could be aminoacylated but the kinetic defect caused an extremely low product release, then "one shot" of aminoacylation would have been detected at these high enzyme concentrations. Thus, the defect is at a step prior to stable aminoacyl-tRNA formation.

The failure to observe aminoacylation of position 3·70 variants after prolonged incubations with excess enzyme at pH 7.5 suggests that, in addition to reduced binding, there could be a severly reduced k_{cat}. This was demonstrated in the following way. At pH 5.5, the binding of synthetases to tRNAs is generally stronger and can be independently measured by the nitrocellulose filter assay. Thus, experiments were attempted at pH 5.5 to determine whether, under these conditions, the binding of the A3·U70 variant might be enhanced so that a complex could be demonstrated and tested for catalytic competence. The rate of aminoacylation of the wild-type $tRNA^{Ala}$ is reduced at pH 5.5, but the K_d measured by the filter assay for the enzyme–$tRNA^{Ala}$ complex (0.28 μM) is in agreement with the K_m measured in the aminoacylation assay under the same conditions (0.22 μM). Binding of A3·U70 $tRNA^{Ala}$ can be detected at pH 5.5 by the nitrocellulose filter assay, and the K_d of 1.2 μM is only 4-fold less than that of the wild-type $tRNA^{Ala}$. However, no aminoacylation of the A3·U70 species can be detected, even though binding was demonstrated. Moreover, the A3·U70 variant is a competitive inhibitor of aminoacylation of wild-type $tRNA^{Ala}$ at pH 5.5. The inhibition constant K_I = 1.5 μM is close to the independently measured K_d. Because the A3·U70 species binds competitively, it is likely that it occupies the same site on the enzyme as wild-type $tRNA^{Ala}$ (Park et al., 1989).

These data show that, even when bound to alanine tRNA synthetase, A3·U70 $tRNA^{Ala}$ cannot be aminoacylated in vitro. The discrimination against this species is a double barrier of both k_{cat} and K_m parameters. Thus, there is a severe reduction in k_{cat}/K_m for the aminoacylation with alanine of tRNAs that have A3·U70 or G3·C70 or other alternatives at position 3·70 (cf. Figure 4). Because the presence of the G3·U70 pair in a number of tRNA sequence frameworks is sufficient to confer aminoacylation with alanine, the enzyme had to develop a rigorous way to distinguish $tRNA^{Ala}$ from those tRNAs that differ by only a single nucleotide at position 3·70. These tRNAs include those for glutamine, glycine, histidine, leucine, lysine, tryptophan, and valine, which have a G3·C70 base pair, and specific isoacceptors of arginine and serine tRNAs, which have an A3·U70 base pair (Sprinzl et al., 1987). The sharply reduced binding at pH 7.5 of those species that differ by only one nucleotide at the 3·70 position, as demonstrated for the A3·U70 variant, also prevents them from being inhibitors of the enzyme under these conditions.

Evidence for Interaction of Alanine Transfer RNA Synthetase with the Amino Acid Acceptor Helix

RNA Footprinting. A three-dimensional structure of alanine tRNA synthetase or of the synthetase–tRNA complex is not available, although diffraction-grade crystals of a fragment of the enzyme have recently been reported (Frederick et al., 1988). In the absence of high-resolution structural information, RNA "footprint" methods can provide a rough model of sites on $tRNA^{Ala}$ that make contact with the bound enzyme. Nuclease digestion of the free and bound tRNA is done under conditions where approximately one cut per molecule is introduced. When end-labeled tRNA is used and the digested species are resolved by gel electrophoresis, the positions of the cleavages can be accurately determined. Because nucleases preferentially cleave at some sites and far less at others, not every position in the molecule can be investigated by these methods.

Figure 5a shows the sequence and cloverleaf structure of $tRNA^{Ala/UGC}$ and indicates by arrows the 32 phosphodiester

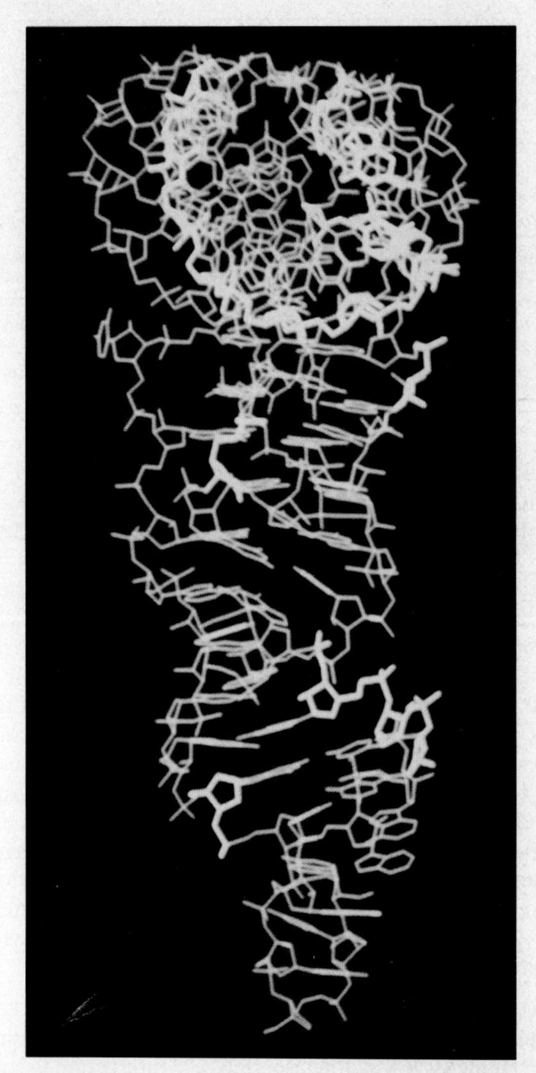

FIGURE 5: (a, left) Sequence and cloverleaf structure of tRNA$^{Ala/UGC}$ and the phosphodiester linkages that are protected from ribonuclease attack by bound alanine tRNA synthetase. The arrows designate the positions that have been examined by use of RNase A or RNase V_1. The arrows with slashed bars are the sites protected by the bound synthetase. The heavy arrow denotes a site of enhanced cleavage in the presence of bound enzyme. Adapted from Park and Schimmel (1988). (b, right) Skeletal model of tRNA^{Ala} as viewed from the 3′ end with sites protected from nuclease cleavage by bound alanine tRNA synthetase. The backbone has been highlighted in white at the places that are protected by the bound enzyme. The protection of the 3′ side of the acceptor helix [see (a)] is seen as a spiral that follows the course of the acceptor–TΨC helix and encompasses the G3·U70 base pair.

linkages that have been probed by the nuclease digestion assay (Park & Schimmel, 1988). For this experiment, RNase A (specially cleaves on the 3′ side of pyrimidines with a preference for single-stranded regions) and cobra venom RNase V_1 (preferentially cleaves on the 5′ side of a purine or pyrimidine base in a double-stranded region) were used. The arrows with cross bars denote those sites that are protected from cleavage by the bound enzyme. The anticodon loop is not protected, and this is consistent with genetic and in vitro kinetic results which show that the anticodon is not important for recognition. However, there is protection (and a site of synthetase-induced hypersensitivity) in the anticodon stem, even though genetic studies have suggested that sequence alterations in this region do not disrupt aminoacylation with alanine in vivo. This simply shows that regions of protection are not necessarily of major consequence for determining amino acid specificity. The greatest concentration of consecutive protected sites is on the 3′ side of the amino acid acceptor helix and extends into the TΨC helix. [These two helices are fused together as one helical branch of the L-shaped tRNA structure (Figure 2).] Both phosphodiester bonds that flank U70 are shielded by the bound enzymes. There is no protection of the 5′ side of the acceptor helix, although the internucleotide phosphodiester linkage between U8 and A9 is protected.

The pattern of protection of the acceptor helix implies that the enzyme spirals around the acceptor–TΨC helix. This spiral is evident by viewing the structure "on end", looking from the 3′ end down the axis of the helix (Figure 5b). The G3·U70 base pair is visible in this projection. If the enzyme recognizes specifically this base pair, and if this pair is arranged in the wobble configuration, then it is formally possible to distinguish G3·U70 from other base pairs. In the wobble configuration, the 4-keto oxygen of U70 is not hydrogen bonded and it projects into the major groove. Introduction of A70 introduces a standard Watson–Crick A·U base pair whereby the 4-keto of uracil is now shifted into a hydrogen bond. On the other hand, change of U70 to C70 introduces an exocyclic amino group at the 4-position which is hydrogen bonded in the standard Watson–Crick configuration with the 6-keto group of guanine. Similar considerations show that the exocyclic 2-amino group of guanine, which is not hydrogen bonded in a G3·U70 wobble pair, can be a site for discrimination. These explanations of specificity are speculative, although they demonstrate how discrimination of subtle structural alterations is possible in principle.

Table III: Apparent Kinetic Parameters for Aminoacylation of $tRNA^{Ala}$ and of Mini- and Microhelices[a]

substrate	K_m (μM)	k_{cat} (s^{-1})	k_{cat}/K_m (M^{-1} s^{-1})
$tRNA^{Ala}$	2.0	0.89	4.4×10^5
$minihelix^{Ala}$	11.4	1.5	1.3×10^5
G3·U70 $minihelix^{Tyr}$	8.8	0.48	5.5×10^4
$microhelix^{Ala}$	35.9	0.28	7.8×10^3

[a] Adapted from Francklyn and Schimmel (1989). Results were obtained at pH 7.5, 37 °C, at a saturating concentration of ATP and a subsaturating concentration (20 μM) of radioactive alanine. Because of the high K_m for alanine, saturation requires prohibitively large amounts of radioactive substrate. Earlier work showed that the K_m for $tRNA^{Ala}$ exhibited little sensitivity to the concentration of alanine (Jasin & Schimmel, 1985).

Recognition of the G3·U70 Base Pair and/or of a Structural Variation? Recognition of $tRNA^{Ala}$ may also be directed, at least in part, at a structural variation in the amino acid acceptor helix that arises from the G3·U70 base pair. McClain et al. (1988) report that other nucleotide combinations including U3·G70 and A3·U70, while weak or inactive suppressors of the alanine-requiring *trpA*(*UAG234*) amber allele, can nonetheless insert alanine at an amber codon of dihydrofolate reductase. The interpretation of this result is unclear because of the poor or nonexistent suppression of the *trpA*(*UAG234*) amber by these $tRNA^{Ala}$ sequence variants and because of the inability to detect significant aminoacylation of the A3·U70 or U70·G3 $tRNA^{Ala}$ species in vitro with substrate levels of enzyme (Figure 4). Possibly the enzyme has a residual activity toward tRNA substrates that have structural irregularities in the acceptor helix and that, in vivo, a minor amount of the aminoacylated A3·U70 and U3·G70 species is produced and is then sequestered by elongation factor Tu and carried to the ribosomes.

Aminoacylation with Alanine of RNA Minihelices

Because of the aforementioned evidence that a major portion of the alanine tRNA synthetase–$tRNA^{Ala}$ interaction is concentrated in the acceptor–TΨC helix, RNA hairpin helices have been designed and synthesized to correspond to this part of the molecule (Francklyn & Schimmel, 1989). These minihelices have been tested as substrates for aminoacylation with alanine tRNA synthetase. In $minihelix^{Ala}$ the acceptor–TΨC helix based on $tRNA^{Ala/GGC}$ has been recreated (Figure 6). This has 12 base pairs with a loop of seven nucleotides and a single-stranded 3′ end that terminates in the sequence $ACCA_{OH}$. This construction effectively deletes the segment from A9 to C48 of $tRNA^{Ala/GGC}$ so that the highly conserved nucleotide U8 has been covalently joined to A49. A G3·C70 $minihelix^{Ala}$ variant has also been constructed.

The wild-type $minihelix^{Ala}$ is efficiently and completely aminoacylated by alanine tRNA synthetase. When compared to $tRNA^{Ala}$, the k_{cat} parameter is similar and K_m is about 6-fold higher for the minihelix (Table III). The elevation in the K_m corresponds to approximately 1 kcal mol^{-1}. This is a small energy and could be due to one or two van der Waals contacts that are missing in the enzyme–$minihelix^{Ala}$ complex.

A G3·C70 variant of $minihelix^{Ala}$ was also synthesized and was found to be inactive for aminoacylation with alanine, even with elevated levels of alanine tRNA synthetase. This behavior is analogous to the effect of G3·U70 in the aminoacylation of $tRNA^{Ala}$. To determine whether the G3·U70 base pair could confer alanine acceptance on an unrelated minihelix, a "tyrosine" minihelix with a G3·U70 variant was synthesized. The $minihelix^{Tyr}$ is based on the acceptor–TΨC sequence of $tRNA^{Tyr}$ and differs at 7 of the 12 base pairs (in the helical region) from that of $minihelix^{Ala}$. The presence of the G3·U70

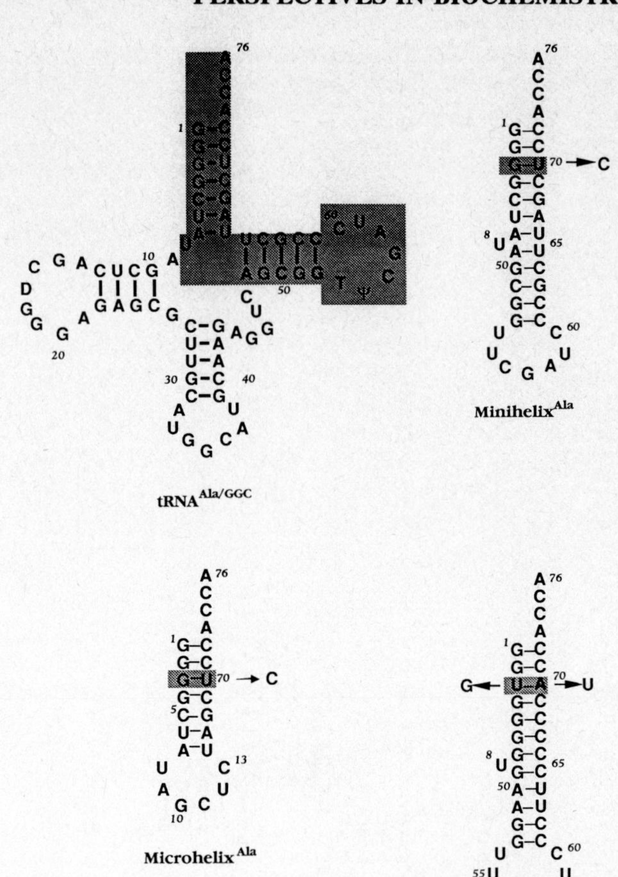

FIGURE 6: RNA minihelices and a microhelix that can be aminoacylated by alanine tRNA synthetase. The alanine helices are based on the sequence of $tRNA^{Ala/GGC}$. $Minihelix^{Ala}$ includes the acceptor–TΨC helix and joins U8 to A49, while $microhelix^{Ala}$ consists only of the acceptor helix (seven base pairs) and joins C13 to U66. In each case, substitution of C70 for U70 eliminates efficient aminoacylation with alanine. The tyrosine minihelix ($minihelix^{Tyr}$) is based on the acceptor–TΨC helix of $tRNA^{Tyr}$; this minihelix becomes an efficient substrate for alanine tRNA synthetase when G3·U70 is substituted for U3·A70 (Francklyn & Schimmel, 1989).

base pair is required for the efficient aminoacylation of this hairpin. Relative to $minihelix^{Ala}$, K_m for G3·U70 $minihelix^{Tyr}$ is comparable and k_{cat} is reduced about 3-fold (Table III). In terms of energy, the reduction in k_{cat} is small and corresponds to less than 1 kcal mol^{-1} (Francklyn & Schimmel, 1989).

Encouraged by the success of these experiments, an even smaller synthetic helix was constructed, which is based on just the amino acid acceptor stem. This consists of seven base pairs connected by a loop of six nucleotides (Figure 6). In this $microhelix^{Ala}$, the sequence of the loop starts at U8 and extends to C13 which is joined to U66. The $microhelix^{Ala}$ is completely aminoacylated by alanine tRNA synthetase at a rate which is reduced relative to $tRNA^{Ala}$. While the K_m is 18-fold higher for $microhelix^{Ala}$, the k_{cat} is reduced by only 3-fold (Table III). On the basis of the relative K_m parameters, the decrease in binding energy is about 1.7 kcal mol^{-1}. The relatively small effect on k_{cat} shows that, once bound to the enzyme, the rate of aminoacylation of the microhelix is comparable to that of intact $tRNA^{Ala}$.

Because the seven base pair $microhelix^{Ala}$ has a k_{cat} for aminoacylation that is within a factor of 3 of that of $tRNA^{Ala}$ (Table III), it is evident that this parameter is not greatly enhanced by sequences which lie outside of the acceptor helix. Moreover, most of the binding energy is derived from interactions with the acceptor–TΨC helix. However, because

$tRNA^{Ala}$ must be discriminated in vivo from the ensemble of noncognate tRNAs, it is possible that there are negative determinants outside of the acceptor–TΨC helix which prevent $tRNA^{Ala}$ from fitting into the tRNA binding sites of other synthetases. In the mapping of contacts between $tRNA^{Ala}$ and its cognate enzyme (Figure 5), sites are protected that are not present in the minihelices which are efficient substrates for the enzyme (Figure 6 and Table III). These could be regions where tRNAs other than $tRNA^{Ala}$ have nucleotides which block their binding to alanine tRNA synthetase and thereby lessen the likelihood that they will inhibit the enzyme.

It has been speculated that, at an early stage in the development of an organized system for protein synthesis, small oligonucleotides may have been aminoacylated and that sequence information in those oligonucleotides could have conferred some specificity for amino acid attachment. The results described above show that, for at least one amino acid, an oligonucleotide can be enzymatically aminoacylated. The proximal location of the enzyme's recognition site relative to the amino acid attachment site was critical to the success of these experiments. It is doubtful that mini- or microhelices analogous to those for alanine (Figure 6) could be aminoacylated in those cases where the major determinants for identity are located in the anticodon or another region (see below) which is distal to the amino acid acceptor end. In those instances it is conceivable that a determinant for identity was at one time proximal to the acceptor end of a small oligonucleotide and, as the tRNA structure became fully elaborated, was translocated to a distal position. Regardless of the origin of the structure of tRNA and of the basis for specific amino acid attachment, however, further investigation of oligonucleotide components of tRNAs will afford a deeper analytical understanding of recognition.

Concluding Remarks

Positive and Negative Determinants for Recognition, the Role of Competition, and the Difficulties of Interpretation. Conceptually, it is necessary to distinguish nucleotides that are recognized by an aminoacyl-tRNA synthetase from those that block or interfere with binding or catalysis. These can be considered as positive and negative determinants, respectively. It is possible for the context of a particular tRNA sequence to inhibit the recognition of an element which is important for identity; that is, a negative determinant may be dominant. For this reason it is necessary to test nucleotides believed important for recognition in more than one sequence framework. A clear example of context effects is shown in the different degree of aminoacylation with alanine in vivo when the G3·U70 base pair is transferred into $tRNA^{Cys/CUA}$ versus into $tRNA^{Phe/CUA}$ (Hou & Schimmel, 1988). In addition, the relative levels of aminoacyl-tRNA synthetases and tRNAs can play a role in determining the recognition of a particular tRNA and whether a given mutant species will be recognized [Yarus et al., 1986; Swanson et al., 1988; see also Yarus (1988)].

There is a clear conceptual distinction between effects due to negative elements and those due to positive elements, and in some cases further experiments need to be done in order to make this distinction. Thus, nucleotide substitutions that allow $tRNA^{Tyr/CUA}$ (Figure 1a) and $tRNA^{Ser/CUA}$ (Rogers & Soll, 1988) to be aminoacylated with glutamine may be effective by blocking the interaction of the mutant tRNAs with the cognate enzyme (tyrosine and serine tRNA synthetases, respectively) rather than by creating a site for the interaction with glutamine tRNA synthetase [that is, in addition to the middle base (U) of the anticodon which is present in the amber suppressor]. Similarly, mutations in the anticodon stem loop of an *E. coli* $tRNA^{Gln/CUA}$ amber suppressor which allow some aminoacylation with tryptophan may weaken the interaction with glutamine tRNA synthetase rather than create specific sites for recognition by tryptophan tRNA synthetase, particularly because some of these substitutions do not recreate nucleotides that are in the analogous locations of $tRNA^{Trp}$ (Yamao et al., 1988a,b). If the nucleotides are acting as blocking elements for glutamine tRNA synthetase, then their transfer into another tRNA sequence framework is not likely to confer tryptophan acceptance.

Limited Nucleotide Constellations Establish the Identities of Some Transfer RNAs. The standards of the field have been raised so that it is no longer sufficient to show only that a specific nucleotide substitution lowers the efficiency of aminoacylation with a particular enzyme. This is because some substitutions that interfere with aminoacylation could create new, unfavorable enzyme–tRNA steric conflicts at sites which are close to but which do not bind to the enzyme through specific atomic interactions in the wild-type complex. The more recent conclusions are based upon the now accepted practice of making nucleotide substitutions into different tRNA sequence frameworks and establishing the effect on amino acid acceptance. This experiment attempts to identify sites that are dominant positive determinants in vivo and in vitro (thorough k_{cat} and/or K_m parameters).

The locations of important nucleotides for some tRNAs are given in Table IV. The listing includes cases where a "transfer" experiment has been done and is provisional because further work may show that, for a particular tRNA, additional nucleotides are required. Moreover, context effects undoubtedly have a major role (see above), and the full extent of these effects will not be known for some time. However, it is now clear that the determinants for identity are idiosyncratic. Also, for at least some tRNAs, a limited constellation of nucleotides is a major determinant of their identities. It is noteworthy that there is evidence that a simple structural feature (base-pair mismatch at the first position of the amino acid acceptor helix) is a major determinant for distinguishing an initiator from an elongator tRNA (Seong & RajBhandary, 1987).

Possible Role for Editing in the Determination of Transfer RNA Identity. An early observation by Baldwin and Berg (1966) indicated that valine could be activated by isoleucine tRNA synthetase and that $tRNA^{Ile}$ then induced hydrolysis of the bound valyl adenylate. There was no evidence that valine was actually transferred to $tRNA^{Ile}$. Schreier and Schimmel (1972) then showed that aminoacyl-tRNA synthetases have a hydrolytic site that removes an amino acid from transfer RNA, in the absence of AMP and pyrophosphate. This was initially discovered as an activity which cleaves the ester bond which links the cognate amino acid to its tRNA, and evidence was presented that the activity is general to synthetases

$$\text{Ile-tRNA}^{\text{Ile}} + \text{Ile-tRNA synthetase} \rightarrow \text{Ile} + \text{tRNA}^{\text{Ile}} + \text{Ile-tRNA synthetase}$$

$$\text{Phe-tRNA}^{\text{Phe}} + \text{Phe-tRNA synthetase} \rightarrow \text{Phe} + \text{tRNA}^{\text{Phe}} + \text{Phe-tRNA synthetase}$$

and, similarly, for other aminoacyl-tRNA synthetases.

Subsequently, Eldred and Schimmel (1972) showed that, for isoleucine tRNA synthetase, this activity is much enhanced when the incorrect amino acid is attached to a tRNA:

$$\text{Val-tRNA}^{\text{Ile}} + \text{Ile-tRNA synthetase} \rightarrow \text{Val} + \text{tRNA}^{\text{Ile}} + \text{Ile-tRNA synthetase}$$

Table IV: Nucleotides Implicated as Important for the Identities of Some Transfer RNAs

tRNA	important positions	evidence
E. coli alanine	G3·U70	in vivo amber suppression with over 30 $tRNA^{Ala}$ sequence variants and with G3·U70 $tRNA^{Cys/CUA}$ and G3·U70 $tRNA^{Phe/CUA}$; in vitro aminoacylation data on several $tRNA^{Ala}$ variants on synthetic minihelix and microhelix substrates [Hou & Schimmel, 1988; Park et al., 1989; Francklyn & Schimmel, 1989; cf. also McClain and Foss (1988a)]
E. coli arginine	A20 and others	preliminary results in vivo based on introducing A20 and A59 into a $tRNA^{Phe}$ amber suppressor; A59 may play a structural role rather than a role in direct recognition; other nucleotides such as those in anticodon may also be important (McClain & Foss, 1988c)
E. coli glutamine	U35	some tRNAs with the amber codon are misacylated with glutamine, and U35 appears responsible (see text); other nucleotides in the acceptor helix may be important as well
E. coli isoleucine	L34	posttranscriptional modification of cytidine 34 to lysidine switches $tRNA^{Ile}{}_2$ from a methionine-accepting to an isoleucine-accepting tRNA in vitro (Muramatsu et al., 1988a)
E. coli methionine	anticodon	in vitro aminoacylation data with anticodon sequence variants and transfer of the CAU anticodon into $tRNA^{Val}$ (Schulman & Pelka, 1988)
yeast phenylalanine	G20, G34, A35, A36, A73	in vitro aminoacylation data with sequence variants of $tRNA^{Phe}$ and of four reconstructed tRNAs (Samson et al., 1989)
E. coli serine	G1·C72, G2·C71, A3·U70, C11·G24	in vivo amber suppression with sequence variants of a $tRNA^{Leu}$ that was converted to a serine-accepting tRNA [Normanly et al., 1986a; cf. Schulman and Abelson (1988)]
E. coli valine	anticodon	in vitro aminoacylation data with the UAC anticodon of $tRNA^{Val}$ transferred into $tRNA^{Met}$ (Schulman & Pelka, 1988)

and Yarus (1972) demonstrated a rapid deacylation of Ile-$tRNA^{Phe}$ by phenylalanine tRNA synthetase:

$$\text{Ile-tRNA}^{\text{Phe}} + \text{Phe-tRNA synthetase} \rightarrow \text{Ile} + \text{tRNA}^{\text{Phe}} + \text{Phe-tRNA synthetase}$$

Considerable investigation of editing reactions was subsequently undertaken [see summaries by Soll and Schimmel (1974), Schimmel and Soll (1979), Yarus (1979), and Fersht (1985)]. Editing can occur by hydrolysis of the aminoacyl adenylate or by charging followed by hydrolysis of the mischarged aminoacyl-tRNA species. The editing activity of a specific enzyme is generally directed toward an amino acid whose steric bulk is not greater than that of the cognate amino acid. Such amino acids can potentially be activated by a specific enzyme (e.g., valine can be activated by isoleucine tRNA synthetase and threonine by the valine enzyme) because they "fit" into the amino acid binding site, albeit with a lower affinity.

The role of editing in vivo is affected by the presence of elongation factor Tu, which tightly binds and sesquesters aminoacyl-tRNA species (Schimmel & Soll, 1979). Thus, if a mischarged tRNA species is released from an enzyme, it can be carried into the ribosomal translation apparatus by Tu and insert a missense substitution in a growing polypeptide chain. The role of editing of charged mutant amber suppressors in vivo, and the influence of this potential reaction on the results obtained with the amber suppression assay, is unknown. When a G3·U70 base pair is transferred into $tRNA^{Cys/CUA}$, the resulting mutant amber suppressor can be aminoacylated (with alanine) in vitro with purified alanine tRNA synthetase (Hou & Schimmel, 1988). The overall yield of aminoacylation of Ala-G3·U70 $tRNA^{Cys/CUA}$ is dependent on the enzyme concentration, however, and may reflect an editing reaction of the alanine tRNA synthetase that is due to enzyme-catalyzed deacylation of Ala-G3·U70 $tRNA^{Cys/CUA}$ and, additionally or alternatively, to some hydrolysis of the enzyme-bound alanyl adenylate when presented with G3·U70 $tRNA^{Cys/CUA}$. Specific editing reactions with "transformed" tRNAs remain to be demonstrated, but these considerations illustrate the role that editing may have in determining the identity of transfer RNAs.

Defining Sites on Aminoacyl-tRNA Synthetases That Determine Transfer RNA Identity. Structural information on aminoacyl-tRNA synthetases has been limited to *Bacillus stearothermophilus* tyrosine and *E. coli* methionine tRNA synthetases (Bhat et al., 1982; Risler et al., 1982; Blow et al., 1983; Brunie et al., 1987). A cocrystal with tRNA has not been obtained in either case. For tyrosine tRNA synthetase, there is a nucleotide fold in the amino-terminal half which is the location of ATP and tyrosine binding sites which have been extensively analyzed by Fersht and co-workers by site-directed mutagenesis (Fersht et al., 1984; Fersht, 1985). The C-terminal half of the protein is required for binding of $tRNA^{Tyr}$ but is disordered in the crystal (Blow et al., 1982; Bedouelle & Winter, 1986). The tRNA binding site has not been definitively located in the structure of methionine tRNA synthetase, although the tRNA cross-linking experiments of Schulman and co-workers have provided evidence that at least a part of the interaction involves the C-terminal half of the protein (Valenzuela et al., 1984; Valenzuela & Schulman, 1986).

Cocrystals of yeast aspartyl-tRNA synthetase with $tRNA^{Asp}$ (Lorber et al., 1983; Podjarny et al., 1987) and of *E. coli* glutamine tRNA synthetase with $tRNA^{Gln}$ (Perona et al., 1988) have been obtained. The latter crystals diffract to high resolution and will provide the first well-defined structure of a complex. The specific binding interactions in the synthetase–tRNA complexes are relatively weak, and the ones most critical will have to be differentiated from weak nonspecific protein–tRNA contacts. It should be possible from the model of a complex to define nucleotide replacements in the bound tRNA that would not be accommodated in the complex. Such nucleotides may constitute the "negative determinants" described above. The possibility of defining the structural basis for discrimination by the k_{cat} parameter is more problematic, because this is influenced by interactions in a transition state and not in the form which is isolated in a cocrystal. The structural interpretation of mutants of glutamine tRNA synthetase that cause misacylation may be instructive (Inokuchi et al., 1984).

For alanine tRNA synthetase, much of the recognition is concentrated in the amino acid acceptor helix and is centered on the G3·U70 base pair. It is not known whether the enzyme recognizes specific atoms in the base pair or a helix variation at this location. It is noteworthy that the U3·G70 $tRNA^{Ala/CUA}$ is not aminoacylated in vitro by substrate levels of the enzyme (Figure 3). In experiments with nucleases that have respectively a strong preference for cleavage in single- or double-stranded regions, the G3·U70 base pair behaves as though it is part of a double-stranded section (Park & Schimmel, 1988).

Alanine tRNA synthetase is a tetramer of identical 875 amino acid polypeptides (Putney et al., 1981a,b). Eighteen fragments of the enzyme have been created and analyzed, and from these studies the locations have been defined for regions important for synthesis of alanyl adenylate, reaction of the adenylate with bound tRNA, binding of tRNA, and tetramer formation (Jasin et al., 1983, 1984, 1985; Regan et al., 1987; Hill & Schimmel, 1989). The first 368 amino acids encode a domain that has the adenylate synthesis activity. Sequences in the 93 amino acid segment from Thr369 to Asp461 have a major influence on both k_{cat} and K_m parameters for the tRNA-dependent step of aminoacylation (Jasin et al., 1983, 1984; Ho et al., 1985; Regan et al., 1987, 1988). Because both parameters are sensitive to the presence of a G3•U70 base pair, the region from Thr369 to Asp461 may encode the structure which recognizes the amino acid acceptor helix of $tRNA^{Ala}$. The isolation of mutations in the enzyme that compensate for substitutions at position 3•70 may further define the location of the critical amino acids and may help to interpret the structural information that is being obtained on this enzyme. The conformational analysis and tRNA interactions of synthetic peptides that recreate at least a part of the binding site may also be instructive.

REFERENCES

Baldwin, A. N., & Berg, P. (1966) *J. Biol. Chem. 241*, 839.

Bedouelle, H., & Winter, G. (1986) *Nature 320*, 371–373.

Bhat, T. N., Blow, D. M., Brick, P., & Nyborg, J. (1982) *J. Mol. Biol. 158*, 699–709.

Blow, D. M., Bhat, T. N., Metclafe, A., Risler, J. L., Brunie, S., & Zelwer, C. (1983) *J. Mol. Biol. 171*, 571–576.

Bruce, A. G., & Uhlenbeck, O. C. (1982) *Biochemistry 21*, 3921.

Brunie, S., Mellot, P., Zelwer, C., Risler, J.-L., Blanquet, S., & Fayat, G. (1987) *J. Mol. Graphics*, 18–21.

Bush, B. L., Jones, T. A., Pflugrath, J. W., & Saper, M. A. (1987) in *PS300 FRODO-Molecular Graphics Program for the PS300* (Version 6.4) (Sack, J. S., Ed.).

Celis, J. E., Hooper, M. L., & Smith, J. D. (1973) *Nature, New Biol. 224*, 261–264.

Celis, J. E., Coulander, C., & Miller, J. H. (1976) *J. Mol. Biol. 104*, 729–734.

de Duve, C. (1988) *Nature 333*, 117–118.

Eldred, E. W., & Schimmel, P. R. (1972) *J. Biol. Chem. 247*, 2961–2964.

Fersht, A. R. (1985) *Enzyme Structure & Mechanism*, 2nd ed., W. H. Freeman, New York.

Fersht, A. R., Shi, J.-P., Wilkinson, A. J., Blow, D. M., Carter, P., Waye, M. M. Y., & Winter, G. P. (1984) *Angew. Chem., Int. Ed. Engl. 23*, 467–473.

Francklyn, C., & Schimmel, P. (1989) *Nature 337*, 478–481.

Frederick, C., Wang, A., Rich, A., Regan, L., & Schimmel, P. (1988) *J. Mol. Biol. 203*, 521–522.

Ghysen, A., & Celis, J. E. (1974) *J. Mol. Biol. 83*, 333–351.

Heck, J. D., & Hatfield, G. W. (1988) *J. Biol. Chem. 263*, 868–877.

Hill, K., & Schimmel, P. (1989) *Biochemistry 28*, 2577–2586.

Ho, C., Jasin, M., & Schimmel, P. (1985) *Science 229*, 389–393.

Hooper, M. L., Russell, R. L., & Smith, J. D. (1972) *FEBS Lett. 22*, 149–155.

Hou, Y.-M., & Schimmel, P. (1988) *Nature 333*, 140–145.

Inokuchi, H., Celis, J. E., & Smith, J. D. (1974) *J. Mol. Biol. 85*, 187–192.

Jasin, M., Regan, L., & Schimmel, P. (1983) *Nature 306*, 441–447.

Jasin, M., Regan, L., & Schimmel, P. (1984) *Cell 36*, 1089–1095.

Jasin, M., Regan, L., & Schimmel, P. (1985) *J. Biol. Chem. 260*, 2226–2230.

Jin, Y., Qiu, M., Li, W., Zeng, K., Bao, J., Gong, P., Wu, R., & Wang, D. (1987) *Anal. Biochem. 161*, 453–459.

Kim, S. H., Suddath, F. L., Quigley, G. J., McPherson, A., Sussman, J. L., Wang, A. H.-J., Seeman, N. C., & Rich, A. (1974) *Science 185*, 435–440.

Lam, S. S. M., & Schimmel, P. R. (1975) *Biochemistry 14*, 2775–2780.

Lorber, B., Giege, R., Ebel, J.-P., Berthet, C., Thierry, J.-C., & Moras, D. (1983) *J. Biol. Chem. 258*, 8429–8435.

Masson, J. M., & Miller, J. H. (1986) *Gene 47*, 179–183.

McClain, W. H., & Foss, K. (1988a) *Science 240*, 793–796.

McClain, W. H., & Foss, K. (1988b) *J. Mol. Biol. 202*, 697–709.

McClain, W. H., & Foss, K. (1988c) *Science 241*, 1804–1807.

McClain, W. H., Chen, Y.-M., Foss, K., & Schneider, J. (1988) *Science 242*, 1681–1684.

Mims, B. H., Prather, N. E., & Murgola, E. J. (1985) *J. Bacteriol. 162*, 837–839.

Muramatsu, T., Nishikawa, K., Nemoto, F., Kuchino, Y., Nishimura, S., Miyazawa, T., & Yokoyama, S. (1988a) *Nature 336*, 179–181.

Muramatsu, T., Yokoyama, S., Horie, N., Matsuda, A., Ueda, T., Yamaizumi, Z., Kuchino, Y., Nishimura, S., & Miyazawa, T. (1988b) *J. Biol. Chem. 263*, 9261–9267.

Murgola, E. J., & Hijazi, K. A. (1983) *Mol. Gen. Genet. 191*, 132–137.

Normanly, J., Ogden, R. C., Horvath, S. J., & Abelson, J. (1986a) *Nature 321*, 213–219.

Normanly, J., Masson, J.-M., Kleina, L., Abelson, J., & Miller, J. H. (1986b) *Proc. Natl. Acad. Sci. U.S.A. 83*, 6548–6552.

Ozeki, H., Inokuchi, H., Yamao, F., Kodaira, M., Sakano, H., Ikemura, T., & Shimura, Y. (1980) in *Transfer RNA: Biological Aspects* (Soll, D., Abelson, J., & Schimmel, P. R., Eds.) pp 341–362, Cold Spring Harbor Laboratory, Cold Spring Harbor, NY.

Park, S. J., & Schimmel, P. (1988) *J. Biol. Chem. 263*, 16527–16530.

Park, S. J., Hou, Y.-M., & Schimmel, P. (1989) *Biochemistry 28*, 2740–2746.

Perona, J. J., Swanson, R., & Steitz, T. A. (1988) *J. Mol. Biol. 202*, 121–126.

Podjarny, A., Rees, B., Thierry, J. C., Cavarelli, J., Jesoir, J. C., Roth, M., Lewitt-Bentley, A., Kahn, R., Lorber, B., Ebel, J. P., Giege, R., & Moras, D. (1987) *J. Biomol. Struct. Dyn. 5*, 187–198.

Prather, N. E., Murgola, E. J., & Mims, B. H. (1984) *J. Mol. Biol. 172*, 177–184.

Putney, S. C., Royal, N. J., Neuman de Vegvar, H., Herlihy, W. C., Biemann, K., & Schimmel, P. (1981) *Science 213*, 1497–1501.

Putney, S. D., Sauer, R., & Schimmel, P. R. (1981) *J. Biol. Chem. 256*, 198–204.

Raftery, L. A., & Yarus, M. (1987) *EMBO J. 6*, 1499–1506.

RajBhandary, U. L. (1988) *Nature 336*, 112–113.

Regan, L., Dignam, J. D., & Schimmel, P. (1986) *J. Biol. Chem. 261*, 5241–5244.

Regan, L., Bowie, J., & Schimmel, P. (1987) *Science 235*, 1651–1653.

Regan, L., Buxbaum, L., Hill, K., & Schimmel, P. (1989) *J. Biol. Chem.* (in press).

Rich, A., & Schimmel, P. R. (1977) *Nucleic Acids Res. 4*, 1649–1655.

Risler, J. L., Zelwer, C., & Brunie, S. (1982) *Nature 292*, 383–386.
Robertus, J. D., Ladner, J. E., Finch, J. T., Rhodes, D., Brown, R. S., Clark, B. F. C., & Klug, A. (1974) *Nature 250*, 546–551.
Rogers, M. J., & Soll, D. (1988) *Proc. Natl. Acad. Sci. U.S.A. 85*, 6627–6631.
Sampson, J., & Uhlenbeck, O. C. (1988) *Proc. Natl. Acad. Sci. U.S.A. 85*, 1033–1037.
Sampson, J. K., DeRenzo, A., Behlen, L., & Uhlenbeck, O. C. (1989) *Science* (in press).
Samuelsson, T., Boren, T., Johansen, T.-I., & Lustig, F. (1988) *J. Biol. Chem. 263*, 13692–13699.
Schimmel, P. R. (1977) *Acc. Chem. Res. 10*, 411–418.
Schimmel, P. (1987) *Annu. Rev. Biochem. 56*, 125–158.
Schimmel, P. (1989) *Adv. Enzymol.* (in press).
Schimmel, P. R., & Soll, D. (1979) *Annu. Rev. Biochem. 48*, 601–648.
Schreier, A. A., & Schimmel, P. R. (1972) *Biochemistry 11*, 1582–1589.
Schulman, L. H., & Abelson, J. (1988) *Science 240*, 1591–1592.
Schulman, L. H., & Pelka, H. (1989) *Science* (in press).
Seong, B. L., & RajBhandary, U. L. (1987) *Proc. Natl. Acad. Sci. U.S.A. 84*, 8859–8863.
Shimura, Y., Aono, H., Ozeki, H., Sarabhai, A., Lamfrom, H., & Abelson, J. (1972) *FEBS Lett. 22*, 144–148.
Smith, J. D., & Celis, J. E. (1973) *Nature, New Biol. 243*, 66–71.
Smith, J. D., Barnett, L., Brenner, S., & Russell, R. L. (1970) *J. Mol. Biol. 54*, 1–14.
Soll, D., & Schimmel, P. R. (1974) *Enzymes 10*, 489–538.
Soll, L. (1974) *J. Mol. Biol. 86*, 233–243.
Sprinzl, M., Hartmann, T., Meissner, F., Moll, H., & Vorderwulbecke, T. (1987) *Nucleic Acids Res. 15*, r53–r188.
Swanson, R., Hoben, P., Sumner-Smith, M., Uemura, H., Watson, L., & Soll, D. (1988) *Science 242*, 1548–1551.
Valenzuela, D., & Schulman, L. H. (1986) *Biochemistry 25*, 4555–4561.
Valenzuela, D., Leon, O., & Schulman, L. H. (1984) *Biochem. Biophys. Res. Commun. 119*, 677–684.
Waye, M. Y., Winter, G., Wilkinson, A. J., & Fersht, A. R. (1983) *EMBO J. 2*, 1827–1829.
Yamo, F., Inokuchi, H., & Ozeki, H. (1988a) *Jpn. J. Genet. 63*, 237–249.
Yamo, F., Inokuchi, H., Normanly, J., Abelson, J., & Ozeki, H. (1988b) *Jpn. J. Genet. 63*, 251–258.
Yaniv, M., Folk, W. R., Berg, P., & Soll, L. (1974) *J. Mol. Biol. 86*, 245–260.
Yarus, M. (1972) *Proc. Natl. Acad. Sci. U.S.A. 69*, 1915–1919.
Yarus, M. (1979) in *Transfer RNA: Structure, Properties, & Recognition* (Schimmel, P. R., Soll, D., & Abelson, J., Eds.) pp 501–515, Cold Spring Harbor Laboratory, Cold Spring Harbor, NY.
Yarus, M. (1986) *J. Mol. Biol. 192*, 235–255.
Yarus, M. (1988) *Cell 55*, 739–741.
Yarus, M., & Berg, P. (1970) *Anal. Biochem. 35*, 450–465.
Yarus, M., Knowlton, R., & Soll, L. (1977) in *Nucleic Acid-Protein Recognition* (Vogel, H. J., Ed.) pp 391–408, Academic Press, New York.

Chapter 20

Mechanisms of Aminoacyl-tRNA Synthetases: A Critical Consideration of Recent Results

Wolfgang Freist

Max-Planck-Institut für experimentelle Medizin, Abteilung Chemie, Göttingen, West Germany

Received March 17, 1989; Revised Manuscript Received April 19, 1989

ABSTRACT: During the last 10 years intensive and detailed studies on mechanisms and specificities of aminoacyl-tRNA synthetases have been carried out. Physical measurements, chemical modification of substrates, site-directed mutagenesis, and determination of kinetic parameters in misacylation reactions with noncognate amino acids have provided extensive knowledge which is now considered critically for its consistency. A common picture emerges: (1) The enzymes work with different catalytic cycles, kinetic constants, and specificities under different assay conditions. (2) Chemical modifications of substrates can have comparable influence on catalysis as can changes in assay conditions. (3) All enzymes show a specificity for the 2′- or 3′-position of the tRNA. (4) Hydrolytic proofreading is achieved in a pre- and a posttransfer process. In most cases pretransfer proofreading is the main step; posttransfer proofreading is often marginal. (5) Initial discrimination of substrates takes place in a two-step binding process. For some investigated enzymes, initial discrimination factors were found to depend on hydrophobic interaction and hydrogen bonds. (6) The overall recognition of amino acids is achieved in a process of at least four steps. At present, only a rough overall picture of aminoacyl-tRNA synthetase action can be given.

Soon after the first aminoacyl-tRNA synthetases were discovered, a general reaction mechanism was postulated by which these enzymes catalyze aminoacylation of tRNAs [for compilations, see Loftfield (1972), Kisselev and Favorova (1974), and Söll and Schimmel (1974)]. According to this mechanism, which is now generally accepted, amino acids are first converted to aminoacyladenylates in the activation step and then attached to tRNA in the transfer step.

$$\mathrm{E + aa + ATP \rightarrow E{\cdot}aa\text{-}AMP + PP_i}$$

$$\mathrm{E{\cdot}aa\text{-}AMP + tRNA \rightarrow E + aa\text{-}tRNA + AMP}$$

In contrast to this general opinion, some workers have insisted on a one-step mechanism (Loftfield, 1972; Loftfield & Eigner, 1969; Lövgren et al., 1975; Deutscher, 1967; Parfait et al., 1972; Thiebe, 1983), and it was also assumed that other unknown reactive intermediates formed with amino acids may play a role in aminoacylation (Thiebe, 1975; Kovaleva et al., 1983).

In spite of the fact that not all questions concerning the chemical mechanism had been solved, the main interest of most research groups turned to the problem of amino acid specificity of the enzymes. Investigations on the reasons for the high enzyme accuracies were carried out with physical methods such as rapid kinetic experiments, equilibrium dialysis, and filtration as well as chemical methods such as modification of substrates, and tests of reactivity of these compounds were applied. Sometimes results obtained by these different methods seemed to be in contradiction and were the subject matter of controversial discussions. Today, regarding these things retrospectively together with new results, important inconsistencies no longer exist.

Influence of Assay Conditions and Chemical Modification. It is the aim of all investigations on aminoacyl-tRNA synthetases to obtain information about the function and the mechanism with which these enzymes work under physiological conditions in the cell. Unfortunately, nobody can simulate the exact physiological conditions in his enzyme assay. Normally, after preparation of an enzyme, biochemists look for assay conditions under which they get the highest enzyme activity or which they just like most and carry out their experiments in this assay. In this way research groups apply to their experiments different temperatures, buffers, and additional compounds. Conclusions on enzyme functions in the cell are difficult, and even comparisons of results reported by different

Table I: Different Orders of Substrate Additions and Product Releases and k_{cat} and K_m Values for Isoleucyl-tRNA Synthetase from Yeast[a]

order	nomenclature	conditions[b]	k_{cat} (s^{-1})	K_m (mM)
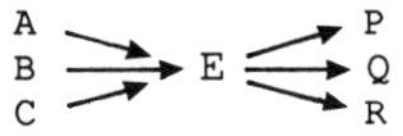 A, B, C → E → P, Q, R	sequential random Ter-Ter	pH 6.5	0.47	0.03
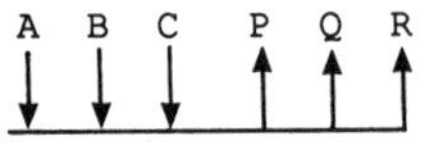 A B C ↓ ↓ ↓ P Q R ↑ ↑ ↑	sequential ordered Ter-Ter	pH 7.65	0.83	0.02
		pH 8.6	0.52	0.02
		pH 7.65 + EF-Tu·GTP	0.28	0.02
		pH 7.65 + EF-Tu·GTP + PPase	0.23	0.05
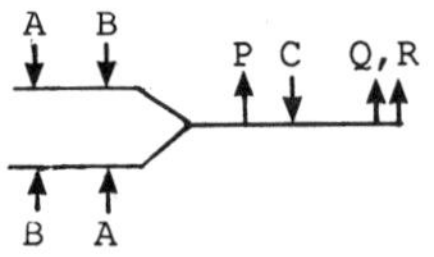 A B ↓ ↓ / B A ↓ ↓; P C Q,R	random Bi-Uni Uni-Bi ping-pong	pH 7.65 + PPase	1.40	0.04
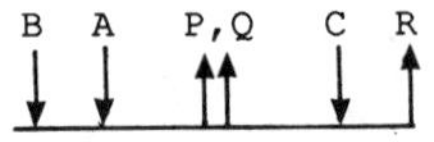 B A ↓ ↓ P,Q ↑↑ C ↓ R ↑	Bi-Bi Uni-Uni ping-pong	pH 7.65 + EF-Tu·GTP + spermine	0.098	0.02
		pH 7.65 + EF-Tu·GTP + spermine (KCl lower)	0.135	0.008

[a] A = ATP, B = isoleucine, C = tRNA^{Ile}-C-C-A, P = pyrophosphate, Q = Ile-tRNA^{Ile}-C-C-A, and R = AMP; nomenclature according to Cleland (1963, 1970). [b] For complete conditions, see Freist and Sternbach (1984) and Freist et al. (1985).

Table II: Order of Substrate Addition and Product Release in Aminoacylation of Modified tRNA^{Ile}-C-C-N for Isoleucyl-tRNA Synthetase from Yeast[a]

tRNA^{Ile}-C-C-N	order	nomenclature	k_{cat} (s^{-1})	K_m (mM)
tRNA^{Ile}-C-C-3′-dA	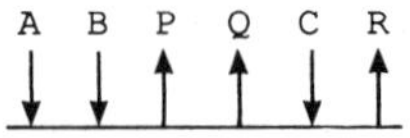 A ↓ B ↓ P ↑ Q ↑ C ↓ R ↑	Bi-Bi Uni-Uni ping-pong	0.2	0.002
tRNA^{Ile}-C-C-A(3′-NH_2)	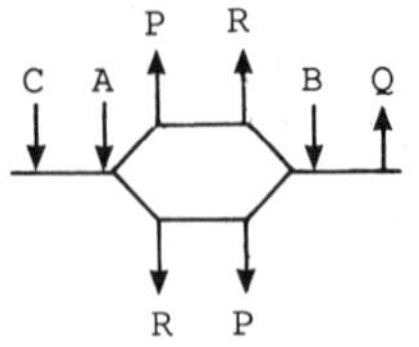 C A ↓ ↓; P R / R P; B ↓ Q ↑	Bi-random-Bi Uni-Uni ping-pong	0.1	0.006

[a] A = ATP, B = isoleucine, C = tRNA^{Ile}-C-C-N, P = pyrophosphate, Q = Ile-tRNA^{Ile}-C-C-N, and R = AMP; nomenclature of orders according to Cleland (1963, 1970); k_{cat} values and K_m values of isoleucine obtained under standard conditions (Freist & Cramer, 1983).

research groups may be impossible.

The influence of assay conditions on the aminoacylation pathway may be illustrated by results obtained with isoleucyl-tRNA synthetase from bakers' yeast, a single-chain enzyme with a molecular weight of 123 000 (Freist & Sternbach, 1984; Freist et al., 1985; Englisch et al., 1987). Four different orders of substrate additions and product releases were found under eight different reaction conditions by initial rate kinetic analyses (Table I). k_{cat} values changed in the range of 0.098–1.4 s^{-1} and K_m values of isoleucine in the range of 0.008–0.05 mM. Addition of inorganic pyrophosphatase, elongation factor EF-Tu·GTP, and spermine can be regarded as a stepwise, but still insufficient, approximation to physiological conditions. The results show that experiments carried out with unmodified natural substrates in the "standard reaction mixture" at pH 7.65 may differ considerably from results obtained under physiological conditions under which additional compounds of the cytosol are present.

It should be emphasized that changes in the order of substrate additions and product releases must not affect the chemical pathway of the reaction. In my opinion the adenylate mechanism must be used by the enzyme in all cases, as will be suggested below in two hypothetical schemes of the catalytic cycle. Even for the curious Bi-Bi Uni-Uni ping-pong order [for nomenclature, see Cleland (1963, 1970)] in which aminoacylated tRNA^{Ile}-C-C-A is released before addition of free tRNA^{Ile}-C-C-A, a catalytic cycle with adenylate formation can be formulated if one assumes that the enzyme has more than one binding site for the substrates and products.

In many mechanistic studies on aminoacyl-tRNA synthetases, modified tRNAs were used as tools [compilations in Sprinzl and Cramer (1978), Chladek and Sprinzl (1985), and Freist (1988)]. As described above, orders of substrate additions and product releases were found to be very sensitive for changes of reaction conditions in the case of isoleucyl-tRNA synthetase from yeast. In Table II are shown the orders of substrate additions and product releases observed in aminoacylation of two modified tRNA^{Ile}s, namely, tRNA^{Ile}-C-C-3′-dA and tRNA^{Ile}-C-C-A(3′-NH_2).

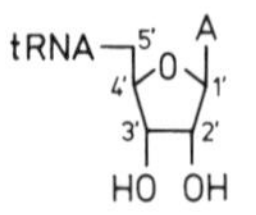

tRNA–5′ A 4′ O 1′ 3′ 2′ HO OH

tRNA-C-C-A

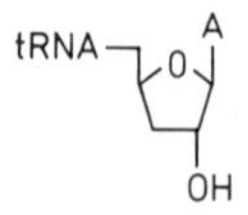

tRNA– O A OH

tRNA-C-C-3′-dA

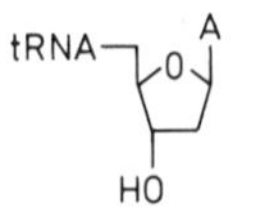

tRNA– O A HO

tRNA-C-C-2′-dA

tRNA– O A H_2N OH

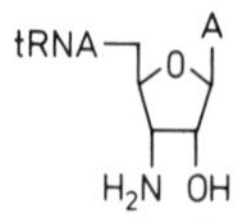

tRNA-C-C-A(3′-NH_2)

These orders deviate from those obtained with natural tRNA^{Ile}-C-C-A under standard conditions. However, these deviations are within the scope of changes that are also observed by variation of buffer conditions or addition of natural compounds of the cytosol such as elongation factor EF-Tu·GTP. In both cases the same care has to be taken in conclusions on enzyme mechanisms occurring in the cytosol.

Enzyme–Substrate Interaction. Substrate analogues have often been tested for their properties as substrates or inhibitors in the aminoacylation reaction [e.g., von der Haar and Cramer (1978) and Freist et al. (1981)]. From inhibition patterns of dead-end inhibitors, hints on the number of substrate binding sites were observed in several cases; further conclusions could be drawn concerning the essential atom groups of the substrates. The results agree in part with investigations obtained by physical methods.

For phenylalanyl-tRNA synthetase from yeast [subunit structure $\alpha_2\beta_2$, M_r = 276 000; compilation of subunit structures and M_r values in Joachimiak and Barciszweski (1980)] it was found by ultracentrifugation analysis, fluorescence titrations, and fast kinetic techniques that the enzyme binds two molecules of tRNA^{Phe}-C-C-A (Krauss et al., 1976, 1975). When chemically modified tRNA^{Phe}s were tested in the aminoacylation reaction, actually some noncompetitive inhibitors of the mixed type were found, indicating a second tRNA binding site (von der Haar & Gaertner, 1975). In this case, results obtained with chemically modified compounds were consistent with results obtained by physical methods.

Differing results were obtained with isoleucyl-tRNA synthetases from yeast and *Escherichia coli* K12 (subunit structures α_1, M_r = 123 000 and 110 000). It was concluded from equilibrium partition or dialysis experiments that one enzyme molecule binds one molecule each of its three substrates (Fersht & Kaethner, 1976a; Berthelot & Yaniv, 1970; Yarus & Berg, 1969; Hustedt & Kula, 1977; Hustedt et al., 1977). Burst experiments with ^{32}P-labeled ATP and filtration of mixtures containing enzyme, labeled ATP, and isoleucine on nitrocellulose filters exhibited also a binding stoichiometry of one molecule per enzyme molecule for isoleucyladenylate, the intermediate after the first reaction step (Fersht & Kaethner, 1976a; Hustedt et al., 1977; Norris & Berg, 1964).

However, when modified tRNAs were tested as inhibitors of the aminoacylation reaction, classical noncompetitive inhibition was observed (von der Haar & Cramer, 1978), results that can only be explained either by the presence of a second binding site for tRNA on one enzyme molecule or by dimerization of the synthetase during catalytic action. In fact, polymerization phenomena could be observed with the enzyme from *E. coli* B (structure α_1, M_r = 112 000) in equilibrium sedimentation experiments (Baldwin & Berg, 1966).

As will be mentioned below, the differing results may perhaps be due to different pathways of aminoacylation depending on assay conditions.

The first opportunity to compare conclusions on essential parts of the ATP molecule made from substrate specificity studies (Freist et al., 1976; Freist & Sternbach, 1988) with results obtained from X-ray analysis was when the structure of the tyrosyl-tRNA synthetase–tyrosyladenylate complex (subunit structure α_2, M_r = 88 000) was solved and several hydrogen bonds between amino acids of the enzyme and tyrosyladenylate were postulated (Irwin et al., 1976; Rubin & Blow, 1981; Fersht et al., 1984; Fersht, 1987). As followed from substrate specificity with regard to ATP analogues, tyrosyl-tRNA synthetases check the hydroxyl groups of the ribose moiety, the conformation at the glycosyl bond, and especially the amino group in position 6 of the adenine part. The results of X-ray analysis agree with these findings except those concerning the adenine base. Whereas no specific interactions between the 6-amino group of the adenine base and the enzyme could be detected in the crystals, tyrosyl-tRNA synthetases do not accept an ATP analogue lacking this group as substrate in the aminoacylation reaction. The specificity for an intact, unmodified adenine moiety shows that during the catalytic cycle the enzyme must check the ATP molecule in a different conformation than found in the crystal, which makes it possible for the enzyme to discriminate ATP from, for example, GTP in the cytosol. Both methods give results that complement one another.

Postulation of Catalytic Cycles. On the basis of binding studies and comparison of the rate constants for the catalytic process, the first catalytic cycle was proposed in 1969 (Yarus & Berg) for isoleucyl-tRNA synthetase from *E. coli* B. In this reaction scheme release of the acylated tRNA is the rate-determining step; for the complete aminoacylation reaction a rate constant of 0.05 s^{-1} was found.

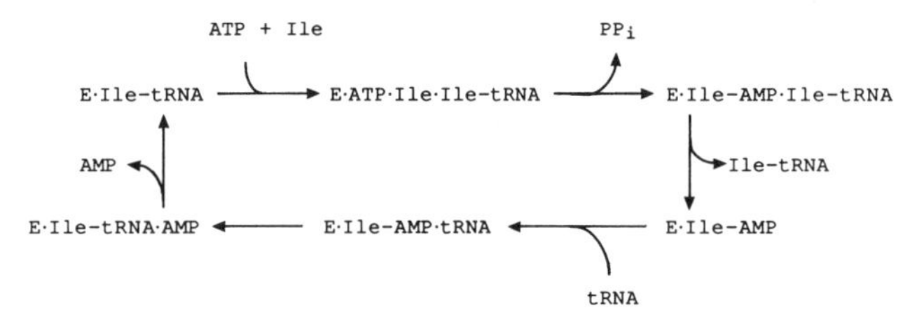

The enzyme from *E. coli* K12, as studied by pulsed quenched-flow techniques, has a different rate-determining step (Fersht & Kaethner, 1976; Fersht & Jakes, 1975): Transfer of the activated amino acid to the tRNA is crucial for the overall rate of the aminoacylation (k = 1.46 s^{-1}).

$$\text{E·tRNA·Ile·ATP} \xrightarrow[\text{fast}]{-\text{PPi}} \text{E·tRNA·Ile-AMP} \xrightarrow[\text{slow}]{} \text{E + Ile-tRNA + AMP}$$

As shown in Table I for isoleucyl-tRNA synthetase from yeast, turnover numbers in the range of 0.1–1.4 s^{-1} were found depending on reaction conditions. To accommodate all the different results that emerged in studies using only physical methods, one can only speculate. For example, the enzyme may act in two types of different catalytic cycles, one rate determined by dissociation of the enzyme–aminoacyl-tRNA complex and the other rate determined by the transfer step (Freist & Sternbach, 1989). Two model cycles that are consistent with a Ter-Ter order of substrate additions and product releases are shown in Scheme I.

The two cycles differ in the number of substrate and product molecules attached to the enzyme. In cycle 1 only one tRNA molecule is complexed with the enzyme; cycle 2 involves complexes in which two tRNA molecules are bound to the enzyme. The protein "E" may be a monomer or a dimer. Similar cycles were also proposed for phenylalanyl- and valyl-tRNA synthetases (Thiebe, 1978; Kern & Gangloff, 1981).

Whereas in cycle 1 a low dissociation rate of about 0.05 s^{-1} may be valid for dissociation of a 1:1 complex of enzyme and tRNA, in cycle 2 from the complex of enzyme and two tRNA molecules tRNA may dissociate at higher rates and the transfer step may become rate determining.

In principle, these hypothetical cycles may be correct, but many details are only suspected because no experiments have been done to investigate the single steps of the cycles under appropriate assay conditions. Most probably, it is impossible to formulate a standard pathway, and the enzymes may function in different pathways according to the growth stages of the cell.

Site of Aminoacylation. An excellent example for elucidation of an important part of the aminoacylation mechanism using modified tRNAs was the determination of the aminoacylation site on the tRNA. It was found that aminoacyl-tRNA synthetases do not attach their cognate amino acids at the same position on tRNA molecules but show specificities for the 2′- or 3′-position (Sprinzl & Cramer, 1975; Fraser & Rich, 1975). Two different types of modified tRNAs were used: tRNA-C-C-2′-dA and tRNA-C-C-3′-dA, which lack one of the terminal hydroxyl groups, and tRNA-C-C-A(2′-

Scheme I

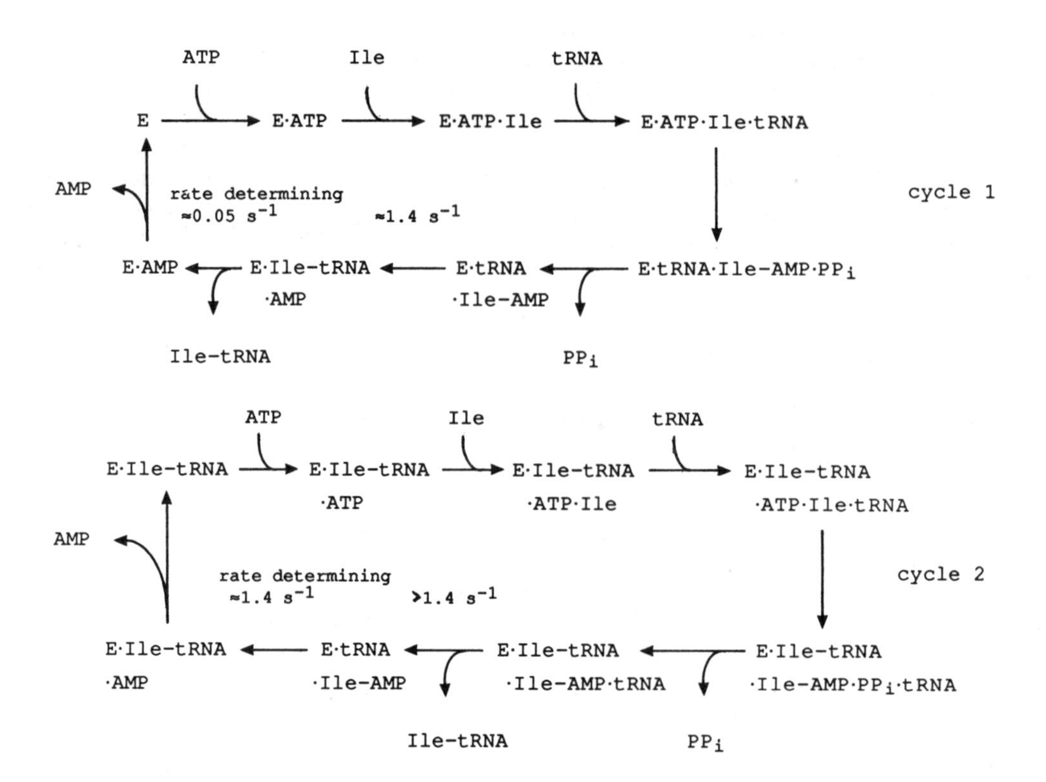

NH_2) and tRNA-C-C-A(3′-NH_2), in which one of the terminal hydroxyl groups is substituted by an amino group. The results obtained with both types of tRNAs were only slightly different and could be combined to a common and general picture (Sprinzl & Cramer, 1978). One could argue that specificity tests carried out with modified tRNAs may not be the same as acylation of natural tRNA under physiological conditions. This objection could be substantiated further because it was shown that orders of substrate additions and product releases are different in aminoacylation of $tRNA^{Ile}$-C-C-A, $tRNA^{Ile}$-C-C-3′-dA, and $tRNA^{Ile}$-C-C-A-(3′-NH_2) (Table II). However, the fact that very similar specificities for the 2′- or 3′-OH group were observed independent of the type of modification may indicate a general property of aminoacyl-tRNA synthetases that may also appear under conditions of protein biosynthesis. Furthermore, it was shown that the 2′,3′ specificity of the enzymes is only lost under extreme conditions at pH 9, at low salt concentrations, and after addition of organic solvents or inorganic pyrophosphatase (Igloi & Cramer, 1979; Freist & Cramer, 1983).

Early Hypotheses on Proofreading Mechanisms. When the amino acid specificities of aminoacyl-tRNA synthetases were determined under in vivo conditions, it was found that they were much higher than would be expected due to differences in binding energies between cognate and noncognate amino acids. For example, the error rate for incorporation of valine instead of isoleucine is 1:3000 in protein biosynthesis (Loftfield & Vanderjagt, 1972). Pauling had calculated a factor of only 4.3 by which incorporation of isoleucine may be favored rather than valine due to higher van der Waals forces (Pauling, 1958). Thus, special proofreading mechanisms were postulated by which the enzymes achieve their high discrimination rates.

The term "chemical proofreading" was created for a correction mechanism by which ester bonds between tRNAs and noncognate amino acids are hydrolyzed by the enzyme before release of the "wrong" product (von der Haar & Cramer, 1976). The hydrolytic capacity of some aminoacyl-tRNA synthetases was measured by incubation of aminoacyl-tRNAs with their cognate enzyme. Val-$tRNA^{Ile}$-C-C-A was hydrolyzed 3–5 times faster than Ile-$tRNA^{Ile}$-C-C-A (von der Haar & Cramer, 1976; Freist & Sternbach, 1989), proposing a "posttransfer proofreading factor" of about 3, a value that was also found later by other methods (Freist et al., 1988). Aminoacyl esters of tRNAs with a 2′- or 3′-deoxyadenosine instead of adenosine as terminal nucleoside were hydrolyzed 10–300 times slower than esters of the natural tRNA. As a rationale for the observation that two hydroxyl groups are essential for effective enzymic hydrolysis of aminoacyl-tRNA esters, it was assumed that hydrolysis takes place only when the aminoacyl moiety has migrated from the OH group to which it is initially attached to the nonaccepting OH. Thus, there would be one OH responsible for acylation and one for hydrolysis (von der Haar & Cramer, 1976; Fersht & Kaethner, 1976b).

According to new investigations (Freist & Sternbach, 1989), it is probable that higher stability of aminoacyldeoxy-tRNAs is due to a simple neighboring group effect of the OH groups as was found for cyclopentanediol esters in chemical hydrolysis (Bruice & Fife, 1962). The nonenzymic hydrolysis rate of cyclopentyl acetate is enhanced by a factor of about 30 by a hydroxyl group in a neighboring position, and this is also the order of magnitude by which aminoacyl-$tRNA^{Ile}$-C-C-A is hydrolyzed faster than aminoacyl-$tRNA^{Ile}$-C-C-dA.

In our recent work (Freist & Sternbach, 1989) we have prepared isoleucyl and valyl esters of $tRNA^{Ile}$-C-C-3′-dA and $tRNA^{Ile}$-C-C-2′-dA. The esters of both tRNAs were stable against enzymic hydrolysis by isoleucyl-tRNA synthetase from bakers' yeast to the same extent, so that at least for this enzyme no preference for one of the two terminal hydroxyl groups in hydrolysis can be postulated.

From kinetic investigations of isoleucyl-tRNA synthetase from *E. coli* K12, it was concluded that rejection of valine in aminoacylation of $tRNA^{Ile}$-C-C-A is mainly achieved by an "editing mechanism" on the pretransfer level (Fersht, 1977a): The enzyme-bound adenylate of noncognate valine is hydrolyzed in a pretransfer proofreading step, and posttransfer hydrolysis of Val-$tRNA^{Ile}$-C-C-A only occurs to a minor extent and is considered to be a "mopping up" step.

At first glance chemical proofreading and the editing mechanism seemed in contrast. However, some years later it was shown more clearly that aminoacyl-tRNA synthetases use both types of error corrections but to different extents (Lin et al., 1984; Freist et al., 1985, 1988; Cramer & Freist, 1987; Freist & Sternbach, 1988).

Two-Step Proofreading. The contribution of proofreading processes to specificity of the aminoacyl-tRNA synthetases can be determined by measuring AMP formation stoichiometry of the aminoacylation reaction. For each misactivated, mistransferred, and rejected noncognate amino acid, one additional molecule of ATP must be hydrolyzed. For example, for mischarging of $tRNA^{Ile}$-C-C-A with valine by isoleucyl-tRNA synthetase from *E. coli* B, a total of 270 AMP molecules formed per molecule of Val-$tRNA^{Ile}$-C-C-A was found in the presence of elongation factor EF-Tu·GTP (Hopfield et al., 1976); a similar value of 192 was obtained for the yeast enzyme under the same conditions (Freist et al., 1985); under standard conditions the numbers were 1490 and 828 (Freist et al., 1988; Cramer et al., 1979). In the first stages of these investigations authors assigned the reaction steps responsible for the high AMP formation rates exclusively either to a pretransfer proofreading event (Hopfield et al., 1976) or to a posttransfer process (Cramer et al., 1979).

After several of those misleading assumptions, distribution of AMP formation to pre- and posttransfer proofreading steps could be established when AMP formation rates in aminoacylation of modified tRNAs, especially $tRNA^{Ile}$-C-C-A(3′-NH_2), were measured (Freist et al., 1987, 1988). When this tRNA species is aminoacylated, the final products of the reaction are not esters but amides because the aminoacyl esters that are initially generated rearrange by spontaneous intramolecular aminolysis to aminoacylamides [compare discussion in Chladek and Sprinzl (1985)].

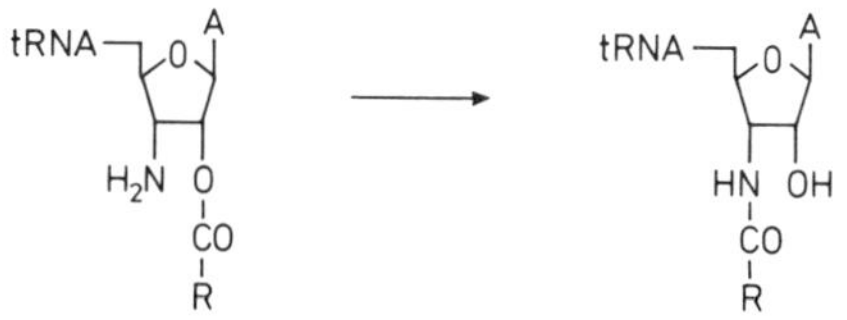

tRNA–O–A (H_2N, O–CO–R) ⟶ tRNA–O–A (HN–CO–R, OH)

These compounds are stable against enzymic hydrolysis (Fraser & Rich, 1975; Freist & Sternbach, 1988b), and thus proofreading can only be achieved on the pretransfer level when this type of modified tRNA is aminoacylated. A falsification of pretransfer AMP formation rates by partial enzymic hydrolysis before acyl migration can be ruled out by considering transacylation rates. For example, $tRNA^{Ile}$-C-C-A is aminoacylated at the 2′-OH group. For transacylation of 2′-*O*-aminoacyladenosines to the 3′-O derivative, rates of 3–11 s^{-1} have been measured, values considerably higher than those of enzymic hydrolysis found with the yeast enzyme (von der Haar & Cramer, 1976; Freist & Sternbach, 1989). Unfortunately, transacylation rates of 2′-*O*-aminoacyl-3′-deoxy-3′-aminoadenosines have not been measured, but according to chemical knowledge this reaction must run much faster, and lack of such measurements may be due to the difficulties in preparing such spontaneously isomerizing compounds. As an analogous reaction, formation of peptide bonds at the ribosome may be mentioned. It has been shown that, without enzymic transferase activity, when the ribosome functions only as template, aminolysis can take place at rates of at least 15–20 s^{-1} (Nierhaus et al., 1980).

As a control that the structure of the tRNA does not influence pretransfer proofreading, in some experiments with isoleucyl-tRNA synthetase from yeast $tRNA^{Ile}$-C-C-A(3′-NH_2) was substituted by $tRNA^{Ile}$-C-C-2′-dA. This latter species lacks the accepting OH group and cannot be aminoacylated; thus AMP formation by proofreading can also only be achieved with the aminoacyladenylate. It turned out that pretransfer proofreading is not dependent on the tRNA structure and that pretransfer proofreading factors determined for one tRNA species are also valid for experiments with other tRNAs (Freist & Sternbach, 1989).

When AMP formation rates were measured in aminoacylation of $tRNA^{Ile}$-C-C-A(3′-NH_2), considerably lower numbers of AMP generated per molecule of Val-$tRNA^{Ile}$-C-C-A(3′-NH_2) were found than in acylation of $tRNA^{Ile}$-C-C-A. These values were, under standard conditions, 111 for the yeast enzyme and 24 for the enzyme from *E. coli* MRE 600 (α_1, M_r = 102 000) (Freist et al., 1987). Divided by the number of AMP molecules generated during aminoacylation with the cognate substrate, pretransfer proofreading factors of Π_1 = 51 and Π_1 = 22 are obtained. These factors show how many times more the noncognate substrate valine is rejected by hydrolysis of the adenylate than the cognate one.

The pretransfer proofreading factor Π_1 is multiplied by the posttransfer factor Π_2 to give the overall proofreading factor Π' (compare Scheme II): $\Pi' = \Pi_1\Pi_2$. Because the overall factor can be determined in aminoacylation of $tRNA^{Ile}$-C-C-A, Π_2 can be calculated as $\Pi_2 = \Pi'/\Pi_1$. In the case of valine posttransfer proofreading factors of 3 and 62 are obtained. In case of the yeast enzyme (Π_1 = 51, Π_2 = 3) pretransfer proofreading is clearly the main correction step, whereas posttransfer proofreading plays a minor role. Π_2 values calculated for the other 18 noncognate amino acids are even lower than 3, confirming the assumption that posttransfer proofreading *can* be only a mopping up step or can even be completely neglected (Fersht, 1977a). However, for rejection of valine by isoleucyl-tRNA synthetase from *E. coli*, the misacylation reaction for which the term mopping up was created, we found more post- than pretransfer proofreading (Π_1 = 22, Π_2 = 62). This observation seems to be consistent with the higher rates of enzymic hydrolysis of Val-$tRNA^{Ile}$-C-C-A by this enzyme (Fersht, 1977a) than found for the yeast synthetase (von der Haar & Cramer, 1976; Freist & Sternbach, 1989). It should be mentioned that under standard assay conditions rejection of valine by isoleucyl-tRNA synthetase from *E. coli* is the only known case in which posttransfer hydrolysis rates are higher than pretransfer proofreading.

Summarizing the results on proofreading, it must be concluded that no general statement concerning distribution to pre- or posttransfer steps can be made. The structure of the misactivated amino acid seems to be decisive for this choice, and in many cases only pretransfer correction takes place. Furthermore, assay conditions also have an influence on the values of proofreading factors as will be discussed later.

Two-Step Initial Discrimination. When Pauling (1958) calculated the probabilities by which errors can occur in protein biosynthesis, he found a factor of 4.3 by which incorporation of alanine instead of glycine or isoleucine instead of valine is favored by higher van der Waals attraction energy. This factor is based on the difference of interaction of a methyl group and a hydrogen atom of 3.78 kJ/mol = 0.90 kcal/mol which causes a discrimination factor of exp (900/*RT*) = 4.3. Similar values were also obtained experimentally from investigation of antigen–antibody interactions (Pauling & Pressman, 1945).

A convenient method for measuring such differences in binding energies $\Delta\Delta G_b$ is given by the equation $(k_{cat}/K_m)_A/(k_{cat}/K_m)_B = \exp(-\Delta\Delta G_b/RT)$, which allows calculation of the difference in Gibbs free energies of binding from the kinetic constants k_{cat} and K_m (Fersht, 1977b).

Because valine is misactivated by isoleucyl-tRNA synthetase, a $\Delta\Delta G$ value of about 13 kJ $\approx$ 3 kcal/mol could be calculated from k_{cat} and K_m values obtained with enzymes from *E. coli* in the pyrophosphate exchange reaction (Loftfield & Eigner, 1966; Fersht, 1977b). Very similar values were also measured later by other authors, and thus an incremental group binding energy of 3.4 kcal/mol was given for a methyl group (Fersht, 1977a; Fersht & Dingwall, 1979a,b; Fersht et al., 1980). As a reason for these higher values, Fersht (1981) discusses higher van der Waals forces due to preformed binding pockets and a dense package in the binding sites of aminoacyl-tRNA synthetases.

Supporting this idea, Hopfield and Yamane (1979) calculated that in an "ideally" arranged binding pocket 6–9 methyl equivalents could be tightly packed around the methyl group by which the isoleucine side chain is longer than the valine side chain and result in a total difference of Gibbs free energy of about 10.5 kJ/mol = 2.5 kcal/mol.

Surprisingly, when $tRNA^{Ile}$-C-C-A(3′-NH_2) was aminoacylated with isoleucine and valine, a difference of $\Delta\Delta G_{I_1}$ = 3.08 kJ/mol = 0.74 kcal/mol in Gibbs free energy was obtained (Freist et al., 1987, 1985), the same value that was calculated and measured by Pauling 30 years ago. For determination of this value it was taken into account that the discrimination factor $D_1 = (k_{cat}/K_m)_{Ile}/(k_{cat}/K_m)_{Val}$ [discrimination factors are called D_1 in aminoacylation of $tRNA^{Ile}$-C-C-A(3′-NH_2); they indicate how many times more the cognate substrate is converted than the noncognate one at the same concentration] must be the product of an initial discrimination factor I_1 and the pretransfer proofreading factor Π_1: $D_1 = I_1\Pi_1$. Thus, I_1 can be calculated as $I_1 = D_1/\Pi_1$ and the difference in Gibbs free energy of binding as $\Delta\Delta G_{I_1} = RT \ln I_1$.

However, in aminoacylation of $tRNA^{Ile}$-C-C-A higher differences in Gibbs free energy of 14.25 kJ/mol = 3.40 kcal/mol were observed (Freist et al., 1987, 1985). Presently, an exact explanation cannot be given for this phenomenon; plausible rationales exist but remain hypothetical. The higher difference in Gibbs free energy obtained in acylation of $tRNA^{Ile}$-C-C-A may indeed by due to a "more dense" package in the amino acid binding pocket. In acylation of $tRNA^{Ile}$-C-C-A(3′-NH_2) this "dense" package may not be achieved.

By rapid kinetic investigations complex formation of aminoacyl-tRNA synthetase and amino acid as well as of enzyme and tRNA has been found to be a two-step process (Holler & Calvin, 1972; Krauss et al., 1979, 1977). With modified tRNAs complex formation is reduced to a one-step process (Krauss et al., 1979, 1977). It seems reasonable to assume that a dense package in the amino acid binding pocket is achieved in a two-step process and that with modified tRNAs the binding site remains uncompleted.

As a working hypothesis one can assume that in the first step of amino acid binding different acids are discriminated by a factor I_1 and after a conformational change of the enzyme in the second step by a factor I_2 (compare Scheme II). Both factors are multiplied to an overall initial discrimination factor I'. The overall difference in Gibbs free energy of binding $\Delta\Delta G_{I'}$ is thus distributed to the two steps of initial discrimination: $\Delta\Delta G_{I'} = \Delta\Delta G_{I_1} + \Delta\Delta G_{I_2} = 3.08 + 11.17 = 14.25$ kJ/mol, or 0.74 + 2.67 = 3.41 kcal/mol (Freist et al., 1987, 1988).

Similarly, as proofreading processes could be assigned to pre- and posttransfer steps by experiments with chemically modified tRNAs, these compounds showed that the initial discrimination of amino acids in their special binding pocket must also be a process that is more complex than thought before. Although no final theory could be postulated, a plausible hypothesis was obtained which is in accordance with physical measurements.

Overall Discrimination of Amino Acids. The factor of physiological relevance is the overall discrimination factor D, valid in aminoacylation of tRNA-C-C-A. As described above, it is determined by four subfactors, two initial discrimination factors and one pre- and one posttransfer proofreading factor:

$$D = I_1 I_2 \Pi_1 \Pi_2$$

Factor D is decisive for the quotient of velocities by which two amino acids are attached to the tRNA by the synthetase; e.g., for discrimination of isoleucine and valine the following equation is valid [see also Fersht (1977b)]:

$$v_{Ile}/v_{Val} = D[\text{Ile}]/[\text{Val}]$$

In detailed studies factors D of 19 noncognate amino acids have been calculated from k_{cat} and K_m values as $D = (k_{cat}/K_m)_{cognate}/(k_{cat}/K_m)_{noncognate}$ for isoleucyl-tRNA synthetases from yeast and *E. coli* MRE 600 as well as for tyrosyl-tRNA synthetase from yeast (Freist et al., 1988; Freist & Sternbach, 1988). Isoleucyl-tRNA synthetases show the highest D values for discrimination of valine (38 000 and 72 000); for most noncognate acids factors are in the range of 10 000–50 000, except for four amino acids which are only rejected with D values between 300 and 3000. D values determined for tyrosyl-tRNA synthetase are considerably higher, in a range from 30 000 to more than 500 000. Whereas isoleucyl-tRNA synthetases show the highest accuracy in rejection of valine, the amino acid most similar to the cognate substrate, tyrosyl-tRNA synthetase shows the lowest specificity in discrimination of the most similar substrate phenylalanine and the highest D values for the acids that deviate most from the tyrosine structure.

It should be mentioned that these D values are in the same order of magnitude as some values given earlier by other authors. For example, for discrimination between isoleucine and valine by isoleucyl-tRNA synthetase a factor of about 18 000 had been estimated (Hopfield et al., 1976); for rejection of tyrosine by phenylalanyl-tRNA synthetase a factor of 20 000 had been found by applying fast kinetic techniques (Lin et al., 1984). As also observed with the isoleucyl-tRNA synthetases from yeast and *E. coli*, tyrosyl-tRNA synthetase from *E. coli* seems to be more specific than the yeast enzyme. For discrimination between phenylalanine and tyrosine a 7-fold higher D value was calculated for the *E. coli* enzyme (Fersht et al., 1980).

Influence of Assay Conditions on Specificity and Energy Consumption. As shown in Table I, k_{cat} and K_m values observed for isoleucine as the cognate substrate are considerably changed by assay conditions. Because these values are decisive for discrimination between cognate and noncognate substrates, factors D should also be changed. This was indeed found when factors D were determined under different assay conditions for discrimination between isoleucine and valine in acylation of $tRNA^{Ile}$-C-C-A by the yeast enzyme (Freist et al., 1985). Applying the same conditions as given in Table I, factors D changed in the range from 2000 to 38 000. Remarkably, these variations were nearly exclusively caused by different k_{cat} and K_m values of the cognate substrate, whereas for unknown reasons kinetic constants of the noncognate substrate valine were nearly constant.

The question concerning which recognition steps are mainly influenced by assay conditions may be answered by comparison

Scheme II: Four-Step Recognition of Amino Acids

```
                                                                    tRNA
                                                                      \
E + aa + ATP ──► E·aa·ATP ──► (E·aa·ATP)' ──► E·aa-AMP ──────── E·aa·tRNA ──► E + aa-tRNA
                                           │          │       │          │
                                           ▼          │       ▼          │
                                          PPi         │      AMP         │
                                                      ▼                  ▼
                                              E + aa + ATP      E + aa + tRNA

              first ini-      second ini-     pretransfer     posttransfer    overall
              tial discri-    tial discri-    proofreading    proofreading    discri-
              mination step   mination step                                   mination
              factor I1       factor I2       factor Π1       factor Π2       factor D =
                                                                              I1·I2·Π1·Π2
```

Table III: Discrimination Factors D and Subfactors I_1, I_2, Π_1, and Π_2 at Different pH Values Obtained for Discrimination of Valine in Aminoacylation of tRNA^{Ile}-C-C-A by Isoleucyl-tRNA Synthetase from Yeast (Freist et al., 1985)

pH	D	I_1	I_2	Π_1	Π_2
6.50	8 000	3.1	31.4	27.7	3.0
7.65	38 000	3.3	76.5	50.5	3.0
8.60	14 000	3.1	53.0	35.3	2.4

of discrimination subfactors calculated from experiments with tRNA^{Ile}-C-C-A and tRNA^{Ile}-C-C-A(3′-NH_2) at different pH values (Freist et al., 1985). In Table III it is shown that for discrimination of valine by isoleucyl-tRNA synthetase initial discrimination factors I_1 and posttransfer proofreading factors Π_2 are approximately constant whereas initial discrimination factors I_2 and pretransfer proofreading factors Π_1 vary up to a factor of about 2. This means that the assumed conformational change of the enzyme which completes the amino acid binding site is pH dependent, as is the following hydrolytic pretransfer correction step. It must be further assumed that these different conformational states of the enzyme are somehow related to the different orders of substrate additions and product releases (Table I).

At present it is not clear whether other aminoacyl-tRNA synthetases exhibit variation of specificity to the same extent depending on assay conditions, but it may be already concluded that in yeast cells isoleucine may be incorporated into proteins with changing accuracy.

Another striking property that has until now only been observed with isoleucyl-tRNA synthetase prepared from commercial bakers' yeast was a very high ATP consumption. According to the overall reaction equation for esterification of the tRNA with the cognate substrate, one molecule of AMP should be generated. For isoleucyl-tRNA synthetase from *E. coli* B an AMP formation stoichiometry of 1.5 (Hopfield et al., 1976) and for that enzyme from *E. coli* MRE 600 an AMP formation stoichiometry of 1.1 (Freist et al., 1988) were found; for tyrosyl- and arginyl-tRNA synthetases from yeast (α_2, M_r = 80 000; α_1, M_r = 73 000) formation stoichiometries of 1.1 and 1.7 AMP/aminoacyl-tRNA have been found (Freist & Sternbach, 1988; Freist et al., in preparation). However, isoleucyl-tRNA synthetase from bakers' yeast needs 5.5 molecules of AMP per molecule of Ile-tRNA^{Ile}-C-C-A under standard conditions (Freist et al., 1985). Obviously, the enzyme prepared from the commercial yeast has lost its ability to recognize exactly its cognate substrate in the hydrolytic correction steps and thus wastes energy by hydrolyzing ATP. However, this high ATP consumption made it possible to observe more clearly ATP consumption dependent on assay conditions in acylation with the cognate substrate. Generally it turned out that high accuracy is connected with high energy consumption of the cognate as well as the noncognate reactions.

Structures of Substrate Binding Pockets. Important progress in understanding the function of single amino acid side chains of the aminoacyladenylate binding site was achieved when the crystal structure of the tyrosyladenylate–tyrosyl-tRNA synthetase complex was solved (Monteilhet & Blow, 1978; Blow & Brick, 1985) and site-directed mutagenesis work was carried out systematically to analyze the function of the substrate binding residues of the enzyme [compilation in Fersht (1987)]. Single amino acids involved in hydrogen bonding could be exchanged and thus Gibbs free energy values for single hydrogen bonds could be determined (Fersht, 1988). The side chain of the substrate tyrosine forms two hydrogen bonds to the enzyme by its hydroxyl group in the para position of the phenyl residue: one to an aspartate side chain and one to a tyrosyl moiety of the enzyme. Changing the tyrosyl residue of the enzyme into a phenylalanyl residue allowed a determination of a contribution of 2.1 kJ = 0.5 kcal/mol of Gibbs free energy to substrate binding due to the tyrosine–tyrosine hydrogen bond. Mutation of the aspartate moiety of the enzyme did result in an inactive enzyme (Fersht, 1987), and Gibbs free energy values of the second hydrogen bond could not be determined directly, but from other experiments a value of 14–18 kJ = 3.5–4.5 kcal/mol could be given for such a hydrogen bond between a charged and an uncharged group (Fersht et al., 1985).

These results could be partly confirmed by changing the substrate instead of single amino acids of the active site of the synthetase. Binding of amino acid side chains is mainly caused by hydrophobic interaction and additionally, as in the case of the cognate substrate tyrosine, by hydrogen bonds. Decisive for hydrophobic interactions are the free accessible surface areas of the amino acids; in the literature a mean energy value of 105 J/mol = 25 cal/mol is given for removal of an accessible surface of 0.01 nm^2 = 1 $Å^2$ from contact with water (Chothia, 1976, 1975, 1974).

When $\Delta\Delta G_{I_1}$ and $\Delta\Delta G_{I_2}$ values (calculated as $\Delta\Delta G_I = RT \ln I$) were plotted against accessible surface areas, linear relationships were observed in which 0.01 nm^2 = 1 $Å^2$ of accessible surface corresponds to 42–150 J/nm^2 or 10–30 cal/$Å^2$, values which are indeed in a range that must be expected for hydrophobic interaction (Freist et al., 1987, 1988).

From X-ray analysis of tyrosyl-tRNA synthetase it was also concluded that, by tyrosyladenylate formation and binding, some amino acid side chains in the vicinity of the binding site are shifted to other positions (Rubin & Blow, 1981; Monteilhet & Blow, 1978). This observation may correspond to some irregularities found in the linear relationships of $\Delta\Delta G$ values to accessible surface areas. If amino acid side chains of the enzyme must be turned aside during substrate binding, parts of the hydrophobic interaction energies may be lost in those processes. A special "stopper model" was created for isoleucyl- and tyrosyl-tRNA synthetases which explains initial recognition of amino acids by hydrophobic interaction and removal of enzyme side chains called "stoppers" (Freist et al., 1988; Freist & Sternbach, 1988). In this stopper model the binding cavity of, e.g., isoleucine is restricted to the length of the isoleucine side chain by a bulky "stopper group". If an amino acid with a longer side chain, such as tryptophan, is bound to this site, the bulky stopper group is shifted to another position. The higher hydrophobic interaction energy gained by the longer side chain of tryptophan is thus compensated by the energy that is necessary for pushing aside the stopper group preventing a favored binding of tryptophan.

From $\Delta\Delta G_I$ plots the Gibbs free energy contributions of the two hydrogen bonds could also be estimated by which the tyrosine side chain is bound to the enzyme. Tyrosine deviates from linear relation of binding energy to its accessible surface area by 12.2 kJ = 2.91 kcal/mol and 6.3 kJ = 1.51 kcal/mol. These two values are in the same range as determined by site-directed mutagenesis methods for hydrogen bonds between an uncharged and a charged group and between two uncharged groups.

Conclusions. Although much progress has been made in the investigation of aminoacyl-tRNA synthetases, we are still far away from a complete understanding of the function and mechanism of these enzymes. As described above for recognition of amino acids, at least a four-step process must be assumed as shown in Scheme II. However, for the complete picture of catalysis this scheme must be combined to a multistep process with the catalytic cycles as shown above for the Ter-Ter substrate additions and product releases. At present this would be far too speculative. Today we have a general and rough knowledge on aminoacyl-tRNA synthetases, but many details are still missing.

Kinetic investigations have shown that the enzymes have many different active conformations and can act in different catalytic cycles. There may be a large population of active enzyme conformations. No conclusions can be made as to the nature of conformational differences at this time. Even if more X-ray structure analysis could be done, it would be difficult to elucidate these processes because in the crystals the molecules are in a "frozen" state favorable for crystallization. Probable action of enzymes like the aminoacyl-tRNA synthetases may reach the borders of "fundamental complex" processes (Cramer, 1979), which are very difficult to understand in all their details.

Registry No. Aminoacyl-tRNA synthetase, 9028-02-8.

References

Baldwin, A. N., & Berg, P. (1966) *J. Biol. Chem. 241*, 831–838, 839–845.

Berthelot, F., & Yaniv, M. (1970) *Eur. J. Biochem. 16*, 123–125.

Blow, D. M., & Brick, P. (1985) in *Biological macromolecules and assemblies. Vol. 2: Nucleic acids and interactive proteins* (Jurnak, F., & McPherson, A., Eds.) pp 443–469, Wiley, New York.

Bruice, T. C., & Fife, T. H. (1962) *J. Am. Chem. Soc. 84*, 1973–1979.

Chladek, S., & Sprinzl, M. (1985) *Angew. Chem. 97*, 377–398; *Angew. Chem., Int. Ed. Engl. 24*, 371–391.

Chothia, C. (1974) *Nature (London) 248*, 338–339.

Chothia, C. (1975) *Nature (London) 254*, 304–308.

Chothia, C. (1976) *J. Mol. Biol. 105*, 1–14.

Cleland, W. W. (1963) *Biochim. Biophys. Acta 67*, 104–137, 173–187, 188–196.

Cleland, W. W. (1970) *Enzymes (3rd Ed.) 2*, 1–65.

Cramer, F. (1979) *Interdiscip. Sci. Rev. 4*, 132–139.

Cramer, F., & Freist, W. (1987) *Acc. Chem. Res. 20*, 79–84.

Cramer, F., von der Haar, F., & Igloi, G. (1979) in *Transfer RNA: Structure, Properties, and Recognition*, pp 267–279, Cold Spring Harbor Laboratory, Cold Spring Harbor, NY.

Deutscher, M. P. (1967) *J. Biol. Chem. 242*, 1132–1139.

Englisch, U., Englisch, S., Markmeyer, P., Schischkoff, J., Sternbach, H., Kratzin, H., & Cramer, F. (1987) *Biol. Chem. Hoppe-Seyler 368*, 971–979.

Fersht, A. R. (1977a) *Biochemistry 16*, 1025–1030.

Fersht, A. R. (1977b) *Enzyme Structure and Mechanism*, Freeman, Reading, MA, and San Francisco, CA.

Fersht, A. R. (1981) *Proc. R. Soc. London, B 212*, 351–379.

Fersht, A. R. (1987) *Biochemistry 26*, 8031–8037.

Fersht, A. R. (1988) *Biochemistry 27*, 1577–1580.

Fersht, A. R., & Jakes, R. (1975) *Biochemistry 14*, 3350–3356.

Fersht, A. R., & Kaerthner, M. M. (1976a) *Biochemistry 15*, 818–823.

Fersht, A. R., & Kaethner, M. M. (1976b) *Biochemistry 15*, 3342–3346.

Fersht, A. R., & Dingwall, C. (1979a) *Biochemistry 18*, 2627–2631.

Fersht, A. R., & Dingwall, C. (1979b) *Biochemistry 18*, 1238–1245.

Fersht, A. R., Shindler, J. S., & Tsui, W.-C. (1980) *Biochemistry 19*, 5520–5524.

Fersht, A. R., Shi, I.-P., Wilkinson, A. I., Blow, D. M., Carter, P., Waye, M. M. Y., & Witner, G. P. (1984) *Angew. Chem. 96*, 455–462; *Angew. Chem., Int. Ed. Engl. 23*, 467–474.

Fersht, A. R., Shi, I.-P., Knill-Jones, I., Lowe, D. M., Wilkinson, A. I., Blow, D. M., Brick, P., Carter, P., Waye, M. M. Y., & Winter, G. (1985) *Nature (London) 314*, 235–238.

Fraser, T. H., & Rich, A. (1975) *Proc. Natl. Acad. Sci. U.S.A. 72*, 3044–3048.

Freist, W. (1988) *Angew. Chem. 100*, 795–811; *Angew. Chem., Int. Ed. Engl. 27*, 773–788.

Freist, W., & Cramer, F. (1983) *Eur. J. Biochem. 131*, 65–80.

Freist, W., & Sternbach, H. (1984) *Biochemistry 23*, 5742–5752.

Freist, W., & Sternbach, H. (1988) *Eur. J. Biochem. 177*, 425–433.

Freist, W., & Sternbach, H. (1989) *Eur. J. Biochem. 178*, 595–602.

Freist, W., von der Haar, F., Faulhammer, F., & Cramer, F. (1976) *Eur. J. Biochem. 66*, 493–497.

Freist, W., Sternbach, H., & Cramer, F. (1981) *Hoppe-Seyler's Z. Physiol. Chem. 362*, 1247–1254.

Freist, W., Sternbach, H., & Cramer, F. (1982) *Eur. J. Biochem. 128*, 315–329.

Freist, W., Pardowitz, I., & Cramer, F. (1985) *Biochemistry 24*, 7014–7023.

Freist, W., Sternbach, H., & Cramer, F. (1987) *Eur. J. Biochem. 169*, 33–39.

Freist, W., Sternbach, H., & Cramer, F. (1988) *Eur. J. Biochem. 173*, 27–34.

Holler, E., & Calvin, M. (1972) *Biochemistry 11*, 3741–3752.

Hopfield, J. J., & Yamane, T. (1979) in *Ribosomes: Structure, function and genetics* (Chambliss, G., Craven, G. R., Davies, I., Davis, K., Kahna, L., & Nomura, M., Eds.) pp

585–591, University Park Press, Baltimore, MD.
Hopfield, J. J., Yamane, T., Yue, V., & Coutts, S. M. (1976) *Proc. Natl. Acad. Sci. U.S.A. 73*, 1164–1168.
Hustedt, H., Flossdorf, J., & Kula, M.-R. (1977) *Eur. J. Biochem. 74*, 199–202.
Igloi, G. L., & Cramer, F. (1978) *FEBS Lett. 90*, 97–102.
Irwin, M. I., Nyborg, I., Reid, B. R., & Blow, D. M. (1976) *J. Mol. Biol. 105*, 577–586.
Joachimiak, A., & Barciszewski, J. (1980) *FEBS Lett. 119*, 201–211.
Kern, D., & Gangloff, I. (1981) *Biochemistry 20*, 2065–2075.
Kisselev, L. L., & Favorova, O. O. (1974) *Adv. Enzymol. Relat. Areas Mol. Biol. 40*, 141–238.
Kovaleva, G. K., Holmuratov, E. G., & Kisselev, L. L. (1983) *FEBS Lett. 151*, 79–82.
Krauss, G., Pingoud, A., Boehme, D., Riesner, D., Peters, F., & Maass, G. (1975) *Eur. J. Biochem. 55*, 517–529.
Krauss, G., Riesner, D., & Maass, G. (1976) *Eur. J. Biochem. 68*, 81–93.
Krauss, G., Riesner, D., & Maass, G. (1977) *Nucleic Acids Res. 4*, 2253–2262.
Krauss, G., von der Haar, F., & Maass, G. (1979) *Biochemistry 18*, 2253–2262.
Lin, S. X., Baltzinger, M., & Remy, P. (1984) *Biochemistry 23*, 4109–4116.
Loftfield, R. B. (1972) *Prog. Nucleic Acid. Res. Mol. Biol. 12*, 87–128.
Loftfield, R. B., & Eigner, E. A. (1966) *Biochim. Biophys. Acta 130*, 426–448.
Loftfield, R. B., & Eigner, E. A. (1969) *J. Biol. Chem. 244*, 1746–1754.
Loftfield, R. B., & Vanderjagt, M. A. (1972) *Biochem. J. 128*, 1353–1356.
Lövgren, T. N. E., Heinonen, J., & Loftfield, R. B. (1975) *J. Biol. Chem. 250*, 3854–3860.
Monteilhet, C., & Blow, D. M. (1978) *J. Mol. Biol. 122*, 407–417.
Nierhaus, K. H., Schulze, H., & Cooperman, B. S. (1980) *Biochem. Int. 1*, 185–192.
Norris, A. T., & Berg, P. (1964) *Proc. Natl. Acad. Sci. U.S.A. 52*, 330–337.
Parfait, R., & Grosjean, H. (1972) *Eur. J. Biochem. 30*, 242–249.
Pauling, L. (1958) *Festschr. Prof. Dr. Arthur Stoll Siebzigsten Geburtstag, 1957*, 597–602.
Pauling, L., & Pressman, D. (1945) *J. Am. Chem. Soc. 67*, 1003–1012.
Rubin, J., & Blow, D. M. (1981) *J. Mol. Biol. 145*, 489–500.
Söll, D., & Schimmel, P. R. (1974) *Enzymes (3rd Ed.) 10*, 489–538.
Sprinzl, M., & Cramer, F. (1975) *Proc. Natl. Acad. Sci. U.S.A. 72*, 3049–3053.
Sprinzl, M., & Cramer, F. (1978) *Prog. Nucleic Acid Res. Mol. Biol. 22*, 1–69.
Thiebe, R. E. (1975) *FEBS Lett. 60*, 342–345.
Thiebe, R. E. (1978) *Nucleic Acids Res. 5*, 2055–2071.
Thiebe, R. E. (1983) *Eur. J. Biochem. 130*, 525–528.
von der Haar, F., & Gaertner, E. (1975) *Proc. Natl. Acad. Sci. U.S.A. 72*, 1378–1382.
von der Haar, F., & Cramer, F. (1976) *Biochemistry 15*, 4131–4138.
von der Haar, F., & Cramer, F. (1978) *Biochemistry 17*, 3139–3145.
Yarus, M., & Berg, P. (1969) *J. Mol. Biol. 42*, 171–189.

Chapter 21

The Allosteric Three-Site Model for the Ribosomal Elongation Cycle: Features and Future

Knud H. Nierhaus

Max-Planck-Institut für Molekulare Genetik, Abteilung Wittmann, Ihnestrasse 73, D-1000 Berlin-Dahlem, West Germany

Received October 25, 1989; Revised Manuscript Received January 4, 1990

ABSTRACT: The ribosome contains three binding sites for tRNA, viz., the A site for aminoacyl-tRNA (decoding site), the P site for peptidyl-tRNA, and the E site for deacylated tRNA (E for exit). The surprising finding of an allosteric linkage between the E and A sites in the sense of a negative cooperativity has three consequences: (a) it improves the proper selection of aminoacyl-tRNAs while preventing interference from noncognate aminoacyl-tRNAs in the decoding process, (b) it provides an explanation for the ribosomal accuracy without having to resort to the proofreading hypothesis, and (c) it has deepened our understanding of the mode of action of some antibiotics.

The elongation cycle of the ribosome is a series of reactions in which the growing peptidyl chain is lengthened by one amino acid. In the early 1960s Watson (1963, 1964) and Lipmann (1963) suggested a model for this elongation cycle in which the ribosome contained two binding sites for tRNA, namely, the P site for the peptidyl-tRNA and the A site for the newly selected aminoacyl-tRNA.

Figure 1A demonstrates the three basic reactions of the elongation cycle in the frame of the two-site model. The left half of the ribosome represents the P site and the right half the A site. In reaction 1 the A site is occupied with an aminoacyl-tRNA cognate to the codon exposed at this site. Next, the peptidyltransferase, an activity associated with the large ribosomal subunit, cleaves off the peptidyl residue from the peptidyl-tRNA and transfers it to the aminoacyl-tRNA at the A site (reaction 2). The result is that the peptidyl-tRNA (extended by one amino acid) is now located at the A site and the deacylated tRNA is at the adjacent P site. Reaction 3 is the translocation step in which the deacylated tRNA leaves the ribosome, the peptidyl-tRNA moves from the A to the P site, and a new codon invades the A site. According to the two-site model, the translocation is necessarily coupled to the release of deacylated tRNA. As a corollary of the two-site model, the posttranslocational ribosome always contains one tRNA, whereas the pretranslocational ribosome contains two tRNAs.

The two-site model became generally accepted (and still survived in the textbooks) when the so-called puromycin reaction was unraveled (Traut & Monro, 1964). Puromycin is a structural analogue of the aminoacyl end of an aminoacyl-tRNA. The drug can bind to the A-site region of the ribosomal active center where the peptide bond is formed and accepts the peptidyl residue from peptidyl-tRNA via formation of a peptide bond. The peptidyl-puromycin then falls off the ribosome.

It is clear that the peptidyl residue can be transferred to puromycin only if the peptidyl-tRNA is present at the P site, whereas no transfer can occur if the peptidyl-tRNA is located at the A site. These facts are the basis for the operational definition of A and P sites; namely, a positive puromycin reaction indicates P-site location of a peptidyl-tRNA, and a negative reaction is evidence for an A-site location. A revised and extended definition of the ribosomal tRNA binding sites taking into account recent observations is given at the end of the next section (Table II).

The puromycin reaction is an exclusive test of the acyl residue of aminoacyl- or peptidyl-tRNA, but the tRNA moiety itself is not tested. Accordingly, we undertook a study of the number of tRNA binding sites on the ribosome. Since three different species of tRNA are found on the ribosome in the course of elongation, i.e., peptidyl-tRNA, aminoacyl-tRNA, and deacylated tRNA (Figure 1A), saturation experiments were performed with all three species. Surprisingly, different answers were found in each of the three cases.

tRNA Binding Sites on the Ribosome

70S ribosomes were saturated with peptidyl-tRNA^{Phe}, aminoacyl-tRNA^{Phe}, or deacylated tRNA^{Phe} in the presence

Table I: tRNA Binding Capacities of 30S and 50S Subunits and of 70S Ribosomes from *E. coli*[a]

			binding sites		
ribosomes	poly(U)	tRNA species	no.	sites	remarks
70S	–	AcPhe-tRNA	1	P	saturates at 0.5
	–	Phe-tRNA	0		
	–	$tRNA^{Phe}$	1	P	
	+	AcPhe-tRNA	1	P or A	exclusion principle
	+	Phe-tRNA	2	P and A	
	+	$tRNA^{Phe}$	3	P, E, and A	binding sequence: first P, then E, then A
30S	–	AcPhe-tRNA	0		
	–	Phe-tRNA	0		
	–	$tRNA^{Phe}$	0		
	+	AcPhe-tRNA	1	P*	P*: prospective P site (becomes part of P site upon 50S association)
	+	Phe-tRNA	1	P*	
	+	$tRNA^{Phe}$	1	P*	
50S	–	AcPhe-tRNA	0		
	–	Phe-tRNA	0		
	–	$tRNA^{Phe}$	1	E*	E*: prospective E site (saturates at 0.2; becomes part of E site upon 30S association)
	+	AcPhe-tRNA	0		
	+	Phe-tRNA	0		
	+	$tRNA^{Phe}$	1	E*	E*: prospective E site (saturates at 0.4; becomes part of E site upon 30S association)

[a] Data are taken from Rheinberger et al. (1981) and Gnirke and Nierhaus (1986).

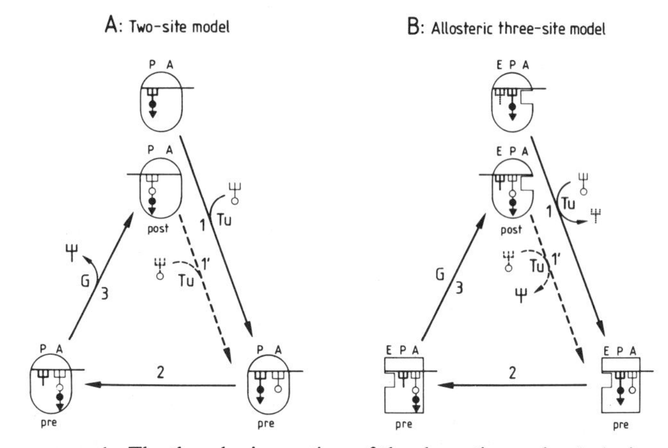

FIGURE 1: The three basic reactions of the elongation cycle: 1, A-site binding; 2, peptidyl transfer; 3, translocation. A, two-site model; B, allosteric three-site model. The pretranslocational conformer of the elongating ribosome is indicated by a rectangular ribosome and the posttranslocational conformer by an oval ribosome. Tu and G mean EF-Tu and EF-G.

or absence of poly(U), *N*-acetyl-Phe-$tRNA^{Phe}$ (AcPhe-$tRNA^{Phe}$) being taken as a simple analogue of peptidyl-$tRNA^{Phe}$. AcPhe-$tRNA^{Phe}$ saturated at one molecule per 70S, Phe-$tRNA^{Phe}$ at two molecules, and—most surprisingly—deacylated $tRNA^{Phe}$ at three molecules per 70S in the presence of poly(U) (Table I, 70S; Rheinberger et al., 1981).

The only not unexpected result was the experiment with Phe-tRNA. This species could obviously occupy the A and P sites of the same ribosome, and formation of dipeptides was observed (the actual saturation value was found to be 1.5 Phe-$tRNA^{Phe}$ molecules per 70S, due to contamination with deacylated $tRNA^{Phe}$, which we could not remove efficiently at that time).

The binding of AcPhe-$tRNA^{Phe}$ leveled off at one molecule per 70S, although AcPhe-$tRNA^{Phe}$ can be present at either the A or P site. This indicates that the binding of AcPhe-tRNA is governed by an "exclusion principle": Two sites (A or P) are available, but if one site is occupied by AcPhe-tRNA, then the second site cannot bind a further AcPhe-$tRNA^{Phe}$ molecule. The tunnel that has been observed in the large ribosomal subunit (Yonath et al., 1987), possibly harboring the growing peptidyl chain, could provide an explanation for this exclusion principle. The peptidyl residue is located in the tunnel regardless as to whether the corresponding tRNA moiety occupies the P or A site. Thus, once the tunnel is occupied, there is no possibility for a second peptidyl-tRNA to bind.

The validity of the exclusion principle has been questioned (Kirillov & Semenkov, 1982; Lill et al., 1984), but it could be verified with our preparations of ribosomes by means of quantitative puromycin reactions: Ribosomes were saturated with purified AcPhe-tRNA (saturation at about one molecule per 70S), and the bound AcPhe-tRNA was found to react quantitatively with puromycin. However, when AcPhe-tRNA was bound to the A site, no reaction took place. It follows that the P sites were exclusively filled and that the adjacent A sites did not bind an additional AcPhe-tRNA (Geigenmüller et al., 1986).

Up to three molecules of deacylated $tRNA^{Phe}$ could be bound per 70S in the presence of poly(U). In the absence of poly(U) only one molecule per 70S was found, and it was known that this molecule was present at the P site. The two additional sites could be occupied only in the presence of poly(U)—but not of poly(A) (Rheinberger, 1984)—indicating that the additional sites can be occupied only if the cognate codons are available. Clearly, codon–anticodon interaction is a prerequisite for a stable occupation of the two additional sites. An "indicator reaction" was developed which demonstrated that in the presence of poly(U) first the P site, then a new site, and finally the A site were occupied (Rheinberger et al., 1981). The new site was termed E site [E for exit according to an early suggestion by Wettstein and Noll (1965)]. The E site exclusively binds deacylated tRNA, and this site was also found and confirmed by other groups (Grajevskaja et al., 1982; Kirillov et al., 1983; Lill et al., 1984, 1988).

The results obtained with *Escherichia coli* ribosomes were so unexpected and strange that we wondered whether or not they could be generalized. We therefore analyzed an unusual organism from another kingdom, the archaebacteria, namely, the extreme halophile *Halobacterium halobium*. This archaebacterium exists with an enormous intracellular concentration of monovalent cations (more than 4 M). For optimal poly(Phe) synthesis even higher amounts (6 M, i.e., saturated salt concentrations) are required. Under these conditions standard nitrocellulose filtration methods cannot be used, since the nonbound tRNA is also trapped quantitatively on the

Table II: Definitions of the Ribosomal tRNA Binding Sites A, P, and E

Site	Definitions
(1) P site	(1.1) the P site is the first site of programmed ribosomes to be occupied at 37 °C
	(1.2) an acylated tRNA at the P site reacts with puromycin
	(1.3) the P site can bind peptidyl-tRNA, aminoacyl-tRNA, and deacylated tRNA
(2) A site	(2.1) a tRNA binds to the A site when its anticodon interacts with the adjacent codon downstream from that at the P site
	(2.2) an acylated tRNA at the A site cannot react with puromycin
	(2.3) the A site can bind peptidyl-tRNA, aminoacyl-tRNA, and deacylated tRNA; the A site exclusively binds the complex of EF-Tu, GTP, and aminoacyl-tRNA
(3) E site	(3.1) a tRNA binds to the E site when its anticodon interacts with the adjacent codon upstream from that at the P site
	(3.2) the E site exclusively binds deacylated tRNA

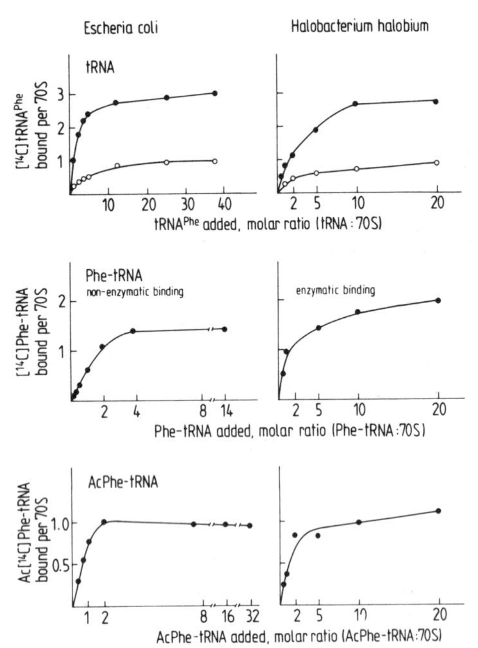

FIGURE 2: Comparison of tRNA-saturation experiments using ribosomes from *E. coli* (left column; Rheinberger et al., 1981) or *H. halobium* (right column; Saruyama & Nierhaus, 1986). Upper panels: ●, plus poly(U); ○, minus poly(U).

Incubation Steps	Expected Binding state	tRNA, bound per 70 S (Nitrocellulose filtration) $tRNA_f^{Met}$	AcLys-tRNA
1. $C_{17}AUGA_4C_{17}$ (10×) $[^{32}P]tRNA_f^{Met}$ (1.5×) Ψ 2. $Ac[^{14}C]$Lys-tRNA (1×)	E P A CCCAUGAAA	**0.94**	0.61 PM: 0.01
3. EF-G + GTP	E P A AUGAAAACC	**0.91** release: 0.03	0.57 PM: 0.49 TL : 0.48
4. 2.5 h / 70 000 × g 5. Resuspension of pelleted 70S · tRNA-complexes	E P A AUGAAAACC	**0.92**	0.56

FIGURE 3: Translocation experiment using the heteropolymeric mRNA $C_{17}AUGA_4C_{17}$. Before translocation $[^{32}P]tRNA^{Met}$ was bound to the P site and $Ac[^{14}C]$Lys-tRNA to the A site (0.94 and 0.61, respectively, per ribosome). The A-site location of AcLys-tRNA is indicated by a negligibly low puromycin reaction (PM; 0.01). Upon EF-G-dependent translocation AcLys-tRNA reacted nearly quantitatively (0.49 out of 0.57), demonstrating a translocation (TL; 0.48 = 0.49 − 0.01) of almost all bound AcLys-tRNA from the A to the P site. During the translocation reaction practically no release of deacylated $tRNA^{Met}$ occurred (0.03 = 0.94 − 0.91); i.e., the $tRNA^{Met}$ was cotranslocated from the P to the E site. From Gnirke et al. (1989).

filters. Instead, centrifugation methods had to be applied.

In spite of the use of different organisms and different analytical techniques, practically the same results for tRNA binding were obtained (Figure 2, right half): deacylated $tRNA^{Phe}$ binding leveled off at three molecules per 70S, Phe-$tRNA^{Phe}$ at two, and AcPhe-$tRNA^{Phe}$ at one, thus confirming the exclusion principle (Saruyama & Nierhaus, 1986). A third binding site specific for deacylated tRNA has also been found in a eukaryotic system (rat liver; Rodnina et al., 1988). It is clear that the results concerning the tRNA binding sites on *E. coli* ribosomes can indeed be generalized. Table I also includes the binding features of the ribosomal subunits. For details and discussion, see Gnirke and Nierhaus (1986).

The features of A, P, and E sites described in this section complement and extend the definitions of the tRNA binding sites mentioned in the introduction. The revised definitions (Table II) rely—in addition to other features—on the position of the codon with respect to that at the P site, allowing a practicable and unequivocal assignment of the tRNA location. However, it would be mistaken to regard the sites as stable and rigid "moulds". We shall see in the next section that the ribosome is a dynamic reactor that oscillates between two major conformers, the pre- and the posttranslocational states (Table III). Therefore, it must be expected that some tRNA–ribosome contacts in one and the same site change during the course of the elongation cycle. Evidence for such changes has been presented already (Abdurashidova et al., 1986; Graifer et al., 1989; Moazed & Noller, 1989a).

This view of the tRNA binding sites, which are dynamic but nonetheless still clearly defined, cannot be easily reconciled with the assumption of stable "hybrid sites" such as an A/P site (Moazed & Noller, 1989b). Furthermore, UV–cross-linking data (Abdurashidova et al., 1986) and affinity-labeling experiments (Graifer et al., 1989) do not support the concept of stable hybrid sites; both groups demonstrate that the A-site environment changes after EF-Tu-dependent GTP cleavage and peptide-bond formation, respectively, but the various A-site patterns are clearly distinguishable from those of the P site.

Features of the Third tRNA Binding Site, the E Site

An important step forward in establishing the properties of the E site and the principles of the elongation cycle was an analysis using a heteropolymeric mRHA, which exposes three different codons at the A, P, and E sites and thus allows an unequivocal assignment of the tRNA binding to any of the three ribosomal binding sites (Gnirke et al., 1989). The mRNA was 41 nucleotides long with the sequence $C_{17}AUGA_4C_{17}$, which contains in its central regions the three codons AUG(Met)-AAA(Lys)-ACC(Thr) or CAU(His)-GAA(Glu)-AAC(Asn). The third possible reading frame was not used here because it contains the stop codon UGA.

In the following we illustrate each of the features of the E site by preferentially describing an experiment using this heteropolymeric mRNA.

Cotranslocation of the Deacylated tRNA from the P to the E Site. The reading frame of the heteropolymeric mRNA is set with deacylated $tRNA^{Met}$, which binds to the P site (Figure 3, step 1). Next, an AcLys-tRNA is bound, for which the puromycin reaction (PM) indicates exclusive A-site binding. Now a translocation reaction is performed with the help of EF-G and GTP (step 3). In this step the amount of bound AcLys-tRNA hardly changes, and the puromycin reaction shows almost quantitative translocation (TL); i.e., AcLys-tRNA has moved from the A to the P site. The point of this experiment is that in the course of translocation the amount of deacylated $tRNA^{Met}$ also remains unaltered; i.e., the $tRNA^{Met}$ does not leave the ribosome from the P site but is rather cotranslocated from the P to the E site. The experiment clearly shows that there is no coupling of translocation and release of deacylated tRNA, in contrast to the prediction of the classical two-site model.

Similar results using the homopolymer poly(U) (Rheinberger & Nierhaus, 1983) were interpreted in a different way, namely, that the deacylated tRNA is released from the P site and rebinds quantitatively to the vacated A site (Baranov & Ryobova, 1988). This interpretation is refuted by the experiment shown in Figure 3, since here the A and P sites harbor different codons, thus preventing any possible rebinding to the A site.

The experiment in Figure 3 shows yet another detail. The posttranslocational complex was pelleted and resuspended, and the amounts of bound tRNAs were determined again. The binding values did not change, indicating that the deacylated tRNA does not easily dissociate from the E site but rather is stably bound at this site and has to be released by an active mechanism.

Wintermeyer and co-workers also observed an uncoupling of translocation and tRNA release, but they observed a labile E-site binding (Robertson et al., 1986; Robertson & Wintermeyer, 1987). The discrepancies are mainly due to technical differences, e.g., the use of different buffer systems and the use of NH_4Cl-washed ribosomes which have partially lost some ribosomal proteins and thus contain a "weak" E site [for details and discussion, see Gnirke et al. (1989)].

Codon–Anticodon Interaction at the Ribosomal E Site. Codon–anticodon interaction at the ribosomal A site provides the signal that is responsible for the selection of the correct amino acid. Therefore, this interaction is the central step in the decoding process and probably also the initial event in aminoacyl-tRNA binding to the ribosome (see next section).

Since a codon is about 10 Å long, and a tRNA has a diameter of about 20 Å, the formation of two adjacent codon–anticodon interactions presents a steric problem. It was thus a surprise when codon–anticodon interaction could also be demonstrated at the P site (Lührmann et al., 1979; Peters & Yarus, 1979; Wurmbach & Nierhaus, 1979; Ofengand & Liou, 1981). Clearly, the universal L-shape of the tRNA molecule has been evolved to solve the steric problem, i.e., to allow adjacent codon–anticodon interactions to occur and to bring peptidyl and aminoacyl residues into juxtaposition at the peptidyltransferase center. One might expect that simultaneous adjacent codon–anticodon interactions are of crucial importance for protein biosynthesis.

The early observation that the E site (just as with the A site) could be occupied with deacylated $tRNA^{Phe}$ only in the presence of poly(U) was taken as evidence that there is also codon–anticodon interaction at the E site (Rheinberger & Nierhaus, 1981). This point has been controversially discussed, and evidence both against (Kirillov et al., 1983; Lill et al., 1984) and in favor of codon–anticodon interaction at this site has been reported (Rheinberger et al., 1986; Lill & Wintermeyer, 1987).

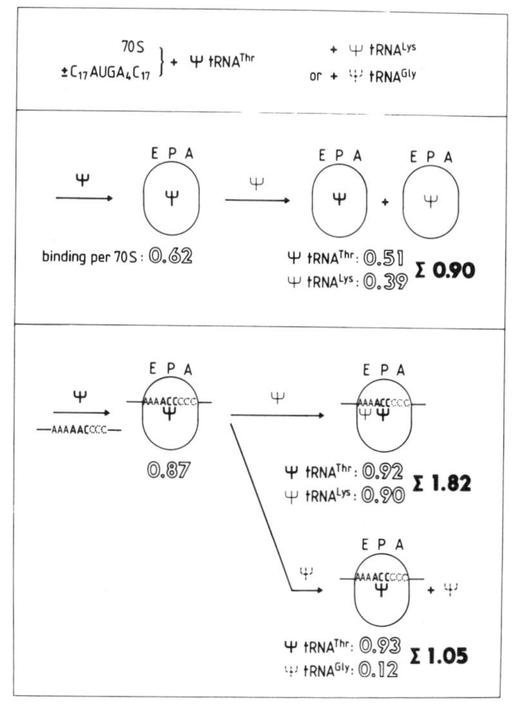

FIGURE 4: Codon-dependent binding to the ribosomal E site. $[^{32}P]tRNA^{Thr}$ was bound to the P site, and then either cognate $[^{3}H]tRNA^{Lys}$ (codon AAA) or $[^{35}S]tRNA^{Gly}$ (codon CCC) was added. In the upper box are shown results from a binding experiment in the absence of mRNA. Both tRNAs added obviously compete for one and the same binding site, since the total binding (0.90) does not exceed one tRNA per 70S (each tRNA was added in a molar ratio 70S:tRNA = 1:4). In the presence of mRNA both tRNAs bind almost stoichiometrically (total binding of 1.82 per 70S), but only if the second deacylated tRNA is cognate to the E-site codon [compare the relatively low binding of $tRNA^{Gly}$ (0.12, codon GGG) with the binding of $tRNA^{Lys}$ (0.90, codon AAA)]. From Gnirke et al. (1989).

The analysis with the heteropolymeric mRNA $C_{17}AUGA_4C_{17}$ gave a decisive answer to the question (Figure 4; Gnirke et al., 1989). In a control experiment without mRNA, two added tRNAs ($tRNA^{Thr}$ and $tRNA^{Lys}$) competed for one and the same binding site, and the total binding did not exceed one molecule per ribosome (0.90). However, when $tRNA^{Thr}$ was first bound to the P site in the presence of mRNA, thus exposing the codon AAA for $tRNA^{Lys}$ at the E site, then $tRNA^{Lys}$ [but not AcLys-$tRNA^{Lys}$; see Gnirke et al. (1989)] could fully bind to the E site, and the total binding approached two molecules per ribosome (1.82). In contrast, when $tRNA^{Gly}$ was added (which is noncognate to the AAA codon), negligibly low binding was found, whereas $tRNA^{Gly}$ bound normally to ribosomes in the presence of its cognate mRNA poly(I). It follows that quantitative filling of the E site depends on the presence of the cognate codon, strongly arguing for codon–anticodon interaction at this site.

Various tRNAs differ in their intrinsic affinities for nonprogrammed ribosomes (Gnirke et al., 1989). For example, $tRNA^{Lys}$ has a relatively low affinity and thus fills the P site but not the E site of nonprogrammed ribosomes. On the other hand, the strong binder $tRNA^{Phe}$ fills the P sites and 20–50% of the E sites of nonprogrammed ribosomes when added in high excess. In the presence of the cognate codons all tRNAs bind strongly to the P and E sites. It follows that the P sites and, in the case of strong binders, the E sites can be occupied in the absence of mRNA. However, in the presence of mRNA codon–anticodon interaction takes place, thereby significantly

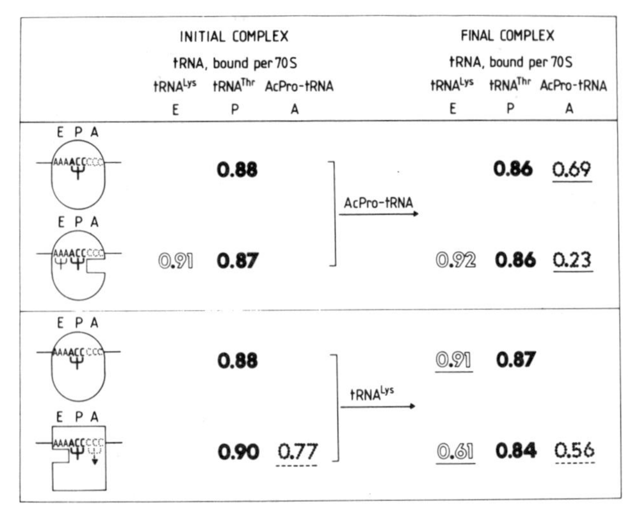

FIGURE 5: Allosteric interactions between A and E sites. When only the P site is occupied, the A site (first line) or the E site (third line) can readily be occupied as the second site. However, if the P and E sites are charged, the binding capacity of the A site is severely impaired (second line, from 0.69 to 0.23). Likewise, occupation of P and A sites reduces the E-site binding (fourth line, from 0.96 to 0.61). Note that in the course of the A-site/E-site interplay the binding at the P site remains unaffected (around 0.87). From Gnirke et al. (1989).

increasing the stability of the tRNA binding.

Allosteric Interactions between the A and E Sites in the Sense of a Negative Cooperativity. Evidence for an allosteric interplay between the A and E sites was first shown in the poly(U) system. When only the P site was filled with deacylated tRNA^{Phe}, the A site could readily be occupied with AcPhe-tRNA at both 0 and 37 °C. However, when the P and E sites were filled, AcPhe-tRNA no longer bound to the A site at 0 °C, but still bound at 37 °C. Clearly, filling of the E site induces a low-affinity A site, and activation energy is required to convert this low-affinity A site into a high-affinity state. Several further observations could be made: (a) when only the P site was filled with deacylated tRNA^{Phe}, charging of the A site with the ternary complex Phe-tRNA·EF-Tu·GTP did not affect the tRNA^{Phe} binding; (b) when both the P and E sites were filled, binding of one ternary complex induced a stoichiometric release of tRNA^{Phe}; and (c) when P, E, and A sites were filled with tRNA^{Phe}, the binding of one ternary complex to the A site induced the release of two molecules of tRHA^{Phe} from A and E sites caused by chasing and allosteric effects, respectively [Figure 1 and Table II in Rheinberger and Nierhaus (1986)]. The interpretation was that occupation of the E site induces a low-affinity A site (posttranslocational state) and vice versa, namely, that occupation of the A site induces a low-affinity E site, thus triggering the release of deacylated tRNA.

This interpretation was questioned, and the observed effects were reinterpreted as chasing effects by tRNA^{Phe} contaminants present in the preparation of the A-site ligand (Robertson & Wintermeyer, 1987). This criticism illustrates a difficulty that is intrinsic to the poly(U) system used both by us and by our colleagues; namely, an assessment of the site location of a deacylated tRNA is often difficult due to the presence of identical codons and tRNAs at each position. Again the use of the heteropolymeric mRNA, which exposes different codons at A, P, and E sites, solved this problem (Figure 5).

When the P site was filled with tRNA^{Thr}, AcPro-tRNA could readily bind to the A site. However, prefilling of the P and E sites with tRNA^{Thr} and tRHA^{Lys}, respectively, severely reduced the AcPro-tRNA binding (from 0.69 to 0.23). The

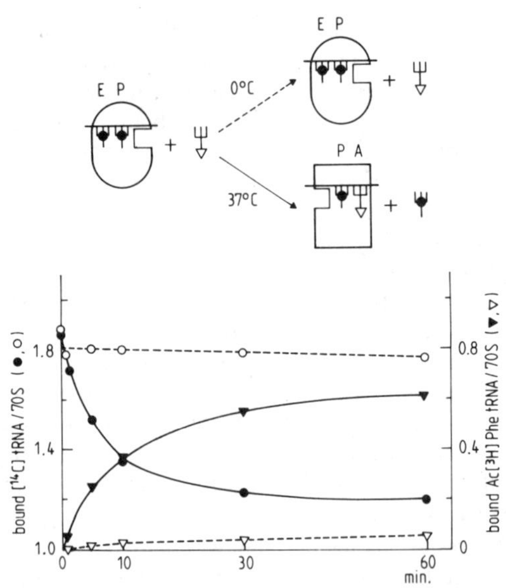

FIGURE 6: Allosteric interactions between A and E sites: kinetics of Ac[^{3}H]Phe-tRNA binding to the A site. When P and E sites are occupied with [^{14}C]tRNA^{Phe} in the presence of poly(U), Ac[^{3}H]-Phe-tRNA does not bind at 0 °C to the A site, indicating its low affinity. When activation energy is provided (incubation at 37 °C) AcPhe-tRNA does bind, and for each molecule of AcPhe-tRNA bound, one molecule of tRNA^{Phe} is released from the E site, demonstrating the negative cooperativity between A and E sites. From Rheinberger and Nierhaus (1986).

opposite is also true; viz., a ribosome with a prefilled P site (tRNA^{Thr}) readily accepts tRNA^{Lys} at the E site, but prefilling of the P and A sites with tRNA^{Thr} and AcPro-tRNA, respectively, impaired the E-site binding (from 0.91 to 0.61). The effect is clear but less pronounced. The reason is that some of the prebound A-site ligand was released (from 0.77 to 0.56) by the occupation of the E site with tRNA^{Lys}, underlining the strength and stability of the tRNA binding at the E site. The essential outcome of the experiment shown in Figure 5 is that occupation of the E site triggers a response at the A site and, vice versa, that occupation of the A site triggers a response at the E site, but *the tRNA at the intervening P site remains unaffected* (at about 0.87). This clearly demonstrates the allosteric coupling between A and E sites in a sense of negative cooperativity.

We believe that the allosteric interplay between A and E sites is of central importance for protein biosynthesis. Therefore, this feature is demonstrated in another way by a simple experiment (Figure 6). The P and E sites of poly(U)-programmed ribosomes are filled with [^{14}C]tRNA^{Phe}, and then the kinetics of Ac[^{3}H]Phe-tRNA binding are followed at 0 and 37 °C. At 0 °C no significant AcPhe-tRNA binding or tRNA^{Phe} release is observed over a period of 60 min, demonstrating the low-affinity A site upon E-site occupation. However, at 37 °C a slow AcPhe-tRNA binding is seen, and deacylated tRNA^{Phe} is released with the identical rate; i.e., for each AcPhe-tRNA molecule bound to the A site, a molecule of deacylated tRNA^{Phe} is released from the E site. It follows that A-site occupation induces a low-affinity E site, thus triggering the release of the deacylated tRNA from this site. Qualitatively the same picture is found when the ternary complex Phe-tRNA·EF-Tu·GTP is used as the A-site ligand instead of AcPhe-tRNA. However, in this case the reaction runs so quickly that a precise assessment of the binding data is difficult (Rheinberger & Nierhaus, 1986).

Recently, a diagnostic assay for the site location of tRNA was proposed, which was based on the protection patterns of specific bases in the rRNA against chemical modification in

Table III: Features of the Allosteric Three-Site Model
(1) the third site, the E site, is allosterically linked to the first site, the A site, in the sense of negative cooperativity (i.e., the oocupation of one site significantly decreases the affinity of the other); this feature has three corollaries
(A) the elongating ribosome oscillates between two states, the pre- and the posttranslocational states: in the pretranslocational state the A and P sites have a high affinity for tRNA binding, whereas the E site has a low affinity; in the posttranslocational state the P and E sites are of high affinity, and the A site now has a low affinity
(B) the elongating ribosome always carries two tRNAs, which are at the A and P sites in the pretranslocational state and at the P and E sites in the posttranslocational state
(C) the deacylated tRNA is not released from the P site during translocation but is cotranslocated from the P to the E site; only occupation of the A site triggers the release of deacylated tRNA from the E site
(2) both tRNAs present on the ribosome before and after translocation contact the mRNA via codon–anticodon interaction (Rheinberger et al., 1986; Rheinberger & Nierhaus, 1986; Gnirke et al., 1989)

the presence of bound tRNA. The E-site pattern was derived from two experiments performed at very different Mg^{2+} concentrations (6–10 and 20–25 mM, respectively). Such a concentration change would affect the ribosomal conformation and also severely change the relative affinities of the three sites (Lill et al., 1986). The resulting complexes can thus hardly be compared. Moreover, they are not comparable for a second reason, namely, that one of the constructed ribosomal complexes represents a state not found during elongation (only one tRNA, an AcPhe-tRNA, present on the ribosome), in contrast to the other complex. An "E-site" pattern was even described in an experiment where it is certain that no E site was occupied [Figure 4; compare Figure 1B in Rheinberger and Nierhaus (1987)]. Furthermore, all the observed strong protections in 23S rRNA specific for P-site binding (as well as for the assumed E-site binding) were dependent on the presence of the 3′-terminal –CA residues of tRNA. A "diagnostic assay" for the location of the relatively rigid tRNA molecules should not rely on the position of the flexible –CCA end of tRNA.

In the same study the A-site assignments were derived from an atypical complex carrying a deacylated tRNA at the P site and a ternary complex at the A site; such a complex again does not occur during elongation. The site assignments are thus dubious and the hybrid-site model (Moazed & Noller, 1989b), which was based on these assignments, is therefore also questionable.

The following point should be noted concerning structural analysis of the tRNA binding sites: In any study of structural changes within elongating ribosomes by means of fluorescence studies, cross-linking or protection experiments, etc., it is a prerequisite to compare the different states of the elongation cycle. It is inappropriate and misleading to compare the structure of a ribosome carrying one tRNA (initation state) with that of a ribosome containing two tRNAs (elongation phase) and then to ascribe the structural differences observed to conformational changes in the elongating ribosome. During elongation the ribosome always contains two tRNAs, either at the A and P sites (pretranslocational state) or at the P and E sites (posttranslocational state).

Recently, it could be shown that the E site is also functional in vivo: Posttranslocational ribosomes in native polysomes contain an occupied E site (Remme et al., 1989).

The properties of the E site outlined in this section have led to the formulation of the allosteric three-site model (Figure 1B). The two key features of the model are summarized in Table III.

Significance of the Allosteric Interplay between A and E Sites for the Accuracy of Translation: An Occupied E Site Prevents the Binding of Noncognate Aminoacyl-tRNAs

Why does the elongation cycle follow the complicated allosteric three-site model, while the simpler two-site model has worked well at least in the textbooks for over 20 years? Two

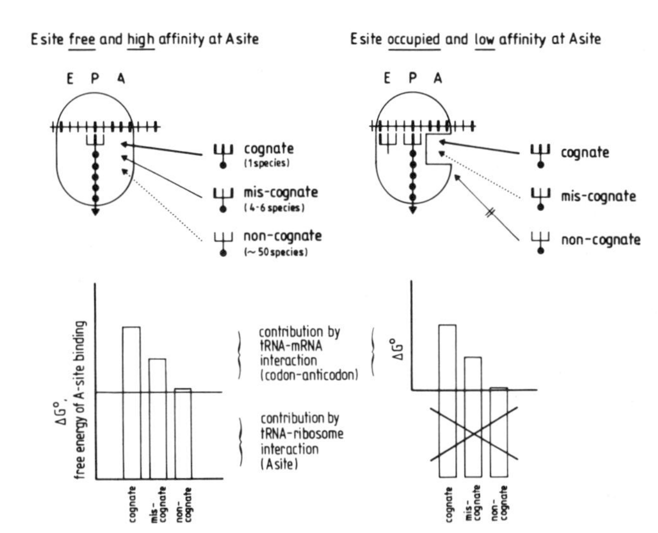

FIGURE 7: Outline of the hypothesis that the E-site-induced lowering of the A-site affinity prevents the binding of noncognate aminoacyl-tRNAs (see text).

possibilities for the physiological significance of the allosteric three-site model can be envisaged, namely, proper translocations and correct A-site binding.

Some observations indicate that the mRNA is pulled through the ribosome via the tRNA "handle", the evidence being that some tRNAs can translocate without mRNA (Belitsina et al., 1982). Furthermore, a tRNA mutant has been described which suppresses (+1) frameshifts via a tetraplet interaction between tRNA and mRNA instead of the usual canonical triplet interaction (Riddle & Carbon, 1973). Likewise, doublet interactions can induce (−1) frameshifts (O'Mahony et al., 1989). The "pulling" mechanism requires a tight coupling between tRNAs and ribosome, i.e., two adjacent codon–anticodon interactions are important for moving the mRNA and for maintaining the frame of the mRNA movement.

The second aspect concerns the accuracy of translation. Evidence will be presented that the most unexpected feature of the allosteric three-site model, namely, the low-affinity state of the A site induced by E-site occupation, plays an important role in discriminating against the binding of noncognate aminoacyl-tRNAs to the A site.

Three different classes of aminoacyl-tRNAs or ternary complexes compete for the programmed A site (Figure 7). The first class is represented by the *cognate* aminoacyl-tRNA, comprising only one species with an anticodon precisely complementary to the codon present at the A site. The second class contains four to six *miscognate* aminoacyl-tRNAs, which bear anticodons similar to that of the cognate tRNA. The third class contains the majority of the tRNA species, comprising about 50 *noncognate* aminoacyl-tRNAs, which contain dissimilar anticodons.

Let us assume that the A site is present in a high-affinity

state (Figure 7, left half). This is realized when only the P site is occupied and the E site is free. The respective free energy of A-site binding can be separated into two terms, one term reflecting tRNA–mRNA contacts (codon–anticodon interaction) and the second term tRNA–ribosome interaction. It is mainly the second term that is responsible for a correct positioning of the tRNA on the ribosome and hence for a successful peptide-bond formation. However, this term does not discriminate between wrong and right tRNA. The discrimination is achieved by the first term, i.e., codon–anticodon interaction, where the three classes of aminoacyl-tRNAs have to be distinguished. In the case of a high-affinity state of the A site where both terms play a part, one would expect significant A-site interaction of even the noncognate tRNAs, thus reducing the rate of protein synthesis and leading to an occasional incorporation of noncognate amino acids.

Now the E site comes in. We assume that E-site occupation induces the low-affinity A site by exclusively abolishing the second term (tRNA–ribosome interaction) but does not impair the effect of the first term (codon–anticodon interaction) and might even improve it (Figure 7, right half). The result is that the ribosome only has to discriminate between cognate and miscognate aminoacyl-tRNAs (which is thought to occur via proofreading mechanisms), whereas for noncognate tRNAs the A site practically does not exist. This mechanism reduces the problem of the selection of the correct aminoacyl-tRNA by an order of magnitude; the ribosome has to select one aminoacyl-tRNA out of five instead of one out of fifty.

In thermodynamic terms the presumed mechanism can be described as follows. It is clear that the discrimination energy $\Delta\Delta G°$ of the binding of cognate vs that of noncognate substrate is not affected by the allosteric shift-down of the A-site affinity. However, the point is that a high-affinity A site (Figure 7, left half) results in significant "sticking times" (reciprocal of the dissociation rate constant) for even noncognate substrates. Therefore, equilibrium between cognate and noncognate substrates at the A site would only be attained after a relatively long time. Furthermore, the increment in $\Delta G°$ of the high-affinity state over that of the low-affinity state is due to tRNA–ribosome interactions, which are an essential prerequisite for peptide-bond formation. Both factors—the slow reaching of the equilibrium and the readiness of the A site to bind the acceptor molecule for the peptidyl transfer—render a premature termination of the binding reaction possible, by peptide-bond formation even with noncognate substrates. In contrast, both of these factors are nonoperative in the low-affinity A-site condition; the reaction of codon–anticodon interaction quickly reaches equilibrium, and the peptidyl-transferase center is not yet ready to accept an aminoacyl-tRNA, thus resulting in a clear separation of the selection process and the peptidyl transfer.

Figure 8 shows a test of this hypothesis (U. Geigenmüller and K. H. Nierhaus, unpublished). AcPhe-tRNA was bound to the P site of poly(U)-programmed ribosomes, and the E site was free, leaving the A site in a high-affinity state. Next, a mixture of ternary complexes containing [^{14}C]Phe-tRNA or the noncognate [^{3}H]Asp-tRNA (codon GAC) was added, and the AcPhe dipeptides formed were analyzed by HPLC techniques. The analysis demonstrates significant AcPhe-Asp formation (about 1% of the total dipeptide formation; Figure 7, left half). In sharp contrast, when the P and E sites were prefilled with AcPhe-tRNA and deacylated tRNA^{Phe}, respectively, the AcPhe-Asp formation was reduced to background values, whereas the formation of the cognate AcPhePhe dipeptide was hardly affected. Clearly, an occupied E site

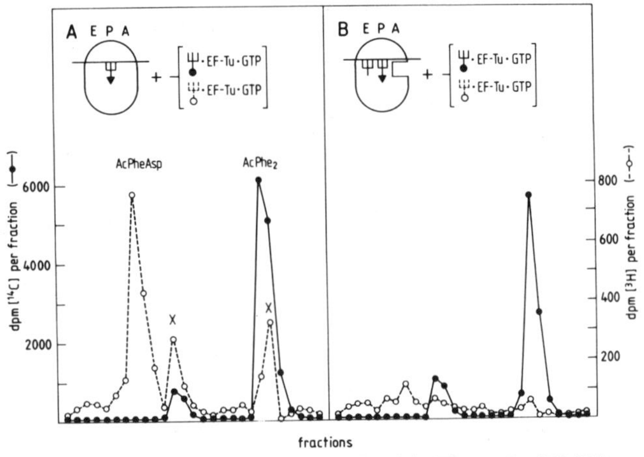

FIGURE 8: Test of the hypothesis outlined in Figure 7. (A) When AcPhe-tRNA is present at the P site in the presence of poly(U) and the E site is free, the A site has a high affinity. Addition of a mixture of cognate and noncognate ternary complexes containing [^{14}C]Phe-tRNA and [^{3}H]Asp-tRNA, respectively, leads to significant formation of "wrong" AcPhe-Asp dipeptides (detected by means of HPLC). (B) When P and E sites are respectively occupied with AcPhe-tRNA and deacyl-tRNA (A site with low affinity), the addition of the mixture of ternary complexes leads exclusively to cognate AcPhe$_2$ formation (U. Geigenmüller and K. H. Nierhaus, unpublished).

prevents a noncognate aminoacyl-tRNA from interacting with the A site, illustrating one important role of the E site during protein synthesis.

When miscognate tRNA^{Leu} was added instead of deacylated tRNA^{Phe} in order to fill the E site, no reduction of AcPhe-Asp formation at all was observed. This shows that only the presence of the cognate tRNA at the E site is a trigger for the reduction of the A-site affinity, underlining the functional importance of codon–anticodon interaction at the E site.

If a mutation were to change a tRNA in such a way that its affinity for the A site was increased without affecting the geometry of the anticodon loop, the allosteric three-site model would predict an increased misreading of this acyl-tRNA. Precisely this has been found. Hirsh (1971) described a mutation that causes a G24 → A change in the D-stem of tRNA^{Trp}, resulting in active misreading. A careful analysis revealed that most probably the geometry of the anticodon loop is not altered, whereas the affinity for the A site seems to be enhanced (Smith & Yarus, 1989a,b). The allosteric three-site model offers a simple explanation; the alteration of tRNA^{Trp} (or its UAG suppressor derivative) in the D-stem increases the affinity for the low-affinity A site, thus counteracting the codon-specificity effect and increasing the frequency of misreading.

The Inhibition Mechanisms of Some Antibiotics Are Better Understood in the Frame of the Three-Site Model

As mentioned in the preceding section, the elongating ribosome always contains two tightly bound tRNAs. During initiation, however, only one tRNA is present, viz., the initiator tRNA at the P site. Therefore, one has to distinguish between the first A-site occupation following initiation when the E site is free (which we refer to as A-site occupation of the i type; i for initiation), and the second and all subsequent A-site occupations where the E site is occupied (A-site occupation of the e type; e for elongation; see Figure 9). The two types of A-site occupation differ thermodynamically. The occupation of the i type occurs well at both 0 and 37 °C, whereas that of the e type requires temperatures above 0 °C; i.e., the allosteric transition (e type) needs higher activation energies.

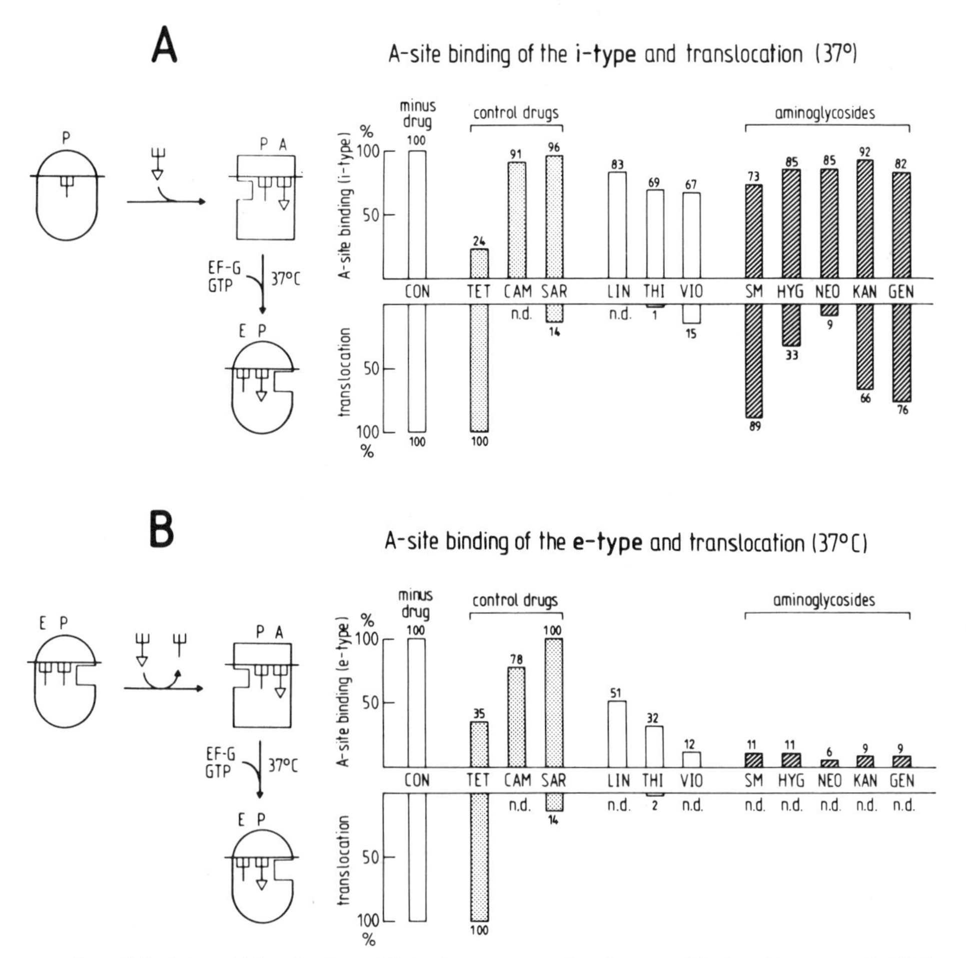

FIGURE 10: A-site occupation of the i type (A) and e type (B) in the presence of various antibiotics: Con, control; TET, tetracycline; CAM, chloramphenicol; SAR, α-sarcin; LIN, lincomycin; THI, thiostrepton; VIO, viomycin; SM, streptomycin; HYG, hygromycin; NEO, neomycin; KAN, kanamycin; GEN, gentamicin. From Hausner et al. (1988).

of the initial selection process (k_1/k_{-1}) is never fully realized". Finally, a lack of correlation has been reported between the efficiencies of a suppressor tRNA in vivo and the predictions of the ribosomal proofreading hypothesis (Faxēn et al., 1988). The experiments were performed in a mutant with hyperaccurate ribosomes, which had earlier been described as excessive ribosomal proofreaders (Ruusala et al., 1984).

We assume in the framework of the allosteric three-site model that a successful codon–anticodon interaction at the A site is the trigger for the allosteric transition from the post- to the pretranslocational state (Gnirke et al., 1989). Thus, the A-site occupation separates into two consecutive steps. In the first step codon–anticodon interaction takes place (complex 2 in Figure 12). Apart from this, no significant interaction between the A site and the ternary complex occurs, since the A site is still in its low-affinity state. A successful codon–anticodon interaction then triggers the second step—the allosteric transition—which leads to a tight binding of the aminoacyl-tRNA (complex 3).

The A-site occupation is a relatively slow process, which in fact is the rate-limiting reaction of the whole elongation cycle (Bilgin et al., 1988). The "lion's share" of the whole process of A-site occupation is most probably represented by the second step, since this step calls for gross structural changes involving the shift from high to low affinity at the E site (tRNA release) and from low to high affinity at the A site. Because this step is slow, and the preceding step of codon–anticodon interaction is quick, the codon–anticodon interaction runs practically under equilibrium conditions during the course of protein biosynthesis (Figure 12). Therefore, the precision of the initial step (k_1/k_{-1}) observed by Thompson and Karim (1982) can become effective during translation. The poor accuracy of ternary complex selection they observed in the presence of GTP is probably caused by the presence of an empty E site under their conditions, and thus the lack of an allosteric transition during A-site occupation.

The EF-Tu-dependent GTPase activity functions in this scenario after the first step of A-site occupation and before tight binding of the aminoacyl-tRNA at the end of the second step (Figure 12). The precise trigger for the factor-dependent GTPase activity is not yet clear; it could be a successful codon–anticodon interaction that is sensed by EF-Tu via a change in the tRNA conformation, or on the other hand, it could be the incipient conformational change at the A site or the contact between EF-Tu and the P-site-bound tRNA (Bosch et al., 1985) during the allosteric transition. In any case, the EF-Tu-dependent GTPase reaction (k_a) is not coupled with the allosteric transition, nor is it a prerequisite for this transition (k_2), since the latter occurs equally well with a ternary complex containing the noncleavable GTP analogue GMPPNP or even with AcPhe-tRNA (Rheinberger & Nierhaus, 1986). Regardless as to what triggers the GTPase reaction, one can easily imagine that in the case of a miscognate ternary complex the signal from the suboptimal codon–anticodon interaction is sufficient to induce the GTPase reaction (and thus the dissociation of EF-Tu·GDP) but is too weak for the allosteric transition (complex 3′ in Figure 12). The result is that the miscognate aminoacyl-tRNA is loosely bound and thus quickly dissociates from the A site. This mechanism explains the increased GTP turnover in the case of miscognate ternary

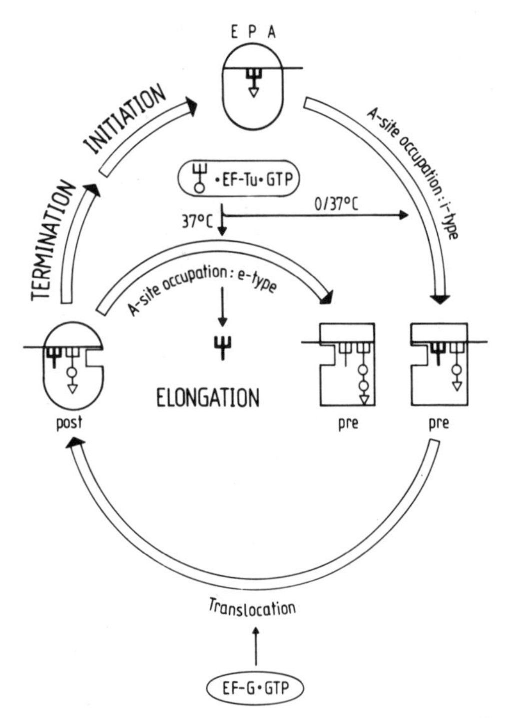

FIGURE 9: Two kinds of A-site occupation in the course of protein biosynthesis. i type: after initiation the E site is free, and therefore, the A site has a high affinity and can be charged even at 0 °C. e type: the second A-site occupation as well as all subsequent ones occur with an occupied P and E site (A site with low affinity). In this case A-site occupation requires a much higher activation energy, and the deacylated tRNA leaves the E site in the course of the allosteric transition from the posttranslocational to the pretranslocational state. From Hausner et al. (1988).

The respective activation energies are 23 kJ/mol (5.4 kcal/mol) and 83 kJ/mol (20 kcal/mol) at 10 mM Mg^{2+} (S. Schilling-Bartetzko and K. H. Nierhaus, unpublished).

The effects of a series of antibiotics were tested on the various reactions of the elongation cycle, in particular on both types of A-site binding and the subsequent EF-G-dependent translocation reaction (Figure 10; Hausner et al., 1988). Some control drugs were included to test the specificity of the system. Tetracycline (TET) inhibited the A-site occupation of both types as expected. Chloramphenicol (CAM) had marginal effects on A-site binding but blocked the puromycin reaction so that the translocation could not be assessed, and α-sarcin [which blocks the binding of elongation factors to the ribosome but has no effect on factor-independent reactions such as spontaneous translocation (Hausner et al., 1987)] did not impair nonenzymatic binding of AcPhe-tRNA to the A site but did block EF-G-dependent translocation.

It can be seen that both thiostrepton (THI) and viomycin (VIO) hardly affect A-site occupation of the i type but severely reduce A-site binding of the e type. In addition, the translocation is blocked by both drugs. It appears that these nonrelated antibiotics are inhibitors of the allosteric transitions in both directions, from the pre- to the posttranslocational state (i.e., translocation reaction) and from the post- to the pretranslocational state (A-site occupation of the e type). The inhibition of the factor-dependent GTPase reaction, which was thought to be the principal inhibition mechanism of thiostrepton (Cundliffe, 1980), is less pronounced (Hausner et al., 1988) and appears to be a consequence rather than the cause of inhibition, since the factor-dependent GTPase reaction follows the allosteric transitions. On the other hand, viomycin is not only an inhibitor of translocation as was hitherto accepted but it equally well blocks the allosteric transition in the reverse reaction [for references and discussion, see Hausner et al. (1988)].

The killing action of aminoglycosides is not caused by their well-documented misreading effect (Fast et al., 1987) but rather results from a not yet identified step shortly after initiation (Davis, 1987). Figure 10 demonstrates that these drugs have practically no effect on A-site occupation of the i type (as was known already) but totally block that of the e type; i.e., they block the second A-site occupation after initiation. Thus the simple picture that arises from the related structure of the aminoglycosides, together with a defined ribosomal region which has been determined at or near their binding sites (Moazed & Noller, 1987) and which harbors rRNA alterations conferring resistance (Cundliffe, 1987), is now complemented with a common point of interference, namely, the e-type occupation of the A site. Figure 11 summarizes the interference points in the elongation cycle of some antibiotics.

Hypothesis: Ribosomes Achieve Accuracy of tRNA Selection without Proofreading

It was shown in a preceding section that the allosteric three-site model explains why noncognate aminoacyl-tRNAs do not impair the rate and accuracy of protein biosynthesis (Figures 7 and 8). In this section the discrimination between cognate versus miscognate aminoacyl-tRNAs is considered. The allosteric three-site model provides a framework according to which the latter discrimination could be achieved via a one-step recognition. If this is so, then the assumption of a proofreading mechanism responsible for the accuracy of aminoacyl-tRNA selection would become unnecessary. A number of observations lend support to this view.

The term "proofreading" is used here in the strict sense that the anticodon of an aminoacyl-tRNA interacts not only once but twice or more with the codon before tight binding of the selected aminoacyl-tRNA at the A site occurs. It is widely accepted that such a proofreading mechanism is a prerequisite for explaining the accuracy values of about 1:1000 (incorporation of one wrong amino acid per 1000 amino acids) observed for protein synthesis in vivo and in vitro [see Bouadloun et al. (1983) and references cited therein]. The increased EF-Tu-dependent GTPase activity observed with a miscognate substrate at the A site is usually taken as evidence for a proofreading mechanism (Thompson & Stone, 1977; Ruusala et al., 1982), and it was suggested that EF-Tu is directly involved in the proofreading mechanism (Tapio & Kurland, 1986).

At least two facts cannot easily be reconciled with the proofreading hypothesis. (1) RNA polymerases without proofreading synthesize RNA with an accuracy of up to 1:1000 (Libby et al., 1989, and references cited therein). If these enzymes are so accurate in checking the pairing precision of only single pairs of nucleotides without proofreading, why should the ribosome not be able to achieve about the same accuracy without proofreading, while checking the pairing precision of the three pairs of nucleotides that comprise the codon–anticodon interaction? (2) Not one mechanism has been proposed up to now that satisfactorily explains a repeated melting and rejoining of codon–anticodon interactions. Even if such a mechanism does exist, the EF-Tu-dependent GTPase activity is not necessarily involved. It has been shown with the hardly cleavable GTP analogue GTPγS that, in principle, the ribosome can select ternary complexes with a precision of better than 1:1000 without needing this GTPase activity (Thompson & Karim, 1982). This precision was due to the dissociation (k_{-1}) of the ternary complexes, and since the rate of the next reaction—the GTP hydrolysis—was comparable to k_{-1}, the authors concluded that "the potential specificity

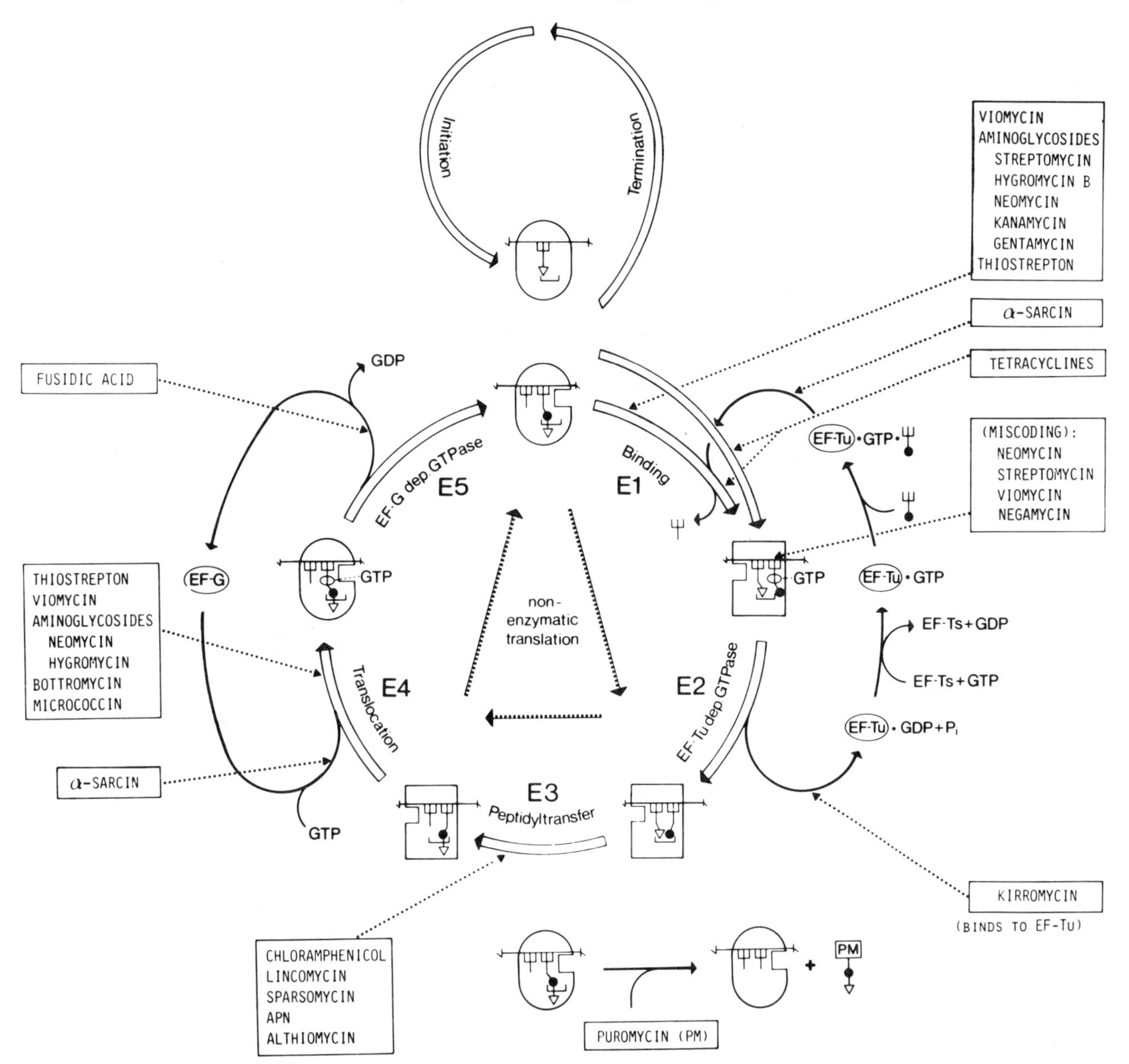

FIGURE 11: Points of interference of various antibiotics inhibiting the ribosomal elongation cycle. From Hausner et al. (1988).

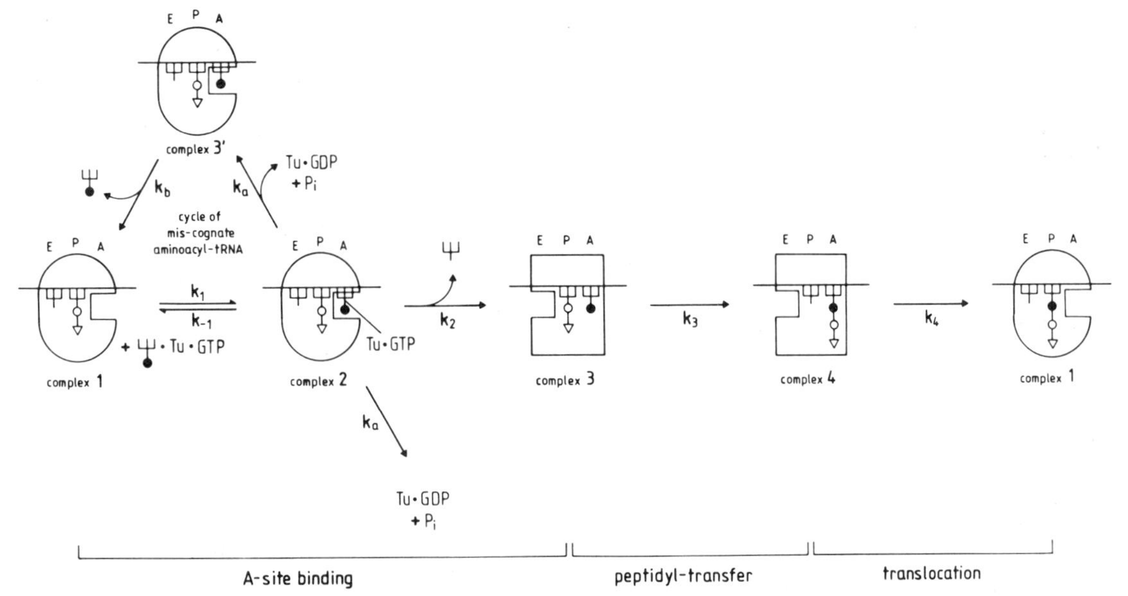

FIGURE 12: Hypothesis for the precision of A-site occupation in the framework of the allosteric three-site model, without proofreading. Complex 3′ indicates an intermediate in the course of binding of a miscognate aminoacyl-tRNA. The aminoacyl-tRNA present in complex 2 and complex 3′ at the A site is loosely bound to the low-affinity A site.

complexes, although codon–anticodon interaction occurs only once.

The mechanism depicted in Figure 12 lacks a proofreading step according to the above-mentioned definition, but it nevertheless represents a branched reaction. Thus, a branched reaction is not indicative of a proofreading mechanism per se, although all proofreading mechanisms necessarily represent branched reactions (Hopfield & Yamane, 1980).

It is not yet clear whether or not proofreading occurs in the ribosomal selection of an aminoacyl-tRNA. Nevertheless, there are good reasons to believe that it does not, and without proofreading ribosomal life would be much easier.

ACKNOWLEDGMENTS

I am grateful to Drs. H. G. Wittmann and R. Brimacombe for help and advice. This work is based on the enthusiastic and dedicated cooperation of my co-workers and colleagues Drs. U. Geigenmüller, A. Gnirke, T.-P. Hausner, J. Remme, H. Saruyama, and S. Schilling-Bartetzko. I thank in particular H.-J. Rheinberger, whose friendly cooperation was important for the development and testing of the allosteric three-site model.

REFERENCES

Abdurashidova, G. G., Baskayeva, I. O., Chernyi, A. A., Kaminir, L. B., & Budowsky, E. I. (1986) *Eur. J. Biochem. 159*, 103–109.

Baranov, V., & Ryabova, A. (1988) *Biochimie 70*, 259–265.

Belitsina, N. V., Tnalina, G. Z., & Spirin, A. S. (1982) *BioSystems 15*, 233–241.

Bilgin, N., Kirsebom, L. A., Ehrenberg, M., & Kurland, C. G. (1988) *Biochimie 70*, 611–618.

Bosch, L., Kraal, B., van Noort, J. M., van Delft, J., Talens, A., & Vijgenboom, E. (1985) *Trends Biochem. Sci. 10*, 313–316.

Bouadloun, F., Donner, D., & Kurland, C. G. (1983) *EMBO J. 2*, 1351–1356.

Cundliffe, E. (1987) *Biochimie 69*, 863–869.

Cundliffe, E. (1980) in *Ribosomes: Structure, Function, and Genetics* (Chambliss, G., Craven, G. R., Davies, J., Davis, K., Kahan, L., & Nomura, M., Eds.) pp 586–604, University Park Press, Baltimore, MD.

Davis, B. D. (1987) *Chem. Rev. 51*, 341–350.

Fast, R., Eberhard, T. H., Ruusala, T., & Kurland, C. G. (1987) *Biochimie 69*, 131–136.

Faxên, M., Kirsebom, L. A., & Isaksson, L. A. (1988) *J. Bacteriol. 170*, 3756–3760.

Geigenmüller, U., Hausner, T.-P., & Nierhaus, K. H. (1986) *Eur. J. Biochem. 161*, 715–721.

Gnirke, A., & Nierhaus, K. H. (1986) *J. Biol. Chem. 261*, 14506–14514.

Gnirke, A., Geigenmüller, U., Rheinberger, H.-J., & Nierhaus, K. H. (1989) *J. Biol. Chem. 264*, 7291–7301.

Graifer, D. M., Babkina, G. T., Matasova, N. B., Vladimirov, S. N., Karpova, G. G., & Vlassov, V. V. (1989) *Biochim. Biophys. Acta 1008*, 146–156.

Grajevskaja, R. A., Ivanov, Yu. V., & Saminsky, E. M. (1982) *Eur. J. Biochem. 128*, 47–52.

Hausner, T.-P., Atmadja, J., & Nierhaus, K. H. (1987) *Biochimie 69*, 911–923.

Hausner, T.-P., Geigenmüller, U., & Nierhaus, K. H. (1988) *J. Biol. Chem. 263*, 13103–13111.

Hirsh, D. (1971) *J. Mol. Biol. 58*, 439–458.

Hopfield, J. J., & Yamanc, T. (1980) in *Ribosomes: Structure, Function, and Genetics* (Chambliss, G., Craven, G. R., Davies, J., Davis, K., Kahan, L., & Nomura, M., Eds.) pp 585–596, University Park Press, Baltimore, MD.

Kirillov, S. V., & Semenkov, Yu. P. (1982) *FEBS Lett. 148*, 235–238.

Kirillov, S. V., Makarov, E. M., & Semenkov, Yu. P. (1983) *FEBS Lett. 157*, 91–94.

Libby, R. T., Nelson, J. L., Calvo, J. M., & Gallant, J. A. (1989) *EMBO J. 8*, 3153–3158.

Lill, R., & Wintermeyer, W. (1987) *J. Mol. Biol. 196*, 137–148.

Lill, R., Robertson, J. M., & Wintermeyer, W. (1984) *Biochemistry 23*, 6710–6717.

Lill, R., Robertson, J. M., & Wintermeyer, W. (1986) *Biochemistry 25*, 3245–3255.

Lill, R., Lepier, A., Schwägele, F., Sprinzl, M., Vogt, H., & Wintermeyer, W. (1988) *J. Mol. Biol. 203*, 699–705.

Lipmann, F. (1963) *Prog. Nucleic Acid Res. 1*, 135–161.

Lührmann, R., Eckard, H., & Stöffler, G. (1979) *Nature 280*, 423–425.

Moazed, D., & Noller, H. F. (1987) *Nature 327*, 389–394.

Moazed, D., & Noller, H. F. (1989a) *Cell 57*, 585–597.

Moazed, D., & Noller, H. F. (1989b) *Nature 342*, 142–148.

Noller, H. F., Moazed, D., Stern, S., Powers, T., Patrick, N. A., Robertson, J. M., Weiser, B., & Triman, K. (1990) in *The Structure, Function and Evolution of Ribosomes* (Hill, W., Ed.) ASM Publications, Washington, DC (in press).

Ofengand, J., & Liou, R. (1981) *Biochemistry 20*, 552–559.

O'Mahony, D. J., Hughes, D., Thompson, S., & Atkins, J. F. (1989) *J. Bacteriol. 171*, 3824–3830.

Peters, M., & Yarus, M. (1979) *J. Mol. Biol. 134*, 471–491.

Remme, J., Margus, T., Villems, R., & Nierhaus, K. H. (1989) *Eur. J. Biochem. 183*, 281–284.

Rheinberger, H.-J. (1982) Thesis at the Freie Universität Berlin.

Rheinberger, H.-J., & Nierhaus, K. H. (1983) *Proc. Natl. Acad. Sci. U.S.A. 80*, 4213–4217.

Rheinberger, H.-J., & Nierhaus, K. H. (1986a) *J. Biol. Chem. 261*, 9133–9139.

Rheinberger, H.-J., & Nierhaus, K. H. (1986b) *FEBS Lett. 204*, 97–99.

Rheinberger, H.-J., & Nierhaus, K. H. (1987) *J. Biomol. Struct. Dyn. 5*, 435–446.

Rheinberger, H.-J., Sternbach, H., & Nierhaus, K. H. (1981) *Proc. Natl. Acad. Sci. U.S.A. 76*, 5310–5314.

Rheinberger, H.-J., Sternbach, H., & Nierhaus, K. H. (1986) *J. Biol. Chem. 261*, 9140–9143.

Riddle, D. L., & Carbon, J. (1973) *Nature, New Biol. 242*, 230–334.

Robertson, J. M., & Wintermeyer, W. (1987) *J. Mol. Biol. 196*, 525–540.

Robertson, J. M., Paulsen, H., & Wintermeyer, W. (1986) *J. Mol. Biol. 192*, 351–360.

Rodnina, M. V., El'skaya, A. V., Semenkov, Yu. P., & Kirillov, S. V. (1988) *FEBS Lett. 231*, 71–74.

Ruusala, T., Ehrenberg, M., & Kurland, C. G. (1982) *EMBO J. 1*, 741–745.

Ruusala, T., Andersson, D., Ehrenberg, M., & Kurland, C. G. (1984) *EMBO J. 3*, 2575–2580.

Saruyama, H., & Nierhaus, K. H. (1986) *Mol. Gen. Genet. 204*, 221–228.

Smith, D., & Yarus, M. (1989a) *J. Mol. Biol. 206*, 489–501.

Smith, D., & Yarus, M. (1989b) *J. Mol. Biol. 206*, 503–511.

Tapio, S., & Kurland, C. G. (1986) *Mol. Gen. Genet. 205*, 186–188.

Thompson, R. C., & Stone, P. J. (1977) *Proc. Natl. Acad. Sci. U.S.A. 74*, 198–202.

Thompson, R. C., & Karim, A. M. (1982) *Proc. Natl. Acad. Sci. U.S.A. 79*, 4922–4926.
Traut, R. R., & Monro, R. E. (1964) *J. Mol. Biol. 10*, 63–72.
Watson, J. D. (1963) *Science 140*, 17–26.
Watson, J. D. (1964) *Bull. Soc. Chim. Biol. 46*, 1399–1425.
Wettstein, F. O., & Noll, H. (1965) *J. Mol. Biol. 11*, 35–53.
Wurmbach, P., & Nierhaus, K. H. (1979) *Proc. Natl. Acad. Sci. U.S.A. 76*, 2143–2147.
Yonath, A., Leonard, K. R., & Wittmann, H. G. (1987) *Science 236*, 813–816.

Chapter 22

Initiation of mRNA Translation in Prokaryotes†

Claudio O. Gualerzi*,‡,§ and Cynthia L. Pon‡

Laboratory of Genetics, Department of Cell Biology, University of Camerino, Camerino (MC), 62032 Italy, and Max-Planck-Institut für Molekulare Genetik, D-1000 Berlin 33, West Germany

Received February 28, 1990

Initiation of protein synthesis consists of several interrelated steps during which the translation initiation region of the messenger RNA (mRNA TIR)[1] and the initiator tRNA (fMet-tRNA^{Metf}) are selected by the ribosome to form a ternary complex. Formation of the first peptide bond between the P-site-bound fMet-tRNA and the A-site-bound aminoacyl-tRNA specified by the second mRNA codon marks the transition from initiation to elongation. Three protein factors (initiation factors IF1, IF2, and IF3) and at least one GTP molecule are required to ensure the efficiency and fidelity of this process.

Initiation is usually the rate-controlling step of translation; under normal conditions (e.g., with the *lacZ* mRNA), elongation proceeds at a rate of ~12 amino acids/s (Sørensen et al., 1989), while the ribosomes load onto the same mRNA at ~3.2-s intervals (Kennell & Riezman, 1977). That initiation is rate-limiting is also indicated by the fact that translational efficiency can often be dramatically affected by changes in the mRNA that alter the TIRs leaving the coding sequence unmodified, while the reverse is generally not true (Gold, 1988, and references cited therein). Initiation also represents the focal point of many posttranscriptional regulatory mechanisms, so that a thorough understanding of its molecular basis and the identification of the determinants of its fidelity and efficiency are prerequisites for understanding regulation of gene expression and for optimizing the production of proteins and peptides from chimeric vectors.

In this paper, we shall review the structural and functional properties of the components involved in translation initiation (with the exception of the ribosomal subunits) and the mechanism of their mutual interactions, which eventually lead to the formation of the 70S initiation complex. We shall also discuss the mechanism of recognition between ribosomes and mRNA during initiation and the molecular determinants of translational efficiency.

Space limitation does not allow the treatment of translational regulation of gene expression. This subject, however, has been reviewed in several recent articles [e.g., Draper (1987), Stormo (1987), Gold (1988), de Smit and van Duin (1990), Dreyfus and Jacques (1990), and McCarthy and Gualerzi (1990)].

The Initiator tRNA

Escherichia coli contains a main and a minor form of initiator tRNA, namely, tRNA^{Metf1} (~75%) and tRNA^{Metf2} (~25%). Both tRNAs contain 77 nucleotides and differ in the presence of either 7-methyl-G or A at position 47 (Figure 1). The two molecules are encoded by *metZ* (tRNA^{Metf1}) and *metY* (tRNA^{Metf2}): *metZ* is present in two tandemly arranged copies at 61′ on the chromosome, while *metY* is the first gene of the *nusA–infB* operon mapping at 69′ (Nagase et al., 1988, and references cited therein). Although no functional difference between these two forms has been reported so far, the presence of two separate genes is likely to serve some regulatory function. In this connection, it is noteworthy that the *metY* promoter is less sensitive to ppGpp than is the *metZ* promoter (25% vs 65% reduction in transcription at 0.1 mM), suggesting that *metY* constitutively supplies tRNA^{Metf} under conditions leading to the stringent response (Nagase et al., 1988). Two forms of initiator tRNA differing by the inversion of a GC base pair in the T-stem have also been found in *Thermus thermophilus* (Watanabe et al., 1979), but only a single gene for tRNA^{Metf} has been found in *Bacillus subtilis*. The latter is present at 275° of the chromosome within a cluster containing 21 tRNA genes (Green & Vold, 1983).

The initiator tRNA has the same CAU anticodon (Figure 1) as the tRNA^{Met} used in elongation and is aminoacylated

† Dedicated to the fond memory of Professor Heinz-Günter Wittmann, to whom we are grateful for the opportunity of having worked for nearly 20 years on the subject of this review and whose support and encouragement we will miss.

* Address correspondence to this author.
‡ Max-Planck-Institut.
§ University of Camerino.

[1] Abbreviations: IF, initiation factor; SD, Shine–Dalgarno; TIR, translation initiation region; DMS, dimethyl sulfate; CMCT, *N*-cyclohexyl-*N′*-[2-(*N*-methylmorpholinio)ethyl]carbodiimide *p*-toluenesulfonate.

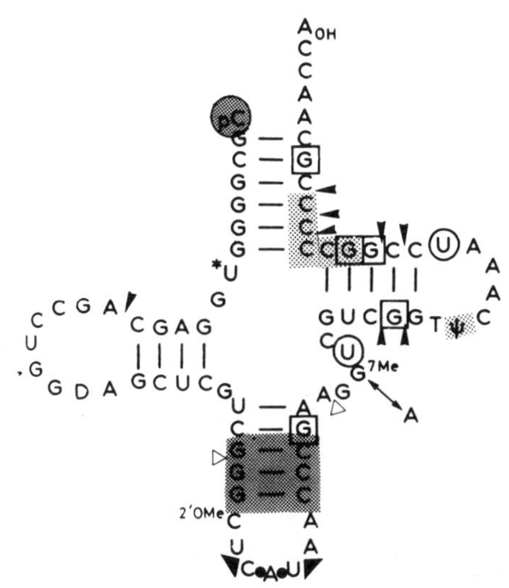

FIGURE 1: Cloverleaf structure of *E. coli* $tRNA^{Metf}$. The symbols for the modified bases are as follows: *U, 4-thiouridylic acid; D, dihydrouridylic acid; 2'OMe, 2'-*O*-methylcytidylic acid; 7MeG, 7-methylguanylic acid; T, thymidylic acid; ψ, pseudouridylic acid. The figure highlights (with dark stippling) two distinct structural features of the initiator tRNA molecule (i.e., the unpaired 5' pC and the three consecutive GC base pairs of the anticodon stem); due to its peculiar anticodon loop conformation, the phosphodiester bonds indicated by dark triangles, which are accessible in elongator tRNAs, are resistant to S1 nuclease in $tRNA^{Metf}$. The phosphodiester bonds of the anticodon loop digested in both initiator and elongator tRNAs are marked by dark circles. The regions of fMet-$tRNA^{Metf}$ affected by its interaction with IF2 are indicated: sites where cleavage by RNase V_1 is prevented are marked with arrowhead and those where cleavage is enhanced, by open triangles; regions where the phosphates are protected from ethylnitrosourea are lightly stippled; guanines whose N7 reactivity toward DMS is increased are boxed; uracils whose reactivity toward CMCT is reduced are encircled. For further details, see the original literature quoted in the text.

by the same synthetase, which recognizes primarily the bases of the anticodon (Schulman, 1979; Pelka & Schulman, 1986, and references cited therein). Although its 3.5-Å crystal structure is overall very similar to that of an elongator tRNA, such as yeast $tRNA^{Phe}$, $tRNA^{Metf}$ is endowed with unique structural features connected with its special role in protein synthesis (Wakao et al., 1989, and references cited therein). One of these is the presence of three consecutive GC base pairs in the anticodon stem that confer rigidity and regularity to the helix and a particular conformation to the anticodon loop (Figure 1), which targets the initiator tRNA to the ribosomal P-site. Indeed, the progressive substitution of the three GC base pairs weakens binding to the ribosomal P-site while increasing accessibility of the anticodon loop to nuclease S1 (Seong & RajBhandary, 1987a).

Furthermore, the special structure of Met-$tRNA^{Metf}$ allows its specific recognition by N^{10}-formyltetrahydrofolate:Met-tRNA transformylase, which modifies the αNH_2 group (Schulman, 1979). This formylation results in a 6–10-fold stimulation of protein synthesis in *E. coli* extracts (Kung et al., 1979). fMet-tRNA interacts specifically with IF2 by virtue of its blocked αNH_2 group, and a portion of the T-stem and loop is protected from chemical and enzymatic attack in the ensuing complex (Figure 1); the binding of IF2 also results (by a long-range effect) in an increased flexibility of the anticodon arm (Wakao et al., 1989).

Finally, the absence of a Watson–Crick base pair at the end of the amino acid acceptor stem (Figure 1) is responsible for the resistance of fMet-tRNA to the action of peptidyl-tRNA hydrolase and, together with the blocking of the αNH_2 group, for its weak interaction with EFTu–GTP (Schulman, 1979; Tanada et al., 1982; Louie et al., 1984). Single base substitutions resulting in mutants containing a base-paired 5' nucleotide enable the initiator tRNA to be used in translation elongation (Seong & RajBhandary, 1987b). The unpaired 5' pC is also likely to be responsible for the restricted mobility of the 3'-terminus observed by EPR spectroscopy (Pscheidt & Wells, 1986).

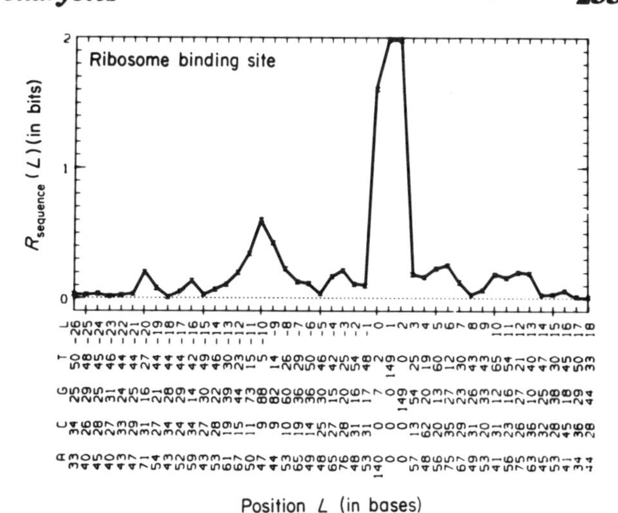

FIGURE 2: Nonrandomness in aligned ribosome-binding sites of mRNAs [from Schneider et al. (1986)].

THE TRANSLATION INITIATION REGION (TIR) OF mRNA AND THE STRUCTURAL BASES FOR TRANSLATIONAL EFFICIENCY

The TIR of most cistrons (≥90%) contains the initiation codon AUG; more rarely GUG (~8%), UUG (~1%), and in only one known case (i.e., *infC*) AUU serve as initiation codons. Almost invariably, the TIRs also contain a 3–9-base-long sequence [the Shine and Dalgarno (SD) sequence] complementary to part or all of the anti-SD sequence 5'-CACCUCCUU-3' found at the 3'-end of 16S rRNA separated by a spacer of variable length (optimally 5–9 bases) from the initiation triplet (Steitz, 1980; Gren, 1984; Schneider et al., 1986).

Other distinguishing features of mRNA TIRs and their relationship to translational efficiency have been searched for by statistical analyses. In addition to the three elements mentioned above, a loose consensus sequence characterized by a strong bias for "nonstructurogenic" nucleotides was identified by Scherer et al. (1980). Subsequently, Schneider et al. (1986) identified at least five additional regions both upstream and downstream from the initiation codon in which the nucleotide sequence of bona fide translational starts is significantly different from "false" starts (Figure 2); the presence of other primary structure elements has also been correlated with efficient expression (McCarthy et al., 1985; Petersen et al., 1988; Gallie & Kado, 1989; Olins & Rangwala, 1989; Thanaraj & Pandit, 1989). None of the above elements, however, seems to be essential per se for initiation, but their interplay and the resulting TIR secondary and tertiary structures probably determine the efficiency of the recognition by ribosomes and the level of translation.

The importance of TIR primary sequence as a determinant of ribosomal recognition, regardless of its context, clearly emerges from the elegant work of Dreyfus (1988), while the importance of secondary structure has been pointed out in numerous reports [e.g., see Ganoza et al. (1987), Gold (1988), and de Smit and van Duin (1990) and references cited therein]. Only in a minority of the cases, however, have the postulated

Table I: Cloned Bacterial Genes Specifying Initiation Factors Whose Primary Structures Are Known

organism	struct gene	map loc	within operon	factor	amino acids	M_r	isoel pt[i]	ref
E. coli	*infA*	20′	no	IF1	71	8118	10.1	*a*
B. subtilis	*infA*	12°	yes	IF1	72	8213	7.8	*b*
E. coli	*infBα*	69′	yes	IF2α	890	97349	5.9	*c*
E. coli	*infBβ*	69′	yes	IF2β	733	79713	5.6	*c*
B. stearothermophilus	*infB*	?	?	IF2	742	82043	7.0	*d*
S. faecium	*infB*	?	?	IF2	785	86415	7.9	*e*
B. subtilis	*infBα*	145°	?	IF2α	716	78600	5.4	*f*
B. subtilis	*infBβ*	145°	?	IF2β	623	68182	4.8	*f*
E. coli	*infC*	38′	yes	IF3	180	20548	10.3	*g*
B. stearothermophilus	*infC*	?	yes	IF3	171	19678	10.7	*h*

[a] Sands et al., 1987. [b] Boylan et al., 1989. [c] Sacerdot et al., 1984. [d] Brombach et al., 1986. [e] Friedrich et al., 1988. [f] Shazand et al., 1990. [g] Sacerdot et al., 1982. [h] Pon et al., 1989. [i] The values were calculated by using the UWGCG program (Devereux et al., 1984).

secondary structures been experimentally verified. At the risk of oversimplifying the issue, it can be said that the primary structure of the TIR has a 2-fold function: to provide sequence-related features for the specific recognition by ribosomal components and to dictate the (presence or absence of) secondary structures that affect (negatively or positively) the efficiency of initiation.

Especially intriguing is the influence that the nature of the initiation triplet (AUG, GUG, UUG, or AUU) has on the level of translational expression. In most (but not all) cases, changing the rare initiation triplet into the more common AUG results in moderate (Reddy et al., 1985; Khudyakov et al., 1988) to large (Brombach & Pon, 1987) increases of expression. Nonetheless, inspection of the catalogue of genes having the rare initiation triplets strongly argues against the idea that the cell uses these codons to attain a substantial reduction in the level of translation. Instead, a likely reason for the cell to use the rare initiation codons is that they serve as targets for regulatory mechanisms aimed at select genes. This concept is best illustrated by the case of the AUU triplet found in *infC* of both *E. coli* (Sacerdot et al., 1982) and *Bacillus stearothermophilus* (Pon et al., 1989). When this AUU is changed to AUG, in vivo expression of both genes is increased about 40-fold because, confirming Gold's intuition (Gold et al., 1984), autorepression by IF3 is lost (Butler et al., 1987). On the other hand, the expression of mRNAs (different from that of IF3) attains comparable levels in vitro regardless of whether the initiation triplet is AUG or AUU. The IF3 autorepression requires a small excess of free IF3 molecules and is specifically aimed at the dissociation of 30S initiation complexes containing mRNAs with an AUU initiation triplet (A. La Teana, C. L. Pon, and C. O. Gualerzi, unpublished results).

The Initiation Factors: Structural, Evolutionary, and Genetic Aspects

Several bacterial genes encoding IFs have been isolated and cloned; their map locations (when available) and some structural characteristics of their products are listed in Table I.

The three IFs do not share any structural homology and appear to be evolutionarily conserved: IF1 of *E. coli* shares 69% identical residues with IF1 of *B. subtilis*, while IF3 of *E. coli* is 50% identical with IF3 of *B. stearothermophilus*. The case of IF2 is more intriguing; all the sequences known so far display a large degree of homology in the C-terminal two-thirds of the molecule, but their N-terminal parts hardly resemble each other (Figure 3), suggesting that this part may have some species-specific regulatory function probably not related to translation. This idea is supported by the finding that even after long deletions from the 5′-terminus (hatched

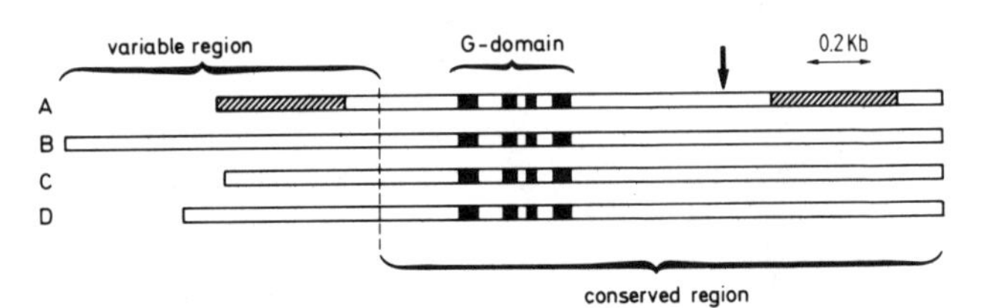

FIGURE 3: Schematic representation of *infB* from (A) *B. stearothermophilus*, (B) *E. coli* (IF2α), (C) *E. coli* (IF2β), and (D) *Streptococcus faecium*. The darkened areas represent the conserved structural elements of the GTP-binding site. The hatched areas represent the deletions introduced by genetic manipulations, and the arrow indicates the point of proteolytic cleavage that separates the G from the C domain (Pon & Gualerzi, 1988; Severini et al., 1990).

area in Figure 3), the *infB* genes encode IF2 fragments fully active in all the basic translation activities, while internal deletions (e.g., like that shown in Figure 3) result in the loss of specific translational functions (Pon & Gualerzi, 1988; Severini et al., 1990). In addition to IF2α, *infB* of *E. coli* (and probably of *B. subtilis*) also expresses a shorter protein (IF2β) resulting from a low-rate initiation event at an in-frame GUG codon downstream from the main initiation start site (Plumbridge et al., 1985). Both the mechanism and the significance of this occurrence remain obscure.

Several attempts to crystallize initiation factors have failed so far. However, the small size of IF1 and its exceptionally well-resolved ^{1}H NMR spectrum have prompted an attempt to determine its solution structure by 2D NMR spectroscopy; IF1 was found to possess a complex structure consisting of extensive β-sheet motifs in parallel and antiparallel orientations, three β-turns, and two short α-helices, one near the N-terminus and the other near the middle of the molecule (M. Paci, R. Boelens, and R. Kaptein, personal communication).

IFs–Ribosome Interaction: Thermodynamic, Structural, and Topographical Aspects

Quantitative data on the association constants and stoichiometry of the interactions between IFs and ribosomes (Table II) have been obtained by fluorescence polarization studies (Weiel & Hershey, 1981, 1982; Zucker & Hershey, 1986) and by Airfuge centrifugation (Pon et al., 1985; Celano et al., 1988). The results are in fairly good agreement and indicate that the 30S subunit has a single high-affinity site for each factor. IF1 and IF3 show negligible affinity for 50S subunits and for 70S monomers (they are actually ejected from the 30S subunit upon subunit association). IF2, on the other hand, binds with a fairly high affinity to the large subunit and also to the 70S monomers from which it is released upon GTP hydrolysis.

The interactions with the 30S subunit display different properties. The binding of IF3 is only slightly affected by IF1

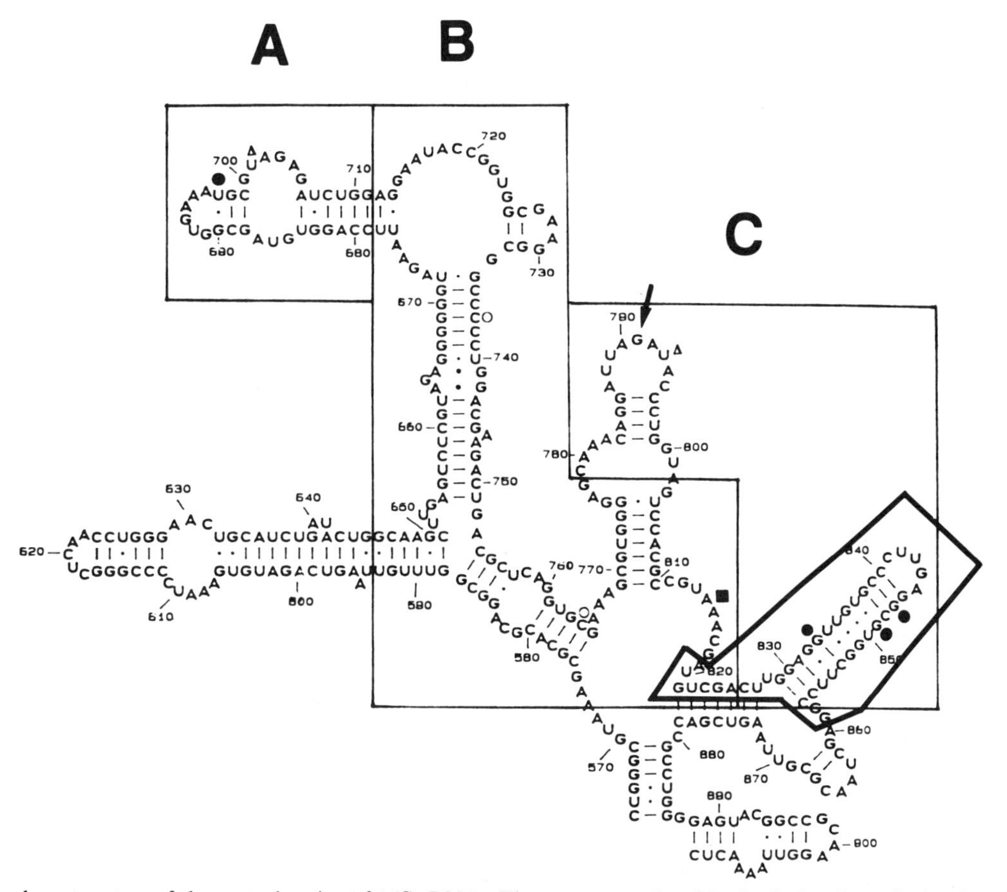

FIGURE 4: Secondary structure of the central region of 16S rRNA. The segment enclosed in the darker frame is that found cross-linked to IF3 by Ehresmann et al. (1986), and the arrow indicates the position of the G → A transition resulting in a decrease of IF3 binding (Tapprich et al., 1989). The sites of enhanced (darkened symbols) or reduced (open symbols) reactivity or cleavage in the presence of IF3 are indicated as follows: hydrolysis by RNase V_1 (○, ●) and reaction with DMS (■) and CMCT (△) [modified from Muralikrishna and Wickstrom (1989)].

and IF2, while the binding of IF1 or IF2 is strongly stimulated by the other two factors. Furthermore, increasing the ionic strength (e.g., from 50 to 150 mM NH_4Cl) reduces 15-fold and 20-fold the K_a of the IF2–30S and IF1–30S complexes, respectively, but decreases only 2-fold the K_a of the IF3–30S complex. All interactions are relatively unaffected by changes in the Mg^{2+} concentration but are drastically weakened by hydrostatic pressure, especially those of IF1 and IF2. It has been suggested that the IF1–30S interaction is an entropy-driven process triggered mainly by the release of counterions from the RNA phosphates and involving a minimum of 2.7–3.6 ion pairs. In the case of IF2 and IF3, ionic and hydrophobic interactions seem to be equally important for ribosomal binding.

Considering the estimated intracellular concentrations of ribosomes (~10 μM) and IFs (~1 μM each) and the K_a's of their interactions, it can be argued that in the cell nearly all the IFs are bound to the native 30S subunits.

The identification of some amino acid residues of the IFs involved in the interactions with the 30S subunit has been obtained by chemical modifications and ^1H NMR spectroscopy (Gualerzi et al., 1986, and references cited therein) and, more recently, by genetic manipulations and site-directed mutagenesis. Thus, Arg 69 plays an important role in the binding of IF1 to the 30S subunit, while the deprotonation of His 29 is apparently involved in the release of the factor from the 30S subunit upon subunit association (Gualerzi et al., 1989). In the case of IF2, two structurally compact and functionally separate domains have been identified: a C-terminal domain containing the fMet-tRNA binding site and a central domain, the G domain, containing the 50S and GTP binding sites as well as the GTPase catalytic center (Spurio et al., 1989).

The topographical localization of the IFs on the ribosome has been investigated by standard protein–protein and protein–RNA cross-linking and, in the case of IF3, visualized by immunoelectron microscopy.

A group of eight ribosomal proteins (S1, S7, S11–13, S18, S19, and S21) has consistently been found cross-linked in significant yield to more than one IF or to the same factor by more than one laboratory, and cross-links between IF1–IF2, IF2–IF3, and IF2–L7/L12 have also been found (Boileau et al., 1983, and references cited therein).

Of the three factors, only IF3 can be efficiently cross-linked to rRNA. The cross-linked factor is found unequally distributed between two 16S rRNA regions that are probably close to each other in the 30S subunit. The major cross-linking site has been localized between residues 819 and 859 in the central domain of the rRNA (Figure 4), while the minor one has been identified between nucleotides 1506 and 1529 (Ehresmann et al., 1986, and references cited therein).

Consistent with this localization is the finding that IF3 reduces the chemical reactivity and enzymatic accessibility of some nucleotides and enhances attack at others in the central (nucleotides 690–850) domain of 16S rRNA. Here three subdomains (A, B, and C of Figure 4) can be distinguished by the effects produced by IF3: domains A and C are either structurally sequestered or stabilized, while domain B is destabilized and more exposed in the presence of the factor (Muralikrishna & Wickstrom, 1989). Finally, mutation of guanine to adenine at position 791 (Figure 4) impairs the association of the 30S subunits with the 50S subunits and increases 10-fold the dissociation rate constant of the IF3–30S complex without altering the on-rate, so that the K_a is reduced by approximately 1 order of magnitude. These results suggest

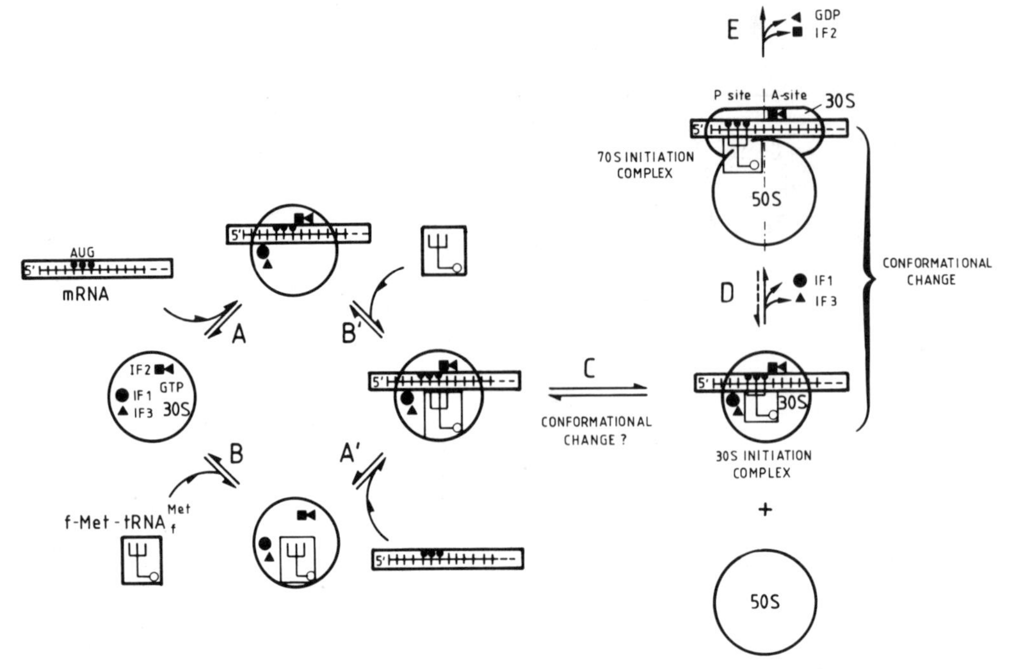

FIGURE 5: Simplified mechanistic model of translation initiation. Steps A, A′, B, and B′ are in rapid equilibrium. Step C represents the first-order, rate-limiting rearrangement of the ternary preinitiation complex kinetically controlled in both directions by the IFs. Step D represents the virtually irreversible subunit association giving rise to the 70S initiation complex [modified from Gualerzi and Pon (1981)].

Table II: Association Constants and Stoichiometry of 30S–Initiation Factor Interaction[a]

complex	K_a (M^{-1}) [stoichiometry (n)] no other factors present	two other factors present
30S–IF1	0.9×10^6 [1.3][b]	1.0×10^8 [0.8][b]
	0.2×10^7 [0.5][c]	9.1×10^7 [0.5][c]
30S–IF2	3.5×10^7[b]	1.8×10^8 [0.75][b]
	1.5×10^7 [0.7][c]	
	3.7×10^7 [0.7][c,d]	
	5.7×10^7 [0.6][e]	
30S–IF3	4.3×10^7 [1.4][b]	

[a] For references, see text. [b] Determined by fluorescence polarization at 10 mM Mg^{2+} and 50 mM NH_4^+. [c] Determined by Airfuge centrifugation at 7.5 mM Mg^{2+} and 50 mM NH_4^+. [d] In the presence of 1 mM GTP. [e] Determined by fluorescence polarization at 5 mM Mg^{2+} and 100 mM NH_4^+.

that IF3 recognizes a specific structure on the 30S ribosomal subunit that includes G791 to stabilize its interaction (Tapprich et al., 1989).

Combining the above information with what is known concerning the topographical organization of the 30S ribosomal subunit [e.g., see Brimacombe (1988) and Stern et al. (1988)], we can draw a relatively unambiguous albeit low-resolution picture of the IFs' binding sites on the ribosome: IF3, having been found cross-linked to elements of both sides, most likely bridges the cleft between head and platform, while IF1 should be located in the neck region and IF2 more toward the head, between IF1 and IF3. A topographical model in which IF2 binds close to both IF1 and IF3 but the latter two factors are not in contact with each other is consistent with the cross-linking data and also with some of the above-mentioned binding studies (Zucker & Hershey, 1986).

The topographical localization of the IFs is also consistent with their role of influencing the kinetics of codon–anticodon interaction at the P-site (see below), which is believed to occur at the bottom of the cleft (Ofengand et al., 1986). Finally, the localization of IF3 at or near bases (e.g., G791) implicated in subunit association (Herr et al., 1979; Tapprich & Hill, 1986) may underlie the well-known subunit antiassociation activity of IF3 [see Hershey (1987) and references cited therein].

The Pathway of Translation Initiation

The probable events occurring during initiation are depicted in Figure 5. This mechanistic model is suggested by kinetic analyses of the formation of 30S and 70S initiation complexes and is compatible with the relevant data available [for review, see Gualerzi et al. (1988)]. Mutatis mutandis, the proposed mechanism is similar to that suggested by McClure (1985) for initiation of transcription.

When the concentrations of mRNA and fMet-tRNA are saturating, the 30S subunit (bearing one molecule each of the three IFs) binds these ligands in random order to form a ternary preinitiation complex in which the two ligands are not yet interacting. A rate-limiting first-order rearrangement, kinetically controlled by the IFs, promotes codon–anticodon interaction at the ribosomal P-site and the formation of the 30S initiation complex. This complex either dissociates into its original components or binds a 50S subunit, giving rise to the 70S initiation complex. Due to the concomitant ejection of IF1 and IF3, formation of the 70S initiation complex is virtually irreversible, while IF2 promotes the positioning of fMet-tRNA in the ribosomal P-site. The ejection of IF2 from the 70S initiation complex is accompanied by the factor- and ribosome-dependent hydrolysis of GTP.

Function of the Initiation Factors

The molecular basis for IF3 activity is probably a change in the conformational dynamics of the 30S subunits induced by the binding of the factor (Pon et al., 1982, and references cited therein), while the basis for IF2 activity probably resides in its capacity to bind fMet-tRNA, to undergo conformational change, and to induce conformational changes of both initiator tRNA and ribosomes (Pon et al., 1985; Zucker & Hershey, 1986; Canonaco et al., 1989; Wakao et al., 1989; our unpublished observations). The mechanism of action of IF1 remains completely obscure.

The main functions of the IFs are summarized in Table III. As mentioned above, the key function of the IFs is to affect

Table III: Main Properties and Functions of Prokaryotic Initiation Factors

factor	properties and function
IF1	binds to 30S subunits, is ejected upon subunit association, increases the affinity of the 30S for IF2 and somewhat for IF3, and stimulates IF2 and IF3 activities
IF2	G-protein: binds to 30S subunits (optimally with GTP), increases the affinity of the 30S subunit for IF1, and contains a binding site for fMet-tRNA; kinetic effector of 30S (and 70S) initiation complex formation and increases on-rate (favoring binding of aminoacyl-tRNAs with blocked αNH_2 groups) and reduces dissociation rate constants; upon subunit association interacts with 50S subunits activating ribosome-dependent GTPase, positions fMet-tRNA in P-site, and is ejected from ribosomes; kinetic proofreading?
IF3	binds to 30S subunits and is ejected upon subunit association; increases the affinity of the 30S for IF1 and IF2; kinetic effector and fidelity factor of 30S initiation complex formation and increases both on- and off-rates of its formation favoring dissociation of noninitiator aminoacyl-tRNAs; subunit antiassociation factor

kinetically the formation and dissociation of the codon–anticodon interaction at the P-site of the 30S subunit and, ultimately, to influence which and how many 30S initiation complexes enter the elongation cycle after association with the 50S subunit. All three factors (but IF1 only in combination with the other two) stimulate the on-rate of 30S initiation complex formation. The effect of IF2 is much larger with aminoacyl-tRNAs having a blocked αNH_2 group but is observable also with aminoacyl-tRNAs that do not form binary complexes with IF2 off the ribosomes; thus the suggestion that this factor promotes the recognition and ribosomal binding of fMet-tRNA through a carrier mechanism analogous to that of EFTu does not seem to be entirely supported by the experimental evidence (Gualerzi et al., 1986, 1988, and references cited therein).

Furthermore, regardless of the type of aminoacyl-tRNA bound, the dissociation rate of the 30S complexes and the rate of exchange between free and 30S-bound aminoacyl-tRNAs are substantially lowered by IF2. IF3, on the other hand, produces a large increase in both dissociation and exchange rates, especially when the bound aminoacyl-tRNA is different from fMet-tRNA. In conclusion, it appears that IF2 and IF3 accelerate the locking and unlocking of the codon–anticodon interaction at the P-site while favoring the formation of the correct over the incorrect and the dissociation of the incorrect over the correct 30S complexes. The role of IF3 as a fidelity factor had been recognized a long time ago and was traced to the recognition by the factor of some specific feature of the initiator tRNA molecule (Gualerzi & Pon, 1981, and references cited therein) but received little attention until Gold's laboratory demonstrated this activity in toeprinting experiments and provided evidence that IF3 "inspects" the correctness of both the anticodon stem of the tRNA and the P-site codon–anticodon interaction (Hartz et al., 1989) (Figure 6).

The IFs also influence in different ways the equilibrium 70S $\rightleftharpoons$ 30S + 50S: by binding to the free 30S subunits, IF3 shifts the equilibrium to the right, thus providing a continuous supply of native 30S subunits to feed the initiation process; IF1 increases the rate of exchange between ribosomal subunits; and IF2 favors subunit association [see Hershey (1987) and references cited therein].

Finally, as mentioned above, IF2 is endowed with an IF2- and ribosome-dependent GTPase activity. At least three nonmutually exclusive physiological functions for this activity can be envisaged: (a) to promote the ejection of IF2 from the ribosome, (b) to position fMet-tRNA in its final transpeptidation-competent ribosomal binding site, and (c) to increase further the fidelity of initiation by driving a kinetic proofreading mechanism.

Ribosome–mRNA Interaction and Initiation Factor Function

The idea that the IFs, IF3 in particular, promote or stimulate (natural) mRNA–ribosome interaction and influence the

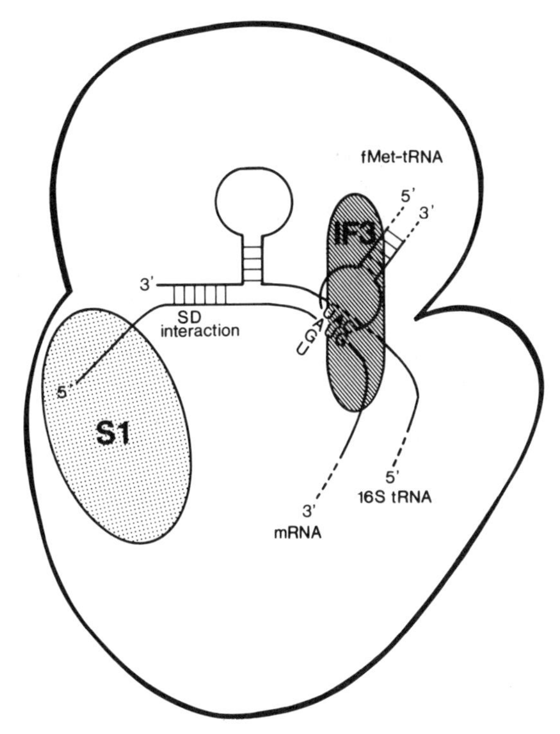

FIGURE 6: mRNA–30S interaction at initiation. This diagram highlights the three main mRNA–ribosome interactions (the SD–anti-SD and the codon–anticodon base pairings and the mRNA–S1 interaction) that dictate translational efficiency. The figure also presents a visual summary of the results obtained in the laboratory of L. Gold concerning the structural elements of initiator tRNA and the portion of the codon–anticodon interaction at the P-site inspected by IF3 to ensure fidelity of translation initiation (see text for details).

SD base pairing remained deeply rooted for a long time, in spite of strong circumstantial evidence to the contrary (Gualerzi & Pon, 1981, and references cited therein).

Direct evidence that neither IF3 nor the other IFs influence the SD interaction and the affinity of the 30S subunit for mRNAs with or without SD sequence has been obtained from several types of binding experiments (Calogero et al., 1988; Canonaco et al., 1989). In agreement with these findings, Laughrea and Tam (1990), using model mDNAs and mRNAs having 4–8-nucleotide-long SD sequences and primary structures similar to the R17 coat protein initiation site, could not detect any effect of the IFs (nor of ribosomal protein S1) on the recognition, binding, or exchange rate of mRNAs on the 30S subunit.

Although the IFs have no detectable effect on the K_a's of the binary complexes between 30S subunits and mRNAs, their presence may well influence the position of the mRNA in its ribosomal binding site. Thus, it has been suggested that, in the absence of IFs, the mRNA preferentially occupies a ribosomal "standby site" corresponding to the region where the SD interaction takes place. In the presence of factors, the mRNA is shifted toward another ribosomal site, possibly closer to the ribosomal P-site where the IFs exert their kinetic influence on codon–anticodon interaction (Canonaco et al., 1989).

Selection of mRNA TIR by Ribosomes

After recalling that the access of ribosomes to noninitiation regions of the mRNA may be restricted by secondary structure constraints [e.g., see Ganoza et al. (1987)], we can ask what are the positive elements governing mRNA–ribosome interaction at initiation and dictating the fidelity and efficiency of the ensuing translational process.

The ribosome is believed to accommodate the mRNA in a "U-shaped" channel or trough (Evstafieva et al., 1983; Olson et al., 1988), probably constituted by structural elements contributed by several ribosomal proteins and stretches of 16S rRNA, so that each mRNA, depending on its structure, binds to the 30S subunit by means of several alternative, more or less specific interactions with both 16S rRNA and ribosomal proteins. Although our knowledge of ribosome structure is still too rudimentary to allow a detailed understanding of these interactions, at least two 30S proteins (i.e., S1 and S21) have consistently been implicated in mRNA binding, and it has recently been demonstrated that the oligo(U)-containing "translational enhancers" found in some mRNA TIRs may be specifically recognized and bound by S1 (I. Boni, USSR Academy of Science, personal communication).

In addition to these interactions, specific base pairing can direct the mRNA TIR to its ribosomal binding site. Base pairing in vivo between the mRNA SD sequence and anti-SD sequence of the 16S rRNA (Figure 6) has been elegantly demonstrated (Hui & de Boer, 1987; Jacob et al., 1987). In vitro, however, mRNAs with and without the SD sequence are translated at the same rate and with the same dependence on IFs, provided that the mRNA concentration is properly selected. The SD interaction is also mechanistically irrelevant for 30S initiation complex formation and for the selection of the mRNA reading frame (Calogero et al., 1988). In agreement with these data, Melançon et al. (1990) showed that 30S mutants lacking the anti-SD sequence are able to initiate translation of natural mRNAs at the correct start sites and concluded that neither translational efficiency (in vitro) nor selection of translational start is solely controlled by the SD interaction. Due to the competition of several mRNA TIRs for a limited number of 30S subunits, however, the influence of the SD interaction on translational efficiency is probably greater in vivo than in vitro. In conclusion, the available data support the suggestion that the in vivo function of the SD interaction is to ensure millimolar concentrations of a potential initiation triplet near the ribosomal P-site (Calogero et al., 1988).

As mentioned above, additional types of base pairing between mRNA and 16S rRNA have also been suggested and correlated with either regulatory mechanisms (Gold et al., 1984) or high levels of translational expression (Petersen et al., 1988; Gallie & Kado, 1989; Olins & Rangwala, 1989; Thanaraj & Pandit, 1989). The occurrence of any other type of base pairing aside from the SD interaction, however, remains to be demonstrated.

Are these thermodynamic-based mechanisms sufficient to explain the specificity of the mRNA selection process and to account for the different levels of translational efficiency encountered in nature? At least in some cases the affinity of the mRNAs for the ribosome either does not correlate (Calogero et al., 1988) or correlates only qualitatively (Lang et al., 1989) with the level of their expression, indicating that an additional mechanism must exist, possibly superimposed on the first. We suggest that this mechanism is based on the kinetic selection of the "best-fit" 30S initiation complexes from among the multitude of 30S ternary complexes that may form in the cell. Thus, the native 30S subunits may be expected, in principle, to bind any aminoacyl-tRNA (or peptidyl-tRNAs, if these are not completely scavenged by peptidyl-tRNA hydrolase) to any triplet (not necessarily cognate) happening to be in the P-site; the stability of these complexes would vary, however, depending on the nature and stereospecificity of the mRNA–ribosome and codon–anticodon interactions (Potapov, 1982). According to the mechanistic scheme presented above (Figure 5), each potential 30S initiation complex can either dissociate into its individual components or become fixed in a 70S initiation complex and enter the elongation cycle. Thus, to be translationally productive, each interaction must pass through the kinetic screen set by the IFs at step C of the initiation pathway. According to this model, both selection of the correct initiation start and translational efficiency depend on the ratio between the (presumably constant) on-rate of 70S initiation complex formation and the (variable) rate of dissociation of the individual 30S complexes.

In addition to offering a unitary explanation for both fidelity and efficiency of translation initiation, this kinetic model provides a simple rationale for why initiation obligatorily begins with the 30S subunit (Blumberg et al., 1979).

References

Blumberg, B. M., Nakamoto, T., & Kezdy, F. (1979) *Proc. Natl. Acad. Sci. U.S.A. 76*, 251–255.

Boileau, G., Butler, P., Hershey, J. W. B., & Traut, R. R. (1983) *Biochemistry 22*, 3162–3170.

Boylan, S. A., Suh, J. W., Thomas, S. M., & Price, C. (1989) *J. Bacteriol. 171*, 2553–2562.

Brimacombe, R. (1988) *Biochemistry 27*, 4207–4214.

Brombach, M., & Pon, C. L. (1987) *Mol. Gen. Genet. 208*, 94–100.

Brombach, M., Gualerzi, C. O., Nakamura, Y., & Pon, C. L. (1986) *Mol. Gen. Genet. 205*, 97–102.

Butler, J. S., Springer, M., & Grunberg-Manago, M. (1987) *Proc. Natl. Acad. Sci. U.S.A. 84*, 4022–4025.

Calogero, R. A., Pon, C. L., Canonaco, M. A., & Gualerzi, C. O. (1988) *Proc. Natl. Acad. Sci. U.S.A. 85*, 6427–6431.

Canonaco, M. A., Gualerzi, C. O., & Pon, C. L. (1989) *Eur. J. Biochem. 182*, 501–506.

Celano, B., Pawlik, R. T., & Gualerzi, C. O. (1988) *Eur. J. Biochem. 178*, 351–355.

de Smit, M. H., & van Duin, J. (1990) *Prog. Nucleic Acids Res. Mol. Biol. 38*, 1–35.

Devereux, J., Haeberli, P., & Smithies, O. (1984) *Nucleic Acids Res. 12*, 387–395.

Draper, D. (1987) in *Translational Regulation of Gene Expression* (Ilan, J., Ed.) pp 1–26, Plenum Press, New York.

Dreyfus, M. (1988) *J. Mol. Biol. 204*, 79–94.

Dreyfus, M., & Jacques, N. (1990) *Mol. Microbiol.* (in press).

Ehresmann, C., Moine, H., Mougel, M., Dondon, J., Grunberg-Manago, M., Ebel, J. P., & Ehresmann, B. (1986) *Nucleic Acids Res. 14*, 4803–4821.

Evstafieva, A. G., Shatsky, I. N., Bogdanov, A. A., Semenkov, Y. P., & Vasiliev, V. D. (1983) *EMBO J. 2*, 799–804.

Friedrich, K., Brombach, M., & Pon, C. L. (1988) *Mol. Gen. Genet. 214*, 595–600.

Gallie, D. R., & Kado, C. I. (1989) *Proc. Natl. Acad. Sci. U.S.A. 86*, 129–132.

Ganoza, M. C., Kofoid, E. C., Marlière, P., & Louis, B. G. (1987) *Nucleic Acids Res. 15*, 345–360.

Gold, L. (1988) *Annu. Rev. Biochem. 57*, 199–233.

Gold, L., Stormo, G., & Saunders, R. (1984) *Proc. Natl. Acad. Sci. U.S.A. 81*, 7061–7065.

Green, C. J., & Vold, B. S. (1983) *Nucleic Acids Res. 11*, 5763–5774.

Gren, E. J. (1984) *Biochimie 66*, 1–29.

Gualerzi, C., & Pon, C. L. (1981) in *Structural Aspects of Recognition and Assembly in Biological Macromolecules* (Balaban, M., Sussman, J. L., Traub, W., & Yonath, A., Eds.) pp 805–826, ISS Press, Rehovot, Israel, and Philadelphia, PA.

Gualerzi, C. O., Pon, C. L., Pawlik, R. T., Canonaco, M. A., Paci, M., & Wintermeyer, W. (1986) in *Structure, Function and Genetics of Ribosomes* (Hardesty, B., & Kramer, G., Eds.) pp 621–641, Springer-Verlag, New York.

Gualerzi, C. O., Calogero, R. A., Canonaco, M. A., Brombach, M., & Pon, C. L. (1988) in *Genetics of Translation* (Tuite, M. F., Picard, M., & Bolotin-Fukuhara, M., Eds.) pp 317–330, Springer-Verlag, Berlin and Heidelberg, West Germany.

Gualerzi, C., Spurio, R., La Teana, A., Calogero, R., Celano, B., & Pon, C. (1989) *Protein Eng. 3*, 133–138.

Hartz, D., McPheeters, D. S., & Gold, L. (1989) *Genes Dev. 3*, 1899–1912.

Herr, W., Chapman, N. M., & Noller, H. F. (1979) *J. Mol. Biol. 30*, 433–449.

Hershey, J. W. B. (1987) in *Escherichia coli and Salmonella typhimurium—Cellular and Molecular Biology* (Neidhardt, F. C., Ingraham, J. L., Low, K. B., Magasanik, B., Schaechter, M., & Umbarger, H. E., Eds.) pp 613–647, American Society for Microbiology, Washington, DC.

Hui, A., & DeBoer, H. A. (1987) *Proc. Natl. Acad. Sci. U.S.A. 84*, 4762–4766.

Jacob, W. F., Santer, M., & Dahlberg, A. E. (1987) *Proc. Natl. Acad. Sci. U.S.A. 84*, 4757–4761.

Kennell, D., & Riezman, H. (1977) *J. Mol. Biol. 114*, 1–21.

Khudyakov, Y. E., Neplyueva, V. S., Kalinina, T. I., & Smirnov, V. D. (1988) *FEBS Lett. 232*, 369–371.

Kung, H. F., Eskin, B., Redfield, B., & Weissbach, H. (1979) *Arch. Biochem. Biophys. 195*, 396–400.

Lang, V., Gualerzi, C., & McCarthy, J. E. G. (1989) *J. Mol. Biol. 210*, 659–663.

Laughrea, M., & Tam, J. (1990) *Biochem. Cell Biol. 67*, 812–817.

Louie, A., Ribeiro, N. S., Reid, B. R., & Jurnak, F. (1984) *J. Biol. Chem. 259*, 5010–5016.

McCarthy, J. E. G., & Gualerzi, C. O. (1990) *Trends Genet. 6*, 78–85.

McCarthy, J. E. G., Schairer, H. U., & Sebald, W. (1985) *EMBO J. 4*, 519–526.

McClure, W. R. (1985) *Annu. Rev. Biochem. 54*, 171–204.

Melançon, P., Leclerc, D., Destroismaisons, N., & Brakier-Gingras, L. (1990) *Biochemistry 29*, 3402–3407.

Muralikrishna, P., & Wickstrom, E. (1989) *Biochemistry 28*, 7505–7510.

Nagase, T., Ishii, S., & Imamoto, F. (1988) *Gene 67*, 49–57.

Ofengand, J., Ciesiolka, J., Denman, R., & Nurse, K. (1986) in *Structure, Function and Genetics of Ribosomes* (Hardesty, B., & Kramer, G., Eds.) pp 472–494, Springer-Verlag, New York.

Olins, P. O., & Rangwala, S. H. (1989) *J. Biol. Chem. 264*, 16973–16976.

Olson, H. M., Lasater, L. S., Cann, P. A., & Glitz, D. G. (1988) *J. Biol. Chem. 263*, 15196–15204.

Pelka, H., & Schulman, L. (1986) *Biochemistry 25*, 4450–4456.

Petersen, G. B., Stockwell, P. A., & Hill, F. D. (1988) *EMBO J. 7*, 3957–3962.

Plumbridge, J. A., Deville, F., Sacerdot, C., Petersen, H. U., Cenatiempo, Y., Cozzone, A., Grunberg-Manago, M., & Hershey, J. W. B. (1985) *EMBO J. 4*, 223–229.

Pon, C. L., & Gualerzi, C. O. (1988) in *Gene Expression and Regulation: The Legacy of Luigi Gorini* (Bissell, M., Dehò, G., Sironi, G., & Torriani, A., Eds.) pp 137–150, Elsevier Science Publishers, Amsterdam, The Netherlands.

Pon, C. L., Pawlik, R. T., & Gualerzi, C. (1982) *FEBS Lett. 137*, 163–167.

Pon, C. L., Paci, M., Pawlik, R. T., & Gualerzi, C. O. (1985) *J. Biol. Chem. 260*, 8918–8924.

Pon, C. L., Brombach, M., Thamm, S., & Gualerzi, C. (1989) *Mol. Gen. Genet. 218*, 355–357.

Potapov, A. P. (1982) *FEBS Lett. 146*, 5–8.

Pscheidt, R. H., & Wells, B. D. (1986) *J. Biol. Chem. 261*, 7253–7256.

Reddy, P., Peterkofsky, A., & McKenney, K. (1985) *Proc. Natl. Acad. Sci. U.S.A. 82*, 5656–5660.

Sacerdot, C., Fayat, G., Dessen, P., Springer, M., Plumbridge, J. A., Grunberg-Manago, M., & Blanquet, S. (1982) *EMBO J. 1*, 311–315.

Sacerdot, C., Dessen, P., Hershey, J. W. B., Plumbridge, J. A., & Grunberg-Manago, M. (1984) *Proc. Natl. Acad. Sci. U.S.A. 81*, 7787–7791.

Sands, J. F., Cummings, H. S., Sacerdot, C., Dondon, L., Grunberg-Manago, M., & Hershey, J. W. B. (1987) *Nucleic Acids Res. 15*, 5157–5168.

Scherer, G. F. E., Walkinshaw, M. D., Arnott, S., & Morre, D. J. (1980) *Nucleic Acids Res. 8*, 3895–3907.

Schneider, T. D., Stormo, G. D., Gold, L., & Ehrenfeucht, A. (1986) *J. Mol. Biol. 188*, 415–431.

Schulman, L. H. (1979) in *Transfer RNA: Structure, Properties and Recognition* (Schimmel, P. R., Söll, D., & Abelson, J. N., Eds.) pp 311–324, Cold Spring Harbor Laboratory, Cold Spring Harbor, NY.

Seong, B. L., & RajBhandary, U. L. (1987a) *Proc. Natl. Acad. Sci. U.S.A. 84*, 334–338.

Seong, B. L., & RajBhandary, U. L. (1987b) *Proc. Natl. Acad. Sci. U.S.A. 84*, 8859–8863.

Severini, M., Choli, T., La Teana, A., Pon, C., & Gualerzi, C. (1990) *J. Protein Chem.* (in press).

Shazand, K., Tucker, J., Chiang, R., Stansmore, K., Sperling-Petersen, H. U., Grunberg-Manago, M., Rabinowitz, J. C., & Leighton, T. (1990) *J. Bacteriol. 172*, 2675–2687.

Sørensen, M. A., Kurland, C. G., & Pedersen, S. (1989) *J. Mol. Biol. 207*, 365–377.

Spurio, R., La Teana, A., Severini, M., Gambacurta, A., Choli, T., Falconi, M., Gualerzi, C., & Pon, C. (1989) *Atti Assoc. Genet. Ital. 35*, 349–350.

Steitz, J. A. (1980) in *Ribosomes: Structure, Function, and Genetics* (Chambliss, G., Craven, G. R., Davies, J., Davis, K., Kahan, L., & Nomura, M., Eds.) pp 479–495, University Park Press, Baltimore, MD.

Stern, S., Weiser, B., & Noller, H. F. (1988) *J. Mol. Biol. 204*, 447–481.

Stormo, G. D. (1987) in *Translational Regulation of Gene Expression* (Ilan, J., Ed.) pp 27–49, Plenum Press, New York.

Tanada, S., Kawakami, M., Nishio, K., & Takemura, S. (1982) *J. Biochem. (Tokyo) 91*, 291–299.

Tapprich, W. E., & Hill, W. E. (1986) *Proc. Natl. Acad. Sci. U.S.A. 83*, 556–560.

Tapprich, W. E., Goss, D., & Dahlberg, A. E. (1989) *Proc. Natl. Acad. Sci. U.S.A. 86*, 4927–4931.

Thanaraj, T. A., & Pandit, M. W. (1989) *Nucleic Acids Res. 17*, 2973–2985.

Wakao, H., Romby, P., Westhof, E., Laalami, S., Grunberg-Manago, M., Ebel, J. P., Ehresmann, C., & Ehresmann, B. (1989) *J. Biol. Chem. 264*, 20363–20371.

Watanabe, K., Kuchino, Y., Yamaizumi, Z., Kato, M., Oshima, T., & Nishimura, S. (1979) *J. Biochem. (Tokyo) 86*, 893–905.

Weiel, J., & Hershey, J. W. B. (1981) *Biochemistry 20*, 5859–5865.

Weiel, J., & Hershey, J. W. B. (1982) *J. Biol. Chem. 257*, 1215–1220.

Zucker, F. H., & Hershey, J. W. B. (1986) *Biochemistry 25*, 3682–3690.

PROTEOLYTIC PROCESSING AND PROTEIN DEGRADATION

Chapter 23

Proteolytic Processing of Polyproteins in the Replication of RNA Viruses†

Christopher U. T. Hellen,* Hans-Georg Kräusslich,‡ and Eckard Wimmer

Department of Microbiology, School of Medicine, State University of New York at Stony Brook, Stony Brook, New York 11794

Received June 28, 1989; Revised Manuscript Received September 12, 1989

Many animal and plant viruses depend on the action of virus-encoded proteinases at various stages in their replication. These enzymes are highly substrate-selective and cleavage-specific, in that they cleave large, virus-specified polypeptides called polyproteins at defined amino acid pairs. Logically, they are potential targets for inhibition of virus-specific proteolytic processing and consequently of viral replication. Inhibitors of viral proteinases may therefore emerge as major antiviral chemotherapeutic agents. Until recently, characterization of viral proteinases has been limited largely to genetic mapping and molecular genetic manipulation, but over the last 2 years several major advances have been made in the understanding of their structure and substrate specificity (Kräusslich et al., 1989b). Several viral proteinases have been purified to homogeneity, and the availability of polyprotein and peptide substrates has facilitated the study of their enzymatic and physical properties. New models relating the structure of viral proteinases to those of their cellular counterparts have been developed, and the three-dimensional structures of two retroviral proteinases have been determined.

A viral polyprotein is a precursor polypeptide that contains several distinct domains and is proteolytically processed to yield diverse structural and nonstructural proteins. Synthesis of such polyproteins is limited to positive-strand RNA viruses and retroviruses (whose genomic RNAs are of the same polarity as the viral mRNAs). Eukaryotic ribosomes normally initiate translation at a single AUG codon; consequently, mRNA species are translated to yield only a single polypeptide. Synthesis and subsequent proteolytic processing of a polyprotein can therefore be regarded as one of a variety of mechanisms by which these viruses express downstream cistrons. In addition, frame-shift or read-through by suppression of leaky termination codons can modify the length of a polyprotein. Other mechanisms include translation from subgenomic RNAs and segmentation of the genome. These strategies allow both for genetic economy and for differential expression of viral proteins. Proteolytic processing enables functionally different domains to be separated and cleavage products to be transported to different cellular compartments. It may also be important in regulating events in viral replication, such as uncoating, activation of replicative enzymes, and morphogenesis (Table II).

We have reviewed the subject recently (Kräusslich & Wimmer, 1988) and in this paper shall mainly discuss the current state of knowledge regarding polyprotein processing in the two best defined viral systems of eukaryotic cells: viz., picornaviruses and their plant virus relatives and retroviruses. Other viral systems will be included as appropriate and will be related to these two main systems.

Proteolysis in Picornavirus Protein Expression

Picornaviridae are a family of small icosahedral viruses that cause a number of important disease syndromes. The family is currently divided into four genera: rhinovirus (the common cold virus), enterovirus (e.g., poliovirus), cardiovirus (e.g., encephalomyocarditis virus), and aphthovirus (foot-and-mouth disease virus). All have a positive-sense single-stranded monopartite genome which contains a single long open reading frame. This is translated to yield a large polyprotein which is subsequently cleaved to yield all structural and nonstructural proteins. Poliovirus was the first eukaryotic RNA virus for which the complete nucleotide seqence was determined, and its polyprotein was the first for which a detailed structure was obtained (Kitamura et al., 1981) [Figure 1; nomenclature according to Rueckert and Wimmer (1984)]. Processing of the polyprotein has subsequently been characterized in detail, and the mechanisms that have been elucidated are typical of many other viruses.

†Work in the authors' laboratories was supported in part by U.S. Public Health Service Grants AI 15122 and CA 28146 to E.W. from the National Institutes of Health and by a grant from Boehringer Ingelheim Pharmaceuticals, Inc.

*Corresponding author.

‡Present address: Deutsches Krebsforschungszentrum, Institut für Virusforschung, Im Neuenheimer Feld 280, D-69 Heidelberg 1, West Germany.

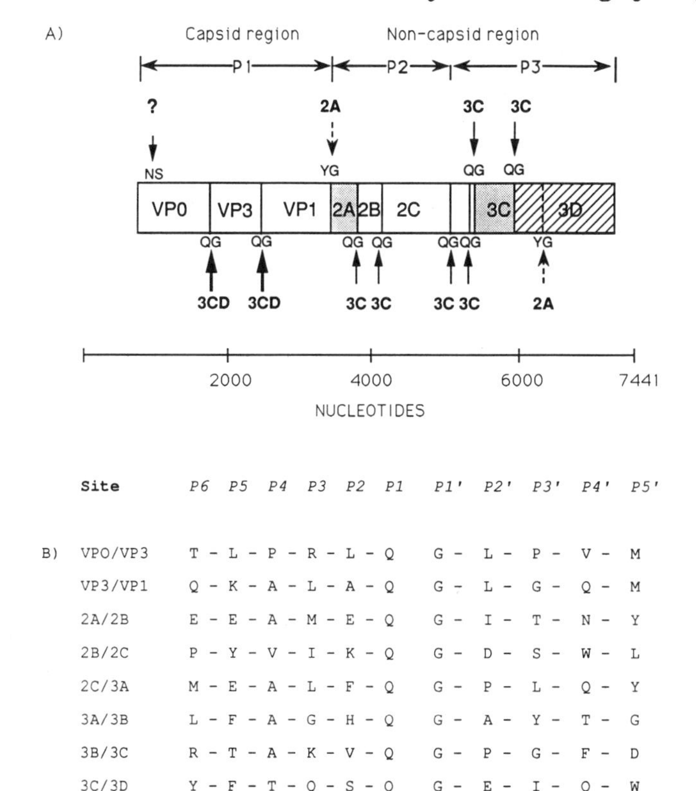

	Site	P6	P5	P4	P3	P2	P1	P1'	P2'	P3'	P4'	P5'
B)	VP0/VP3	T	L	P	R	L	Q	G	L	P	V	M
	VP3/VP1	Q	K	A	L	A	Q	G	L	G	Q	M
	2A/2B	E	E	A	M	E	Q	G	I	T	N	Y
	2B/2C	P	Y	V	I	K	Q	G	D	S	W	L
	2C/3A	M	E	A	L	F	Q	G	P	L	Q	Y
	3A/3B	L	F	A	G	H	Q	G	A	Y	T	G
	3B/3C	R	T	A	K	V	Q	G	P	G	F	D
	3C/3D	Y	F	T	Q	S	Q	G	E	I	Q	W
C)	VP1/2A	K	D	L	T	T	Y	G	F	G	H	Q
	3C'/3D'	K	L	L	D	T	Y	G	I	N	L	P

FIGURE 1: Gene organization, processing scheme, and cleavage sites of the poliovirus polyprotein. (A) Proteolytic cleavages of the polyproteins occur between amino acid pairs indicated by standard single-letter code. Arrows above and below the polyprotein indicate sites that are cleaved in inter- and intramolecular reactions, respectively, by proteinases as indicated. The question mark indicates that the mechanism of cleavage at this site is not known. The positions of virus-encoded proteinases within the polyprotein are indicated by shaded boxes. The nomenclature of poliovirus proteins is according to Rueckert and Wimmer (1984). The amino acid residues at sites cleaved (B) by $3C^{pro}$ and $3CD^{pro}$ and (C) by $2A^{pro}$ are indicated by standard single-letter code and are described according to the nomenclature of Berger and Schechter (1970). The newly generated carboxy terminus, after cleavage of the peptide bond, is designated P1, preceded by the P2 residue, etc. The newly generated amino terminus is designated P1′, followed by the P2′ residue, etc.

Picornaviral capsids are composed of essentially equimolar amounts of four nonidentical polypeptides. They are synthesized as a precursor P1 whose N-terminus is myristoylated (Chow et al., 1987). In poliovirus (Toyoda et al., 1986), P1 is separated from the nascent polypeptide in a reaction catalyzed by the adjacent polypeptide $2A^{pro}$, a proteinase that hydrolyzes a Y–G bond at its own amino terminus and cleaves a second Y–G dipeptide, within the RNA polymerase 3D. The initial cleavage is probably an intramolecular event and must occur before the capsid precursor can be processed to yield capsomer proteins (Nicklin et al., 1987). The second cleavage is not essential for viral proliferation (Lee & Wimmer, 1988), so that the main function of $2A^{pro}$ is to sever the structural proteins of the viral nucleocapsid from the nonstructural proteins, allowing respectively assembly of particles and generation of a functional replication complex. A likely second function of $2A^{pro}$ is the rapid "shut off" of host cell protein synthesis which occurs on infection with poliovirus. This is associated with proteolytic cleavage of a component of the cap-binding complex, which is induced but not directly catalyzed by $2A^{pro}$ in vitro (Kräusslich et al., 1987) and in vivo (Sun & Baltimore, 1989). Amino acid sequence determination of the termini of poliovirus proteins established that all but one of the remaining cleavages of the polyprotein occurred at QG dipeptides (Nicklin et al., 1986, and references cited therein; Figure 1). Cleavage at QG sites in the P2 region of the polyprotein can probably be catalyzed by $3C^{pro}$ alone, but processing of the P1 capsid precursor requires additional sequences from the P3 region (Ypma-Wong & Semler, 1987). Protein 3CD (which contains the sequence of the polymerase 3D in addition to the 3C sequence) was able to cleave the P1 precursor efficiently at both sites in vitro (Jore et al., 1988; Ypma-Wong et al., 1988a; Figure 1). Similar results were obtained with proteins isolated from infected HeLa cells (Nicklin et al., 1988). Larger precursors (3BCD or P3) can also cleave P1 (Kuhn et al., 1988). It is therefore likely that the 3D sequence of $3CD^{pro}$ interacts with P1 in such a way that the QG sites in P1 can be recognized and cleaved. Cleavage by $3CD^{pro}$ of mutated P1 lacking a myristoylation signal is defective, which indicates that the interaction between P1 and $3CD^{pro}$ probably involves the myristic acid moiety (Kräusslich et al., 1989c). There are likely to be additional specific contacts between $3CD^{pro}$ and the structural domains of the capsid precursor, since even minor perturbations to the integrity of the P1 precursor are sufficient to impair cleavage (Ypma-Wong et al., 1988b, and references cited therein). These contacts may stabilize the precursor–proteinase complex, thus enabling $3CD^{pro}$ to cleave sites in the P1 precursor much more efficiently in vitro than $3C^{pro}$. $3C^{pro}$ is part of the nonstructural precursor and can catalyze all cleavages within it. However, the interaction between the P1 structural precursor and an appropriate proteinase may require a higher K_m, since they may have become spatially separated following the initial cleavage of structural from nonstructural precursors.

Processing of the poliovirus polyprotein serves three different functions, which can be correlated with three different proteolytic activities. First, structural and nonstructural precursors are separated from the nascent polyprotein by $2A^{pro}$, and second, release of different functional proteins is catalyzed by $3C^{pro}$ and $3CD^{pro}$. The last processing step in the formation of infectious particles is cleavage of VP0 to VP4 and VP2 at an N-S dipeptide (Figure 1), which occurs on encapsidation of viral RNA. It is not catalyzed by either $2A^{pro}$ or $3C^{pro}$. The three-dimensional structures of polio-, rhino-, and mengovirus each show the close proximity of a serine residue in VP2 to the carboxy end of VP4. It may act as a nucleophile, as in serine proteases; viral RNA could then act as a proton-abstracting base since there is no suitable His residue (Arnold et al., 1987). The ordered RNA in bean pod mottle virus (a plant virus relative of picornaviruses) is in a position and orientation that would be ideal to participate in this catalysis (Chen et al., 1989). However, the role of Ser10 in VP2 of poliovirus in the maturation cleavage has recently become questionable, since mutation of this serine to alanine or to cysteine yielded viable virus in which VP0 was cleaved (J. Harber and E. Wimmer, unpublished results).

All picornaviruses encode active and closely related 3C proteinases, but there is little similarity between the sequences of enzymatically active 2A proteinases of rhino- and enteroviruses and their counterparts in cardioviruses. In aphthoviruses, 2A consists of only 16 amino acid residues. A rapid primary cleavage reaction separates the structural precursor from the growing polyprotein of the latter two groups of viruses, but the mechanism by which it occurs has not been elucidated. Structural precursors can be further processed by $3C^{pro}$ in vitro to yield capsid proteins, but larger precursors of 3C^pro are probably responsible for cleavage in vivo. For example, 3ABC is probably responsible for processing of the encephalomyocarditis virus capsid protein precursor (Jackson, 1986). This cleavage normally occurs in a defined stepwise manner, resulting in sequential release of the various capsid proteins (Shih & Shih, 1981). The mechanism by which this cascade is ordered is not known [although substitution of residues at one site has been shown to result in premature cleavage at another (Parks & Palmenberg, 1987)], but it is thought to be important in the formation of virion capsid structures.

Determinants of Picornavirus Substrate Recognition

Cleavage site recognition by $3C^{pro}$ of poliovirus is unusually stringent, for no corresponding picornaviral proteinase cleaves exclusively at one dipeptide (Figure 1). The residues occurring at cleavage sites within the polyproteins of other picornaviruses have been determined or predicted with reasonable certainty. It is clear that cleavage occurs within a small subset of dipeptides comprising Q-G, -S, -T, -V, -A, and -M and E-G and -S. Four of thirteen QG dipeptides and eight of ten YG dipeptides in the poliovirus polyprotein are not cleaved by $3C^{pro}$ and by $2A^{pro}$, respectively, so there must be additional determinants of cleavage site recognition. Moreover, there are clear differences in the efficiency of cleavage at different sites, resulting, for example, in the relative stability of 3AB and 3CD of poliovirus, and in the ordered stepwise cleavage of the encephalomyocarditis virus capsid precursor.

The specificity of $2A^{pro}$ and $3C^{pro}$ has been addressed by mutation of cleavage sites and by cleavage of peptide substrates. Lee and Wimmer (1988) noted that both bona fide $2A^{pro}$ cleavage sites are preceded by a Thr residue and have a Leu residue at position P4; this residue and other hydrophobic residues (I, V, and M) occupy the P4 position of VP1–2A cleavage sites in all other entero- and rhinoviruses. Substitution of a Thr by an Ala residue at position P2 of the 3C′/3D′ site abolished cleavage in vivo. Novel QA and naturally occurring QS and QG sites appear to be processed by encephalomyocarditis virus $3C^{pro}$ at normal rates in vitro [e.g., Parks et al. (1989)], but ten other mutated P1–P1′ sites were not cleaved, confirming the importance of residues at these positions. Interactions at positions other than P1 and P1′ are required since a QG dipeptide was not cleaved by purified poliovirus $3C^{pro}$ (Pallai et al., 1989). The P4 position may be important in cleavage site recognition: Nicklin et al. (1986) noted the prevalence of alanine or other aliphatic residues at this position, and Kuhn et al. (1988) found that some substitutions at this position impaired cleavage at the 3B/3C site. It is apparent that residues surrounding the QG dipeptide strongly influence cleavage efficiency, and this therefore appears to be a novel mechanism of regulation of gene expression. For example, the rate of cleavage of a peptide that corresponded to the 3C/3D site was increased by ca. 2 orders of magnitude by substitution of Thr by Ala at the P4 position (Pallai et al., 1989). Flanking Pro residues occur at most sites cleaved by $3C^{pro}$ within the encephalomyocarditis virus polyprotein, and it is interesting to note that polio peptides that are cleaved efficiently have Pro at the P2′ position, which may favor formation of a β-turn.

Determinants of Catalytic Activity

Inhibitor studies [references in Kräusslich and Wimmer (1988) and Sommergruber et al. (1989)] indicate that picornaviral 3C and 2A proteinases contain an active-site thiol group, suggesting that these enzymes constitute a class of cysteine proteinases. They are not related to the papain superfamily, since they are not inhibited by the characteristic inhibitor E-64 [e.g., Nicklin et al. (1988)] and show no sequence similarity with this family. Analysis of aligned sequences suggested that the viral proteinases are structurally related to trypsin-like serine proteinases and, more specifically, that His40, Asp85, and Cys147 of $3C^{pro}$ and His20, Asp38, and Cys109 of $2A^{pro}$ form the catalytic triad (Bazan & Fletterick, 1988). An evolutionary relationship has been proposed between a broad class of viral 3C-like Cys proteinases and the family of cellular trypsin-like Ser proteinases (Gorbalenya et al., 1989a,b; Bazan & Fletterick, 1989a,b); particularly notable is the replacement of the nucleophilic Ser by a spatially and functionally equivalent Cys residue in the catalytic triad of the viral enzyme. When a similar Ser-Cys substitution was introduced into the active sites of trypsin (Higaki et al., 1987) and subtilisin (Polgar & Bender, 1969; Neet et al., 1969), large reductions in activity were observed. However, additional changes may have a compensatory effect in viral proteinases, since substitution of the active-site Cys by Ser does not increase their activity as might have been predicted. Substitution of Cys147 by Ser was reported to inactivate $3C^{pro}$ of poliovirus in a bacterial expression system (Ivanoff et al., 1986); a similar Cys106-Ser substitution in polio $2A^{pro}$ reduced its activity by over 90% (M. Fäcke, H.-G. Kräusslich, and E. Wimmer, unpublished data). Viral proteinases function in controlled and limited proteolysis rather than in digestion, so that a reduction in turnover number coupled with a highly restricted substrate specificity may be a desirable property. Bazan and Fletterick (1988) suggested that the 3C and 2A subclasses of viral enzymes are homologous to the large (e.g., trypsin) and small (e.g., α-lytic protease) subclasses of trypsin-like proteases. Their model (derived by assignment of viral sequences which form the 12 components β-strands of the trypsin fold) enables prediction of structural features to be made. These are currently being investigated by X-ray crystallography of purified $3C^{pro}$ of poliovirus. Similarities between small bacterial and 2A-like proteinases may also involve the substrate-binding pocket. The primary specificity of α-lytic protease was relaxed and became much more permissive to large P1 residues as a result of substitutions that increased the size of the substrate binding pocket (Bone et al., 1989). Small residues occur naturally at equivalent positions in viral 2A proteinases, which may explain their ability to accommodate Tyr residues and to tolerate a number of substitutions at the P1 position (C. U. T. Hellen, C.-K. Lee, and E. Wimmer, unpublished results).

Proteolytic Processing of Plant Virus Polyproteins

There is a surprising degree of similarity between 3C and other nonstructural proteins of picornaviruses and proteins encoded by cowpea mosaic virus (Argos et al., 1974) and tobacco etch virus (Domier et al., 1987), which are members respectively of the comovirus and potyvirus groups of plant viruses. Moreover, the same order of nonstructural domains (putative helicase, VPg, proteinase, polymerase) is found in genomic RNAs of all three families of viruses. Although sequence similarity between como- and picornaviruses in the capsid regions is not apparent, capsid proteins of four picornaviruses and two comoviruses (cowpea mosaic virus and bean pod mottle virus) do have a common eight-stranded antiparallel β-barrel motif (Chen et al., 1989).

Comoviruses have a genome consisting of two positive-sense RNA molecules, both of which contain single large open reading frames; one encodes the capsid proteins while the other

Table I: Proteolytic Cleavage Sites in the Polyproteins Encoded (A) by Cowpea Mosaic Virus B-RNA, (B) by Cowpea Mosaic Virus M-RNA, (C) by Red Clover Mottle Virus M-RNA, and (D) by Bean Pod Mottle Virus M-RNA[a]

		sequence									
	site	P5	P4	P3	P2	P1	P1′	P2′	P3′	P4′	P5′
A	32K:58K	K	D	N	A	Q	S	S	P	V	I
	58K:VPg	S/G	A	E	P	Q	S	R	K	P	N
	VPg:24K	W	A	D	A	Q	M	S	L	D	Q
	24K:87K	I	A	Q	A	Q	G	A	E	E	Y
B	48/58:60	V	A	F	P	Q	M	E	Q	N	L
	VP37:VP23	G	A	I	A	Q	G	P	V	C	A
C	48/58:60	F	A	N	P	Q	T	D	T	D	L
	VP37:VP23	Q	A	E	A	Q	G	G	V	V	R
D	48/58:60	E	V	Q	A	Q	M	E	T	N	L
	VP37:VP23	G	T	I	P	Q	S	I	S	Q	Q

[a]Amino acid sequence data (A, B) from Wellink et al. (1986) and (D) from Chen et al. (1989) and G. Lomonossoff (personal communication); deduced amino acid sequence (C) from Shanks et al. (1986). The amino acid residues at cleavage sites are indicated by standard single-letter code and are described according to the nomenclature of Berger and Schechter (1970).

encodes nonstructural proteins. In contrast to picornaviruses, comoviruses synthesize their structual and nonstructural proteins as part of different polyproteins and therefore do not need a 2A^{pro}-like activity. Potyviruses resemble picornaviruses in having monopartite positive-sense RNA genomes, which contain single large open reading frames (Figure 2). These are translated to yield polyproteins that are proteolytically processed to yield eight mature proteins (Dougherty & Carrington, 1988). The 24- and 49-kDa polypeptides of como- and potyviruses, respectively, are proteinases that resemble 3C^{pro} of picornaviruses. The cowpea mosaic virus proteinase can cleave all sites within both polyproteins in vitro. Remarkably, it also requires a cofactor for processing of the capsid precursor, but this is not the viral polymerase, as for poliovirus (Vos et al., 1988). The 49-kDa proteinase encoded by tobacco etch virus mediates its own autocatalytic release (probably in cis) and cleaves at three other positions in the polyprotein (Carrington & Dougherty, 1988; Carrington et al., 1988, and references cited therein). Potyviruses encode a second proteinase that is not related to the trypsin-like cysteine proteinases. It cleaves at a dipeptide (G-G) which is surrounded by residues that differ considerably from those at sites cleaved by the 49-kDa proteinase (Carrington et al., 1989; Figure 2). Cleavage by the 49-kDa proteinase of ten different potyviruses occurs at the dipeptide Q/(A, G, or S); all but one of these sites have H or F at the P2 position, and all but one have aliphatic residues at the P4 position. These potyvirus cleavage sites therefore resemble those of picornaviruses at P4, P1, and P1′ positions, but differ in that they extend over seven residues (Dougherty et al., 1989). The five sites recognized by the 49-kDa proteinase of tobacco etch virus are defined by the seven amino acid sequence: ExxYxQ/(G or S) (Figure 2). The role of these residues in defining a cleavage site has been investigated by substitution of residues at positions P7–P2′ (Dougherty et al., 1989, and references cited therein). Substitutions at absolutely conserved residues (P6, P3, P1) significantly reduced or even eliminated cleavage, while substitutions at other positions had little effect. Differences in nonconserved P4 and P2 positions of natural cleavage sites modulate their rate of cleavage and may therefore regulate the kinetics of formation of different gene products (Dougherty & Parks, 1989). Such a mechanism, which has been alluded to as a likely feature of processing of picornaviral polyproteins, would overcome the limitation that expression of a single polyprotein imposes on temporal regulation of expression. Cleavage sites within comovirus polyproteins resemble the 3C cleavage sites of picornaviruses at P4, P1, and P1′ positions (Table I). The trypsin/3C^{pro} sequence/structure profiles developed by Bazan and Fletterick

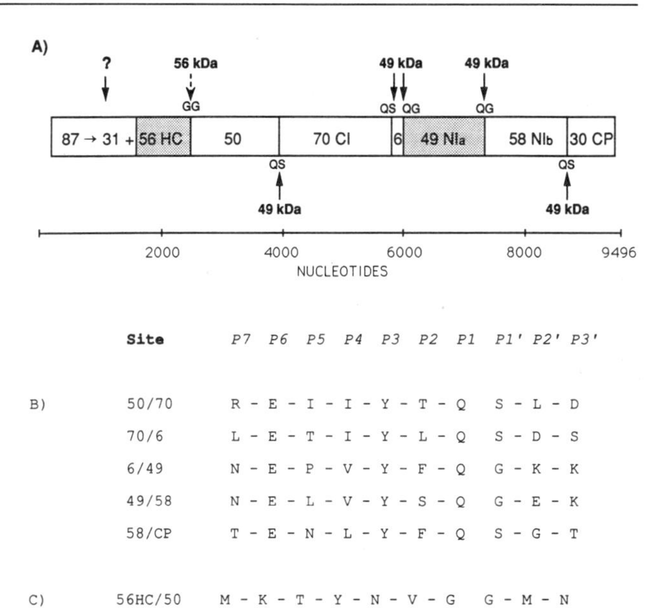

FIGURE 2: Gene organization, processing scheme, and cleavage sites of the tobacco etch virus polyprotein. (A) Proteolytic cleavages of the polyproteins occur between amino acid pairs indicated by standard single-letter code. Arrows above and below the polyprotein indicate sites that are cleaved in inter- and intramolecular reactions, respectively, by proteinases as indicated. The question mark indicates that the site and mechanism of cleavage are not known. The nomenclature of potyvirus proteins is as described by Domier et al. (1987), and the molecular weights are as described by Dougherty and Carrington (1988). The amino acid residues at sites cleaved (B) by the 49-kDa and (C) by the 56-kDa proteinases are described according to the nomenclature of Berger and Schechter (1970).

(1988) and by Gorbalenya and colleagues have been used to identify proteinase domains within polyproteins encoded by further families of plant (nepo-, sobemo-, and luteovirus) and animal (flavi-, pesti- and coronavirus) viruses (Table II; Bazan & Fletterick, 1989a,b; Gorbalenya et al., 1989a,b). All have genomes which encode long polyproteins and which have been shown to be processed to yield mature proteins. The predicted spatial disposition of active-site residues in these putative proteinases is similar to that of 3C^{pro}, but surprisingly, proteinase domains in four of these virus groups have Ser-active centers.

RETROVIRIDAE

Retroviridae are a family of enveloped RNA viruses that have common morphological, biochemical, and physical properties (Teich, 1984) and differ from plus-strand RNA viruses in replicating via an obligatory DNA intermediate. The genomes of all replication-competent retroviruses consist of three major genetic elements that are arranged in the order

Table II: Viral Proteinases[a]

family	virus	enzyme	enzyme type	catalytic class	role in viral replication
picornavirus	polio	2A	trypsin-like (small subclass)	Cys-proteinase	separation of functionally different domains; mediates cleavage of cellular proteins
	polio	3C	trypsin-like	Cys-proteinase	release of nonstructural components from polyprotein
		3CD	trypsin-like	Cys-proteinase	release of capsid proteins from precursor
	FMDV	3C	trypsin/3C-like	Cys-proteinase	release of individual components from polyprotein; cleavage of cellular proteins
	FMDV	L	?	?	autocatalytic release from precursor; mediates cleavage of cellular proteins
	FMDV	?	?	?	separation of domains at 2A/2B junction
	polio	?	?	base-catalyzed?	maturation cleavage of VP0 to VP4 and VP2
comovirus	CPMV	24 kDa	trypsin/3C-like	Cys-proteinase	release of individual components from polyprotein
potyvirus	TEV	49 kDa	trypsin/3C-like	Cys-proteinase	release of individual proteins from polyprotein
	TEV	87 kDa	?	?	autocatalytic release from polyprotein
nepovirus	TBRV	23 kDa	trypsin/3C-like	Cys-proteinase	release of individual components from polyprotein
sobemovirus	SBMV	105 kDa	trypsin/3C-like	Ser-proteinase	release of individual components from 105-kDa P1 polyprotein
luteovirus	PLRV	70 kDa	trypsin/3C-like	Ser-proteinase	release of individual components from 115-kDa polyprotein?
flavivirus	YFV	ns3	trypsin/3C-like	Ser-proteinase	separation of individual components from polyprotein
pestivirus	HCV	p80	trypsin/3C-like	Ser-proteinase	separation of individual components from polyprotein
coronavirus	IBV	?	trypsin/3C-like	Cys-proteinase	release of individual components from polyprotein?
			trypsin-like	Cys-proteinase?	release of individual components from polyprotein?
togavirus	SV	capsid	?	Ser-proteinase	separation of functionally different domains
		ns2	?	?	separation of individual components from polyprotein destined for different subcellular compartments
retrovirus	HIV-1	PR	pepsin-like	Asp-proteinase	separation of individual components from polyprotein; early events in infection?
nodavirus	FHV	α-protein	?	?	maturation cleavage of capsid protein
adenovirus		19 kDa	?	Ser-proteinase	maturation cleavage of precursor proteins

[a] Abbreviations: polio, poliovirus; FMDV, foot-and-mouth disease virus; CPMV, cowpea mosaic virus; TEV, tobacco etch virus; TBRV, tobacco black ring virus; SBMV, southern bean mosaic virus; PLRV, potato leaf roll virus; YFV, yellow fever virus; HCV, hog cholera virus; IBV, infectious bronchitis virus; SV, Sindbis virus; HIV-1 human immunodeficiency virus, type 1; FHV, flockhouse virus. Proteinases encoded by specific viruses are not necessarily encoded by all members of a virus family. Questions marks indicate that the nature or identity of a specific viral proteinase has not been identified. Proteinases encoded by SBMV, PLRV, and IBV have been identified by sequence and structural-pattern analysis, but their sizes have not been determined. Processing of viral precursor proteins has been demonstrated in other RNA viruses, including birna-, reo-, calici-, and tymoviruses [see Kräusslich and Wimmer (1988) for references]. It has been suggested that DNA viruses, such as vaccinia virus and adeno- (see above) and hepadaviruses, encode proteinases that cleave virus-encoded proteins; recent experimental evidence does not support the assignment of proteinase activity to a sequence found within hepadaviruses, which was made on the basis of weak similarity between it and aspartic proteinases.

5′-gag-pol-env-3′. The *gag* (**g**roup-specific **a**nti**g**en) region encodes up to six structural proteins, which form the retroviral nucleocapsid and which are translated as a precursor from 35S mRNA. The *pol* region encodes the viral replication enzymes [reverse transcriptase (RT), integrase (IN), and in most instances, proteinase (PR), in the order PR-RT-IN] and is also translated from 35S mRNA as part of a *gag-pol* fusion polyprotein [Figure 3; nomenclature according to Leis et al. (1988)]. Synthesis of the *gag-pol* polyprotein is achieved either by suppression of an *amber* termination codon at the end of the *gag* gene (e.g., murine leukemia virus) or by ribosomal frame-shifting at one or two sites at the end of or within the *gag* gene (e.g., Rous sarcoma virus and HIV-1) [references in Kräusslich and Wimmer (1988); Figure 3). The infrequency of frame-shifting and suppression of termination codons leads to an overproduction of structural proteins compared to replicative enzymes.

The gene order of the *gag-pol* polyprotein is reminiscent of the poliovirus polyprotein; a proteinase maps to a location between structural and nonstructural proteins and severs these two regions. Proteolytic processing by the retrovirus-encoded proteinase is essential for a productive infectious cycle. Defective avian retroviruses which lack PR do not process the *gag* polyprotein (Hayman et al., 1979); cleavage was similarly abolished on mutation of the coding sequence of proteinases of avian, murine, and human retroviruses. PR-deficient mutants of murine leukemia virus (Crawford & Goff, 1985) and HIV-1 (Kohl et al., 1988) produced noninfectious virions containing unprocessed polyprotein, which indicates that processing of the polyprotein is not required for particle formation but is necessary to render these particles infectious. These observations are crucial if inhibitors of retroviral proteinases are to serve as antiviral drugs.

Structural and Biochemical Characterization of Retroviral Proteinases

The amino acid sequences of a number of purified retroviral PRs have been either wholly or partially determined by direct analysis. Comparison of these sequences with the deduced amino acid sequences of other retroviruses and related genetic elements revealed a conserved Asp-Thr(Ser)-Gly sequence, which is homologous with a sequence in the active site of the aspartic proteinase family (Toh et al., 1985). Aspartic proteinases have two homologous domains and may have evolved by gene duplication and fusion, whereas retroviral PRs are less than half as long and could thus correspond to only one such domain. Refined consensus templates for cellular and putative viral aspartic proteinases (established on the basis of patterns of residue conservation and known or predicted secondary structure) implied strong similarities between these domains. It was therefore proposed that retrovirus proteinases are dimers of identical PR polypeptide chains, with a fold similar to other aspartic proteinases; viral domains were predicted to differ principally in having shortened connecting loops between the strands of β-sheet that formed the core (Pearl & Taylor, 1987b). Similar parsimony was also noted when viral and cellular trypsin-like enzymes were compared (Bazan & Fletterick, 1988). Many of these proposals were confirmed following determination of the crystal structures of Rous sarcoma and HIV-1 proteinases.

The Rous sarcoma virus PR monomer is 124 amino acids long; it has an approximate intramolecular 2-fold symmetry and consists of two helices and of several β-strands connected by loops and turns (Figure 4). The strands in the core of the

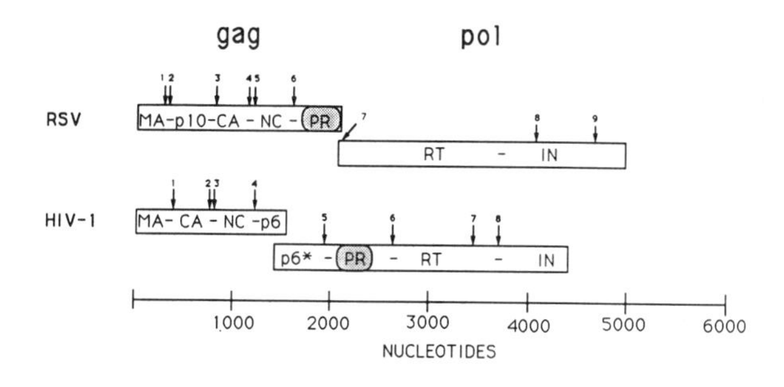

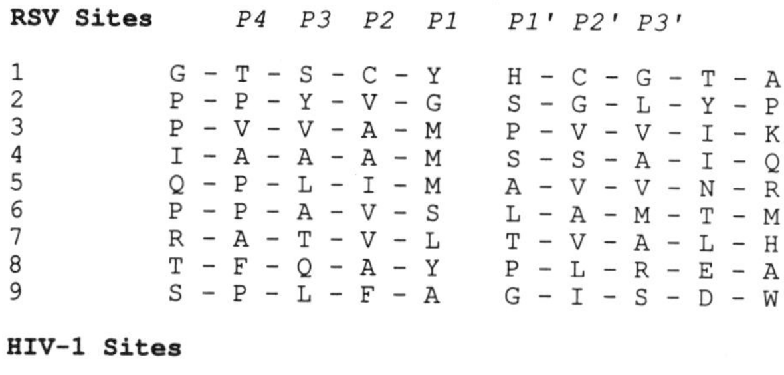

RSV Sites

	P4	P3	P2	P1	P1'	P2'	P3'		
1	G	T	S	C	Y	H	C	G	T	A
2	P	P	Y	V	G	S	G	L	Y	P
3	P	V	V	A	M	P	V	V	I	K
4	I	A	A	A	M	S	S	A	I	Q
5	Q	P	L	I	M	A	V	V	N	R
6	P	P	A	V	S	L	A	M	T	M
7	R	A	T	V	L	T	V	A	L	H
8	T	F	Q	A	Y	P	L	R	E	A
9	S	P	L	F	A	G	I	S	D	W

HIV-1 Sites

1	V	S	Q	N	Y	P	I	V	Q	N
2	K	A	R	V	L	A	E	A	M	S
3	T	A	T	I	M	M	Q	R	G	N
4	R	P	G	N	F	L	Q	S	R	P
5	V	S	F	N	F	P	Q	I	T	L
6	C	T	L	N	F	P	I	S	P	I
7	G	A	Q	T	F	Y	V	N	L	R
8	I	R	K	I	L	F	L	D	G	I

FIGURE 3: Coding regions from the genomes of Rous sarcoma virus (RSV) and human immunodeficiency virus, type 1 (HIV-1). Only the *gag* and *pol* regions have been shown, the latter being oriented relative to the former region, which has arbitrarily been placed in the first reading frame. Solid lines indicate start and end positions of open reading frames and stop codons within a functional reading frame. The nomenclature of retrovirus proteins is according to Leis et al. (1988). The amino-terminal product of the *pol* reading frame of the HIV-1 genome is designated p6*, and the positions of the proteinases within the polyproteins are indicated by stippled boxes. Cleavage sites within the RSV and HIV-1 polyproteins are numbered and indicated by arrows; amino acid residues at these positions are indicated by standard single-letter code and are described accoridng to the nomenclature of Berger and Schechter (1970).

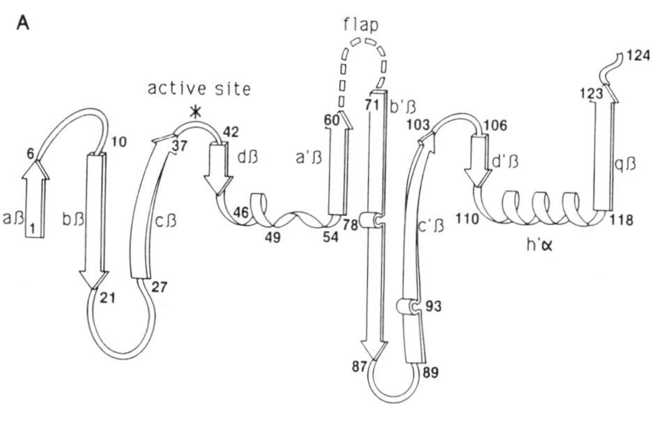

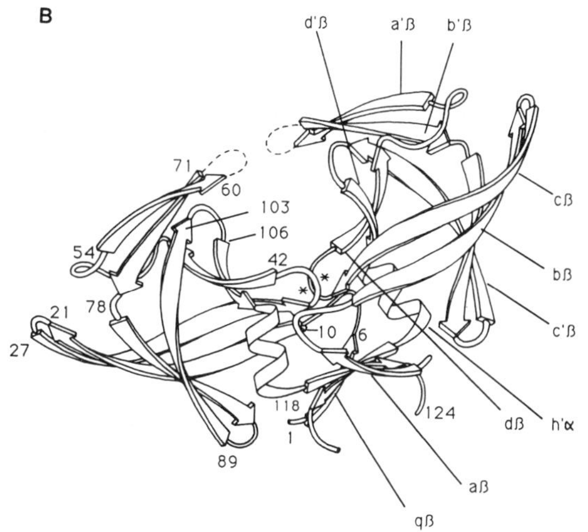

FIGURE 4: Schematic representation of (A) the secondary and (B) the tertiary structures of Rous sarcoma virus proteinase. Amino acid residues are numbered from the amino terminus; α-helices are depicted as helical ribbons and β-strands as arrows (labeled a–d, a′–d′, and q). The active site of the proteinase (Asp37–Ser38–Gly39) is located between cβ and dβ strands and is indicated by an asterisk. The disordered flap region (residues 61–70 in each monomer) is represented by a dashed line. Aspartic proteinases have a helix between d and a′ β-strands, which in Rous sarcoma virus proteinase is reduced to a single turn followed by a distorted loop (residues 50–55). Part B was drawn by J. Richardson, Duke University, and is used with permission.

monomer are organized into a sandwich of two four-stranded β-sheets; one is composed of antiparallel β-chains whereas the other comprises two superimposed ψ structures. The terminal strands of two monomers form a common four-stranded β-sheet (the dimer interface), which therefore differs from the six-stranded interface regions of pepsin-like proteinases. The active-site triplet of each subunit (residues 37–39) is located between cβ and dβ chains. Mutation of the Asp37 putative active-site residue confirmed its importance for catalytic activity (Kotler et al., 1988). The apposing residues of the active site are hydrogen-bonded in a "fireman's grip" configuration, which is characteristic of aspartic proteinases. A disordered "flap" region from each monomer (residues 61–70) projects over the active site, forming a large substrate-binding cleft; by contrast, pepsin-like proteinases have only one, N-terminal, flap. The subunits of the dimer are related by a nearly perfect 2-fold axis of symmetry which passes between the flaps, the active-site Asp residues, and the two C-terminal strands of the interface sheet. The resulting symmetrical active site contrasts with the asymmetric nature of retroviral substrates.

The HIV-1 PR is 99 amino acids long and is thus 25 residues shorter than the Rous sarcoma virus PR. The predicted similarity between the two PRs has recently been confirmed by analysis of the crystal structure (Wlodawer et al., 1989); they have the same fold, and there is an rms deviation of only 1.5 Å for 86 common C_α atoms; deletion of very few residues is necessary for proper alignment of the two PRs. Similar results were reported previously in another determination of the HIV-1 PR crystal structure, but the exact topology of the terminal strands could not be assigned accurately because of the lower resolution of that study (Navia et al., 1989). The structure determined by Wlodawer et al. (1989) indicates that Weber et al. (1989) accurately predicted protein subsites for each substrate residue (positions P4–P3′). Large regions of HIV-1 and Rous sarcoma virus PRs have structural analogues in microbial aspartic proteinases; over half the residues in an HIV-1 PR monomer could be fitted closely with the N- and C-terminal domains of microbial aspartic proteinases (Wlodawer et al., 1989).

Mutations have been made at each residue within HIV-1 PR, and their effects can now be related to the enzyme's structure. The importance of the active-site residue was confirmed by substitution of Asp25 [e.g., Kohl et al. (1988)]. Loeb et al. (1989b) found that three regions were highly sensitive to mutation. The first region comprises residues Ala22–Leu33, which surround the sequence noted by Toh et al. (1985). These residues form parts of the cβ and dβ strands and contain the active-site residue and the highly conserved Thr26 residue, which is believed to contribute toward dimer stabilization (Wlodawer et al., 1989). The second region comprises residues Ile47–Gly52 and forms part of the flap, which probably interacts with the substrate (Weber et al., 1989). The third region comprises residues Thr74–Arg87, which form part of c′β and d′β strands. The flap and the superimposed ψ structures formed by the first and third regions are presumably the main structural and functional determinants of the enzyme.

Studies with proteinase inhibitors are consistent with the classification of retroviral proteinases as aspartic enzymes. Pepstatin A (a transition-state analogue inhibitor of aspartic proteinases) inhibited polyprotein processing (Katoh et al., 1987; Kräusslich et al., 1988; Roberts & Oroszlan, 1989) and also inhibited cleavage of synthetic peptide substrates by purified PR at inhibitory concentrations similar to those observed for renin (Kräusslich et al., 1989a; Meek et al., 1989; Richards et al., 1989). Characteristic inhibitors of aspartic proteinases supported the classification of HIV-1 PR as a member of this class of enzyme (Katoh et al., 1989; Meek et al., 1989). Biochemical experiments supported predictions from molecular modeling and structure determination that only dimers of retroviral PRs would be catalytically competent (Yoshinaka & Luftig, 1980; Nutt et al., 1988; Meek et al., 1989; Katoh et al., 1989). Results which suggested that the concentration of the PR-containing precursor is an important determinant of activity (Kräusslich et al., 1988) are consistent with the requirement for dimerization to generate the functional enzyme. A minimum cleavage concentration of PR (related to the dissociation constant of the dimer) may be required for activity. The pH optimum for cleavage of synthetic peptides by HIV-1 and RSV PR was 4.5–5.5 (Nutt et al., 1988; Billich et al., 1988; Kotler et al., 1989). Higher pH inhibited enzyme activity strongly (Richards et al., 1989), and the cytoplasmic neutral pH is therefore nonoptimal and may inhibit premature processing, an important point considering that cleavage of the retrovirus polyprotein normally only occurs in immature particles. The pH optimum of retroviral proteinases is considerably higher than that of cellular aspartic proteinases (e.g., pepsin, for which the optimum is ca. pH 3.5), with the sole exception of renin (whose optimum is pH 5.5–7.5). The different pH optimum of renin is due in part to substitution by Ala218 of the Thr or Ser residue common to most aspartic proteinases (Sielecki et al., 1989). It is therefore noteworthy that substitution of Ala40 by Ser in Rous sarcoma virus PR (which corresponds to Ala218 in renin) lowered the pH optimum by one pH unit (J. Leis, personal communication).

Determinants of Retroviral Cleavage Site Recognition

Retroviral proteinases are highly substrate-specific, since they correctly cleave homologous or closely related precursors (Yoshinaka et al., 1985), but not *gag* precursors, or peptides corresponding to *gag* or *pol* cleavage sites of distantly related retroviruses (Yoshinaka et al., 1986; Kräusslich et al., 1989a). Cleavage sites recognized by a number of different proteinases have been determined (Figure 3; Katoh et al., 1985; Henderson et al., 1988a,b; Hizi et al., 1989, and references cited therein), but surprisingly, no consistent pattern is apparent that would account for the high specificity for their natural substrates that distinguishes retroviral PRs from many other aspartic proteinases. However, consensus sequences and binding requirements at PR subsites have been suggested (Pearl & Taylor, 1987a; Henderson et al., 1988b; Weber et al., 1989). The cleavage site is defined by residues between P4 and P3′ positions; significant cleavage by HIV-1 PR was only observed for peptides spanning these residues [e.g., Darke et al. (1988) and Billich et al. (1988)] and did not increase as substrate length increased to include P5, P6, P4′, or P5′ residues (Moore et al., 1989). Cleavage of the HIV-1 MA/CA site was impaired by substitutions made between but not outside these positions (Partin et al., 1989). Tyr/Pro and Phe/Pro dipeptides appear frequently at P1 and P1′ positions, but the variety of other residues in these positions is remarkable: Leu/Ala, Leu/Phe, Met/Met, Met/Ser, and Phe/Leu dipeptides occur at HIV-1 cleavage sites (Figure 3). Most substitutions made at these positions of three HIV-1 *pol* sites impaired processing (Loeb et al., 1989a). The residues in the P2 and P4 positions are additional determinants of HIV-1 substrate specificity. The P2 position is commonly occupied by Asn residues, and even substitution of this residue by Gln at the MA/CA site abolished cleavage (K. Partin, C. Carter, and E. Wimmer, unpublished results). Structural analysis suggested that a small residue would be required to bind to the S4 site of HIV-1 PR. Ala, Ser, and Thr occur frequently at this position in natural cleavage sites, and a peptide that contained Phe at this position was not cleaved (Kräusslich et al., 1989a). Substitution of the Ser residue at the P4 position of the MA/CA cleavage site by Arg reduced the efficiency of cleavage, whereas substitution by Thr had no effect (K. Partin, C. Carter, and E. Wimmer, unpublished results). Pepstatin had a bulky isovaleryl group at the P4 position and has a K_i for HIV-1 PR that is 200 times worse than that of acetylpepstatin (Richards et al., 1989). The specificity of HIV-1 PR is broad enough for it to be able to accommodate peptide substrates in vitro that correspond to eight known *gag* and *pol* cleavage sites. Cleavage was efficient enough to suggest that these sites are also processed by HIV-1 PR in vivo (Darke et al., 1988). The variation in susceptibility to cleavage of peptides that correspond to different sites (Kräusslich et al., 1989a; Darke et al., 1988) suggests that residues which form cleavage sites may determine the order of cleavage in the processing cascade.

Proteolytic Processing Is a Determinant of Retroviral Morphogenesis

Retroviral polyproteins are not cleaved immediately after synthesis but are instead transported to the plasma membrane and are normally only processed once budding of immature particles occurs. The N-terminal glycine residues of many retroviral *gag* and *gag-pol* polyproteins are covalently linked to myristic acid (Schultz et al., 1988), and this modification directs the polyproteins to the membrane (Rhee & Hunter, 1987; Schultz & Rein, 1989). Membrane association of *gag* and *gag-pol* precursors probably concentrates and aligns them, promoting intermolecular interactions that may result in some self-assembly of *gag* molecules and dimerization of PR domains. Myristoylation is required for proteolytic processing of the *gag* precursors of murine leukemia and Mason–Pfizer monkey viruses (Rein et al., 1986; Rhee & Hunter, 1987). Göttlinger et al. (1989) found that myristoylation of HIV-1 *gag* and *gag-pol* polyproteins was not necessary for their cleavage in vivo when they were expressed at a high level (which may obviate the need for concentration by membrane association). However, myristoylation is necessary for HIV-1 assembly and consequently for production of infectious virus particles. Cleavage of the *gag* polyprotein is not required for budding of viral particles since immature particles released from retrovirus-transformed cells contain solely or largely uncleaved *gag* precursor (Yoshinaka & Luftig, 1977). Cleavage of polyproteins occurs largely after the release of virions from cells (Witte & Baltimore, 1978). Characteristic morphological changes associated with virion maturation (Gonda et al., 1985) are caused directly by cleavage of *gag* precursor (Yoshinaka & Luftig, 1977), but the mechanism by which the processing cascade and budding of viral complexes from cells are initiated is not understood. Since HIV-1 PR is only active in dimeric form, it is possible that concentration of *gag-pol* precursor proteins at the plasma membrane is required to achieve a "minimal cleavage concentration". Premature processing, on the other hand, is undesirable because the liberated polyprotein domains would be devoid of their myristate "anchor" and would therefore diffuse back into the cytoplasm. Cleavage of the retrovirus polyprotein could occur as an intramolecular event in which two *gag-pol* polyproteins dimerize to cleave at their N- and C-termini or as an intermolecular event in which two such dimers cleave each other. It is difficult to envisage the major structural rearrangement of strands necessary for intramolecular cleavage of HIV-1 and RSV PRs, since the termini of both are ordered (Miller et al., 1989; Wlodawer et al., 1989). It is therefore likely that intermolecular cleavage, probably involving two polyprotein dimers, leads to the initial release of PR. Once released, PR would be confined within the budding particle; it could thus initiate a cascade of proteolytic cleavage events resulting in maturation of virus particles. Active enzyme can be isolated from mature, extracellular retrovirions. The packaged proteinase of equine infectious anemia virus is able to cleave the p11 NC protein to p6 and to p4 proteins when capsids are isolated at pH 7.6 (Roberts & Oroszlan, 1989). This step may be important in the early stages of the replication cycle, possibly including integration of viral DNA. It is not known whether it occurs in other retroviruses.

Conclusion

Numerous proteinases that are encoded by RNA viruses have been identified, and many have been well characterized (Table II). Most fall into one of two groups, related by sequence similarity and active-site topology to either trypsin-like serine or pepsin-like aspartic proteinases. $3C^{pro}$-like proteinases encoded by picorna-, como-, and potyvirus fall into the former category, whereas retroviral proteinases are of the latter type. Proteinases encoded by members of a number of other virus families have been identified, and although many have not been thoroughly characterized, it is likely that most will prove to be related to one or the other of these groups. However, there are viral proteinases, such as the nonstructural proteinase of Sindbis virus (Ding & Schlesinger, 1989), that are apparently not related to either of these groups.

Viral enzymes are much smaller than equivalent polypeptides of their cellular counterparts and also differ from most in having a lower turnover number and an unusually high degree of substrate specificity. This probably reflects the different functions that viral proteinases have evolved to fulfill but may fortuitously aid the design of specific inhibitors. Functions assigned to or suggested for virus-encoded proteinases include the separation of functionally different domains, the separation of components destined for different cellular components, maturation, and morphogenesis (Table II).

Cleavage sites have been identified and are defined by a hierarchy of parameters. The first to be identified was the importance of residues at the scissile bond, such as QG residues at P1–P1′ positions of poliovirus cleavage sites and hydrophobic residues in these positions for retroviruses. The importance of conserved residues in other positions indicates that viral cleavage sites are defined by an extended sequence of up to seven amino acids. Obvious additional determinants include the accessibility and flexibility of amino acid sequences surrounding cleavage sites. Attempts to identify all residues which determine a cleavage site by comparison of cleavage sites with a polyprotein are complicated by the heterogeneity at the same position within various sites. It is likely that this variation determines that rate and order of cleavage at the various sites, so that differential proteolytic release of specific precursors and products can be achieved despite the fact that synthesis of a polyprotein results in equimolar production of different domains. Different functions for cleavage intermediates and end products have been described (e.g., the different proteolytic specificities of $3C^{pro}$ and $3CD^{pro}$), and it is likely that other examples will be found in the future. Proteinases would thus appear to function in a regulatory capacity during the viral replication cycle.

The high degree of cleavage specificity of different viral proteinases raises the interesting possibility that they could be used as tools in molecular biological research, in an analogous manner to restriction enzymes. A number of examples have been described in which bona fide cleavage sites inserted into novel positions within unrelated proteins have been recognized and accurately cleaved by appropriate proteinases. For example, a $3C^{pro}$ site that can be cleaved by $3C^{pro}$ has been inserted into a surface loop of a viable poliovirus isolate (C. Mirzayan and E. Wimmer, unpublished results), and a site recognized by the tobacco etch potyvirus 49-kDa proteinase has similarly been introduced into a novel position within its polyprotein (Carrington & Dougherty, 1988).

Over the past decade viral proteinases have emerged as viable targets for chemotherapy, creating the exciting prospect that effective drugs will be developed to treat numerous viral diseases for which vaccines do not currently exist. The exquisite specificity of viral proteinases makes it unlikely that such drugs will interfere with normal cellular proteolytic processes, but the similarities between many viral proteinases make it likely that derivatives of an effective chemotherapeutic agent will be effective in inhibiting a whole family of viruses. Increasing appreciation of this potential has been parallelled by increasing awareness of the subtle interactions between viral substrates and proteinases and the consequent opportunities for sophisticated regulation of viral expression, replication, and morphogenesis that this affords.

Acknowledgments

We thank Alexander Wlodawer, Mariusz Jaskólski, and Kathryn Partin for critical reading of the manuscript and Alexander Wlodawer/Cold Spring Harbor Press for permission to use their figures. We are grateful to W. G. Dougherty and to the many other colleagues who provided unpublished information and preprints of their work.

References

Argos, P., Kamer, P., Nicklin, M. J. H., & Wimmer, E. (1984) *Nucleic Acids Res. 12*, 7251–7267.

Arnold, E., Luo, M., Vriend, G., Rossman, M. G., Palmenberg, A. C., Parks, G. D., Nicklin, M. J. H., & Wimmer, E. (1987) *Proc. Natl. Acad. Sci. U.S.A. 84*, 21–25.

Bazan, J. F., & Fletterick, R. J. (1988) *Proc. Natl. Acad. Sci. U.S.A. 85*, 7872–7876.

Bazan, J. F., & Fletterick, R. J. (1989a) *FEBS Lett. 249*, 5–7.

Bazan, J. F., & Fletterick R. J. (1989b) *Virology 171*, 637–639.

Berger, A., & Schechter, I. (1970) *Philos. Trans. R. Soc. London, B 257*, 249–264.

Billich, S., Knoop, M. T., Hansen, J., Strop, P., Sedlacek, J., Mertz, R., & Moelling, K. (1988) *J. Biol. Chem. 263*, 17905–17908.

Bone, R., Silen, J. L., & Agard, D. A. (1989) *Nature 339*, 191–195.

Carrington, J. C., & Dougherty, W. G. (1988) *Proc. Natl. Acad. Sci. U.S.A. 85*, 3391–3395.

Carrington, J. C., Cary, S. M., & Dougherty, W. G. (1988) *J. Virol. 62*, 2313–2320.

Carrington, J. C., Cary, S. M., Parks, T. D., & Dougherty, W. G. (1989) *EMBO J. 8*, 365–370.

Chen, Z., Stauffacher, C., Li, Y., Schmidt, T., Bomu, W., Kramer, G., Shanks, M., Lomonossoff, G. & Johnson, J. E. (1989) *Science 245*, 154–159.

Chow, M., Newman, J. F. E., Filman, D. J., Hogle, J. M., Rowlands, D. J., & Brown, F. (1987) *Nature 327*, 482–486.

Crawford, S., & Goff, S. P. (1985) *J. Virol. 53*, 899–907.

Darke, P. L., Nutt, R. F., Brady, S. F., Garsky, V. M., Ciccarone, T. M., Leu, C.-T., Lumma, P. K., Freidinger, R. M., Veber, D. F., & Sigal, I. S. (1988) *Biochem. Biophys. Res. Commun. 156*, 297–303.

Ding, M., & Schlesinger, M. J. (1989) *Virology 171*, 280–284.

Domier, L. L., Shaw, J. G., & Rhoads, R. E. (1987) *Virology 158*, 20–27.

Dougherty, W. G., & Carrington, J. C. (1988) *Annu. Rev. Phytopathol. 26*, 123–143.

Dougherty, W. G., & Parks, T. D. (1989) *Virology 172*, 145–155.

Dougherty, W. G., Cary, S. M., & Parks, T. D. (1989) *Virology 171*, 356–364.

Gonda, M. A., Wong-Staal, F., Gallo, R. C., Clements, J. E., Narayan, O., & Gilden, R. V. (1985) *Science 227*, 173–177.

Gorbalenya, A. E., Donchenko, A. P., Blinov, V. M., & Koonin, E. V. (1989a) *FEBS Lett. 243*, 103–114.

Gorbalenya, A. E., Koonin, E. V., Donchenko, A. P., & Blinov, V. M. (1989b) *Nucleic Acids Res. 17*, 4847–4861.

Göttlinger, H. G., Sodroski, J. G., & Haseltine, W. A. (1989) *Proc. Natl. Acad. Sci. U.S.A. 86*, 5781–5785.

Hayman, M. J., Royer-Pokora, B., & Graf, T. (1979) *Virology 92*, 31–45.

Henderson, L. E., Copeland, T. D., Sowder, R. C., Schultz, A. M., & Oroszlan, S. (1988a) in *Human Retroviruses, Cancer and AIDS: Approaches to Prevention and Therapy*, pp 135–147, Alan R. Liss, New York.

Henderson, L. E., Benveniste, R. E., Sowder, R. C., Copeland, T. D., Schultz, A. M., & Oroszlan, S. (1988b) *J. Virol. 62*, 2587–2595.

Higaki, J. N., Gibson, B. W., & Craik, C. S. (1987) *Cold Spring Harbor Symp. Quant. Biol. 52*, 615–621.

Hizi, A., Henderson, L. E., Copeland, T. D., Sowder, R. C., Krutzsch, H. C., & Oroszlan, S. (1989) *J. Virol. 63*, 2543–2549.

Ivanoff, L. A., Towatari, T., Ray, J., Korant, B. D., & Petteway, S. R., Jr. (1986) *Proc. Natl. Acad. Sci. U.S.A. 83*, 5392–5396.

Jackson, R. J. (1986) *Virology 149*, 114–127.

Jore, J., de Geus, B., Jackson, R. J., Pouwels, P. H., & Enger-Valk, B. E. (1988) *J. Gen. Virol. 69*, 1627–1636.

Katoh, I., Yoshinaka, Y., Rein, A., Shibuya, M., Odaka, T., & Oroszlan, S. (1985) *Virology 145*, 280–292.

Katoh, I., Yasunaga, T., Ikawa, Y., & Yoshinaka, Y. (1987) *Nature 329*, 654–656.

Katoh, I., Ikawa, Y., & Yoshinaka, Y. (1989) *J. Virol. 63*, 2226–2232.

Kitamura, N., Semler, B. L., Rothberg, P. G., Larsen, G. R., Adler, C. J., Dorner, A. J., Emini, E. A., Hanecak, R., Lee, J. J., van der Werf, S., Anderson, C. W., & Wimmer, E. (1981) *Naturc 291*, 547–553.

Kohl, N. E., Emini, E. A., Schleif, W. A., Davis, L. J., Heimbach, J. C., Dixon, R. A. F., Scolnick, E. M., & Sigal, I. S. (1988) *Proc. Natl. Acad. Sci. U.S.A. 85*, 4686–4690.

Kotler, M., Katz, R. A., Danho, W., Leis, J., & Skalka, A. M. (1988) *Proc. Natl. Acad. Sci. U.S.A. 85*, 4185–4189.

Kräusslich, H.-G., & Wimmer, E. (1988) *Annu. Rev. Biochem. 57*, 701–754.

Kräusslich, H.-G., Nicklin, M. J. H., Toyoda, H., Etchison, D., & Wimmer, E. (1987) *J. Virol. 61*, 2711–2718.

Kräusslich, H.-G., Schneider, H., Zybarth, G., Carter, C. A., & Wimmer, E. (1988) *J. Virol. 62*, 4393–4397.

Kräusslich, H.-G., Ingraham, R. H., Skoog, M. T., Wimmer, E., Pallai, P. V., & Carter, C. A. (1989a) *Proc. Natl. Acad. Sci. U.S.A. 86*, 807–811.

Kräusslich, H.-G., Oroszlan, S., & Wimmer, E., Eds. (1989b) *Viral proteinases as targets for chemotherapy*, Cold Spring Harbor Laboratory, Cold Spring Harbor, NY.

Kräusslich, H.-G., Hölscher, C., Reuer, Q., Harber, J., & Wimmer, E. (1989c) (submitted for publication).

Kuhn, R. J., Tada, H., Ypma-Wong, M.-F., Semler, B. L., & Wimmer, E. (1988) *J. Virol. 62*, 4207–4215.

Lee, C.-K., & Wimmer, E. (1988) *Virology 166*, 405–414.

Leis, J., Baltimore, D., Bishop, J. M., Coffin, J., Fleissner, E., Goff, S. P., Oroszlan, S., Robinson, H., Skalka, A. M., Temin, H. M., & Vogt, V. (1988) *J. Virol. 62*, 1808–1809.

Loeb, D. D., Hutchison, C. A., Edgell, M. H., Farmerie, W. G., & Swanstrom, R. (1989a) *J. Virol. 63*, 111–121.

Loeb, D. D., Swanstrom, R., Everitt, L., Manchester, M., Stamper, S. E., & Hutchinson, C. A. (1989b) *Nature 340*, 397–400.

Meek, T. D., Dayton, B. D., Metcalf, B. W., Dreyer, G. B., Strickler, J. E., Gorniak, J. G., Rosenberg, M., Moore, M. L., Magaard, V. W., & Debouck, C. (1989) *Proc. Natl. Acad. Sci. U.S.A. 86*, 1841–1845.

Miller, M., Jaskolski, M., Rao, J. K. M., Leis, J., & Wlodawer, A. (1989) *Nature 337*, 576–579.

Moore, M. L., Bryan, W. M., Fakhoury, S. A., Magaard, V. W., Huffman, W. F., Dayton, B. D., Meek, T. D., Hyland, L., Dreyer, G. B., Metcalf, B. W., Strickler, J. E., Gorniask, J. G., & Debouck, C. (1989) *Biochem. Biophys. Res. Commun. 159*, 420–425.

Navia, M. A., Fitzgerald, P. M. D., McKeever, B. M., Leu, C.-T., Heimbach, J. C., Herber, W. K., Sigal, I. S., Darke, P. L., & Springer, J. P. (1989) *Nature 337*, 615–620.

Neet, K. E., Nanci, A., & Koshland, D. E., Jr. (1968) *J. Biol. Chem. 243*, 6392–6401.

Nicklin, M. J. H., Toyoda, H., Murray, M. G., & Wimmer, E. (1986) *Bio/Technology 4*, 33–42.

Nicklin, M. J. H., Kräusslich, H.-G., Toyoda, H., Dunn, J. J., & Wimmer, E. (1987) *Proc. Natl. Acad. Sci. U.S.A. 84*, 4002–4006.

Nicklin, M. J. H., Harris, K. S., Pallai, P. V., & Wimmer, E. (1988) *J. Virol. 62*, 4586–4593.

Nutt, R. F., Brady, S. F., Darke, P. L., Ciccarone, T. M., Colton, C. D., Nutt, E. M., Rodkey, J. A., Bennett, C. D., Waxman, L. H., Sigal, I. S., Anderson, P. S., & Veber, D. F. (1988) *Proc. Natl. Acad. Sci. U.S.A. 85*, 7129–7133.

Pallai, P. V., Burckhardt, F., Skoog, M. T., Schreiner, K., Bax, P., Cohen, K., Hansen, G., Palladino, D. E. H., Harris, K. S., Nicklin, M. J. H., & Wimmer, E. (1989) *J. Biol. Chem. 264*, 9738–9741.

Parks, G. D., & Palmenberg, A. C. (1987) *J. Virol. 61*, 3680–3687.

Parks, G. D., Baker, J. C., & Palmenberg, A. C. (1989) *J. Virol. 63*, 1054–1058.

Partin, K., Kräusslich, H.-G., Bradley, J., Wimmer. E., & Carter, C. (1989) in *Modern Approaches to new Vaccines*

including Prevention of AIDS (Lerner, R., Ginsberg, H., Chanock, R. M., & Brown, F., Eds.) Cold Spring Harbor Laboratory, Cold Spring Harbor, NY.
Pearl, L. H., & Taylor, W. R. (1987a) *Nature 328*, 482.
Pearl, L., H., & Taylor, W. R. (1987b) *Nature 329*, 351–354.
Polgar, L., & Bender, M. L. (1969) *Biochemistry 8*, 136–141.
Rein, A., McClure, M. R., Rice, N. R., Luftig, R. B., & Schultz, A. M. (1986) *Proc. Natl. Acad. Sci. U.S.A. 83*, 7246–7250.
Rhee, S. S., & Hunter, E. (1987) *J. Virol. 61*, 1045–1053.
Richards, A. D., Roberts, R., Dunn, B. M., Graves, M. C., & Kay, J. (1989) *FEBS Lett. 247*, 113–117.
Roberts, M. M., & Oroszlan, S. (1989) *Biochem. Biophys. Res. Commun. 160*, 486–494.
Rueckert, R. R., & Wimmer, E. (1984) *J. Virol. 50*, 957–959.
Schultz, A. M., & Rein, A. (1989) *J. Virol. 63*, 2370–2373.
Schultz, A. M., Henderson, L. E., & Oroszlan, S. (1988) *Annu. Rev. Cell Biol. 4*, 611–647.
Shanks, M., Stanley, J., & Lomonossoff, G. P. (1986) *Virology 155*, 697–706.
Shih, C. T., & Shih, D. S. (1981) *J. Virol. 40*, 942–945.
Sielecki, A. R., Hayakawa, K., Fujinaga, M., Murphy, M. E. P., Fraser, M., Muir, A. K., Carilli, C. T., Lewicki, J. A., Baxter, J. D., & James, M. N. G. (1989) *Science 243*, 1346–1351.
Sommergruber, W., Zorn, M., Blaas, D., Fessl, F., Volkmann, P., Maurer-Fogy, I., Pallai, P., Merluzzi, V., Matteo, M., Skern, T., & Kuechler, E. (1989) *Virology 169*, 68–77.
Sun, X.-H., & Baltimore, D. (1989) *Proc. Natl. Acad. Sci. U.S.A. 86*, 2143–2146.
Teich, N. (1984) in *RNA Tumor Viruses* (Weiss, R., Teich, N., Varmus, H., & Coffin, J., Eds.) pp 25–207, Cold Spring Harbor Laboratory, Cold Spring Harbor, NY.
Toh, H., Ono, M., Saigo, K., & Miyata, T. (1985) *Nature 315*, 691.
Toyoda, H., Nicklin, M. J. H., Murray, M. G., Anderson, C. W., Dunn, J. J., Studier, F. W., & Wimmer, E. (1986) *Cell 45*, 761–770.
Vos, P., Verver, J., Jaegle, M., Wellink, J., van Kammen, A., & Goldbach, R. (1988) *Nucleic Acids Res. 16*, 1967–1985.
Weber, I. T., Miller, M., Jaskolski, M., Leis, J., Skalka, A. M., & Wlodawer, A. (1989) *Science 243*, 928–931.
Wellink, J., Rezelman, G., Goldbach, R., & Beyreuther, K. (1986) *J. Virol. 59*, 50–58.
Witte, O. N., & Baltimore, D. (1978) *J. Virol. 26*, 750–761.
Wlodawer, A., Miller, M., Jaskôlski, M., Sathyanarayana, B. K., Baldwin, E., Weber, I. T., Selk, L. M., Clawson, L., Schneider, J., & Kent, S. B. H. (1989) *Science 245*, 616–621.
Yoshinaka, Y., & Luftig, R. B. (1977) *Proc. Natl. Acad. Sci. U.S.A. 74*, 3446–3450.
Yoshinaka, Y., & Luftig, R. B. (1980) *J. Gen. Virol. 48*, 329–340.
Yoshinaka, Y., Katoh, I., Copeland, T. D., & Oroszlan, S. (1985) *J. Virol. 55*, 870–873.
Yoshinaka, Y., Katoh, I., Copeland, T. D., Smythers, G. W., & Oroszlan, S. (1986) *J. Virol. 57*, 826–832.
Ypma-Wong, M.-F., & Semler, B. L. (1987) *Nucleic Acids Res. 15*, 2069–2088.
Ypma-Wong, M.-F., Dewalt, P. G., Johnson, V. H., Lamb, J. G., & Semler, B. L. (1988a) *Virology 66*, 265–270.
Ypma-Wong, M.-F., Filman, D. J., Hogle, J. M., & Semler, B. L. (1988b) *J. Biol. Chem. 263*, 17846–17856.

Chapter 24

Covalent Modification Reactions Are Marking Steps in Protein Turnover

Earl R. Stadtman*

Laboratory of Biochemistry, National Heart, Lung, and Blood Institute, National Institutes of Health, Bethesda, Maryland 20892

Received February 14, 1990; Revised Manuscript Received March 27, 1990

In pioneering studies Schoenheimer (1946) discovered that intracellular proteins undergo continuous synthesis and degradation. In the meantime, it has been demonstrated that various enzymes turnover at different rates and that the turnover of a particular enzyme may be selectively influenced by nutritional factors, cellular differentiation, organelle distribution, environmental conditions, and metabolic abnormalities of genetic or nongenetic origins. Attempts to identify structural features of proteins that dictate turnover rates have on the whole been disappointing. However, rates of turnover in vivo vary with respect to molecular size, isoelectric point, hydrophobicity, the N-terminus amino acid, thermal stability, conformational changes provoked by protein–protein and protein–lipid interactions, and the concentrations of coenzymes, allosteric effectors, and enzyme substrates. For general reviews, see Schimke and Doyle (1970), Pine (1972), Goldberg and Dice (1974), Schimke and Bradley (1975), Goldberg and St. John (1976), Holzer and Heinrick (1980), and Rivett (1986). Nevertheless, exceptions to all of these generalizations have frustrated attempts to correlate protein turnover with unique structural features. Likewise, attempts to identify the proteolytic systems involved in protein turnover have led to the realization that no single system in responsible. Indeed, it is now evident that ATP-dependent and ATP-independent, lysosomal and nonlysosomal, and ubiquitin-dependent and ubiquitin-independent pathways are all involved in protein turnover at one stage or another.

It is particularly significant that the rates of turnover of various proteins in vivo correlate reasonably well with their susceptibilities to degradation by endopeptidases in vitro (Dice et al., 1973; Segal et al., 1974), and there is a reasonable correlation between the rates of degradation of various enzymes in vivo and their thermal stabilities (McLendon & Radany, 1978); moreover, abnormal and denatured proteins are degraded more rapidly than their native counterparts. Thus, proteins containing amino acid substitutions resulting from site-specific mutations or the incorporation of amino acid analogues (viz., canavalin for arginine, azetitinecarboxylic acid for cysteine, fluorophenylalanine for phenylalanine) are degraded more rapidly than their normal counterparts.

Because the posttranslational modification of a given amino acid residue in a protein is equivalent to the substitution of that residue by an analogue, it has been proposed that the covalent modifications of amino acid residues may serve as "marking" steps for protein degradation (Hood et al., 1977; Holzer & Heinrich, 1980; Oliver et al., 1980; Levine et al., 1981).

This concept is confirmed by the results of in vitro experiments showing that several kinds of covalent modification increase the sensitivity of various enzymes to degradation by purified proteases [for reviews, see Rivett (1986) and Stadtman (1986, 1988a,b)]. Such modifications include the oxidation of amino acid residues by activated oxygen species, the esterification of hydroxyl groups (adenylylation of tyrosine residues, phosphorylations of serine or threonine residues), the acylation (carbamylation, acetylation, ubiquitination) of N^{ϵ}-lysine residues, the oxidation of non-heme iron-sulfur centers, the amino acylation of N-terminal glutamate or aspartate residues, the deamidation of glutamine and asparagine residues, and the glycosylation of amide or hydroxyl groups [for specific examples, see Rivett (1986), Stadtman (1986, 1988a), Holzer and Heinrick (1980), and Gonda et al. (1989)]. Other modifications that are probable marking steps for proteolytic degradation but that have not been directly tested for this activity include the glycation of N^{ϵ}-lysyl groups, proline isomerization, aspartate racemization, the addition of α,β-unsaturated aldehydes to cysteine sulfhydryl groups (thiol ester formation), the reaction of aldehydes with N^{ϵ}-lysyl groups, and various enzymic and nonenzymic reactions leading to inter- and intraprotein cross-linking or to peptide bond cleavage. Because the intracellular accumulation of proteins with these modifications is associated with a loss of endogenous protease activities (viz., with aging), it seems likely that such modifi-

* Address correspondence to the author at 9000 Rockville Pike, Building 3, Room 222, Bethesda, MD 20892.

cations serve as marking steps for protein degradation.

Oxidative Denaturation Marks Enzymes for Proteolysis. Studies on the regulation of enzyme turnover in *Escherichia coli* and *Klebsiella aerogenes* led to the discovery that the intracellular degradation of several enzymes is preceded by oxidative modifications catalyzed by a mixed-function oxidation (MFO) system comprised of NAD(P)H, O_2, Fe(III) or Cu(II), and endogenous enzymes [presumably NAD(P)H oxidases]. Subsequently, it was found that a variety of enzymic and nonenzymic MFO systems, now referred to as metal-ion-catalyzed oxidation (MCO) systems (Amici et al., 1989), have the capacity to catalyze oxidative modifications of proteins (Oliver et al., 1980; Levine et al., 1981; Fucci et al., 1983; Stadtman & Wittenberger, 1985). From these studies, it was concluded that the H_2O_2 and Fe(II) or Cu(I) generated by the MCO systems interact at metal binding sites on proteins to form activated oxygen species ($\dot{O}H$, O_2^-, ferryl ions), which oxidize the side chains of amino acid residues at the metal binding sites. It was proposed that the oxidation is one modification that "marks" an enzyme for proteolytic degradation (Oliver et al., 1980; Levine et al., 1981; Rivett et al., 1985). In the meantime, this concept gained support from studies showing that neutral proteases with high specificity for the oxidized forms of proteins are widely distributed (Rivett et al., 1985; Rivett, 1985a,b; Roseman & Levine, 1987; Davies, 1987; Davies & Goldberg, 1987; Lee et al., 1988; Davies & Lin, 1988; Marcillat et al., 1988) and that the oxidized forms of purified enzymes are degraded more rapidly than their native counterparts by any one of several common proteases, viz., trypsin, pepsin, subtilisin, calpain, and cathepsin D (Farber & Levine, 1982; Rivett, 1985b; Dean & Pollak, 1985; Hunt et al., 1988).

Mechanisms of Oxygen-Radical-Mediated Modifications. Histidine, proline, arginine, lysine, methionine, and cysteine residues are particularly sensitive to site-specific metal-ion-catalyzed oxidation. Levine (1983) showed that some amino acid residues are converted to carbonyl derivatives. It is evident from studies from homopolymers of amino acids that carbonyl derivatives are formed in the metal-ion-catalyzed oxidation of proline, arginine, lysine, and histidine residues (Oliver et al., 1985; Amici et al., 1989). Glutamic semialdehyde residues have been identified as products of proline and arginine oxidation (Amici et al., 1988; Climent et al., 1989), and an adipic semialdehyde residue is tentatively identified as the product of lysine oxidation (Miller, Berlett, and Stadtman, unpublished data). Presumptive evidence that threonine residues may be oxidized to 2-amino-3-ketobutyric acid residues has also been presented (Taborsky, 1973). In addition to carbonyl derivatives, proline residues are converted also to glutamic acid and/or pyroglutamic acid residues, and histidine residues are converted to asparagine and/or aspartic acid residues (Farber & Levine, 1986; Creeth et al., 1983; Cooper et al., 1985).

Methionine residues are oxidized to methionine sulfoxide derivatives (Maier et al., 1989; Auroma & Halliwell, 1989), and cysteine residues are converted to mixed disulfides and to inter- and intra-disulfide cross-linkages (Ballard & Hopgood, 1976; Francis & Ballard, 1980; Bond & Offermann, 1981; Offerman et al., 1984).

It is noteworthy that tryptophan, phenylalanine, and tyrosine residues in proteins are not major sites of oxidation by metal-ion-catalyzed oxidation systems. In contrast, all amino acid residues are modified by radicals produced during γ-radiolysis, but in this case the aromatic amino acids are the preferred targets [for review, see Swallow (1960)]. These differences in target specificity are probably attributable to the site-specific nature of the metal-ion-catalyzed reactions as noted above. In the absence of O_2, $\dot{O}H$ radicals produced by radiolysis lead to extensive protein–protein cross-linkage via tyrosine–tyrosine bonding, and possibly other amino acid cross-links as well (viz., –S–S– cross-links). In the presence of O_2, the cross-linking reactions are suppressed, and considerable peptide bond cleavage occurs, with concomitant formation of protein carbonyl groups (Garrison et al., 1962; Schuessler & Schilling, 1984; Wolff et al., 1986; Davies et al., 1987). On the basis of results from studies with amino acids and amines, it is generally agreed that peptide bond cleavage probably involves $\dot{O}H$-mediated hydrogen abstraction from the α-carbon atom to form carbon-centered radicals; these in the presence of O_2^- will form peroxy radical intermediates, which can facilitate peptide bond cleavage to peptide amide and carbonyl peptide derivatives. One of several mechanisms proposed by Garrison et al. (1962) is

$$H_2O + O_2 \rightarrow \dot{O}H + H\dot{O}_2 \quad (1)$$

$$RCONHC(R^2)HCONHR^3 + \dot{O}H \rightarrow RCONH\dot{C}(R^2)CONHR^3 + H_2O \quad (2)$$

$$RCONH\dot{C}(R^2)CONHR^3 + O_2 \rightarrow RCONHC(R^2)(\dot{O}_2)CONHR^3 \quad (3)$$

$$RCONHC(R^2)(\dot{O}_2)CONHR^3 + H\dot{O}_2 \rightarrow RCONHC(R^2)(OOH)CONHR^3 + O_2 \quad (4)$$

$$RCONHC(R^2)(OOH)CONHR^3 \rightarrow RCON{=}C(R^2)CONHR^3 + H_2O_2 \quad (5)$$

$$RCON{=}C(R^2)CONHR^3 + H_2O \rightarrow RCONH_2 + R^2COCONHR^3 \quad (6)$$

The overall reaction is formally analogous to the oxygen-dependent, site-specific metal-ion-catalyzed reaction described by Bateman et al. (1985) and the oxidative cleavage of peptide–copper complexes described by Levitzki et al. (1967).

Subsequently, Schuessler and Schilling (1984) noted that the number of peptide fragments obtained upon radiolysis of bovine serum albumin is approximately equal to the number of prolyl residues in the protein and postulated that a preferred target of radiation-induced chain scission "may be the amino acyl–proline bond, because tertiary amide bonds are easier to oxidize than secondary amide bonds." As noted by Wolff et al. (1986), the possibility that peptide bond cleavage is due to oxidation of the proline moiety itself must also be considered, but see below.

Deamidation of Asparaginyl and Glutaminyl Residues. Asparaginyl and glutaminyl residues undergo spontaneous deamidation at rates determined by the amino acid sequences. Robinson and co-workers [see Robinson and Rudd (1974)] examined 70 different pentapeptides containing either an asparaginyl or a glutaminyl residue in the central position and showed that their rates of deamidation varied from 6 days to 9 years. They proposed that sequences around asparaginyl and glutaminyl residues may have been selected through evolutionary pressures to serve as biological clocks that specify the rates of protein turnover in vivo. There is in fact a positive correlation between the rates of enzyme turnover in vitro and their contents of asparaginyl plus glutaminyl residues (Robinson & Rudd, 1974). Nevertheless, direct evidence that deamidation renders an enzyme more susceptible to proteolytic degradation, though reasonable, is still lacking. A role of deamidation in enzyme turnover is suggested by the fact that the accumulation of deamidated forms of several enzymes varies reciprocally with respect to the intracellular levels of proteases, viz., during aging and in individuals with premature aging disease [for review, see Gracy et al. (1985a), Dreyfus et al. (1978), and McKerrow (1979)]. The most convincing evidence that deamidation contributes to the pool of altered enzymes comes from studies showing that the age-related accumulation of four more acidic, more labile isoenzyme forms of triosephosphate isomerase in several test systems (Yuan et al., 1981; Gracy et al., 1985a,b) is due to the sequential deamidation of two asparaginyl residues (Asn 71 and Asn 15). The observation that deamidation destabilizes subunit interactions and that the monomeric subunit is more susceptible to proteolytic degradation (Yuan et al., 1981) supports the thesis that deamidation "marks" enzymes for degradation. This thesis is also supported by the fact that the rate of deamidation (8 days) of rabbit muscle aldolase is about the same as the half-life of the enzyme in vivo (Midelfort & Mehler, 1972) and by the studies of Gonda et al. (1989) showing that enzymes in reticulocyte lysates and in yeast catalyze the deamidation of N-terminal asparaginyl and glutaminyl residues of mutant forms of β-galactosidase. The resulting N-terminal Asp and Glu are then susceptible to aminoacylation by Arg-tRNA- or Lys-tRNA-specific transferases to generate protein derivatives with destabilizing N-terminal Arg or Lys residues. Thus modified, the β-galactosidase becomes susceptible to ubiquitin conjugation and subsequent degradation by the ATP-dependent ubiquitin pathway (see below). The generality of the N-terminal Glu or Asp aminoacylation pathway for the degradation of other proteins remains to be determined.

Dean (1987) and Wolff et al. (1986) proposed that N-terminal glutamyl and aspartyl residues generated in the oxidative cleavage of prolyl and histidyl residues, respectively, might be targeted for degradation by the above Arg-modification mechanism. It is noteworthy, however, that neither glutamyl nor aspartyl residues have been shown to be primary products of the metal-ion-catalyzed oxidation reactions, nor have they been identified as N-terminal residues of fragments obtained by oxygen-radical-mediated peptide bond scission. The primary products of oxidation are more likely pyroglutamyl (Amici et al., 1989) and asparaginyl (Farber & Levine, 1986) residues, which are converted to glutamic and aspartic acids during the hydrolysis procedures used in their identification. Whereas polyhistidine has been shown to undergo fragmentation by metal-ion-catalyzed reactions and aspartate has been identified in the acid hydrolysates of the oxidized polymer (Cooper et al., 1985), the oxidation of histidyl to asparaginyl residues in proteins, at least in the case of glutamine synthetase, is not associated with peptide bond cleavage (Farber & Levine, 1986). The N-terminal pyroglutamyl peptide that could be formed in the oxidation of prolyl residues in protein (Amici et al., 1989) might resist degradation by the ubiquitin pathway because of a blocked *N*-amino group (Herschko et al., 1984). Moreover, scission of prolyl peptide bonds by Garrison's mechanism, reactions 1–6, would not lead to an N-terminal glutamate residue.

Racemization and Isomerization of Aspartyl and Asparaginyl Residues. Proteins containing D-aspartyl and/or isoaspartyl residues have been shown to accumulate in eye lens (Masters et al., 1977), in teeth (Helfman & Bada, 1975, 1976), and in human erythrocyte membrane during aging (Brunauer & Clarke, 1986). Geiger and Clark (1987) demonstrated that these derivatives as well as the deamidation of asparaginyl residues likely occur via a common succinimide intermediate, as shown in Figure 1. The carboxyl groups of these abnormal aspartyl derivatives are readily methylated by highly specific *S*-adenosylmethionine-dependent carboxyl methyltransferases. Because L-aspartyl residues are not substrates, methylation of protein in the presence of the transferase and [*methyl*-^{14}C]-*S*-adenosylmethionine provides a highly sensitive measure of abnormal aspartyl residues in protein (Clarke, 1985). Johnson et al. (1987) have shown that cycles of carboxylmethylation can facilitate the conversion of the D-aspartyl and isoaspartyl residues back to the L-aspartyl configuration.

Prolyl Isomerization. Amino acid residues in proteins and peptides exist mainly in the trans configuration. However, depending upon amino acid composition and sequence, the prolyl residues of naturally occurring proteins may be stabilized in the cis configuration [for review, see Stadtman (1988a)]. The potential importance of cis–trans prolyl isomerization in protein degradation is suggested by the fact that chymotrypsin and other proline-specific endopeptidases are able to split prolyl bonds only if the prolyl residue is in the trans configuration (Fischer et al., 1983, 1984; Bachinger, 1987); the trans configuration is essential also to enable a protein to assume a triple-helix configuration. The biological significance of cis–trans isomerization in the proper folding of proteins is highlighted by the discovery of a highly specific prolyl isomerase (Fischer & Bang, 1984; Bachinger, 1987; Lang et al., 1987) and the identification of this isomerase as the cyclosporin receptor protein (Takahashi et al., 1989).

Ubiquitin-Dependent Proteolysis. The conjugation of ubiquitin to the ε-amino group of lysyl residues of some denatured proteins facilitates their degradation (Ciechanover et al., 1980, 1984; Wilkinson et al., 1980; Chin et al., 1982). The ubiquitination is a multistep process in which the C-terminal glycine carboxyl group of ubiquitin (Ub–COOH) is activated by an activating enzyme (E1–SH) to form an enzyme-bound Ub–C(O)–AMP (reaction 7). This is followed by thiolytic

$$\text{E1–SH} + \text{Ub–COOH} + \text{ATP} \rightarrow \text{Ub–C(O)–AMP·E1–SH} + \text{PP}_i \quad (7)$$

$$\text{Ub–C(O)–AMP·E1–SH} \rightarrow \text{E1–S–C(O)–Ub} + \text{AMP} \quad (8)$$

$$\text{E1–S–C(O)–Ub} + \text{E2–SH} \rightarrow \text{E2–S–C(O)–Ub} \rightarrow \text{E1–SH} \quad (9)$$

$$\text{E2–S–C(O)–Ub} + \text{Pr–NH}_2 \xrightarrow{\text{E3}} \text{Pr–NH–C(O)–Ub} + \text{E2–SH} \quad (10)$$

$$\text{Pr–NHC(O)–Ub–NH}_2 + n[\text{E2–S–C(O)–Ub}] \xrightarrow{\text{E3}} \text{Pr–NHC(O)–Ub–NH–[C(O)–Ub]}_n \quad (11)$$

cleavage of the Ub–C(O)–AMP bond by a sulfhydryl group on E1–SH to form a ubiquitin thiolester, E1–S–C(O)–Ub (reaction 8). The ubiquitin moiety of E1–S–C(O)–Ub is then transferred to the sulfhydryl group of a small ubiquitin carrier protein (E2–SH), reaction 9, and then to a particular ε-amino group (Pr–NH_2) of the target protein (reaction 10); this reaction is catalyzed by a specific ubiquitin transferase, E3. Whereas monoubiquitination of proteins is implicated in diverse physiological processes [see Cook and Chock (1988), Rechsteiner (1988), Schlesinger and Herschko (1988), and Ciechanover and Schwartz (1989)], it is insufficient to target denatured proteins for degradation. To be recognized by the ubiquitin–ATP-dependent proteolytic system, several ubiquitin molecules must be conjugated to the target protein. Using genetically engineered constructs of β-galactosidase, Chau et al. (1989) established that only one lysyl residue (residue 15 or 17) of β-galactosidase is modified. Multiply ubiquitinated conjugates are formed by the subsequent addition of up to 20 ubiquitin peptides to the ubiquitin moiety of the monoconjugate (reaction 11). This involves the generation of an ordered

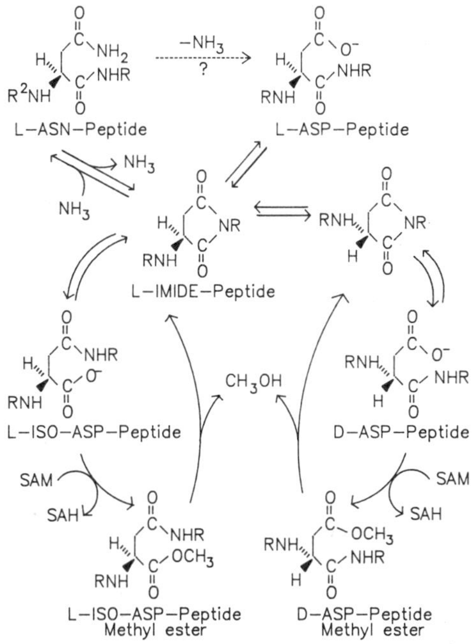

FIGURE 1: Role of succinimide peptide intermediate and carboxyl-methylation in the deamidation, racemization, and isomerization of asparaginyl- and aspartyl-containing peptides. Abbreviations: SAM, *S*-adenosylmethionine; SAH, *S*-adenosylhomocysteine.

chain of branched ubiquitin–ubiquitin conjugates by a mechanism in which the carboxyl-terminal Gly 76 of one ubiquitin moiety is joined in isopeptide linkage to the internal Lys 48 of an adjacent ubiquitin moiety. In this way, the conjugation of a single Lys residue on the target protein by a highly branched polyubiquitin tree-like structure serves as the secondary "marking" step for proteolytic degradation. The "primary" marking steps, i.e., the denaturing steps that target proteins for ubiquitination, are still poorly defined. In the case of RNase, however, the oxidation of a methionine residue to the sulfoximine derivative renders the enzyme highly susceptible to degradation by the ubiquitin-dependent pathway (Herschko et al., 1986).

Acetylation of the α-amino group of some proteins inhibits their conjugation to ubiquitin and protects them from degradation by the ubiquitin-dependent pathway (Herschko et al., 1984). This focused attention on the importance of an unmodified, exposed amino acid residue at the N-terminus of proteins as a recognition site for the ubiquitin transferase. By means of site-directed mutagenesis, Bachmiar et al. (1986) replaced the N-terminal Met residue of *E. coli* β-galactosidase with other amino acids and found that enzyme species with either Met, Ser, Ala, Thr, Val, Gly, Pro, or Cys at the N-terminus had in vivo half-lives of 20 h or more, whereas the constructs with either Glu, Tyr, Gln, His, Arg, Lys, Phe, Trp, Leu, Asn, or Asp at the N-terminus had very short half-lives (2–30 min). On the basis of these results and correlations with the N-terminal amino acid compositions of the more stable enzymes in vivo, these workers proposed that the half-lives of proteins are specified by the identity of the N-terminal amino acid residue, and they formalized this concept in terms of the so-called "N-end rule". In the meantime, evidence both consistent with and inconsistent with the N-end rule has been obtained [see, for example, Ciechanover and Schwartz (1989), Rogers et al. (1986), Rogers and Rechsteiner (1988), Ciechanover et al. (1984), Bachmair and Varschavsky (1989), Ghoda et al. (1989), Speiser and Etlinger (1983), Saus et al. (1982), and Rote et al. (1989)]. Clearly, structural features of proteins downstream from the N-terminus ("body-type" features) are also important for the recognition of proteins as substrates for ubiquitination and degradation. In fact, body-type proteins appear to be more abundant than N-end-type proteins (Ciechanover & Schwartz, 1989). It is noteworthy, however, that single amino acid substitutions of internal amino acid residues of a protein may alter its thermal stability and susceptibility to proteolytic degradation (Ogasahara et al., 1985; McLendon & Radnay, 1978; Yutani et al., 1977, 1984). It is therefore possible that the effects of N-terminal amino acid substitutions are merely the expression of a more general principle, namely, that the substitution of one amino acid residue in a protein by another amino acid may alter sensitivity of the protein to degradation. A systematic study of the substitution of various amino acids for a particular residue in the body of a protein might disclose correlations similar to those observed for N-end group substitutions [for example, see Yutani et al. (1977)]. If so, the enunciation of a body-rule would be no less justified than the enunciation of an N-end rule.

Schiff Base Formation. The reaction of amino groups of proteins with aldehydes to form Schiff base adducts is well-known. Until now, nonenzymic reactions of malondialdehyde and glucose with proteins are the only ones known to occur in vivo. The possible implication of these reactions in atherogenesis, diabetes, aging, and cataractogenesis has prompted considerable interest in their metabolism.

The potential role of such reactions in targeting proteins for proteolysis is inferred by the fact that macrophages and sinusoidal liver cells possess "scavenger" receptors for the binding, endocytosis, and degradation of these aldehyde-derived adducts. Two kinds of receptors have been identified. One type recognizes malondialdehyde-modified proteins, e.g., the low-density lipoproteins (Fogleman et al., 1980). The other type is implicated in the endocytosis of advanced glycosylation end products (AGE) formed in vivo (Vlassara et al., 1985, 1987), as well as a number of chemically synthesized aliphatic aldehyde–protein conjugates (Horiuchi et al., 1986; Takata et al., 1988). In model studies with ribonuclease A, Chio and Tappel (1969) showed that malondialdehyde formed in the peroxidation of polyunsaturated fatty acids (viz., arachidonic or linolenic acid) react with the ϵ-amino groups of lysyl residues in the enzyme (Enz) to form highly fluorescent, catalytically inactive intra- or interprotein conjugated imine adducts (reactions 12 and 13).

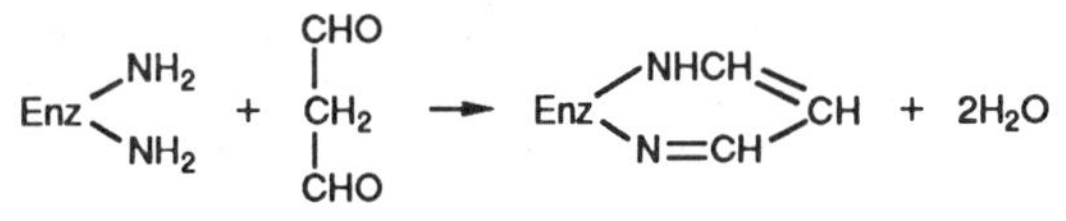

$$\mathrm{Enz}(NH_2)_2 + CHO\text{-}CH_2\text{-}CHO \rightarrow \mathrm{Enz}\langle^{NHCH=}_{N=CH}\rangle CH + 2H_2O \quad (12)$$

$$2\mathrm{Enz{-}NH_2} + CHO\text{-}CH_2\text{-}CHO \rightarrow \mathrm{Enz{-}NHCH{=}CHCH{=}N{-}Enz} + 2H_2O \quad (13)$$

Glycation. Glycation of proteins involves the nonenzymic reaction of reducing sugars with the N-terminal α-amino group or, more often, with the ϵ-amino groups of lysyl residues to form Schiff base adducts that undergo rapid Amadori rearrangements to relatively stable ketoamines (Figure 2). These ketoamines are slowly converted to brown, fluorescent polymeric derivatives that are referred to as advanced glycosylation end products (AGE). The overall process is known as the browning reaction or the Maillard reaction. It is responsible for the nonenzymic deterioration of foods during storage and has been implicated in several physiological disorders, viz., diabetes, cataractogenesis, and aging. Moreover, glycated

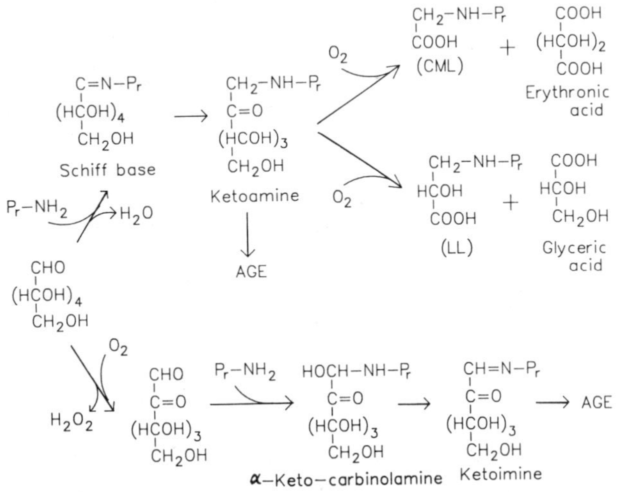

FIGURE 2: Mechanisms of protein glycation. Abbreviations: CML, N^{ϵ}-(carboxymethyl)lysyl residue; LL, 3-(N^{ϵ}-lysino)lactic acid; $PrNH_2$, ϵ-amino group of a protein lysyl residue; AGE, advanced glycosylation end products.

derivatives of several enzymes/proteins have been shown to occur in vivo [for reviews, see Harding (1985), Cerami and Crabbe (1986), and Cerami et al. (1987, 1986)].

A nonbrowning pathway of glycation was disclosed by the studies of Ahmed et al. (1986, 1988) showing that oxidative cleavage of fructosyllysyl residues in protein is catalyzed by an oxygen-dependent metal ion free radical mechanism, both in vivo and in vitro (Figure 2). Cleavage between C_2 and C_3 of the carbohydrate moiety yields an N^{ϵ}-(carboxymethyl)lysyl (CML) residue and D-erythronic acid, whereas cleavage between C_3 and C_4 yields the 3-(N^{ϵ}-lysino)lactic acid (LL) derivative and D-glyceric acid; both CML and LL were detected in human urine and in glycated protein in vivo.

Wolff and Dean and their co-workers have presented evidence that glucose-derived fluorescent derivatives of proteins as well as fragmentation of the polypeptide chains is mediated by oxygen free radicals generated in the metal-ion-catalyzed oxidation of glucose (Wolff & Dean, 1987; Hunt et al., 1988). They proposed that oxidation of free glucose yields a ketoaldehyde, which subsequently reacts with the protein amino groups to yield an α-ketocarbinolamine derivative. Upon enolization, this would form a ketoimine, which could undergo further reactions to form brown fluorescent products (Figure 2). The associated fragmentation of the protein is presumably due to side reactions involving random attack of the protein by $\dot{O}H$ generated during glucose oxidation. The data presented by Dean and his colleagues are generally consistent with this hypothesis. However, the relative contributions of the oxidative and nonoxidative pathways of protein glycation are under current debate (Harding & Besnick, 1988; Wolff & Dean, 1988).

Because there is a correlation between advanced glycosylation end product (AGE) formation and the protein–protein cross-linking that occurs during aging, during cataract formation, and in tissue degeneration in diabetes, it is believed that glycation facilitates protein cross-linkage. On the basis of their isolation and characterization of 2-furoyl-4(5)-(2-furanyl)-1*H*-imidazole (FFI) from acid hydrolysates of AGE-containing proteins, Cerami et al. (1986, 1987) proposed an attractive mechanism for glycation-dependent cross-linking. However, a role of FFI in protein cross-linking is discounted by results of studies showing that FFI is not a component of glycated proteins but is an artifactual product of the procedures used in its isolation (Horiuchi et al., 1988; Njoroge et al., 1988). A plausible mechanism for the formation of FFI from unlinked glycated proteins has been presented (Njoroge et al., 1988). It is noteworthy that, in addition to glucose, reactions with fructose (Suárez et al., 1989), mannose and frucose (Davis et al., 1989), and glucose 6-phosphate (Beswick & Harding, 1987) have been shown to glycate proteins. Moreover, glycated proteins and adducts formed by the reactions of protein with simple aliphatic aldehydes and ketones (dihydroxyacetone etc.), formaldehyde, glycolaldehyde, and glyceraldehyde all appear to bind at the f-Alb receptor (so-called because it recognizes the formaldehyde-modified albumin) of rat liver sinusoidal cells (Horiuchi et al., 1986). The Schiff base adduct formed in the reactions of glycolaldehyde with protein lysyl residues also undergoes Amadori rearrangement to form an aldoamine, which reacts readily with the lysyl groups of a second molecule of protein to form protein–protein cross-links (Acharya & Manning, 1983).

Other Modifications. There is reason to believe that the protein modifications described in the foregoing sections mark enzymes for degradation. There are a number of other protein modifications that may also render protein susceptible to degradation, but until now their effects on degradation have not been investigated. These include thiolether derivatization of protein sulfhydryl groups, the methylation of carboxyl and amino acid groups, and a variety of enzymic and nonenzymic protein–protein cross-linking reactions.

(*a*) *Thiolether Formation*. Metal-catalyzed oxidation of polyunsaturated lipids leads to numerous products among which are α,β-unsaturated aldehydes and 4-hydroxyalkenals. These unsaturated aldehydes have been shown to react with the sulfhydryl groups of proteins to form thiolether adducts (reactions 14 and 15); the thiolether derivatives produced from

$$RCH_2OH{=}CHCHO + PrSH \rightarrow RCH_2CH(SPr)CH_2CHO \quad (14)$$

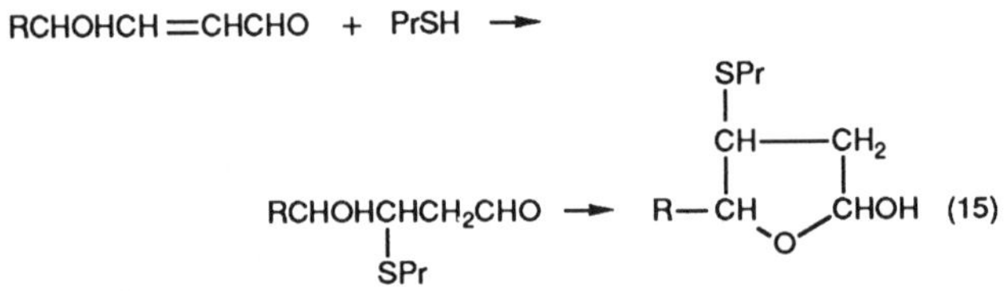

$$RCHOHCH{=}CHCHO + PrSH \rightarrow RCHOHCH(SPr)CH_2CHO \rightarrow R{-}CH{-}CH(SPr){-}CH_2{-}CHOH{-}O \text{ (ring)} \quad (15)$$

4-hydroxyalkenals may undergo cyclization to form hemiacetals. For reviews, see Esterbauer and Zolner (1989). Because the protein thiolethers contain a reactive aldehydic function, their derivatization with 2,4-dinitrophenylhydrazine has been used as a measure of protein-bound alkenals in tissue extracts (Benedetti et al., 1982). However, protein carbonyl groups are found in the metal-catalyzed oxidation of side chains of amino acid residues, in the oxidative cleavage of proteins by α-amido transfer mechanisms, and in the glycation of proteins (see above). Therefore, more specific assay procedures are needed in order to assess the origins of protein carbonyl groups.

Thiolether adducts of proteins are also formed by the addition of mevalonic acid derived isoprenyl groups to cysteine residues. These are the result of highly specific enzyme-catalyzed posttranslational modifications that have been shown to occur in yeast, Chinese hamster ovary cells, and HeLa cells (Farnsworth et al., 1989; Rilling et al., 1989). They are presumably involved as mating factors but have been implicated also in the modification of the $P21^{ras}$ gene product and in the modification of human lamin B.

(*b*) *Protein–Protein Cross-Linking*. A number of enzymic and nonenzymic posttranslational modifications of proteins lead to intra- and/or interprotein cross-links. As already noted, the exposure of proteins to oxygen free radicals can lead to intermolecular cross-linkage via tyrosyl–tyrosyl or –S–S bonds. Moreover, malondialdehyde produced in the free radical mediated oxidation of unsaturated fatty acids can facilitate protein–protein cross-links via Schiff base formation (reaction 13). Also, thiolether conjugates obtained by reaction of lipid-derived 4-hydroxyalkenals with protein sulfhydryl groups (reaction 15) can react further with ϵ-amino groups of lysyl residues ($PrNH_2$) of another protein to form Schiff base protein–protein cross-links (reaction 16). Likewise, Schiff

$$RCHOHC(SPr)HCH_2CHO + PrNH_2 \rightarrow RCHOHC(SPr)HCH_2CH{=}NPr + H_2 \quad (16)$$

base cross-links can arise by reaction of the ketoamine adducts generated in the glycation of proteins (Figure 2) with lysyl residues of another protein molecule (reaction 17) or by re-

$$PrNHCH_2C(O)(CHOH)_nCH_2OH + PrNH_2 \rightarrow PrNHCH_2C(NPr)(CHOH)_nCH_2OH + H_2O \quad (17)$$

action of protein carbonyl groups formed by oxygen free radical mediated oxidation of amino acid side chains with the ϵ-amino group of another protein molecule (reaction 18).

$$PrCHO + PrNH_2 \rightarrow PrCH{=}NPr + H_2O \quad (18)$$

Whereas Schiff base cross-links may be relatively labile, subsequent reduction of the Schiff base [viz., by ascorbate; see Tuma et al. (1984)] will yield stable secondary amine-linked conjugates.

In addition to the above nonenzymic protein–protein cross-linking reactions, one should consider also enzyme-catalyzed posttranslational cross-linking reactions, such as those catalyzed by transglutaminase (TG). This enzyme catalyzes substitution of the amide moiety of a protein glutaminyl residue [$PrC(O)NH_2$] with a primary amine (reaction 19) or with the ϵ-amino group of another protein to yield

$$PrC(O)NH_2 + RNH_2 \xrightarrow{TG} PrC(O)NHR + NH_3 \quad (19)$$

$$PrC(O)NH_2 + PrNH_2 \xrightarrow{TG} PrC(O)NHPr + NH_3 \quad (20)$$

$$2PrC(O)NH_2 + NH_2CH_2(CH_2)_nCH_2NH_2 \xrightarrow{TG} PrC(O)NHCH_2(CH_2)_nCH_2NHC(O)Pr + 2\ NH_3 \quad (21)$$

protein–protein conjugates in which the glutaminyl residue of one protein is linked to the lysyl residue of another protein (reaction 20). Moreover, when a simple primary amine is replaced by a polyamine (e.g., putrescine on spermidine), then protein cross-linking can occur by the polyamine bridge (reaction 21). The reactions catalyzed by transglutaminase presumably serve important physiological functions [for reviews, see Placentini et al. (1988), Fesus and Thomazy (1988), Birckbichler et al. (1988), and Lorand (1988)]; however, functionally incompetent protein complexes produced as a consequence of enzymic nonspecificity would be likely candidates for proteolytic degradation. Indeed, cross-linked protein aggregates produced by one ore more of the above cross-linking reactions accumulate during aging and in various disease states, i.e., under conditions where the levels of proteases are known to decline (Starke-Reed & Oliver, 1989; Taylor & Davies, 1989; Gracy et al., 1985a).

General Considerations. (*a*) *Regulatory Role of Covalent Modification Dependent Proteolysis.* As noted by Schimke (1970), the intracellular concentration of an enzyme is determined by the balance between its rates of synthesis and degradation. Whereas various mechanisms for the repression and depression of enzyme synthesis at both transcriptional and translational levels are well recognized, the mechanisms by which cells regulate the rates of degradation of one enzyme with respect to another are not well understood. That such regulation exists is evident from the fact that there are large differences in the rates of turnover of various enzymes and that the rate of degradation of one enzyme can vary independently of another by variations in the nutritional state of the organism. Because the covalent modification of an enzyme can target it for degradation, it is tacitly accepted that the rate of degradation and hence the intracellular level of an enzyme are regulated by covalent modification reactions. The fact that phosphorylation of proteins constitutes a major mechanism for the regulation of diverse biological functions leads to the proposition that phosphorylation might serve to regulate also the degradation of proteins. This suggestion is supported by the demonstration that the phosphorylated forms of some enzymes are more susceptible to proteolysis than their nonphosphorylated counterparts [for review, see Rivett (1986) and Holzer and Heinrich (1980)]; however, earlier evidence that protein phosphorylation is a critical step in the glucose-induced "proteolytic catabolite inactivation" of some enzymes in yeast is being reassessed in light of more recent findings (Holzer, 1989).

The possibility that metal-ion-catalyzed oxidative modification of enzymes can be used for the selective regulation of enzyme degradation is suggested by the demonstration that the substrates of enzymes can protect them from oxidative modification (Levine et al., 1981; Fucci et al., 1983; Stadtman & Wittenberger, 1985). Such metabolite effects could account for the differential responses of various enzyme levels to nutritional deficiencies. For example, starvation for nitrogen results in a decrease in the intracellular level of glutamate, one of the substrates that protects glutamine synthetase against oxidative damage and subsequent degradation. Teleologically, the regulation of an enzyme's susceptibility to degradation by the concentrations of its substrates makes good sense. In the absence of its substrate, an enzyme is biologically inactive, therefore, its selective degradation can have little effect on its biological functions, but by degradation it can yield amino acids and other products needed for the synthesis of other proteins that are essential for survival during this period of stress.

The metal-catalyzed oxidation of enzymes is definitely the basis of a regulatory mechanism in *Klebsiella aerogenes.* This organism uses two different metabolic pathways for growth on glycerol. Under anaerobic conditions, glycerol is metabolized via the pathway glycerol → dehydroxyacetone → dihydroxyacetone-P. When shifted for anaerobic to aerobic condition, the glycerol dehydrogenase that catalyzes the first step in this sequence is inactivated by a hydrogen peroxide dependent mechanism and is subsequently rapidly degraded (Chevalier et al., 1990). Concomitantly, a new set of enzymes is induced that catalyzes the reaction sequence glycerol → glycerol-3-P → dihydroxyacetone-P. In miroorganisms, the switch from anaerobic to aerobic environment often leads to repression of anaerobic genes, the degradation of the anaerobic gene products, and the derepression of a set of genes coding for enzymes needed for aerobic metabolism. The mechanism elucidated in the studies with *K. aerogenes* may therefore be a prototype for the regulation of shifts from anaerobic to aerobic metabolism.

(*b*) *Housekeeping Function.* A substantial fraction of the covalent modification of protein is the result of uncontrolled, spontaneous changes in protein structure or is due to chemical insults (viz., oxygen radical damage) they defy biological defenses. The selective degradation of these nonfunctional modified proteins is therefore the responsibility of intracellular proteases that can distinguish between "good" and "bad" proteins. The importance of this housekeeping function is evident from the results of studies showing that the intracellular accumulation of catalytically inactive or less active forms of several enzymes which occurs during aging is correlated with an age-dependent decrease in the intracellular levels of the neutral-alkaline proteinase (Starke-Reed & Oliver, 1989). As judged by the carbonyl content of the total cellular protein, it is evident that a considerable portion of the modified proteins that accumulate with age is due to metal-ion-catalyzed reactions. With the assumption that there is on the average one carbonyl group per 50-kDa polypeptide chain, it can be calculated that 40–50% of the cellular protein in old animals is oxidized. This agrees with the finding that the specific activity of several enzymes in cells from old animals is only about half that measured in cells from young animals. Obviously, still unanswered questions are as follows: (a) How much damaged protein (cellular garbage) can a cell tolerate? (b) Is the age-dependent intracellular accumulation of modified protein due to an increase in the rate of covalent modifications or to a decrease in the capacity to degrade the damaged protein? (c) Is the age-dependent decrease in neutral protease activity due to a deficiency in the rate of protease synthesis or to posttranslational modifications (viz., oxidation) of the proteases themselves?

Finally, it is evident that there is a need to develop sensitive techniques for the quantitation ov various kinds of covalently modified proteins in order to evaluate the contributions of each to the overall pool of damaged protein.

(c) How Are Modified Proteins Recognized as Targets for Degradation? In view of the extraordinary diversity of posttranslational modifications that target proteins for degradation and the fact that the modifications involve alteration of only one or a few amino acid residues in a given protein, it is surprising that so many different proteinases can distinguish between the modified and native forms of proteins. The suggestion that all modified proteins have in common a structural feature that serves as a recognition signal for proteolytic attack is not eaisly reconciled with the marked differences in peptide bond specificites of the many proteinases that preferentially degrade modified proteins. Perhaps even subtle perturbations of protein structure, as occurs with most posttranslational modification reactions, can lead to a multiplicity of recognition parameters and thus specify interactions with more than one type of proteinase. This appears to be the case with *E. coli* glutamine synthetase. The native form of this enzyme is almost completely resistant to attack by highly purified preparations of the multicatalytic proteinase from rat liver (Rivett, 1985a) and a neutral proteinase from *E. coli* (Roseman & Levine, 1987); however, following its modification by the ascorbate/Fe(III)/O_2 MCO system, the enzyme is readily degraded by both proteinases and is more susceptible to degradation by several secretory proteinases (Farber & Levine, 1982; Rivett, 1985b). Studies of the time course of the changes that occur during exposure to the MFO system (Rivett & Levine, 1990) revealed that the susceptibility to degradation by the multicatalytic proteinase is associated with the loss of two histidine residues per subunit and is independent of changes in the hydrophobicity of the enzyme. In contrast, the susceptibility to degradation by the *E. coli* proteinase is correlated with the conversion of the enzyme to a more hydrophobic form (Cervera & Levine, 1987; Levine, 1989).

REFERENCES

Acharya, A. S., & Manning, J. M. (1983) *Proc. Natl. Acad. Sci. U.S.A. 80*, 3590–3594.

Ahmed, M. U., Thorpe, S. R., & Baynes, J. W. (1986) *J. Biol. Chem. 261*, 4889–4894.

Ahmed, M. U., Dunn, J. A., Walla, M. D., Thorpe, S. R., & Baynes, J. W. (1988) *J. Biol. Chem. 263*, 8816–8821.

Amici, A., Levine, R. L., Tsai, L., & Stadtman, E. R. (1989) *J. Biol. Chem. 264*, 3341–3346.

Aruoma, O. I., & Halliwell, B. (1989) *FEBS Lett. 244*, 76–80.

Bachinger, H. P. (1987) *J. Biol. Chem. 262*, 17144–17148.

Bachmair, A., & Varshavsky, A. (1989) *Cell 56*, 1019–1032.

Bachmair, A., Finley, D., & Varshavsky, A. (1986) *Science 234*, 179–186.

Ballard, F. J., & Hopgood, M. F. (1976) *Biochem. J. 154*, 717–724.

Bateman, R. C., Jr., Youngblood, W. W., Busby, W. J., Jr., & Kiser, J. S. (1985) *J. Biol. Chem. 260*, 9088–9091.

Benedetti, A., Esterbauer, H., Ferrali, M., Fulceri, R., & Comporti, M. (1982) *Biochim. Biophys. Acta 711*, 345–356.

Beswick, H., & Harding, J. J. (1987) *Biochem. J. 246*, 761–769.

Birckbichler, P. J., Anderson, L. E., & Dell'orco, R. T. (1988) in *Advances in Post-Translational Modifications of Proteins and Aging* (Zappia, V., Galletti, P., Porta, R., & Wold, F., Eds.) pp 109–117, Plenum Press, New York.

Bond, J. S., & Offermann, M. K. (1981) *Acta Biol. Med. Ger. 40*, 1365–1374.

Brunauer, L. S., & Clarke, S. (1986) *J. Biol. Chem. 261*, 12538–12543.

Cerami, A., & Crabbe, M. J. C. (1986) *Trends Pharmacol. Sci. 7*, 271–274.

Cerami, A., Vlassara, H., & Brownlee, M. (1986) *J. Cell. Biochem. 30*, 111–120.

Cerami, A. C., Vlassara, H., & Brownlee, M. (1987) *Sci. Am. 256*, 90–96.

Cervera, J., & Levine, R. L. (1987) *FASEB J. 2*, 2591–2595.

Chau, V., Tobias, J. W., Bachmair, A., Marriott, D., Ecker, D. J., Gonda, D. K., & Varshavsky, A. (1989) *Science 243*, 1576–1583.

Chevalier, M., Lin, E. C. C., & Levine, R. L. (1990) *J. Biol. Chem. 265*, 40–46.

Chin, D. T., Kuehl, L., & Rechsteiner, M. (1982) *Proc. Natl. Acad. Sci. U.S.A. 79*, 5857–5861.

Chio, K. S., & Tappel, A. L. (1969) *Biochemistry 8*, 2827–2833.

Ciechanover, A., & Schwartz, A. L. (1989) *Trends Biochem. Sci. 14*, 483–488.

Ciechanover, A., Heller, H., Elias, S., Hass, A. L., & Hershko, A. (1980) *Proc. Natl. Acad. Sci. U.S.A. 77*, 1365–1368.

Ciechanover, A., Finley, D., & Varshavsky, A. (1984) *Cell 37*, 57–66.

Clarke, S. (1985) *Annu. Rev. Biochem. 54*, 479–506.

Climent, I., Tsai, L., & Levine, R. L. (1989) *Anal. Biochem. 182*, 226–232.

Cook, J., & Chock, P. B. (1988) *BioFactors 1*, 133–146.

Cooper, B., Creeth, M., & Donald, A. S. R. (1985) *Biochem. J. 228*, 615–626.

Creeth, J. M., Cooper, B., Donald, A. S. R., & Clamp, J. R. (1983) *Biochem. J. 211*, 323–332.

Davies, K. J. A. (1987) *J. Biol. Chem. 262*, 9895–9901.

Davies, K. J. A., & Goldberg, A. L. (1987) *J. Biol. Chem. 262*, 8227–8234.

Davies, K. J. A., & Lin, S. W. (1988) *Free Radical Biol. Med. 5*, 215–223.

Davies, K. J. A., Delsignore, M. E., & Lin, S. W. (1987) *J. Biol. Chem. 282*, 9902–9907.

Davis, L. J., Hakim, G., & Rossi, C. A. (1989) *Biochem. Biophys. Res. Commun. 160*, 362–366.

Dean, R. T. (1987) *FEBS Lett. 220*, 278–282.

Dean, R. T., & Pollak, J. K. (1985) *Biochem. Biophys. Res. Commun. 126*, 1082–1089.

Dice, J. F., Dehlinger, P. J., & Schimke, R. T. (1973) *J. Biol. Chem. 248*, 4220–4228.

Dreyfus, J. C., Kahn, A., & Schapira, F. (1978) *Curr. Top. Cell. Regul. 14*, 143–297.

Esterbauer, H., & Zollner, H. (1989) *Free Radical Biol. Med. 7*, 197–203.

Farber, J. M., & Levine, R. L. (1982) *Fed. Proc., Fed. Am. Soc. Exp. Biol. 41*, 865.

Farber, J. M., & Levine, R. L. (1986) *J. Biol. Chem. 261*, 4574–4578.

Farnsworth, C. C., Wolda, S. L., Gelb, M. H., & Glomset, J. A. (1989) *J. Biol. Chem. 264*, 20422–20429.

Fesus, L., & Thomazy, V. (1988) in *Advances in Post-Translational Modifications of Proteins and Aging* (Zappia, V., Galletti, P., Porta, R., & Wold, F., Eds.) pp 119–134, Plenum Press, New York.

Fischer, G., & Bang, H. (1984) *Biomed. Biochim. Acta 43*, 1101–1111.

Fischer, G., Heins, J., & Barth, A. (1983) *Biochim. Biophys. Acta 742*, 452–462.

Fischer, G., Bang, H., Berger, E., & Schellenberger, A. (1984) *Biochim. Biophys. Acta 791*, 87–97.

Fogelman, A. M., Shechter, I. S., Hokom, M., Child, J. S., & Edwards, P. A. (1980) *Proc. Natl. Acad. Sci. U.S.A. 77*, 2214–2218.

Francis, G. L., & Ballard, F. J. (1980) *Biochem. J. 186*, 581–590.

Fucci, L., Oliver, C. N., Coon, M. J., & Stadtman, E. R. (1983) *Proc. Natl. Acad. Sci. U.S.A. 80*, 1521–1525.

Garrison, W. M., Jayko, M. E., & Bennett, W. (1962) *Radiat. Res. 16*, 483–502.

Geiger, T., & Clarke, S. (1987) *J. Biol. Chem. 262*, 785–794.

Ghoda, L., Van Daalen Wetters, T., Macrae, M., Ascherman, D., & Coffino, P. (1989) *Science 242*, 1493–1495.

Goldberg, A. L., & Dice, J. F. (1974) *Annu. Rev. Biochem. 48*, 835–869.

Goldberg, A. L., & St. John, A. C. (1976) *Annu. Rev. Biochem. 45*, 747–803.

Gonda, D. K., Bachmair, A., Wünning, I., Tobias, J. W., Lane, W. S., & Varshavsky, A. (1989) *J. Biol. Chem. 264*, 16700–16712.

Gracy, R. W., Yüksel, K. Ü., Chapman, M. L., Cini, J. K., Jahani, M., Lu, H. S., Gray, B., & Talent, J. M. (1985a) in *Modifications of Proteins during Aging*, pp 1–18, Alan R. Liss, New York.

Gracy, R. W., Chapman, M. L., Cini, J. E., Jahani, M., Tollefsbol, T. O., & Yüksel, K. Ü. (1985b) in *Molecular Biology of Aging* (Woodhead, A. D., Blackett, A. D., & Hollaender, A., Eds.) pp 427–442, Plenum Press, New York.

Harding, J. J. (1985) *Adv. Protein Chem. 37*, 247–334.

Harding, J. J., & Beswick, H. T. (1988) *Biochem. J. 249*, 617–618.

Helfman, P. M., & Bada, J. L. (1975) *Proc. Natl. Acad. Sci. U.S.A. 72*, 2891–2894.

Helfman, P. M., & Bada, J. L. (1976) *Nature (London) 262*, 279–281.

Hershko, A., Heller, H., Eytan, E., Kaklij, G., & Rose, I. A. (1984) *Proc. Natl. Acad. Sci. U.S.A. 81*, 7201–7205.

Hershko, A., Heller, H., Eytan, E., & Reiss, Y. (1986) *J. Biol. Chem. 261*, 11992–11999.

Holzer, H. (1989) *Cell. Biol. Rev. 21*, 306–319.

Holzer, H., & Heinrich, P. C. (1980) *Annu. Rev. Biochem. 49*, 63–91.

Hood, W., DeLa Morena, E., & Grisolia, S. (1977) *Acta Biol. Med. Ger. 36*, 1667–1672.

Horiuchi, S., Murakami, M., Takata, K., & Morino, Y. (1986) *J. Biol. Chem. 261*, 4962–4966.

Horiuchi, S., Shiga, M., Araki, N., Takata, K., Saitoh, M., & Morino, Y. (1988) *J. Biol. Chem. 263*, 18821–18826.

Hunt, J. W., Simpson, J. A., & Dean, R. T. (1988) *Biochem. J. 250*, 87–93.

Johnson, B. A., Murray, E. D., Clarke, S., Glass, D. B., & Aswad, D. W. (1987) *J. Biol. Chem. 262*, 5622–5629.

Lang, K., Schmid, F. X., & Fischer, G. (1987) *Nature (London) 329*, 268–270.

Lee, Y. S., Park, S. C., Goldberg, A. L., & Chung, C. H. (1988) *J. Biol. Chem. 263*, 6643–6646.

Levine, R. L. (1983) *J. Biol. Chem. 258*, 11823–11827.

Levine, R. L. (1989) *Cell Biol. Rev. 21*, 347–360.

Levine, R. L., Oliver, C. N., Fulks, R. M., & Stadtman, E. R. (1981) *Proc. Natl. Acad. Sci. U.S.A. 78*, 2120–2124.

Levitzky, A., Anbar, M., & Berger, A. (1967) *Biochemistry 6*, 3757–3765.

Lorand, L. (1988) in *Advances in Post-Translational Modifications of Proteins and Aging* (Zappia, V., Galletti, P., Porta, R., & Wold, F., Eds.) pp 79–94, Plenum Press, New York.

Maier, K. L., Matejkova, E., Hinze, H., Leuschel, L., Weber, H., & Beck-Speier, I. (1989) *FEBS Lett. 250*, 221–226.

Marcillat, O., Zhang, Y., Lin, S. W., & Davies, K. J. A. (1988) *Biochem. J. 254*, 677–683.

Masters, P. M., Bada, J. L., & Zigler, J. S., Jr. (1978) *Proc. Natl. Acad. Sci. U.S.A. 75*, 1204–1208.

McKerrow, J. H. (1979) *Mech. Ageing Dev. 10*, 371–377.

McLendon, G., & Radany, E. (1978) *J. Biol. Chem. 253*, 6335–6337.

Midelfort, C. F., & Mehler, A. H. (1972) *Proc. Natl. Acad. Sci. U.S.A. 69*, 1816–1819.

Njoroge, F. G., Fernandes, A. A., & Monnier, V. M. (1988) *J. Biol. Chem. 263*, 10646–10652.

Offerman, M. K., McDay, M. J., Marsh, M. W., & Bond, J. S. (1984) *J. Biol. Chem. 259*, 8886–8891.

Ogasahara, K., Tsunasawa, S., Soda, Y., Yutani, K., & Sugino, Y. (1985) *Eur. J. Biochem. 150*, 17–21.

Oliver, C. N., Levine, R. L., & Stadtman, E. R. (1980) in *Metabolic Interconversion of Enzymes* (Holzer, H., Ed.) pp 259–268, Springer-Verlag, Berlin.

Oliver, C. N., Levine, R. L., & Stadtman, E. R. (1982) in *Experiences in Biochemical Perception* (Ornston, L. N., & Sligar, S. G., Eds.) pp 233–249, Academic Press, New York.

Oliver, C. N., Ahn, B., Wittenberger, M. E., & Stadtman, E. R. (1985) in *Cellular Regulation and Malignant Growth* (Ebasi, S., Ed.) pp 320–331, Springer-Verlag, Berlin.

Oliver, C. N., Levine, R. L., & Stadtman, E. R. (1987) *J. Am. Geriat. Soc. 35*, 947–956.

Pine, M. J. (1972) *Annu. Rev. Microbiol. 26*, 103–126.

Placentini, M., Cerű-Argento, M. P., Farrace, M. G., & Autuori, F. (1988) in *Advances in Post-Translational Modifications of Proteins and Aging* (Zappia, V., Galletti, P., Porta, R., & Wold, F., Eds.) pp 185–196, Plenum Press, New York.

Rechsteiner, M. (1988) *Ubiquitin*, Plenum Press, New York.
Rilling, H. C., Bruenger, E., Epstein, W. W., & Kandutsch, A. A. (1989) *Biochem. Biophys. Res. Commun. 163*, 143–148.
Rivett, A. J. (1985a) *J. Biol. Chem. 260*, 12600–12606.
Rivett, A. J. (1985b) *Arch. Biochem. Biophys. 243*, 624–632.
Rivett, A. J. (1986) *Curr. Top. Cell. Regul. 28*, 291–337.
Rivett, A. J., & Levine, R. L. (1990) *Arch. Biochem. Biophys.* (in press).
Rivett, A. J., Roseman, J. A., Oliver, C. N., Levine, R. L., & Stadtman, E. R. (1985) in *Intracellular Protein Catabolism* (Khairallah, E. A., Bond, J. S., & Bird, J. W. C., Eds.) pp 317–328, Alan R. Liss, New York.
Robinson, A. B., & Rudd, C. J. (1974) *Curr. Top. Cell. Regul. 8*, 247–294.
Rogers, S., Wells, R., & Rechsteiner, M. (1986) *Science 234*, 179–186.
Rogers, S. W., & Rechsteiner, M. (1988) *J. Biol. Chem. 263*, 19850–19862.
Roseman, J. E., & Levine, R. L. (1987) *J. Biol. Chem. 262*, 2101–2110.
Rote, K., Rogers, S., Pratt, G., & Rechsteiner, M. (1989) *J. Biol. Chem. 264*, 9772–9779.
Saus, J., Timoneda, J., Hernando-Yago, J., & Grisolia, S. (1982) *FEBS Lett. 143*, 225–227.
Schimke, R. T., & Doyle, D. (1970) *Annu. Rev. Biochem. 39*, 929–976.
Schimke, R. T., & Bradley, M. O. (1975) in *Proteases and Biological Control* (Reich, E., Rifkin, D. B., & Shaw, E., Eds.) pp 515–530, Cold Spring Harbor Laboratory, New York.
Schlesinger, M. J., & Hershko, A. (1988) *The Ubiquitin System*, Cold Spring Harbor Laboratory, New York.
Schuessler, H., & Schilling, K. (1984) *Int. J. Radiat. Biol. 45*, 267–281.
Segal, H. L., Winkler, J. R., & Miyagi, M. P. (1974) *J. Biol. Chem. 249*, 6364–6365.
Speiser, S., & Etlinger, J. D. (1982) *J. Biol. Chem. 257*, 14122–14127.
Stadtman, E. R. (1986) *Trends Biochem. Sci. 11*, 11–12.
Stadtman, E. R. (1988a) *J. Gerontol. Biol. Sci. 43*, B112–B120.
Stadtman, E. R. (1988b) *Exp. Gerontol. 23*, 237–347.
Stadtman, E. R., & Wittenberger, M. E. (1985) *Arch. Biochem. Biophys. 239*, 379–387.
Starke-Reed, P., & Oliver, C. N. (1989) *Arch. Biochem. Biophys. 275*, 559–567.
Suárez, G., Rajaram, R., Oronsky, A. L., & Gawinowicz, M. A. (1989) *J. Biol. Chem. 264*, 3674–3679.
Swallow, A. J. (1960) in *Radiation Chemistry of Organic Compounds* (Swallow, A. J., Ed.) pp 211–224, Pergamon Press, New York.
Taborsky, G. (1973) *Biochemistry 12*, 1341–1348.
Takahashi, N., Hayano, T., & Suzuki, M. (1989) *Nature (London) 337*, 473–475.
Takata, K., Horiuchi, S., Araki, N., Shiga, M., Saitoh, M., & Morino, Y. (1988) *J. Biol. Chem. 263*, 14819–14825.
Taylor, A., & Davies, K. J. A. (1987) *Free Radical Biol. Med. 3*, 371–377.
Tuma, D. J., Donohue, T. M., Jr., Medina, V. A., & Sorrell, M. F. (1984) *Arch. Biochem. Biophys. 234*, 377–381.
Vlassara, H., Brownlee, M., & Cerami, A. (1986) *Proc. Natl. Acad. Sci. U.S.A. 82*, 5588–5592.
Vlassara, H., Valinsky, J., Brownlee, M., Cerami, C., Nishimoto, S., & Cerami, A. (1987) *J. Exp. Med. 166*, 539–549.
Wilkinson, K. D., Urban, M. K., & Haas, A. L. (1980) *J. Biol. Chem. 255*, 7529–7532.
Wolff, S. P., & Dean, R. T. (1987) *Biochem. J. 245*, 243–250.
Wolff, S. P., & Dean, R. T. (1988) *Biochem. J. 249*, 618–619.
Wolff, S. P., Garner, A., & Dean, R. T. (1986) *Trends Biochem. Sci. 1*, 28–31.
Yuan, P. M., Talent, J. M., & Gracy, R. W. (1981) *Mech. Ageing Dev. 17*, 151–162.
Yutani, K., Ogasahara, K., Sugino, Y., & Matsushiro, A. (1977) *Nature (London) 267*, 274–275.
Yutani, K., Ogasahara, K., Aoki, K., Kakuno, T., & Sugino, Y. (1984) *J. Biol. Chem. 259*, 14076–14081.

CELL GROWTH AND REGULATION

Chapter 25

Positive and Negative Controls on Cell Growth[†]

Robert A. Weinberg

Whitehead Institute for Biomedical Research, Nine Cambridge Center, Cambridge, Massachusetts 02142, and Department of Biology, Massachusetts Institute of Technology, Cambridge, Massachusetts 02139

Received May 15, 1989; Revised Manuscript Received June 12, 1989

The oncogenes that have attracted great attention over the past decade would seem to provide us with much of the explanation at the molecular level of the origins of cancer. We now count as many as 50 distinct cellular oncogenes, many of which are able to induce at least some of the cell phenotypes associated with the neoplastic state. Of these, about a dozen are directly implicated in the etiology of human cancer. The remainder have been found associated with a variety of mammalian and avian retroviruses, having been abstracted from the cellular genome by these transducing viruses. Yet others in this large group have been uncovered by virtue of nucleic acid homologies with previously known oncogenes (Bishop, 1983; Varmus, 1984). Taken together, these genes leave an impression of a richly complex cellular growth regulatory circuitry that can be perturbed at many points by the actions of oncogenes and their encoded proteins.

In spite of this already well-documented complexity, there are reasons to believe that oncogenes can provide at best only part of the explanation of the molecular and genetic mechanisms of cancer. These reasons stem from two attributes that are shared by all these genes. First, oncogenes act in one way or another as positive effectors of cell growth. As such, the oncogene paradigm overlooks the possible existence of an equally complex and important cellular growth regulatory network that is dedicated to suppressing or constraining cell growth. Such negative regulatory genes, to the extent that they exist, could become involved in cancer when they are lost or inactivated; such loss would remove a normally existing brake or constraint on the cell's growth, thereby triggering the runaway growth of neoplasia.

Second, oncogenes invariably arise as consequences of somatic alterations of the target cell genome. Thus, in human tumors, the cellular oncogenes studied have been found to arise as a consequence of well-defined somatic mutations of normal growth regulating genes often termed protooncogenes. A minority of human tumors acquire viral oncogenes following infectious events occurring in one or another target organ; such infections also represent somatic changes in the cell genome. Yet it is clear that inborn genetic determinants can also strongly affect the probability of tumor formation during an organism's lifetime. The existence of inherited cancer genes that are passed through the germline is also not addressed by the oncogene paradigm.

A class of genes is described here that promises to address both deficiencies. Altered forms of one of these genes are indeed transmitted through the germline and can strongly affect tumor incidence. Moreover, these genes behave as if they were negative regulators of growth, thus acting in a fashion directly opposite to that of the well-studied oncogenes. Genes in this new class have been termed "tumor-suppressor" genes, "recessive oncogenes", "antioncogenes", "emerogenes", and "growth-suppressor" genes (Klein, 1987). None of these terms is ideal; indeed, because we do not understand the normal functions of these genes, use of any specific term is a bit presumptuous. Most acceptable is the term growth-suppressor gene, which will be used here.

Evidence of Genetic Loss during Tumor Formation

Two lines of work have converged on the idea that loss of genetic functions underlies at least some of the steps of carcinogenesis. The first of these derived from the use of somatic cell hybridization, which was achieved by fusion of cells in monolayer culture. The resulting genetic hybrids provided the first indication that many tumor cells lack critical genetic elements that are present and functional in their normal counterparts. Upon fusion of tumor cells with normal cells, the great majority of resulting hybrids are found to be non-tumorigenic. Such loss of tumorigenicity has been observed for a wide range of inter- and intraspecific cell hybrids (Harris, 1988), including hybrids of tumor cells fused with normal cells from the same tissue. These results imply that the normal parent used in the hybridization process is contributing genes

[†]Some of the work described here was supported by Outstanding Investigator Grant 5-R35-CA39826 of the National Cancer Institute and by American Cancer Society Grant CD355 to R.A.W., who is an American Cancer Research Professor.

to the tumorigenic partner that impose normal growth control on the latter. One presumes that these normal genes replace similar or identical genes that were previously lost from the tumor cell during its progression from normalcy to malignancy.

Results of this sort are difficult to interpret with precision, as they involve the comingling of two entire cell genomes. More focused conclusions come from work showing that only a small number of genes derived from the normal parent are required for the reimposition of normal growth control on the tumorigenic partner. Such observations are made possible by the karyotypic instability of certain cell hybrids, which tend to lose preferentially the chromosomes of one or another parent. In hybrids in which the normal parent's chromosomes are preferentially lost, one may often observe the reemergence of tumorigenic cell clones. This reacquisition of a tumorigenic phenotype can be correlated with the loss of one or another specific chromosome originating with the normal parent. This suggests the presence of a gene or small number of genes on this chromosome which act(s) to impose normal growth control on the tumor cell (Stanbridge et al., 1981).

Even more persuasive than these correlative experiments are those involving microcell fusions that enable an investigator to introduce selectively a specific normal chromosome into the tumorigenic cell. In the case of human HeLa cells, loss of tumorigenicity is observed upon introduction of a normal chromosome 11. Indeed, a single copy of this normal chromosome suffices to restore more normal growth control (Saxon et al., 1986). The nature of the chromosome 11 associated growth controlling gene(s) remains obscure.

The second line of work underscoring the importance of genetic loss during tumor pathogenesis has involved study of a small set of relatively rare tumors—retinoblastoma and Wilms' tumor. Both are tumors of children, and their biology suggests that the tumor cells represent embryonic cell types that have not succeeded in their normal progression to a differentiated state.

Here the indications of genetic loss are even more direct: karyological analysis of certain tumors of each type has revealed the presence of interstitial deletions that repeatedly affect specific chromosomal bands. In the case of retinoblastoma, deletion of 13q14 is occasionally observed (Yunis & Ramsay, 1978); Wilms' tumors often show deletion of chromosomal material around band 11q13 (Riccardi et al., 1978). These occasionally observed karyotypic abnormalities must underrepresent the true frequency of genetic loss. After all, millions of DNA base pairs need be lost before the microscopic morphology of a chromosome is affected. Many other deletions involving loss of smaller DNA segments may occur much more frequently but escape detection by the karyologist. In the case of retinoblastoma, these various chromosome 13q14 alterations have been presumed to involve a specific locus termed Rb.

The Case of Retinoblastoma

Retinoblastoma tumors are relatively rare, occurring in only one of 20 000 children, and are almost always seen from birth up to the age of 5. The disease occurs as two distinct clinical entities. The "familial" form of retinoblastoma is seen in children having a similarly afflicted parent. These children usually show multiple independent foci of tumor formation affecting both eyes. In "sporadic" retinoblastoma, only a single focus of tumor formation is observed, and then in a child having no family history of this disease.

Alfred Knudson provided a unifying genetic explanation of these two forms of disease by proposing that both depend upon a common genetic mechanism involving two critical mutations (Knudson, 1971). He argued that the appearance of sporadic retinoblastomas depends upon two somatic mutations sustained by a single retinal stem cell, the descendants of which then proliferate to form the tumorigenic cell clone. In the case of familial retinoblastoma, he proposed that one of the required mutations is already present in the conceptus, having been acquired from sperm or egg. This mutation, by necessity, is implanted in all cells of the developing retina. Knudson suggested that the second of the required mutations is then sustained in one or another retinal cell, creating the doubly mutated cell whose genotype now favors neoplastic growth. The occurrence of doubly mutated retinal cells is obviously strongly favored in those children who already carry one mutant allele in each of their retinal cells. In this sense, one can say that the presence of such an allele from the moment of conception strongly predisposes to tumor onset.

We now realize that the two target genes postulated by Knudson are in fact the two copies of the Rb gene on chromosome 13 (Sparkes et al., 1983; Godbout et al., 1983; Cavenee et al., 1983; Dryja et al., 1984). The mutations leading to retinoblastoma involve molecular changes that cause loss of Rb gene function. This was already apparent from the aforementioned karyological studies that showed frequent deletion of the chromosomal region carrying the Rb locus. Because loss of both Rb gene copies favors malignant cell transformation, this would suggest, as mentioned earlier, that the intact Rb gene functions normally to suppress the runaway growth of retinal cells. Indeed, a single intact copy would appear to suffice to orchestrate normal cell growth, since the bulk of the retina in a child afflicted with familial retinoblastoma appears to be perfectly normal.

The Rb gene has been isolated by molecular cloning (Friend et al., 1986; Lee et al., 1987; Fung et al., 1987; Bookstein et al., 1988). It is a large gene of 190 kilobases that is affected by gross structural alterations in as many as 30% of tumors. These estimates of the frequency of Rb gene alterations in retinoblastomas have been obtained by Southern blotting analysis of genomic DNAs, a procedure that provides a relatively insensitive assay of mutation. Recent studies of Rb alleles found in tumor cells and having a grossly normal structure have revealed subtle changes in gene sequences including point mutations that affect splicing of Rb mRNA precursors and result in the deletion of exons from the Rb mRNA (Horowitz et al., 1989). The most sensitive assays of gene inactivation are those designed to analyze the Rb protein through use of immunoprecipitation. One recent study has revealed the absence of the Rb protein from 18 out of 18 retinoblastoma cell lysates although this protein is readily demonstrable in normal retinoblasts, the apparent precursors of the tumor cells (J. Horowitz and R. A. Weinberg, unpublished results). Taken together, these studies suggest that inactivation of both Rb alleles occurs during the genesis of virtually all retinoblastomas. A functional proof that the cloned DNA indeed represents the Rb gene has been produced recently through the introduction of the cloned gene into retinoblastoma and osteosarcoma cells (Huang et al., 1988). These cells lose their tumorigenicity following reacquisition of the gene; this in turn strongly supports the conclusion that display of the tumor phenotype depends upon loss of this gene from these cells.

Puzzles and Paradoxes Surrounding the Rb Gene

The Rb gene is involved in far more than the genesis of retinoblastomas. This was already apparent from long-term followups of children cured of the familial type of retinoblastoma early in life. These children suffer various types of sarcoma later in life at rates greatly above normal (Draper et al., 1986; Toguchida et al., 1988). Accordingly, Rb hemizygosity predisposes to tumors in at least two developmental lineages. Moreover, use of the cloned Rb gene as probe has also shown RB inactivation occurring frequently in osteosarcomas and soft tissue sarcomas that appear to arise purely from somatic mutational accidents (Friend et al., 1987; Weichselbaum et al., 1988).

This picture is further complicated by the more recent findings that Rb inactivation is also found in a very high proportion of small cell lung carcinomas, in about one-third of bladder carcinomas, and in a smaller proportion of mammary carcinomas (Harbour et al., 1988; Lee et al., 1988; T'Ang et al., 1988). These gene losses all appear due to somatic mutation. This creates an interesting paradox since these particular tumor types are not known to occur with increased rates in children born with defective Rb alleles. Thus, we have no idea why the loss of the surviving Rb allele in these children, which must occur with substantial frequency throughout their bodies, does not cause them to contract these other types of tumors at unusually high rates.

The Rb gene is expressed in a wide range of tissues throughout the body (Bernards et al., 1989). Such expression would suggest an important role of Rb in regulating the proliferation of a wide variety of cell types. Once again a puzzle arises, in that Rb inactivation triggers tumor formation in only a relatively narrow range of tissue types. This might suggest that intactness of the Rb gene is not critical to the growth regulation of many of the cell types in which it is expressed.

Yet a third puzzle surrounding the Rb gene concerns the species distribution of the retinoblastomas: it has been seen only in human beings even though there have been extensive opportunities to observe the disease in many other mammals. Not unexpectedly, the Rb gene is represented in the genomes of all other mammals and even more distantly related chordates. The mouse Rb homologue encodes a protein that is 90% identical in its amino acid sequence to its human counterpart (Bernards et al., 1989). Clearly Rb is an important regulator of growth in other mammals which must also sustain occasional homozygous losses of the Rb gene. Perhaps some idiosyncrasy of human embryogenesis makes our retinas particularly susceptible to tumor formation; alternatively, the molecular biology of our retinal cells may cause them to be especially dependent on Rb function for their growth regulation.

The Rb-Encoded Protein

The gene product of Rb is a 105-kilodalton nuclear phosphoprotein that has DNA binding properties (Lee et al., 1987; Whyte et al., 1988). These facts focus one's thinking on mechanisms of action involving regulation of transcription. Although plausible, such mechanistic speculations are still premature, since no specific transcriptional factor or target gene is known to be affected by this protein, which is termed p105-Rb.

One year ago, insight into the functioning of this protein came in a dramatic fashion and from a totally unexpected quarter. The work providing this insight originated in laboratories studying the molecular mechanisms of cell transformation by DNA tumor viruses, specifically human adenovirus type 5. This virus induces upper respiratory tract infections in humans, but in certain rodents it can act as a potently tumorigenic agent. Its ability to induce tumors is traceable to two oncogenes carried in the viral genome, termed E1A and E1B.

The E1A oncogene is the most intensively studied of these by virtue of its multiple effects on cells. When introduced in rodent embryo cells, it has an "immortalizing" power that enables these cells to grow indefinitely in culture, in this way avoiding the senescence that is the fate of normal embryo cells. The E1A oncogene can also collaborate with a *ras* oncogene in converting a fully normal embryo cell into a tumorigenic one (Ruley, 1983). Finally, E1A-encoded proteins can act as transcriptional regulators, inducing the expression of some viral and cellular genes and repressing yet others (Berk, 1986).

These multiple powers of the E1A oncogene required a molecular explanation. Two laboratories studying E1A function found an important clue in the observation that the E1A-encoded proteins are able to form stable complexes with as many as six distinct cellular proteins (Yee & Branton, 1985; Harlow et al., 1986). Thus, antibodies specifically reactive with the viral oncoproteins precipitate both these proteins and cellular proteins from adenovirus-transformed cells. The presence of these six cellular proteins in the immunoprecipitates is not due to some adventitious cross-reactivity of the precipitating antibodies, which recognizes none of these cellular proteins in lysates of cells not transformed by adenovirus.

Such observations are not without precedent in the field of DNA tumor virology. The SV40 large T antigen oncoprotein is known to complex with the host cell protein, p53 (Lane & Crawford, 1979; McCormick & Harlow, 1980). Similarly, the middle T oncoprotein of polyomavirus complexes with and activates the cellular pp60^{src} protein (Courtneidge & Smith, 1983; Bolen et al., 1984). Clearly, tumor viruses have evolved an ability to alter host cell metabolism by developing proteins that are able to complex with specific target proteins of host cell origin. One assumes that the functioning of these host cell targets is perturbed following complex formation with the viral oncoproteins.

In the case of the adenovirus E1A protein, its wide range of activities may be attributable to its ability to complex with six or more cellular proteins, each of which may serve as a controller of one or another important cellular regulatory pathway. The results of last summer revealed that one of these host cell targets is p105-Rb, the same protein that is lost from retinoblastomas and osteosarcomas through genetic alteration of the Rb gene (Whyte et al., 1988).

This confluence of two apparently unrelated areas of research revealed a fully unexpected confrontation between the gene product of an oncogene and a gene product ostensibly involved in contraining normal cellular growth. The first of these acts as a growth agonist while the second would seem to act as a growth antagonist. One simple model would state that, by complexing with p105-Rb, the E1A oncoprotein is able to neutralize its growth-suppressing effects and, in so doing, liberate the cell from previously operative constraints on growth. All this encourages those who would call Rb an "antioncogene", but this term still falls short, since it would not seem that the normal function of the Rb gene is to counteract oncogene action.

Coming close on the heels of the discovery of the E1A·p105-Rb complex, findings from two other groups extended this paradigm by demonstrating that this p105-Rb also forms complexes with two other, distinct viral oncoproteins. The SV40 large T oncoprotein (SV40 LT) forms a complex with p105-Rb in addition to its previously mentioned ability to complex with the cellular p53 protein (De Caprio et al., 1988). In addition, the E7 oncoprotein of human papillomavirus type 16 also has binding affinity for p105-Rb (Dyson et al., 1989). This latter virus is implicated as an etiologic agent in a substantial proportion of cervical carcinomas.

These more recent findings attach even greater significance to p105-Rb. The three oncogene-encoded proteins are structurally very distinct and are specified by viruses that otherwise have little in common. This ability of their respective oncoproteins to complex with p105-Rb would seem to have arisen through some process of convergent evolution governed by the central position of p105-Rb in regulating cell growth and the importance of its deregulation to various viral growth cycles. Thus, by altering p105-Rb function, these viral oncoproteins may allow a quiescent cell to progress to a more active growth state that results in a more hospitable intracellular milieu for virus replication.

Biological tests had previously shown that these three viral oncogenes function analogously in cell transformation. Like E1A, the other viral oncogenes act to immortalize embryo cells. In addition, like E1A, the SV40 LT and HPV E7 oncogenes can each collaborate with a *ras* oncogene to effect the conversion of embryo fibroblasts to a tumorigenic state (Land et al., 1983; Phelps et al., 1988). Hence, the shared physiological functions of these three oncogenes can now be traced to a common biochemical substrate.

One final set of data underscores the relevance and importance of these complexes to the transforming powers of the oncogene proteins. Mutations of the E1A and SV40 LT oncogenes that knock out their transforming abilities simultaneously ablate their ability to complex with p105-Rb (De Caprio et al., 1988; Whyte et al., 1989). This connection between complex formation and transforming ability is highlighted most dramatically by a simple point mutation in SV40 LT that wipes out both traits with one blow (De Caprio et al., 1988). Accordingly, these complexes cannot be dismissed as adventitious aggregation artifacts; instead, they clearly reflect important functional interactions.

Models of p105-Rb Function

These various transforming mechanisms involving either alterations of the Rb gene or perturbation of its encoded protein force one to confront the role of gene and protein in the normal cell. How indeed do these act to constrain normal cell proliferation? And what are the precise consequences of removing p105-Rb from the cell or altering its function through complex formation?

Two quite different models come to mind. The first holds that, in nongrowing cells, p105-Rb acts as a suppressor of growth, perhaps by repressing the expression of genes essential for proliferation. Upon receiving mitogenic stimuli, signals are conveyed to p105-Rb that neutralize its function and thus allow the derepression of functions essential for growth. These signals impinging upon p105-Rb might include cellular proteins that act analogously to the viral oncoproteins described earlier. For example, increased levels of the cellular *myc* protein are induced by mitogens, and the *myc* protein is said to have certain structural and functional similarities with the viral oncoproteins described earlier (Phelps et al., 1988; Moran, 1988; Ralston & Bishop, 1983; Figge et al., 1988; Vousden & Jat, 1989). Thus, a *myc*-like protein could interact with p105-Rb and act to depress growth functions.

Following an alternative model, p105-Rb sits in an inactive state in growing cells. When these cells encounter growth-inhibitory signals in their environment, such signals may be transduced to the nucleus where they activate p105-Rb molecules that then proceed to down-regulate cellular functions including the transcription of genes essential for continued growth. Currently available evidence does not allow us to choose one of these models over the other.

Two other biochemical/molecular phenomena merit mention. First, no evidence obtained to date directly implicates p105-Rb as a transcription regulator. The E1A protein that interacts with p105-Rb is undeniably a regulator of transcription, but site-directed mutagenesis indicates that this activity is assignable to domains of the E1A protein that are distinct from the regions involved in transformation and p105-Rb binding (Lillie et al., 1987). Second, p105-Rb represents a collection of distinct molecular species that differ in their state of phosphorylation. Since SV40 LT oncoprotein binds only to the unphosphorylated form of p105-Rb (Ludlow et al., 1989), this might suggest that this subfraction of the p105-Rb pool represents its biologically active form that mediates growth regulation.

Growth Suppressor Genes Masquerading as Oncogenes

The Rb gene would appear to be only one example of a large class of cellular genes that act analogously to down-regulate cell growth. Most of these genes are known only from indirect types of genetic analysis. However, it now appears that, unbeknownst to us, two of these antigrowth genes have been in the hands of molecular biologists for many years, each masquerading as a growth-stimulating oncogene.

The first of these is *erb*A, an oncogene known originally from its presence in the genome of avian erythroblastosis virus. The *erb*A oncogene resides in this viral genome with a second viral oncogene, *erb*B, which like *erb*A is a transduced cellular gene that has acquired oncogenic potential following its mobilization by the retrovirus. Each of these viral oncogenes stems from a well-known cellular progenitor: the normal cellular version (i.e., the protooncogene) of *erb*A encodes the thyroid hormone receptor while the cellular progenitor of *erb*B specifies the epidermal growth factor (EGF) receptor (Sap et al., 1986; Weinberger et al., 1986; Downward et al., 1984).

Altered versions of the two receptor proteins conspire with one another to induce malignant conversion of avian erythroid precursors. The *erb*B oncoprotein acts as a mitogen to drive the proliferation of undifferentiated precursors of the avian erythrocytes. However, these erythroblasts tend to differentiate with high frequency into erythrocytes, which lack proliferative potential. Because of this, erythroid precursors transformed by *erb*B alone are poorly tumorigenic. *erb*A acts in a distinct and synergistic way on these erythroid cells by blocking their ability to differentiate. As such, it traps these cells in a pool of undifferentiated red cell precursors whose size is continuously expanded by the growth-stimulatory effects of the *erb*B oncoprotein (Graf & Beug, 1983).

Recent biochemical studies have generated a provocative model of how the *erb*A oncoprotein acts to block differentiation: it actively blocks the functioning of its normal counterpart, the thyroid hormone (triiodothyronine, T3) receptor (K. Damm, C. C. Thompson, and R. M. Evans, submitted for publication). It would appear that, during normal avian erythropoiesis, ambient levels of the T3 hormone cause the receptor to activate expression of a bank of differentiation-specific genes. In this way, the normal T3 receptor acts as a growth suppressor since the resulting differentiated cells, in this instance red cells, irreversibly lose proliferative potential. In the presence of the aberrantly functioning *erb*A oncoprotein, normal T3 receptor present in the same cell is unable to interact productively with its bank of responder genes, and thus the path to differentiation is blocked. In genetic terms, the oncogenic (i.e., *erb*A) allele of the T3 receptor gene would appear to act as a dominant negative, blocking the functioning of its wild-type counterpart.

The *p53* gene, mentioned previously in this discussion, represents a second growth-suppressing gene that until recently was viewed as a growth agonist. The *p53* gene is named after its encoded protein. This protein is known largely from its association with the SV40 LT oncoprotein in virus-transformed rodent cells. SV40 LT avidly binds this host cell protein (Lane & Crawford, 1979; McCormick & Harlow, 1980). In so doing, it increases the half-life of p53 from ca. 15 min to more than 24 h. As a consequence, the steady-state level of p53 is increased by as much as 2 orders of magnitude (Oren et al., 1981).

This large increase in p53 protein levels has parallels to the deregulation of expression levels seen upon activation of oncogenes like *myc* and *fos* (Adams et al., 1985; Coppola & Cole, 1986; Baumbach et al., 1986; Ruther et al., 1987; Verma & Sassone-Corsi, 1987). These oncogenic activations are achieved through deregulation of transcript levels of the respective genes. SV40 LT achieves a similar end result, but through the expedient of stabilizing the gene product. The increased level of gene product was presumed in turn to have a growth-promoting effect on the cell, resulting ultimately in cell transformation. Moreover, *p53* mRNA levels increase in the cell in response to mitogens (Reich, 1984), further strengthening the analogy with the *myc* and *fos* genes, which are also turned on by serum factors.

An apparently conclusive proof of *p53*'s growth-agonistic function was provided several years ago when three groups showed that an oncogene can be created by fusing *p53* cDNA with a strong constitutive transcriptional promoter. The resulting chimeric construct behaved like clones of the *myc*, E1A, SV40 LT, and HPV E7 oncogenes: it could immortalize cells and could collaborate with a *ras* oncogene in malignant transformation (Eliyahu et al., 1984; Parada et al., 1984; Jenkins et al., 1984).

A discordant note soon came from reports that the *p53* gene is deleted or grossly damaged in certain tumor cell genomes (Mowat et al., 1985). This is hardly consistent with the proposed role of *p53* as a growth-stimulating gene. More to the point were detailed characterizations of the cDNAs used in the earlier experiments, which were found unexpectedly to carry point mutations (Finlay et al., 1988; Eliyahu et al., 1988, 1989; Hinds et al., 1989). Moreover, when p53 cDNAs carrying the bona fide wild-type sequence were used in attempts to construct oncogenes, these were found to be actively inhibitory for growth (Eliyahu et al., 1989; Finlay et al., 1989).

These data suggest that the normal *p53* gene acts as a growth-suppressing element in the cell and that mutant forms of its encoded protein act in a quite opposite way to stimulate growth. In strong analogy with *erb*A, discussed above, it now appears that mutant alleles of *p53* act as dominant negatives to inhibit functioning of wild-type *p53* alleles. How this is achieved on a molecular level is not yet totally clear. One clue may be provided by the observation that p53 molecules form stable oligomers (Kraiss et al., 1988). In summary, two genetic mechanisms can involve *p53* in tumorigenesis: reading frame mutations that create dominant negative alleles that neutralize function of the surviving wild-type allele, and mutations that inactivate both copies of this gene. In either event, the consequence of this is the physiological inactivation of p53 function, resulting in turn in a release from its growth-constraining effects.

Genetic Mechanisms That Involve Growth-Suppressor Genes in the Cancer Process

We have described two distinct genetic mechanisms involving growth-suppressor genes that favor the malignant outgrowth of tumor cells: the inactivation of both gene copies and the creation of dominant negative alleles that succeed in neutralizing the activity of the surviving wild-type allele. In the case of the Rb gene, the inactivation of gene copies creates recessively acting null alleles. When both gene copies are reduced to this state, then runaway cell growth is unleashed. To date, dominant negative alleles of Rb have not been reported although they may well exist. In the case of *p53*, the neutralization of gene function can be achieved both through gene inactivation and through the creation of dominant negative alleles. A dominant negative allele of *erb*A has been documented; null alleles have not yet been seen in tumor cell genomes and presumably await discovery. The end result of all these various genetic alterations is the same—removal of a previously operative barrier to cell proliferation.

As suggested earlier, these three genes appear as representatives of a much larger class of cellular genes that normally function as growth suppressors. The indications for the existence of these genes rest largely on indirect genetic arguments that are worth describing in outline. Many of these genes appear to behave like the Rb in tumorigenesis: conversion of both of their copies to null alleles triggers cancer in one or another organ.

Clues pointing to the existence of these genes derive from the specific genetic mechanisms believed responsible for the elimination of the two functional copies of these genes. To repeat the pattern described for sporadic retinoblastoma, we note the following: an initial random somatic mutation inactivates one Rb copy; subsequently, a second somatic event removes the surviving wild-type allele. Critical to our discussion is the mechanism responsible for loss of the *second* allele. It could in principle depend upon a mutational event that creates a novel null allele unrelated to the inactive allele generated earlier on the other chromosome. The probability of each of these two mutational events would be equal and very low (e.g., 10^{-6} per cell generation), and the probability of nullizygosity of the cell would be the product of these two small probabilities.

A far easier route to losing the second gene copy is provided by mechanisms involving chromosomal loss or mitotic recombination. Here the chromosomal region carrying the surviving wild-type allele is lost wholesale and replaced by a duplicated copy of the corresponding region on the homologous, already mutated chromosome. Such events may occur as frequently as 10^{-3} per cell generation and now result in two identical copies of the initially mutated, defective allele. Importantly, any previous heterozygosity that existed in the chromosomal region surrounding the growth-suppressing gene is lost, since all chromosomal markers in this region are now present in two identical copies. Thus, this chromosomal region has suffered a "reduction to homozygosity".

All this leads to a powerful and simple genetic analysis that involves searching tumor cell genomes for chromosomal regions that repeatedly suffer reductions to homozygosity (Hansen & Cavenee, 1987). Repeated observation of chromosome-specific loss of heterozygosity in a series of tumors represents prima facie evidence for the presence of a growth-suppressor gene on this chromosome, recessive alleles of which are able to affect the neoplastic phenotype when the dominant wild-type alleles are lost. The chromosomal markers most readily used to detect loss of heterozygosity are restriction fragment length polymorphisms (RFLPs), detected through Southern blotting and use of auspiciously chosen probes. In this fashion, growth-suppressor genes involved in acoustic neuromas have been assigned to chromosome 22 and genes involved in colon car-

Table I: Chromosomal Locations of Putative Growth Suppressor Loci Based on Tumor-Specific Reduction to Homozygosity

neuroblastoma	1 p
small cell lung carcinoma	3 p
colon carcinoma	5 q
Wilms' kidney tumor	11 p
bladder carcinoma	11 p
retinoblastoma, osteosarcoma	13 q
ductal breast carcinoma	13 q
astrocytoma, colon carcinoma	17 p
meningioma, acoustic neuroma	22 q
colon carcinoma	18

cinoma assigned to chromosomes 17 and 18 (see Table I for a larger list). These molecular analyses are powerful in being able to scan the whole genome for chromosomal regions carrying possible suppressor genes. Nonetheless, these results remain at best indirect proofs of the existence of such genes. Only molecular cloning of these genes, which will require enormous experimental effort, will provide thoroughly convincing proof that all these genes exist and function as hypothesized.

Precis and Prospect

Various lines of evidence have converged in the last years on the idea that our genome carries a group of genes that is normally involved in negatively regulating cellular proliferation. Each of these genes would seem to act as a critical guardian of normal growth in only a subset of the body's many cell types. We suspect this because loss of both gene copies is associated in each case with a particular, idiosyncratic subset of tumor types.

The true size of this class of genes is unknown and at present unknowable, since we have only imperfect means for discovering them. Stated simply, we know about most of these genes only when the cell is deprived of their functions. Experimentally, the bulk of these genes has become apparent through somatic cell hybridization, karyologic studies, and RFLP analysis. In addition, two well-studied oncogenes, *p53* and *erb*A, are now realized to be aberrant versions of growth-suppressing genes. Taken together, this class still composes only a dozen or so genes. Many others may escape detection by the admittedly limited types of analysis currently available to us.

Granting our imperfect understanding of growth-suppressor genes, it must be said that these genes may ultimately be seen as more important in tumorigenesis than the intensively studied oncogenes. After all, it is far easier to inactivate gene function than it is to create the hyperactive forms of genes that we study as oncogenes. Creating oncogenes often involves precise and subtle tinkering with specific nucleotide sequences. Contrast the limited number of sites at which point mutations (and these mutations alone) can create a *ras* oncogene (Barbacid, 1987) with the almost unlimited number of genetic changes that can act to cripple the 190-kb-long Rb gene.

Most interesting are the biochemical and cell physiological mechanisms that enable these genes and their encoded proteins to constrain cell growth. Perhaps it is more than mere coincidence that the three suppressor genes isolated to date, *Rb*, *erb*A, and *p53*, all encode nuclear proteins that appear as candidates for transcriptional regulators (*Rb*, *p53*) or are known to function in this manner (*erb*A).

One might argue that cell growth is governed by separate, parallel growth-agonistic and growth-antagonistic signaling pathways that converge upon the ultimate decision determining cell growth or quiescence. But the recent work demonstrating direct physical interactions between oncoproteins and growth-suppressing proteins suggests a different scenario—that these two types of proteins are interwoven in a common signaling pathway that is responsible for making the big decisions about proliferation and differentiation, the decisions that lie at the heart of morphogenesis and the dysmorphogenesis that we know as cancer.

Acknowledgments

I thank Ehry Anderson for excellent help in the preparation of the manuscript.

References

Adams, J. M., Harris, A. W., Pinkert, C. A., Corcoran, L. M., Alexander, W. S., Cory, S., Palmiter, R. D., & Brinster, R. L. (1985) *Nature 318*, 533–538.

Barbacid, M. (1987) *Annu. Rev. Biochem. 56* 779–827.

Baumbach, W. R., Keath, E. J., & Cole, M. D. (1986) *Mol. Cell. Biol. 59*, 276–283.

Berk, A. J. (1986) *Annu. Rev. Genet. 20*, 45–79.

Bernards, R., Schackleford, G. M., Gerber, M. R., Horowitz, J. M., Friend, S. H., Schartl, M., Bogenmann, E., Rapaport, J. M., McGee, T., Dryja, T. P., & Weinberg, R. A. (1989) *Proc. Natl. Acad. Sci. U.S.A.* (in press).

Bishop, J. M. (1983) *Annu. Rev. Biochem. 52*, 301–354.

Bolen, J. B., Thiele, C. J., Israel, M. A., Yonemoto, W., Lipsich, L. A., & Brugge, J. S. (1984) *Cell 38*, 767–777.

Bookstein, R., Lee, E. Y.-H. P., To, H., Young, L. J., Sery, T. W., Hayes, R. C., Friedmann, T., & Lee, W.-H. (1988) *Proc. Natl. Acad. Sci. U.S.A. 85*, 2210–2214.

Cavenee, W. K., Dryja, T. P., Phillips, R. A., Benedict, W. F., Godbout, R., Gallie, B. L., Murphree, A. L., Strong, L. C., & White, R. L. (1983) *Nature 305*, 779–784.

Coppola, J. A., & Cole, M. D. (1986) *Nature 320*, 760–763.

Courtneidge, S. A., & Smith, A. E. (1983) *Nature 303*, 435–439.

De Caprio, J. A., Ludlow, J. W., Figge, J., Shew, J.-Y., Huang, C. M., Lee, W. H., Marsilio, E., Paucha, E., & Livingston, D. M. (1988) *Cell 54*, 275–283.

Downward, J., Yarden, Y., Mayes, E., Scrace, G., Totty, N., Stockwell, P., Ullrich, A., Schlessinger, J., & Waterfield, M. D. (1984) *Nature 307*, 521–527.

Draper, G. J., Sanders, B. M., & Kingston, J. E. (1986) *Br. J. Cancer 53*, 661–671.

Dryja, T. P., Cavenee, W., White, R., Rapaport, J. M., Petersen, R., Albert, D. M., & Bruns, G. A. P. (1984) *N. Engl. J. Med. 310*, 550–553.

Dyson, N., Howley, P. M., Munger, K., & Harlow, E. (1989) *Science 243*, 934–937.

Eliyahu, D., Raz, A., Gruss, P., Givol, D., & Oren, M. (1984) *Nature 312*, 646–649.

Eliyahu, D., Goldfinger, N., Pinhasi-Kimhi, O., Shaulski, G., Skurnik, Y., Arai, N., Rotter, V., & Oren, M. (1988) *Oncogene 3*, 313–321.

Eliyahu, D., Michalovitz, D., Eliyahu, S., Pinhasi-Kimhi, O., & Oren, M. (1989) (in press).

Figge, J., Webster, T., Smith, T. F., & Paucha, E. (1988) *J. Virol. 62*, 1814–1818.

Finlay, C., Hinds, P., Tan, T.-H., Eliyahu, D., Oren, M., & Levine, A. J. (1988) *Mol. Cell. Biol. 8*, 531–539.

Finlay, C. A., Hinds, P. W., & Levine, A. J. (1989) *Cell* (in press).

Friend, S. H., Bernards, R., Rogelj, S., Weinberg, R. A., Rapaport, J., Albert, D., & Dryja, T. P. (1986) *Nature 323*, 643–646.

Friend, S. H., Horowitz, J. M., Gerber, M. R., Wang, X.-F., Bogenmann, E., Li, F. D., & Weinberg, R. A. (1987) *Proc.*

Natl. Acad. Sci. U.S.A. 84, 9059–9063.
Fung, Y.-K. T., Murphree, A. L., T'Ang, A., Qian, J., Hinrichs, S. H., & Benedict, W. F. (1987) *Science 236*, 1657–1661.
Godbout, R., Dryja, T. P., Squire, J., Gallie, B. L., & Phillips, R. A. (1983) *Nature 304*, 451–453.
Graf, T., & Beug, H. (1983) *Cell 34*, 7–9.
Hansen, M. F., & Cavenee, W. K. (1987) *Cancer Res. 47*, 5518–5527.
Harbour, J. W., Lai, S.-L., Whang-Peng, J., Gazdar, A. F., Minna, J. D., & Kaye, F. J. (1988) *Science 241*, 353–357.
Harlow, E., Whyte, P., Franza, B. R., & Schley, C. (1986) *Mol. Cell. Biol. 6*, 1579–1589.
Harris, H. (1988) *Cancer Res. 48*, 3302–3306.
Hinds, P., Finlay, C., & Levine, A. J. (1989) *J. Virol. 63*, 739–746.
Horowitz, J. M., Yandell, D. W., Park, S.-H., Canning, S., Whyte, P., Buchkovich, K., Harlow, E., Weinberg, R. A., & Dryja, T. P. (1989) *Science 243*, 937–940.
Huang, H.-J. S., Yee, J.-K., Shew, J.-Y., Chen, P.-L., Bookstein, R., Friedmann, T., Lee, E. Y.-H. P., & Lee, W.-H. (1988) *Science 242*, 1563, 1566.
Jenkins, J. R., Rudge, K., & Currie, G. A. (1984) *Nature 312*, 651–654.
Klein, G. (1987) *Science 238*, 1539–1544.
Knudson, A. G. (1971) *Proc. Natl. Acad. Sci. U.S.A. 68*, 820–823.
Kraiss, S., Quaiser, A., Oren, M., & Monternarh, M. (1988) *J. Virol. 62*, 4737–4744.
Land, H., Parada, L., & Weinberg, R. A. (1983) *Nature 304*, 596–602.
Lane, D. P., & Crawford, L. V. (1979) *Nature 278*, 261–263.
Lee, E. Y.-H. P., To, H., Shew, J.-Y., Bookstein, R., Scully, P., & Lee, W.-H. (1988) *Science 241*, 218–221.
Lee, J.-Y., Shew, J.-Y., & Hong, F. D. (1987) *Nature 329*, 642–645.
Lee, W. H., Bookstein, R., Hong, F., Young, L.-J., Shew, J.-Y., & Lee, E. Y.-H. P. (1987) *Science 235*, 1394–1399.
Lillie, J. W., Loewenstein, P. M., Green, M. R., & Green, M. (1987) *Cell 50*, 1091–1100.
Ludlow, J. W., De Caprio, J. A., Huang, C.-M., Lee, W.-H., Paucha, E., & Livingston, D. M. (1989) *Cell 56*, 57–65.
McCormick, F., & Harlow, E. (1980) *J. Virol. 34*, 213–224.
Moran, E. (1988) *Nature 334*, 168–170.
Mowat, M., Cheng, A., Kimura, N., Bernstein, A., & Benchimol, S. (1985) *Nature 314*, 633–636.
Oren, M., Maltzman, W., & Levine, A. J. (1981) *Mol. Cell. Biol. 1*, 101–110.
Parada, L. F., Land, H., Weinberg, R. A., Wolf, D., & Rotter, V. (1984) *Nature 312*, 649–651.
Phelps, W. C., Yee, C.-L., Münger, K., & Howley, P. M. (1988) *Cell 53*, 539–547.
Ralston, R., & Bishop, J. M. (1983) *Nature 306*, 803–806.
Reich, N. C., & Levine, A. J. (1984) *Nature 308*, 199–201.
Riccardi, V. M., Sujanski, E., Smith, A. C., & Francke, U. (1978) *Pediatrics 61*, 604–610.
Ruley, H. E. (1983) *Nature 304*, 602–606.
Rüther, V., Garber, C., Komitowski, D., Müller, R., & Wagner, E. F. (1987) *Nature 325*, 412–416.
Sap, J., Munoz, A., Damm, K., Goldberg, Y., Ghysdael, J., Leutz, A., Berg, H., & Vennström, B. (1986) *Nature*, 635–640.
Saxon, P. J., Srivatsan, E. S., & Stanbridge, E. J. (1986) *EMBO J. 5*, 3461–3466.
Skurnik, Y., Arai, N., Rotter, V., & Oren, M. (1988) *Oncogene 3*, 313–321.
Sparkes, R. S., Murphree, A. L., Lingua, R. W., Sparkes, M. C., Field, L. L., Funderburk, S. J., & Benedict, W. F. (1983) *Science 219*, 971–973.
Stanbridge, E. J., Flandermeyer, R. R., Daniels, D. W., & Nelson-Rees, W. A. (1981) *Somatic Cell Mol. Genet. 7*, 699–712.
T'Ang, A., Varley, J. M., Chakraborty, S., Murphree, A. L., & Fung, Y.-K. T. (1988) *Science 242*, 263–266.
Toguchida, J., Ishizaki, K., Sasaki, S., Ikenaga, M., Sugimoto, M., Kotoura, Y., & Yamamuro, T. (1988) *Cancer Res. 48*, 3939–3943.
Varmus, H. E. (1984) *Annu. Rev. Genet. 18*, 53–612.
Verma, I. M., & Sassone-Corsi, P. (1987) *Cell 51*, 513–514.
Vousden, K. H., & Jat, P. S. (1989) *Oncogene 4*, 153–158.
Weichselbaum, R. R., Beckett, M., & Diamond, A. (1988) *Proc. Natl. Acad. Sci. U.S.A. 85*, 2106–2109.
Weinberger, C., Thompson, C., Ong, E. S., Lebo, R., Gruol, D. J., & Evans, R. M. (1986) *Nature 324*, 641–646.
Whyte, P., Buchkovich, K. J., Horowitz, J. M., Friend, S. H., Raybuck, M., Weinberg, R. A., & Harlow, E. (1988) *Nature 334*, 124–129.
Whyte, P., Williamson, N. M., & Harlow, E. (1989) *Cell 56*, 67–75.
Yee, S. P., & Branton, P. E. (1985) *Virology 147*, 142–153.
Yunis, J. J., & Ramsay, N. (1978) *Am. J. Dis. Child. 132*, 161–163.

Chapter 26

Xenopus Oocytes and the Biochemistry of Cell Division†

James L. Maller
Department of Pharmacology, University of Colorado School of Medicine, Denver, Colorado 80262
Received September 19, 1989; Revised Manuscript Received October 27, 1989

ABSTRACT: The control of cell proliferation involves both regulatory events initiated at the plasma membrane that control reentry into the cell cycle and intracellular biochemical changes that direct the process of cell division itself. Both of these aspects of cell growth control can be studied in *Xenopus* oocytes undergoing meiotic maturation in response to mitogenic stimulation. All mitogenic signaling pathways so far identified lead to the phosphorylation of ribosomal protein S6 on serine residues, and the biochemistry of this event has been investigated. Insulin and other mitogens activate ribosomal protein S6 kinase II, which has been cloned and sequenced in oocytes and other cells. This enzyme is activated by phosphorylation on serine and threonine residues by an insulin-stimulated protein kinase known as MAP-2 kinase. MAP kinase itself is also activated by direct phosphorylation on threonine and tyrosine residues in vivo. These results reconstitute one step of the insulin signaling pathway evident shortly after insulin receptor binding at the membrane. Several hours after mitogenic stimulation, a cell cycle cytoplasmic control element is activated that is sufficient to cause entry into M phase. This control element, known as maturation-promoting factor or MPF, has been purified to near homogeneity and shown to consist of a complex between $p34^{cdc2}$ protein kinase and cyclin B2. In addition to apparent phosphorylation of cyclin, regulation of MPF activity involves synthesis of the cyclin subunit and its periodic degradation at the metaphase → anaphase transition. The $p34^{cdc2}$ kinase subunit is regulated by phosphorylation/dephosphorylation on threonine and tyrosine residues, being inactive when phosphorylated and active when dephosphorylated. Analysis of phosphorylation sites in histone H1 for $p34^{cdc2}$ has revealed a consensus sequence of $(^{K}/_{R})^{S}/_{T}P(X)^{K}/_{R}$, where the elements in parentheses are present in some but not all sites. Sites with such a consensus are specifically phosphorylated in mitosis and by MPF in the protooncogene $pp60^{c\text{-}src}$. These results provide a link between cell cycle control and cell growth control and suggest that changes in cell adhesion and the cytoskeleton in mitosis may be regulated indirectly by MPF via protooncogene activation. S6 kinase II is also activated upon expression of MPF in cells, indicating that MPF is upstream of S6 kinase on the mitogenic signaling pathway. Further study both of the signaling events that lead to MPF activation and of the substrates for phosphorylation by MPF should lead to a comprehensive understanding of the biochemistry of cell division.

Cell proliferation as a process encompasses a vast array of biological phenomena. Included in the process are the transduction of extracellular signals into intracellular biochemical signals, the complex structural reorganization of the cellular architecture to prepare for a division cycle, and finally the reprogramming of metabolism to actually carry out cell division. Each of these aspects usually is studied as an isolated process with a particular cell type or model system. However, it seems clear that in order to gain a comprehensive understanding of cell proliferation, it will be necessary to study all of these phenomena as an integrated process in a single cell type. The purpose of this paper is to review the progress that has been made in each of these areas in the *Xenopus* oocyte system and to propose that the oocyte is uniquely suited to provide an understanding of how extracellular signals are transduced into intracellular biochemical signals that activate regulatory elements of cell cycle control, which reorganize cellular metabolism for division. The central process that underlies all aspects of cell proliferation analysis is the cell cycle. Therefore, in the first part of this paper, I will review briefly the general concept of cell cycle control that has emerged from studies in somatic cells and then consider the

† Work in this laboratory is supported by grants from the NIH (GM26743 and DK28353) and the American Cancer Society (NP-517E).

oocyte system in relation to this general picture.

Cell Cycle Restriction Points. Two general restriction points have been identified in the cell cycle, one in the G_0/G_1 period governing entry into S phase and the other in the G_2 phase governing entry into mitosis [for reviews, see Pardee et al. (1978), Marcus et al. (1985), and Baserga (1981)]. Since both DNA synthesis and mitosis are hallmarks of cellular reproduction, analysis of both of these restriction points is valid for cellular control processes. Although the G_0 state is believed to exist in vivo in various types of stem cells, experimentally it has been defined in relation to in vitro culture of somatic cells. Generally, cultured mammalian cells require the addition of external growth factors, present in serum, in order to undergo a round of DNA synthesis and cell division. By removal of serum leading to cell cycle arrest in a G_0 state, the events that accompany reentry into the cell cycle can be studied upon readdition of defined growth factors. The particular growth factors required appear to depend on the cell type and the exact stage of G_0 arrest. In some cases growth factors make cells competent to reenter the cell cycle upon treatment with other factors, while in other cases cell cycle progression is directly stimulated. Analysis of growth factor action has been carried out at several levels. The receptors to which growth factors bind have been studied intensively, and the most progress has been made with receptors that express a tyrosine protein kinase activity, as in the case of EGF, PDGF, insulin, and IGF_1 receptors [see Yarden and Ullrich (1988) for a review]. Binding of these growth factors has several immediate consequences. The first is activation of the tyrosine protein kinase activity of the receptor itself, which has been shown in most cases to correlate with the appearance of new phosphotyrosyl proteins. In several cases, this phosphorylation is closely correlated with changes in phospholipid metabolism. Within the next few minutes, several immediate early events occur that appear to be of general significance. One is the transcription of several genes, notably c-*fos*, c-*myc*, and c-*jun*, whose protein products are transcription factors that probably activate as yet unidentified genes for events later in the cell cycle. At the biochemical level, two immediate early events are seen in most cell types. One is an increase in intracellular pH due to Na/H exchange, and the other is the phosphorylation on serine residues of ribosomal protein S6. Although both these events are very highly correlated with mitogenic stimuli, the exact function of the pH increase or of S6 phosphorylation has not been established. In the case of S6 phosphorylation, it has not yet been possible to artificially increase or block S6 phosphorylation, which would be necessary to evaluate its potential importance. However, it seems unlikely that S6 phosphorylation itself is sufficient to increase the rate of protein synthesis. In spite of the uncertain regulatory significance of these events, the high conservation of their expression encourages analysis of their regulation as models for the biochemistry of early mitogenic activation, especially by stimuli that work through tyrosine kinase activation.

It is evident from the analysis of regulation at the G_0/G_1 transition that cell cycle progression is characterized by both transcriptional and translational regulation and is most likely concerned with the immediate reprogramming of the cellular metabolism for DNA synthesis. It is probable that different changes are needed for cells to pass through the G_2 restriction point.

G_2 Restriction Point. Like the G_0/G_1 restriction point, it appears that in vivo some cell populations are physiologically arrested in G_2 phase (Pedersen & Gelfant, 1970; Melchers &

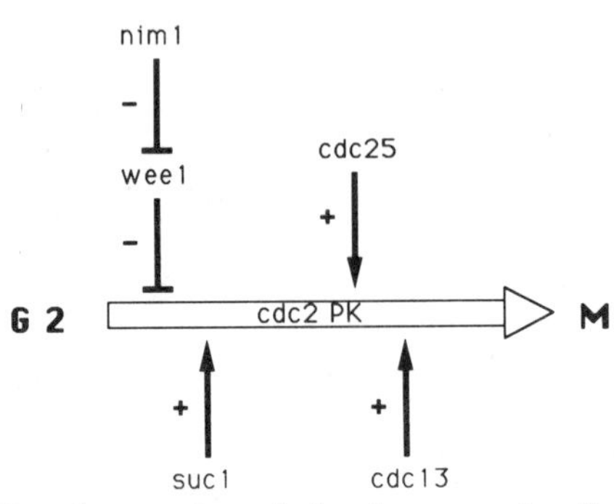

FIGURE 1: Genetic network regulating the eucaryotic cell cycle. The figure depicts the *cdc2*$^+$ gene prodct as the central regulator of the $G_2 \rightarrow M$ transition. Other genes indicated are known to be required either for activating *cdc2* kinase or for specifying the timing of *cdc2* kinase activation in the cell cycle, as described in the text. Modified from Russell and Nurse (1987b).

Lernhardt, 1985). A number of cultured cells can be artificially blocked in G_2 by elevating the level of cAMP (Nose & Katsuta, 1975; Stambrook & Velez, 1976), suggesting that protein kinase A exerts an inhibitory effect on cell division at this point in the cell cycle. Biochemical support for a G_2 block has been particularly convincing from studies on various types of oocytes, where a physiological G_2 block is released by a reduction in cAMP levels (Maller et al., 1977; Meijer & Zarutskie, 1987; Bornslaeger et al., 1986).

In both the budding yeast *Saccharomyces cerevisiae* and the fission yeast *Schizosaccharomyces pombe*, elegant genetic analysis has shown that mutations in several genes lead to an arrest at the G_2/M border of the cell cycle [see Lee and Nurse (1988) for a review]. The analysis has been more extensive in *S. pombe* because it has a cell cycle more like that of vertebrate cells, with well-defined G_1 and G_2 periods. Figure 1 shows a scheme proposed by Nurse's laboratory that takes into account the genetic analysis to date (Russell & Nurse, 1987b). The central control element regulating the $G_2 \rightarrow M$ transition is cell division cycle gene 2 (*cdc2*$^+$). This gene encodes a 34-kDa serine/threonine protein kinase. While $p34^{cdc2}$ has all the usual consensus sequences for protein kinases, the remaining sequence is not closely related to any other member of the protein kinase family that has been sequenced.

Mutations in *cdc2*$^+$ that cause arrest in G_2 have been intensely studied, although in some cases mutants arrest in G_1, indicating that this protein kinase is involved in both G_1/S and G_2/M control points in the cell cycle [for reviews, see Nurse (1985) and Lee and Nurse (1988)]. Suppressors of mutations in *cdc2*$^+$ have fallen into three classes. About half of the suppressors involve a gene known as *sucI*, which encodes a protein that binds to $p34^{cdc2}$ in vitro and stimulates the kinase activity in extracts of thermolabile mutants (Brizuela et al., 1987; Moreno et al., 1989; Dunphy et al., 1988). The exact function of the *sucI* gene is unknown at present, but deletion of *sucI* leads to mitotic arrest, suggesting a role in exit from mitosis (Moreno et al., 1989). This role is likely to be important inasmuch as *sucI* homologues have been identified in human cells (Draetta et al., 1987). Over- or underexpression of *sucI* does not appear to affect the timing of mitosis.

A positive stimulus regulating the timing of *cdc2*$^+$ activation is evident after expression of the *cdc25* gene, whose sequence is unrelated to any other sequence in the gene bank. Other timing genes include *weeI*$^+$ and *nimI*$^+$, both of which encode proteins with consensus sequences for protein kinases (Russell & Nurse, 1987a,b). *weeI*$^+$ inhibits the timing of *cdc2*$^+$ activation, while *nimI*$^+$ stimulates *cdc2*$^+$ by a mechanism that involves *weeI*$^+$. The timing controls on *cdc2*$^+$ by *cdc25* and *wee1*$^+$ are independent, but they can be affected by mutations in other genes that encode homologues of protein phosphatase 1 (Booher & Beach, 1989). Fission yeast *wee1*$^+$ has been shown to exert mitotic control when expressed in the distantly related budding yeast, *S. cerevisiae*, and its action is opposed by the *MIH1* gene, a budding yeast homologue of *cdc25*$^+$ (Russell et al., 1989). Another element required for *cdc2* kinase activation is the product of the *cdc13*$^+$ gene, which is required for activation of *cdc2* in mitosis and may also be necessary for spindle microtubule organization (Moreno et al., 1989; Hagan et al., 1988; Booher & Beach, 1988). Sequence analysis indicates the *cdc13*$^+$ gene product is similar to the cyclin proteins implicated in mitotic control in the eggs of various marine invertebrates (Evans et al., 1983; Solomon et al., 1988; Goebl & Byers, 1988). These results indicate that the *cdc2*$^+$ gene product is regulated by a complex network of other proteins, including other protein kinases, and suggest that protein phosphorylation will prove to be the fundamental biochemical process underlying the molecular basis of the $G_2 \rightarrow M$ transition in the cell cycle.

The function of *cdc2*$^+$, and presumably its kinase specificity, has been highly conserved in evolution. *cdc2*$^+$ is a homologue of a gene in budding yeast known as *CDC28*$^+$, which has been cloned and sequenced and found to also encode a 34-kDa serine/threonine protein kinase. Each homologue is able to complement mutations in the other gene when expressed in the heterologous yeast (Beach et al., 1982). More specifically, the human homologue of *cdc2*$^+$ was also cloned by complementation (Lee & Nurse, 1987), indicating that the function of this gene in cell cycle control is conserved from yeast to man. Despite the clear importance of *cdc2*$^+$ and the delineation of a genetic regulatory network for the $G_2 \rightarrow M$ transition, little biochemical information has been forthcoming from this system. Instead, most biochemical information on the $G_2 \rightarrow M$ transition has come from the *Xenopus* oocyte system, as discussed in the next section.

Xenopus Oocyte System. Like most vertebrates, oocytes from *Xenopus* are physiologically arrested at the G_2/M border in first meiotic prophase. In amphibians like *Xenopus*, release from this prophase block occurs in response to mitogenic stimulation by progesterone, insulin, or IGF_1. Several hours later entry into meiosis I is signaled by breakdown of the germinal vesicle (nucleus), termed GVBD. Because of the perfect synchrony of the oocyte population prior to mitogenic stimulation, biochemical changes are more easily monitored even if they have transient kinetics. Many of the changes seen in oocytes undergoing maturation are similar to those described for cultured cells traversing the G_0/G_1 restriction point after growth factor addition. For analysis of phosphorylation events involved in the $G_2 \rightarrow M$ transition, we have focused on the activation of ribosomal protein S6 as a defined biochemical end point in the signaling pathway used by all mitogens so far studied, with the objective of working backward up a pathway, ultimately reaching the mitogen receptor itself. The particular mitogen we have chosen to focus on is insulin, in part because the question of signaling by tyrosine kinases is a major unresolved problem in cell growth and cell cycle control. One reason for working backward up the pathway is the disappointing lack of progress in any cell type in characterizing initial substrates or initial steps in the insulin signaling pathway beyond activation of the tyrosine kinase activity of the insulin receptor. Oocytes have genuine insulin receptors as judged by insulin binding, anti-insulin receptor antibody binding, and inhibition of insulin effects by microinjected antibodies against the tyrosine kinase domain of the β subunit (Maller & Koontz,

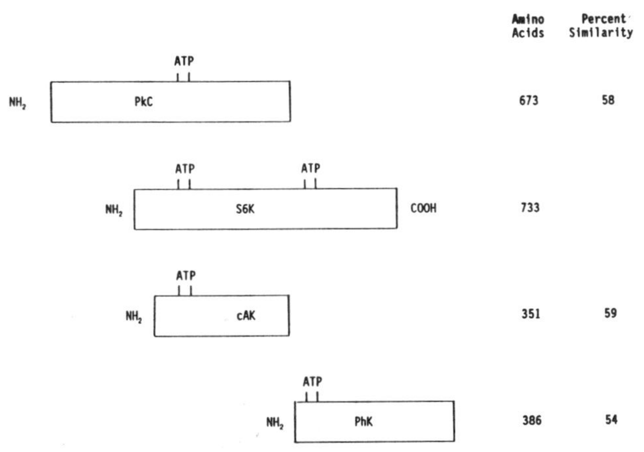

FIGURE 2: Sequence similarity of kinase domains in S6KII. The figure shows schematically that S6KII contains two apparent consensus sequences for ATP-binding sites in protein kinases, one in each half of the molecule. The kinase domain in the amino-terminal half is 58% similar at the amino acid level to protein kinases A and C, while the domain in the C-terminal half is more similar to phosphorylase *b* kinase. Whether both domains are catalytically active is unknown at present.

1980; Morgan et al., 1986). However, oocytes also possess structurally similar IGF_1 receptors, and it is possible all of these criteria actually measured IGF_1 receptors. Because of the very close structural and functional relationship between insulin and IGF_1 receptors, it is likely that information about signaling pathways for either receptor will have general significance.

Upon stimulation of oocytes with insulin, there is a rapid activation of an S6 kinase activity within 15–30 min that returns to near basal by 1 h post insulin (Stefanovic et al., 1986; Stefanovic & Maller, 1988). From 2 to 3 h after insulin, S6 kinase activity begins to increase again, reaching a maximal rate just prior to GVBD at 6–7 h post insulin. This maximal rate of increase in S6 kinase activity correlates with a large 3–5-fold increase in protein phosphorylation that has been called the "burst" of phosphorylation (Maller et al., 1977; Doree et al., 1983). This event is a hallmark of biochemical reprogramming of the cell for division and is evident about 1 h prior to nuclear breakdown. The early increase in S6 kinase activation is unaffected by treatment of oocytes with cycloheximide, while the burst of phosphorylation is blocked by pretreatment with cycloheximide (Stefanovic & Maller, 1988). Because S6 kinase activation is maximal at the completion of maturation, metaphase II arrested unfertilized eggs, which have undergone maturation in vivo, were used as a source of material for purification. DEAE-Sephacel profiles of S6 kinase activity in eggs revealed two peaks of non-cAMP-dependent activity, termed S6KI and S6KII in order of their elution from the column. Purification of S6KII to homogeneity yielded a single polypeptide chain of M_r 92000 on Laemmli gels, while S6KI is an M_r 90000 protein (Erikson & Maller, 1986, 1989a). Both kinases are relatively specific for S6 although S6KI can also weakly phosphorylate histone H1 or α-casein, common substrates for broad specificity kinases. One hundred micrograms of S6KII was purified from 1.2 kg of eggs, tryptic peptides were sequenced, and oligonucleotide probes were constructed for screening of a *Xenopus* ovary cDNA library in λgt10. Two clones that were sequenced showed 96% identity with each other, and antibodies against the bacterially expressed protein reacted with purified S6KII (Jones et al., 1988). The sequence of S6KII proved to be rather interesting, since there were apparently complete consensus sequences for protein kinase domains in both the N-terminal and C-terminal halves of the molecule (Figure 2). The N-terminal domain is 58% similar in sequence to the sequences of protein kinases A and C, while the C-terminal half is much more similar to the catalytic domain of phosphorylase *b* kinase. Whether both catalytic domains are active has not yet been established. However, the presence of two domains encouraged additional examination of the substrate specificity of S6KII, and it was found to also phosphorylate glycogen synthase, tyrosine hydroxylase, troponin I, and lamin C (Erikson & Maller, 1988). These phosphorylation sites were found to exhibit a consensus motif of RXXS. The phosphorylation of lamin C occurs at a site increased in mitosis when disassembly of the nuclear envelope occurs, suggesting that S6KII may be directly involved in disassembly of the envelope at GVBD (Ward & Kirschner, 1988).

Recent data indicate that S6 kinases of similar size and sequence are present in mouse and chicken cells, react with antibodies against bacterially expressed S6KII, and are activated in response to mitogenic stimulation (Alcorta et al., 1989). In cells transformed by a temperature-sensitive mutant of $pp60^{v\text{-}src}$, an S6KII homologue is activated in a temperature-sensitive fashion, suggesting that S6KII is a target for oncogene action (Erikson et al., 1987). These results suggest S6KII and its homologues are a family of highly conserved kinases involved in mitogenic stimulation. While S6KII is clearly a major partner in the mechanism of mitogenic stimulation of S6 phosphorylation, these data do not exclude that other S6 kinases are also activated by mitogens. Enzymes of lower molecular weight have been described in frogs (S6KI), chicken (Blenis et al., 1987), mouse (Jeno et al., 1989), and rat (Price et al., 1989) that may be members of a distinct group of S6 kinases, possibly more involved in secondary waves of S6 kinase activation after initial mitogen treatment.

Interest in this laboratory is focused on the pathway of activation of S6 kinase rather than on detailed molecular characterization of the enzyme. Thus we were excited by the initial finding that incubation of S6 kinase II with either protein phosphatase 1 or 2A caused up to a 98% inactivation of the enzymatic activity (Maller, 1987; Sturgill et al., 1988). Since these two phosphatases are major serine/threonine-specific phosphatases in cells (Ingebritsen & Cohen, 1983), it suggested that the next step back in the signaling pathway was another serine/threonine protein kinase. It was reported that Swiss 3T3 cell S6 kinase could also be inactivated by phosphatase (Ballou et al., 1988), suggesting that S6 kinases in general are activated by direct phosphorylation rather than be second messengers. A screening of a variety of broad specificity kinases, such as protein kinases A and C, demonstrated no reactivation of S6 kinase activity under phosphorylation conditions. However, a purified, highly specific, insulin-stimulated protein kinase from 3T3-L1 cells able to phosphorylate microtubule-associated protein 2 (MAP-2) phosphorylated and reactivated dephospho-S6KII (Sturgill et al., 1988). Phosphopeptide mapping demonstrated that two new phosphopeptides were present in S6KII after MAP kinase phosphorylation and, significantly, reactivation was accompanied by the de novo appearance of phosphothreonine in S6KII. Recently S6KII has also been shown to be phosphorylated in vivo on serine and threonine residues as it becomes activated during maturation induced by a variety of mitogens (Erikson & Maller, 1989a). The phosphopeptide map of in vivo activated S6KII is more complex than the in vitro pattern for S6KII activated with MAP kinase induced by a variety of stimuli. This is not unexpected, since phosphatase 2A causes 98% inactivation of S6KII and MAP kinase can only restore 30% of the activity, implying the existence of other S6KII kinases able to phosphorylate other sites in S6KII evident in in vivo mapping. Activation of S6 kinase by phosphorylation is correlated with a decrease in electrophoretic mobility on SDS gels, permitting Western blotting to be used for assessing phosphorylation and dephosphorylation. Analysis by this method has shown that all S6 kinase molecules become phosphorylated during maturation and all become dephosphorylated upon egg activation (Erikson & Maller, 1989a).

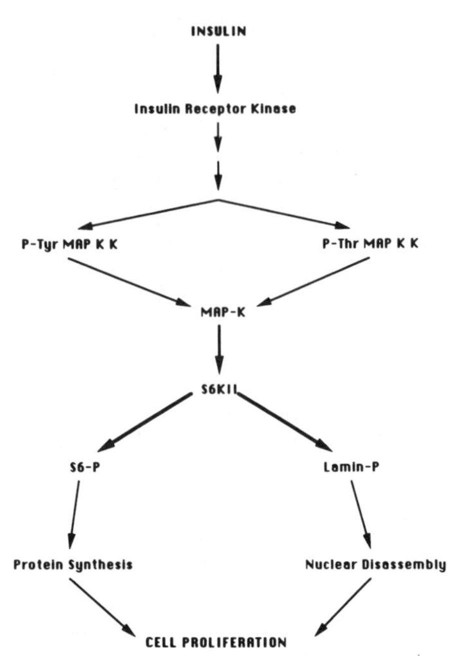

FIGURE 3: Possible signaling pathway for activation of S6KII. The diagram shows S6KII activation by MAP kinase, which is itself activated by two phosphorylation events mediated by different kinases, one tyrosine and one serine/threonine. Several additional intermediates are likely to exist between the MAP kinase kinases and the insulin receptor itself. These results suggest two types of signaling pathways are required for activation of MAP kinase and thus S6KII.

These results demonstrate the reconstitution of one step in the insulin signaling pathway for the $G_2 \rightarrow M$ transition and support the concept that working backward up the signaling pathway is a productive approach. Although the concept of kinase cascades dominates the literature of protein phosphorylation, this S6KII system is only the third example of one serine/threonine kinase phosphorylating and activating another serine/threonine kinase, the classic case being the phosphorylation of phosphorylase kinase by cAMP-dependent protein kinase (Walsh et al., 1968). Ultimately, these results lead us to consider the activation of MAP kinase as the next step back in the signaling pathway. Since MAP kinase was known to be phosphorylated on threonine and tyrosine residues in vivo after insulin stimulation (Sturgill & Ray, 1988), MAP kinase was incubated with phosphatases 1 and 2A and the activity of MAP kinase assessed. Phosphatase 1 had no effect on MAP kinase, but phosphatase 2A caused over an 80% inactivation of MAP kinase activity (Sturgill et al., 1988). Phosphoamino acid analysis of dephosphorylated MAP kinase confirmed that inactivation correlated with removal of phosphothreonine (T. Sturgill, personal communication). Although the function of tyrosine phosphorylation of MAP kinase is unknown at present, it is clearly not sufficient to activate the enzyme, and attempts to phosphorylate MAP kinase in vitro with purified insulin receptor kinase have so far been unsuccessful. However, these results show that MAP kinase is also regulated by direct phosphorylation and point to the existence of a serine/threonine MAP kinase kinase as being important in the signaling pathway. The pathway depicted in Figure 3 summarizes this discussion. It should be emphasized that a large number of steps are likely to be present between activation of the insulin receptor tyrosine kinase activity and activation of MAP kinase. In addition, as mentioned earlier, other mitogenic stimuli also cause activation of oocyte S6KII by phosphorylation. It is not yet clear whether these other stimuli represent completely independent but parallel stimuli or whether they intersect at MAP kinase or at a point further upstream in the insulin signaling pathway.

Analysis of Maturation-Promoting Factor. As mentioned earlier, S6 phosphorylation is a small component of the burst of phosphorylation observed to occur in oocytes 30–60 min prior to GVBD. This burst of phosphorylation represents a pleiotropic response to activation of maturation-promoting factor (MPF). Thus in working backward up the late pathway of S6 kinase activation, it could be predicted that MPF will be an intermediate between the receptor and MAP kinase kinase. MPF was first identified by cytoplasmic transfer experiments in which metaphase cytoplasm induced precocious maturation when injected into resting oocytes in the absence of mitogenic factors or protein synthesis (Masui & Markert, 1971; Smith & Ecker, 1971). Subsequently, an activity with similar properties able to induce GVBD in *Xenopus* was described in oocytes of other species and in mitotic cells from yeast to man, suggesting MPF was likely to be of fundamental importance in cell division and might have evolutionarily conserved components. Little was known about the molecular nature of MPF from studies using the microinjection assay. However, it was shown that oocytes contain a latent store of MPF, since a large (>100-fold) amplification of MPF activity was evident within 2 h in injected oocytes, even in the presence of protein synthesis inhibitors (Reynhout & Smith, 1974; Gerhart et al., 1984; Cyert & Kirschner, 1988). Correlated with the amplification was the large burst of phosphorylation also evident just prior to GVBD in mitogen-stimulated oocytes undergoing maturation (Maller et al., 1977). This led to the hypothesis that MPF was a kinase or an activator of a kinase. Other work using the oocyte injection assay showed that MPF was already present in latent form in small oocytes unable to respond to external mitogenic administration, suggesting that the activation mechanism for MPF was under acute developmental control (Hanocq-Quertier et al., 1976; Sadler & Maller, 1983; Wasserman et al., 1984; Taylor & Smith, 1985). Wasserman and Smith (1978) were the first to show that MPF activity oscillated during the mitotic cell cycle after fertilization, and Gerhart et al. (1984) showed that MPF activity also oscillated between meiosis 1 and 2 in *Xenopus*. Oscillations in the total level of phosphorylation were also found to occur in various species in concert with changes in MPF activity (Doree et al., 1983; Peaucellier et al., 1984).

For nearly 15 years, little progress was made in MPF purification using the oocyte injection assay, although several attempts were reported (Wu & Gerhart, 1980; Adlakha et al., 1985; Nguyen-Gia et al., 1986). In retrospect, two main reasons for difficulty are evident. One is the instability of MPF, and the other is the requirement for activity to be highly concentrated in order to give GVBD in oocytes. Additionally, since the oocyte only has one nucleus, the end point is all or none. This situation changed because of two key developments—improved stabilization procedures and a cell-free assay for MPF. Extraction of MPF was markedly improved by the inclusion of EGTA and β-glycerophosphate in the medium, and stability was further enhanced by the inclusion of γ-S-ATP in the extraction. Since thiophosphorylated proteins are often poor substrates for protein phosphatases, this suggested that maintenance of a phosphoprotein was important in stabilizing MPF activity.

Xenopus Cell-Free System. The most significant development in the characterization of MPF was the generation by Manfred Lohka of a cell-free system from unfertilized, metaphase-arrested frog eggs that could carry out early mitotic events in vitro in response to MPF (Lohka & Masui, 1984; Lohka & Maller, 1985; Miake-Lye & Kirschner, 1985). These events include nuclear envelope breakdown, chromosome condensation, and spindle formation. If eggs are in interphase at the time of extraction, the preparation induces nuclear assembly and total semiconservative DNA replication instead of mitotic events (Blow & Laskey, 1986). Interconversion from M phase to interphase could be accomplished by addition of calcium, and conversion from interphase to M phase could be accomplished by addition of MPF even in the presence of protein synthesis inhibitors (Lohka & Maller, 1985). The development of a cell-free system able to carrry out a single cell cycle is a breakthrough in the biochemistry of cell division, and the system has been rapidly adopted by a number of workers. For the first time, the control mechanisms regulating entry into and exit out of M phase can be studied functionally in vitro, and the individual component processes of mitosis such as nuclear breakdown, chromosome condensation, and spindle formation are now able to be studied in a functional, reversible in vitro system. This cell-free system promises to have a profound impact on cell biology. Already, Murray and Kirschner have shown that if much more concentrated interphase egg extracts are prepared, multiple cell cycles of alternating M phase and DNA synthesis occur in vitro spontaneously, permitting study of the entire cell cycle in a dynamic in vitro system (Murray & Kirschner, 1989; Murray et al., 1989).

As a first use of the MPF-dependent single cell cycle system, Lohka et al. (1988) purified MPF 3500-fold to near homogeneity, using as an assay breakdown of synthetic nuclei in the cell-free system. Purified MPF was also able to induce GVBD when injected into oocytes in the presence of cycloheximide. The purified preparation consisted largely of two polypeptide chains of M_r 32 000–34 000 and 45 000. Upon incubation with [γ-^{32}P]ATP, the 45-kDa component became phosphorylated on serine and threonine residues, but only in fractions that also contained the 34-kDa component. This indicated MPF was a complex of a 34-kDa serine/threonine protein kinase and a 45-kDa subunit.

Identification of $p34^{cdc2}$ *in MPF.* The finding that MPF contained a 34-kDa protein kinase was striking because of the extensive genetic work described earlier in the fission yeast *S. pombe* that had identified the *cdc2*$^+$ gene as a central regulator of the $G_2 \rightarrow M$ transition in the cell cycle; the *cdc2*$^+$ gene encodes a 34-kDa serine/threonine protein kinase.

Work by Nurse and his colleagues has shown that while there is only 63% sequence identity between the functionally interchangeable human and fission yeast *cdc2* products, all *cdc2* homologues show perfect conservation of a 16 amino acid PSTAIR region unrelated to any normal kinase consensus sequence (Lee & Nurse, 1987, 1988). Antibodies raised against a synthetic peptide corresponding to this sequence detected a single protein in Western blots of extracts of a wide variety of cells, which suggested they were specific for $p34^{cdc2}$. The PSTAIR sequence itself is likely to be a functionally important domain for *cdc2*, inasmuch as a synthetic peptide encoding the sequence has biological effects in the cell-free system or in injected oocytes (Gautier et al., 1988; Labbe et al., 1989a). However, recently other genes were reported to contain sequences containing 14 of the 16 amino acids, indicating that some caution must be exercised in the use of this antibody (Toh-e et al., 1988). However, this antibody was able to detect by Western blot and to immunoprecipitate the 34-kDa component of MPF, providing direct evidence that $p34^{cdc2}$ was the kinase component of MPF (Gautier et al., 1988). Indirect evidence supporting the same conclusion was obtained at the same time by Dunphy et al. (1988), who showed that the product of the *sucl* gene, known as p13, which binds the product of *cdc2*, could inhibit MPF activity in the cell-free system and deplete MPF activity from extracts. A large number of other proteins also bound to $p13^{sucl}$–Sepharose beads in extracts, including a 42-kDa protein (Dunphy et al., 1988) that is unrelated to the p45 component seen in purified MPF or in PSTAIR immunoprecipitates (Gautier et al., 1988). These results were exciting because the same protein had been identified as a central regulator of the cell cycle and M phase by two independent approaches, genetics in yeast and biochemistry in *Xenopus*. It also indicated the mitotic role of $p34^{cdc2}$ was as a subunit of the MPF complex. In addition, it provided an immediate potential regulatory system for MPF in the light of the extensive network of other genetically identified regulatory elements affecting *cdc2* (Lee & Nurse, 1988), some of which were described earlier.

Properties and Regulation of cdc2 Protein Kinase. Biochemically, *cdc2* protein kinase has several unusual properties. It is only the second serine/threonine protein kinase able to utilize both ATP and GTP, and γ-S-ATP is also a substrate. Although β-glycerophosphate was important for extraction and stabilization of MPF activity, this compound is the most potent inhibitor yet found of the kinase activity (Erikson & Maller, 1989b). The kinase can utilize as divalent cations both Mg^{2+} and Mn^{2+} with optima of 10 and 0.2 mM, respectively. Analysis of MPF has revealed a very narrow substrate specificity, with histone H1 being the best substrate so far identified. α-Casein and phosphatase inhibitor 1 are phosphorylated at less than 5% the rate of H1, and MAP-2 is phosphorylated at 40% the rate of H1. *cdc2* protein kinase is unable to phosphorylate glycogen synthase, unlike most previously characterized serine/threonine protein kinases. Analysis of the sites in histone H1 phosphorylated by *cdc2* kinase has revealed that they are the so-called growth-associated sites involved in chromatin condensation, which become stoichiometrically phosphorylated in mitosis (Allan et al., 1982; Langan, 1982). A growth-associated kinase purified from chromatin fractions of Novikoff heptoma cells on the basis of phosphorylation of these sites has been found to contain $p34^{cdc2}$ complexed with a 62-kDa protein (Langan et al., 1989). Several other investigators have also reported identification of an "M phase specific" histone H1 kinase that contains $p34^{cdc2}$ with one or more additional components (Arion et al., 1988; Meijer et al., 1989; Brizuela et al., 1989). Whether these cytosol-derived kinase complexes exhibit MPF activity is not certain at present, but they have appreciable kinase activity in interphase, as originally reported for chromatin-derived mammalian growth-associated kinase (Langan, 1982), indicating they are not really M phase "specific" but do undergo a severalfold activation in M phase.

Analysis of the phosphorylation sites in H1 histone that are phosphorylated by *cdc2* kinase has revealed a consensus sequence of $(^{K}/_{R})^{S}/_{T}P(X)^{K}/_{R}$, where the elements in parentheses are present in some but not all sites (Langan et al., 1980). Given the rather narrow substrate specificity of *cdc2* kinase, a computer search was carried out of 5251 known sequences, which identified 7% as having at least one consensus site. Many of these proteins turn out to be protooncogenes, including $pp60^{c\text{-}src}$, c-*myc*, c-*myb*, and retinoblastoma-associated protein, and many are also DNA-binding proteins (Table I). Some of these putative substrates are more likely to be involved in DNA synthesis and thus the G_1/S function of $p34^{cdc2}$. The likelihood that MPF carries out phosphorylation at these sites has been shown directly in the case of $pp60^{c\text{-}src}$. In vitro, both immunoprecipitated and baculovirus-produced $pp60^{c\text{-}src}$ are phosphorylated by *cdc2* kinase at the consensus sites (Shenoy et al., 1989; Morgan et al., 1989). These are the same sites specifically phosphorylate in mitosis in fibroblasts, when $pp60^{c\text{-}src}$ kinase activity is 4–7-fold activated and the protein undergoes a pronounced shift in electrophoretic mobility (Chackalaparampil & Shalloway, 1988). At present, however, it has not been shown that phosphorylation by *cdc2* kinase is sufficient to activate $pp60^{c\text{-}src}$ in vitro. The phosphorylation of sites in other substrates not containing this consensus is likely, inasmuch as RNA polymerase II has been reported to be phosphorylated by a homologue of $p34^{cdc2}$ at sites remote from any basic residue, although the motif of proline C-terminal to serine or threonine is conserved (Cisek & Gordon, 1989). Clearly the identification of additional substrates, particularly protooncogenes, merits urgent attention. These results are exciting because they provide a direct link between cell cycle control and cell growth control by protooncogenes. One could speculate that the changes in cell/cell interactions and cytoskeletal structure in mitosis may be mediated by protooncogenes activated by MPF, inasmuch as several structural features of normal mitotic cells resemble those seen in transformed cells (Cooper, 1989; Warren & Nelson, 1987). These studies also raise the question of when in the cell cycle oncogenes exert their regulatory roles. It has been assumed such roles are most evident in the G_0/G_1 transition, but examination of mitotic roles now seems warranted.

Table I: Subset of Proteins with *cdc2* Kinase Consensus Sites

$pp60^{src}$	RB-associated protein	*c-*myc*
D-*src28*	polyoma mT	*c-*myb*
E1b	*Hep B DNA polym[a]	*polyoma lgT
E2A	*HIV polym	*SV40 lgT
mos	*HTLV-II trans-activating protein	*JC lgT

[a] An asterisk indicates a DNA-binding protein.

As mentioned earlier, when MPF is injected into oocytes, a large burst in protein phosphorylation occurs, encompassing a 3–5-fold increase in the total amount of protein phosphate (Maller et al., 1977). This pleiotropic increase is likely to be reflective of the metabolic and structural reprogramming of the cellular biochemistry for mitosis. At present only three substrates have been identified as undergoing phosphorylation—lamins, nucleoplasmin, and ribosomal protein S6 (Miake-Lye & Kirschner, 1985; Neilsen et al., 1982; Cotten et al., 1986). In the case of S6, it has been shown that, upon MPF injection, S6KII is activated by phosphorylation on serine and threonine residues to the same extent as seen with progesterone (Erikson & Maller, 1989a). This activation may represent part of the mechanism by which MPF increases protein synthesis 2-fold (Wasserman et al., 1982). However, S6KII is also able to phosphorylate lamins at mitotic site(s), suggesting S6KII activation could be important for disassembly of the nuclear envelope (Erikson & Maller, 1988; Ward & Kirschner, 1989). Extended analysis shows that S6KII activity against both S6 and lamins oscillates in the cell cycle with kinetics similar to those of MPF. Whether the pathway for S6KII activation is via the MPF-dependent phosphorylation and activation of $pp60^{c\text{-}src}$ remains to be determined. However, in vitro MPF is unable to phosphorylate either MAP kinase or S6KII, indicating that there are additional intermediates between MPF and MAP kinase in the pathway.

Identification of p45 as a B-Type Cyclin. An important question concerns the nature of the other subunit of purified MPF, p45. A leading candidate was one of the cyclins. Cyclins were first identified in sea urchin eggs as proteins that accumulated continuously during interphase but were quantitatively degraded at the metaphase → anaphase transition during mitosis (Evans et al., 1983; Standart et al., 1987; Murray & Kirschner, 1989). Sequence analysis of cyclins from several species has shown they fall into two classes, A and B, on the basis of sequence similarity. Direct evidence that cyclins played a role in mitotic regulation came from the finding that synthetic mRNA for cyclins from various species could induce maturation when injected into *Xenopus* oocytes (Swenson et al., 1986; Pines & Hunt, 1987). Suggestive evidence for interaction of cyclins with *cdc2* protein came from the findings that small amounts of cyclin could be found in $p34^{cdc2}$ immunoprecipitates and that cyclin antibodies coprecipitated a histone H1 kinase activity (Draetta et al., 1989). In general, cyclins from different species are divergent in sequence except for a prototypic "cyclin box" motif found in the middle third of many cyclin molecules. Hunt and coworkers used oligonucleotide probes against this region to isolate *Xenopus* cDNAs for two B-type cyclins (Minshull et al., 1989) termed B1 and B2. Ablation of these mRNAs in *Xenopus* egg cell-free extracts caused arrest of the cell cycle. Western blotting and immunoprecipitation experiments with antibodies to cyclin B2 showed that it corresponded to p45 in purified MPF, and cyclin B1 was also present in similar amounts (Gautier et al., 1990). A B-type cyclin was also found to be the second subunit of MPF purified from starfish oocytes (Labbe et al., 1989b; Meijer et al., 1989). Both cyclins B1 and B2 could be phosphorylated in vitro by purified MPF, and cell cycle analysis showed the kinase activity against cyclin oscillated with kinetics identical with those of $p34^{cdc2}$. These results are exciting because they complete the molecular identification of MPF and provide an immediate control mechanism to regulate MPF activity via degradation of the cyclin component at the metaphase → anaphase transition. As mentioned earlier, in oocytes injected with catalytic amounts of MPF, a several hundred fold activation of MPF occurs. This result implies that cyclin B2 should be already present in resting G_2 phase oocytes, and Western blotting experiments show that cyclin B2 is indeed present in resting oocytes (Gautier et al., in preparation). A similar situation in meiosis I has also been described in surf clam eggs (Swenson et al., 1986). Although amounts of *cdc2* protein do not change during the cell cycle in various cell types, it thus appears meiosis I is unusual among cell cycle phases in containing a stored reservoir of cyclin as well. This result raises the question of why protein synthesis is needed for meiosis I in *Xenopus*, since both subunits of MPF are present and already associated as "pre-MPF" in resting oocytes (Gautier et al., in preparation). Recent evidence indicates the synthesis of the protooncogene c-*mos* increases severalfold early after mitogenic stimulation (Sagata et al., 1988, 1989; Freeman et al., 1989). Antisense oligonucleotides directed against c-*mos* block activation of MPF and S6 kinase (Barrett et al., 1990), and injection of synthetic c-*mos* mRNA induces maturation in the absence of mitogenic stimuli (Sagata et al., 1989). These results suggest the hypothesis that c-*mos* is involved in the activation of pre-MPF in the oocyte.

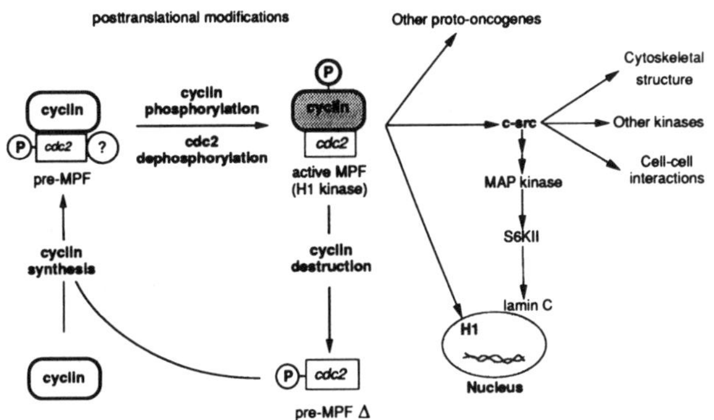

FIGURE 4: Working model of cell cycle control by MPF. The model shows that MPF activity is regulated by synthesis and degradation of the cyclin subunit and phosphorylation/dephosphorylation of the $p34^{cdc2}$ subunit. Other possible regulatory features are phosphorylation of cyclin and association with other components, but this is not yet clear. The mechanism of action of MPF to cause nuclear events can be accounted for in part by direct phosphorylation and activation of other kinases such as $pp60^{c\text{-}src}$. Since S6KII is regulated indirectly by $pp60^{v\text{-}src}$ in transformed cells, it is possible but not yet proven that the MPF-dependent but indirect activation of S6KII is via phosphorylation and activation of the c-*src* protooncogene.

Regulation of $p34^{cdc2}$ Kinase Activity. It seems clear that the cyclin subunit of MPF is regulated by its synthesis and degradation. It is likely also to be regulated by phosphorylation by $p34^{cdc2}$ and other kinases, perhaps including c-*mos*. In the case of $p34^{cdc2}$ kinase, two general mechanisms have been identified for regulating its activity (Figure 4). One is phosphorylation/dephosphorylation, and the other is formation of complexes with other proteins. Draetta and Beach (1988) showed that during G_2 phase in HeLa cells a small fraction of *cdc2* protein (5%) entered into a 170-kDa complex with a 62-kDa protein (most likely a cyclin) and the 13-kDa product (p13) of the *sucI* gene. However, only a small fraction of *sucI* protein in the cell is bound to *cdc2* protein (Draetta & Beach, 1988). Experimentally, the heat-stable *sucI* protein is valuable because when bound to beads it can remove all $p34^{cdc2}$ from extracts and does not appear to markedly inhibit kinase activity of preactivated $p34^{cdc2}$ (Dunphy & Newport, 1989). Thus, although many proteins in addition to $p34^{cdc2}$ are brought down by $p13^{sucI}$ beads (Dunphy et al., 1988), the beads can be used to monitor changes in total *cdc2* kinase activity more conveniently than immunoprecipitation. However, since only a small fraction of $p34^{cdc2}$ is present in various complexes, the *cdc2* protein bound to $p13^{sucI}$ beads may be derived from more than one complex or contain both inactive monomer and active complexed $p34^{cdc2}$, complicating the interpretation of the results. Elution of *cdc2* protein from $p13^{sucI}$ beads requires either SDS or soluble *sucI* protein (Dunphy et al., 1988; Labbe et al., 1989b); elution with the latter protein has been proposed as an aid in the purification of MPF.

Several different complexes of $p34^{cdc2}$ have been identified in cells by either purification or coprecipitation with antibodies or $p13^{sucI}$ beads (Lohka et al., 1988; Gautier et al., 1988; Brizuela et al., 1987; Draetta & Beach, 1988, Wittenberg & Reed, 1988). In most of those cases there is evidence to suggest at least one of the components is cyclin. However, it is not certain if all of these complexes represent MPF activity. At present MPF has been purified from two sources, and in both cases it consists only of a B-type cyclin complexed with $p34^{cdc2}$ with no detectable *sucI* protein (Lohka et al., 1988; Labbe et al., 1989b). This suggests that $p13^{sucI}$-containing complexes identified by immunoprecipitation have another function, although it is possible that *sucI* protein participates in early steps of MPF activation and is then removed.

The other major mechanism regulating *cdc2* protein kinase activity is phosphorylation/dephosphorylation. Draetta and Beach (1988) first reported that human *cdc2* protein immunoprecipitated in a high molecular weight complex was phosphorylated, and they suggested phosphorylation underlay the activation process. However, more careful studies with more highly synchronized cells showed that in fact just prior to entry into M phase mouse *cdc2* protein became *de*phosphorylated (Morla et al., 1989). There is now a very good correlation in several cell systems between phosphorylation and kinase inactivation of *cdc2* protein upon exit from M phase and between dephosphorylation and activation of kinase activity during entry into M phase (Gautier et al., 1989; Dunphy & Newport, 1989; Morla et al., 1989). This biochemical finding is exciting because of the previous genetic work in *S. pombe* showing that the timing of mitosis-specific *cdc2* kinase functions was delayed by the expression of the $wee1^+$ gene, which encodes a serine/threonine protein kinase (Russell & Nurse, 1987a). This is another example of the convergence of genetics and biochemistry in the studies of cell cycle regulation.

$p34^{cdc2}$ is only the second example of a serine/threonine protein kinase that is inactivated by phosphorylation, the classic case being the phosphorylation and inactivation of myosin light chain kinase by cAMP-dependent protein kinase (Conti & Adelstein, 1981). Phosphoamino acid analysis of *cdc2* protein from interphase has shown it is phosphorylated on both tyrosine and threonine residues (Morla et al., 1989), and anti-phosphotyrosine antibodies have also been used to monitor changes in phosphotyrosine content (Dunphy & Newport, 1989). This suggests two different signaling systems are likely to be involved in $p34^{cdc2}$ inactivation, one a tyrosine kinase pathway and the other a threonine kinase pathway. However, the significance of the tyrosine phosphorylation of $p34^{cdc2}$ in vertebrate cells is unclear at present. Beach and collaborators (Draetta et al., 1988) originally reported that human *cdc2* protein was a substrate for $pp60^{c\text{-}src}$, a particularly interesting possibility in light of the phosphorylation of *src* protein by *cdc2* kinase (Shenoy et al., 1989; Morgan et al., 1989). However, the study by Draetta et al. (1988) used inactive denatured *cdc2* protein, and native $p34^{cdc2}$ does not appear to be a substrate for $pp60^{c\text{-}src}$ (Shenoy et al., 1989; Morgan et al., 1989), although after denaturation it does become a substrate (J. L. Maller, unpublished). Inactivation by phosphorylation is likely to be a conserved feature since $p34^{cdc2}$ is also tyrosyl- and threonyl-phosphorylated in *S. pombe* (Gould & Nurse, 1989), where tyrosyl-phosphorylated proteins have been difficult to find. Dunphy and Newport (1989) showed that the *sucI* protein blocked the tyrosyl dephosphorylation of pre-MPF from oocytes and prevented activation of its kinase activity and ability to cause nuclear breakdown. Once $p34^{cdc2}$ was activated, however, *sucI* protein bound equally well to $p34^{cdc2}$ but had no effect on kinase activity, a property that makes $p13^{sucI}$ beads a particularly valuable tool for monitoring changes in $p34^{cdc2}$ kinase activity during the cell cycle. Whether the *sucI* protein also prevents threonine dephosphorylation was not investigated, since all analysis was by anti-phosphotyrosine antibody blotting. However, it seems likely that *sucI* protein also blocks threonine removal, since tyrosine dephosphorylation of mouse $p34^{cdc2}$ does not cause activation (Morla et al., 1989) and both phosphorylated residues are adjacent. In spite of the very strong correlation between phosphorylation and inactivation of *cdc2* kinase, isolation of the kinases and phosphatases that act on $p34^{cdc2}$ is required to establish that phosphorylation/dephosphorylation is causally responsible for changes in enzyme activity. Identification of the tyrosine kinase involved may be most efficient in yeast, due to the great scarcity of phosphotyrosyl proteins and presumably kinases in this cell and to the power of genetic analysis.

Figure 4 presents a pictorial view of the current level of understanding of MPF regulation and action. This picture is likely to be outdated rapidly because of the remarkable rate of progress recently in characterizing MPF. The model shows a cycle in which MPF activity is regulated by the synthesis and degradation of the cyclin subunit as well as phosphorylation of cyclin by *cdc2* kinase and possibly other kinases. The model also suggests that the association of $p34^{cdc2}$ with cyclin, and possibly other (?) proteins, is a regulated event, although no evidence is currently available on this point. $p34^{cdc2}$ is depicted as undergoing regulation by phosphorylation/dephosphorylation on both threonine and tyrosine residues. Targets of active MPF are shown as histone H1, at sites believed to be crucial for chromosome condensation, and $pp60^{c\text{-}src}$, which may be involved in the pleiotropic change in the cytoskeleton and cell/cell interactions that occur in mitosis. The bulk of the MPF-induced burst of phosphorylation occurs on proteins that are not direct substrates for MPF, such as ribosomal protein S6 and lamins. The figure suggests S6KII is a downstream kinase activated by MPF for both S6 and lamin phosphorylation, possibly via activation of $pp60^{c\text{-}src}$, MAP kinase kinase, and MAP kinase, although this has not yet been established. Presumably the early increase in S6KII activity occurring immediately after inactivation of the insulin receptor kinase does not involve MPF and represents only those elements of the signaling pathway distal to tyrosine kinase activation.

In summary, the *Xenopus* oocyte and egg system has been instrumental in defining steps in tyrosine kinase signaling pathways and in developing a cell-free system able to carry out cell cycle traverse in vitro. Use of this system has led to molecular elucidation of MPF and identified links between cell cycle control and cell growth control by protooncogenes. It is likely that further insights into the biochemistry of cell proliferation will come from analysis of oocyte maturation and of MPF in the cell-free system.

ACKNOWLEDGMENTS

I thank David Shalloway for identifying proteins in the gene bank with consensus sites for *cdc2* kinase and Karen Eckart for secretarial support.

REFERENCES

Adlakha, R. C., Wright, D. A., Sahasrabuddhe, C. G., Davis, F. M., Prashad, N., Bigs, H., & Rao, P. N. (1985) *Exp. Cell Res. 160*, 471–482.

Alcorta, D., Crews, C. M., Sweet, L., Bankston, L., Jones, S. W., & Erikson, R. L. (1989) *Mol. Cell. Biol. 9*, 3850–3859.

Allan, J., Hartman, P. G., Crane-Robinson, C., & Aricles, F. X. (1980) *Nature 288*, 675–679.

Arion, D., Meijer, L., Brizuela, L., & Beach, D. (1988) *Cell 55*, 371–378.

Ballou, L. M., Jeno, P., & Thomas, G. (1988) *J. Biol. Chem. 263*, 1188–1194.

Barrett, C. B., Schroetke, R. M., Van der Hoorn, F., Nordeen, S. K., & Maller, J. L. (1990) *Mol. Cell. Biol. 10*, 310–315.

Baserga, R. (1981) in *Tissue Growth Factors, Handbook of Experimental Pharmacology* (Baserga, R., Ed.) Vol. 57, pp 1–12, Springer, New York.

Beach, D., Durkacz, B., & Nurse, P. (1982) *Nature 300*, 706–709.

Blenis, J., Kuo, C. J., & Erikson, R. L. (1987) *J. Biol. Chem. 262*, 14373–14376.

Blow, J. J., & Laskey, R. A. (1986) *Cell 47*, 577–587.

Booher, R., & Beach, D. (1988) *EMBO J. 7*, 2321–2327.

Booher, R., & Beach, D. (1989) *Cell 57*, 1009–1016.

Bornslaeger, F. A., Mattei, P., & Schultz, R. M. (1986) *Dev. Biol. 114*, 453–462.

Brizuela, L., Draetta, G., & Beach, D. (1987) *EMBO J. 6*, 3507–3514.

Brizuela, L., Draetta, G., & Beach, D. (1989) *Proc. Natl. Acad. Sci. U.S.A. 86*, 4362–4366.

Chackalaparampil, I., & Shalloway, D. (1988) *Cell 52*, 801–810.

Cisek, L. J., & Corden, J. L. (1989) *Nature 339*, 679–684.

Cooper, J. A. (1989) in *Peptides and Protein Phosphorylation* (Kemp, B. E., & Alwood, P. F., Eds.) CRC Press, Boca Raton, FL (in press).

Cotten, M., Sealy, L., & Chalkley, R. (1986) *Biochemistry 25*, 5063–5069.

Cyert, M. S., & Kirschner, M. W. (1988) *Cell 53*, 185–195.

Doree, M., Peaucellier, G., & Picard, A. (1983) *Dev. Biol. 99*, 489–501.

Draetta, G., & Beach, D. (1988) *Cell 54*, 17–26.

Draetta, G., Brizuela, L., Potaskin, J., & Beach, D. (1987) *Cell 50*, 319–325.

Draetta, G., Piwnica-Worms, H., Morrison, D. K., Druker, B., Roberts, T., & Beach, D. (1988) *Nature 336*, 738–744.

Draetta, G., Luca, F., Westendorf, J., Brizuela, L., Ruderman, J., & Beach, D. (1989) *Cell 56*, 829–838.

Dunphy, W. G., & Newport, J. W. (1989) *Cell 58*, 181–191.

Dunphy, W. G., Brizuela, L., Beach, D., & Newport, J. (1988) *Cell 54*, 423–431.

Erikson, E., & Maller, J. L. (1986) *J. Biol. Chem. 261*, 350–355.

Erikson, E., & Maller, J. L. (1988) *Second Messengers Phosphoproteins 12*, 135–143.

Erikson, E., & Maller, J. L. (1989a) *J. Biol. Chem. 264*, 13711–13717.

Erikson, E., & Maller, J. L. (1989b) *J. Biol. Chem. 264*, 19577–19582.

Erikson, E., Stefanovic, D., Blenis, J., Erikson, R. L., & Maller, J. L. (1987) *Mol. Cell. Biol. 7*, 3147–3155.

Evans, T., Rosenthal, E., Youngblom, J., Distel, D., & Hunt, T. (1983) *Cell 33*, 389–396.

Freeman, R. S., Pickham, K. M., Kanko, J. P., Lee, B. A., Pena, S. V., & Donoghue, D. J. (1989) *Proc. Natl. Acad. Sci. U.S.A. 86*, 5805–5809.

Gautier, J., Norbury, C., Lohka, M., Nurse, P., & Maller, J. L. (1988) *Cell 53*, 185–195.

Gautier, J., Matsukawa, T., Nurse, P., & Maller, J. L. (1989) *Nature 339*, 626–629.

Gautier, J., Minshull, J., Lohka, M., Glotzer, M., Hunt, T., & Maller, J. L. (1990) *Cell* (in press).

Gerhart, J. C., Wu, M., & Kirschner, M. W. (1984) *J. Cell Biol. 98*, 1247–1255.

Goebl, M., & Byers, B. (1988) *Cell 54*, 739–740.

Gould, K. L., & Nurse, P. (1989) *Nature 342*, 39–45.

Hagan, I. M., Hayles, J., & Nurse, P. (1988) *J. Cell Sci. 91*, 587–595.

Hanocq-Quertier, J., Baltus, E., & Brachet, J. (1976) *Proc. Natl. Acad. Sci. U.S.A. 73*, 2028–2032.

Ingebritsen, T. S., & Cohen, P. (1983) *Science 221*, 331–338.

Jeno, P., Jaggi, N., Luther, H., Siegmann, M., & Thomas, G. (1989) *J. Biol. Chem. 264*, 1293–1297.

Jones, S. W., Erikson, E., Blenis, J., Maller, J. L., & Erikson, R. L. (1988) *Proc. Natl. Acad. Sci. U.S.A. 85*, 3377–3381.

Labbe, J. C., Picard, J. C., Picard, A., Peaucellier, G., Cavadore, J. C., Nurse, P., & Doree, M. (1989a) *Cell 57*, 253–263.

Labbe, J.-C., Capon, J.-P., Caput, D., Cavadore, J.-C., Derancourt, J., Kaghad, M., Lebois, J.-M., Picard, A., & Doree, M. (1989b) *EMBO J. 8*, 3053–3058.

Langan, T. A. (1982) *J. Biol. Chem. 257*, 14835–14836.

Langan, T. A., Zeilig, C. E., & Leichtling, B. E. (1980) in *Protein Phosphorylation and Bio Regulation* (Thomas, G., Podesta, E. J., & Gordon, J., Eds.) pp 70–82, S. Karger, Basel, Switzerland.

Langan, T. A., Gautier, J., Lohka, M., Hollingsworth, R., Moreno, S., Nurse, P., Maller, J., & Sclafani, R. A. (1989) *Mol. Cell. Biol. 9*, 3860–3868.

Lee, M., & Nurse, P. (1987) *Nature 327*, 31–35.

Lee, M., & Nurse, P. (1988) *Trends Genet. 4*, 287–295.

Lohka, M. J., & Masui, Y. (1984) *Dev. Biol. 101*, 518–523.

Lohka, M. J., & Maller, J. L. (1985) *J. Cell Biol. 101*, 518–523.

Lohka, M. J., Hayes, M. K., & Maller, J. L. (1988) *Proc. Natl. Acad. Sci. U.S.A. 85*, 3009–3113.

Maller, J. L. (1987) *J. Cyclic Nucleotide Protein Phosphorylation Res. 11*, 543–555.

Maller, J. L., & Koontz, J. W. (1980) *Dev. Biol. 85*, 309–316.

Maller, J. L., Wu, M., & Gerhart, J. C. (1977) *Dev. Biol. 58*, 295–312.

Marcus, M., Fainsod, A., & Diamond, G. (1985) *Annu. Rev. Genet. 19*, 389–422.

Masui, Y., & Markert, C. L. (1971) *J. Exp. Zool. 177*, 129–146.

Meijer, L., & Zarutskie, P. (1987) *Dev. Biol. 121*, 306–315.

Meijer, L., Arion, D., Goldsteyn, R., Pines, J., Brizuela, L., Hunt, T., & Beach, D. (1989) *EMBO J. 8*, 2275–2282.

Melchers, F., & Lernhardt, W. (1985) *Proc. Natl. Acad. Sci. U.S.A. 82*, 7681–7685.

Miake-Lye, R., & Kirschner, M. W. (1985) *Cell 41*, 165–175.

Minshull, J., Blow, J., & Hunt, T. (1989) *Cell 56*, 947–956.

Moreno, S., Hayles, J., & Nurse, P. (1989) *Cell 58*, 361–372.

Morgan, D. O., Ho, L., Korn, L. J., & Roth, R. A. (1986) *Proc. Natl. Acad. Sci. U.S.A. 83*, 328–332.

Morgan, D. O., Kaplan, J. R., Bishop, J. M., & Varmus, H. E. (1989) *Cell 57*, 775–786.

Morla, A., Draetta, G., Beach, D., & Wang, J. Y. J. (1989) *Cell 58*, 193–203.

Murray, A. W., & Kirschner, M. W. (1989) *Nature 339*, 275–280.

Murray, A. W., Solomon, M. J., & Kirschner, M. W. (1989) *Nature 339*, 280–286.

Nguyen-Gia, P., Bomsel, M., Lahrousse, J. P., Gallien, C. L., & Weintraub, H. (1986) *Eur. J. Biochem. 161*, 771–777.

Nielsen, P. J., Thomas, G., & Maller, J. L. (1982) *Proc. Natl. Acad. Sci. U.S.A. 79*, 2937–2941.

Nose, K., & Katsuta, H. (1975) *Biochem. Biophys. Res. Commun. 64*, 983–988.

Nurse, P. (1985) *Trends Genet. 1*, 51–55.

Pardee, A. B., Dubrow, R., Hamlin, J. L., & Kletzien, R. F. (1978) *Annu. Rev. Biochem. 47*, 715–750.

Peaucellier, G., Doree, M., & Picard, A. (1984) *Dev. Biol. 106*, 267–274.

Pedersen, T., & Gelfant, S. (1970) *Exp. Cell Res. 59*, 32–36.

Pines, J., & Hunt, T. (1987) *EMBO J. 5*, 157–160.

Price, D. J., Nemenoff, R. A., & Avruch, J. (1989) *J. Biol. Chem. 264*, 13825–13833.

Reynhout, J. K., & Smith, L. D. (1974) *Dev. Biol. 25*, 232–247.
Russell, P., & Nurse, P. (1987a) *Cell 49*, 559–567.
Russell, P., & Nurse, P. (1987b) *Cell 49*, 569–576.
Russell, P., Moreno, S., & Reed, S. I. (1989) *Cell 57*, 295–303.
Sadler, S. E., & Maller, J. L. (1983) *Dev. Biol. 98*, 165–177.
Sagata, N., Oskarsson, M., Copeland, T., Brumbaugh, J., & Vande Woude, G. F. (1988) *Nature 335*, 519–525.
Sagata, N., Daar, I., Oskarsson, M., Showalter, S. D., & Vande Woude, G. (1989) *Science 245*, 643–646.
Shenoy, S., Choi, J.-K., Bagrodia, S., Copeland, T. D., Maller, J. L., & Shalloway, D. (1989) *Cell 57*, 763–774.
Smith, L. D., & Ecker, R. E. (1971) *Dev. Biol. 25*, 233–247.
Solomon, M., Booher, R., Kirschner, M., & Beach, D. (1988) *Cell 54*, 738–739.
Stambrook, P., & Velez, C. (1976) *Exp. Cell Res. 99*, 57–62.
Standart, N., Minshull, J., Pines, J., & Hunt, T. (1987) *Dev. Biol. 124*, 248–258.
Stefanovic, D., & Maller, J. L. (1988) *Exp. Cell Res. 179*, 104–114.
Stefanovic, D., Erikson, E., Pike, L. J., & Maller, J. L. (1986) *EMBO J. 5*, 157–160.
Sturgill, T. W., & Ray, L. B. (1988) *Proc. Natl. Acad. Sci. U.S.A. 85*, 3753–3757.
Sturgill, T. W., Ray, L. B., Erikson, E., & Maller, J. L. (1988) *Nature 334*, 715–718.
Swenson, K. I., Farrell, K. M., & Ruderman, J. V. (1986) *Cell 47*, 861–870.
Taylor, M. H., & Smith, L. D. (1987) *Dev. Biol. 121*, 111–118.
Toh-e, A., Tanaka, K., Vesono, Y., & Wickner, R. B. (1988) *MGG, Mol. Gen. Genet. 214*, 162–164.
Walsh, D. A., Perkins, J. P., & Krebs, E. G. (1968) *J. Biol. Chem. 243*, 3763–3765.
Ward, G. E., & Kirschner, M. W. (1988) *J. Cell Biol. 107*, 524a.
Warren, S. L., & Nelson, W. J. (1987) *Mol. Cell. Biol. 7*, 1326–1337.
Wasserman, W. J., & Smith, L. D. (1978) *J. Cell Biol. 78*, R15–R22.
Wasserman, W. J., Richter, J. D., & Smith, L. D. (1982) *Dev. Biol. 89*, 152–158.
Wasserman, W. J., Houle, J. G., & Samuel, D. (1984) *Dev. Biol. 105*, 315–324.
Wittenberg, C., & Reed, S. I. (1988) *Cell 54*, 1061–1072.
Wu, M., & Gerhart, J. C. (1980) *Dev. Biol. 79*, 465–477.
Yarden, Y., & Ullrich, A. (1988) *Annu. Rev. Biochem. 57*, 443–478.

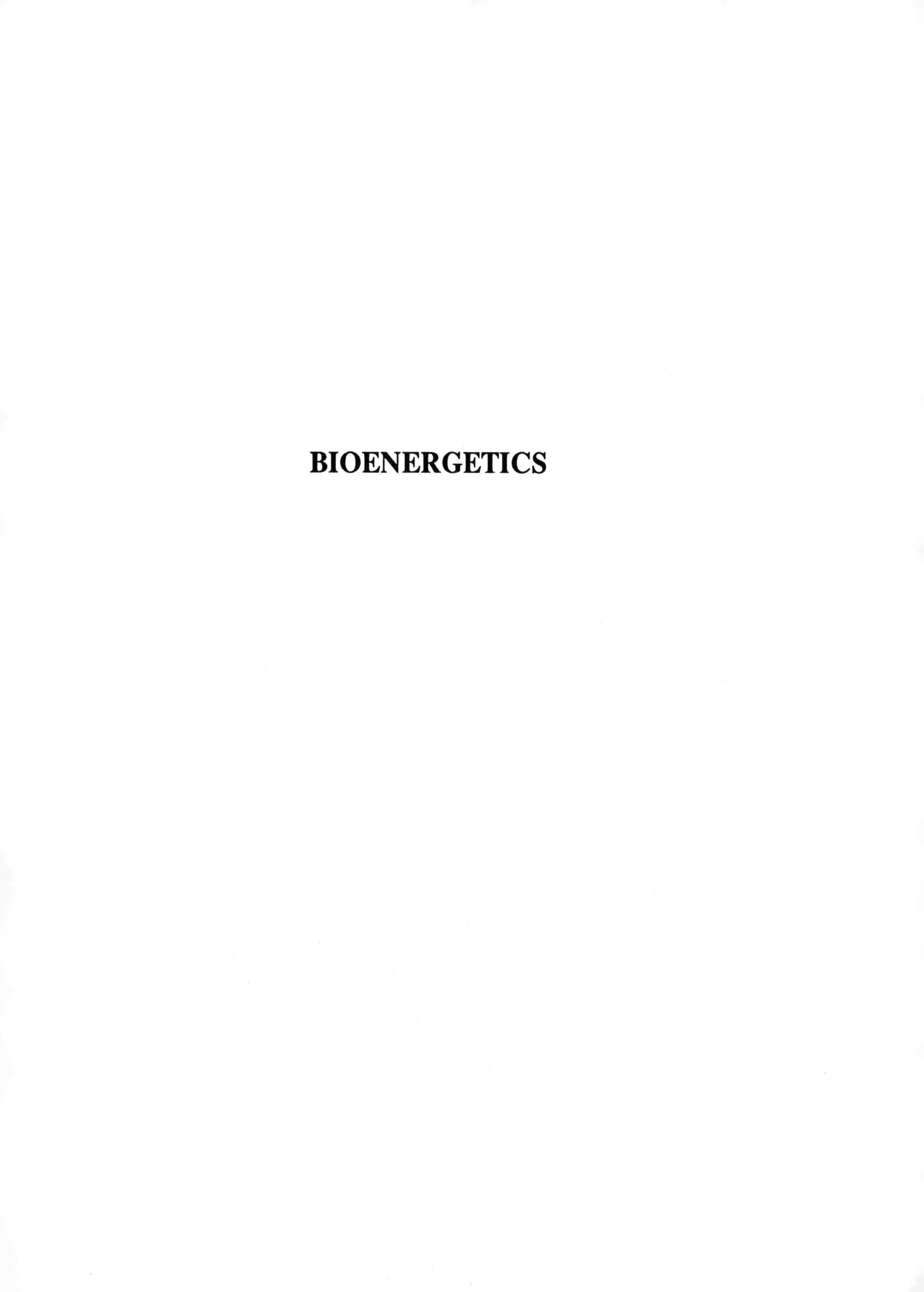

BIOENERGETICS

Chapter 27

Water Oxidation in Photosystem II: From Radical Chemistry to Multielectron Chemistry[†]

G. T. Babcock,[*,‡] B. A. Barry,[‡,§,||] R. J. Debus,[§,⊥] C. W. Hoganson,[‡,#] M. Atamian,[‡] L. McIntosh,[§] I. Sithole,[§] and C. F. Yocum[○]

Department of Chemistry, Michigan State University, East Lansing, Michigan 48824-1322, Plant Research Laboratory, Department of Energy, Michigan State University, East Lansing, Michigan 48824, and Departments of Biology and Chemistry, The University of Michigan, Ann Arbor, Michigan 48109-1055

Received June 13, 1989; Revised Manuscript Received July 6, 1989

The basic reactions in photosynthesis involve the light-driven disruption of stable chemical bonds and transfer of the electrons in these bonds to form reactive intermediate species. Ultimately, the free energy trapped in this process is used by the organism to drive the synthesis of proteins, nucleic acids, and other complex biomolecules. Schematically, these events can be represented as

low energy electrons (stable substrate) $\xrightarrow{h\nu}$ high energy electrons (reactive intermediates) $\xrightarrow[\text{biochemical reactions}]{\text{spontaneous}}$ plant growth and reproduction

In higher plants, algae, and cyanobacteria the low-energy electron source is water, which is oxidized to molecular oxygen with the injection of four electrons into the photosynthetic system:

$$2H_2O \rightarrow O_2 + 4H^+ + 4e^-$$

In developing the capacity to metabolize water, these organisms overcame two major difficulties. First, water is a very stable molecule, hence its abundance, and a reactive, unstable species must be created within the organism to oxidize it. Access to this highly oxidizing species must be controlled so that it is directed at water and not at components vital to the organism itself. Second, the photochemistry that occurs in photosynthesis is one-electron photochemistry; that is, one absorbed photon generates one oxidizing equivalent. Water chemistry, as indicated above, is a four-electron process, and a means by which to combine the oxidizing power of four absorbed photons is necessary. Moreover, this has to be done in such a way that reactive intermediates in the water oxidation process, such as hydroxyl radical or peroxide, are not released into the cell.

The photochemical and electron-transfer reactions that occur in the membrane-bound structure that evolved to overcome these obstacles, the photosystem II/oxygen-evolving complex (PSII/OEC),[1] can be represented as

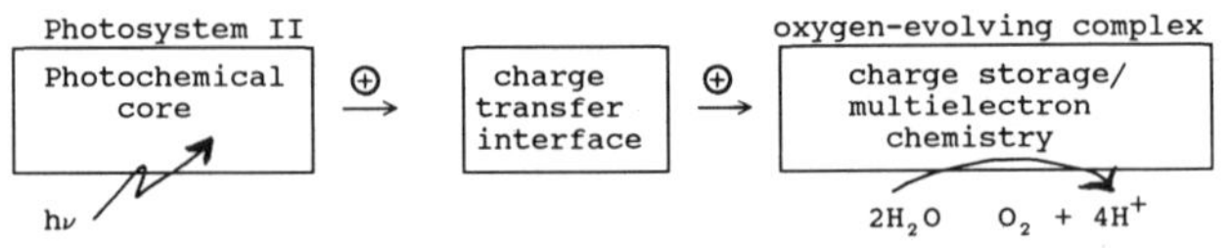

Photosystem II forms the photochemical core of the system: upon light absorption PSII generates oxidizing equivalents or holes at sufficiently high potential to oxidize water. These oxidizing equivalents, designated as ⊕, are transferred one at a time through the charge-transfer interface to the oxygen-evolving complex where the four equivalents required in the water oxidation process are accumulated. Accordingly, the OEC can exist in five redox states depending on the number of stored oxidizing equivalents. These redox states are designated S_0–S_4, where the subscript denotes the charge storage state, with S_4 being the most oxidizing and capable of oxidizing water. In many respects the OEC acts as a capacitor, and the charge-transfer reactions that precede water splitting have a marked dependence on the S state of the complex.

[†] Research carried out in the authors' laboratories was supported by the USDA Competitive Research Grants Office, Photosynthesis Program (G.T.B. and C.F.Y.), by the National Science Foundation (C.F.Y.), by the U.S. DOE (L.M.), and by the NIH (GM 37300 to G.T.B.).

[‡] Department of Chemistry, Michigan State University.

[§] Plant Research Laboratory, DOE, Michigan State University.

[||] Present address: Department of Biochemistry, University of Minnesota, St. Paul, MN 55108.

[⊥] Present address: Department of Biochemistry, University of California, Riverside, Riverside, CA 92521.

[#] Present address: Department of Biochemistry and Biophysics, Chalmers Institute of Technology, S-41296 Goteborg, Sweden.

[○] Departments of Biology and Chemistry, The University of Michigan.

[1] Abbreviations: PSII, photosystem II; OEC, oxygen-evolving complex.

The PSII/OEC exists as a discrete, isolable entity in the photosynthetic membrane [for reviews see Ghanotakis and Yocum (1985), Babcock (1987), Andréasson and Vänngård (1988), Renger (1988), and Brudvig et al. (1989)], and there is no indication of charge-transfer communication between different PSII/OEC units. Thus, a remarkable aspect of this complex, which has emerged as a general theme in the organization of photosynthetic electron-transfer assemblies, is the rapid departure from free radical chemistry. That is, the charge separation that occurs in the photochemical core necessarily generates one-electron oxidized and reduced species. The oxidizing and reducing equivalents carried by these radical intermediates are quickly combined at staging components or "gates", and subsequent reactions occur from these multielectron oxidized or reduced species. This gating activity usually takes place within the complex that contains the photochemical core and provides catalytic flexibility, as well as protection from reactive radical species. In the water-splitting process, this function is particularly apparent. The OEC acts as a four-electron gate, and radical reactions in water oxidation are avoided. This perspective aims at using recent results on the structure and function of the PSII/OEC to provide insight into the underlying principles of its operation.

Biochemical Organization of the Photosystem II/Oxygen-Evolving Complex

The close association of the photochemistry that produces the oxidizing equivalents necessary for water oxidation and the catalytic site at which water splitting occurs was suggested by the flashing light, oxygen-evolution measurements of Joliot, Kok, and their respective co-workers (Joliot & Kok, 1975). These experiments led to the concept of the charge storage center with its S_n valence states and to the tight physical coupling between PSII photochemistry and water oxidation at the charge storage center. As the structure of the photosynthetic membrane was elucidated, the expectation developed that both functions are incorporated in a single membrane-bound complex. This expectation was confirmed when Stewart and Bendall (1979) isolated an oxygen-evolving PSII complex from a thermophilic bacterium; Yocum and co-workers (Berthold et al., 1981) provided an extremely convenient, high-yield procedure by which to disintegrate the photosynthetic membrane and recover O_2-evolving PSII complexes. This latter development provided the starting material for a range of biochemical and biophysical experiments that have generated detailed insight into the organization and function of the PSII/OEC.

From a biochemical perspective, the PSII/OEC complex has been remarkably easy to resolve. Thus, a variety of smaller assemblies has been isolated following selective detergent solubilization of individual polypeptides. This situation has allowed rapid progress to be made in understanding the subunit composition of the complex and the functions of the specific polypeptides in the light absorption, charge separation, charge storage, and water oxidation reactions. While a considerable effort is needed to complete these assignments and, in particular, to understand the molecular basis for various activities associated with the PSII/OEC, there is enough information available to construct reasonable working models for the polypeptide and cofactor organization. Figure 1 shows such a model. The redox cofactors that are required for photochemistry and oxygen evolution are bound into a membrane-spanning complex that contains both extrinsic and intrinsic polypeptides. Several of these cofactors are indicated in Figure 1 and include the photochemically active chlorophyll P_{680}, a pheophytin initial electron acceptor, the quinone

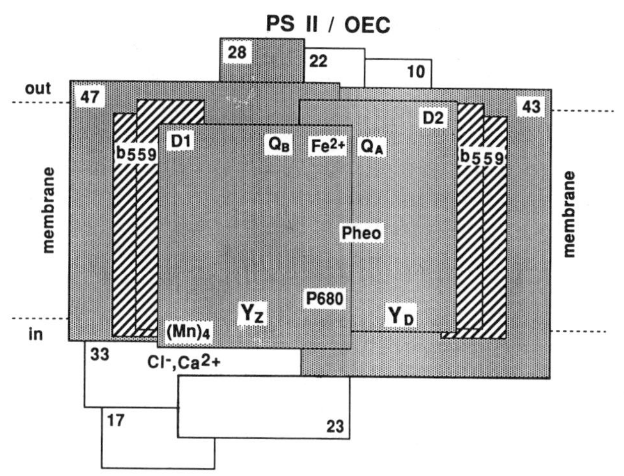

FIGURE 1: Model for the structure of the photosystem II/oxygen-evolving complex. Masses (in kilodaltons) are indicated for each polypeptide except for D1 and D2, which have masses of approximately 34 and 32 kDa, respectively, and cytochrome *b*-559, which occurs as a heterodimer of 4- and 9-kDa polypeptides. Chlorophyll binding subunits are stippled, and cytochrome *b*-559, which probably has two copies per PSII, is dashed. The 17-, 23-, and 33-kDa polypeptides are peripheral to the membrane; all other subunits are membrane-spanning. P_{680} is the reaction center chlorophyll, Pheo is an intermediate pheophytin electron acceptor, Q_A and Q_B are quinone acceptors, and Y_Z and Y_D are redox-active tyrosines. Other chlorophyll and pheophytin chromophores, which are associated with the PSII/OEC, are not shown.

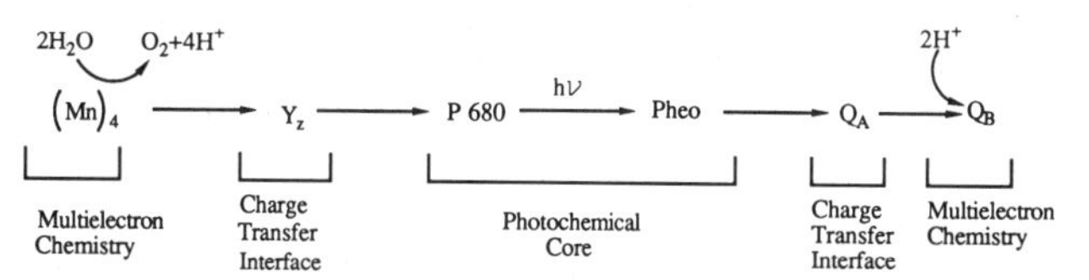

FIGURE 2: Sequence of electron transfers that occur in the PSII/OEC. On the acceptor side, the two-electron reduced species Q_BH_2 dissociates from its D1 binding site and diffuses to the cytochrome b_6f complex where further redox chemistry and proton translocation occur.

electron acceptors Q_A and Q_B, and the ferrous ion that is associated with these species. Four manganese atoms are involved in the charge storage/water oxidation site. This center is linked to P_{680} by a redox-active tyrosine, Y_Z, which functions as the charge-transfer interface. A second redox-active tyrosine, Y_D, and cytochrome *b*-559 are also present but do not appear to operate intimately in the main PSII/OEC functions. In terms of electron-transfer sequence and the transition from radical to multielectron chemistry, a simple linear chain best describes the situation (Figure 2). The kinetics of these reactions are considered in more detail below; before doing this, we discuss briefly the polypeptides and non-redox-active cofactors, Ca^{2+} and Cl^-, shown in Figure 1.

Approximately 12 different polypeptides are involved in the PSII/OEC; at least 3 of these are peripheral, the other 9 are membrane-spanning. The extrinsic polypeptides—the 17-, 23-, and 33-kDa subunits in Figure 1—appear to serve at least two functions. The first is to facilitate binding of the inorganic cofactors that are necessary for water-splitting chemistry. These include Cl^- and Ca^{2+}, in addition to the manganese ions at the site of charge storage and water oxidation chemistry [for review, see Homann (1987) and Babcock (1987)]. The function of Cl^- is not well understood, although the fact that amines, which bind to manganese (Beck & Brudvig, 1988; Britt et al., 1989) and inhibit O_2 evolution, are competitive with Cl^- provides support for the idea that the anion ligates manganese in some of the S states in a mechanistically important way (Sandusky & Yocum, 1984, 1986). The role of

Ca^{2+} is even more unclear. That this species is required for stable charge storage in the Mn ensemble is apparent, but the mechanism by which Ca^{2+} operates and the specific S-state transitions that are affected by its depletion are matters of considerable controversy [e.g., Ono & Inoue (1989), Boussac and Rutherford (1988), and dePaula et al. (1986)]. Interesting possibilities include models in which Ca^{2+} is either directly involved with manganese in a heteronuclear cluster (Bonadies et al., 1989; Penner-Hahn et al., 1989) or occurs with a unique coordination sphere but in close proximity to the manganese ensemble (Rutherford, 1989). The second function of the extrinsic subunits, and particularly of the 33-kDa polypeptide (Ghanotakis et al., 1984), is to isolate the endogenous redox-active species in the complex from the aqueous milieu. Owing to the likelihood of rearrangements occurring in the intrinsic polypeptide core upon removal of the soluble subunits [e.g., Hoganson et al. (1989)], this function has been more difficult to study quantitatively. Nonetheless, achieving such isolation was the key to the successful construction of the water oxidation system, and the erection of a diffusion barrier to limit access of unwanted reductants to the intermediate S states in water splitting seems to be part of the strategy.

Turning to the intrinsic polypeptides in Figure 1, the D1/D2 heterodimer forms the photochemical core by binding the chlorophylls (P_{680}), the pheophytins, and quinones that mediate the light-driven charge separation reaction (Nanba & Satoh, 1987). The sequence and secondary structure analogies that exist between D1/D2 and the corresponding L/M complex that comprises the bacterial reaction center core have been well documented [e.g., Hearst (1986) and Michel and Deisenhofer (1988)]. The pseudo-C_2 symmetry in protein structure and chromophore arrangement that occurs in the bacterial reaction center appears to be preserved in D1/D2 and has been shown to be a useful tool in designing experiments to study PSII (Debus et al., 1988a,b; Vermaas et al., 1988). This topic and the related issue of whether and where the pseudo-C_2 symmetry breaks down in the O_2-evolving system are considered in detail below.

At least three of the other intrinsic polypeptides in Figure 1, the 47-, 43-, and 28-kDa polypeptides, are also known to bind chlorophyll. The 47-kDa polypeptide can be isolated along with D1, D2, and cytochrome *b*-559 in a discrete unit. Recent work has shown that these particles from two-dimensional lattices can be studied by electron microscopy and that, despite the fact that this preparation lacks endogenous quinone [Petersen et al., 1989; see also Diner et al. (1988b)], a tyrosine radical (presumably either Y_Z^+ and Y_D^+) can be generated by light in the presence of an exogenous acceptor. This latter finding contrasts with the situation in more highly resolved D1/D2/*b*-559 preparations for which tyrosine radical generation has not been reported. That the 47-kDa polypeptide may play additional roles in PSII beyond a simple light-harvesting function is also suggested by reports that the 33-kDa peripheral polypeptide can be cross-linked to it with retention of both manganese and O_2-evolution activity [Enami et al., 1989; Bricker et al., 1988; see also Gounaris et al. (1988)]. The 28-kDa polypeptide binds chlorophyll *a* and had been isolated in pure form (Ghanotakis et al., 1987). It has also been implicated in maintaining the native Q_B binding site on the D1 polypeptide (Bowlby et al., 1988). The 22- and 10-kDa polypeptides, which apparently do not bind cofactors, also appear to influence events in the vicinity of the quinones as their removal from PSII preparations correlates with increased accessibility of Q_A^- to exogenous acceptors. The other intrinsic polypeptides in PSII, the 4 and 9 kDa, provide histidine ligands, one from each polypeptide, to the heme of cyt *b*-559 (Babcock et al., 1985). Recent work has shown that both *b*-559 hemes are likely to be oriented to the outside of the membrane (Tae et al., 1988).

ELECTRON-TRANSFER KINETICS IN THE PSII/OEC

The simple linear scheme for PSII electron transfer shown in Figure 2 is now generally accepted. In this section we focus on the kinetics of these reactions as they occur under conditions that are close to physiological. At low temperatures, the S-state oxidation reactions become slow reactive to those of competing pathways, and interesting alterations to the scheme in Figure 2 occur. In one of these alternative pathways, cytochrome *b*-559 is oxidized by way of a redox-active chlorophyll through Y_Z^+ and/or P_{680}^+. This pathway is postulated to have physiological significance in protecting PSII from photoinhibition, to which it is especially sensitive owing to the highly oxidizing conditions that are necessary for water oxidation (Thompson & Brudvig, 1988). This model is well suited to experimental test; the details of this pathway, however, will not be considered further here.

Primary Charge Separation and Electron Removal in PSII. The analogy to the photochemical reactions and structures that occur in the photosynthetic bacteria has led to rapid progress in understanding the primary charge separation in PSII (Michel & Deisenhofer, 1988). Schatz et al. (1988) predicted a formation time of ~3 ps for P_{680}^+ $Pheo^-$ in open reaction centers, which has recently been observed by Wasielewski et al. (1989). An interesting aspect to this reaction is that this primary charge-separated state is close in energy to the precursor P_{680} electronic excited state (Van Gorkom, 1985; Schatz et al., 1988). This circumstance leads to several consequences: the primary photochemistry is reversible, with P_{680} serving as a shallow trapping state for excitons in the antenna pigment bed; the exciton decay kinetics are trap-limited rather than diffusion-limited, and the trapping time is a function of the antenna size. Mathis and co-workers (Hansson et al., 1988) have considered similar effects in discussing their observation that the P_{680}^+ $Pheo^-$ lifetime and yield are functions of the antenna size in various PSII preparations [see also Schlodder and Brettel (1988)]. Schatz et al. (1988) note that the interrelationship between antenna size and trapping time allows PSII to use the antenna system for exciton storage; that is, the antenna serves as a buffer for the slower electron-transfer steps that follow charge separation. This strategy was cited as increasing the overall efficiency of charge separation. As noted below, the use of kinetic buffers to increase stable charge separation efficiencies is also apparent in the reactions that precede O_2 evolution.

The electron transfer from $Pheo^-$ to Q_A occurs in the 300–600-ps range and has been resolved by studying both the oxidation of $Pheo^-$ (Nuijs et al., 1986; Schatz & Holzwarth, 1987) and the reduction of Q_A (Eckert et al., 1988). This time appears to be slightly slower than the 200-ps $BPheo^-$ to Q_A transfer in bacterial centers (Michel & Deisenhofer, 1988). This difference has also been observed recently in direct, time-resolved photovoltage measurements of the primary charge separation event (Trissl & Leibl, 1989). In the same study, the dielectric distance between Pheo and Q_A was found to be shorter in PSII than in the bacterial reaction centers. The slowing down of the $Pheo^-/Q_A$ reaction, despite the shorter distance, was attributed to a larger reorganization energy accompanying Q_A reduction in the oxygen-evolving organisms relative to the bacteria.

The final electron-transfer event in the shuttling of electrons out of PSII involves the transition to multielectron chemistry

Table I: Stoichiometries and Kinetics for Reactions Preceding O_2 Evolution

reaction	S state prior to photon absorption				ref
	S_0	S_1	S_2	S_3	
$Y_ZP_{680}^+ \rightarrow Y_Z^+P_{680}$ (ns)	20	20	40, 280	40, 280	Meyer et al., 1989
$Y_Z^+S_n \rightarrow Y_ZS_{n+1}$	30 μs	100 μs	300 μs	1.0–1.2 ms	Babcock et al., 1976; Dekker et al., 1984b
H^+ release	1	0	1	2	Saphon & Crofts, 1977
H^+ appearance kinetics	250 μs		220 μs	1.2 ms	Forster & Junge, 1985
formal change[a]	0	0	+1	+1	
$S_n \rightarrow S_{n-1}$ (min)		stable	3–3.5	3.5–4	Styring & Rutherford, 1988
T_{freeze} (K)[b]	220–225	135–140	230	235	Styring & Rutherford, 1988
E_a (kJ/mol)[c]		9.6	26.8	15.5 (T >289 K), 59.4 (T <289 K)	Koike et al., 1987

[a] The formal charge is determined by considering proton release and the number of oxidizing equivalents stored in the OEC; below pH 5.3 protonation of an endogenous group occurs that also influences the formal charge; see Meyer et al. 1989). [b] T_{freeze} is defined as the temperature at which the indicated S-state transition is half-inhibited. [c] Determined in a thermophilic cyanobacterium; the exact kinetics in this system vary slightly from those in higher plant preparations, but the general characteristics of the various reactions appear to be maintained.

in the $Q_A^- \rightarrow Q_B$ transfer. The Q_B^- species remains tightly bound in its binding site and is reduced to the Q_B^{2-} species in a second electron transfer. The kinetics of the oxidation of the Q_A^- depend on the redox state of Q_B: Q_B^- is formed in ~200 μs whereas Q_B^{2-} is formed at about half this rate [see Crofts and Wraight (1983) for a review of this topic and of the fate of Q_BH_2 after it leaves its binding site].

Hole Transfer and Accumulation Prior to Water Oxidation. The analogy to electron transfer in bacterial reaction centers clearly breaks down in the reactions that transfer holes out of the oxidized reaction center chlorophyll. In PSII, P_{680}^+ is reduced by a tyrosine residue (Y_Z in Figure 2; Barry & Babcock, 1987; Debus et al., 1988a,b; Gerken et al., 1988) whereas a cytochrome is the electron source in the non-oxygen-evolving systems. This difference clearly reflects the difference in function. In PSII, the oxidizing power generated in producing P_{680}^+ ($E_m^{0\prime} \approx 1.2$ V) must be maintained and directed eventually at water. Tyrosine oxidation, which produces the highly oxidizing tyrosine radical ($E_m^{0\prime} \approx 1.0$ V), is a means by which to achieve this. The reduction potentials generated photochemically in the bacterial reaction center are significantly lower, and cytochrome oxidation provides an energetically economic means by which to achieve stabilization of the primary charge separation.

Table I summarizes current data on the half-times and, where available, the temperature dependencies of the various electron-transfer and proton release events that occur as the PSII/OEC builds the oxidizing potential required for H_2O decomposition. Examination of these data reveals that the electron-transfer kinetics depend on both the protonation state in PSII and the oxidation state of the manganese complex. This is summarized in Table I under the "formal charge" heading; for the $S_0 \rightarrow S_1$ transition, for example, the oxidizing equivalent stored in the OEC is neutralized by the proton released during this S-state advance, and thus the formal charge in the PSII/OEC is the same in the S_0 and S_1 states. Witt and co-workers have recently carried out an examination of the pH dependence of the $Y_ZP_{680}^+ \rightarrow Y_Z^+P_{680}$ reaction (Meyer et al., 1989). Their results indicate that the protonation state of an acid/base group (pK_a = 5.3) also influences the kinetics of this reaction: at pH values below the pK_a of this group the $Y_ZP_{680}^+$ electron transfer is retarded compared to its kinetics at pH values above the pK_a. The biphasic time course of the $Y_ZP_{680}^+$ reaction for complexes in the S_2 and S_3 states (Table I) was attributed to heterogeneity in the sample with respect to the protonation state of this group. Witt and co-workers consider that the formal charge dependence of the redox reactions in PSII reflects direct electrostatic effects; protein rearrangements that accompany an S-state advance also provide an explanation for this behavior.

The rate of the $Y_ZP_{680}^+ \rightarrow Y_Z^+P_{680}$ reaction correlates well with the formal charge concept. The $S_nY_Z^+ \rightarrow S_{n+1}Y_Z$ reactions, however, depend more strongly simply on the oxidation state of the cluster, slowing down by roughly a factor of 3 for each additional oxidizing equivalent stored in the S ensemble. Babcock et al. (1976) noted early on that the half-time of the reduction of Y_Z^+ during the $S_3 \rightarrow S_0$ transition corresponded to the release kinetics of the O_2 formed on this transition and concluded that the electron-transfer, rather than water chemistry, rate limits the oxygen-evolution process. This conclusion accorded well with the lack of a significant deuterium isotope effect in the water oxidation process (Sinclair & Arnason, 1974). Recent work by Van Gorkom and co-workers (Plijter et al., 1988), however, has suggested that the early measurements of O_2 release kinetics may be in error and that O_2 remains bound for approximately 50 ms after its formation on the $S_3 \rightarrow S_0$ transition. Resolution of this issue is necessary.

The nature of the individual S-state transitions and the factors that influence their kinetics are beginning to be understood more clearly. X-ray absorption spectroscopy indicates a relatively high average manganese oxidation state (~+3) for the relaxed, low S-state ensemble (Sauer et al., 1988; Penner-Hahn et al., 1989). Data from a variety of spectroscopic approaches including optical absorption, EPR, X-ray absorption, and NMR [for reviews see Babcock (1987), Andreasson and Vanngard (1988), and Brudvig et al. (1989)] suggest that manganese oxidation is involved in most, if not all, of the S-state transitions. The optical data are probably the most direct in this context and, on the basis of near-UV absorption changes that are similar for each S-state advance [with the possible exception of the $S_0 \rightarrow S_1$ transition; see Lavergne (1989)], have been interpreted to indicate that the manganese ensemble acts as a linear charge accumulator as the OEC moves to the S_4 state (Kretschmann et al., 1988; Lavergne, 1989; Dekker, 1989). The other spectroscopies generally support this notion [see, for example, Evelo et al. (1989a)], with the important exception of $S_2 \rightarrow S_3$. For this transition, there is no evidence for a manganese valence change in the X-ray absorption edge, and both NMR and EPR fail to detect alterations in manganese valence as judged by relaxation measurements. These results are somewhat less direct than the optical work, however, and this situation, coupled with the fact that the $S = 1/2$ multiline EPR signal of the S_2 state disappears upon formation of S_3, suggests that manganese oxidation, perhaps accompanied by structural rearrangement, does occur on this transition.

Insight into the extent and functional implication of conformational change during the various S-state transitions is becoming available. Koike et al. (1987) and Styring and Rutherford (1988) have recently measured the temperature dependencies of the individual S-state transitions. The former group used a thermophilic cyanobacteria for their work and obtained activation energies for three of the four S-state transitions as indicated in Table I. They found that the $S_1 \rightarrow S_2$ transition had a relatively small activation energy, which accords well with the observation that this transition persists to much lower temperatures than the other S-state changes (Table I). Styring and Rutherford interpreted the low inhibition temperature as reflecting the fact that $S_1 \rightarrow S_2$ is the only S-state advance that is not accompanied by proton release, suggesting that conformation rearrangements on this transition will be minimal. This interpretation, however, is at odds with the observation that the activation entropy is significantly larger for $S_1 \rightarrow S_2$ [-140 J/(mol·K)] than for $S_2 \rightarrow S_3$ [-85 J/(mol·K)], which indicates that conformational rearrangements to accommodate the increase in formal charge on the $S_1 \rightarrow S_2$ transition are of consequence (Koike et al., 1987). In agreement with this interpretation, de Paula et al. (1987) have noted that the conversion between the two EPR-detectable forms of the S_2 state, the *g* 4.1 and the multiline signals, proceeds with a significant negative activation entropy (~-90 J/mol·K) when the multiline is the product state.

The view of the S-state transitions that emerges from these considerations can be summarized as follows. The valence states of the four manganese that are involved in the OEC are primarily +3 and +4, with +3 occurring as the average value in the lower S states and advancing sequentially toward the +4 average in the higher S states. Water oxidation is probably a concerted event in the $S_3 \rightarrow S_0$ transition, although recent suggestions of H_2O_2 involvement in the OEC (Johansen, 1988; Wydrzynski et al., 1989) may indicate this species as an intermediate. Despite the apparent metal-centered oxidation character of the S-state transitions, however, there appear to be significant protein and/or cluster nuclear rearrangements on each of the transitions. Proton release is likely to be involved, and S-state-dependent Cl^- ligation to the manganese ensemble (Sandusky & Yocum, 1984; Rutherford, 1989) may also play a role in these nuclear activities.

Electron-Hole Recombination in PSII and the Necessity of Y_Z. At first glance, it would appear that water oxidation could proceed effectively with only the reaction center chlorophyll and the charge-accumulating/water-splitting manganese ensemble, that is, that the intermediacy of Y_Z is unnecessary in this process. The kinetic considerations above, however, combined with information on charge recombination rates in PSII, show clearly that Y_Z is vital to preserving high quantum yield in the O_2-evolving process.

Efficient photosynthesis requires that the initial products of primary charge separation, P^+ and Q^-, be engaged in productive electron transfer before deleterious P^+Q^- charge recombination can occur. This can be accomplished by removing the electron from Q^- or by filling the hole in P^+. In PSII the Q_A^- to Q_B electron transfer, like the analogous bacterial reaction, proceeds with a half-time of ~200 μs (Crofts & Wraight, 1983). In the oxygen-evolving system this time is essentially the same as the $P_{680}^+Q_A^-$ recombination time (Mathis & Rutherford, 1987; Hoganson & Babcock, 1989), which indicates that the net quantum yield would drop to 50% if the system relied on this reaction to preserve charge separation.

The relative sluggishness of Q_A reoxidation indicates that to achieve high quantum yield the hole on P_{680}^+ must be filled rapidly. From the discussion above, it appears that fairly substantial nuclear rearrangements are likely to accompany the S-state transitions, however, which suggests that the reorganization energies for these electron transfers are relatively large. Moreover, the average driving force for these transfers is small: the redox potential for P_{680}^+ is ~1.2 V, and the water/oxygen couple has a potential of 0.92 V at pH 5, the relevant pH in the photosynthetic water-splitting process. This combination, i.e., small driving force and large reorganization energy, is likely to limit the rates of hole transfer into the manganese ensemble, and indeed, Table I shows that these reactions proceed with times comparable to the $P_{680}^+Q_A^-$ recombination time. Even if these rates were increased by a factor of 10 by eliminating Y_Z, a significant departure from unit quantum yield in charge separation would occur. By introducing Y_Z, however, the photosynthetic system has avoided this pitfall. The forward rates of electron transfer from the tyrosine into P_{680}^+ are 3–4 orders of magnitude faster than $P_{680}^+Q_A^-$ charge recombination, which provides a quantum yield approaching unity. As importantly, the $Q_A^-Y_Z^+$ recombination reaction proceeds with a half-time that is approximately 2 orders of magnitude slower (~60 ms; Dekker et al., 1984a) than the $P_{680}^+Q_A^-$ recombination. Relative to the $Q_A^-Y_Z^+$ reaction, the productive $Y_Z^+S_n$ reactions are now fast, which ensures that high quantum yield is preserved all the way to the water-splitting site. Thus, the necessity of Y_Z^+ is clear: it functions as a hole storage tank to allow productive forward electron-transfer reactions to compete successfully with wasteful, but thermodynamically favored, charge recombination reactions.

C_2 Symmetry in Photosystem II and Identification of Y_Z and Y_D

In addition to Y_Z, photosystem II contains a second tyrosine free radical, Y_D^+ (Barry & Babcock, 1987). This species is stable in its free radical form and is not involved in the electron-transfer reactions that lead to water oxidation. The EPR line shapes of Y_Z^+ and Y_D^+ are identical, which indicates that the unpaired electron spin density distribution and the orientation of the tyrosine phenol ring with respect to the polypeptide backbone are identical for both species (Barry & Babcock, 1988). The orientation of the phenol ring, in particular, influences the EPR line shape, and differences in the orientation account for the fact that the spectra of Y_D^+ and Y_Z^+ are distinct from those of the tyrosine radicals that occur in ribonucleotide reductase and prostaglandin synthase (Bender et al., 1989).

This unusual and surprising circumstance, i.e., the occurrence in PSII of two tyrosine radicals, one active and the other inactive in electron transport, has been the key to their assignment to specific residues in PSII and to providing insight into the organization of the reaction center.

For Y_Z^+ and Y_D^+ the similar EPR line shapes indicate that the polypeptide sequences in the vicinity of the two radicals are similar. Debus et al. (1988a,b) used this information plus the idea that analogies exist between PSII and the bacterial reaction center (Michel & Deisenhofer, 1988) to design directed mutagenesis experiments aimed at engineering out one or the other radical. The striking lessons apparent from the bacteria were, first, the near C_2 symmetric arrangement of both the L and M polypeptides and of the chromophores involved in charge separation and, second, the inactivity of one of the C_2 symmetric branches in charge separation. The occurrence of Y_Z/Y_D in PSII echoes this situation and suggested that the two tyrosines may occur in symmetry-related amino acid clusters on D1 and D2. Inspection of the sequences of these two polypeptides with these considerations in mind showed that only the tyrosine pair at the 160 (D2) and 161 (D1) positions fulfilled these criteria: in the folding scheme of D1 and D2 proposed by Trebst (1986) [see also Sayre et al. (1986)], these two tyrosines occur in symmetry-related positions in the C membrane-spanning helices of the two polypeptides; moreover, the four upstream and four downstream amino acids adjacent to the tyrosines are essentially conserved between D1 and D2, which is important in rationalizing the nearly identical spectral properties of the two radicals. Mutagenesis of these tyrosines to phenylalanine and subsequent spectral and functional analysis showed that Y_D corresponds to Tyr^{160} of D2 and that Y_Z is Tyr^{161} of D1 (Debus et al., 1988a,b). Vermaas and co-workers (1988) have obtained the same result for Y_D, and Metz et al. (1989) have recently mutagenized the D1 Tyr^{161} and agree with its assignment as Y_Z. These results also accord with iodination data that suggest that Y_D occurs in D2 and Y_Z in D1 [e.g., Ikeuchi and Inoue (1988) and Takashi and Satoh (1989)].

Two conclusions follow from these data. First, the analogy between PSII and the bacterial reaction center, in terms of C_2 symmetry considerations and in terms of active and inactive electron-transport branches, is considerably strengthened by the Y_Z/Y_D assignments. In fact, they show that experiments designed with these considerations in mind are a powerful tool in analyzing PSII. A caveat to this approach, which is also considered in the following section, is that the analogy should be used with prudence: PSII splits water whereas the bacterial reaction center does not; in addition, there are likely to be evolutionary embellishments, for example, in pigment composition (Dekker et al., 1989), that have occurred in PSII that are absent in the bacteria. Second, when Y_Z is mutagenized to phenylalanine, photosynthetic growth is absent, whereas when Y_D is deleted, photosynthetic growth continues (Debus et al., 1988,b). Thus, Y_D cannot substitute for Y_Z, and despite the apparent C_2 symmetry in PSII, the Y_D branch is functionally incompetent. Within the context of reaction center design and function these results indicate that the symmetry-related branches are not functionally redundant; i.e., one does not serve as a "backup" for the other.

That Y_D^+ is stable in its radical form is a very surprising observation. Its redox potential is expected to be extremely oxidizing—the corresponding Y_Z^+ species is sufficiently unstable to be able to drive the oxidation of water—and consequently one would expect a very short lifetime of the free radical. Recent experiments have provided a basis to interpret this behavior. Rodriguez et al. (1987) used ENDOR spectroscopy to show that the phenol oxygen in the radical is involved in a hydrogen bond interaction and that the hydrogen-bonded proton is extremely resistant to exchange with bulk solvent. Evelo et al. (1989b) have recently confirmed this observation in a series of spin-echo EPR measurements. The hydrogen bond in Y_D^+ contrasts in an interesting way with the situation in ribonucleotide reductase where the phenol oxygen of the neutral radical is clearly not involved in a hydrogen bond (Bender et al., 1989). Y_D^+ is thus well isolated from solvent, which explains its insensitivity to added reductants, and suggests that the phenol hydrogen may undergo a rocking motion upon redox change as indicated in Figure 3 [see also Eckert and Renger (1988)]. Whether such a situation occurs for Y_Z upon redox change is unknown, but it will be interesting to study the effect of deuterium exchange upon the redox kinetics of this species. These results support the idea, which can also be inferred from D1/D2 folding patterns, that Y_Z and Y_D are buried in the membrane and that the stability of Y_D^+ arises from simple isolation from redox-active species. Seclusion of Y_D^+ is also indicated in EPR relaxation measurements reported by Isogai et al. (1988) and by Innes and

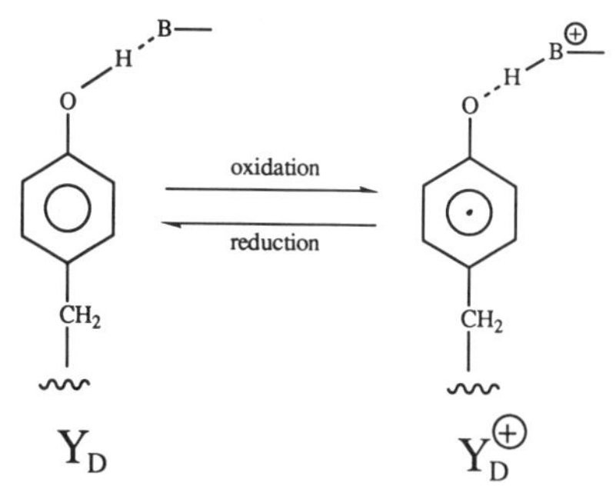

FIGURE 3: Proton rocking motion model for redox transitions in Y_D. In its reduced form (Y_D) the phenol proton is postulated to be hydrogen bonded to a base (e.g., histidine, amide N) in the vicinity of the tyrosine. Proton transfer from tyrosine to the base does not occur owing to the high pK_a ($\sim$10) of the tyrosine. Upon oxidation of the residue to form Y_D^+, the phenol become strongly acidic ($pK_a \approx -1$), and proton transfer to the base occurs to form the $H{-}B^+$ species. The proton back hydrogen bonds to the oxygen of the neutral radical. Evidence for the occurrence of the hydrogen-bonded proton has been found by both ENDOR (Rodriguez et al., 1987) and ESE (Evelo et al., 1989b) spectroscopies. The H_2O/D_2O exchange rate of the hydrogen-bonded proton is remarkably slow (Rodriguez et al., 1987), which leads to a model in which the proton rocks back and forth, depending on tyrosine redox state, between the phenol oxygen and the base.

Brudvig (1989). The latter authors were able to estimate the distance between the paramagnet and the solution phase and found that Y_D^+ is centered essentially in the middle of the PSII membrane at least 25 Å from the nearer solution phase.

WHERE DOES PSEUDO-C_2 SYMMETRY IN PSII BREAK DOWN—LOCATION OF THE MANGANESE ENSEMBLE

Like the bacterial reaction center subunits L and M, the PSII subunits D1 and D2 are each likely to contain five membrane-spanning helices (Sayre et al., 1986). The mutagenesis results above have established that Y_D and Y_Z occur in the third membrane-spanning helix, the C helix of the two polypeptides. On the basis of conserved amino acid residues and the bacterial analogy, most researchers identify the fourth and fifth membrane-spanning helices, the D and E helices of the polypeptides, as forming the binding site for P_{680}, the intermediate pheophytin acceptor, Q_A, Q_B, and the associated Fe^{2+} [e.g., Trebst (1986) and Michel and Deisenhofer (1988)]. Such a structure is consistent with a P_{680} Y_Z separation distance of 8–12 Å that has been estimated recently (Hoganson & Babcock, 1989).

Within this context two extreme models for the location of the manganese ensemble can be constructed as shown in Figure 4. In the first (Figure 4a), pseudo-C_2 symmetry is preserved in PSII as the manganese ensemble is incorporated roughly along the C_2 symmetry axis and derives ligands from amino acids near the carboxy termini of both D1 and D2. Dismukes (1988) and Rutherford (1989), among others, have argued in favor of such a structure. Consideration of the requirements for effective manganese function in its water-oxidizing capacity leads to the conclusion that the most likely ligands are carboxyl groups (Pecoraro, 1988), a conjecture supported by the failure to observe nitrogen hyperfine coupling in the S_2 multiline EPR signal (Andréasson, 1989; Britt et al., 1989). Consequently, most speculation as to appropriate ligands centers on conserved aspartates and glutamates, and indeed, in the carboxy-terminal regions of D1 and D2 a number of these occur that are absent in the bacterial reaction center. In the second model (Figure 4b), the pseudo-C_2 symmetry is broken by manganese incorporation, which is located off the C_2 symmetry axis. Inspection of the amino acid sequences of D1 and D2 reveals that several

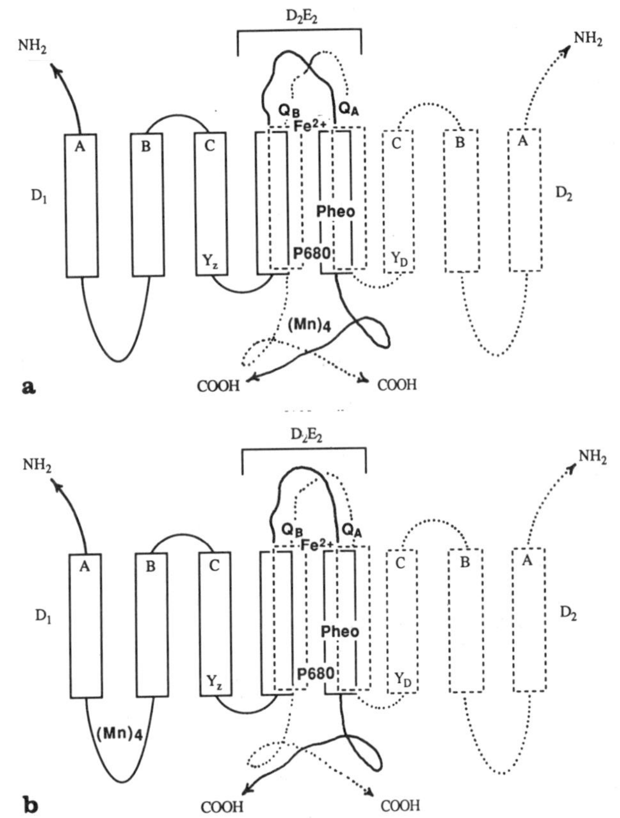

FIGURE 4: Models for the disposition of the Mn binding site in the PSII/OEC. The reaction center polypeptides D1 and D2, each with five membrane spanning helices (A–E), are shown. In analogy with the bacterial reaction center, the chromophores involved with primary charge are associated with the 4-helix D_2E_2 structure. The pseudo-C_2 symmetry axis is along the $Fe^{2+}\cdots P_{680}$ direction. In (a), the placement of the manganese cluster preserves pseudo-C_2 symmetry as it is postulated to occur near the –COOH termini of D1 and D2. In (b), incorporation of Mn is envisioned to occur in the AB interhelical region of the D1 polypeptide only, which disrupts pseudo-C_2 symmetry in the reaction center. An interesting speculation, not shown explicitly in (b), is that Ca^{2+} binds to the corresponding AB interhelical region in D2.

unique, conserved glutamates and aspartates occur in the AB interhelical loop in D1, which suggests this locus as a plausible Mn binding site in the model of Figure 4b.

Before examining the sparse experimental data that are useful in arguing for one or the other of these two extreme models, two points can be made. First, the organization of the manganese ensemble is not known. Most of the discussion on this issue suggests that all four metals are in close proximity and multinuclear clusters (n = 2–4) are considered as attractive structures for the ensemble [for reviews see Christou and Vincent (1987) Pecoraro (1988), and Brudvig et al. (1989)]. Recent EXAFS data appear to support this idea (George et al., 1989; Penner-Hahn et al., 1989). Nonetheless, other structures remain viable, particularly given the Ca^{2+} requirement for O_2 evolution (Bonadies et al., 1989; Rutherford, 1989), and this area is one of intense research activity. Second, the data supporting an association of manganese with only D1 and D2 are indirect [see, for example, Dismukes (1988)], and other PSII polypeptides, in particular the integral 47-kDa chlorophyll binding polypeptide and 33-kDa extrinsic polypeptide, may also be involved in supplying ligands to the metal ions.

As indicated above, there is little experimental data available that bear directly on the two extreme models in Figure 4. A non-O_2-evolving, low fluorescent mutant of *Scenedesmus*, LF1, has been extensively characterized and shown to be defective

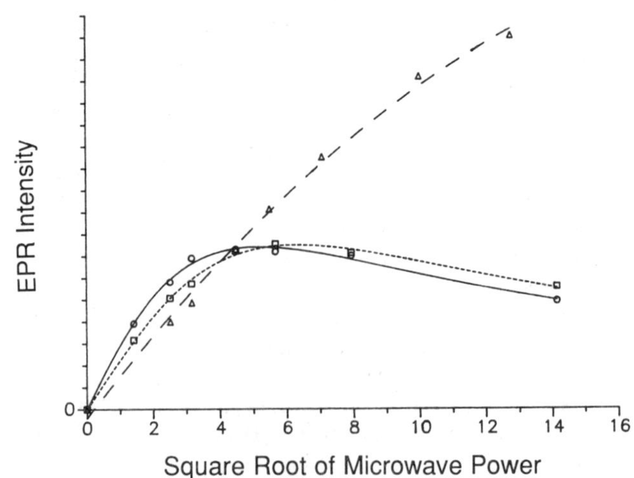

FIGURE 5: Effect of microwave power on the amplitude of the EPR signals of Y_Z^+ and Y_D^+ in O_2-evolving PSII membranes. The Y_Z^+ (△) and Y_D^+ (□) measurements were made by using a 1-Hz flash repetition rate; the Y_D^+ (○) measurement was performed in the dark. The curves have been normalized at 20 mW.

in assembling the manganese complex [see Rutherford et al. (1988)]. This has been linked to a failure to process the carboxy terminus of the D1 polypeptide (Diner et al., 1988a). As the authors point out, however, this effect, while supporting an association of Mn with D1, may well be indirect. Similarly, recent chemical modification studies have implicated histidine residues on D1 in manganese binding (Tamura et al., 1989). Although the suggestion of histidine ligation runs counter to the spectroscopic data discussed above, it may be rationalized if the multiline EPR signal arises from only two or three of the four Mn atoms in the complex. The authors in this study have not yet identified the specific histidines involved and thus discussed both the symmetric and asymmetric models in Figure 4 as viable alternatives.

Some insight into resolving the question posed in Figure 4 is available from EPR power saturation studies. Warden et al. (1976) showed that Y_Z^+ was much more strongly relaxed than Y_D^+ and attributed the effect to an interaction with the manganese ensemble that occurred for the former but not for the latter radical. This observation suggests an asymmetric disposition for the water-splitting site relative to the two tyrosines and has been studied in detail recently. The occurrence of different magnetic interactions between manganese and the two tyrosines has been confirmed by measuring the power saturation of the two organic radicals under identical flashing light conditions [Figure 5; see also Rutherford (1989)]. A weaker, S-state-dependent magnetic interaction has been observed for Y_D^+, and Evelo et al. (1989a) used this to estimate a distance of 30–40 Å between this radical and the Mn cluster. The small magnitude of the relaxation enhancement in these experiments, however, suggests that this estimate should be viewed with some caution. Hoganson and Babcock (1988) showed that the stronger Y_Z^+/Mn interaction does not broaden the Y_Z^+ spectrum appreciably, despite its increased relaxation rate, and suggested a distance somewhat greater than 15 Å between the two paramagnents. A more detailed analysis of this interaction is not possible without more specific information on the magnetic properties of the cluster. Taken together, these considerations provide some evidence, albeit weak, in favor of a departure from C_2 symmetry at the Mn cluster. Site-directed mutagenesis, which can now be carried out on both D1 and D2 (Debus et al., 1988a,b) and chemical modification studies are likely to provide a resolution to this question.

Conclusions

In the past decade the PSII/OEC has gone from being mysterious to being one of the most well-understood membrane proteins in biology. Several factors have facilitated this rapid progress. First, oxygen evolution has proven to be remarkably stable to biochemical manipulation, despite early indications to the contrary. This stability, coupled with the relative ease by which individual polypeptides can be removed from the complex, has allowed functional assignments for the various subunits to be made. It has also provided samples for biophysical analysis in which the PSII concentrations can approach 1 mM. This has significantly improved signal to noise ratios in some spectroscopies [e.g., Penner-Hahn et al. (1989)] and made others possible [e.g., the first magnetic susceptibility measurement on O_2-evolving PSII preparations has recently appeared (Sivaraja et al., 1989)]. Second, the crystallization of the bacterial reaction center, the realization of the analogies between it and PSII (Michel & Deisenhofer, 1988), and the demonstration of the relevance of constructing experiments on the basis of the anology (Debus et al., 1988a,b) have provided a "shortcut" to dissecting the oxygen-evolving system. Third, because the water-splitting process, like all photosynthetic reactions, can be initiated by light, kinetic studies over the picosecond to second time regime have become possible. Rate constants and their temperature dependencies are becoming available for the charge-transfer processes that lead from photochemical electron-hole separation to the actual water-splitting event. The relevant redox partners in these events have been identified, and the factors that influence the rate constants are beginning to emerge.

Despite this rapid evolution in our understanding of the PSII/OEC, many of its most interesting and novel aspects are nebulous. The physical arrangement of the chromophores remains uncertain, particularly with regard to the actual site of water oxidation. In a broader sense, the issue is in what areas the bacterial analogy will break down. This pertains not only to the location of the manganese cluster but also to the pigment composition and organization. Even more fundamentally, the PSII/OEC provides a vexing paradox. The arrangement of the redox components and the kinetics of the electron-transfer reactions that are involved in producing O_2 efficiently provide an excellent example of the economy of nature. The occurrence of Y_D and the demonstration that it is not a backup to the active charge-transfer pathway, on the other hand, seem to be counterexamples. Some insight into the resolution of this paradox comes from the observation that the Y_D-less mutant, while competent in O_2 evolution, grows at about one-third the rate of the wild type (Debus et al., 1988a). This suggests that the inactive branch in PSII, and perhaps in the bacteria as well, is involved in dynamic interactions that facilitate assembly (perhaps in a template role) or stabilize the reaction center once it is formed. Finally, with regard to water-splitting chemistry itself, we have little knowledge of mechanism. The majority of current work, both on the natural system itself and in new and stimulating inorganic model complexes (e.g., Pecoraro, 1988), is aimed at determining the structure of the complex. As progress in this effort continues, an appreciation of how the complex actually works should grow.

Added in Proof

Although Y_D and Y_Z occur as neutral radicals in their oxidized states as discussed in the text, we prefer to designate the radicals as Y_D^+ and Y_Z^+. The basis for this preference derives partly from the fact that the proton does not leave the radical site (see Figure 3) and partly from the convenience that the "+" notation provides in designating the redox state of the tyrosine.

Registry No. H_2O, 7732-18-5.

References

Andréasson, L.-E. (1989) *Biochim. Biophys. Acta 973*, 456–467.

Andréasson, L.-E., & Vänngård, T. (1988) *Annu. Rev. Plant Physiol. Plant Mol. Biol. 39*, 379–411.

Babcock, G. T. (1987) *Photosynth.: New Compr. Biochem. 15*, 125.

Babcock, G. T., Blankenship, R. E., & Sauer, K. (1976) *FEBS Lett. 61*, 286–289.

Babcock, G. T., Widger, W. R., Cramer, W. A., Oertling, W. A., & Metz, J. G. (1985) *Biochemistry 24*, 3638–3645.

Barry, B. A., & Babcock, G. T. (1987) *Proc. Natl. Acad. Sci. U.S.A. 84*, 7099–7103.

Barry, B. A., & Babcock, G. T. (1988) *Chem. Scr. 28A*, 117–122.

Beck, W. F., & Brudvig, G. W. (1988) *Chem. Scr. 28A*, 93–98.

Bender, C. J., Sahlin, M., Babcock, G. T., Barry, B. A., Chandrashekar, T. K., Salowe, S. P., Stubbe, J., Lindstrom, B., Petersson, L., Ehrenberg, A., & Sjoberg, B.-M. (1989) *J. Am. Chem. Soc.* (in press).

Berthold, D. A., Babcock, G. T., & Yocum, C. F. (1981) *FEBS Lett. 134*, 231–234.

Bonadies, J. A., Maroney, M., & Pecoraro, V. L. (1989) *Inorg. Chem. 28*, 2044–2050.

Boussac, A., & Rutherford, A. W. (1988) *Biochemistry 27*, 3476–3483.

Bowlby, N. R., Ghanotakis, D. F., Yocum, C. F., Petersen, J., & Babcock, G. T. (1988) *Light-Energy Transduction in Photosynthesis: Higher Plant and Bacterial Models* (Stevens, S. E., Jr., & Bryant, D. A., Eds.) p 215, The American Society of Plant Physiology Publishers, Baltimore, MD.

Bricker, T. M., Odom, W. R., & Queirolo, C. B. (1988) *Biochim. Biophys. Acta 933*, 358–364.

Britt, R. D., Zimmermann, J. L., Sauer, K., & Klein, M. P. (1989) *J. Am. Chem. Soc. 111*, 3522–3532.

Brudvig, G. W., Beck, W. F., & dePaula, J. C. (1989) *Annu. Rev. Biophys. Biophys. Chem. 18*, 25–46.

Christou, G., & Vincent, J. B. (1987) *Biochim. Biophys. Acta 895*, 259–274.

Crofts, A. R., & Wraight, C. A. (1983) *Biochim. Biophys. Acta 726*, 149–185.

Debus, R. J., Barry, B. A., Babcock, G. T., & McIntosh, L. (1988a) *Proc. Natl. Acad. Sci. U.S.A. 85*, 427–430.

Debus, R. J., Barry, B. A., Sithole, I., Babcock, G. T., & McIntosh, L. (1988b) *Biochemistry 27*, 9071–9074.

Dekker, J. P. (1989) in *Manganese Redox Enzymes* (Pecoraro, V. L., Ed.) VDL Publications, New York (in press).

Dekker, J. P., van Gorkom, H. J., Brok, M., & Ouwehand, L. (1984a) *Biochim. Biophys. Acta 764*, 301–309.

Dekker, J. P., Plijter, J. J., Ouwehand, L., & van Gorkom, H. J. (1984b) *Biochim. Biophys. Acta 767*, 176–179.

Dekker, J. P., Bowlby, N. R., & Yocum, C. F. (1989) *FEBS Lett.* (in press).

dePaula, J. C., Li, P. M., Miller, A.-F., Wu, B. W., & Brudvig, C. W. (1986) *Biochemistry 25*, 6487–6494.

dePaula, J. C., Beck, W. F., Miller, A.-F., Wilson, R. B., & Brudvig, G. W. (1987) *J. Chem. Soc., Faraday Trans. 1 83*, 3635–3651.

Diner, B. A., Ries, D. F., Cohen, B. N., & Metz, J. G. (1988a) *J. Biol. Chem. 263*, 8972–8980.

Diner, B. A., de Vitry, C., & Popot, J.-L. (1988b) *Biochim. Biophys. Acta 934*, 47–54.
Dismukes, C. (1988) *Chem. Scr. 28A*, 99–104.
Eckert, H.-J., & Renger, G. (1988) *FEBS Lett. 236*, 425–431.
Eckert, H.-J., Wiese, N., Bernarding, J., Eichler, H.-J., & Renger, G. (1988) *FEBS Lett. 240*, 153–158.
Enami, I., Miyaoka, T., Mochizuki, Y., Shen, J.-R., Satoh, K., & Katoh, S. (1989) *Biochim. Biophys. Acta 973*, 35–40.
Evelo, R. G., Styring, S., Rutherford, A. W., & Hoff, A. J. (1989a) *Biochim. Biophys. Acta 973*, 428–442.
Evelo, R. G., Dikanov, S. A., & Hoff, A. J. (1989b) *Chem. Phys. Lett. 159*, 25–30.
Forster, V., & Junge, W. (1985) *Photochem. Photobiol. 41*, 183–190.
George, G. N., Prince, R. C., & Cramer, S. P. (1989) *Science 243*, 789–791.
Gerken, S., Brettel, K., Schlodder, E., & Witt, H. T. (1988) *FEBS Lett. 237*, 69–75.
Ghanotakis, D. F., & Yocum, C. F. (1985) *Photosynth. Res. 7*, 97–114.
Ghanotakis, D. F., Babcock, G. T., & Yocum, C. F. (1984) *Biochim. Biophys. Acta 765*, 388–398.
Ghanotakis, D. F., Demetriou, D. M., & Yocum, C. F. (1987) *Biochim. Biophys. Acta 891*, 15–21.
Gounaris, K., Chapman, D. J., & Barber, J. (1988) *FEBS Lett. 234*, 374–378.
Hansson, O., Duranton, J., & Mathis, P. (1988) *Biochim. Biophys. Acta 932*, 91–96.
Hearst, J. E. (1986) *Encycl. Plant Physiol., New Ser. 19*, 382.
Hoganson, C. W., & Babcock, G. T. (1988) *Biochemistry 27*, 5848–5855.
Hoganson, C. W., & Babcock, G. T. (1989) *Biochemistry 28*, 1448–1454.
Hoganson, C. W., Babcock, G. T., & Yocum, C. F. (1989) *Photosynth. Res.* (in press).
Homann, P. H. (1987) *J. Bioenerg. Biomembr. 19*, 10–20.
Ikeuchi, M., & Inoue, Y. (1988) *Plant Cell Physiol. 29*, 695–705.
Innes, J. B., & Brudvig, G. W. (1989) *Biochemistry 28*, 1116–1125.
Isogai, Y., Nishimura, M., Iwaki, M., & Itoh, S. (1988) *Biochim. Biophys. Acta 936*, 259–268.
Johansen, J. (1988) *Biochim. Biophys. Acta 933*, 406–412.
Joliot, P., & Kok, B. (1975) in *Bioenergetics of Photosynthesis: Oxygen Evolution in Photosynthesis* (Govindjee, Ed.) p 387, Academic Press, New York.
Koike, H., Hanssum, B., Inoue, Y., & Renger, G. (1987) *Biochim. Biophys. Acta 893*, 524–533.
Kretschmann, H., Dekker, J. P., Saygin, O., & Witt, H. T. (1988) *Biochim. Biophys. Acta 932*, 358–361.
Lavergne, J. (1989) *Photochem. Photobiol. 50*, 235–241.
Mathis, P., & Rutherford, A. W. (1987) *Photosynth.: New Compr. Biochem. 15*, 63.
Metz, J. G., Nixon, P. J., Rogner, M., Brudvig, G. W., & Diner, B. A. (1989) *Biochemistry 28*, 6960–6969.
Meyer, B., Scholdder, E., Dekker, J. P., & Witt, H. T. (1989) *Biochim. Biophys. Acta 974*, 36–43.
Michel, H., & Diesenhofer, J. (1988) *Biochemistry 27*, 1–7.
Nanba, O., & Satoh, K. (1987) *Proc. Natl. Acad. Sci. U.S.A. 84*, 109–112.
Nuijs, A. M., Van Gorkom, H. J., Plijter, J. J., & Duysens, L. N. M. (1986) *Biochim. Biophys. Acta 848*, 167–175.
Ono, T., & Inoue, Y. (1989) *Biochim. Biophys. Acta 973*, 443–449.
Pecoraro, V. L. (1988) *Photochem. Photobiol. 48*, 249–264.
Penner-Hahn, J. E., Fronko, R. M., Pecoraro, V. L., Yocum, C. F., Betts, S. D., & Bowlby, N. R. (1989) *J. Am. Chem. Soc.* (in press).
Petersen, J., Dekker, J. P., Bowlby, N. R., Ghanotakis, D. F., Yocum, C. F., & Babcock, G. T. (1989) *Biochemistry* (submitted for publication).
Plijter, J. J., Aalbers, S. E., Barends, J. P. F., Vos, M. H., & Van Gorkom, H. J. (1988) *Biochim. Biophys. Acta 935*, 299–306.
Renger, G. (1988) *ISI Atlas Sci.: Biochem.*, 41–47.
Rodriguez, I. D., Chandrashekar, T. K., & Babcock, G. T. (1987) *Progress in Photosynthesis Research* (Biggins, J., Ed.) Vol. 1, p 479, Martinus Nijhoff Publishers, The Hague, The Netherlands.
Rutherford, A. W. (1989) *Trends Biochem. Sci. 14(6)*, 227–232.
Rutherford, A. W., Seibert, M., & Metz, J. G. (1988) *Biochim. Biophys. Acta 932*, 171–176.
Sandusky, P. O., & Yocum, C. F. (1984) *Biochim. Biophys. Acta 766*, 603–611.
Sandusky, P. O., & Yocum, C. F. (1986) *Biochim. Biophys. Acta 849*, 85–93.
Saphon, S., & Crofts, A. R. (1977) *Z. Naturforsch. 32C*, 617–626.
Sauer, K., Guiles, R. D., McDermott, A. E., Cole, J. L., Yachandra, V. K., Zimmermann, J. L., Klein, M. P., Dexheimer, S. L., & Britt, R. D. (1988) *Chem. Scr. 28A*, 87–91.
Sayre, R. T., Andersson, B., & Bogorad, L. (1986) *Cell 47*, 601–608.
Schatz, G. H., & Holzwarth, A. R. (1987) *Progress in Photosynthesis Research* (Biggins, J., Ed.) Vol. I, p 67, Martinus Nijhoff Publishers, The Hague, The Netherlands.
Schatz, G. H., Brock, H., & Holzwarth, A. (1988) *Biophys. J. 54*, 397–405.
Schlodder, E., & Brettel, K. (1988) *Biochim. Biophys. Acta 933*, 22–34.
Sinclair, J., & Arnason, T. (1974) *Biochim. Biophys. Acta 368*, 393–400.
Sivaraja, M., Philo, J. S., Lary, J., & Dismukes, G. C. (1989) *J. Am. Chem. Soc. 111*, 3221–3225.
Stewart, A. C., & Bendall, D. S. (1979) *FEBS Lett. 107*, 308–312.
Styring, S., & Rutherford, A. W. (1988) *Biochim. Biophys. Acta 933*, 378–387.
Tae, G.-S., Black, M. T., Cramer, W. A., Vallon, O., & Bogorad, L. (1988) *Biochemistry 27*, 9075–9080.
Takahasi, Y., & Satoh, K. (1989) *Biochim. Biophys. Acta 973*, 138–146.
Tamura, N., Ikeuchi, M., & Inoue, Y. (1989) *Biochim. Biophys. Acta 973*, 281–289.
Thompson, L. K., & Brudvig, G. W. (1988) *Biochemistry 27*, 6653–6658.
Trebst, A. (1986) *Z. Naturforsch. 41C*, 240–245.
Trissl, H.-W., & Leibl, W. (1989) *FEBS Lett. 244*, 85–88.
Van Gorkom, H. J. (1985) *Photosynth. Res. 6*, 97–112.
Vermaas, W. F. J., Rutherford, A. W., & Hansson, O. (1988) *Proc. Natl. Acad. Sci. U.S.A. 85*, 8477–8481.
Warden, J. T., Blankenship, R. E., & Sauer, K. (1976) *Biochim. Biophys. Acta 423*, 462–478.
Wasielewski, M. R., Johnson, D. G., Seibert, M., & Govindjee (1989) *Proc. Natl. Acad. Sci. U.S.A. 86*, 524–528.
Wydrzynski, T., Ångström, J., & Vänngård, T. (1989) *Biochim. Biophys. Acta 973*, 23–28.

Chapter 28

Cytochrome c Oxidase: Understanding Nature's Design of a Proton Pump†

Sunney I. Chan* and Peter Mark Li‡

A. A. Noyes Laboratory of Chemical Physics, California Institute of Technology, Pasadena, California 91125

Received June 16, 1989; Revised Manuscript Received July 17, 1989

It has been estimated that nearly 90% of the O_2 consumed by aerobic organisms participates in the dioxygen chemistry of cytochrome c oxidase and becomes reduced to water in the terminal step of respiration. Cytochrome oxidases of the aa_3 type (having two a-type cytochromes) are found in a wide variety of aerobic organisms including bacteria, fungi, single-celled eukaryotes, plants, and animals. It is an integral membrane protein complex comprised of 2 or 3 subunits in the simplest bacterial systems and as many as 13 dissimilar subunits in mammals [for a review, see Wikström et al. (1981)].

All aa_3-type oxidases contain four redox-active metal centers (two iron hemes and two copper ions) and catalyze the four-electron reduction of molecular oxygen to water with reducing equivalents derived from cytochrome c:

$$4\text{Cyt } c^{2+} + O_2 + 4H^+ \rightarrow 4\text{Cyt } c^{3+} + 2H_2O$$

The electrons enter the protein from the cytosol side of the mitochondrial inner membrane, and the protons consumed in the dioxygen reduction reaction are taken up from the matrix. In this manner, the sidedness of the membrane is exploited to convert redox free energy into a proton electrochemical gradient across the inner mitochondrial membrane. In addition, cytochrome oxidase is also an electrogenic proton pump capable of transporting up to four protons from the matrix side of the mitochondrial membrane to the cytosol side for every dioxygen molecule reduced. This "vectorial" proton pumping activity augments the "scalar" proton consumption associated with dioxygen reduction, increasing the efficiency of the free energy conversion from redox energy to the synthesis of ATP [for a review, see Krab and Wikström (1987)].

Cytochrome oxidase has been extensively studied for nearly 5 decades with quite reasonable success. However, our understanding of the enzyme has now reached a level where further real progress will require an improved molecular definition of the problem. Cytochrome oxidase is an extremely complex enzyme, both structurally and functionally. In fact, it is quite impressive that we have come this far in our understanding of the biochemistry of this enzyme without a detailed picture of the assembly of the subunit composition and three-dimensional structure of the protein complex.

On an elementary level, there is still uncertainty regarding the polypeptide composition and the minimal molecular mass, which has been reported to range from 70 to 120 kDa (Brunori et al., 1987a). Moreover, we do not yet have an unequivocal definition of the functional unit. As suggested by Brunori et al. (1987a), the simplest view of the enzyme is that the basic unit is comprised of all the polypeptides and prosthetic groups which copurify with the electron transfer and dioxygen reduction activity. On the other hand, since the enzyme is also a proton pump, this definition of the functional unit must be revised to include the minimum number of additional polypeptides that are necessary to carry out redox-linked proton translocation, if any. As these authors point out, a clear-cut distinction between these two viewpoints is difficult because our knowledge of the electron transfer processes and the mechanism of the redox-linked proton pumping activities is still quite rudimentary.

Finally, cytochrome oxidase represents a distinct class of proton translocation devices whose principles of operation are not well understood. In simplest of terms, it is a molecular machine, capable of existing in a large number of conformational states, but which must operate according to an ordered sequence of conformational transitions to achieve kinetic competence as electrons flow from one metal center to another and dioxygen is reduced to give a series of intermediates at the dioxygen reduction site. Whether the problem is sufficiently tractable and amenable to description in terms of a

†Contribution No. 7961 from the Arthur Amos Noyes Laboratory of Chemical Physics, California Institute of Technology. This work was supported by Grant GM 22432 from the National Institute of General Medical Sciences, U.S. Public Health Service. Acknowledgement is made to the donors of the Petroleum Research Fund, administered by the American Chemical Society, for partial support.

*To whom correspondence should be addressed.

‡Recipient of predoctoral fellowships from the National Science Foundation, the Department of Education, and the Josephine de Karmān Trust.

small subset of these conformational states remains to be seen.

The purpose of this paper is to summarize our current understanding of the structure and function of cytochrome oxidase, particularly those aspects that bear on the proton pumping function. We begin with a general introduction to the structural biochemistry of cytochrome oxidase and review our current knowledge of the structures of the redox-active metal centers. Since the proton pumping function of the enzyme is linked to the electron transfer from ferrocytochrome *c* to dioxygen, we also discuss the chemistry of dioxygen reduction at the binuclear center as well as the role of the other redox-active metals in regulating the electron flow. Finally, we attempt to illustrate how information about the structure and function of the enzyme limits the choice of mechanisms to describe the redox-linked proton translocation.

Structural Biochemistry

Metal Centers. Cytochrome oxidase contains four redox-active metal centers which are important to its catalytic activity. These four centers may be distinguished on the basis of function. One pair, cytochrome a_3 and Cu_B, forms a binuclear cluster where dioxygen is bound and reduced during the catalytic cycle. The other pair, cytochrome *a* and Cu_A, mediates the flow of electrons from ferrocytochrome *c* to the binuclear center. Spectroscopically, cytochrome *a* and Cu_A interact only weakly and are often treated as independent electron acceptors. On the basis of spectroscopic evidence as well as sequence homology data, it is now generally agreed that these redox-active metal centers are located in subunits I and II (Mueller et al., 1988). Specifically, it has been suggested that cytochrome *a*, cytochrome a_3, and Cu_B reside in subunit I while Cu_A is associated with subunit II [for a model, see Holm et al. (1987)].

Recently, there has also been a report that cytochrome oxidase contains three copper ions (Steffens et al., 1987; Yewey et al., 1988). This additional copper (Cu_X) does not appear to be redox active and has been suggested to be copurified adventitious copper (Li et al., 1989). In addition, there is evidence that cytochrome oxidase binds zinc and magnesium (Yewey et al., 1988). The zinc has been identified as a Zn^{2+} ion ligated almost exclusively by the cysteine sulfurs of subunit V (Naqui et al., 1988). Naqui et al. have suggested that the zinc ion plays a structural role in the enzyme. The role of magnesium is not yet well understood.

Cytochrome *a* is a six-coordinate low-spin heme A, axially ligated by the nitrogens from two neutral imidazoles in both the oxidized and reduced states (Babcock & Callahan, 1983; Martin et al., 1985). This metal center is generally assumed to be the primary acceptor of electrons from ferrocytochrome *c*. However, this position is not unambiguous and must be reevaluated in light of the rapid electron equilibration between cytochrome *a* and Cu_A recently reported by Morgan et al. (1989). It has been argued that if this center is the primary point of entry for electrons, then the degradation of the redox free energy should be minimized in this electron transfer step. However, there is now compelling evidence that the cytochrome *a* redox potential does vary as the enzyme is turning over (Thörnström et al., 1988; Wikström et al., 1981). In the resting form of the enzyme, cytochrome *a* has a fairly high midpoint potential (ca. 350 mV), but it decreases to ca. 280 mV when the dioxygen binding site becomes reduced. The redox potential of cytochrome *a* also displays a moderate pH dependence of ca. 30 mV/pH unit. The source of this pH dependence has been localized to the titration of a protonatable group on the inner side of the mitochondrial membrane (Artazabanov et al., 1978). This pH-dependent midpoint potential has led to the suggestion that cytochrome *a* is the site of redox linkage to proton translocation. However, it must be noted that the pH dependence of the cytochrome *a* midpoint potential appears to change in response to the state of the enzyme. For example, in the CO mixed-valence form of the enzyme, the pH dependence decreases to ca. 9 mV/pH unit (Blair et al., 1986a).

Copper A (Cu_A) is the low-potential copper of cytochrome oxidase, having a midpoint potential of ca. 285 mV. It has been suggested that the role of this metal center is to transfer electrons from cytochrome *a* to the dioxygen reduction site. However, the possibility that Cu_A is the primary electron acceptor from ferrocytochrome *c* has been implicated recently. It has been noted that Cu_A is the metal cofactor most exposed to the cytosolic side of the membrane, and the putative metal binding site is near a patch of negatively charged amino acid residues which could serve as the docking site for ferrocytochrome *c* (Millet et al., 1982; Holm et al., 1987).

Irrespective of the resolution of these issues, Cu_A is an unusual metal center as evidenced by its enigmatic spectroscopic signatures. In the oxidized state, Cu_A exhibits a weak optical transition in the near infrared (830 nm) which has been assigned to a charge-transfer transition between the copper ion and a sulfur ligand (Beinert et al., 1962). In addition, Cu_A displays an EPR spectrum atypical of Cu complexes and copper sites in proteins. There is no resolvable hyperfine splitting when the spectrum is recorded at X-band ($g = 2.18$, 2.03, and 1.99). Of particular interest is that one *g* value is below the free electron *g* value, a situation atypical of simple Cu^{2+} centers (Assa et al., 1976). X-ray absorption spectroscopy indicates that the oxidized Cu_A site is in a highly covalent ligand environment and that there is considerable charge transfer from the ligands to the copper ion (Hu et al., 1977; Powers et al., 1981).

ENDOR studies have identified hyperfine and superhyperfine interactions between the unpaired electron of Cu_A and various nuclei. The copper hyperfine interaction is unusually small and isotropic [Stevens et al. (1982) and references cited therein]. EPR and ENDOR studies using [^2H]Cys- and [^{15}N]His-substituted yeast cytochrome oxidase have implicated at least one histidine and at least one cysteine ligand to Cu_A. The proton hyperfine couplings from the cysteine β-CH_2's are unusually strong (12 and 19 MHz) whereas the corresponding His ^{14}N superhyperfine interaction is significantly smaller than in blue copper proteins (Stevens et al., 1982; Martin et al., 1988). More recently, EXAFS measurements comparing native, chemically modified, and Cu_A-depleted cytochrome oxidases have shown that two cysteine sulfurs are probably involved in the Cu_A ligation structure (Li et al., 1987). Accordingly, we have proposed that Cu_A is ligated by two cysteine sulfur atoms and two histidine nitrogen atoms. Comparison of amino acid sequences for cytochrome oxidases across a wide variety of organisms does show the presence of two highly conserved cysteine residues in subunit II (Steffens et al., 1987; Hall et al., 1988), and it has been surmised that these are the two cysteine ligands to Cu_A. No other cysteine residues are conserved.

The structure of the binuclear center has been studied for many years by a variety of spectroscopic techniques, particularly in conjunction with the binding of externally added ligands. The binuclear center coordinates a variety of ligands, including F^-, CN^-, formate, and peroxide in the oxidized state and O_2, CO, and NO in the reduced state. Cytochrome a_3 is a high-spin ferric heme in both the oxidized and reduced forms of the enzyme, with one histidine nitrogen ligand in the axial position distal to Cu_B (Stevens & Chan, 1981). The other axial ligand is variable depending on the state of the enzyme. In the resting enzyme, the ferric heme of cytochrome a_3 is strongly antiferromagnetically coupled to Cu_B yielding a net $S = 2$ paramagnetic species (Brudvig et al., 1986). The bridging ligand is not known, but it is possibly a μ-oxo, μ-hydroxyl, or μ-chloro species. In the reduced state, cytochrome a_3 is a high-spin ferrous heme (the Cu_B center is d^{10} and hence diamagnetic) and also $S = 2$. The Cu_B ion has been less well characterized because there is no visible absorption from this metal center. However, ENDOR and EPR studies on intermediates trapped during the turnover cycle indicate that Cu_B is ligated by at least three histidine nitrogens in a fashion similar to the type 3 copper centers (Cline et al., 1983; Reinhammar et al., 1980).

Interactions among the Metal Centers. There is considerable evidence that all four metal centers interact with one another to varying degrees. These interactions are either magnetic or electrostatic by virtue of their spatial proximity or conformational by virtue of their spatial and conformational linkage. Both types of interactions occur in the binuclear center. The electrostatic interactions are strong here and manifest themselves in terms of an exchange interaction between the cytochrome a_3 and Cu_B spins, which behave as a magnetic unit. The other two centers can be treated as isolated centers, although they do interact magnetically with each other and with the binuclear center (as a unit), albeit weakly. These magnetic interactions have been used to infer the spatial distribution of the metal centers (Leigh et al., 1974; Brudvig et al., 1984).

The conformational interactions are reflected in the redox behavior of the metal centers. It is well-known that cytochrome a exhibits an anticooperative interaction with cytochrome a_3 and Cu_B. By contrast, this type of redox interaction has not been observed between Cu_A and the binuclear center. However, Cu_A does interact *allosterically* with the binuclear center, specifically cytochrome a_3. When Cu_A is chemically modified, the iron–histidine stretching and formyl stretching modes of cytochrome a_3 change in frequency as observed by resonance Raman spectroscopy (Larsen et al., 1989). Also, the rate of cyanide binding to cytochrome a_3 is accelerated when Cu_A is modified by heat treatment (Li et al., 1988). Recently, spectroelectrochemical experiments have also uncovered an anticooperative interaction of $\sim$40 mV between cytochrome a and Cu_A which may be either electrostatic or conformational in nature (Blair et al., 1986a). In support of this, Brudvig et al. (1984) and Scholes et al. (1984) have reported that the reduction of cytochrome a affects the EPR and ENDOR spectra of Cu_A. Interactions of this type between cytochrome a and Cu_A and between the dioxygen reduction site and the other metal centers in the enzyme are of considerable interest because allosteric coupling is undoubtedly involved in the regulation of intramolecular electron transfer in the proton pumping reaction. Figure 1 shows our current view of the redox-active metal centers in cytochrome oxidase.

Subunits. Mammalian cytochrome oxidase has been shown to contain at least 13 inequivalent subunits: 3 are coded by the mitochondrial DNA and synthesized on the mitochondrial ribosomes (subunits I–III), and the remaining subunits (subunits IV–XIII) are coded for by the nuclear DNA and synthesized in the cytosol. Most eukaryotic organisms have multisubunit cytochrome oxidases containing more than the three mitochondrially coded polypeptides, while the prokaryotes tend to have simpler oxidases with two or three subunits which are homologous to the mitochondrially coded subunits

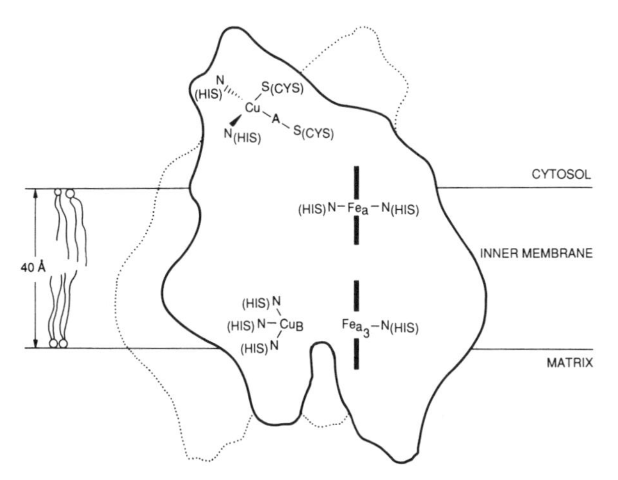

FIGURE 1: Pictorial representation of the cytochrome oxidase dimer in the inner mitochondrial membrane, including the approximate relative positions of the redox-active metal centers. The non-redox-active metal centers (Zn, Mg) are not shown. It should be noted that the intersite distances are not drawn exactly to scale.

I–III. The remainder of the subunits presumably perform a regulatory function.

Since all the redox-active metal centers are contained in subunits I and II, there has been some ambiguity as to the function of subunit III. At one time subunits III was proposed to be involved in the proton translocation process because it is a membrane-spanning polypeptide which is sensitive to the carboxyl reagent dicyclohexylcarbodiimide (DCCD). DCCD has been shown to inhibit proton translocation in the ATP synthase system, and it also inhibits proton translocation in cytochrome oxidase, albeit to a lesser extent. For this reason, it has been argued that subunit III plays a role in proton translocation. Finel and Wikström (1988) recently suggested that subunit III plays a role in the oligomerization of the protein, which is perhaps important to the proton pumping function of the enzyme. However, Moody and Rich (1989) have recently carried out experiments which suggest that the monomer form of the enzyme is competent in proton translocation.

Several methods of subunit III removal have now been developed. These subunit III-less enzymes often display no proton pumping activity when assayed with a pH-sensitive glass electrode. However, it was found that these enzyme species yielded poorly coupled phospholipid vesicles, and the loss of proton pumping activity could be correlated with a lowered respiratory control ratio (RCR). Fast kinetics methods have determined that the subunit III-less enzyme retains 47% of the proton pumping activity, more than observed in the glass electrode experiments [for extensive reviews, see Prochaska and Fink (1987) and Brunori et al. (1987b)]. In our laboratory, we also find that subunit III depletion does not abolish proton pumping activity, so long as the Cu_A site is not modified during the subunit III depletion procedures. In these experiments, the H^+/e^- stoichiometry is reduced to ca. 0.5 (Li et al., 1988). Recently, a bacterial cytochrome oxidase with only two subunits has been purified from *Paracoccus denitrificans*. Reconstitution of this enzyme into phospholipid vesicles resulted in coupled proteopliposomes that displayed proton pumping behavior, albeit with a lowered stoichiometry of ca. 0.5 (Solioz et al., 1982). This result has led to the suggestion that subunit III is not part of either the essential proton pumping or the electron transfer machinery but may be involved in a regulatory role of some kind. However, this interpretation may be oversimplified (see below).

Bacterial Oxidases. With the advent of new DNA technologies, the DNA sequences for a number of bacterial oxidases have been obtained. A few of these bacterial oxidases have also been isolated in quantities large enough for spectroscopic studies. These include aa_3-type cytochrome oxidases from *P. denitrificans* (Ludwig & Schatz, 1980), *Thiobacillus novellus* (Yamanaka & Fujii, 1980), *Thermus thermophilus* (Fee et al., 1980), and PS3 (a thermophile) (Sone et al., 1979). As isolated, all of these oxidases contain either two or three subunits which seem to be homologous to the three mitochondrially coded mammalian oxidases. Of these bacterial enzymes, the one from *P. denitrificans* is the most thoroughly studied. It also most closely resembles the mitochondrially coded bovine heart subunits, as does the *T. novellus* enzyme. The structure of the cytochrome oxidases from the thermophilic bacteria, however, seem to be different. The *T. thermophilus* and PS3 enzymes have only one major polypeptide, and some oxidases of this type have an intrinsic cytochrome *c* associated with the oxidase as well. However, all these bacterial aa_3-type oxidases exhibit similar functional characteristics and have spectroscopic properties almost indistinguishable from those of the bovine enzyme. It is clear from these bacterial studies that the catalytic core of cytochrome oxidase is comprised of at most three major subunits.

The isolation of two- and three-subunit bacterial cytochrome oxidases provides a simpler and more versatile system for studying redox-linked proton pumping in cytochrome oxidase. Because these oxidases are bacterial in nature, one can take advantage of techniques not accessible to the mammalian oxidase system. Recently, Yamanaka and co-workers reported the purification of copper-deficient aa_3-type oxidases from *Pseudomonas* AM1 (Fukomori et al., 1985), *Nitrosomonas europaea* (Numata et al., 1989), and *Halobacterium halobium* (Fujiwara et al., 1989). Using bacterial growth under copper-deficient conditions, they were able to isolate a Cu_A-deficient enzyme from the *Pseudomonas* AM1 and *N. europa* strains capable of oxidizing ferrocytochrome *c* and reducing dioxygen. The cytochrome oxidase from *H. halobium* grown under copper-deficient conditions contains no copper as isolated. This enzyme was found to be devoid of electron transfer and dioxygen reduction activity when ferrocytochrome *c* was used as the substrate. These experiments show that the electron transfer between cytochrome *a* and the binuclear center is viable, and that electrons can enter the enzyme via cytochrome *a*. Similar conclusions have been derived from the chemical modification experiments on the bovine enzyme (Gelles & Chan, 1985). Unfortunately, it is not known whether these Cu_A-less enzymes are capable of pumping protons. Finally, a cytochrome oxidase from *T. thermophilus* has recently been purified which contains a *b*-type cytochrome instead of the cytochrome *a* (Zimmermann et al., 1988). This oxidase has a copper site spectroscopically similar to Cu_A, but it contains only one cysteine residue. Again, it is not known whether this enzyme is capable of pumping protons at this time.

Another new development in the bacterial oxidase field is the cloning of the cytochrome oxidase genes for *P. denitrificans* (Raito et al., 1987) and PS3 (Sone et al., 1988). These studies have already proven useful in identifying a subunit III in *P. denitrificans* that was not observed in the initial purification. The advent of an expression system and site-directed mutagenesis techniques will allow for specific perturbations of any of the active sites in the protein complex and the opportunity to more clearly delineate structure–function questions that are not amenable to traditional biochemical and biophysical techniques.

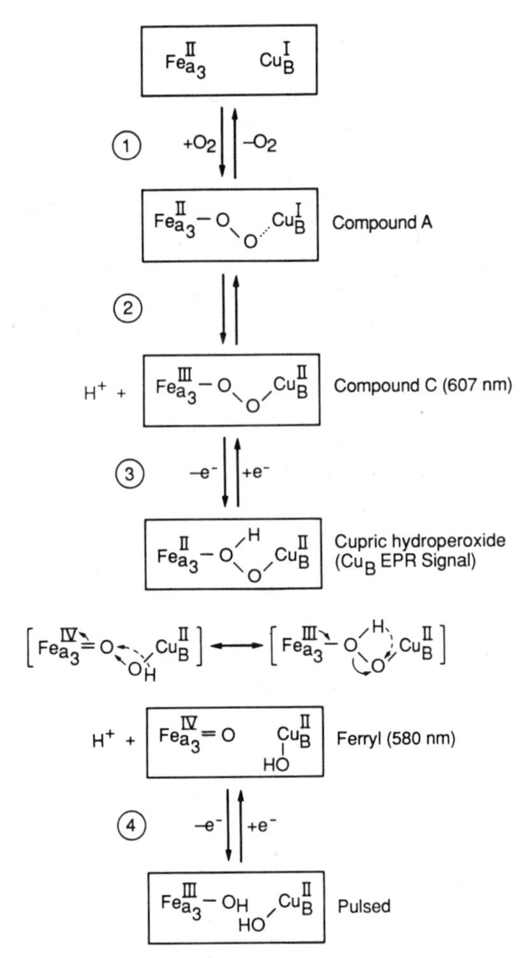

FIGURE 2: Scheme depicting the four electron reducing steps of the dioxygen reduction reaction on the basis of the available chemical and spectroscopic evidence. (See text for details.)

Dioxygen Chemistry

In cytochrome oxidase, the proton pumping reaction of the enzyme is coupled to the highly exergonic enzyme-mediated reduction of dioxygen at the binuclear center. Specifically, we expect the details of the electron transfers which drive the proton pump to be controlled by the chemistry of dioxygen reduction. Accordingly, the chemistry of the dioxygen reduction reaction is of great importance, and this subject has been a major focus of research for the past decade. In this section, we highlight some of the more recent work in this area. The chemistry of dioxygen reduction has been reviewed in detail recently (Hill et al., 1986; Chan et al., 1988b), so we will be brief here and refer the reader to these more extensive treatises.

Our current view of the dioxygen reduction mechanism is summarized in Figure 2. In step 1, dioxygen binds to ferrous cytochrome a_3, forming a dioxygen adduct called compound A (Chance et al., 1975a,b). This species is short-lived, lasting no more than 10 μs according to a recent resonance Raman investigation (G. Babcock, personal communication). The frequency of the O–O stretch observed in the Raman experiment suggests some electron redistribution from the iron toward the bound dioxygen. Further electron redistribution at the binuclear cluster, particularly from Cu_B, converts compound A to the peroxidic adduct, compound C (step 2). This species is characterized by an optical difference spectrum (compound C minus resting) with a distinctive maximum feature at 607 nm. Compound C is stable in the absence of the transfer of additional reducing equivalents from the low-potential centers. Further reduction by one electron (step 3) generates two intermediates at the three-electron level of dioxygen reduction. The first is cupric hydroperoxide species in which the O–O bond is still intact. The second intermediate is an O–O bond cleavage product of the first (Blair et al., 1985). There is now compelling experimental evidence that the second intermediate is an oxyferryl cytochrome a_3/cupric Cu_B species. It exhibits an optical spectrum with difference features (ferryl minus resting) at 580 and 428 nm. In addition, the EPR (Witt & Chan, 1987; Witt, 1988), resonance Raman (Witt, 1988), Mössbauer (Fee et al., 1988), and EXAFS (Chance & Powers, 1985; Kumar et al., 1988) experiments are consistent with an oxyferryl structure. Finally, in step 4, further reduction by the fourth electron yields the pulsed enzyme and water.

Confirming evidence for some of these intermediates has come from the activation of the enzyme by H_2O_2. Here, the chemistry is less heterogeneous and can be controlled to obtain compound C and the oxyferryl intermediates in high yields (Chan et al., 1988b). In particular, the reaction of the pulsed enzyme with stoichiometric amounts of H_2O_2 gives compound C, and the reaction with excess H_2O_2 produces nearly quantitative yields of the oxyferryl intermediate. This technique has allowed for the spectroscopic characterization of the oxyferryl species mentioned above.

Finally, Wikström (1981) has shown that the catalytic cycle of cytochrome *c* oxidase can also be partially reversed in mitochondria. When mitochondria are poised in a highly oxidizing environment (in the presence of high ferricyanide/ferrocyanide), the addition of high concentrations of ATP can generate a sufficient membrane potential to induce the sequential formation of two optically distinct water oxidation intermediates. These are identical with the two intermediates generated at the two- and three-electron-reduced levels of dioxygen reduction in the experiments which proceed in the forward direction. The first intermediate observed upon charging the membrane is a species with the 580-nm absorption band, spectroscopically identical with the oxyferryl species generated by Chan et al. (1988b) at the three-electron level of dioxygen reduction. In the reverse reaction, this species corresponds to a one-electron oxidation product of the pulsed enzyme. The second intermediate exhibits an intense absorption difference band at 607 nm and is spectroscopically indistinguishable from the peroxidic intermediate compound C. These experiments are particularly important because they are performed at equilibrium and allow the midpoint potentials of these intermediates to be measured: 939 mV for the oxyferryl intermediate and 801 mV for compound C (Wikström, 1988).

From these studies, it is clear that the chemistry of dioxygen reduction is quite complicated and that each electron level of dioxygen reduction yields chemically distinct oxygen intermediates at the binuclear center, each with a different chemical reactivity and affinity for reducing equivalents. It is therefore improbable that one of the binuclear center metals is involved in redox linkage. Nevertheless, it is important to emphasize the important role that the dioxygen chemistry plays in the proton pumping reaction. At the very least, formation of the peroxide and oxyferryl intermediates increases the redox potential of the binuclear cluster by ca. 500 mV. This can have profound effects on the intramolecular electron transfer reaction rates. In addition, the formation of these intermediates can trigger conformational changes in the protein that can result in electron gating and proton gating. Consistent with these ideas, Wikström has recently reported that only the electron transfers from the low-potential centers to the highly oxidizing compound C and oxyferryl intermediates are involved in translocating protons (Wikström, 1989). This result is supported by the observation that dioxygen binds only to ferrous cytochrome a_3, implying that at least one step in the cycle involves an electron transfer to the "unloaded" oxidized binuclear center. Such an electron transfer does not contain enough free energy to pump protons against an electrochemical gradient and cannot be expected to be involved in proton pumping. However, although some electron transfer steps may not be involved in proton pumping, it is clear that the details of all the electron transfer reactions leading to dioxygen binding and reduction must play a major role in the catalytic cycle.

Kinetics and Pathways of Electron Transfer

The available body of literature on the kinetics of cytochrome oxidase is copious, and reviews on this subject are available [see Brunori et al. (1981), Wikström et al. (1981), and Hill & Greenwood (1984a,b)]. In this section, we merely point out the major questions which we feel remain unanswered about the kinetics of the enzyme, focusing on the kinetics of the intramolecular electron transfer events. Understanding the details of these electron transfers has become particularly important in relating electron transfer to proton pumping, since it is most probably the intramolecular electron transfer from either cytochrome *a* or Cu_A to the dioxygen intermediates that is linked to proton translocation.

Under typical experimental conditions, the cytochrome *c* oxidase molecule can catalyze the oxidation of 30–600 molecules of cytochrome *c* per second [see Wikström et al. (1981) and Brunori et al. (1987a) for a review]. The oxidation of ferrocytochrome *c* can be monitored with optical spectroscopy (Smith, 1955), and the consumption of dioxygen can be monitored polarographically (Ferguson-Miller et al., 1976). In addition the redox states of the hemes and Cu_A can be monitored during turnover by optical spectroscopy (Gibson & Greenwood, 1963).

Two classes of experiment have been used to investigate the turnover cycle of the enzyme. One is the steady-state experiment, which is designed to study the rate-limiting step of the enzyme and the events involved in the approach toward the steady state. The second class involves transient kinetics, designed to follow the enzyme through one turnover cycle in real time. These latter experiments have been particularly valuable in the study of the intramolecular electron transfers between the low-potential centers and dioxygen bound at the binuclear center.

Steady-State Kinetics. The steady-state behavior of cytochrome oxidase is complicated. It is well-known that the oxidation of ferrocytochrome *c* by oxidase exhibits nonhyperbolic kinetics [Malmstöm and Andrẽasson (1985) and reference cited therein]. When the cytochrome *c* concentration is varied in these experiments, two distinct kinetic phases of cytochrome *c* oxidation are often observed, each with a characteristic turnover number and K_m. These values also depend on ionic strength, pH, and the detergent used to solubilize the enzyme (Singorjo et al., 1986). The biphasic kinetics were initially used to argue for the presence of two catalytically competent cytochrome *c* binding sites (Ferguson-Miller et al., 1976). Recently, however, it has been suggested that there may be only one catalytically competent cytochrome *c* binding site but that there are two conformations of the enzyme (E_1 and E_2) which can accept electrons (Malmström & Andrẽasson, 1985; Thörnström et al., 1988). Malmström and co-workers have proposed that the enzyme can pump protons only in one of these two conformations (E_2)

(Brzezinski & Malmström, 1987). They further note that the existence of two conformations is an intrinsic property of ion pumps displaying alternating access.

One of the newer developments in the steady-state kinetics of cytochrome oxidase has been a set of experiments in which the reduction levels of cytochrome *a* and Cu_A were monitored during turnover. It has been shown that under turnover cytochrome *a* is significantly more reduced than Cu_A (Brzezinski et al., 1986; Thörnström et al., 1988). Furthermore, after complete oxidation of the cytochrome *c* at the end of the reaction, Cu_A appeared to be oxidized completely while cytochrome *a* remained partially reduced. To account for these observations, these authors have proposed a model advocating that the enzyme switches from the E_1 (nonpumping) to the E_2 (proton pumping) conformation only when cytochrome *a* and Cu_A are reduced and that intramolecular electron transfer from cytochrome *a* and Cu_A to the "unloaded" binuclear center is a concerted two-electron process. However, since it now appears that protons are pumped only when dioxygen is bound to the enzyme, some of these ideas may need to be revised.

Transient Kinetics. When stopped-flow techniques are used to follow the concomitant reduction of cytochrome *a* and Cu_A by ferrocytochrome *c*, biphasic kinetics are also observed. Most of the available literature suggests that these two metal centers are reduced synchronously, in two distinct phases (Antalis & Palmer, 1982; Andréasson et al., 1982), with some reports indicating that Cu_A may lag slightly (Wilson et al., 1975). To date, all of the experiments which measure the electron input have been limited by the binding rate of cytochrome *c*. The best estimates for this rate are between 10^6 and 3×10^7 M^{-1} s^{-1} (Antalis & Palmer, 1982; Andréasson et al., 1982; Wilson et al., 1975).

Following entry into the enzyme at cytochrome *a* and/or Cu_A, electrons are transferred to the binuclear center (cytochrome a_3 and Cu_B) intramolecularly. Most of the measurements on the kinetics of this electron transfer reaction have come from flow–flash experiments which monitor the reoxidation of cytochrome *a* and Cu_A in the presence of dioxygen. Accordingly, these electron transfer events are relevant to the proton pumping reaction. These studies have shown that the reoxidation of both cytochrome *a* and Cu_A is multiphasic. Hill and Greenwood (1984a,b) reported that 40% of cytochrome *a* is reoxidized simultaneously with cytochrome a_3 at nearly 30 000 s^{-1}. Following this phase, 60% of Cu_A is reoxidized at 7000 s^{-1}. Finally, the remainder of reduced cytochrome *a* and Cu_A is reoxidized at 700 s^{-1}. These results clearly indicate that the downhill electron transfer events are heterogeneous, depending on the conformation state of the enzyme and the nature of the intermediate at the binuclear site.

Another method used to infer the rate of electron transfer between the low-potential centers and the dioxygen reduction site is based on the measurement of the rate of the reverse electron transfer from the reduced dioxygen binding site to the oxidized low-potential centers following CO photodissociation from the CO mixed-valence enzyme. In these experiments cytochrome a_3 and Cu_B are reduced initially. Boelens et al. (1982) have reported that, following CO photodissociation, a rapid backflow of electrons from cytochrome a_3 to Cu_A occurs in ~5% of the enzyme molecules. On the basis of these results, they suggested that the electron transfer in the forward direction proceeded from Cu_A to the binuclear center at ~10 000 s^{-1}. Recently, Brzezinski and Malmström (1987) confirmed these observations and obtained a rate of 14 000 s^{-1} for the electron transfer rate from Cu_A to cytochrome a_3. On this basis, these authors argued that Cu_A is the primary electron donor to the oxygen binding site. In addition, they also observed a slower electron transfer from Cu_A to cytochrome *a* (~700 s^{-1}). It should be noted, however, that these electron transfer rates pertain only to the enzyme with an unloaded oxygen binding site, where the redox potential of the binuclear center is at most marginally (~100 mV) more positive than that of the low-potential metal centers, and thus may not be relevant to the proton pumping forms of the enzyme.

The rate of electron transfer between cytochrome *a* and Cu_A has also received attention. This is an important issue because if the rate is fast compared to the turnover rate of the enzyme, then the issue of one vs two cytochrome *c* binding sites, or the issue of one vs two electron input sites, becomes moot. In addition, a knowledge of this rate under a wide variety of circumstances would facilitate the interpretation of data from the flow–flash experiments alluded to earlier. Toward addressing this question, Morgan et al. (1989) recently studied the electron equilibration between cytochrome *a* and Cu_A in a partially reduced, CO-inhibited form of the enzyme (where the low-potential centers are reduced on the average by one electron) using the perturbed equilibrium method. These workers obtained a value of 17 000 s^{-1} for the sum of the forward and reverse rate constants (cytochrome $a^{2+}/Cu_A{}^{2+} \rightleftharpoons$ cytochrome $a^{3+}/Cu_A{}^{1+}$). Thus, the electron equilibration is extremely rapid compared to the turnover rate of the enzyme (30–600 s^{-1}), at least in this form of the enzyme. In this experiment, the binuclear site is reduced. It would be of interest to verify that the electron equilibration between the low-potential centers is indeed significantly slower when the binuclear center is oxidized, as suggested by a number of stopped-flow experiments on the resting enzyme (Wilson, 1975).

It is evident that many questions regarding the electron flow remain unanswered. In particular, it would be important to know whether the protein shuttles electrons from the low-potential centers to the dioxygen reduction site through a different pathway depending upon whether the binuclear site is activated by dioxygen or not. The answer to this question has taken on an increased significance and urgency as we attempt to formulate molecular mechanisms to describe the proton pumping process.

PROTON PUMPING

Redox Loops and Proton Pumps. Cytochrome oxidase links the electron transfer reaction between ferrocytochrome *c* and dioxygen to a net translocation of proton from the mitochondrial matrix to the cytosol. The concept of linkage between electron transfer and proton translocation in mitochondria was first proposed by Mitchell (1966) as part of the chemiosmotic hypothesis. However, for some time, there was disagreement as to whether the vectorial electron transfer mediated by cytochrome oxidase was coupled to proton translocation (Moyle & Mitchell, 1978). The question now is not whether cytochrome oxidase is involved with linking an electron transfer reaction to proton translocation but rather how this linkage occurs.

For many years, it was thought that cytochrome oxidase was the electron transfer arm of a redox loop. In this model, the electron transfers were catalyzed by the cytochrome oxidase enzyme, with no vectorial translocation of protons (Mitchell, 1966). However, on the basis of proton ejection experiments in coupled mitochondria, Wikström (1977) proposed that cytochrome oxidase is a "proton pump". Wikström argued that cytochrome oxidase uses the free energy of electron transfer to translocate protons vectorially in the opposite direction to the electron transfers. Since that time, Wikström and many others have come to view cytochrome oxidase as a proton pump which translocates proton via a mechanism other than a "redox loop". However, Mitchell (1988) has maintained that the proton-carrying function of the substrates has been overlooked. As an alternative to Wikström, Mitchell has offered two ligand-based redox loop mechanisms to account for the proton ejections based on the dioxygen substrate and its intermediates as carriers of oxidizing equivalents and the protons. In the first model, H_2O_2 formed during dioxygen reduction is the ligand which accepts the electrons and delivers the protons to and from the cytosol, respectively (Mitchell et al., 1985). The second model is a Cu_A-based mechanism in which oxidoreduction of the Cu_A is linked to the translocation of a hydroxide from the cytosol to the matrix (Mitchell, 1987). These schemes are ingenious. Unfortunately, ligand-based redox loops necessarily have unity H^+/e^- stoichiometry, and it has been shown that the H^+/e^- stoichiometry may be variable (Papa et al., 1989).

The distinction between a redox loop and a proton pump has caused some confusion in the past, and much discussion has ensued concerning the definition of these two terms (Mitchell, 1988; Malmström, 1988). We take a proton pump to describe an enzyme which actively translocates a proton via a mechanism other than a substrate-based redox loop, i.e., the coupling between electron transfer and proton trannsfer does not involve the association and dissociation of protons to and from a redox-active substrate molecule. As a result, the proton pumping function of cytochrome oxidiase has gained wide attention because it represents another class of active proton translocation enzymes (probably belonging to the class of active ion pumps) and has general mechanistic implications on how electron transfer can be linked to proton translocation in the respiratory and photosynthetic electron transport chains.

Basic Requirements of a Proton Pump. Any enzyme that couples two reactions must somehow catalyze both reactions in such a way that the "uphill" reaction does not occur in the absence of the "downhill" reaction. In cytochrome oxidase, protons are pumped by use of energy derived from the exergonic transfer of electrons from cytochrome *c* to molecular oxygen. Therefore, the driving reaction is electron transfer, and the driven reaction is the energetically unfavorable translocation of a proton against an electrochemical gradient. To ensure that the proton pump does not act as a passive proton transporter, three general requirements must be met for redox-linked proton translocation [Wikström et al., 1981; Malmström, 1985; Blair et al., 1986b; for a review of theoretical proton pumping models, see Krab and Wikström (1987)]. We refer to these as (1) linkage, (2) electron gating, and (3) proton gating. In this section we review these concepts and attempt to clarify the nomenclature which exists in the literature.

(*A*) *Linkage.* In order for electron transfer to be linked to proton transfer, one requirement is that these two activities be linked by some common intermediate. Many schemes for linkage have been proposed. The coupling can be direct, with the redox center also being the proton translocator. Or the coupling may be indirect, with the redox element being in conformational contact with the proton translocating element. In both cases, there are two distinct states of the redox center (reduced and oxidized), as well as two distinct states of the proton translocating element (protonated and deprotonated). When the redox and proton translocating elements are linked in a model which includes the sidedness of the membrane, one can envision an eight-state "cubic" formalism as proposed by

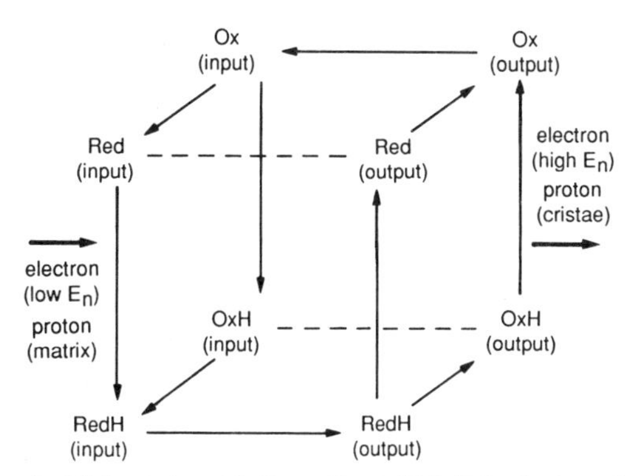

FIGURE 3: Eight-state cubic formalism which links the two states of the electron translocating element, the two states of the proton translocating element, and the two sides of the membrane, as proposed by Wikström (1981). (See text for a more detailed description.)

Wikström (1981) and shown in Figure 3.

It is important to note that the cubic formalism is not itself a mechanistic model for proton pumping. It is, rather, a formal way of describing the eight possible states in which the transducer of redox free energy to protonmotive energy can exist. As described by Krab and Wikström (1987), the eight states arise from separating the possible redox, protonation, and sidedness states of the transducer along Cartesian coordinates: the *x* axis describes the redox state of the pump element; the *y* axis describes the sidedness of the pump element with the input (I) state denoting protonic contact with the matrix space and electronic contact with the electron donor and the output (O) state denoting protonic contact with the cytosol and electronic contact with the electron acceptor; the *z* axis describes the protonation state of the pump element. There has recently been some discussion on the generality of this treatment. As pointed out by Blair et al. (1986b) and Krab and Wikström (1987), the eight-state model arises from requiring that the electronic and protonic specificities be symmetric. It is not necessary for the electronic sidedness to be the same as the protonic sidedness in any given state (for example, separate proton input and electron input states), in which case there would be 16, and not 8 states. However, it is argued that the eight-state scheme is the simplest one in which electron transfer, proton transfer, and I/O reactions may be discussed independently, although this scheme seems unnecessarily restrictive (Blair et al., 1986b).

To rectify this, Krab and Wikström (1987) recently introduced "specificity of the first kind" to denote the existence of distinct electronic and protonic states on the input and output sides of the cubic pump description. These authors attribute the existence of these states to be a consequence of "electron gating". The importance of electron gating in a redox-linked proton pump has recently been emphasized by Chan and coworkers. We assume that specificity of the first kind is implicitly built into the eight-state cubic scheme.

(*B*) *Electron Gating, Proton Gating, and Specificity of the Second Kind.* In order for a redox-linked proton pump to attain a maximum H^+/e^- stoichiometry, the electron transfer reaction, which is highly exergonic, must not take place in the absence of proton transfer. Most models for proton pumping argue that this requirement is manifested in a need for two distinct redox states of the pump site. In one, the redox element is in an "electron input state" which is ready to accept electrons. The other state is an "electron output state" with all of the associated proton movements completed, which is ready to pass electrons onto another acceptor, presumably the dioxygen reduction site. And the process by which the protein ensures that electron transfer occurs only into and out of these two states is electron gating (Blair et al., 1986b), or specificity of the first kind (Krab & Wikström, 1987). If electron gating is absent, electron transfer may occur in states of the enzyme which are not competent in proton translocation (i.e., states in which the proton translocating element has not responded to the redox element). These reactions are futile and lead to electron "leaks" which are not described by the conventional cubic scheme. It is important to distinguish electron leaks from electron slippage (backward movement of electrons) which is represented by the cycle inscribing the left face of the cube. The latter is a nonproductive event and does not bear on the stoichiometry.

Implicit in the discussion above is that, in addition to electron gating, the proton flow must be similarly gated. It goes without saying that there must not be proton leaks through the protein which dissipate the transmembrane protonmotive force, even as the enzyme is turning over. The enzyme must also ensure that proton slippage does not occur, such as the cycle that is inscribed on the bottom face of the cube.

Krab and Wikström (1987) have argued that electron and proton gating alone do not ensure a viable catalytic cycle and that specificity of a second kind is needed. This specificity of the second kind kinetically controls the four input to output conversions so that only two are viable pathways. In our view, the purpose of electron gating and proton gating is to ensure that the coupled processes are kinetically enhanced at the expense of the uncoupled events. In this connection it bears pointing out that a viable pumping model requires that electron and proton movements are synchronized, meaning that both electron and proton gating must be present. Having both types of gating specifies that only two of the four I/O conversions are allowed because both electron and proton transfers must be completed before any I/O conversion may proceed. In other words, only one state on the input side may be I/O competent, and one state on the output side may be I/O competent. Therefore, we feel that electron and proton gating together correspond to "specificity of the second kind".

Site of Redox Linkage. The largest free energy change of the cytochrome oxidase redox reaction is associated with the electron transfer from the primary acceptors (Cu_A and cytochrome *a*) to the dioxygen anchored at the binuclear center. Since this free energy is expended once the electron has reached the oxygen binding site, these "low-potential" centers have been considered the most natural candidates for the site of redox linkage. Until recently, the argument that either cytochrome *a* or Cu_A should be the site of redox linkage was based solely on this idea without experimental support. However, the recent experiments of Wikström and Casey (1985) on whole mitochondria appear to confirm that these two sites are the most likely sites for redox linkage to proton translocation.

Of the two low-potential centers, cytochrome *a* has received the most attention as the site of redox linkage. Three main arguments have been advanced to support cytochrome *a* as the site of linkage. First, the cytochrome *a* midpoint potential exhibits a dependence on the pH of the mitochondrial matrix of ca. 30 mV/pH unit (Arzatbanov et al., 1978). Second, the reoxidation kinetics of cytochrome *a* following flash photolysis of the CO-inhibited enzyme has been shown to be heterogeneous (Hill et al., 1984a,b). Third, Moroney et al., (1984) have observed that the steady-state reduction level of cytochrome *a* is dependent on both the pH and the transmembrane potential. The latter two observations have been rationalized in terms of two different states of cytochrome *a*, both electron transfer competent, that are presumably the input and output states of the pump site. In support of cytochrome *a*, Babcock and co-workers have obtained evidence from resonance Raman experiments that the heme A formyl group of cytochrome *a* is hydrogen bonded to a hydrogen-bond donor in the protein and that the strength of this hydrogen bond increases as the heme iron is reduced (Babcock & Callahan, 1983). This group has offered a proposal for the mechanism of redox linkage based on this result.

More recently, however, some circumstantial evidence from this laboratory seems to implicate Cu_A as the site of redox-linked proton translocation. In these experiments, Cu_A was chemically modified by *p*-(hydroxymercurio)benzoate (pHMB) to produce a structurally altered type 2 Cu_A site (Gelles & Chan, 1985). The resultant enzyme exhibited a rapid extravesicular alkalinization when it was reconstituted into membrane vesicles and assayed for proton pumping activity (Nilsson et al., 1988). These authors attribute this behavior to the formation of a facile transmembrane proton conduction pathway through the protein upon Cu_A modification. In subsequent work, Li et al. (1988) showed that heating cytochrome oxidase at 43 °C in the nonionic detergent lauryl maltoside also results in the structural modification of the Cu_A site. This heat-modified enzyme was shown to contain a mixture of type 1 and type 2 Cu_A sites in addition to native Cu_A. When assayed for proton pumping activity, this modified enzyme preparation either displayed no proton pumping activity (Sone & Nicholls, 1984) or revealed a proton conduction pathway through the protein similar to that of the pHMB-modified enzyme (Li et al., 1988). On the other hand, when the Cu_A site was protected from heat-induced modification by reduction of the enzyme or by ligand binding to the binuclear center, proton pumping activity was retained (Li et al., 1988). These results strongly implicate Cu_A as an important part in the proton pumping machinery of the enzyme.

At this time, the biochemical evidence supporting either of the two low-potential metal centers as the site of redox linkage is circumstantial at best. Nevertheless, these studies have provided impetus for the development of molecularly based models for redox linkages. In such exercises it seems important to be as explicit about the details of the proposal as possible, so that the various ingredients of the models can be subjected to critical experimental testing. Ultimately, the fate of a particular model must rest on how well predictions match the experimental facts. Depending on the outcome, a model will have to be either abandoned or refined for further assessment.

Models for Redox Linkage. Although the basic requirements for a redox-linked proton pump have been discussed, there are few mechanistic models of redox linkage that attempt to build in these requirements at the molecular level. Two such models exist in the literature. Here, we present each model and discuss the extent to which each incorporates the requirements of a redox-linked proton pump.

(*A*) *Babcock Model.* Babcock and Callahan (1983) observed that the strength of hydrogen bonding between the formyl oxygen of cytochrome *a* and some proton donor(s) in the protein varies between the oxidized and reduced states of the heme center. From the formyl C=O stretching frequency measured by resonance Raman spectroscopy, it was calculated that the hydrogen-bond strength differs by 110 mV between the ferric and ferrous forms of cytochrome *a*. Babcock and Callahan proposed that this energy contributes to the total free energy required to drive a proton against the electrochemical gradient (~200 mV) across the inner mitochondrial membrane. The details of the mechanism proposed by Babcock

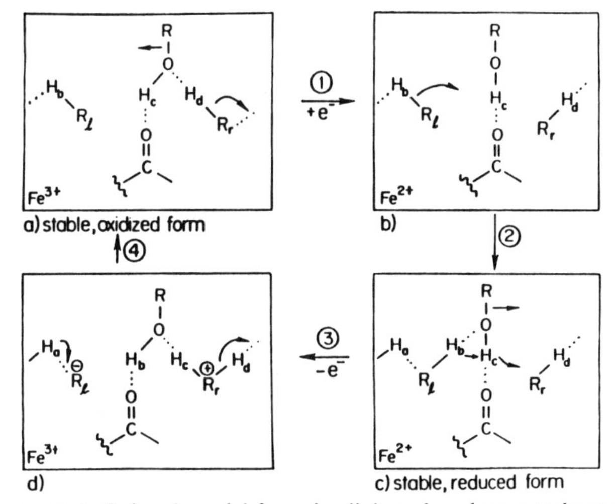

FIGURE 4: Babcock model for redox linkage based on cytochrome *a*. (See text for details.)

and Callahan (1983) are shown in Figure 4. When cytochrome *a* is oxidized, the formyl oxygen is hydrogen bonded to a protein proton donor lying between two hydrogen-bonded channels. One of these is connected to the matrix side of the membrane and the other to the cytosolic side. Upon reduction of the site, the hydrogen-bond strength increases between the now electron-rich formyl oxygen and the proton of the donor group. The change in hydrogen bond strength is proposed to cause a change in the geometry of the conjugate base, allowing it to also interact with the proton of an adjacent acidic residue in contact with the matrix-facing hydrogen-bonding chain. As the cycle continues, the hydrogen-bond strength increases between the conjugate base and the matrix-derived proton at the expense of the proton hydrogen bonded to the formyl group. Eventually, the latter proton leaves to occupy a place on the cytosol-facing hydrogen-bonded channel as it is replaced by the proton from the matrix that is hydrogen bonded to the conjugate base. The proton hole left in the matrix-facing hydrogen-bonded channel is eventually replenished by a tandem proton migration along the channel toward the pump element, followed by the uptake of another proton into the channel from the matrix space. In this scheme, the redox center is linked directly to the proton binding steps via a direct mechanism in which there is alternating access of the "pump site" to the two sides of the membrane. The cytochrome *a* formyl group serves to gate the proton flow in response to a change in the redox state of the center. Although the elements of redox linkage and proton gating are clearly evident here, unfortunately this model makes no provisions for the gating of the electron flow to obviate futile cycles.

Since a number of treatments of the enzyme have been reported to disrupt proton pumping while leaving the environment around cytochrome *a* intact, Babcock and Callahan have recently offered a revised version of this model in which they allow for the possibility of a redox linkage of the cytochrome *a* formyl group to a more global conformational change linking the redox element to a distant proton binding and transport element.

(*B*) *Chan Model.* Gelles et al. (1986) have proposed an alternate model based on Cu_A as the site of redox coupling to proton translocation. In this model, the redox element is also linked directly to the proton transfer element. However, these authors have explicitly included the gating of electron flow. Gelles et al. (1986) argue that the enzyme must be able to control the rate constants of the possible electron transfers in order to enhance the coupled processes and suppress all of

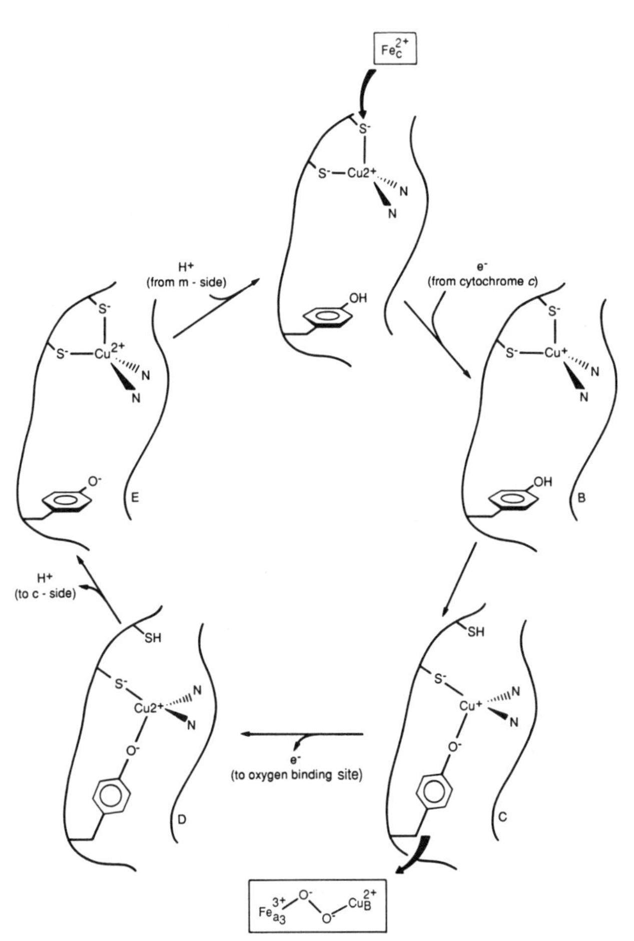

FIGURE 5: Chan model for redox linkage based on Cu_A. (See text for details.)

the leak pathways. Conformational switching is proposed as a means of achieving this electron gating. The electron enters the Cu_A site in one conformation (the "input state"), and facile electron transfer out of the site occurs only after the enzyme has switched conformations to the "output state". The redox linkage actually occurs during this switching process.

The Chan model is outlined in Figure 5. Two hydrogen-bonded channels connect the pump element with the matrix and cytosol. In the oxidized state, the copper ion is ligated by two histidine and two cysteine ligands in a distorted tetrahedral geometry. This is the electron input state. When the site becomes reduced, the bis(dithiolate) coordination becomes asymmetric, and one cysteine bond lengthens relative to the other (Chan et al., 1988a). At this point, a tyrosine (or another residue with a similar pK_a) interacts with the copper ion, displacing one cysteine ligand. The change in pK_a's of the incoming tyrosine ligand and outgoing cysteine ligand leads to a proton transfer from the tyrosine to the cysteine. In this way, part of the redox energy from the reduction of the Cu_A site is expended in moving the proton from the matrix side of the pump site to the cytosol side. Following this ligand exchange (or rearrangement) and proton transfer, the reduced Cu_A site is in the electron output state and transfers the electron to the dioxygen reduction site. Since this is a reduced type 1 site, Gelles et al. (1986) proposed that electron transfer to dioxygen intermediates at the binuclear center is extremely facile. Thus, the process is kinetically driven. When the Cu_A site becomes oxidized, reverse ligand exchange (or rearrangement) occurs, and eventually, the tyrosinate is returned to the matrix side of the pump to be reprotonated following the tandem migration of the protons in the matrix-facing hydrogen-bonding channel toward the pump site. It is this last step that is proposed to be the rate limiting in the proton pumping cycle.

While there is no direct experimental evidence to support the above model, the Cu_A modification experiments discussed earlier do suggest a central role for Cu_A in proton pumping. In formulating this model, Gelles et al. (1986) have attempted to incorporate all three requirements of a redox-linked proton pump as well as the available information on the ligand structure and electronic structure of the Cu_A site.

Identification of Redox Linkage Site: Search for a Crossover Point. With the general acceptance of proton pumping in cytochrome oxidase, recent research efforts have been directed toward identifying the site of redox linkage and unraveling the molecular mehanism of the redox-linked proton translocation reaction. The identification of this site is of particular importance because the nature of the redox center(s) involved will dictate the kinds of mechanisms at work. Because there are essentially no other examples of a redox-linked proton pump and because the theoretical considerations of such a pump are relatively undeveloped, it is unclear what special properties such a center(s) should possess.

One of the most popular approaches toward identifying the site of redox linkage is the search for a classical "crossover point" in cytochrome oxidase (Rich, 1988). The concept of a crossover point was originally established during the effort to identify the energy coupling sites in the respiratory chain. This approach hinges on the assumption that any element in a linear electron transport chain which is involved with proton translocation will necessarily have a steady-state turnover rate that is sensitive to a membrane potential. Therefore, when the system is at steady state and a membrane potential is applied, the proton translocating element will slow down in response, causing the electron carrier upstream in the chain to become more reduced while causing the electron carrier downstream to become more oxidized. Extended to cytochrome oxidase, this approach makes several assumptions. First, this approach relies on a linear sequence of intramolecular electron transfer events, or that all electron transfers pass through the site of redox linkage. If there is a branched pathway in which one arm is not coupled to proton translocation, one would not necessarily expect the classical crossover behavior. Second, it assumes that the electron transfer events are rate limiting. For a proton pump, it seems more likely that the electron transfer rates are facile and that the turnover numbers are limited by conformational events subsequent to the electron transfer events. When the electron transfer rates become sufficiently retarded by the membrane potential, the electron transfers can become rate limiting. Third, the existence of a crossover point assumes that the redox elements are noninteractive. If the redox element which is coupled to proton translocation interacts with the upstream and downstream redox elements in such a way as to modulate the rates of electron transfer depending on the magnitude of the membrane potential, then one will not observe a standard crossover point behavior. Fourth, many of these experiments utilize cyanide to slow down the steady-state turnover rate. Under this circumstance, the majority of the cytochrome oxidase molecules are inhibited with cyanide, and it is unclear whether the steady-state situation is representative of normal turnover. Finally, the interpretation of the steady-state kinetics of cytochrome oxidase is complicated, as the analysis involves no fewer than four electron acceptors at the dioxygen reduction site, each with a different affinity for reducing equivalents and different chemical reactivities. Thus, while a crossover point may exist in cytochrome oxidase, the complex behavior of the

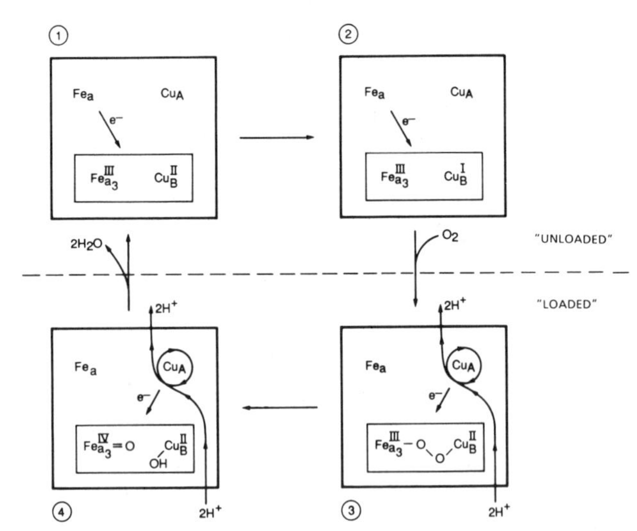

FIGURE 6: Proposal for the complete turnover cycle of cytochrome *c* oxidase which integrates the intramolecular electron transfer, dioxygen reduction, and proton pumping events. (See text for a detailed description.)

redox element linked to proton translocation in response to a transmembrane potential may obscure its identification, and the absence of such behavior does not necessarily disprove the involvement of a redox center in proton translocation.

Complete Turnover Cycle: A Hypothesis

We conclude this paper with a proposal for the complete turnover cycle of the enzyme, wherein we attempt to integrate both the electron transfer and proton pumping events with the dioxygen chemistry.

Any chemically acceptable proton pumping model must be consistent with all of the available biochemical evidence at hand. Hence, one of the challenges behind the model-building exercise is integrating the proton pump into an already complicated turnover cycle which involves four electron transfers to at least four different dioxygen reduction intermediates. Once again, we emphasize the crucial role that dioxygen plays in proton pumping because it is the reduction of dioxygen to water that provides the free energy for proton pumping. Accordingly, we assert that it is the electron transfer from Cu_A (or cytochrome *a*) to the dioxygen intermediates bound at the binuclear center that is linked to the translocation of the protons. Such a model is consistent with the recent results of Wikström (1989) that suggest that protons are pumped during only two of the four electron transfer steps. However, a proton pump which relies on two of the four electron transfer steps in the catalytic cycle does require a level of complexity which has not been considered in most models up until this point. In this section, we describe how a model based on Cu_A as the site of redox linkage can be integrated into a pumping cycle that derives energy from only two of the four electron transfer steps.

If the proton pump derives energy from only two of the four electron transfers to dioxygen, the pumping site (or the site of linkage) must be able to distinguish between the energetically productive and nonproductive electron transfers, *or* alternatively, electrons must pass through the pump site only when protons are pumped. We propose that allosteric coupling between the binuclear center and the low-potential centers modulates the downhill electron transfer pathways. There is some circumstantial evidence for an allosteric interaction between Cu_A and the binuclear center (Li et al., 1988), so we propose that Cu_A is the site of redox linkage.

As shown in Figure 6, the first two electrons enter the oxidized binuclear center in the absence of dioxygen. Since cytochrome *a* and the "unloaded" binuclear center have similar potentials, these electron transfer steps do not contain sufficient free energy to pump a proton. However, these electron transfer events initiate the dioxygen chemistry and therefore must occur with reasonable rapidity. We suggest that these two electron transfers occur via cytochrome *a*.

When dioxygen binds to the reduced binuclear center and becomes reduced to the peroxy intermediate, the driving force for the electron transfers from the low-potential centers increases dramatically. The electron transfer steps which follow must be linked to proton pumping, and we propose that they occur via the Cu_A site. The suggestion is that dioxygen binding and its subsequent reduction to the peroxy intermediate trigger a conformational change that directs the electron flow through the pumping site (Cu_A). In studies of the single turnover of the enzyme, Cu_A is always reoxidized before cytochrome *a*, suggesting that electron transfer from Cu_A to the binuclear center is more facile than from cytochrome *a* when the dioxygen reduction site is activated.

One consequence of having two distinct electron transfer paths is that the maximum level of reduction at steady state for cytochrome *a* and Cu_A cannot exceed 50%. Assuming that the highly driven electron transfer steps from Cu_A to the dioxygen intermediates are fast, the enzyme population at steady state will always contain a large contribution from states of the enzyme in which both cytochrome *a* and Cu_A are oxidized. This is consistent with the long-standing observation that the steady-state levels of reduction for cytochrome *a* and Cu_A never exceed 50% (Moroney et al., 1984; Gregory & Ferguson-Miller, 1989).

An important associated question is how the transfer of one electron can be coupled to two proton transfers. Two possible mechanisms are possible. The pump site may transport two directly linked protons per electron or, alternatively, one proton may be translocated via a directly linked process while a second proton is translocated via an indirect mechanism at another site. Our model of redox linkage proposes that the Cu_A site pumps one proton for every electron transfer that passes through the pump site. Accordingly, we favor a picture that includes two sites for proton translocation. The alternate site could reside in subunit III, where DCCD binding and subunit III depletion consistently diminish proton pumping activity by 50%. These results are consistent with subunit III being one of the two proton pumping sites. It should be noted that if this model is correct, Cu_A and subunit III must be in conformational contact. Finally, the possibility of two distinct electron transfer pathways to the unloaded and loaded dioxygen reduction site makes the issue of electron transfer pathways especially significant. Specifically, it becomes important to determine whether two pathways exist, and if so, how the enzyme switches between the two. Two possible switching mechanisms exist. First, the enzyme may control the equilibration between cytochrome *a* and Cu_A such that electron transfer through the pump site (presumably Cu_A) does not occur in species of the enzyme with an "unloaded" dioxygen reduction site. Second, the enzyme may directly control the rate of electron transfer from the low-potential centers to the dioxygen reduction site, depending on the state of the enzyme. In this case, the cytochrome *a* to Cu_A electron equilibration rate may remain fast in all forms of the enzyme. These possibilities provide an exciting framework for the next generation of experiments on this fascinating enzyme.

Registry No. H^+, 12408-02-5; cytochrome *c* oxidase, 9001-16-5.

References

Aasa, R., Albracht, S. P. J., Falk, K. E., Lanne, B., & Vänngård, T. (1976) *Biochim. Biophys. Acta 422*, 260–272.

Andréasson, L. E., Malmström, B. G., Strömberg, B. G., & Vänngård, T. (1982) *FEBS Lett. 28*, 297–301.

Antalis, T. M., & Palmer, G. (1982) *J. Biol. Chem. 257*, 6194–6206.

Aratzabanov, V. Y., Konstantinov, A. A., & Skulachev, V. P. (1978) *FEBS Lett. 87*, 188–195.

Babcock, G. T., & Callahan, P. M. (1983) *Biochemistry 22*, 2314–2319.

Beinert, H., Griffith, D. E., Wharton, D. C., & Sands, R. H. (1962) *J. Biol. Chem. 237*, 2337–2346.

Blair, D. F., Witt, S. N., & Chan, S. I. (1985) *J. Am. Chem. Soc. 107*, 7389–7399.

Blair, D. F., Ellis, W. R., Wang, H., Gray, H. B., & Chan, S. I. (1986a) *J. Biol. Chem. 261*, 11524–11537.

Blair, D. F., Gelles, J., & Chan, S. I. (1986b) *Biophys. J. 50*, 713–733.

Boelens, R., Wever, R., & Van Gelder, B. F. (1982) *Biochim. Biophys. Acta 682*, 264–272.

Brudvig, G. W., Blair, D. F., & Chan, S. I. (1984) *J. Biol. Chem. 259*, 11001–11009.

Brudvig, G. W., Morse, R. ., & Chan, S. I. (1986) *J. Magn. Reson. 67*, 198–201.

Brunori, M., Antonini, G., & Wilson, M. T. (1981) *Met. Ions Biol. Syst. 13*, 187–228.

Brunori, M., Antonini, G., Malatesta, F., Sarti, P., & Wilson, M. T. (1987a) *Adv. Inorg. Biochem. 7*, 93–154.

Brunori, M., Antonini, G., Malatesta, F., & Sarti, P. (1987b) *Eur. J. Biochem. 169*, 1–8.

Brzezinski, P., & Malmström, B. G. (1987) *Biochim. Biophys. Acta 894*, 29–38.

Brzezinski, P., Thörnström, P.-E., & Malmström, B. G. (1986) *FEBS Lett. 194*, 1–5.

Chan, S. I., Li, P. M., Nilsson, T., Gelles, J., Blair, D. F., & Martin, C. T. (1988a) in *Oxidases and Related Redox Systems* (Mason, H., Ed.) Alan R. Liss, New York.

Chan, S. I., Witt, S. N., & Blair, D. F. (1988b) *Chem. Scr. 28A*, 51–56.

Chance, B., & Powers, L. (1985) *Curr. Top. Bioenerget. 14*, 1–19.

Chance, B., Saronio, C., & Leigh, J. S., Jr. (1975a) *Proc. Natl. Acad. Sci. U.S.A. 72*, 1635–1640.

Chance, B., Saronio, C., & Leigh, J. S., Jr. (1975b) *J. Biol. Chem. 250*, 9226–9237.

Cline, J., Reinhammar, B., Jensen, P., Venters, R., & Hoffman, B. M. (1983) *J. Biol. Chem. 258*, 5124–5128.

Fee, J. A., Choc, M. G., Findling, K. L., Lorence, R., & Yoshida, T. (1980) *Proc. Natl. Acad. Sci. U.S.A. 77*, 141–151.

Fee, J. A., Zimmerman, B. H., Nitsche, C. I., Rusnak, F., & Münck, E. (1988) *Chem. Scr. 28A*, 75–79.

Ferguson-Miller, S., Brautigan, D. L., & Margoliash, E. (1976) *J. Biol. Chem. 251*, 1639–1650.

Finel, M., & Wikström, M. (1988) *Eur. J. Biochem. 176*, 125–129.

Fujiwara, T., Fukomori, Y., & Yamanaka, T. (1989) *J. Biochem. 105*, 287–292.

Fukumori, Y., Nakayama, K., & Yamanaka, T. (1985) *J. Biochem. 98*, 1719–1722.

Gelles, J., & Chan, S. I. (1985) *Biochemistry 24*, 3963–3972.
Gelles, J., Blair, D. F., & Chan, S. I. (1986) *Biochim. Biophys. Acta 853*, 205–236.
Gibson, Q. H., & Greenwood, C. (1963) *Biochem. J. 86*, 541–554.
Gregory, L., & Ferguson-Miller, S. (1989) *Biochemistry 28*, 2655–2662.
Hall, J., Moubarak, A., O'Brien, P., Pan, L. P., Choi, I., & Millet, F. (1988) *J. Biol. Chem. 263*, 8142–8149.
Hill, B. C., & Greenwood, C. (1984a) *Biochem. J. 218*, 913–921.
Hill, B. C., & Greenwood, C. (1984b) *FEBS Lett. 166*, 362–366.
Hill, B. C., Greenwood, C., & Nicholls, P. (1986) *Biochim. Biophys. Acta 853*, 91–113.
Holm, L., Saraste, M., & Wikström, M. (1987) *EMBO J. 6*, 2819–2823.
Hu, V. W., Chan, S. I., & Brown, G. S. (1977) *Proc. Natl. Acad. Sci. U.S.A. 74*, 3821–3825.
Krab, K., & Wikström, M. (1987) *Biochim. Biophys. Acta 895*, 25–29.
Kumar, C., Naqui, A., Powers, L., Ching, Y., & Chance, B. (1988) *J. Biol. Chem. 263*, 7159–7163.
Larsen, R. W., Ondrias, M. R., Copeland, R. A., Li, P. M., & Chan, S. I. (1989) *Biochemistry 28*, 6418–6422.
Leigh, J. S., Jr., Wilson, D. F., Owen, C. S., & King, T. E. (1974) *Arch. Biochem. Biophys. 160*, 476–486.
Li, P. M., Gelles, J., Chan, S. I., Sullivan, R. J., & Scott, R. A. (1987) *Biochemistry 26*, 2091–2095.
Li, P. M., Morgan, J. E., Nilsson, T., Ma, M., & Chan, S. I. (1988) *Biochemistry 27*, 2091–2095.
Li, P. M., Malmström, B. G., & Chan, S. I. (1989) *FEBS Lett. 248*, 210–211.
Ludwig, B., & Schatz, G. (1980) *Proc. Natl. Acad. Sci. U.S.A. 77*, 196–200.
Malmström, B. G. (1985) *Biochim. Biophys. Acta 811*, 1–12.
Malmström, B. G. (1988) *FEBS Lett. 231*, 268–269.
Malmström, B. G., & Andréasson, L.-E. (1985) *J. Inorg. Biochem. 23*, 233–242.
Martin, C. T., Scholes, C. P., & Chan, S. I. (1985) *J. Biol. Chem. 260*, 2857–2861.
Martin, C. T., Scholes, C. P., & Chan, S. I. (1988) *J. Biol. Chem. 263*, 8420–8429.
Millet, F., Darley-Usmar, V. M., & Capaldi, R. A. (1982) *Biochemistry 21*, 3857–3862.
Mitchell, P. (1966) *Chemiosmotic Coupling in Oxidative and Photosynthetic Phosphorylation*, Glynn Research Institute, Bodmin, Cornwall.
Mitchell, P. (1987) *FEBS Lett. 222*, 235–245.
Mitchell, P. (1988) *FEBS Lett. 231*, 270–271.
Mitchell, P., Mitchell, R., Moody, A. J., West, I. C., Baum, H., & Wrigglesworth, J. M. (1985) *FEBS Lett. 188*, 1–7.
Moody, A. J., & Rich, P. (1989) *Biochim. Biophys. Acta 973*, 29–34.
Morgan, J. E., Li, P. M., Jang, D. J., El-Sayed, M. A., & Chan, S. I. (1989) *Biochemistry 28*, 6975–6983.
Moroney, P. M., Scholes, T. A., & Hinkle, P. C. (1984) *Biochemistry 23*, 4991–4997.
Moyle, J., & Mitchell, P. (1978) *FEBS Lett. 88*, 268–272.
Mueller, M., Schlapfer, B., & Azzi, A. (1988) *Proc. Natl. Acad. Sci. U.S.A. 85*, 6647–6651.
Naqui, A., Powers, L., Lundeen, M., Constantinescu, A., & Chance, B. (1988) *J. Biol. Chem. 263*, 12342–12345.
Nilsson, T., Gelles, J., Li, P. M., & Chan, S. I. (1988) *Biochemistry 27*, 296–301.
Numata, M., Yamazaki, T., Fukumori, Y., & Yamanaka, T. (1989) *J. Biochem. 105*, 245–248.
Papa, S., Capitano, N., & Steverding, L. (1989) *Ann. N.Y. Acad. Sci. 550*, 238–259.
Powers, L., Chance, B., Ching, Y., & Angiolillo, P. (1981) *Biophys. J. 34*, 465–498.
Prochaska, L. J., & Fink, P. S. (1987) *J. Bioenerg. Biomembr. 19*, 143–166.
Raito, M., Jalli, T., & Saraste, M. (1987) *EMBO J. 6*, 2825–2833.
Reinhammar, B., Malkin, R., Jensen, P., Karlsson, B., Andréasson, L.-E., Aasa, R., Vänngård, T., & Malmström, B. G. (1980) *J. Biol. Chem. 255*, 5000–5004.
Rich, P. (1988) *Ann. N.Y. Acad. Sci. 550*, 254–259.
Scholes, C. P., Janakiraman, R., Taylor, H., & King, T. E. (1984) *Biophys. J. 45*, 1027–1030.
Sinjorgo, K. M. C., Steinbach, O. M., Dekker, H. L., & Muijsers, A. O. (1986) *Biochim. Biophys. Acta 850*, 108–115.
Smith, L. (1955) in *Methods in Biochemical Analysis* (Glick, D., Ed.) Vol. 2, Wiley, New York.
Solioz, M., Carafoli, E., & Ludwig, B. (1982) *J. Biol. Chem. 257*, 1579–1582.
Sone, N., & Nicholls, P. (1984) *Biochemistry 23*, 6550–6554.
Sone, N., Ohyama, K. T., & Kagawa, Y. (1979) *FEBS Lett. 106*, 39–42.
Sone, N., Yoki, F., Fu, T., Ohta, S., Metso, T., Raito, M., & Saraste, M. (1988) *J. Biochem. 103*, 606–610.
Steffens, G. C. M., Biewald, E., & Buse, G. (1987) *Eur. J. Biochem. 164*, 295–300.
Stevens, T. H., & Chan, S. I. (1981) *J. Biol. Chem. 256*, 1069–1071.
Stevens, T. H., Martin, C. T., Wang, H., Brudvig, G. W., Scholes, C. P., & Chan, S. I. (1982) *J. Biol. Chem. 257*, 12106–12133.
Thörnström, P.-E., Brzezinski, P., Fredriksson, P.-O., & Malmström, B. (1988) *Biochemistry 27*, 5441–5447.
Wikström, M. K. F. (1977) *Nature 266*, 271–273.
Wikström, M. (1981) *Proc. Natl. Acad. Sci. U.S.A. 78*, 4051–4054.
Wikström, M. (1988) *Chem. Scr. 28A*, 71–74.
Wikström, M. (1989) *Nature 338*, 776–778.
Wikström, M., & Casey, R. P. (1985) *J. Inorg. Biochem. 23*, 327–334.
Wikström, M., Krab, K., & Saraste, M. (1981) *Cytochrome Oxidase: A Synthesis*, Academic Press, New York.
Wilson, M. T., Greenwood, C., Brunori, M., & Antonini, E. (1975) *Biochem. J. 147*, 145–153.
Witt, S. N. (1988) Ph.D. Thesis, California Institute of Technology, Pasadena, CA.
Witt, S. N., & Chan, S. I. (1987) *J. Biol. Chem. 262*, 1446–1448.
Yamanaka, T., & Fujii, K. (1980) *Biochim. Biophys. Acta 591*, 63–62.
Yewey, G. L., & Caughey, W. S. (1988) *Ann N.Y. Acad. Sci. 550*, 22–32.
Zimmermann, B. H., Nitsche, C. I., Fee, J. A., Rusnak, F., & Münck, E. (1988) *Proc. Natl. Acad. Sci. U.S.A. 85*, 5779–5783.

Chapter 29

Long-Range Electron Transfer in Multisite Metalloproteins†

Harry B. Gray* and Bo G. Malmström‡

Arthur Amos Noyes Laboratory, California Institute of Technology, Pasadena, California 91125

Received March 1, 1989; Revised Manuscript Received April 27, 1989

Electron-transfer (ET)[1] reactions play major roles in respiration and photosynthesis. In both of these processes, there are electron-transport chains associated with specific cell organelles. These chains consist of a number of ET complexes firmly bound to biological membranes. ET between the complexes is mediated by small, diffusible molecules. The ET components within the complexes are flavins, iron–sulfur clusters, heme-bound iron, and manganese and copper ions. The complexes span the membranes, and ET through them is coupled to the translocation of protons across the membranes. This leads to the creation of an electrochemical potential, which drives the synthesis of ATP according to the chemiosmotic mechanism of Mitchell (1966).

Biological ET reactions have a number of characteristics distinguishing them from most redox processes involving small metal complexes. First, they occur rapidly over large molecular distances (>10 Å). Second, the ET event is often accompanied by only minor changes in the structure of the redox site. Finally, many biological ET complexes are molecular ion pumps, whose operation requires a structural control of the electron flow.

A good deal is known about the factors governing short-range ET in metal complexes (Taube, 1984; Marcus & Sutin, 1985), whereas the understanding is less complete concerning long-range ET in biological systems. In recent years, there has, however, been real progress in the understanding of ET involving small metalloproteins such as plastocyanin and cytochrome *c* (Sykes, 1985; Scott et al., 1985). In addition, there has been a considerable amount of experimental work with artificial multisite metalloproteins, aimed at illuminating the effects of the driving force, the reorganization energy, the distance, and the intervening medium on the rate of long-range ET. In this paper, we will review the current status of such model studies. To set the scene, this review will be preceded by a short discussion of biological ET, in particular the different properties needed in proteins, like cytochrome *c*, which mediate ET alone compared to proteins in which the ET is coupled to the pumping of protons. In the concluding section of the paper, the lessons learned from the model studies will be applied to one particular proton pump, cytochrome oxidase.

Electron Transfer in Biology

There are great similarities in the types of ET complexes encountered in respiration and photosynthesis, as illustrated in Figure 1. In both cases, ET between complexes is mediated by a quinone at one point and by a small, water-soluble metalloprotein at another. In particular, the composition of complex III in respiration is nearly identical with that of the b_6f complex in photosynthesis. Proton translocation occurs at these complexes and also at complexes I and IV (cytochrome oxidase) in respiration (Wikström & Saraste, 1984).

Mitchell (1966) suggested the redox loop as a device for achieving proton translocation. [The Q-cycle (Mitchell, 1975), believed to operate in the bc_1 (complex III) and b_6f complexes, is a variation of the redox loop concept.] The redox loop is defined as a system that translocates hydrogen atoms one way across the membrane and electrons the other way, leaving the proton on one side of the membrane. Thus, a net translocation of protons is accomplished with a H^+/e^- stoichiometry of 1 only. There is, however, good experimental evidence that, in complexes I and IV of respiration, the stoichiometry is actually 2 H^+/e^- (Wikström & Saraste, 1984; Brown & Brand, 1988). With cytochrome oxidase (complex IV), it has been established that 1 H^+/e^- is pumped across the membrane (Wikström, 1977) and, in addition, that the protons consumed in the reduction of dioxygen to water are specifically taken up from the matrix side of the inner mitochondrial membrane

† Our research on protein electron transfer is supported by grants from the National Science Foundation, the National Institutes of Health (H.B.G.), and the Swedish Natural Science Research Council (B.G.M.). This is Contribution No. 7913 from the Arthur Amos Noyes Laboratory.

‡ Visiting Associate of the Beckman Institute, California Institute of Technology. Permanent address: Department of Biochemistry and Biophysics, Chalmers University of Technology, S-41296 Göteborg, Sweden.

[1] Abbreviations: a, ammine; cyt, cytochrome; Mb, sperm whale myoglobin; cyt *c*-Zn, zinc derivative of horse heart cytochrome *c*; Mb-ZnP, zinc mesoporphyrin IX derivative of sperm whale myoglobin; ET, electron transfer; isn, isonicotinamide; P, mesoporphyrin IX; py, pyridine.

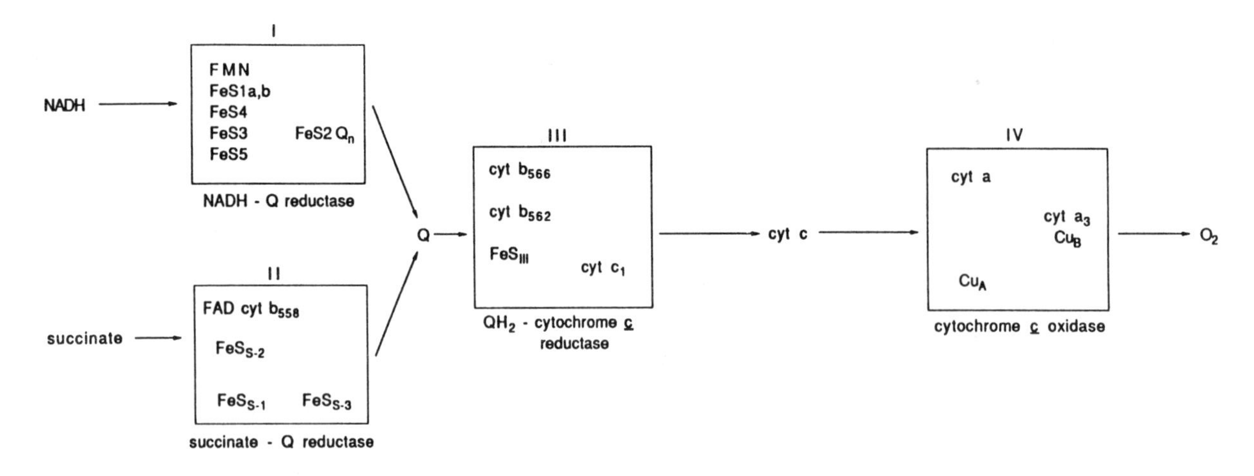

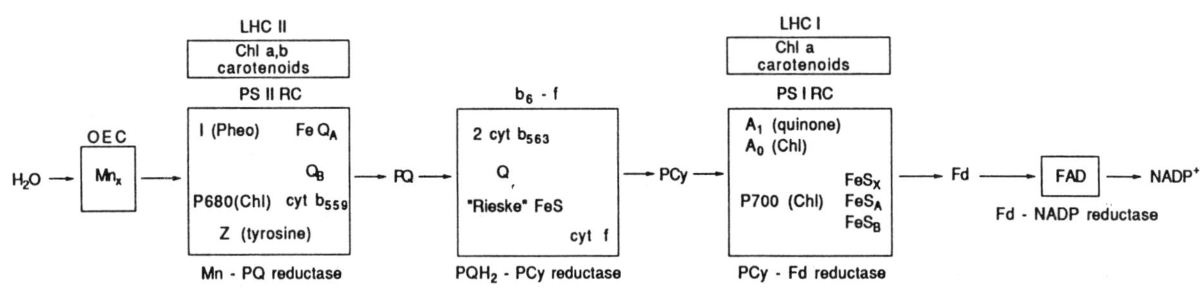

FIGURE 1: ET chains of respiration and photosynthesis. Abbreviations: Q, ubiquinone; PQ, plastoquinone; cyt, cytochrome; FeS, iron–sulfur cluster; PCy, plastocyanin, Fd, ferredoxin; Chl, chlorophyll; Pheo, pheophytin; OEC, oxygen-evolving complex, LHC, light-harvesting complex; PS, photosystem; RC, reaction center.

(Wikström, 1988). In this way, an effective H^+/e^- stoichiometry of 2 is obtained.

Any redox-linked proton pump must follow certain basic principles, as summarized in the cubic reaction scheme of Wikström and Krab (1979). One important feature of this scheme is the existence of separate input and output states for H^+ and e^-. In this way, the ET from donor to acceptor cannot be completed unless the transducer undergoes a transition from the input to the output state. If the e^- input–output states are also the input–output states for H^+, and the transition between states can only occur when the pump is protonated, then the ET reactions will drive proton pumping.

From what has just been said, it is obvious that in redox-linked proton pumps there must be redox-induced conformatonal changes that drastically alter the ET properties of specific redox sites, as will be illustrated later with cytochrome oxidase as an example. Small ET proteins with a single redox site, on the other hand, display minimal structural changes on reduction and reoxidation. Thus, the reorganization energy, which, as will be discussed shortly, is a major determinant of the ET rate, is minimized, and electron transfer is rapid. Good examples are cytochrome *c* and plastocyanin, the proteins that transfer electrons between membrane-bound complexes in respiration and photosynthesis, respectively.

It has been shown (Gray & Malmström, 1983) that the facile ET kinetic properties of blue copper proteins, such as plastocyanin, can be explained in terms of a "rack mechanism" (Lumry & Eyring, 1954; Eyring et al., 1956). According to this hypothesis, functional groups in a protein are in a strained configuration induced by the overall conformation. In plastocyanin, this results in the unusual situation that the metal ligands have almost the same positions in the Cu(II) and the Cu(I) state as well as in the apoprotein (Freeman, 1981). The strained configuration is also evidenced by the high reduction potentials and unusual spectroscopic properties of blue copper proteins (Gray & Malmström, 1983; Ainscough et al., 1987).

MODIFIED PROTEINS: INTRAMOLECULAR ELECTRON TRANSFER AT FIXED DISTANCES

Modification of a single-site metalloprotein with a redox-active complex produces a two-site (donor–acceptor) system that allows long-range ET through a protein to be studied in a systematic manner. Ruthenium ammines have been used as modification agents primarily because they can be attached specifically to surface histidines, and the resulting complexes are kinetically stable in solution in the relevant oxidation states (Gray, 1986; Mayo et al., 1986; Osvath et al., 1988). Crystal structure analyses are available for the proteins that have been investigated, and so the distances and intervening media between the donor and acceptor sites in the modified derivatives are known.

Distance Dependence of ET Rates in Heme Proteins. In the analyses that follow, we will assume that the intramolecular ET rate constant is the product of a nuclear frequency factor, ν_n, an electronic factor, Γ, and an activation energy term, $\exp(-\Delta G^*/RT)$ (Lieber et al., 1987):

$$k_{ET} = \nu_n \Gamma \exp(-\Delta G^*/RT) \quad (1)$$

The frequency factor, ν_n, is generally taken to be 10^{13} s^{-1}. The electronic factor, Γ, is unity when the donor and acceptor are strongly coupled, but at long donor–acceptor distances in protein ET reactions the coupling will be much weaker. For such weakly coupled systems, Γ is expected to fall off exponentially with distance (d):

$$\Gamma = \Gamma(d_0) \exp[-\beta(d - d_0)] \quad (2)$$

In eq 2, d_0 is the van der Waals contact distance, usually assumed to be 3 Å (Marcus & Sutin, 1985). The rate of the decrease in electronic coupling with increasing distance is given by β, which depends on the nature of the medium.

Table I: ET Distances and Rates for Ruthenium-Labeled Heme Proteins

derivative	distance (Å)[a]	k_{ET} (s^{-1})[b]
a_5Ru(His-33)cyt *c*-Zn	10.8–11.7 (11.6)	7.7×10^5
a_5Ru(His-48)Mb-ZnP	11.8–16.6 (12.7)	7.2×10^4
a_5Ru(His-81)Mb-ZnP	18.8–19.3 (19.3)	1.5×10^2
a_5Ru(His-116)Mb-ZnP	19.8–20.4 (20.1)	3.0×10^1
a_5Ru(His-12)Mb-ZnP	21.5–22.3 (22.0)	1.4×10^2

[a] The donor–acceptor edge–edge distances are lower and upper values estimated by Lieber et al. (1987). The value for the lowest energy conformation is in parentheses. [b] Rate for Mb-ZnP* → Ru^{III} (Axup et al., 1988; R. K. Upmacis, S. S. Kim, F.-D. Tsay, D. E. Malerba, J. R. Winkler, and H. B. Gray, unpublished results) or cyt *c*-Zn* → Ru^{III} ET (Elias et al., 1988).

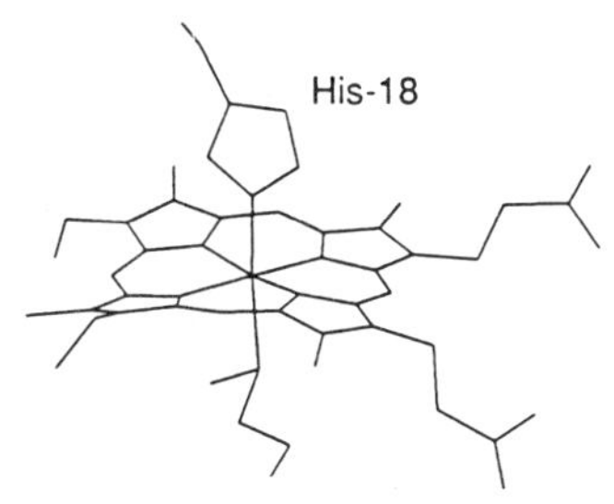

FIGURE 2: View of the ET centers in a_5Ru(His-33)cyt *c*. The distance between the two redox units is 11.6 Å (see Table I).

Sperm whale myoglobin (Mb) has four surface histidines (His-48, His-81, His-116, His-12) that can be modified with ruthenium ammines. Pentaammineruthenium(III) (a_5Ru) has been covalently attached to these histidines. After separation and purification of the reaction products, four singly labeled myoglobin species are obtained (Crutchley et al., 1986; Axup et al., 1988; Karas, 1989). In contrast to the myoglobin system, horse heart cytochrome *c* has only one surface histidine (His-33) that readily binds ruthenium (Yocom et al., 1982). The edge-to-edge distance from the histidine to the heme in each of the a_5RuMb and a_5Rucyt *c* derivatives is given in Table I. The two ET units in a_5Ru(His-33)cyt *c* are shown in Figure 2.

ET rates in photoactive zinc derivatives of a_5RuMb and a_5Ru(His-33)cyt *c* have been measured by transient absorption spectroscopy. In Mb, the heme group can be replaced by zinc mesoporphyrin IX (ZnP) (Axup et al., 1988); in cyt *c*, Zn for Fe substitution gives cyt *c*-Zn (Elias et al., 1988). A laser flash generates triplet ZnP* (or triplet cyt *c*-Zn*), which subsequently returns to the ground state by ET as well as by the usual decay pathways:

$$\mathrm{ZnP^{*}\text{-}Ru^{III}} \xrightarrow{k_{ET}} \mathrm{ZnP^{+}\text{-}Ru^{II}}$$

$$h\nu \uparrow\downarrow k_D \qquad k_b \swarrow$$

$$\mathrm{ZnP\text{-}Ru^{III}}$$

Rates for the a_5RuMb-ZnP and a_5Ru(His-33)cyt *c*-Zn ET reactions are given in Table I. These data are incorporated in two ln k_{ET} versus distance plots shown in Figure 3. A fit

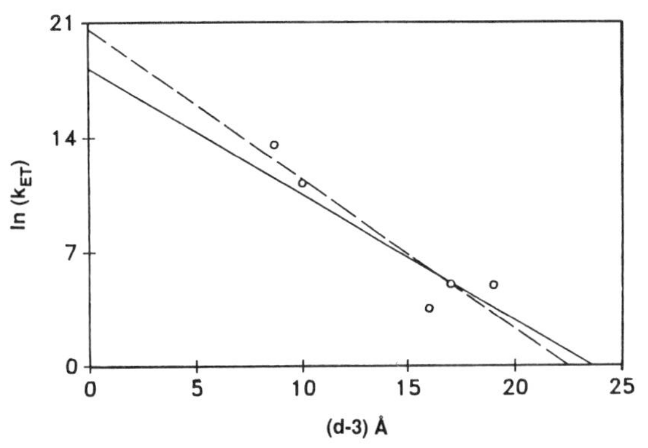

FIGURE 3: ET rate versus distance for Ru-labeled Mb and cyt *c* derivatives. Data are given in Table I. Solid line, $\beta = 0.78$ Å^{-1}; dashed line, $\beta = 0.91$ Å^{-1}.

to all five points gives $\beta = 0.91$ Å^{-1}; restricting the fit to the Mb data yields $\beta = 0.78$ Å^{-1}.

An important conclusion can be drawn immediately from each of the plots: namely, that the edge–edge distance or some through-protein bond pathway related to it is a reasonable indicator of the long-range electronic coupling. An exclusively peptide backbone pathway for Mb ET can be ruled out, because the through-backbone distances are of the order 210, 56, 110, and 380 Å from the proximal histidine to His-48, -81, -116, and -12, respectively, and show no systematic variation with ET rate (Cowan et al., 1989).

In the particular case of a_5Ru(His-12)Mb-ZnP, which displays a slightly higher rate than would be expected from the edge–edge distance, the possibility of a medium effect involving the Trp-14 aromatic group, which lies in a parallel orientation directly along the ET pathway, has been mentioned (Axup et al., 1988). Work on MgP diacid and diester substituted a_5RuMb derivatives also indicates that intervening aromatic residues increase the rate of ET in proteins (Cowan & Gray, 1988; Cowan et al., 1989).

It might be expected that β would be larger for proteins, since direct through-bond pathways are not available. Possibly related to this point is the observation that the maximum rates for intramolecular ET in organic donor–acceptor molecules with rigid spacers are significantly larger than those for Ru-labeled protein systems at similar distances (Mayo et al., 1986; Closs & Miller, 1988). It has been suggested that this difference arises because there are always weakly (noncovalently) coupled units (van der Waals and H-bonded atoms) along the long-range pathway for protein ET reactions (Cowan et al., 1989).

MARCUS THEORY: ROLE OF THE REORGANIZATION ENERGY

According to Marcus, the activation free energy for ET (ΔG^*, eq 1) depends on the reaction free energy, $\Delta G°$, and the nuclear reorganization energy, λ (Marcus & Sutin, 1985):

$$\Delta G^* = (\Delta G° + \lambda)^2/4\lambda \qquad (3)$$

As the reaction free energy is increased, the ET rate is predicted to increase, reach a maximum when $-\Delta G° = \lambda$, and then fall off (Figure 4). The highly exoergic region where the ET rate is predicted to decrease is called the "inverted region".

Both driving force and temperature effects on ET rates have been investigated in attempts to evaluate the reorganization energy accompanying protein ET. Measurements of ET in protein–protein complexes have been analyzed in terms of λ values ranging from 0.8 eV for the cyt *c*/cyt b_5 complex (McLendon, 1988) to 2.1 eV for a Zn,Fe hybrid hemoglobin

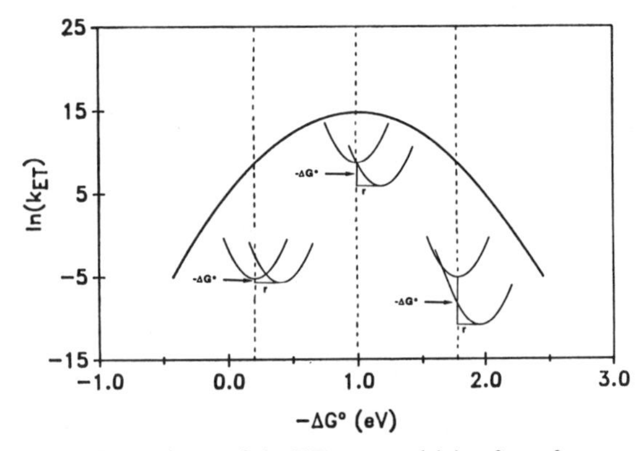

FIGURE 4: Dependence of the ET rate on driving force for a system following eq 3 in which the reorganization energy is 1 eV. Shown under the Marcus curve are reactant–product energy surfaces at driving forces of 0.2, 1.0, and 1.8 eV. The product $\nu_n\Gamma$ is 2.8×10^6 s^{-1} in all three cases.

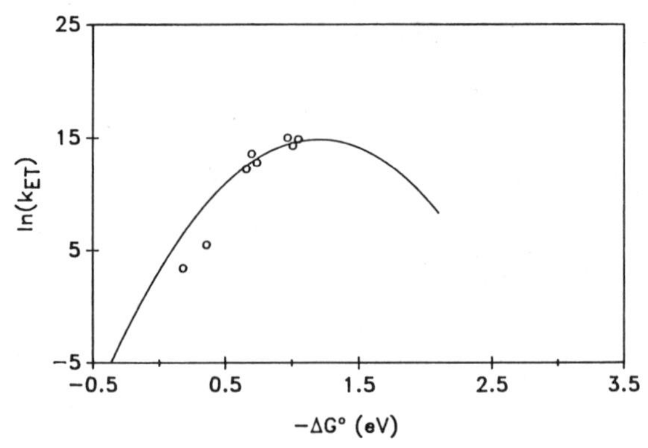

FIGURE 5: ET rate as a function of the free energy change for Rucyt *c* (see Table II). The Marcus curve is for $\lambda = 1.2$ eV.

Table II: ET Rates in Ru(His-33)cyt *c* Derivatives[a]

donor	acceptor	$-\Delta G°$ (eV)	k_{ET} (s^{-1})
a_5Ru^{II}	cyt *c*-Fe^{III}	0.18	3.0×10^1
a_5Ru^{II}	cyt *c*-Zn*	0.36	2.4×10^2
a_4isnRu^{II}	cyt *c*-Zn^+	0.66	2.0×10^5 (k_b)
cyt *c*-Zn*	a_5Ru^{III}	0.70	7.7×10^5
a_4pyRu^{II}	cyt *c*-Zn^+	0.74	3.5×10^5 (k_b)
cyt *c*-Zn*	a_4pyRu^{III}	0.97	3.3×10^6
a_5Ru^{II}	cyt *c*-Zn^+	1.01	1.6×10^6 (k_b)
cyt *c*-Zn*	a_4isnRu^{III}	1.05	2.9×10^6

[a] From Meade et al. (1989).

(Peterson-Kennedy et al., 1986), and reorganization energies of 0.7 and 1.2 eV have been estimated from the kinetics of the cyt *c* and cyt b_5 ET self-exchange reactions (Dixon et al., 1989). We shall concentrate on two Ru-modified protein systems, a_4LRu(His-33)cyt *c* and a_4LRu(His-48)Mb (L = NH_3, py, isn), because the data available in these cases are quite extensive and the ET distances and intervening media are fixed and known.

Ruthenated Cytochrome c. The ET rates for a_4LRu(His-33)cyt *c*-M (M = Fe, Zn) vary from 3.0×10^1 to 3.3×10^6 s^{-1} over a 0.18–1.05-eV range of driving forces (Table II). The cyt *c*-Zn* → Ru^{III} and Ru^{II} → ZnP^+ reactions with comparable driving forces proceed at comparable rates in Rucyt *c*-Zn derivatives, thereby indicating that these ET processes have similar values of λ and Γ. It is not likely, however, that the same holds for the Rucyt *c*-Fe (Ru^{II} → Fe^{III}) ET reaction. If just the Rucyt *c*-Zn reactions are considered, analysis of the rates gives $\lambda = 1.2$ eV and $\nu_n\Gamma = 2.8 \times 10^6$ s^{-1} (Meade et al., 1989). As can be seen from the solid curve in Figure 5, the Ru^{II} → Fe^{III} ET reaction proceeds somewhat more slowly than expected, suggesting weaker electronic coupling between the Fe porphyrin and the Ru complex.

Configurational changes of both the inner coordination spheres of the metal centers (λ_i) and the surrounding medium (λ_o) contribute to the total nuclear reorganization energy (λ) in Rucyt *c* ($\lambda = \lambda_i + \lambda_o$). The inner-sphere contribution can be estimated from the structural changes that are known to accompany oxidation of Ru(II) ammines and Zn porphyrins. The Ru–N bond lengths generally change by less than 0.05 Å upon one-electron oxidation of Ru(II) ammine complexes (Gress et al., 1981), and estimates of the inner-sphere reorganization energies are on the order of 0.05 eV (Brown & Sutin, 1979; Siders & Marcus, 1981). An X-ray crystal structure of a Zn porphyrin radical cation reveals fairly small changes in bond lengths and angles compared to the neutral precursor (Collins & Hoard, 1970; Spaulding et al., 1974), and spectroscopic evidence indicates relatively minor distortions of the Zn porphyrin upon excitation to the lowest-lying triplet state (Meade et al., 1989). Hence, the contribution of the inner-sphere configurational changes about the Zn porphyrin is not likely to be more than 0.15 eV, leading to an estimated upper limit of 0.2 eV for the total inner-sphere reorganization barrier in the Rucyt *c* intramolecular ET reactions.

The remaining contribution to λ in Rucyt *c* must arise from configurational changes of the solvent and polypeptide backbone of the protein (λ_o). Calculations of solvent reorganization energies typically treat the solvent as a dielectric continuum. The original Marcus model, which represents the reactants as two conducting spheres embedded in a dielectric continuum, yields an estimate of 1.1 eV for the solvent reorganization energy in Rucyt *c* (Meade et al., 1989). More sophisticated treatments describe the reorganization energy associated with transferring a single charge from one point to another within spherical or ellipsoidal cavities of low dielectric constant embedded in dielectric continua (Brunschwig et al., 1986). Taking the Rucyt *c* system as a single sphere of 32-Å diameter leads to an estimate of 0.63 eV for the solvent reorganization energy (Meade et al., 1989). Neither solvent model accounts for reorganization of the protein matrix in response to the charge transfer. A calculation of the protein-only reorganization energy accompanying the cyt *c* ET self-exchange reaction indicates a contribution of 0.15 eV from protein dielectric relaxation to λ_o (Churg et al., 1983). The total λ for Rucyt *c* would then be predicted to fall in the 1.0–1.4-eV range, which encompasses the best experimentally derived values.

Ruthenated Myoglobin. The ET rates for a_5Ru(His-48)- and $a_4$pyRu(His-48)-labeled Mb's have been determined with flash photolysis (Karas et al., 1988). The observed ET rate constants are 0.058 s^{-1} for a_5RuMb-Fe and 2.5 s^{-1} for $a_4$pyRuMb-Fe. The Fe^{II}/Ru^{III} ET rate for a_5Ru(His-48)-Mb-Fe is within experimental error of the ET rate previously determined by generating the Ru^{II}/Fe^{III} mixed-valence species. The observation that the rate constant is independent of the initial $[Ru^{III}Fe^{II}]$:$[Ru^{II}Fe^{III}]$ ratio demonstrates that long-range ET in a_5Ru(His-48)Mb-Fe is reversible. At 25 °C the forward and reverse ET rates are calculated to be 0.04 and 0.02 s^{-1}, respectively.

In order to extend the driving force dependence to large values of $-\Delta G°$ (~1 eV), Mb derivatives were prepared by removing the heme in Ru(His-48)Mb and replacing it with a photoactive porphyrin (MP). Rate constants and driving forces are given in Table III.

The donor orbital of a photoactive porphyrin MP* is more delocalized than the orbital of an iron porphyrin, and therefore, the MP* to Ru electronic coupling should be larger than that

Table III: ET Rates in Ru(His-48)Mb Derivatives[a]

donor	acceptor	$-\Delta G^\circ$ (eV)	k_{ET} (s^{-1})
$Fe^{II}P$	a_5Ru^{III}	0.02	0.040
$Fe^{II}P$	a_4pyRu^{III}	0.28	2.5
H_2P^*	a_5Ru^{III}	0.53	7.6×10^2
PdP*	a_5Ru^{III}	0.70	9.1×10^3
PtP*	a_5Ru^{III}	0.73	1.2×10^4
CdP*	a_5Ru^{III}	0.85	6.3×10^4
MgP*	a_5Ru^{III}	0.87	5.7×10^4
ZnP*	a_5Ru^{III}	0.88	7.0×10^4
PdP*	a_4pyRu^{III}	0.96	9.0×10^4

[a] From Karas et al. (1988) and Cowan et al. (1989).

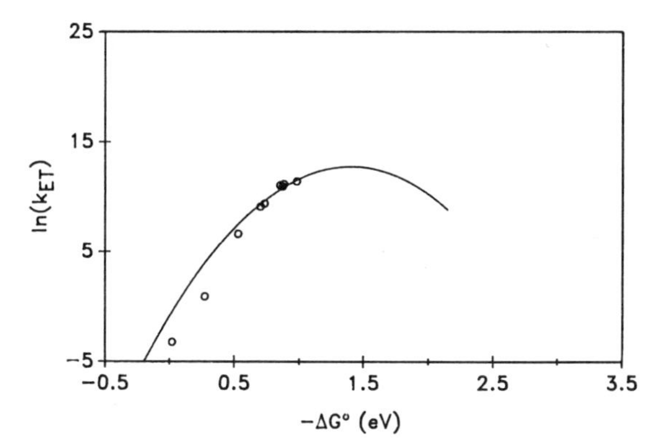

FIGURE 6: Plot of ln k_{ET} versus $-\Delta G^\circ$ for the RuMb ET reactions (see Table III). The Marcus curve is for λ = 1.4 eV.

between FeP and Ru (Cowan et al., 1989). The best fit to the data for the MP* → Ru^{III} ET reactions yields λ = 1.4 eV and $\nu_n\Gamma = 3.5 \times 10^5$ s^{-1} (Figure 6). The $Fe^{II} \rightarrow Ru^{III}$ rates fall well below the $\lambda \sim 1.4$ eV curve, consistent with the argument that the FeP to Ru electronic coupling is weaker. Another factor in the low $Fe^{II} \rightarrow Ru^{III}$ rates could well be a larger λ associated with the binding of water to the Fe^{III} proteins. Inclusion of the FeP points in the Marcus plot indicates a very high λ (~2 eV), and the strong temperature dependence of the $Fe^{II} \rightarrow Ru^{III}$ ET rate in a_5Ru(His-48)Mb also is consistent with a relatively large λ (Karas et al., 1988; Cowan & Gray, 1988; Crutchley et al., 1986).

Reorganization Energy Control of Long-Range Protein ET. Work on Ru-modified cytochrome *c* and myoglobin derivatives has shown that long-range ET reactions can be controlled through tradeoffs between driving force and reorganization energy. Since rearrangements of water molecules are usually associated with large reorganization energies, the exclusion of water in either the inner or outer sphere of a donor or acceptor would be expected to lead to much higher ET rates if other factors—notably the driving force—were not changed very much. Alternatively, λ could be reduced by tightening the polypeptide pockets in which the donor and acceptor reside. Reductions in λ by 1 eV or more conceivably could be achieved by a combination of lowering the polarity of the ET medium and tightening its hold on the donor and acceptor, thereby drastically limiting nuclear rearrangements accompanying the reaction. Calculations based on the parameters derived from work on ruthenated proteins show that a protein ET reaction at 0.1-V driving force would increase in rate from 5×10^{-2} to 2×10^4 s^{-1} if λ were reduced from 1.5 to 0.5 eV at a donor–acceptor edge–edge distance of 15 Å (Figure 7).

CYTOCHROME OXIDASE

Cytochrome oxidase contains four redox-active metal centers. Two of them, cytochrome *a* and a_3, contain a heme A group in separate environments, whereas the other two, Cu_A and Cu_B, are copper ions coordinated in different ways to the

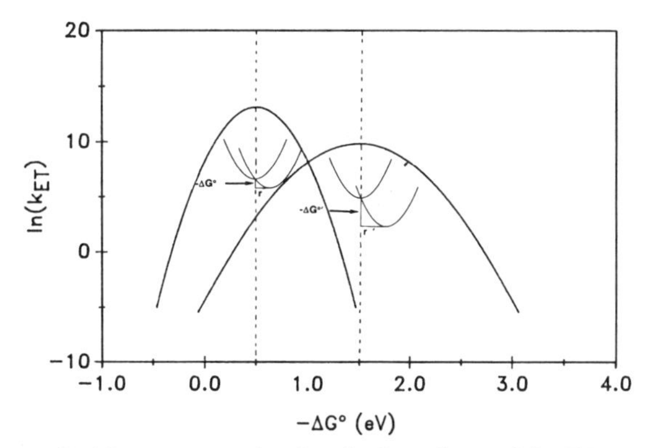

FIGURE 7: Marcus curves showing the dependence of the ET rate on $-\Delta G^\circ$ for λ values of 1.5 and 0.5 eV. These two λ values could be associated with two different enzyme conformations, E_1 (λ = 1.5 eV) and E_2 (λ = 0.5 eV). Below the curves are reactant–product energy surfaces for activationless ET ($-\Delta G^\circ = \lambda$). The two curves were calculated for ET at d = 15 Å, using parameters derived from work on ruthenated proteins (M. Therien and H. B. Gray, unpublished results).

protein (Wikström et al., 1981). All centers are more than 15 Å from each other, except cytochrome a_3 and Cu_B, which form a binuclear unit. Electrons from cytochrome *c* first reduce Cu_A and cytochrome *a*, which are in rapid redox equilibrium (Antalis & Palmer, 1982). From these reduced centers, they are transferred intramolecularly to the cytochrome a_3–Cu_B center, which is the dioxygen-binding and -reducing site.

The intramolecular ETs from cytochrome *a* and Cu_A to the binuclear site have been thought to limit k_{cat}, the catalytic constant (Wilson et al., 1981; Brzezinski et al., 1986). Both transfers are, however, rapid compared to k_{cat} in the mixed-valence state (Boelens et al., 1982; Brzezinski & Malmström, 1987). Cytochrome *a* and Cu_A are also rapidly oxidized when the fully reduced enzyme is mixed with dioxygen (Hill et al., 1986; Orii, 1988). This suggests that the enzyme can exist in two states, in which the ET properties are different. Since this is a requirement of a proton pump (Blair et al., 1986), it is attractive to consider that these two forms represent the separate input and output states for electrons in the pump redox sites.

There are strong redox interactions between the metal centers (Wikström et al., 1981; Blair et al., 1986). This may also reflect the existence of two conformational states with different ET properties, as the interactions can hardly be electrostatic in origin in view of the large metal–metal distances [cf. Nocera et al. (1984)]. In addition, conformational changes induced by reduction of Cu_A and/or cytochrome *a* have been detected by fluorescence measurements (Copeland et al., 1987) and by drastic changes in the rate of cyanide binding to oxidized cytochrome a_3 (Jones et al., 1984). It has recently been demonstrated that the rate of the conformational change detected by cyanide binding varies with pH in the same way as k_{cat} (Thörnström et al., 1988). This suggests that it is this conformational transition rather than the internal ET which is the rate-limiting step. It also indicates that this transition corresponds to the switch between the input and output states, because it has already been pointed out that the internal ET appears slow only when starting from the fully oxidized enzyme. Finally, it should be mentioned that in the output state the rate of ET between Cu_A and the binuclear unit is almost independent of temperature, the enthalpy of activation being as low as 2.1 kJ·mol^{-1} (Brzezinski & Malmström, 1987).

Gating by Changes in Reorganization Energy. The experimental results summarized earlier on the effects of distance, driving force, and reorganization energy on the rates of ET in model systems strongly suggest that the most efficient way to gate the electron flow in a redox-linked proton pump is by variations in reorganization energy. The same decrease in rate as is given by an increase in the distance of 10 Å is achieved at relatively low driving forces (≤0.5 eV) by an increase in the reorganization energy of 1 eV (96 kJ·mol^{-1}) (Lieber et al., 1987). Driving force variations by themselves would not change the rate substantially, because it is known that the difference in driving force between the input and output states in cytochrome oxidase is less than 0.2 V (Brzezinski & Malmström, 1987).

The argument just given, together with an analysis of the temperature dependence of the ET from Cu_A to the binuclear site in the mixed-valence cytochrome oxidase (Brzezinski & Malmström, 1987), can be used to formulate a mechanism of electron gating. The enzyme has two conformations, E_1 and E_2, corresponding to the input and output states for e^- and H^+. In E_1, there is a large barrier to ET from Cu_A to the binuclear site. This barrier is assumed to be associated mainly with the electron acceptor, cytochrome a_3–Cu_B, because Cu_A has the same structure in the fully oxidized enzyme as in the mixed-valence state, in which electron equilibration with the binuclear site is rapid, as evidenced by EPR (Clore et al., 1980). Reduction of Cu_A and cytochrome *a*, followed by protonation of the proton-translocating group, leads to a transition from E_1 to E_2. The structure of the oxidized cytochrome a_3–Cu_B site is changed in E_2, as evidenced by the large change in the rate of cyanide binding, in such a way that the reorganization energy is lowered. This could be accomplished by a structural change leading to the dissociation of ligated water molecules, or a change in the coordination geometries of the metal ions themselves, or a combination of both. In any case, the structural change should be such that the oxidized binuclear site in E_2 adopts a structure very close to that of the reduced form, thereby minimizing the reorganization energy. This change would be a type of rack mechanism, first suggested to operate in mitochondrial "energy conservation" by Lumry (1963). To gate the electron flow effectively, λ in E_1 should probably be 1 eV higher than in E_2, as depicted in Figure 7. This is a reasonable rack energy in a multisite metalloprotein [cf. Gray and Malmström (1983)].

ACKNOWLEDGMENTS

We thank David Smith and Mike Therien for assistance in preparing the figures and for several helpful discussions.

Registry No. Cytochrome oxidase, 9001-16-5.

REFERENCES

Ainscough, E. W., Bingham, A. G., Brodie, A. M., Ellis, W. R., Gray, H. B., Loehr, T. M., Plowman, J. E., Norris, G. E., & Baker, E. N. (1987) *Biochemistry 26*, 71–82.

Antalis, T. M., & Palmer, G. (1982) *J. Biol. Chem. 257*, 6194–6206.

Axup, A. W., Albin, M., Mayo, S. L., Crutchley, R. J., & Gray, H. B. (1988) *J. Am. Chem. Soc. 110*, 435–439.

Blair, D. F., Gelles, J., & Chan, S. I. (1986) *Biophys. J. 50*, 713–733.

Boelens, R., Wever, R., & Van Gelder, B. F. (1982) *Biochim. Biophys. Acta 682*, 264–272.

Brown, G. C., & Brand, M. D. (1988) *Biochem. J. 252*, 473–479.

Brown, G. M., & Sutin, N. (1979) *J. Am. Chem. Soc. 101*, 883–892.

Brunschwig, B. S., Ehrenson, S., & Sutin, N. (1986) *J. Phys. Chem. 90*, 3657–3668.

Brzezinski, P., & Malmström, B. G. (1987) *Biochim. Biophys. Acta 894*, 29–38.

Brzezinski, P., Thörnström, P.-E., & Malmström, B. G. (1986) *FEBS Lett. 194*, 1–5.

Churg, A. K., Weiss, R. M., Warshel, A., & Takano, T. (1983) *J. Phys. Chem. 87*, 1683–1694.

Clore, M. G., Andréasson, L.-E., Karlsson, B., Aasa, R., & Malmström, B. G. (1980) *Biochem. J. 185*, 155–167.

Closs, G. L., & Miller, J. R. (1988) *Science 240*, 440–447.

Collins, D. M., & Hoard, J. L. (1970) *J. Am. Chem. Soc. 92*, 3761–3771.

Copeland, R. A., Smith, P. A., & Chan, S. I. (1987) *Biochemistry 26*, 7311–7316.

Cowan, J. A., & Gray, H. B. (1988) *Chem. Scr. 28A*, 21–26.

Cowan, J. A., Upmacis, R. K., Beratan, D. N., Onuchic, J. N., & Gray, H. B. (1988) *Ann. N.Y. Acad. Sci. 550*, 68–84.

Crutchley, R. J., Ellis, Jr., W. R., & Gray, H. B. (1986) *J. Am. Chem. Soc. 107*, 5002–5004.

Dixon, D. W., Xiaole, H., Woehler, S. E., Mauk, A. G., & Sishta, B. P. (1989) *J. Am. Chem. Soc.* (in press).

Elias, H., Chou, M. H., & Winkler, J. R. (1988) *J. Am. Chem. Soc. 110*, 429–434.

Eyring, H., Lumry, R., & Spikes, J. D. (1954) in *The Mechanism of Enzyme Action* (McElroy, W. D., & Glass, B., Eds.) pp 123–136, The John Hopkins Press, Baltimore, MD.

Freeman, H. C. (1981) in *Coordination Chemistry-21* (Laurent, J. P., Ed.) pp 29–51, Pergamon Press, Oxford.

Gray, H. B. (1986) *Chem. Soc. Rev. 15*, 17–30.

Gray, H. B., & Malmström, B. G. (1983) *Comments Inorg. Chem. 2*, 203–209.

Gress, M. E., Creutz, C., & Quicksall, C. O. (1981) *Inorg. Chem. 20*, 1522–1528.

Hill, B. C., Greenwood, C., & Nicholls, P. (1986) *Biochim. Biophys. Acta 853*, 91–113.

Jones, M. G., Bickar, D., Wilson, M. T., Brunori, M., Colosimo, A., & Sarti, P. (1984) *Biochem. J. 220*, 57–66.

Karas, J. L. (1989) Ph.D. Thesis, California Institute of Technology, Pasadena, CA.

Karas, J. L., Lieber, C. M., & Gray, H. B. (1988) *J. Am. Chem. Soc. 110*, 599–600.

Lieber, C. M., Karas, J. L., Mayo, S. L., Albin, M., & Gray, H. B. (1987) *Proceedings of the Robert A. Welch Conference on Chemical Research. Design of Enzymes and Enzyme Models*, Nov 2–4, 1987, pp 9–33, Robert A. Welch Foundation, Houston, TX.

Lumry, R. (1963) in *Photosynthesis Mechanisms of Green Plants*, pp 625–634, Publication No. 1145, National Academy of Sciences, Washington, DC.

Lumry, R., & Eyring, H. (1954) *J. Phys. Chem. 58*, 110–120.

Marcus, R. A., & Sutin, N. (1985) *Biochim. Biophys. Acta 811*, 265–322.

Mayo, S. L., Ellis, W. R., Jr., Crutchley, R. J., & Gray, H. B. (1986) *Science 233*, 948–952.

McLendon, G. (1988) *Acc. Chem. Res. 21*, 160–167.

Meade, T. J., Gray, H. B., & Winkler, J. R. (1989) *J. Am. Chem. Soc. 111*, 4353–4356.

Mitchell, P. (1966) *Chemiosmotic Coupling in Oxidative and Photosynthetic Phosphorylation*, Glynn Research, Bodmin, England.

Mitchell, P. (1975) *FEBS Lett. 59*, 137–139.

Nocera, D. G., Winkler, J. R., Yocom, K. M., Bordignon, E., & Gray, H. B. (1984) *J. Am. Chem. Soc. 106*, 5145–5150.

Orii, Y. (1988) *Chem. Scr. 28A*, 63–69.

Osvath, P., Salmon, G. A., & Sykes, A. G. (1988) *J. Am. Chem. Soc. 110*, 7114–7118.

Peterson-Kennedy, S. E., McGourty, J. L., Kalweit, J. A., & Hoffman, B. M. (1986) *J. Am. Chem. Soc. 108*, 1739–1746.

Scott, R. A., Mauk, A. G., & Gray, H. B. (1985) *J. Chem. Ed. 62*, 932–938.

Siders, P., & Marcus, R. A. (1981) *J. Am. Chem. Soc. 103*, 741–747.

Spaulding, L. D., Eller, P. G., Bertrand, J. A., & Felton, R. H. (1974) *J. Am. Chem. Soc. 96*, 982–987.

Sykes, A. G. (1985) *Chem. Soc. Rev. 14*, 283–315.

Taube, H. (1984) *Angew. Chem., Int. Ed. Engl. 23*, 329–339.

Thörnström, P.-E., Nilsson, T., & Malmström, B. G. (1988) *Biochim. Biophys. Acta 935*, 103–108.

Wikström, M. (1977) *Nature 266*, 271–273.

Wikström, M. (1988) *FEBS Lett. 231*, 247–252.

Wikström, M., & Krab, K. (1979) *Biochim. Biophys. Acta 549*, 177–222.

Wikström, M., & Saraste, M. (1984) in *Bioenergetics* (Ernster, L., Ed.) pp 49–94, Elsevier, Amsterdam.

Wikström, M., Krab, K., & Saraste, M. (1981) *Cytochrome Oxidase—A Synthesis*, Academic Press, London.

Wilson, M. T., Peterson, J., Antonini, E., Brunori, J., Colosimo, A., & Wyman, J. (1981) *Proc. Natl. Acad. Sci. U.S.A. 78*, 7115–7118.

Yocom, K. M., Shelton, J. B., Shelton, J. R., Schroeder, W. A., Worosila, G., Isied, S. S., Bordignon, E., & Gray, H. B. (1982) *Proc. Natl. Acad. Sci. U.S.A. 79*, 7052–7055.

Author Index

Affiliation Index

Subject Index

D

E

Production: Kurt Schaub
Indexing: Deborah Steiner
Acquisition: Robin Giroux

Books printed and bound by Maple Press, York, PA

Paper meets minimum requirements of American National Standard for Information Sciences—Permanence of Paper for Printed Library Materials, ANSI Z39.48–1984 ∞

Other ACS Books

Chemical Structure Software for Personal Computers
Edited by Daniel E. Meyer, Wendy A. Warr, and Richard A. Love
ACS Professional Reference Book; 107 pp;
clothbound, ISBN 0–8412–1538–3; paperback, ISBN 0–8412–1539–1

Personal Computers for Scientists: A Byte at a Time
By Glenn I. Ouchi
276 pp; clothbound, ISBN 0–8412–1000–4; paperback, ISBN 0–8412–1001–2

Biotechnology and Materials Science: Chemistry for the Future
Edited by Mary L. Good
160 pp; clothbound, ISBN 0–8412–1472–7; paperback, ISBN 0–8412–1473–5

Polymeric Materials: Chemistry for the Future
By Joseph Alper and Gordon L. Nelson
110 pp; clothbound, ISBN 0–8412–1622–3; paperback, ISBN 0–8412–1613–4

The Language of Biotechnology: A Dictionary of Terms
By John M. Walker and Michael Cox
ACS Professional Reference Book; 256 pp;
clothbound, ISBN 0–8412–1489–1; paperback, ISBN 0–8412–1490–5

Cancer: The Outlaw Cell, Second Edition
Edited by Richard E. LaFond
274 pp; clothbound, ISBN 0–8412–1419–0; paperback, ISBN 0–8412–1420–4

Practical Statistics for the Physical Sciences
By Larry L. Havlicek
ACS Professional Reference Book; 198 pp; clothbound; ISBN 0–8412–1453–0

The Basics of Technical Communicating
By B. Edward Cain
ACS Professional Reference Book; 198 pp;
clothbound, ISBN 0–8412–1451–4; paperback, ISBN 0–8412–1452–2

The ACS Style Guide: A Manual for Authors and Editors
Edited by Janet S. Dodd
264 pp; clothbound, ISBN 0–8412–0917–0; paperback, ISBN 0–8412–0943–X

Chemistry and Crime: From Sherlock Holmes to Today's Courtroom
Edited by Samuel M. Gerber
135 pp; clothbound, ISBN 0–8412–0784–4; paperback, ISBN 0–8412–0785–2

For further information and a free catalog of ACS books, contact:
American Chemical Society
Distribution Office, Department 225
1155 16th Street, NW, Washington, DC 20036
Telephone 800–227–5558